CCTV Surveillance

CCTV Surveillance

Video Practices and Technology

by Herman Kruegle

Butterworth-Heinemann

Boston • Oxford • Melbourne • Singapore • Toronto • Munich • New Delhi • Tokyo

Library of Congress Cataloging-in-Publication Data
Kruegle, Herman.
  CCTV surveillance : video practices and technology / by Herman Kruegle.
      p.   cm.
  Includes bibliographical references and index.
  ISBN 0-7506-9028-3
    1. Closed-circuit television—Design and construction.
  2. Television in security systems.   I. Title.
  TK6680.K78 1995   94-32729
  621.389′28–dc20   CIP

British Library Cataloguing-in-Publication Data
A catalogue record for this book is available from the British Library.

Butterworth–Heinemann
313 Washington Street
Newton, MA 02158

10 9 8 7 6 5 4 3 2 1

Printed in the United States of America

CCTV Systems Application & Design

...... a *Generic* program that teaches you to professionally design and layout CCTV Video Surveillance Systems

...... a *must* for Consultants, Engineers, Facilities Managers and CCTV Sales Personnel

APPLICATION/DESIGN VIDEO TAPE TRAINING SERIES

- 13 Video Tapes 40 – 60 minutes each :

  I . . . .Video Theory
  II . . . .Cameras and Design
  III . . .Visible Lighting Considerations
  IV . . .Infrared Lighting
  V . . . .Lenses & Tools
  VI . . .Coax and Fibre Options
  VII . . .Microwave, Infrared and other Transmission Options
  VIII . .System Monitoring – Ground Fault Loops
  IX . . .System Power
  X . . . .Switchers, Multiplexing and Alarm Interfacing
  XI . . .Time Lapse & Event Video Recorders
  XII . . .Pan/Tilt Housings and Domes
  XIII . .Site Assessment & Design

APPLICATION & DESIGN MANUAL

- 250 pages of the most concise, easy to read CCTV Application and Design Information in the industry

APPLICATION & DESIGN CERTIFICATION TEST

- A Self Administered test you complete at home and return to LTC for grading

FRAMED LTC CERTIFICATE

- Delivered to you upon successful program completion

CCTV Systems Installation & Field Service

...... a *Generic* program designed specifically for Field Personnel

...... a *must* for Technicians and Service Managers

...... the Installation/Field Service Manual is *indispensable* and has been referred to as the "Field Person's Bible"

INSTALLATION/FIELD SERVICE VIDEO TAPE TRAINING SERIES

- 14 Video Tapes 20 – 40 minutes each :

  I . . . .What is a Video System
  IIA . . .Cameras : Tubes
  IIB . . .Cameras : Chips
  III . . .Lighting and IR Enhancement
  IV . . .Lenses
  V . . . .Coax and Connectors
  VI . . .Non-Coaxial Video Transmission Options
  VII . .Monitors
  VIII . .Ground Faults/Loops
  IX . . .Switchers
  X . . . .Pan/Tilt Housings and Domes
  XI . . .Oscilloscopes
  XII . . .System Documentation and Troubleshooting

INSTALLATION & FIELD SERVICE MANUAL

- 300 pages of the most concise, easy to read CCTV Installation and Service Information in the industry

INSTALLATION & FIELD SERVICE CERTIFICATION TEST

- A Self Administered test you complete at home and return to LTC for grading

FRAMED LTC CERTIFICATE

- Delivered to you upon successful program completion

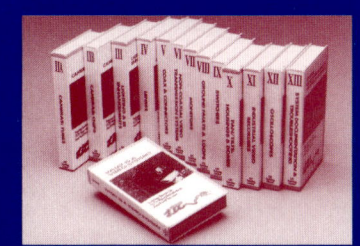

LTC Training Courses certified by :
NAAA
NFBA
ASIS

Learn How to Sell and Install CCTV Today!

Order by :   Phone **800-331-7412**

Fax **613-544-1235**

Mail . . . . . .DSF INTERNATIONAL INC.
P.O. Box 1504
Kingston, Ontario K7L 5C7

Item	Regular	CANASA Member	Amount
Application & Design			
Complete Program	$565	$485	_____
incl. 13 Video Tapes, Manual LTC Certification Test Certificate on successful completion			
Manual only	$105	$105	_____
Inst. & Field Service			
Complete Program	$595	$495	_____
incl. 14 Video Tapes, Manual LTC Certification Test Certificate on successful completion			
Manual only	$115	$115	_____
Encyclopedia of CCTV*	$200	$200	_____
		Sub-Total	_____
		7% GST	_____
	PST (where applicable)		_____
	Canadian Order Shipping & Handling		$18.50
		Total	_____

*   one comprehensive book combining the Application & Design and
the Installation & Field Service manuals

❑ I am a current CANASA Member

Name : _____

Company : _____

Address : _____

City : _____ Prov : _____ Postal Code :_____

Tel : _____ Fax : _____

❑ Cheque enclosed - *payable to DSF International Inc.*

❑ Charge to :   ❑ Visa   ❑ Mastercard

Card # : _____ Exp : _____

Signature : _____

100% Satisfaction Guaranteed!

CCTV TRAINING

Design

Application

Installation

Field Service

Canadian Alarm and Security • L'Association canadienne de l'alarme et de la sécurité

CANASA

UP TO **$200.00 OFF** for CANASA MEMBERS

Limited Time Offer!

**The most sought after training
in the CCTV industry!**

For Carol

Add Picture Power With Multiplexers

Frame multiplexers have myriad features and functions that dealers can easily sell to clients who require high-quality video documentation.

By Allan B. Colombo
Associate Editor, SDM

When documentation is critical, the need to view and record video images independent of one another is integral to all well-engineered multiple-camera, closed circuit television (CCTV) systems. One way that security dealers can accomplish independent image recording is with frame multiplexing technology.

Unlike sequential switchers, which leave gaps in video coverage because only one image at a time can be recorded, frame multiplexing technology provides for entire image frames to be stored on videotape in sequential fashion, at a rate of up to 60 per second using digital technology. This makes it possible to capture a large number of images per camera in near real-time fashion and do so with little or no interruption of coverage. Frame multiplexers also record full frame images, which means higher resolutions on playback.

To appreciate frame multiplex technology, you must understand its predecessor, the quad multiplexer. Quads, as they often are called, will display up to four camera images on a single monitor. Not only does this save money by eliminating the need for additional monitors, but it also makes it easier for officers to observe four different areas at once.

Quads are generally considered a step up from sequential switchers, but they have a disadvantage that affects playback. Quads digitize the images and then display them on a screen in four quadrants. These four images are then recorded as a whole, just as they are seen on the monitor.

Multiplexers in Action

Convenience stores are common targets for crime in the United States. However, many stores don't have sufficient security equipment. Without it, criminals never may be caught.

In February 1997, three women working together stole 16 cartons of cigarettes from Quick Change #3, one of a chain of 50 stores located primarily in Georgia. When the cashier came up short during shift-change inventory, she reviewed the surveillance tape, saw the theft, and called police.

The police were able to make still pictures of the women from the tape, because the picture was so clear. This allowed them to identify the suspects, which assisted in their arrest.

"Our older surveillance system gave us poor resolution. Our pictures weren't clear and our sound was poor," says Jeri Tanner, president of

A good recording system can quickly resolve false liability suits, says security consultant Kim Sellers, who recommends using multiplexers over quads, especially for stores such as Quick Change #3, pictured here.

Quick Change Food Marts. "We purchased a new system in November 1996 from Capitol Business Equipment (CBE), Montgomery, Ala., that allows us to monitor and hear anything going on in the store including every transaction at the registers. Without this new system, we may not have caught our offenders."

CBE's recommendation combined high-density recording with a multiplexer to get the highest quality images. Kim Sellers, point of sale and security consultant for CBE, chose the DigiScan DS9 digital multiplexer, from Gyyr, Anaheim, Calif.

Although quads are commonly used in convenience stores, Sellers says multiplexers are more well-suited for this application.

"I almost refuse to sell people quads with today's technology. With quads you miss so much. When you blow up images, the reso-

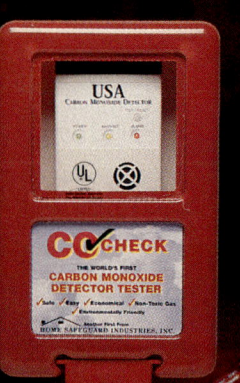

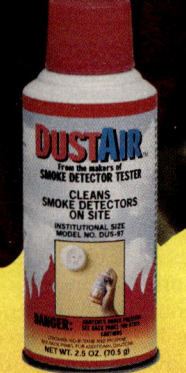

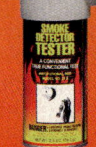

The drawback is that the resolution of each image is about a fourth of what it was in its original form. Even though each image within each quadrant is digitally marked so a user can view one image at a time on the monitor screen, the resolution only will be about 25 percent of what it could have been if full-frame recording were used. The result on playback is an image that might not hold up in court.

However, in situations where observation—not documentation—is the primary mission, quads perform very well for this task.

Frame Multiplexers in Action

By now you may have realized that behind every good frame multiplexer there is a micro-controller or microprocessor. Which one of these powerful, solid-state devices is used depends on how the multiplexer is packaged.

For example, some frame multiplexers are integral to a computer-based camera system. In this case, a microprocessor is at work. Where the multiplexer is used in stand-alone form (black box, so to speak), you will find a microcontroller. No matter which platform you select the basic principle of operation is essentially the same.

Frame multiplexers operate by recording full-frame, full-resolution images, one after the other in rapid succession. Like cameras, these devices are designed to process and record up to 60 frames of video every second to simulate real-time motion to the human eye.

This is a tremendous improvement over the more than 1 second refresh times that are common with conventional sequential camera switchers. However, the number of recorded images from each camera each second—called refresh time—depends on the number of cameras connected to the CCTV system.

For example, in a two-camera system, you can expect to record up to 30 frames from each camera every second. In a four-camera system, the number of frames per camera is reduced to 15; 7.5 frames per second in an eight-camera system; and 3.75 frames per second in a 16-camera system. Thus, the more cameras you have in a frame multiplexed system, the fewer number of frames the multiplexer will capture and record from each camera per unit time.

One factor that can affect refresh times of recorded images is the use of a time-lapse video cassette recorder. Multiplexers must be programmed to match a VCR's time-lapse mode. If not done, performance will suffer.

"The thing you have to do is decide on what speed you are going to record at so you can program the multiplexer to match the VCR," says Steve Wilber, sales application manager, Gyyr, Anaheim, Calif. "If this is not done, you will either skip pictures on recording or you may take unnecessary time sequencing, which means you could take two or three times as long to record the same information."

Programming a multiplexer to match a time-lapse VCR is usually accomplished by looking up the time-lapse mode on a chart and then programming an assigned value in the multiplexer's programming.

Product Enhancements

Today's frame multiplexers come with a host of enhancements that make them even more powerful than simple recording devices in the hands of knowledgeable security dealers and end users.

There are two basic types of frame multiplexers on the market: simplex and duplex. Simplex-type units are used when end-users either record or play back recorded information—never both at the same time. Duplex-type units allow end users to do both simultaneously, although two VCRs are necessary to

---

lution is poor and the images are distorted. A multiplexer is higher priced, but when our customers need a clear picture, they get it."

Sellers' customer agrees: "Our new Gyyr multiplexer gives us the opportunity to record and see virtually all cameras at the same time. Our older [system] switched back and forth, seldom recording what we needed at the time," Tanner says.

"With approximately 750 customers entering the store every day, we wanted to eliminate as many opportunities for employee, vendor, or customer theft as we could by installing the new equipment. We were looking to improve the bottom line with a complete system that would allow us to add cameras that would interface with our point of sale equipment. Our new system gives us good, clear pictures compared to our old black and white system, and the mul-

This multiplexer, installed at a convenience store, overlooks the cash register and is integrated with the point of sale system.

tiplexer allows us to view more areas of the store," Tanner continues.

"We were recently given a counter-feit $100 bill and were able to review the tape and turn it over to the Georgia Bureau of Investigation. Any disputes that might occur with customers or vendors can be resolved quickly by viewing the tape and being able to hear and see the situation. We feel this system greatly reduces cash shortages and opportunities for shrink," Tanner says.

The Gyyr recorder and multiplexer work together with five color cameras from Ultrak Inc., Westminster, Colo., two interfaces from American Video Equipment, Houston, Texas, an audio kit from Louroe Electronics, Van Nuys, Calif., and smoked domes from Wren Associates Ltd., Jefferson City, Mo., at Quick Change #3.

"The smoked domes add a touch of class and don't bother our honest customers, but they make shoplifters wonder which way the camera is facing," Tanner says.

Frame multiplexers can record up to 60 video images per second on videotape in sequential fashion.

take advantage of this feature.

"Multiplexers offer many options for display, like picture in a picture. You can do what some manufacturers call cameo modes, which means you can mix multiple display formats on the same screen. For example, you might have a quad display with another quad display within one of the quadrants," Wilber says. "It's even possible to have a ring of eight images along the outside of a larger one in the middle."

Some models allow dealers to designate certain camera scenes for priority recording, which is referred to by some manufacturers as interleaving. Interleaving means that the number of frames per second from select cameras can be increased over that of other lower-priority scenes (cameras) for recording purposes. One model allows the user to program camera scene recording by time-of-day and day-of-week criteria, which is consistent with the capabilities inherent with microcon-

trollers and processors.

Another handy feature built into current frame multiplexers is digital motion sensing. This feature may be used to trigger priority recording (interleaving), eliminating the need to give recording preference to select cameras until it is needed.

Another recent enhancement is remote diagnostics. It is now possible to diagnose maintenance problems with multiplexers, and possibly correct them, without a dealer having to leave his office, says Lore Pierson, marketing manager, Dedicated Micros, Reston, Va.

Using a modem and a specific communications protocol, data can be sent to Dedicated Micros' technical staff over POTS (plain old telephone system), who will then report what they find to the dealer. This diagnostic tool may someday find its way into dealers' offices; but at this time, according to Pierson, there are no plans to do this. **SDM**

FOR MORE INFORMATION...

Use this list of selected manufacturers to obtain more information about video frame multiplexers. Simply circle the corresponding number on the Reader Service Card, located elsewhere in this issue.

◆ ACI Int'l Inc. . . . . .Circle 231
◆ ATV Research Inc. . . . . .Circle 232
◆ Advanced Technology Video Circle 233
◆ American Fibertek Inc. . . . . .Circle 234
◆ American Video Equipment .Circle 235
◆ Appro Technology Inc. . . . . .Circle 236
◆ Asian World Ltd. . . . . .Circle 237
◆ Audio Video Supply . . . . . .Circle 238
◆ Automation Displays Inc. . . .Circle 239
◆ CCTV Corp. . . . . .Circle 240
◆ CSI/Speco . . . . .Circle 241
◆ Camtron Electronics Int'l . . .Circle 242
◆ Crest Electronics Inc. . . . . . .Circle 243
◆ Colorado Video Inc. . . . . . . .Circle 244
◆ Costar Video Systems . . . . .Circle 245
◆ Cybermation Systems Inc. . .Circle 246
◆ Dedicated Micros Inc. . . . . .Circle 247
◆ Elbex America . . . . .Circle 248
◆ Electronics Line USA Inc. . .Circle 249
◆ Elmo Mfg. Corp. . . . . .Circle 250
◆ FM Systems . . . . .Circle 251
◆ Fiber & Wireless Inc. . . . . . .Circle 252
◆ Fiberlign . . . . .Circle 253
◆ For-A Corp. of America . . . .Circle 254
◆ GPS Standard USA . . . . . . .Circle 255
◆ Gyyr . . . . .Circle 256
◆ Impath Networks Inc. . . . . . .Circle 257
◆ Javelin Electronics . . . . . . .Circle 258
◆ Knogo North America Inc. . .Circle 259
◆ Koyo Int'l Inc. of America . . .Circle 260

◆ Larscom Inc. . . . . .Circle 261
◆ Litton Poly-Scientific . . . . . .Circle 262
◆ Math Associates Inc. . . . . . .Circle 263
◆ Meret Optical Comm. . . . . . .Circle 264
◆ Metronet Protection Svcs. . .Circle 265
◆ Microwave Filter Co. Inc. . . .Circle 266
◆ Mitsubishi Electronics
      America Inc. . . . . .Circle 267
◆ Multiplex Technology Inc. . .Circle 268
◆ Navco Security Systems . . .Circle 269
◆ Northern Computers Inc. . . .Circle 270
◆ Navco Security Systems . . .Circle 271
◆ Paragon Electric Co. Inc. . . .Circle 272
◆ Pelco Sales Inc. . . . . .Circle 273
◆ Philips Comm. & Security . .Circle 274
◆ Q Enterprises . . . . .Circle 275
◆ QSI Systems . . . . .Circle 276
◆ Quark Digital Systems Inc. . .Circle 277
◆ RGB Spectrum . . . . .Circle 278
◆ RNJ Electronics Inc. . . . . . .Circle 279
◆ Richardson Electronics Ltd. .Circle 280
◆ Ronan Engineering . . . . .Circle 281
◆ Senstar . . . . .Circle 282
◆ Sensormatic Electronics . . .Circle 283
◆ Shepherd Surveillance . . . .Circle 284
◆ Silent Witness . . . . .Circle 285
◆ Superscope Technology . . .Circle 286
◆ TVS Inc. . . . . .Circle 287
◆ Tellabs . . . . .Circle 288
◆ 3M Telecom Systems Div. . .Circle 289
◆ Tie Security Systems . . . . . .Circle 290
◆ UHF Associates . . . . .Circle 291
◆ Ultrak Inc. . . . . .Circle 292
◆ Vicon Industries Inc. . . . . . .Circle 293
◆ Vision Factory . . . . .Circle 294
◆ Vision Research Corp. . . . . .Circle 295

Contents

Foreword

In recent years, the applications of electronic systems for security purposes have changed dramatically. Until the early 1980s, the predominant electronic systems were alarm systems dedicated to monitoring and annunciating building or site intrusions. CCTV was used primarily in outdoor intrusion-detection applications for the purpose of alarm assessment.

A major breakthrough in CCTV applications occurred when solid-state cameras were introduced in the mid-1980s. Such cameras require minimal maintenance and very quickly became competitive with their tube counterparts in performance and price. Further advances in signal processing and advanced communications found their way into CCTV products, resulting in an industry that is rich in computer and communications technology. For example, remote building surveillance can now be performed using compressed video transmission over dial-up telephone lines, with picture quality and transmission speeds that begin to rival coaxial systems. All of that capability is now available at costs that allow a convenience store to utilize the technology to deter crime and reduce operating costs.

The CCTV industry is now maturing, since one can define a consistent and comprehensive view of its inner workings that is shared by many. To take that view and write a thorough reference manual about the CCTV industry is a major achievement. Such is the case with *CCTV Surveillance*. Previously, the only sources of information about CCTV have been manufacturers' data sheets, articles published in industry periodicals, and a few training manuals intended primarily for CCTV dealers and installers. At last, there is a single source of information covering the theory, design, and application of CCTV systems that can be used by the everyone involved in security.

The release of this book comes at a time when the security profession is undergoing significant change. The challenge to "do more with less," the result of downsizing trends in the military and corporate worlds, coincides with a change in the security threat scenario—from a focus on external threat (terrorism, burglary, asset protection, and so on) to a focus on internal threat (workplace violence, information security, and asset management). The security director can no longer afford to be merely reactive to security threats as they occur but must be proactive and preemptive. CCTV systems provide a tool that, when properly applied, can help the security director reduce manpower requirements and respond to a variety of threats.

In this book, Herman Kruegle creates a very complete picture of CCTV, starting with its role in the security plan and then covering each component of a CCTV system, from theory to practice. Abstract and concrete descriptions alike are very clear, and the numerous illustrations will facilitate understanding of each concept.

Herman Kruegle's personal knowledge of CCTV and electro-optics, accumulated over a 30-year career, enables him to create *the* definitive reference book, a book that should be in the library of every person involved in any security-related function. From security design engineer to facility manager to security director, this book offers a wealth of information that will serve the reader well.

Ronald C. Thomas
President and CEO
Securacom, Inc.
Woodcliff Lake, N.J.
Fall 1994

Preface

WHY WRITE A CCTV SURVEILLANCE BOOK?

In the past decade the security industry has grown from a small, immature one, with a helter-skelter conglomeration of companies and disciplines, into a large, well-organized association of companies, with better defined product areas and solutions to customer-asset-protection problems. Closed-circuit television (CCTV) technology has emerged as a distinct entity. To implement it successfully, one must understand diverse technical disciplines, such as scene illumination, signal transmission, and display and recording technology. Until now, there has been no book available that teaches all aspects of CCTV security. This book attempts to inform managers, security and safety directors, consultants and designers, and security dealers and installers, so that they can make intelligent decisions about when and how to use CCTV–and how it integrates into the overall security system. From a practical viewpoint, the book explains how to design a system, specify functions and features, and choose hardware.

This book is intended for those professionals associated with safety and/or security. For managers, security personnel, indeed for entire institutions, CCTV has become one of the most important tools available for protecting employees and assets. Significant technical advances in lenses, television cameras, and electronics have provided equipment for meeting almost all remote vision needs, whether real-time or recorded.

New solid-state image sensors and microelectronics have led to complete CCTV cameras smaller than a deck of cards. Special systems using existing telephone networks allow a viewer to see a still television picture from anywhere on earth for the price of a long-distance telephone call.

This book covers the theory, hardware, and design of CCTV surveillance systems for security and safety applications. A second companion book, *CCTV Access Control*, covers the theory, hardware, design, and application of CCTV for personnel access control in walk-up and vehicle entry applications. *CCTV Access Control* describes the use of CCTV, electronic card readers, personal descriptor, and video image storage and retrieval applications.

GUIDE TO READING THIS BOOK

CCTV Surveillance can be used by nontechnical managers and security directors as well as technically oriented personnel, such as security consultants, architects, engineers, and security dealers and installers. Each chapter contains useful nontechnical and technical information, presented in a form suitable for technical personnel who need details about specialized aspects of the equipment and physics of security.

To make the book more helpful to this wide audience of nontechnical managers and technical practitioners, each chapter begins with an overview for nontechnical readers, which summarizes the main points of the chapter. The remainder of the chapter provides information needed to understand the concept, analyze and design some specific applications, and choose the best CCTV hardware.

Chapter Overviews

Part 1

Chapter 1, CCTV's Role in the Security Plan, introduces the asset protection problems to be solved and how CCTV fits into the security plan. An effective plan needs an integrated security system, with CCTV playing a vital role.

Chapter 2, CCTV Technology Overview, touches broadly on all aspects of CCTV security surveillance hardware. From lighting sources, cameras, and accessories to transmission media, to switching, monitoring, and recording systems, this chapter provides an understanding of all the components of a modern CCTV security system.

Part 2

Chapter 3 serves as a tutorial on CCTV lighting requirements for effective indoor and outdoor use of camera sensors. The characteristics of natural and artificial light sources are explained, such as spectral content, intensity, beam pattern, life expectancy, and cost, as well as their applicability to monochrome and color CCTV systems.

Chapter 4 analyzes the large variety of lenses available for CCTV systems. These include the workhorses of the industry: fixed-focal-length lenses and zoom, pinhole, and specialty lenses. Lenses are the eyes of a CCTV system and deserve special attention for any system, overt or covert.

Chapter 5 reviews CCTV camera technology and the many types of cameras available. Significant technological advances and lower costs for solid-state cameras largely account for the increased use of monochrome and color cameras in security systems.

Chapter 6 discusses the important subject of signal transmission, one of the most neglected, but crucial, parts of any CCTV system. It covers video, audio, and control signal transmission and the hardware available. Transmission over coaxial and fiber-optic cable is described, as well as wireless transmission, including microwave, radio frequency, infrared, and slow-scan (non—real-time) transmission.

Chapter 7 covers the monitors, display terminals, and accessories used for displaying the CCTV picture. Included is an analysis of the human engineering considerations in designing the control console.

Chapter 8 reviews the available videocassette recorders (VCRs), other magnetic and optical devices for permanently storing video images, and video image printers.

Chapter 9 explains the operation of all the different types of camera switchers, from the simple manual switcher to the complex-matrix, computer-controlled switcher.

Chapters 10 and 11 describe the camera pan/tilt mechanisms, which are used to point the camera/lens from a remote console location, as well as all types of indoor and outdoor camera housings.

Chapter 12 describes the video motion detector, an important component used to alert security personnel to movement within the field of view of the CCTV camera. These smart devices can discriminate between human targets and other animals and objects and can even track an intruder.

Chapter 13 presents some special CCTV devices and how they are used to produce particular effects on displayed or recorded video pictures. Such components include image splitters, combiners, and alphanumeric annotators.

Chapter 14, on covert CCTV, provides a comprehensive analysis of hidden CCTV camera and lens equipment and its surveillance applications to detect unlawful activity without the suspect's knowledge. Unique covert optics including pinhole, mini, and fiber-optic lenses and their installation are covered.

Chapter 15 presents some of the latest information on new low-light-level cameras and systems, including tube and microchannel plate light-intensifying (amplification) devices and thermal infrared sensing devices.

Chapter 16 provides the theory and guidelines for choosing the proper electrical power source and backup equipment necessary for CCTV security equipment. This chapter also analyzes the critical importance of uninterruptible power supplies to maintain CCTV equipment on-line during a power failure.

Part 3

Chapter 17 presents specific applications to illustrate how CCTV equipment is used to solve particular security problems. The examples include an office environment, a parking lot, a building's front lobby, elevators and a lobby, a retail store, a correctional institution, and a banking surveillance system. Each case includes an analysis, a block diagram, and a bill of materials of the CCTV equipment needed.

Chapter 18 describes new electronics and CCTV technology that is expected in the near future.

Chapter 19 provides a checklist to serve as a starting point for preparing design specifications, responding to requests for quotation or proposals, and conducting facility site surveys. The aspects covered include environmental factors for outdoor systems, lighting requirements for normal and low-light-level conditions, and hardware considerations.

Chapter 20 is a listing of some of the older and newer tests and recommended safety standards issued by Underwriters Laboratories Inc. and equipment standards recommended by the Electronic Industries Association (EIA). The standards serve as guidelines for CCTV equipment manufacturers, consultants, and system designers.

Glossary and Bibliography

For both nontechnical and technical personnel, a comprehensive glossary is provided to give a better understanding of common terms used in lighting, television optics, and electronics. Many of these terms are unfamiliar to practitioners in the CCTV and security industries because they come from many diverse disciplines, including physics, lighting, optics, and electronics. Finally, the bibliography will help the reader find more extensive information on a particular subject.

Acknowledgments

Over the years I have had the opportunity to speak with many individuals who have provided technical insight in electro-optics and CCTV technology. I particularly appreciate the many discussions with Stanley Dolin and Lee Gallagher, from whom I received many insights into optics and the physics of lighting, lenses, and optical sensors. Many discussions with Victor Houk, Ralph Ward, Ron Thomas, and Ed Henkel have likewise been very useful.

I also acknowledge the initial encouragement of Kevin Kopp at Butterworth in the formative stages of the book, and Greg Franklin for his excellent editorial advice, motivation, and recommendations in planning the layout of the book. I also want to recognize the professional copyediting by Chris Keane.

I likewise appreciate the insight given by David Bittner and the editorial advice offered by Jane Stokes, former editor at *CCTV Magazine,* over the years.

I would also like to thank several colleagues who reviewed sections of the manuscript: H. Eugene Crow—Chapters 3–5; Terry McGhee—Chapter 3; Robert DeLia—Sections 6.1–6.4; Thomas Christ—Section 6.5; and Sal Raia—Chapter 9.

I gratefully acknowledge the dedication, patience, and skill of my wife, Carol, in preparing a manuscript of this size.

I would like to thank the following companies for supplying photographs or other illustrations.

American Dynamics	Figures 3-15, 9-11
American Fibertech	Figure 6-26
American Laser	Figure 6-40
Applied Engineering Products	Figure 10-24
Burle Industries	Figures 2-8, 9-11
Chinon	Figures 8-9, 8-12, 14-19
Computar	Figures 4-14, 4-19, 14-19
Dellstar	Figure 6-35
Dynalens	Figure 4-38
Elbex	Figure 10-22
Elmo	Figure 2-8
EMI	Figures 11-2, 11-13
FOR-A	Figure 13-9
Fujinon	Figure 4-19
General Electric	Figure 3-10
HDS	Figure 6-32
Hitachi	Figure 5-7
Holobeam	Figure 3-10
Ikegami	Figures 2-8, 4-29
ITT	Figure 15-6
Interoptics	Figure 6-26
Javelin	Figure 11-1
Kenall Lighting	Figures 3-7, 3-17
Math Associates	Figure 6-27
Metal Arc	Figure 3-10
Mitsubishi	Figure 2-23
OCLI	Figure 7-9
Panasonic	Figures 2-21, 2-22, 7-4, 10-19, 14-20
Pelco	Figures 10-2, 10-19, 11-8, 11-11 11-12
Pentax	Figure 2-6, 4-12, 4-13
Portac	Figure 13-10
Questar	Figure 15-20
Sanyo	Figure 8-6

Schwem   Figure 4-38
Seagate   Figure 8-9
Sensormatic   Figure 10-14
Solarex   Figure 16-11
Sony   Figures 2-21, 2-23, 7-11,
                            8-4, 8-15
Southwest Microwave   Figure 6-32
Sovonics   Figure 16-11
Star Tech   Figure 14-26
Startron   Figures 2-6, 4-16, 15-20
Telesite   Figure 6-11
Thomas and Betts   Figure 3-7
Unisolar   Figure 16-12
United Technology   Figure 17-20
Vicon   Figures 2-25, 2-26, 4-27,
                            6-7, 10-1, 10-2, 10-11, 10-12,
                            11-1, 11-11, 11-12

Video Alarm   Figures 2-25, 10-12, 11-4,
                            11-5, 11-6, 11-7, 11-13
Videor   Figures 2-25, 2-26, 14-5,
                            14-22
VMI   Figures 2-6, 2-14, 2-15, 2-24,
                            4-24, 4-28, 4-32, 4-33, 4-34,
                            4-36, 10-5, 10-7, 11-5, 11-6,
                            13-12, 14-5, 14-7, 14-8, 14-11,
                            14-13, 14-14, 14-20, 14-24,
                            17-18
Watec   Figures 2-28, 4-19, 10-23,
                            14-19
Winsted   Figures 7-2, 7-3

PART I

Chapter 1
CCTV's Role in the Security Plan

CONTENTS

1.1 PROTECTION OF ASSETS: AN OVERVIEW

The application and integration of closed-circuit television (CCTV) to safety and security has come of age. CCTV is a reliable, cost-effective deterrent and a means for apprehending and prosecuting offenders. Today, most safety and security applications require several different types of systems, such as alarm, fire detection, intrusion prevention and detection, access control—and now CCTV.

Security personnel today are responsible for multifaceted security and safety systems, in which CCTV plays an important role. With today's increasing labor costs, CCTV more than ever before has earned its place as a cost-effective means for improving security and safety, while reducing security budgets.

Loss of assets and time due to theft is a growing cancer on our society that eats away at the profits of every organization or business, be it government, retail, service, or manufacturing. The size of the organization makes no difference to the thief. The larger the company, the more theft occurs, and the greater the opportunity for losses. The more valuable the product, the easier it is for a thief to dispose of it, and thus the greater the temptation is to steal it. A properly designed and applied CCTV system can be an extremely profitable investment for an institution. The main objective of the CCTV system should be not the apprehension of thieves but rather increased deterrence through security. A successful thief needs privacy; a television system can deny that privacy.

As a security by-product, CCTV has emerged as an effective training tool for managers and security personnel. The use of CCTV systems has resulted in improved employee efficiency and increased productivity.

The public at large has accepted the use of CCTV systems in public and industrial facilities, and workers' resistance to it is steadily decreasing. With the belt-tightening under way in today's businesses, employees and others begin looking for other ways—such as theft or industrial espionage—to increase their income and pay the bills. CCTV is being applied to counteract these losses and increase corporate profits. In many case histories, after CCTV was installed, shoplifting and employee thefts dropped sharply. The number of thefts cannot be counted exactly, but the reduction in shrinkage can be measured, and it has been shown that CCTV is an effective psychological deterrent to crime.

Theft is not only the unauthorized removal of valuable property but also the removal of information, such as computer software, magnetic tape and disks, optical disks, microfilm, and data on paper. CCTV surveillance systems provide a means for successfully deterring such thievery and/or detecting or apprehending offenders. Another form of loss that CCTV prevents is the willful destruction of property, for example vandalizing buildings, defacing elevator interiors, painting graffiti on art objects and facilities, demolishing furniture or other valuable equipment, and destroying computer rooms. CCTV offers the greatest potential benefit when integrated with other sensing systems (e.g., alarms) and used to view remote areas. For example, when combined with smoke detectors, CCTV cameras in inaccessible areas can be used to give advance warning of a fire.

But CCTV is only a link in the overall security of a facility. Organizations must develop a complete security plan rather than adopt piecemeal protection measures and react to problems only as they occur. To make the best use of CCTV technology, the practitioner and end user must understand all of its aspects—from light sources to video monitors. The capabilities and limitations of CCTV during daytime and nighttime operation must also be understood.

The protection of assets is a management function. Three key factors govern the planning of an assets protection program: (1) an adequate plan designed to prevent losses from occurring, (2) adequate countermeasures to limit losses and unpreventable losses, and (3) support of the protection plan by top management.

1.1.1 History

Throughout history, humans have valued above all else their own life and the lives of their loved ones. Next in value has been their property. Over the centuries many techniques have been developed to protect property against invaders or aggressors threatening to take or destroy it.

More recently, manufacturing, industrial, and government organizations have hired "watchmen" to protect their facilities. These private police, wearing uniforms and using equipment much like the police do, sought to prevent crime—primarily theft—on the protected premises. Contract protection organizations, typified by Pinkerton's and Burns, provided a new and usually less expensive guard force, which industrial employers began to use widely.

World War II supplied the single most important impetus to the growing protection of industrial premises. Private corporations obtained such protection through contract agencies to guard classified facilities and work.

As technology advanced, alarm systems and eventually CCTV were introduced in the early 1960s, when companies such as the Radio Corporation of America (RCA) began introducing vacuum-tube television cameras.

Today's state-of-the-art security system includes CCTV as a key component. There are many applications for CCTV equipment to provide general surveillance, security, and safety.

Having begun in the 1960s, the CCTV industry grew rapidly throughout the 1970s, because of increased reliability and technological improvements in the tube-type camera. In the 1980s growth continued at a more modest level, with further improvements in functions and other accessories for television security systems. The most significant advance during the 1980s was the introduction of the solid-state CCTV camera; by the early 1990s, it had replaced most tube cameras of the past 30 years.

The most significant driving force behind this CCTV explosion has been the worldwide increase in theft and terrorism and the commensurate need to more adequately protect personnel and assets. Another factor has been the rapid improvement in equipment capability at affordable prices, resulting from the widespread use of solid-state CCTV for consumer use (made possible through technological breakthroughs) and the availability of low-cost videocassette recorders (VCRs) and associated camera equipment. These two driving forces have accelerated the development and implementation of the excellent CCTV equipment available today.

In the past, the camera—in particular, the vidicon sensor tube—was the critical item in system design. The camera determined the overall performance, quantity, and quality of visual intelligence obtainable from the security system, because the camera's image tube, subject to degradation with age and usage, was the weakest link in the system. The complexity and variability of the image tube and its analog electrical nature made it less reliable than the other solid-state components. Performance varied considerably between different camera models and camera manufacturers, and as a function of temperature and age. Today the situation is considerably different. Solid-state charged coupled device (CCD) and metal-oxide semiconductor cameras are now available. While the various solid-state cameras have different features, they are reliable, with modest variations in sensitivity and resolution depending on the manufacturer—rather than inherent generic differences as in tube cameras. Systems are more reliable and stable, since the tube has been replaced with a solid-state device that does not wear out. This innovation and popular consumer use of camcorders has resulted in the widespread use of solid-state monochrome and color cameras in security applications.

1.2 THE ROLE OF CCTV IN ASSET PROTECTION

In one phase of asset protection, CCTV is used to detect unwanted entry into a facility, beginning at the perimeter location, and continuing by following the intruder throughout the facility (Figure 1-1).

In a perimeter protection role, CCTV can be used with intrusion-detection devices to alert the guard at the security console that an intrusion has occurred. If an intrusion occurs, multiple CCTV cameras located throughout the facility follow the intruder so that there is a proper response by guard personnel or designated employees. Management must determine whether specific guard reaction is required and what the response will be.

Obviously, CCTV allows the guard to be more effective, but it also improves security by permitting the camera scene to be documented via a VCR and/or printed out on a hard copy video printer. In the relatively short history of CCTV, there have been great innovations in the permanent recording of video images for later use, brought about primarily by the consumer demand and availability of video camcorders and VCRs. The ability to record video provides CCTV security with a new dimension, i.e., going beyond real-time camera surveillance. The specialized time-lapse

recorders and video printers as well as magnetic storage of video images on magnetic and optical hard disks now allow management to present hard evidence for prosecution of criminals. This ability of CCTV is of prime importance to those protecting assets, since it permits permanent identification of wrongdoing.

Most CCTV security is accomplished with monochrome equipment, but the solid-state camera has now made color security practical. Tube-type color cameras were unreliable, they had short lifespans and high maintenance costs, and their color balance could not be maintained over even short periods of time. The development of color CCD cameras for the consumer VCR market accelerated the availability of these reliable, stable, long-life cameras for the security industry. Likewise, availability of VCR technology, also resulting from consumer demand, made possible the excellent time-lapse VCR, providing permanent documentation for CCTV security applications. While monochrome cameras are still specified in most major security applications, the trend is toward the use of color in security. As the sensitivity and resolution of color cameras increase and cost decreases, color cameras will replace most monochrome types.

Along with the introduction of the solid-state camera has come a decrease in the size of ancillary equipment, such as

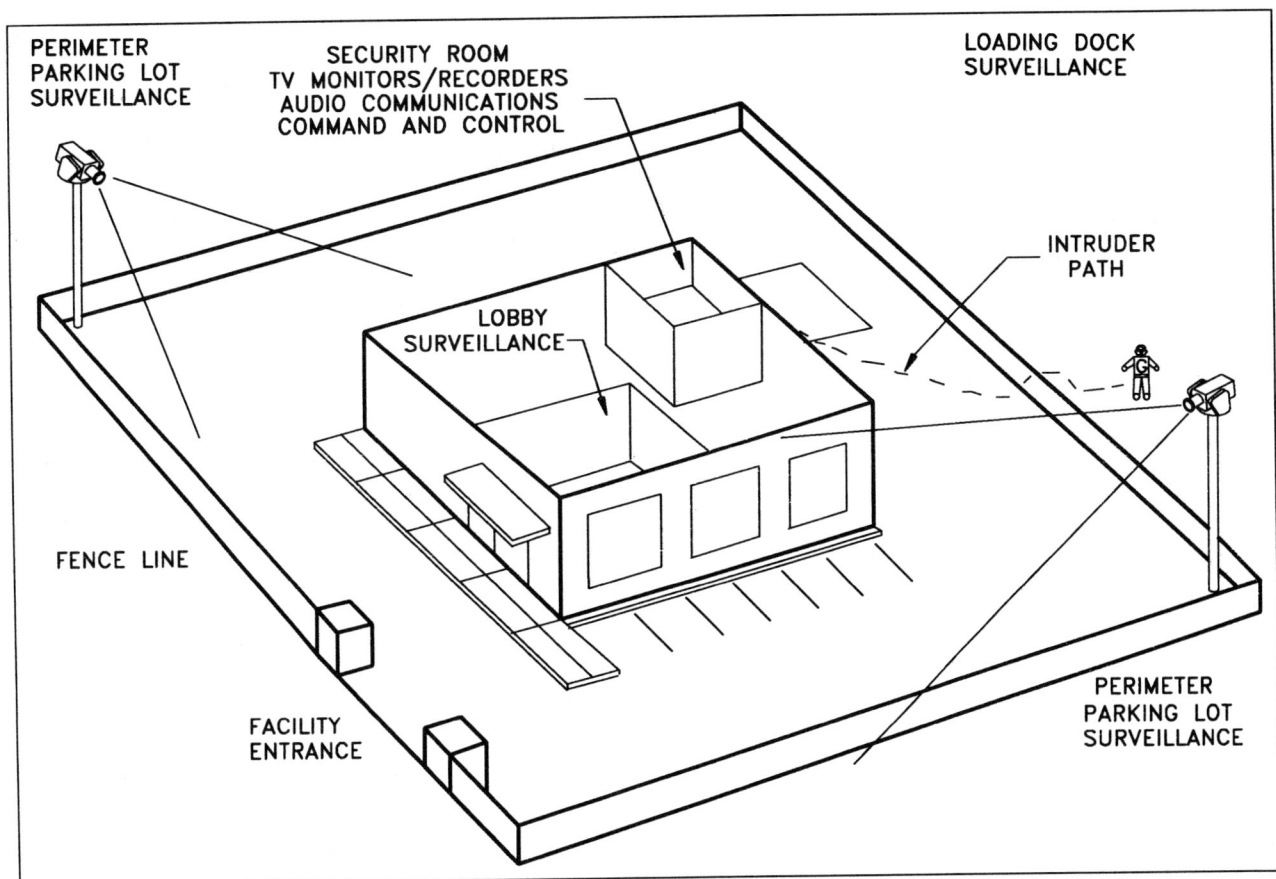

FIGURE 1-1 CCTV security system

lenses, housings, pan/tilt mechanisms, and brackets, which lower costs and provide more aesthetic installations. For covert CCTV applications, the small cameras and lenses are easier to conceal.

The potential importance of color in surveillance applications can be illustrated very clearly by looking at a color television scene on a television monitor, be it surveillance or other, and turning off the color to make it a monochrome scene. It becomes quite obvious how much information is lost when the colors in the scene change to shades of gray. Objects easily identified in the color scene become difficult to identify in the monochrome scene. It is much easier to pick out a person with red socks in the color scene than in the monochrome scene.

The security industry has long recognized the value of color to enhance personnel and article identification in video surveillance and access control. One reason why we can identify subjects more easily in color is that we are used to seeing color, both in the real world and on our TV at home. When we see a monochrome scene, we have to make an additional effort to recognize certain information, besides the actual missing colors, thereby decreasing the intelligence available. Providing more accurate identification of personnel and objects leads to a higher degree of apprehension and conviction for crimes.

1.2.1 CCTV as Part of the Emergency and Disaster Plan

Every organization, regardless of size, should have an emergency and disaster control plan, which should include CCTV as a critical component. Part of the plan should be a procedure for succession of personnel in the event one or more members of top management are unavailable when disaster strikes. In large organizations the plan should include the designation of alternate headquarters if possible, a safe document-storage facility, and remote CCTV operations capability. The plan must provide for medical aid and assure the welfare of all employees. Using CCTV as a source of information, there should be a method to alert employees in the event of a dangerous condition and a plan to provide for quick police and emergency response. There should be an emergency shutdown plan and restoration procedures, with designated employees acting as leaders. There should be CCTV cameras stationed along evacuation routes and instructions for practice tests. The evacuation plan should be prepared in advance and tested.

A logical and effective disaster control plan should do the following:

1. define emergencies and disasters that could occur as they relate to the particular organization
2. establish an organization and specific tasks with personnel designated to carry out the plan immediately before, during, and immediately following a disaster

3. establish a method for utilizing the organization's resources, in particular CCTV, to analyze the disaster situation and bring to bear all other available resources.
4. recognize a plan to change from normal operations into and out of the disaster emergency mode as soon as possible.

CCTV plays a very important role in any emergency and disaster plan:

1. CCTV helps protect human life by enabling security or safety officials to see remote locations and view first-hand what is happening, where it is happening, what is most critical, and what areas must be attended to first.
2. CCTV aids in minimizing personal injury by permitting "remote eyes" to get to those people who require immediate attention, or to send personnel to the area being hit hardest to remove them from the area, or to bring in equipment to protect them.
3. CCTV reduces the exposure of physical assets to oncoming disaster, such as fire or flood, and prevents or at least assesses document removal (of assets) by intruders or any unauthorized personnel.
4. CCTV documents equipment and assets that were in place prior to the disaster, recording them on VCR, hard disk, or other medium to be compared to the remaining assets after the disaster has occurred. It also documents personnel and their activities before, during, and after an incident.
5. Probably more so than any other part of a security system, CCTV will aid management and the security force in minimizing any disaster or emergency. CCTV is useful in restoring an organization to normal operation by determining that no additional emergencies are in progress and that procedures and traffic flow are normal in those restored areas.

1.2.1.1 Protecting Life and Minimizing Injury

Through the intelligence gathered from CCTV, security and disaster control personnel should move all personnel to places of safety and shelter. Personnel assigned to disaster control and remaining in a threatened area should be protected by using CCTV to monitor their safety as well as access to these locations. By such monitoring, advance notice is available if a means of support and assistance for those persons is required or if injured personnel must be rescued or relieved.

1.2.1.2 Reducing Exposure of Physical Assets and Optimizing Loss Control

Assets should be stored or secured properly before an emergency, so they will be less vulnerable to theft or loss. CCTV is an important tool for continually monitoring safe areas during and after a disaster to ensure that material is not removed. In an emergency or disaster, the well-documented plan will call for specific personnel to locate highly valued assets, secure them, and evacuate personnel.

1.2.1.3 Restoring Normal Operations Quickly

After the emergency situation has been brought under control, CCTV and security personnel can monitor and maintain the security of assets. They can also help determine that employees are safe and have returned to normal work.

1.2.1.4 Documenting an Emergency

For purposes of planning, insurance, and evaluation by management and security, CCTV coverage of critical areas and operations during an emergency can save an organization considerable money. CCTV recordings of assets lost or stolen or personnel injured or killed can support a company's claim that it was not negligent and that it initiated a prudent emergency and disaster plan prior to the event. Although CCTV can provide crucial documentation of an event, it should be supplemented with high-resolution photographs of specific instances or events.

Moreover, if fences or walls are destroyed or damaged in a disaster, CCTV can help prevent and document intrusion or looting by employees, spectators, or other outsiders.

1.2.1.5 Emergency Shutdown and Restoration

In the overall disaster plan, shutting down equipment such as machinery, utilities, processes, and so on, must be considered. If furnaces, gas generators, electrical power equipment, boilers, high-pressure air or oil systems, chemical equipment, or rapidly rotating machinery could cause damage if left unattended, they should be shut down as soon as possible. CCTV can help determine if the equipment has been shut down properly, if personnel must enter the area to do so, or if it must be shut down by other means.

1.2.1.6 Testing the Plan

While a good emergency plan is essential, it should not be tested for the first time in an actual disaster situation. Deficiencies are always discovered during testing. Also, a test serves to train the personnel who will carry out the plan if necessary. CCTV can help evaluate the plan to identify shortcomings and show personnel what they did right and wrong. Through such peer review a practical and efficient plan can be put in place to minimize losses to the organization.

1.2.1.7 Standby Power and Communications

During any emergency or disaster, primary power and communications between locations will probably be disrupted. Therefore, a standby power-generation system should be provided for emergency monitoring and response. This standby power, comprised of a backup gas-powered generator or an uninterruptible power supply, with DC batteries to extend backup operation time, will keep emergency lighting, communications, and strategic CCTV equipment on-line as needed. Most installations use a power sensing device that monitors the normal supply of power at various locations. When the device senses that power has been lost, the various backup equipments automatically switch to the emergency power source.

A prudent security plan anticipating an emergency will include a means to power vital CCTV, audio, and other sensor equipment to ensure its operation during the event. Since emergency CCTV and audio communications must be maintained over remote distances, alternative communication pathways should be supplied, in the form of either auxiliary hard-wired cable or a wireless (RF, microwave, infrared) system. It is usually practical to provide a backup path to only the critical cameras, not all of them. The standby generator supplying power to the CCTV, safety, and emergency equipment must be sized properly. For equipment that normally operates on 120-volt AC, for example, inverters are used to convert the low voltage from the DC batteries (typically 12 or 24 volts DC) to the required 120 volts AC.

1.2.2 Security Investigations

Security investigators have used CCTV very successfully regarding company assets and theft, negligence, outside intrusion, and so on. Using covert CCTV, that is, using a hidden camera and lens, it is easy to positively identify a person or to document an event. Better video image quality, smaller lenses and cameras, and easier installation and removal of such equipment have led to this high success. Many lenses and cameras are available today that can be hidden in rooms, hallways, or stationary objects. Equipment to provide such surveillance is available for locations indoors or outdoors, in bright sunlight or in no light.

1.2.3 Safety

CCTV equipment is not always installed for security reasons alone. For safety purposes as well, security personnel can be alerted to unsafe practices or accidents that require immediate attention. An attentive guard can use CCTV cameras distributed throughout a facility, in stairwells and loading docks and around machinery, to observe and document immediately any safety violations or incidents.

1.2.4 The Role of the Guard

Although, historically, guards have been used primarily for plant protection, today they are also used for asset protection. Management is now more aware that guards are only

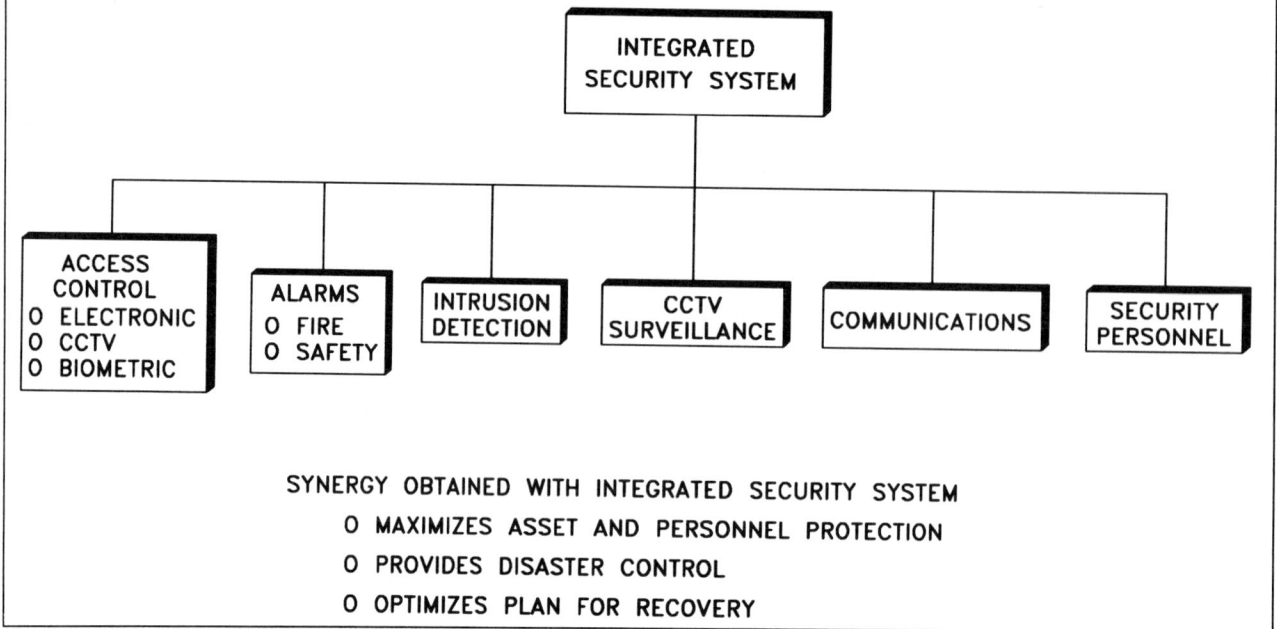

FIGURE 1-2 Integrated security system

one element of an organization's complete security plan. As such, the guard force's cost and its ability to protect are analyzed in relation to the costs and roles of other security plan functions. In this respect, CCTV has much to contribute: increased security for relatively low capital investment, and low operating cost as compared to a guard. Using CCTV, guards can increase the security coverage or protection of a facility. Alternatively, installing new CCTV equipment enables guards to monitor remote sites, allowing guard count and security costs to be reduced significantly.

1.2.5 Employee Training and Education

CCTV is a powerful training tool. It is used widely in education because it can demonstrate lessons and examples vividly and conveniently to the trainee. Example procedures of all types can be shown conveniently in a short time period, and with instructions given during the presentation. Videotaped real-life situations (not simulations or performances) can demonstrate the consequences of misapplied procedures—and the benefits of proper planning and execution by trained and knowledgeable personnel.

Every organization can supplement live training with either professional training videos or actual scenes from their own video system, demonstrating good and poor practices as well as proper guard reaction in real cases of intrusion, unacceptable employee behavior, and so on. Such internal video systems can also be used in training exercises: trainees may take part in videotaped simulations, which are later critiqued by their supervisor. Trainees can

then observe their own actions to find ways to improve and become more effective. Finally, such internal video systems are very important tools during rehearsals or tests of an emergency or disaster plan. After the run-through, all team members can monitor their own reactions, and managers or other professionals can critique them.

1.3 SYNERGY THROUGH INTEGRATION

CCTV equipment is most effective when integrated with other security hardware and procedures to form a coherent security system. Such an integrated security system is more than a combination of sensing security equipment. The hardware used in synergy with CCTV is electronic access control, fire and safety alarms, intrusion detection alarms, communications, and security personnel (Figure 1-2).

Functionally, the integrated security system is designed as a coordinated combination of equipment, personnel, and procedures that (a) uses each component in a way that enhances the use of every other component and (b) optimally achieves the system's stated objective.

In designing a security system, one must consider each element's potential contribution to loss prevention, asset protection, or personnel safety. For example, if an intrusion occurs, where and when should it be detected, what should be the response, and how should it be reported and recorded? If the intruder has violated a barrier or fence, the intrusion detection system should be able to determine that a person—not an animal, bird, insect, leaf, or other object—passed through the barrier. CCTV provides the most positive means for establishing this. Next, this information must be communicated by some means to a security personnel reaction force, with enough detail to permit a guard to respond directly to the intrusion location.

Another example: if material is being removed by an unauthorized person in an interior location, a CCTV surveillance system activated by a video motion detector alarm should alert a guard and transmit the video information to security personnel for appropriate action. In both cases a guard force would be dispatched or some other security measure would be taken to respond to the act, and the event would be recorded on a VCR and/or printed on hard copy for efficient response, documentation, and prosecution. In these examples and many others, the combination of sensors, intelligence communication devices, a guard force, and documentation equipment provide a synergy that maximizes the security function. The integration of CCTV, access control, alarms, intrusion detection, and security guards increases a facility's overall security and maximizes asset protection and employee safety.

Because a complete CCTV system may be assembled from components manufactured by different companies, all equipment must be compatible. The CCTV equipment should be specified by one consulting or architecture/engineering firm, and the system and service should be purchased, installed, and maintained through a single dealer/installer or general contractor. If a major supplier provides a turnkey system, including all equipment, training, and maintenance, the responsibility of system operation lies with one vendor, which is easier to control. Buying from one source also permits management to go back to one installer or general contractor if there are any problems, instead of having to point fingers or negotiate for service among several vendors.

Choosing a single supplier obviously requires thorough analysis to determine that the supplier (1) will provide a good system, (2) will be available for maintenance when required, and (3) will still be in business in 5 or 10 years. Unlike the situation in the 1970s and mid-1980s, today there are multiple sources for pan/tilt mechanisms, lenses, time-lapse recorders, housings, and other equipment required for a sophisticated CCTV system. Though such variety gives the end user a choice for each component, it nevertheless can result in incompatible equipment. The system designer and installer who integrate the system must be aware of the differences and interface the equipment properly.

If the security plan calls for a simple system with potential for later expansion, the equipment should be modular and ready to accept new technology as it becomes available. Many larger manufacturers of security equipment anticipate this integration and expansion requirement and design their products accordingly.

Service is a key ingredient in successful system operation. If one component fails, repair or replacement must be done quickly, so that the system is not shut down. Near-continuous operation is accomplished by the direct replacement method: immediate maintenance by an in-house service organization or quick-response service calls from the installer/contractor. Service considerations should be addressed during the planning and initial design stages, as they affect choice of manufacturer and service provider. Most vendors use the replacement technique to maintain and service equipment. If part of the system fails, the vendor replaces the defective equipment and sends it to the factory for repair. This service policy decreases security system downtime.

The key to a successful security plan is to choose the right equipment and service company, one that is customer oriented and knowledgeable about reliable, technologically superior products that satisfy its customers' needs.

1.4 CCTV'S ROLE AND ITS APPLICATIONS

In its broadest sense, the purpose of CCTV in any security plan is to provide remote eyes for a security operator: to create live-action displays from a distance. The CCTV system should have recording means—either a VCR or other storage media—to maintain permanent records for training or evidence. Following are some applications for which CCTV provides an effective solution:

1. For security purposes, an overt visual observation of a scene or activity is required from a remote location.
2. An area to be observed contains hazardous material or some action that may kill or injure personnel. Such areas may have toxic chemicals, radioactive material, substances with high potential for fire or explosion, or items that may emit X-ray radiation or other nuclear radiation.
3. Visual observation of a scene must be covert. It is much easier to hide a small camera and lens in a concealable location than to station a person in the area.
4. There is little activity to watch in an area, as in an intrusion detection location or a storage room, but significant events must be recorded when they occur. Integration of CCTV with alarm sensors and a time-lapse/real-time VCR would provide an extremely powerful solution.
5. Many locations must be observed simultaneously by one person from a central security position. For example, tracing a person or vehicle from an entrance into a facility to a final destination, where the person or vehicle will be interdicted by a security force. Often a guard or security officer must only periodically review a scene for activity. The use of CCTV eliminates the need for a guard to make rounds to remote locations, which would have been wasteful of the guard's time and unlikely to detect a trespasser.

6. When a crime has been committed, it is important to have a hard-copy printout of the activity and event, which requires a television or photographic system. The proliferation of high-quality printed images from VCR equipment has clearly made the case for using CCTV for creating permanent records.

1.4.1 Problems Solved by CCTV

The most effective way to determine that a theft has occurred, when, where, and by whom, is to use CCTV for detection and recording. The particular event can be identified, stored, and later reproduced for display or hard copy. Personnel can be identified on monochrome or color CCTV monitors. Most security installations to date use monochrome CCTV cameras, which provide sufficient information to document the activity and event or identify personnel or articles. Many newer installations use color CCTV, which permits easier identification of personnel or objects.

If there is an emergency or disaster and security personnel must see if personnel are in a particular area, CCTV can provide an instantaneous assessment of personnel location and availability.

In many cases during normal operations, CCTV can help ensure the safety of personnel in a facility, determine that personnel have not entered the facility, or confirm that personnel have exited the facility. Such functions are used, for example where dangerous jobs are performed or hazardous material is handled.

The synergistic combination of audio and CCTV information from a remote site provides an effective source for security. Several camera manufacturers and installers combine video and audio (one way or duplex) using an external microphone or one installed directly into the camera. The video and audio signals are transmitted over the same coaxial, shielded two-wire, or fiber-optic cable, to the security monitoring location, where they are watched live and/or recorded on a VCR. When there is activity in the camera area, the video and audio signals are switched onto the monitor, the guard sees and hears the scene, and initiates a response.

1.4.2 Choosing Overt or Covert CCTV

Most CCTV installations use both overt and covert CCTV cameras, with more cameras overt than covert. Overt installations are designed to deter crime and provide general surveillance of remote areas, such as parking lots, perimeter fence lines, warehouses, entrance lobbies, hallways, or production areas. When CCTV cameras and lenses are exposed, all managers, employees, and visitors realize that the premises are under constant television surveillance. When the need arises, covert installations are used to detect and observe clandestine activity. Although overt video equipment is often large and not meant to be concealed, covert equipment is usually small and designed to be hidden in objects in the environment or behind a ceiling or wall. Overt CCTV is often installed permanently, whereas covert CCTV is often designed to be installed quickly, left in place for a few hours, days, or weeks, and then removed. Since minimizing installation time is desirable when installing covert CCTV, video signal transmission often is wireless.

1.4.3 Security Surveillance Applications

CCTV applications fall broadly into two types, indoor and outdoor. This division sets a natural boundary between equipment types: those suitable for controlled indoor environments and those suitable for harsher outdoor environments. The two primary parameters are environmental factors and lighting factors. The indoor system requires artificial lighting, which may or may not be augmented by daylight. The indoor system is subject to only mild indoor temperature and humidity variations, dirt, dust, and smoke. The outdoor system must withstand extreme temperatures, precipitation (fog, rain, snow), wind, dirt, dust, sand, and smoke.

1.4.4 Safety Applications

In public, government, industrial, and other facilities, a safety, security, and personnel protection plan must guard personnel from harm caused by accident, human error, sabotage, or terrorism. Security forces are expected to know the conditions at all locations in the facility through the use of CCTV.

In a hospital room or hallway, the television cameras may serve a dual function: monitoring patients while also determining the status and location of employees, visitors, and others. A guard can watch entrance and exit doors, hallways, operating rooms, drug dispensaries, and other vital areas.

Safety personnnel can use CCTV for evacuation and to determine if all personnel have left the area and are safe. Security personnel can use CCTV for remote traffic monitoring and control and to ascertain high-traffic locations and how best to control them. CCTV plays a critical role in public safety, as a tool for monitoring vehicular traffic on highways and city streets, in truck and bus depots, and at public rail and subway facilities and airports.

1.4.5 CCTV Access Control

As security systems become more complex and necessary, CCTV access control and electronic access control equipments are being combined to work synergistically with each other. For medium- to low-level access control security requirements, electronic card reading systems are adequate after a person has first been identified at some exterior perimeter location. For higher security, biometric descriptors of a person and/or CCTV identification are necessary.

CCTV surveillance is often used with electronic or CCTV access control equipment. CCTV access control uses television to identify remotely a person requesting access, on foot or in a vehicle. A guard can compare the live image and the photo ID carried by the person and then either allow or deny entry. For the highest level of access control security, the guard uses a system to compare the live image to an image of the person retrieved from a video image data bank. The two images are displayed side by side on a split-screen monitor, perhaps along with other pertinent information. At present, different companies manufacture CCTV and electronic access control equipment; there is no single supplier of all the equipment.

The CCTV access control system can be combined with an electronic access control system to increase security and provide a means to track all attempted entries.

This book provides a brief description of CCTV access control equipment and applications. A complete description of specific techniques, equipment, and applications integrating the CCTV access control function with electronic access control is covered in a companion book, *CCTV Access Control*.

1.5 THE BOTTOM LINE

The synergy of a CCTV security system implies the following functional scenario:

1. Some alarm sensor will detect immediately an unauthorized intrusion or entry or attempt to remove equipment.
2. A CCTV camera located somewhere in the alarm area will be fixed on the location or may be pointed manually or automatically (from the guard site) to view the alarm area.
3. The information from the alarm sensor and CCTV camera is transmitted immediately to the security console, monitored by personnel, and/or recorded for permanent documentation.
4. The security operator receiving the alarm information has a plan to dispatch personnel to the location or to take some other appropriate action.
5. After dispatching a security person to the alarm area, the guard resumes normal security duties to view any future event.
6. If after a reasonable amount of time the person dispatched does not neutralize the intrusion or other event, the security guard resumes monitoring that situation to bring it to a successful conclusion.

Use of CCTV plays a crucial role in the overall system plan. During an intrusion or theft, the CCTV system (when signaled by the intrusion alarm) provides information to the guard, who must make some identification of the perpetrator, assess the problem, and respond appropriately. An installation containing suitable and sufficient alarm sensors and CCTV cameras permits the guard to follow the progress of the event and assist the response team in countering the attack.

The use of CCTV to track an intruder is most effective. With an intrusion alarm and visual CCTV information, all the elements are in place for a timely, reliable transfer of information to the security officers. For maximum effectiveness, all parts of the security system must work properly; for total success, each part must rely on the others. If an intrusion alarm fails, the command post cannot see the CCTV image at the right location and the right time. If the CCTV fails, the guard cannot identify the perpetrator even though he may know that an intrusion has occurred. If a security officer is not alert or misinterprets the alarm and CCTV input, the data from either or both are not processed and acted upon and the system fails.

In an emergency, such as a fire, flood, malfunctioning machinery, burst utility pipeline, and so on, the operation of CCTV, safety sensors, and human response at the console are all required. CCTV is an inexpensive investment for preventing accidents and minimizing damage when an accident occurs. In the case of a fire, CCTV cameras act as real-time eyes at the emergency location, permitting security and safety personnel to send the appropriate reaction force with adequate equipment to provide optimum response. Since the reaction time to a fire or other disaster is critical, having various cameras on the location before personnel arrive is very important. In the case of a fire, while a sprinkler may activate or a fire sensor may produce an alarm, a CCTV camera can quickly ascertain whether the event is a false alarm, a minor alarm, or a major event. The automatic sprinkler and fire alarm system might alert the guard to the event, but the CCTV "eyes" viewing the actual scene prior to the emergency team's dispatch often save lives and reduce asset losses.

In a security violation, if a sensor detects an intrusion, the guard monitoring the CCTV camera can determine if the intrusion requires the dispatch of personnel or some other response. In the event of a major, well-planned attack on a facility by a terrorist organization or other intruder, a diversionary tactic such as a false alarm can quickly be discovered through the use of CCTV, thereby preventing an inappropriate response.

To justify expenditures on security and safety equipment, an organization must expect a positive return on investment; that is, the value of assets protected must be greater than the amount spent on security, and the security system must adequately protect personnel and visitors. An effective security system reduces theft, saves money, and saves lives.

Chapter 2
CCTV Technology Overview

CONTENTS

2.1 OVERVIEW

Although today's security system hardware is based on new technology, it has followed the same basic concepts over the past 20 to 30 years. In the 1950s, American and foreign governments agreed upon standards for video signal parameters. These characteristics and specifications have by and large remained the same, as detailed in Chapter 5. The current chapter describes security CCTV system components and the theory of operation.

The primary function of any CCTV system in a security or safety application is to provide remote eyes for the

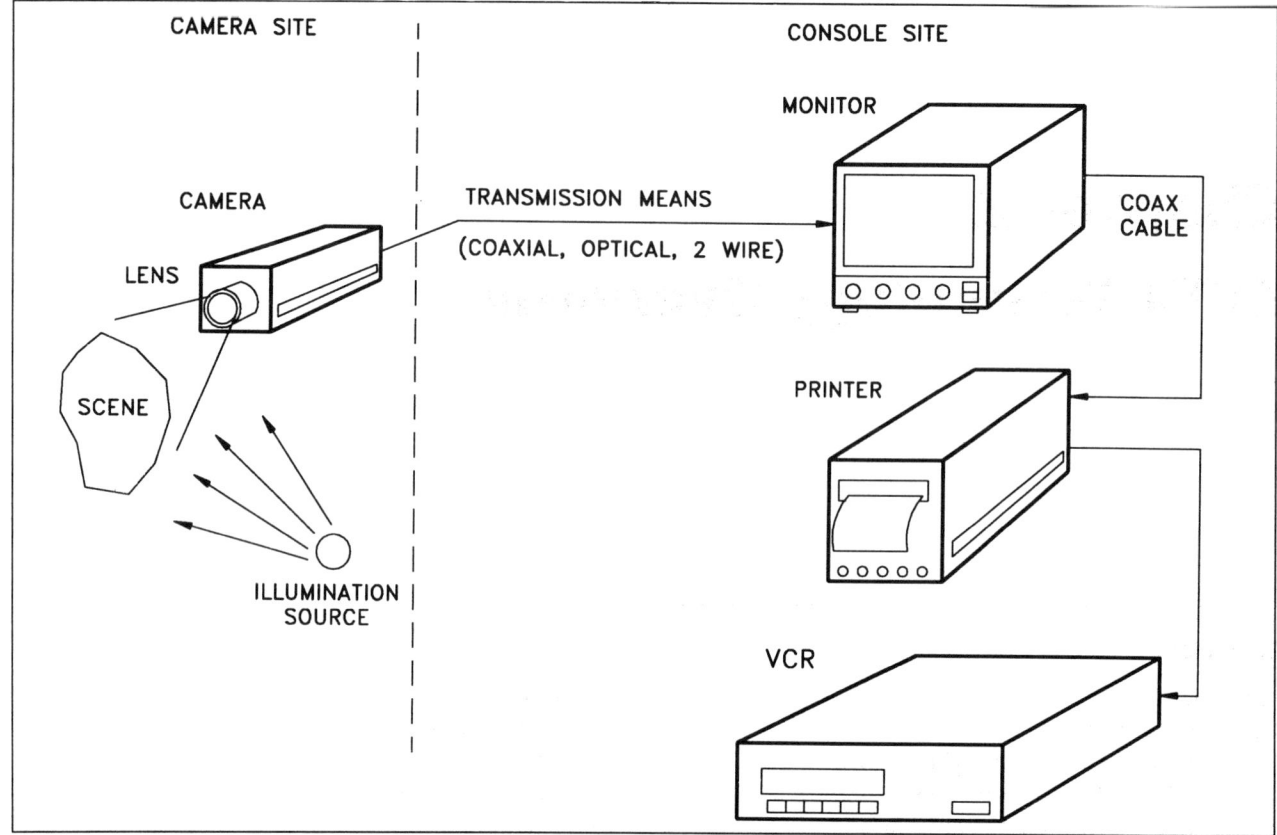

FIGURE 2-1 Single-camera CCTV system

security force located at a central control console. The CCTV system is a complete entity, from the illumination source, the scene to be viewed, the camera lens, the camera, and the means of transmission to the remote monitoring and recording equipment. Other equipments often necessary to complete the system include video switchers, motion detectors, housings, scene combiners and splitters, and character generators.

This chapter describes the technology used to (1) capture the visual image, (2) convert it to a video signal, (3) transmit it to and receive it at a remote location, (4) display it on a television monitor, and (5) record and print it for permanent record.

Figure 2-1 shows the simplest CCTV application, requiring only one television camera and a monitor. The printer and VCR are optional.

The camera may be used to monitor employees, visitors, people entering or leaving a front reception area of a building, or other single scenes from a remote location. The camera could be located in the lobby ceiling and pointed at the reception area, the front door, or an internal access door. The monitor might be located hundreds or thousands of feet away, in another building or another country, with the security personnel viewing that same

lobby, front door, or reception area. The television camera–monitor system effectively extends the eyes, reaching from the location of the observer to the observed location. The basic one-camera system shown in Figure 2-1 includes the following hardware components.

Lens. The lens collects the light from the scene and forms an image of the scene on the light-sensitive camera sensor.

Camera. The camera sensor converts the visible scene formed by the lens into an electrical signal suitable for transmission to the remote monitor, recorder, or printer.

Transmission link. The transmission media carries the electrical video signal from the camera to the remote monitor. Media choices include (a) electrical coaxial or two-wire cable; (b) fiber-optic cable; or (c) radio frequency (RF), microwave, or infrared (IR) wireless transmitter-receiver system.

Monitor. The video monitor displays the camera picture by converting the electrical video signal back into a visible picture on the monitor screen.

These first four items are required to make a simple CCTV system work. Figure 2-2 shows a block diagram of a

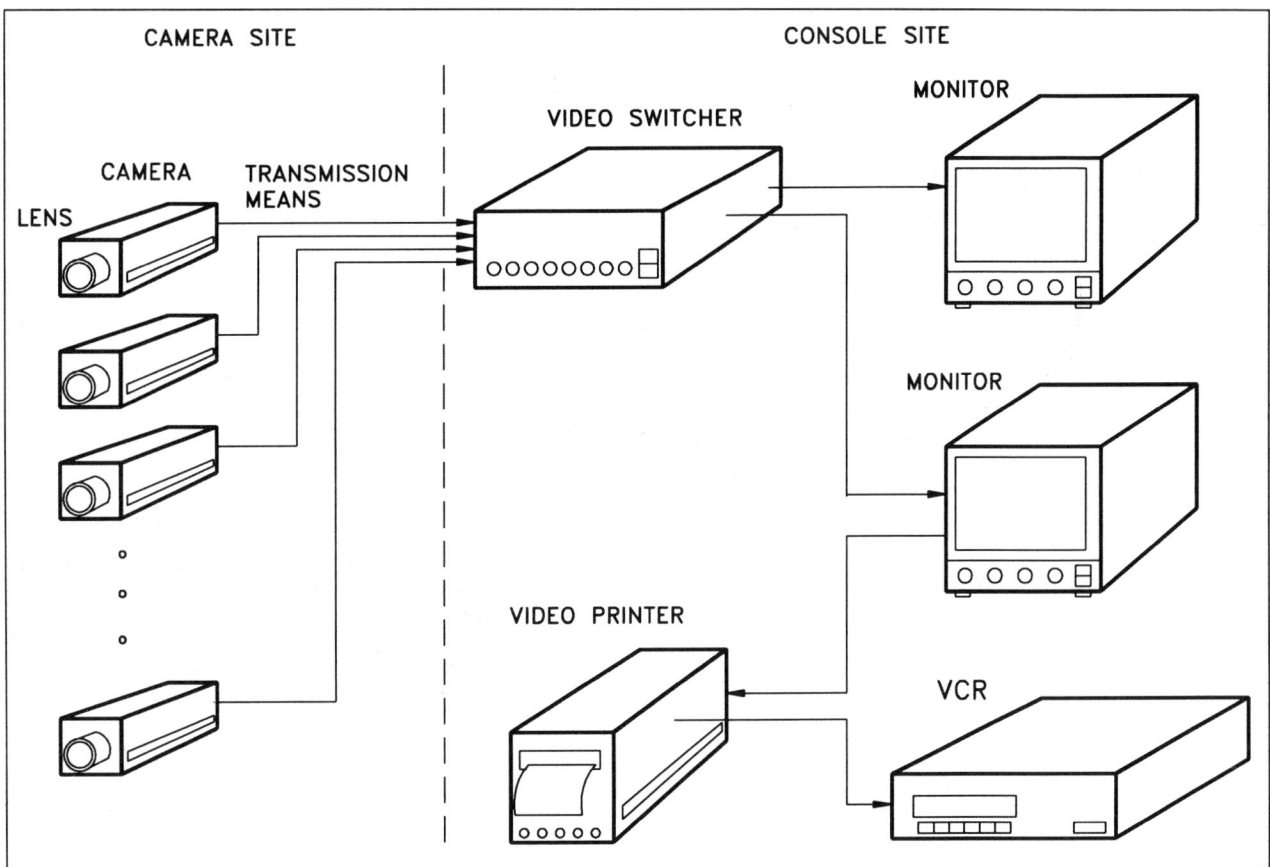

FIGURE 2-2 Comprehensive CCTV security system

more comprehensive CCTV security system using these components plus four additional hardware and options (listed below) to expand the capability of the single-camera system to multiple cameras, monitors, recorders, and so on, into a more complex CCTV security system.

Camera switcher. When a CCTV security system has multiple cameras, an electronic switcher is used to select different cameras automatically or manually, displaying their pictures on a single or multiple monitors, a recorder, or a printer.

Recorder. The picture from the camera is permanently recorded by a VCR onto a magnetic tape cassette.

Hard-copy printer. The video printer produces a hard-copy printout of any live or recorded video scene, using thermal or other sensitized paper.

Ancillary equipment. Supporting equipment required for outdoor or more complex systems includes environmental camera housings, camera pan and tilt mechanisms, image combiners and splitters, and scene annotators.

Housing. The many varieties of camera/lens housings fall into two categories: indoor and outdoor. Indoor housings protect the camera and lens from tampering and usually are made from lightweight materials. Outdoor housings protect the camera and lens from the environment, such as precipitation, extremes of heat and cold, dust, and dirt.

Pan/tilt mechanism. When a camera must view a large area, a pan and tilt mount is used to rotate it horizontally (called panning) and tilt it, providing a large angular coverage.

Splitter/combiner. To display more than one camera scene on a single monitor, an optical or electronic image combiner or splitter is used. From 2 to 32 scenes can be displayed on a single monitor.

Annotator. A time and date generator can annotate the video scene with chronological information. A camera identifier puts a number on the monitor screen to identify the camera scene being displayed.

2.2 THE CCTV SYSTEM

2.2.1 The Role of Light and Reflection

Figure 2-3 shows the essentials of the CCTV camera environment: illumination source, camera, lens, and its combined field of view (FOV), that is, the scene the camera-lens combination sees.

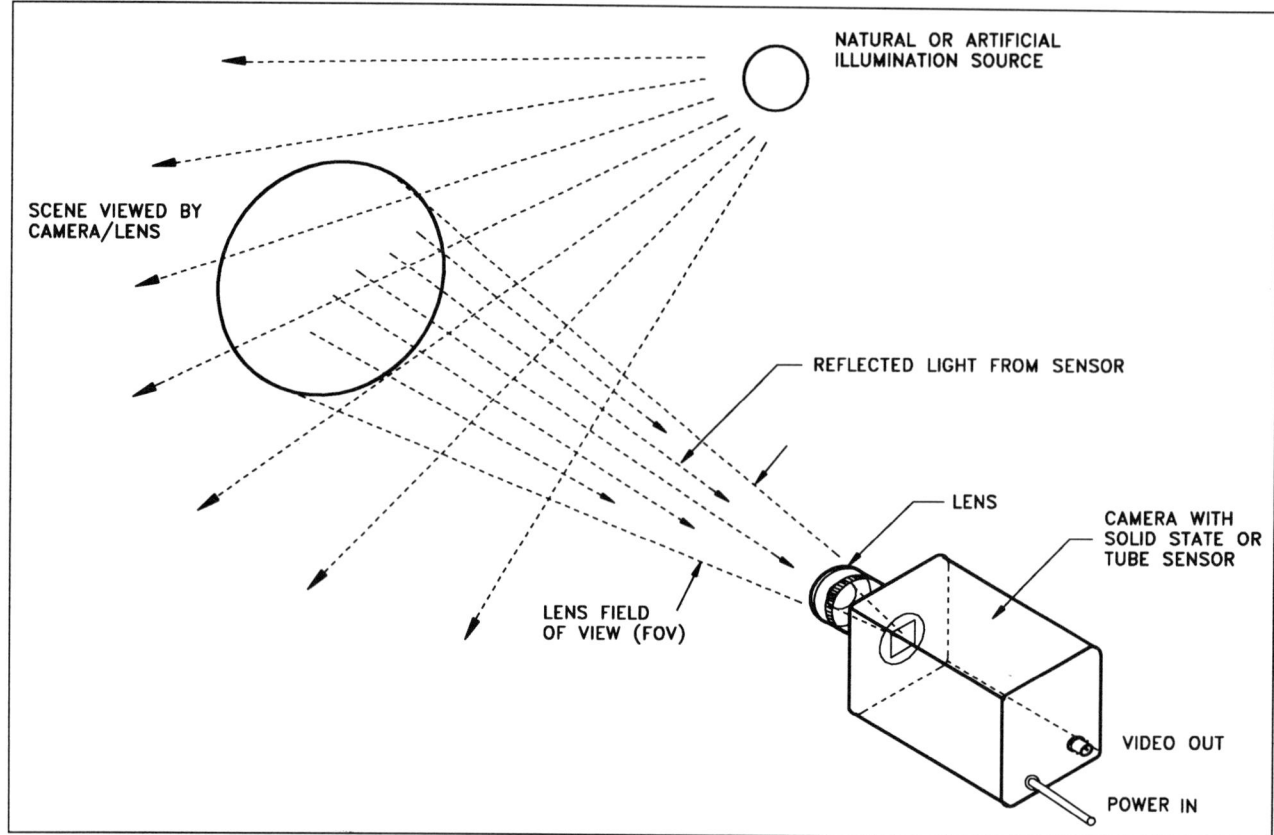

FIGURE 2-3 CCTV camera, scene, and source illumination

A scene or target area to be viewed is illuminated by natural or artificial light sources. Natural sources include the sun, the moon (reflected sunlight), and starlight. Artificial sources include incandescent, sodium, metal arc, fluorescent, infrared, and other man-made lights. Chapter 3 describes all of these light sources.

The camera lens receives the light reflected from the scene. Depending on the scene to be viewed, the amount of light reflected from objects in the scene can vary from 5 or 10% to 80 or 90% of the light incident on the scene. Typical values of reflected light for normal scenes such as foliage, automobiles, personnel, and streets fall in the range from about 25 to 65%. Snow-covered scenes may reach 90%.

The amount of light received by the lens is a function of the brightness of the source, the reflectivity of the scene, and the transmission characteristics of the intervening atmosphere. In outdoor applications, there is usually a considerable optical path from the source to the scene, and back to the camera; hence the transmission through the atmosphere must be considered. When atmospheric conditions are clear, there is generally little or no attenuation of the reflected light from the scene. However, when there is precipitation (rain, snow, or sleet, or when fog intervenes) or in dusty or sand-blown environments, this attenuation must be considered. Likewise in hot climates thermal effects (heat waves) and humidity can cause severe attenuation and/or distortions of the scene. Complete attenuation of the reflected light from the scene

(zero visibility) can occur, in which case no scene image is formed.

Since most tube and solid-state cameras operate in the visible and near-infrared wavelength energy region, the general rule of thumb with respect to visibility is that if the human eye cannot see the scene, neither will the camera. Under this situation, no amount of increased lighting will help; however, if the visible light can be filtered out of the scene and only the IR portion used, scene visibility might be increased somewhat.

Figure 2-4 illustrates the relationship between the viewed scene and the scene image on the camera sensor.

The lens located on the camera forms an image of the scene and focuses it onto the sensor. All television systems used in security have a 4-by-3 aspect ratio (4 units wide by 3 units high) for both the image sensor and the field of view. The width parameter is h, the vertical is v.

2.2.2 The Lens Function

The function of the lens is to collect light reflected from the scene and focus it into an image onto the CCTV camera sensor. A fraction of the light reaching the scene from the natural or artificial illumination source is reflected toward the camera and intercepted and collected by the camera

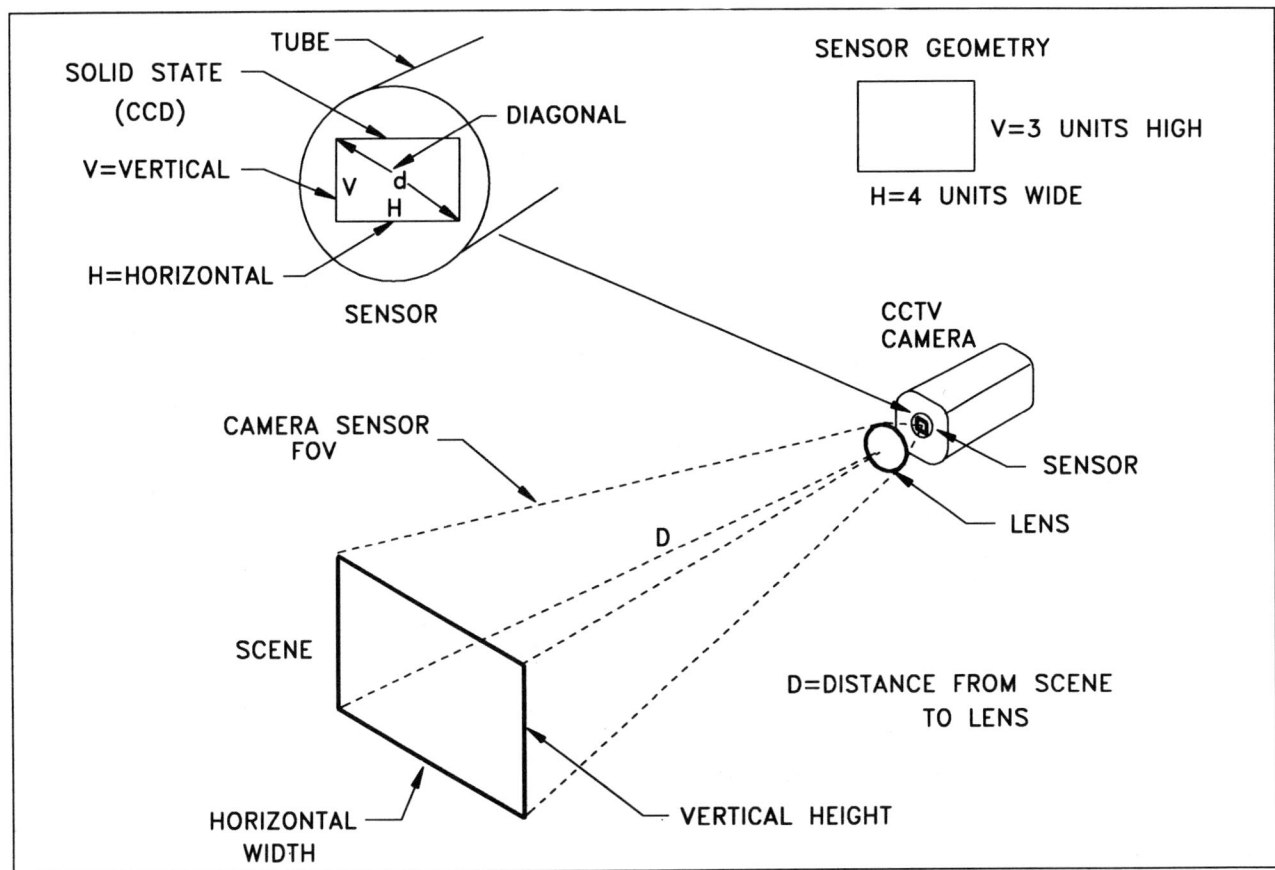

FIGURE 2-4 CCTV scene and sensor geometry

lens. The larger the lens diameter, the more light will be gathered, the brighter the image on the sensor, and the better the final image on the monitor will be. This is why larger-aperture (diameter) lenses, having a higher optical throughput, are better (and more expensive) than smaller-diameter lenses, which collect less light. Under good lighting conditions—bright indoor lighting, outdoors under sunlight—the large-aperture lenses are not required; there is sufficient light to form a bright image on the sensor by using small-diameter lenses.

The camera lens is analogous to the lens of the human eye (Figure 2-5) and collects the reflected radiation from the scene much like the lens of your eye or a film camera.

Most CCTV applications use a fixed-focal-length (FFL) lens, which, like the eye's lens, covers a constant angular field of view (FOV). That is, the FFL images a scene with constant magnification. A large variety of CCTV camera lenses are available with different focal lengths (FLs), which will provide different FOVs. Wide-angle, medium-angle, or narrow-angle telephoto lenses produce different magnifications and FOVs.

Most CCTV lenses have an iris diaphragm (as does the human eye) to adjust the open area of the lens, and hence change the amount of light passing through it. Manual or automatic irises are used, depending on the application. In an automatic-iris CCTV lens, as in a human eye lens, the iris

closes automatically when the illumination is too high and opens automatically when it is too low, thereby maintaining the optimum illumination on the sensor at all times. Figure 2-6 shows representative samples of CCTV lenses, including FFL, zoom, pinhole, and large catadioptric (which combine both mirror and glass optical elements). Chapters 4 and 14 describe CCTV lens characteristics and applications in detail.

2.2.3 The Camera Function

The CCTV lens focuses the scene onto the television image sensor, which acts like the retina of the eye or the film in a photo camera. The CCTV camera sensor and electronics convert the visible image into an equivalent electrical signal suitable for transmission to a remote monitor. Figure 2-7 is a block diagram of a typical CCTV camera.

The camera converts the optical image produced by the lens into a time-varying electric signal, modulated in accordance with the light-intensity distribution throughout the scene. Other camera electronic circuits produce synchronizing pulses so that the time-varying video signal can later be displayed on a monitor, recorded on a VCR, or printed out as hard copy on a video printer. While cameras may differ in size and shape depending on specific type and

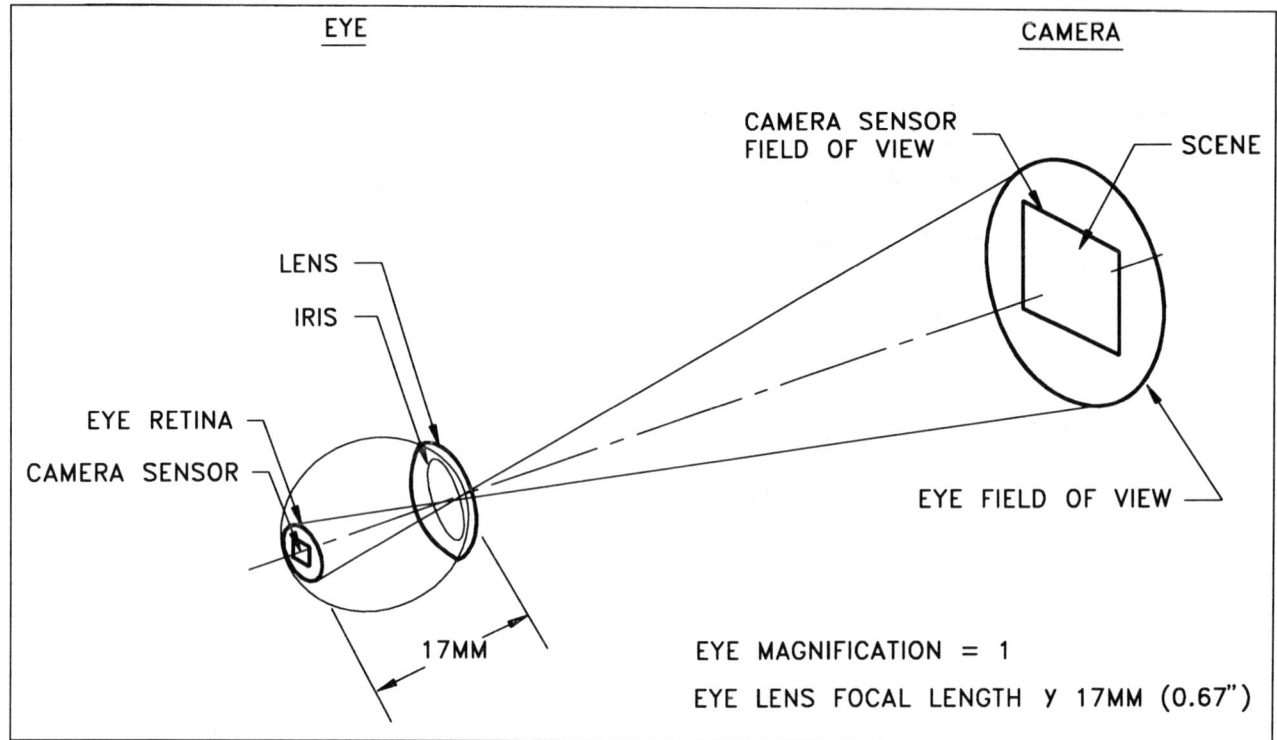

FIGURE 2-5 Comparing the human eye to the CCTV camera lens

capability, the scanning process used by most cameras is essentially the same. Almost all cameras must scan the scene, point by point, as a function of time. (An exception is the image intensifier, as discussed in Chapter 16.) Solid-state charged coupled devices (CCD) are used in standard applications. In scenes with very low illumination, low light level (LLL) devices such as intensified CCD (ICCD), silicon intensified target (SIT), or intensified SIT (ISIT) CCTV cameras are used. These cameras are complex and expensive.

Most CCTV security cameras in use are monochrome, that is, black and white. Until solid-state color CCTV cameras became available, color did not find much use in security applications. The nonbroadcast, tube-type color cameras available for security applications lacked long-term stability, sensitivity, and high resolution. With the development of solid-state sensor technology and widespread use of consumer color CCD cameras, VCRs, and camcorders, color cameras came into use regularly in security systems. Figure 2-8 shows representative CCTV cameras, including monochrome and color solid-state CCD and metal-oxide semiconductor (MOS) miniature and remote head types. Chapters 5 and 15 describe standard and LLL security CCTV cameras in detail.

2.2.4 The Transmission Function

Once the camera has generated an electrical video signal representing the scene image, the signal is transmitted to a remote security monitoring site via some transmission means (by coaxial, fiber-optic, or two-wire cable, or by wireless technique). The choice of transmission media depends on factors such as distance, environment, and facility layout.

If the distance between the camera and the monitor is short (10–500 feet), coaxial cable is used. For longer distances (1000 to several thousand feet) or where there are electrical disturbances, fiber-optic cable is preferred. In applications where the camera and monitor are separated by roadways or where there is no right-of-way, wireless systems are used. Chapter 6 describes all of these video transmission media.

2.2.5 The Monitor Function

At the monitor site the video signal is electronically converted back into a visual image on the monitor face via electronic circuitry similar but inverse to that in the camera. The final scene is produced by a scanning electron beam in the cathode-ray tube (CRT) in the video monitor. This beam activates the phosphor on the cathode-ray tube, thereby producing a representation of the original scene (as seen by the camera) on the faceplate of the monitor. Chapter 7 describes monitor technology and hardware. A permanent record of the monitor video scene can be made with a VCR or hard-disk magnetic storage media, and a permanent hard copy can be printed on a video printer.

FIGURE 2-6 Representative CCTV lenses: (a) motorized zoom, (b) catadioptric long FFL, (c) narrow FOV (telephoto) FFL, (d) wide FOV FFL, (e) straight and (f) right-angle pinhole lenses

2.3 SCENE ILLUMINATION

As previously stated, illumination of the scene is produced via natural or artificial illumination. Monochrome cameras can operate with any type of light source, but color cameras need light that contains all the colors in the visible spectrum. To produce a satisfactory color image, color cameras also need light with a reasonable balance of all the colors.

2.3.1 Natural Light

During the day the amount of illumination and spectral distribution of light (color) reaching a scene depends on the time of day and atmospheric conditions. The color spectrum of the light reaching the scene is important if color CCTV is being used. Direct sunlight produces the highest-contrast scene, allowing maximum identification of objects. On a cloudy or overcast day, less light is received by the objects in the scene, resulting in less contrast. To produce an optimum camera picture under the wide variation in light level (such as occurs when the sun is obscured by clouds), an automatic-iris camera system is required.

Figure 2-9 shows the light levels for outdoor illumination under bright sun, partial clouds, overcast clouds, rain, snow, dust, sand, and fog.

Typically, scene illumination measured in foot-candles (fc) can vary over a range of 10,000 to 1 (or more), which exceeds the operating range of the camera and image tube for producing a good-quality video image. After the sun has gone below the horizon and the moon is overhead, reflected sunlight from the moon illuminates the scene and may be detected by a sensitive camera. Detection of information in a scene under this condition requires an LLL camera, since there is very little light reflected into the camera lens from the scene. As an extreme, when the moon is not overhead, the only light received is ambient light from (1) local man-made lighting sources; (2) night-glow caused by distant ground lighting reflecting off particulate (pollution), clouds, and aerosols in the lower atmosphere; and (3) direct light caused by starlight. This is the most severe lighting condition and requires the most sensitive ICCD, SIT, or ISIT cameras. Table 2-1 summarizes the light levels occurring under daylight and these LLL conditions. The equivalent metric measure of light level (lux) compared with the English (fc) is given.

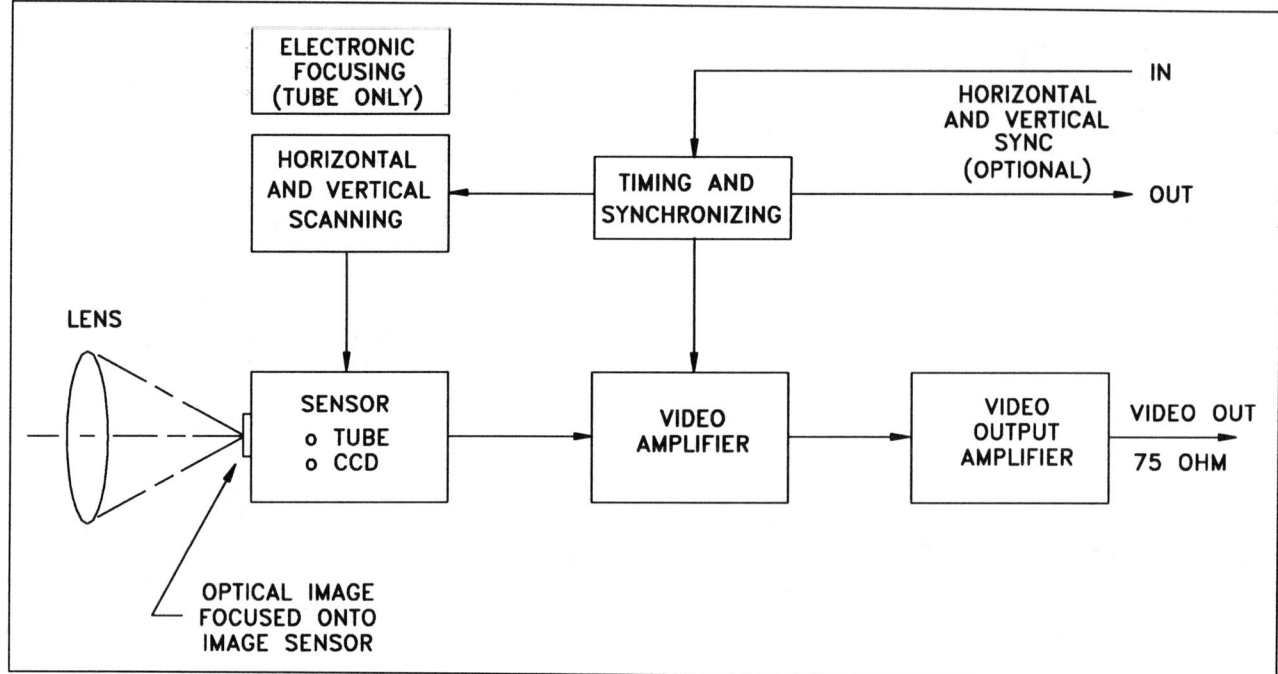

FIGURE 2-7 CCTV camera block diagram

2.3.2 Artificial Light

Artificial illumination is often used to augment outdoor lighting to obtain adequate CCTV security information at night. These light sources consist of tungsten, tungsten-halogen, metal-arc, mercury, sodium, xenon, and IR lamps. Figure 2-10 illustrates several examples of these lamps.

The type of lighting chosen depends on architectural requirements and the specific function. Often a particular lighting design is used for safety reasons, so that personnel at the scene can see better, as well as for CCTV security. By far the most efficient visual outdoor light types are the low- and high-pressure sodium-vapor lamps, to which the human eye is most sensitive. However, metal-arc and mercury lamps provide sufficient illumination to produce excellent-quality video images. Long-arc and xenon lamps are often used in outdoor sports arenas and large parking areas.

Artificial indoor illumination is similar to outdoor, with fluorescent lighting used in addition to high-pressure sodium, metal-arc and mercury lamps. Since indoor lighting has a relatively constant light level, automatic-iris lenses are often unnecessary. However, if the CCTV camera views a scene near an outside window, where additional light comes in during the day, or if the indoor lighting changes between daytime and nighttime operation, then an automatic-iris lens or electronically shuttered camera is required. The illumination level from most indoor lighting is significantly lower (by several orders of magnitude) than that of sunlight. Chapter 3 describes outdoor natural and artificial lighting and indoor man-made lighting systems, as well as the different types of light sources available for CCTV use.

2.4 SCENE CHARACTERISTICS

The quality of the CCTV picture depends on various scene characteristics, including (1) the sharpness and contrast of objects relative to the scene background, (2) whether objects are in a simple, uncluttered background or a complicated scene, and (3) whether objects are stationary or in motion.

Several factors, though not directly parts of the CCTV system, play very important roles in determining the final monitor picture: (1) the scene lighting, (2) the contrast of objects in the scene, and (3) the system resolution. These factors will determine whether the system will be able to detect, determine orientation, recognize, or identify objects and personnel. As will be seen later, the scene illumination—via sunlight, moonlight, or artificial sources—and the actual scene contrast play important roles in the types of lens and camera necessary and the quality of the resultant scene on the monitor.

2.4.1 Target Size

Besides the scene's illumination level and the object's contrast with respect to the scene background, the object's apparent size—that is, its angular FOV as seen by the camera—influences a person's ability to detect it. (For example, try to find a football referee with a striped shirt in a field of zebras.)

The performance of a television system depends on the job's requirements, such as (1) detection of the object or

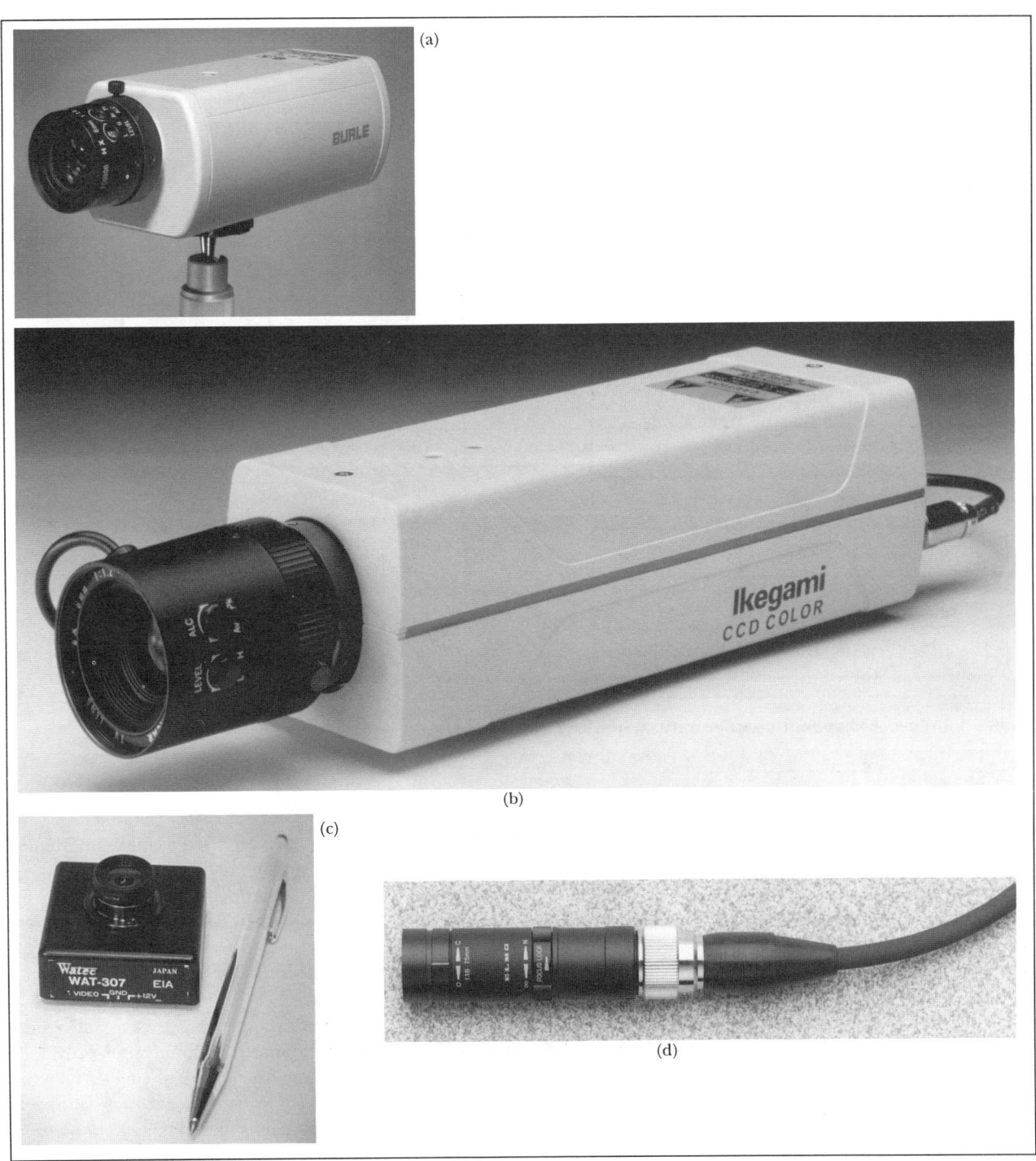

FIGURE 2-8 Representative CCTV cameras: (a) ½-inch format solid-state CCD camera, (b) ⅔-inch format color camera, (c) ⅓-inch format miniature camera, (d) ½-inch format remote head color camera

movement in the scene; (2) determination of the object's orientation; (3) recognition of the type of object in the scene, that is, adult or child, car or truck; or (4) identification of the object (Who is the person? Exactly what kind of truck?). Making these distinctions depends on the system's resolution, contrast, and signal-to-noise ratio (S/N). In a typical scene, the average observer can detect a target about one-tenth of a degree in angle. Relating this information to a standard television picture, which has about 525 horizontal lines and about 350 TV line vertical and 500 TV line horizontal resolution, Figure 2-11 and Table 2-2 summarize the number of lines required to detect, orient, recognize, or identify an object in a television picture. The number of TV lines required will increase for conditions of poor lighting, highly complex backgrounds, reduced contrast, or fast movement of the camera or target.

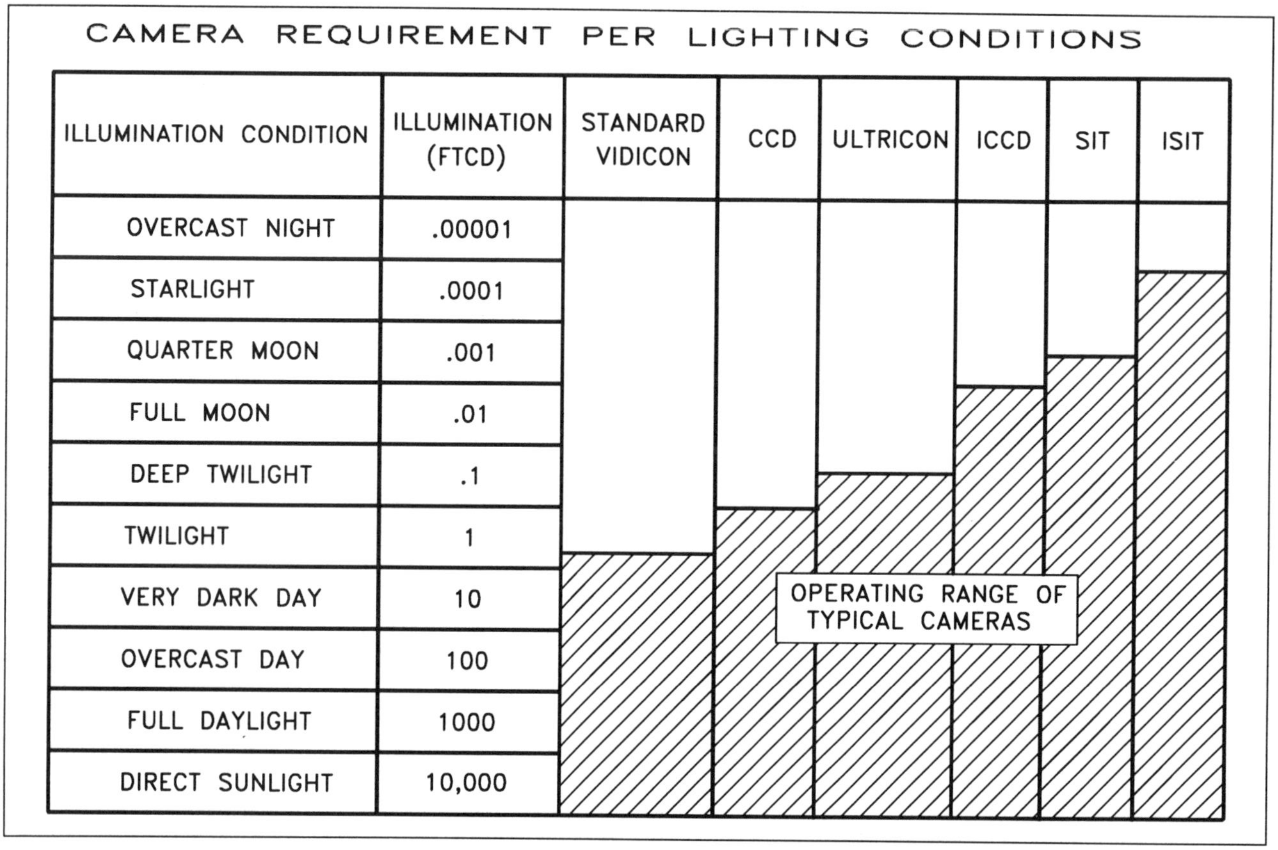

CAMERA REQUIREMENT PER LIGHTING CONDITIONS

ILLUMINATION CONDITION	ILLUMINATION (FTCD)	STANDARD VIDICON	CCD	ULTRICON	ICCD	SIT	ISIT
OVERCAST NIGHT	.00001						
STARLIGHT	.0001						
QUARTER MOON	.001						
FULL MOON	.01						
DEEP TWILIGHT	.1						
TWILIGHT	1						
VERY DARK DAY	10						
OVERCAST DAY	100		OPERATING RANGE OF TYPICAL CAMERAS				
FULL DAYLIGHT	1000						
DIRECT SUNLIGHT	10,000						

FIGURE 2-9 Camera capability under natural lighting conditions

2.4.2 Reflectivity

The reflectivity of different materials varies greatly depending on their composition and surface texture. Table 2-3 gives some examples of materials and objects viewed by television cameras and their respective reflectivities.

Since the camera responds to the amount of light reflected from the scene, one should be aware that most common objects have a large range of reflectivities. The objects with the highest reflectivities produce the brightest images. To detect one object located within the area of another, the objects must differ in either reflectivity, color, or texture. Therefore, if a red box is in front of a green wall, and both have the same reflectivity and texture, the box will not be seen on a monochrome CCTV system. In this case, the total reflectivity in the visible spectrum is the same for the green wall and the red box.

The case of a color scene is more complex. While the reflectivity of the red box and the green wall may be the same as averaged over the entire visible spectrum from blue to red, the color camera can distinguish between green and red.

It is easier to identify a scene characteristic by a difference in color in a color scene than it is to identify it by a difference in gray scale (intensity) in a monochrome scene.

For this reason the target size required to make an identification in a color scene is generally less than it is to make the same identification in a monochrome scene.

2.4.3 Effects of Motion

A moving object in a television scene is easier to detect but more difficult to recognize than a stationary one, provided that the camera can respond to it. Tube-type cameras produce sharp images for stationary scenes but smeared images for moving targets. This is caused by a phenomenon called tube lag. Solid-state sensors (CCD, MOS, and others) do not exhibit smear or lag and can therefore produce sharp images of both stationary and moving scenes. When the target in the scene moves very fast, the inherent camera scan rate (30 frames per second) causes a blurred image of this moving target in both the tube and solid-state sensors. This is analogous to the blurred image in a photograph when the shutter speed is too slow for the action. There is no cure for this as long as the standard NTSC television scan rate (30 frames per second) is used. However, CCTV snapshots can be taken without any blurring, using fast-shuttered CCD cameras.

CONDITION	ILLUMINATION		COMMENTS
	(FTCD)	(LUX)	
DIRECT SUNLIGHT	10,000	107,527	DAYLIGHT RANGE
FULL DAYLIGHT	1,000	10,752.7	
OVERCAST DAY	100	1,075.3	
VERY DARK DAY	10	107.53	
TWILIGHT	1	10.75	
DEEP TWILIGHT	.1	1.08	
FULL MOON	.01	.108	LOW LIGHT LEVEL RANGE
QUARTER MOON	.001	.0108	
STARLIGHT	.0001	.0011	
OVERCAST NIGHT	.00001	.0001	

NOTE: 1 LUX=.093 FTCD

Table 2-1 Light Levels under Daytime and Nighttime Conditions

2.5 LENSES

Many different lens types are used for CCTV security and safety applications. They range from the simplest FFL manual-iris lenses to the more complex variable-focal-length zoom lenses, with the automatic iris being an option for all types.

In addition, pinhole lenses are available for covert applications, split-image lenses for viewing multiple scenes on one camera, right-angle lenses for viewing a scene perpendicular to the camera axis, and fiber-optic lenses for viewing through thick walls.

A lens collects reflected light from the scene and focuses it onto the camera image sensor, analogous to the lens of the human eye focusing a scene onto the retina at the back of the eye (Figure 2-5). As in the human eye, the camera lens inverts the scene image on the image sensor, but the eye and camera electronics compensate to perceive an upright scene. The retina of the human eye differs from any CCTV lens in that it focuses a sharp image only in the central 10% of its total 160-degree FOV. All vision outside the central focused scene is out of focus. This central imaging part of the human eye can be characterized as a medium FL lens: 16 to 25 mm. In principle, Figure 2-5 represents the function of any lens in a television system. Lens variations consist primarily of different physical sizes and optical speeds, different fields of view or variable FL (zoom), and special lenses to produce unique results.

2.5.1 Fixed-Focal-Length Lens

Figure 2-12 illustrates three FFL or fixed FOV lenses with narrow, medium, and wide FOVs and the corresponding FOV obtained when used with a half-inch camera sensor format.

Wide-FOV (short FL) lenses permit viewing a very large scene with low magnification, and therefore provide low resolution and low identification capabilities. Narrow-FOV or telephoto lenses have high magnification, with high resolution and high identification capabilities.

2.5.2 Zoom Lens

The zoom lens is more versatile and complex than the FFL lens. The FL is variable from wide angle to narrow (telephoto) FOV (Figure 2-13). The lens angular FOV depends on the camera sensor size, as shown in the table in Figure 2-13.

Zoom lenses consist of multiple lens groups that are moved within the lens barrel by means of an external zooming ring, thereby changing the lens FL and angular FOV without having to switch lenses.

2.5.3 Covert Pinhole Lens

This special security lens class is used when the lens and CCTV camera must be hidden. The front lens element or aperture is small (from $1/16$ to $5/16$ of an inch in diameter). While this

FIGURE 2-10 Representative artificial light sources: (a) tungsten-halogen, (b) fluorescent (straight and U), (c) high-pressure sodium, (d) parabolic aluminized reflector and standard tungsten bulb, (e) high-intensity-discharge mercury and metal arc, (f) infrared

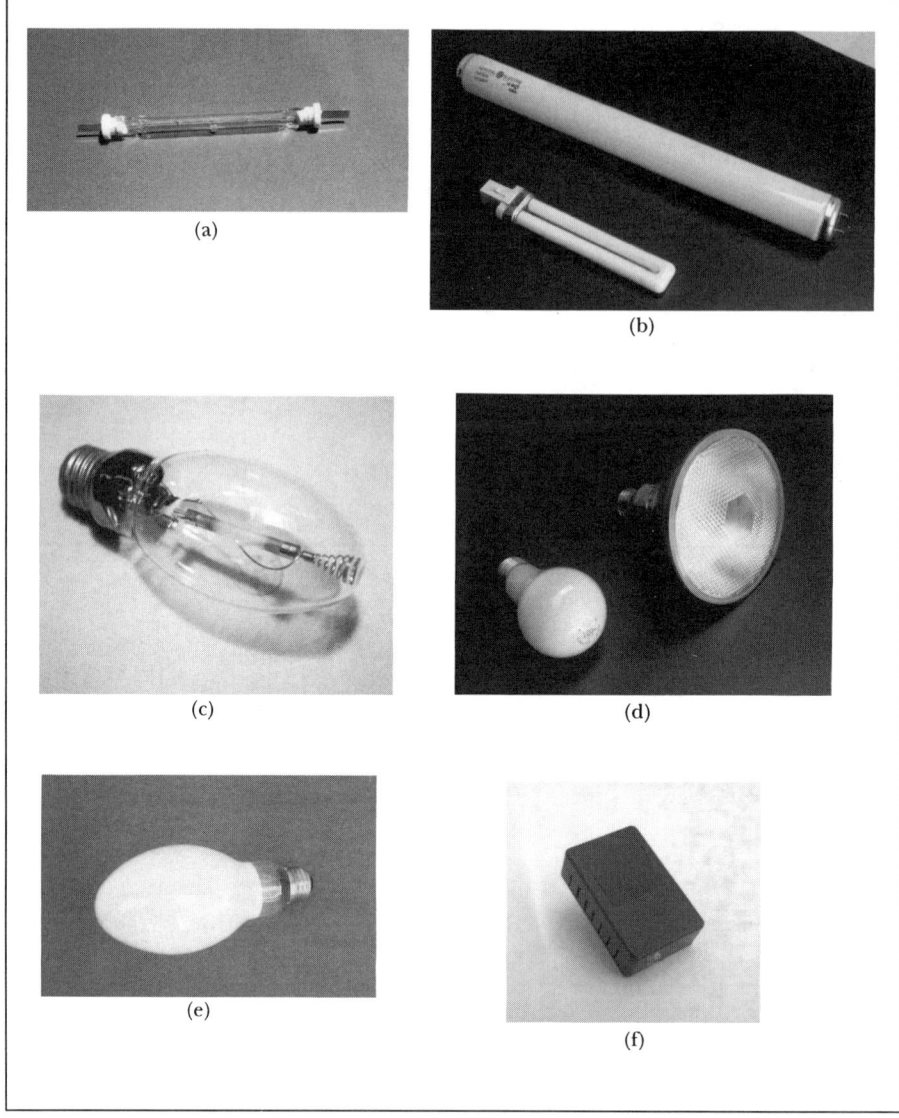

is not the size of a pinhead, it nevertheless has been labeled a pinhole lens. Figure 2-14 shows examples of straight, right-angle, and mini-pinhole lenses.

2.5.4 Special Lenses

Some special lenses useful in security applications include split-image, right-angle, relay, and fiber-optic (Figure 2-15).

The dual-split and tri-split lenses use only one camera to produce multiple scenes. These are useful for viewing the same scene with different magnifications or different scenes with the same or different magnifications. Using only one camera reduces cost and increases reliability.

The right-angle lens permits the camera to view scenes that are perpendicular to the camera's optical axis. There are no restrictions on the focal lengths, so they can be used in wide- or narrow-angle applications.

The flexible and rigid coherent fiber-optic lenses are used to mount a camera several inches to several feet away from the front lens, as might be required to view the opposite side of a wall. The primary function of the fiber-optic bundle is to transfer the image from one location to another. This may be useful (1) for protecting the camera; (2) for locating the lens in one environment (outdoors) and the camera in another (indoors); or (3) if the camera is too large to fit in a place where the lens does.

2.6 CAMERAS

The lens focuses the scene image onto the image sensor area and the camera electronics transform the visible image to an electrical signal, point by point. The camera video signal (containing all picture information) is made up of frequencies from 30 cycles per second, or hertz (Hz), to 4.2

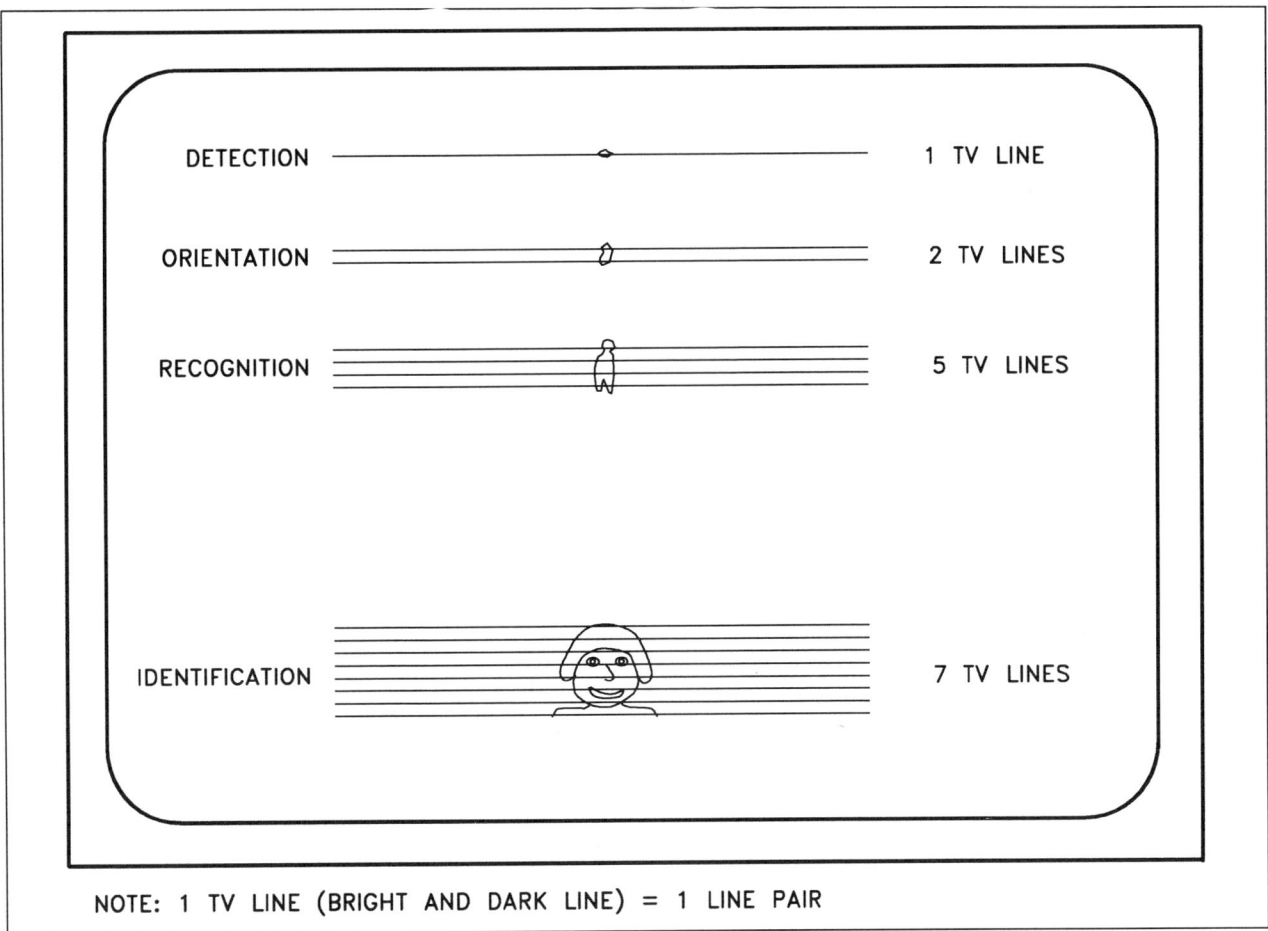

FIGURE 2-11 Object size vs. intelligence obtained

million cycles per second, or megahertz (MHz), which are transmitted via the coaxial cable. Most camera controls are internal, though mechanical focus (sensor position) may be either internal or external.

Significant technology advancements have been made in the last few years in security CCTV cameras. All security cameras made betweeen the 1950s and 1980s have been of the vacuum tube type, either vidicon, silicon, Newvicon (a trademark of Matsushita Corp.), or LLL types using SIT and ISIT (see Section 2.2.3). In the 1980s came the CCD solid-state television image sensor. Increased consumer demand for VCRs using CCD sensors in camcorder applications has caused a technology explosion and the availability of mono-chrome and color solid-state cameras, replacing tube cam-eras in security systems. While many of the cameras in use today are tube type, the highly reliable, long-lasting, high-performance solid-state cameras are fast replacing tube cameras. Solid-state camera costs have steadily decreased and are competitive with most tube types. Considering initial hardware and maintenance costs and camera life, the

solid-state camera is the clear choice. With respect to LLL capability, vidicon, silicon, Newvicon, and CCD cameras do not achieve the sensitivity of the ICCD, SIT, and ISIT cameras. Because the poor performance of tube color cam-eras restricted their use in security applications, solid-state CCD color cameras, which have good stability, sensitivity, and resolution, are now used.

INTELLIGENCE	MINIMUM TV LINES
DETECTION	1 ± 0.25
ORIENTATION	1.4 ± 0.35
RECOGNITION	4 ± 0.8
IDENTIFICATION	6.4 ± 1.5

Table 2-2 TV Lines vs. Intelligence Obtained

Table 2-3 Reflectivity of Common Materials

MATERIAL	REFLECTIVITY (%) *
SNOW	85–95
ASPHALT	5
PLASTER (WHITE)	90
SAND	40–60
TREES	20
GRASS	40
CLOTHES	15–30
CONCRETE–NEW	40
CONCRETE–OLD	25
CLEAR WINDOWS	70
HUMAN FACE	15–25
WOOD	10–20
PAINTED WALL (WHITE)	75–90
RED BRICK	25–35
PARKING LOT AND AUTOMOBILES	40
ALUMINUM BUILDING (DIFFUSE)	65–70

* VISIBLE SPECTRUM: 400–700 NANOMETERS

2.6.1 The Scanning Process

All tube and solid-state cameras use some form of scanning to generate the television picture. A block diagram of the CCTV camera and a brief description of the scanning process and video signal are shown in Figures 2-7 and 2-16, respectively.

The television sensor converts the optical image from the lens into an electrical signal. The camera electronics process the video signal and generate a composite video signal containing the picture information (luminance and color) and horizontal and vertical synchronizing pulses. Signals are transmitted in what is called a frame of picture video, made up of two fields of information. Each field is transmitted in $1/60$ of a second, the entire frame in $1/30$ of a second, for a repetition rate of 30 frames per second. In the United States, this format is the Electronic Industries Association (EIA) standard, called the NTSC (National Television System Committee) system. The European standard uses 625 horizontal lines with a field taking $1/50$ of a second, a frame $1/25$ of a second, and a repetition rate of 25 frames per second.

In the NTSC system, the first picture field is created by scanning $262\frac{1}{2}$ horizontal lines. The second field of the

frame contains the second $262\frac{1}{2}$ lines, which are synchronized so that they fall between the gaps of the first field lines, thus producing one completely interlaced picture frame containing 525 lines. If the scan lines of the second field fall *exactly* halfway between the lines of the first field, a 2-to-1 *interlace* system results.

As shown in Figure 2-16, the first field starts at the upper-left corner (of the tube camera sensor or the CRT monitor) and progresses down the sensor (or screen), line by line, until it ends at the bottom center of the scan. Likewise the second field starts at the top center of the screen and ends at the lower-right corner. Each time one line in the field traverses from the left side of the scan to the right, it corresponds to one horizontal line (scan), as shown in the video waveform at the bottom of Figure 2-16. The video waveform consists of negative synchronization pulses and positive picture information. The horizontal and vertical synchronization pulses are used by the television monitor (and VCR or video printer) to synchronize the video picture and paint an exact replica in time and intensity of the camera scanning function on the monitor face. Black picture information is indicated on the waveform at the bottom (approximately 0 volts) and white picture infor-

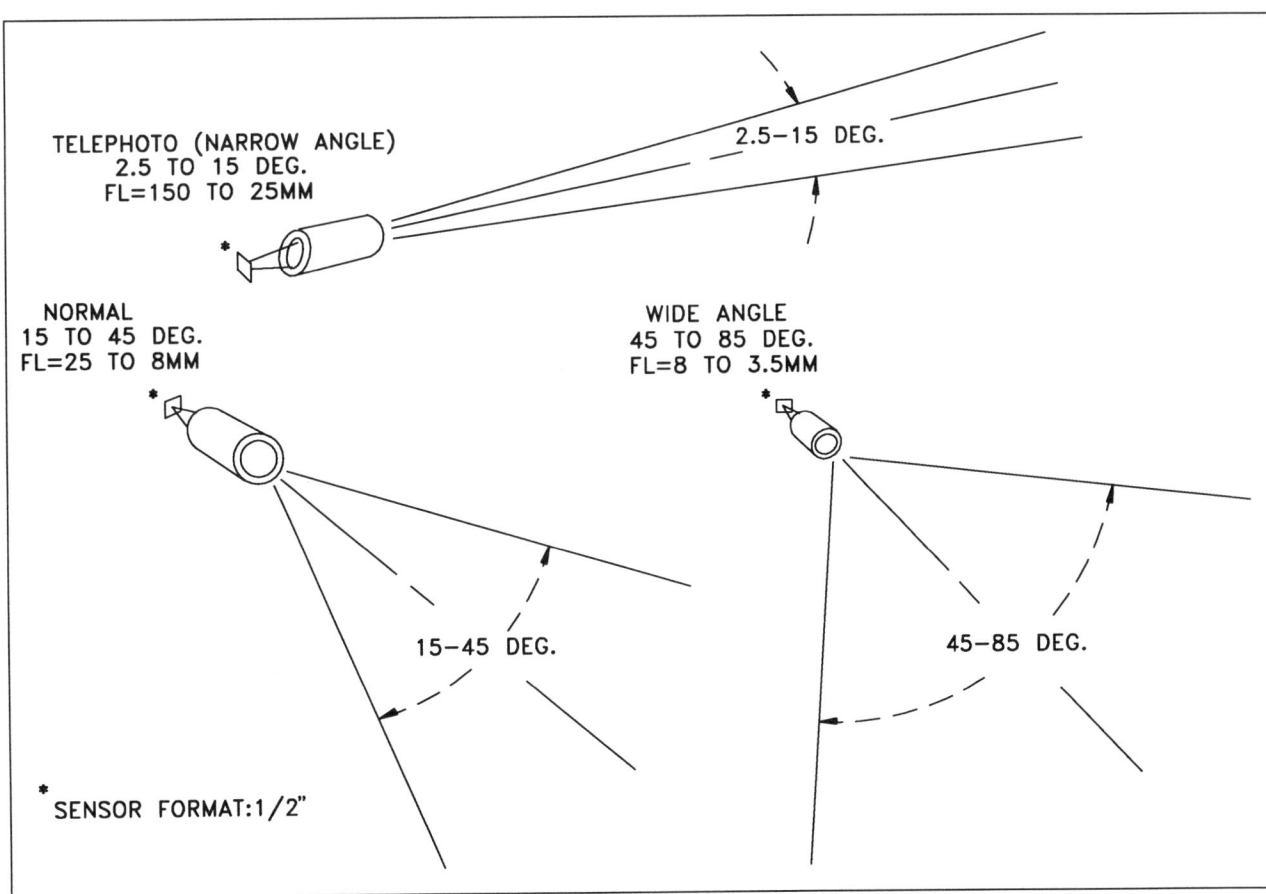

FIGURE 2-12 Representative FFL lenses and their FOVs

mation at the top (1 volt). The amplitude of a standard NTSC signal is 1.4 volts peak to peak. In the 525-line system, the frame of picture information consists of approximately 512 lines. The lines with no picture information are necessary for vertical blanking, which is the time when the camera electronics or the beam in the monitor CRT tube moves from the bottom to the top to start a new field.

Random-interlace cameras do not provide complete synchronization between the first and second fields. The horizontal and vertical scan frequencies are not locked together; therefore, fields do not interlace exactly. This condition, however, results in an acceptable picture, and the asynchronous condition is difficult to detect. The 2-to-1 interlace system has an advantage when multiple cameras are used with multiple monitors and/or VCRs in that they prevent jump or jitter when switching from one camera to the next.

The scanning process for solid-state cameras is different. The solid-state sensor consists of an array of very small picture elements (pixels) that are read out serially (sequentially) by the camera electronics to produce the same

NTSC format—525 TV lines in $1/30$ of a second (30 frames per second)—shown in Figure 2-16.

2.6.2 Tube and Solid-State Cameras

CCTV security cameras have gone through rapid technological change during the last half of the 1980s. For decades the vidicon tube camera was the only security camera available. In the 1980s silicon-diode and Newvicon tube cameras with increased sensitivity lowered the light-level requirements for a good picture. The solid-state sensor camera has now replaced the tube camera. Both types are summarized briefly in the following sections.

2.6.2.1 Tube Camera

Most tube-type security cameras have a $2/3$- or 1-inch sensor format, and use a standard antimony trisulfide photocathode target in the tube. One-inch vidicon cameras are used when higher resolution is required. All vidicon cameras use automatic light control (ALC), which electronically

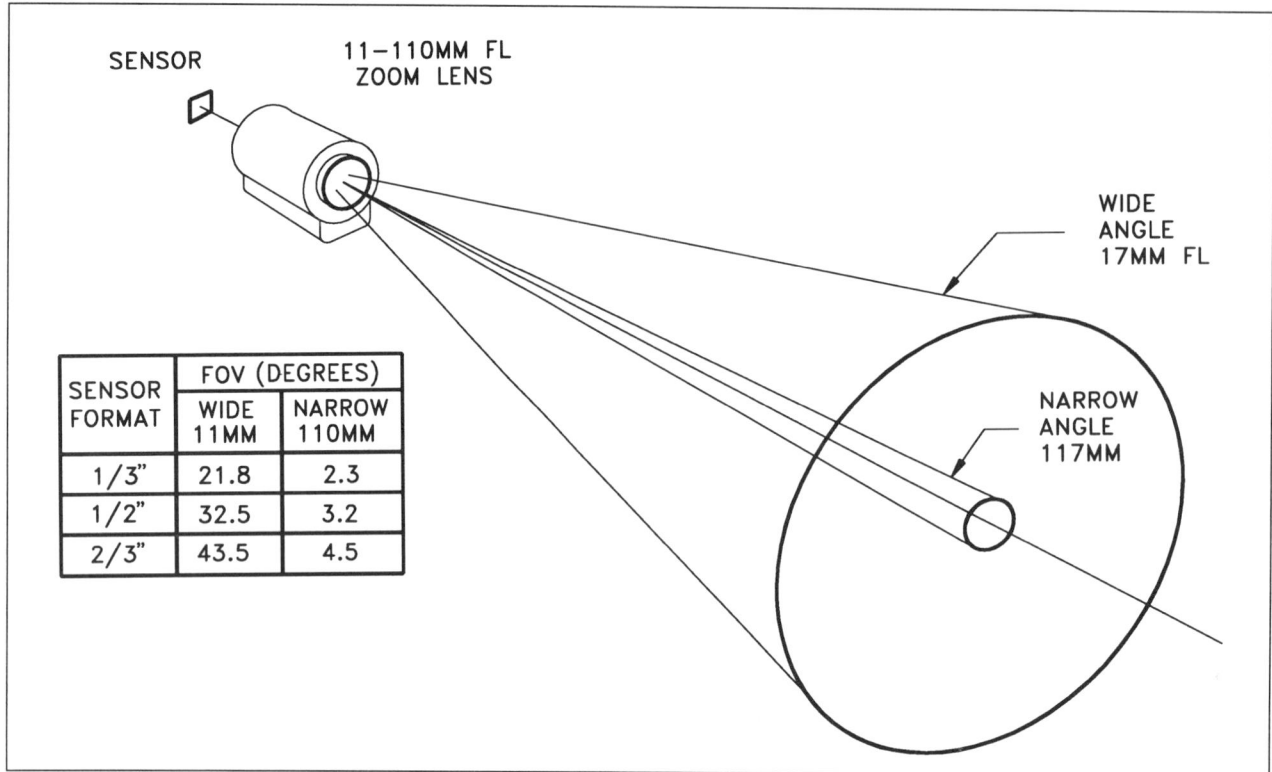

SENSOR FORMAT	FOV (DEGREES)	
	WIDE 11MM	NARROW 110MM
1/3"	21.8	2.3
1/2"	32.5	3.2
2/3"	43.5	4.5

FIGURE 2-13 Zoom CCTV lens FOV

changes the camera amplification to adapt to large variations in light levels. ALC compensation over a 10,000:1 range is standard, although cameras having a 100,000:1 ALC capability are available. Vidicon tube cameras can be damaged if they view bright targets such as the sun. Dark spots or lines will appear permanently on the camera output image.

In poorly lighted scenes when standard vidicon cameras do not produce an adequate picture, more-sensitive Newvicon, silicon tube, or solid-state CCD sensors are used. These cameras have a sensitivity between 10 and 100 times better than the standard vidicon. All three are relatively immune to image burn when they are pointed at bright lights or the sun or pointed continuously at the same scene; however, the Newvicon can be damaged at high light levels. At low light levels these cameras cannot produce an adequate picture, so ICCD, SIT, or ISIT cameras are required. These cameras are extremely sensitive and must use lenses having special automatic-iris characteristics. These LLL cameras are expensive and only justified when raising the light level is not practical.

2.6.2.2 Solid-State Camera

CCD solid-state cameras are available in three different image formats: ⅔-inch, ½-inch, and ⅓-inch. The first generation of solid-state cameras available from most manufacturers was ⅔-inch. While these provided good performance, the second-generation ⅔-inch and smaller ½-inch format produced higher resolution and sensitivity.

The ½-inch permitted the design of smaller, less expensive lenses to accomplish the same functions as their ⅔-inch counterpart. Many manufacturers now produce ⅓-inch format cameras with medium resolution. The solid-state sensor cameras have important attributes, including (1) precise, repeatable pixel geometry; (2) low power requirements; (3) small size; (4) excellent color rendition and stability; and (5) ruggedness and long life expectancy.

2.6.3 Low-Light-Level Intensified Camera

When a security application requires viewing during nighttime conditions where available light is moonlight, starlight, or other residual reflected light, LLL intensified cameras are necessary. The ICCD, SIT, and ISIT cameras have sensitivities between 100 and 1000 times more than the best solid-state or tube cameras. Camera costs are between 10 and 20 times more than CCD cameras. Chapter 15 describes the characteristics of these cameras.

2.7 TRANSMISSION

The function of CCTV is to be the remote eyes of the security guard. By definition, the camera must be remotely located from the monitor, and therefore the television signal must

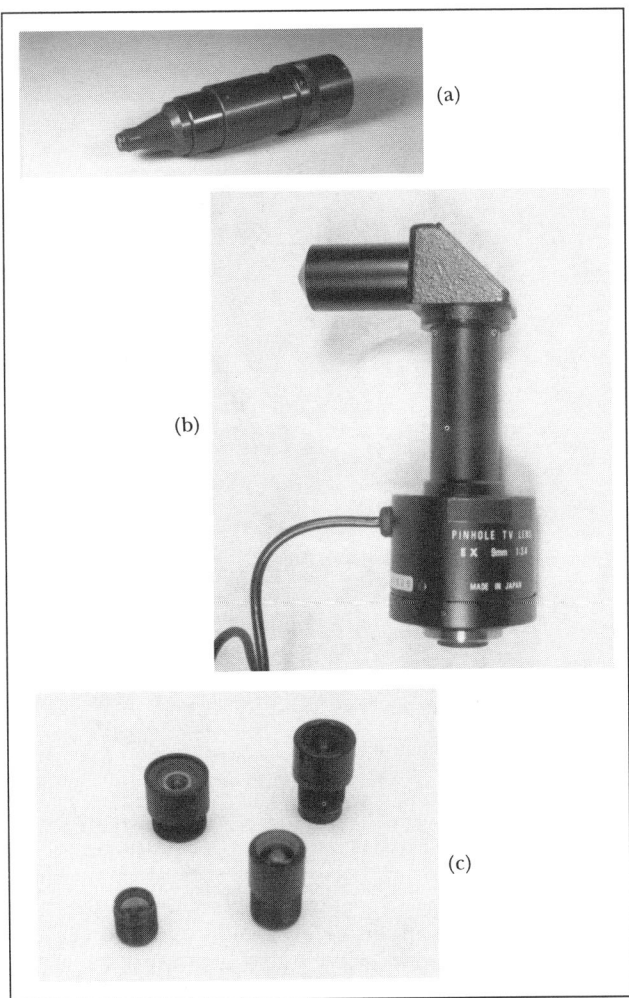

FIGURE 2-14 Pinhole and mini-pinhole lenses

be transmitted by some means. In security applications, the distance between the camera and the monitor is from tens of feet to many miles or perhaps completely around the globe. The transmission cable or path travel may be inside buildings, outside buildings, or through the atmosphere, and in almost any environment imaginable. For this reason the transmission means must be carefully assessed and an optimum choice of hardware made to satisfactorily transmit the video signal over such distances. Chapter 6 describes and analyzes the characteristics, advantages, and disadvantages of all of these transmission means, and the hardware available to transmit the video signal.

2.7.1 Hard-Wired

There are currently many hard-wired means for transmitting a video signal, including coaxial, two-wire (such as intercom or telephone wire), and fiber-optic cable. Figure 2-17 shows some examples of transmission cables and hardware.

The most common means are coaxial and fiber-optic cable. Other techniques, such as two-wire telephone cable, are used as alternatives. Wired cable systems offer a more reliable signal at the monitor, straightforward installation and maintenance, and lower cost. Chapter 6 describes in detail all hard-wired transmission techniques.

2.7.2 Wireless

Sometimes it is more economical or beneficial to transmit the video signal without cable—wireless—from the camera to the monitor using a microwave, RF, or IR atmospheric link. Two broad applications for wireless transmission are (1) covert CCTV installations and (2) building-to-building transmission over a roadway. However, the Federal Communications Commission (FCC) restricts some wireless transmitting devices using microwave or RF frequencies to government and law enforcement use. Only recently have some RF transmitters been given FCC approval for general use. These devices operate above the normal television frequency bands at approximately 920 MHz. The atmospheric IR link, which requires no FCC approval, transmits a video image over a narrow beam of light. Figure 2-18 illustrates some of the wireless transmission techniques available today.

When transmitting a color signal over microwave, IR (optical), or RF, the specifications for the transmitter and receiver units must be suitable for color transmission.

2.7.3 Slow-Scan

The techniques mentioned so far provide a means for real-time transmission of a video signal, requiring a full 4.2-MHz bandwidth to reproduce real-time motion. When these techniques cannot be used for real-time transmission, alternative delayed or slow-scan techniques are used. In these systems, a non–real-time video transmission takes place, so that full scene action is lost; only snapshots of the video image are transmitted.

2.7.4 Fiber Optics

Fiber-optic-transmission technology has advanced significantly in the last 5 to 10 years and now represents the present and future transmission means of choice. Fiber-optic transmission holds several significant advantages over other hard-wired systems: (1) very long transmission paths, up to many miles, without any significant degradation in the video signal with monochrome or color; (2) immunity to external electrical disturbances from weather or equipment; (3) very wide bandwidth, permitting one or more video, control, and audio signals to be multiplexed on a single fiber; (4) resistance to tapping (eavesdropping) and therefore a very secure transmission means.

FIGURE 2-15 Special CCTV lenses: (a) dual-split-image lens, (b) tri-split-image lens, (c) right-angle lens, (d) rigid fiber optics, (e) relay lens, (f) flexible fiber optics

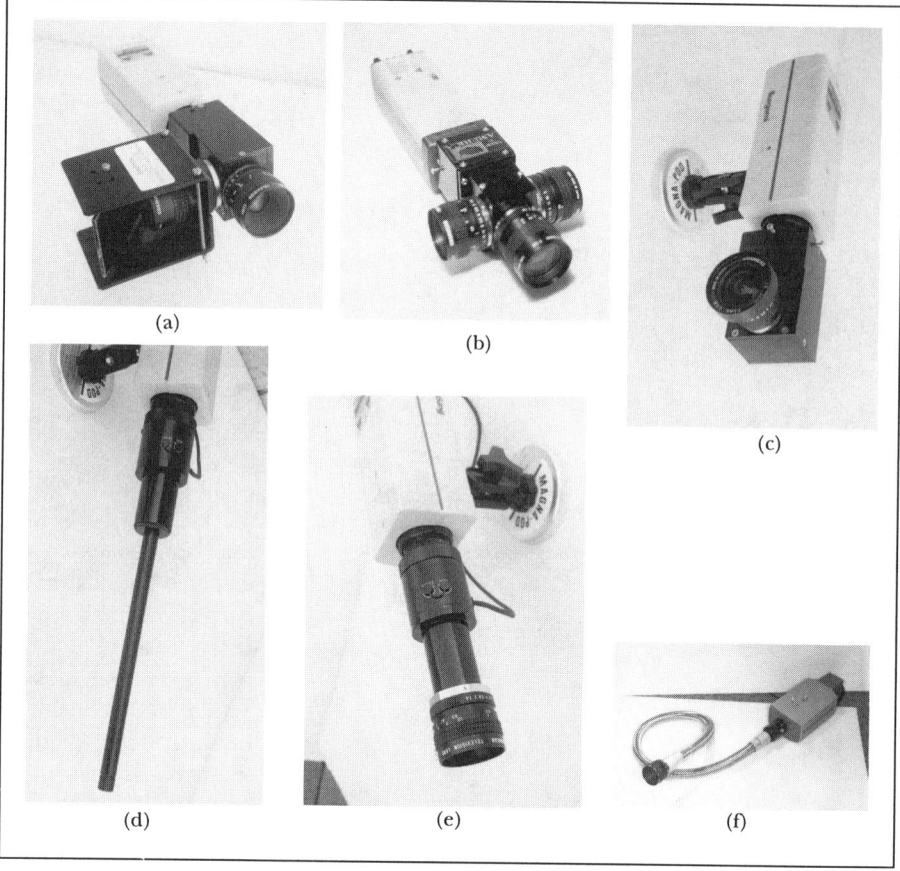

(a)   (b)   (c)

(d)   (e)   (f)

While the installation and termination of fiber-optic cable requires somewhat more skilled technicians, it is well within the capability of qualified security professionals. Particular attention should be paid to transmission means when transmitting color CCTV signals, since the color signal is significantly more complex and susceptible to distortion than monochrome. Almost all hard-wired installations involving color use fiber-optic cable.

2.8 SWITCHERS

The video switcher, an important component in many CCTV security systems, accepts video signals from many different television cameras. Using manual or automatic activation or alarming signal input, the switcher selects one or more of the cameras and directs its video signal to a specified monitor, recorder, or some other device or location.

2.8.1 Standard

There are four basic switcher types: manual, sequential, homing, and alarming. Figure 2-19 shows how these are connected into the CCTV security system.

The manual switcher connects one camera at a time to the monitor or other recording or printing device. The sequential switcher automatically switches the cameras in sequence to the output device. The operator can override the automatic sequence with the homing sequential switcher. The alarming switcher connects the alarmed camera to the output device automatically, when an alarm is received.

2.8.2 Microprocessor-Controlled

When the security system requires many cameras with pan/tilt camera mountings, in various locations, and with multiple monitors and other alarm input functions, a microprocessor-controlled switcher and keyboard can be used to manage these additional requirements (Figure 2-20).

In large security systems, the switcher is microprocessor controlled and can switch hundreds of cameras to dozens of monitors, recorders, or video printers via an RS-232 communications control link. Numerous manufacturers make comprehensive keyboard-operated, computer-controlled consoles that integrate the functions of the switcher, pan/tilt pointing, automatic scanning, automatic preset pointing for pan/tilt systems, and many other functions. The power of the software-programmable console resides in its flexibility, expandability, and ability to accommodate a large variety of applications and changes in facility design. In place of a dedicated hardware system built for each

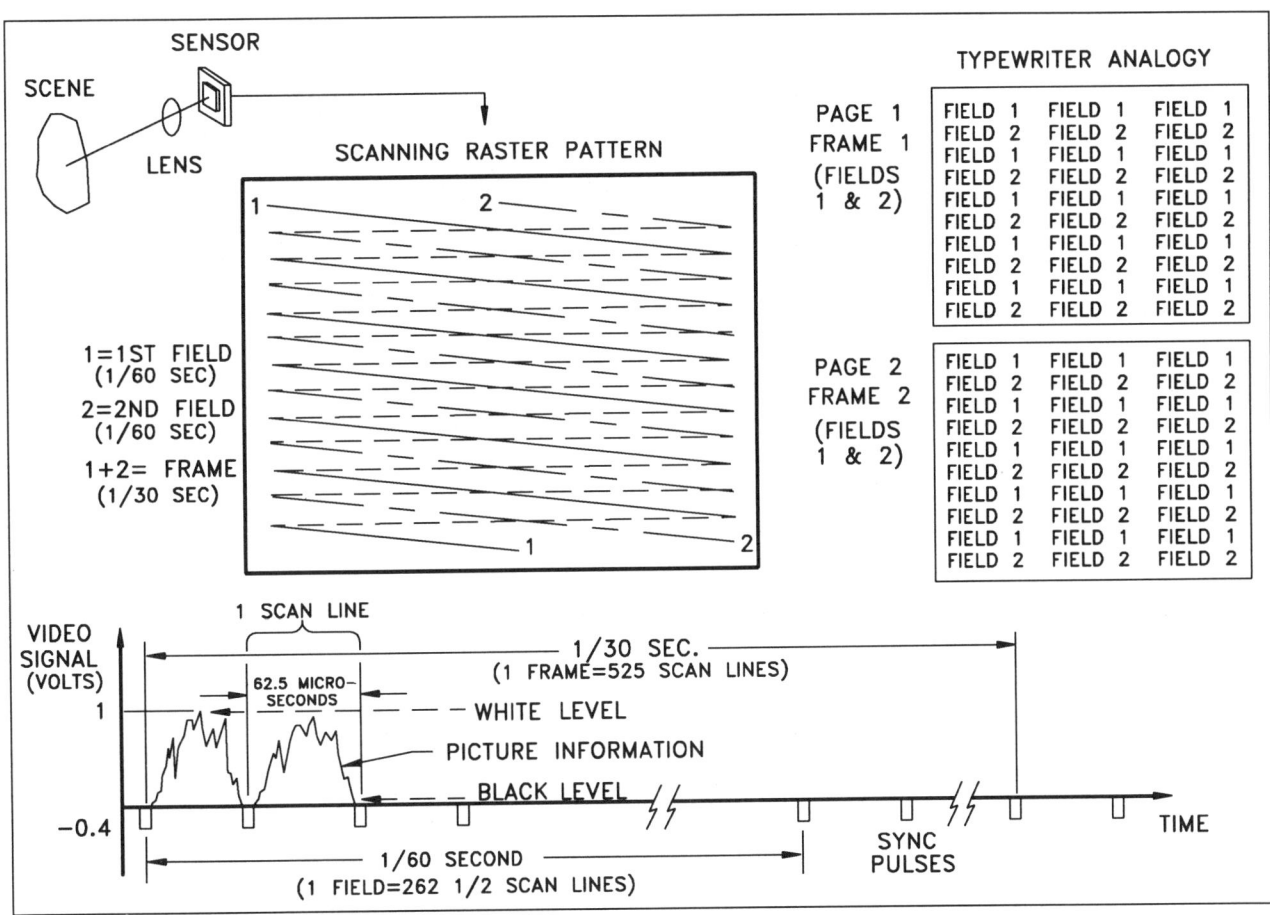

FIGURE 2-16 CCTV scanning process and video signal

specific application, this computer-controlled system can be configured via software to match almost all the required applications. Chapter 9 describes types of switchers and their functions and applications.

2.9 MONITORS

Television monitors can be divided into several categories: (1) monochrome, (2) color, (3) video graphics and text, and (4) audio and video. Almost all monitors used for CCTV applications are video monitors, not the graphics monitors used for computer displays. Because scenes displayed on security monitors are mostly visual, sometimes annotated with alphanumeric characters, the monitor must be capable of displaying a continuous range of light intensities. That is, the monitor needs to display a large number of shades of gray, or halftones. Video monitors also display color hues with higher purity. Computer graphics monitors, on the other hand, lack good gray-scale capabilities, since most of the information displayed is generally made up of alphanumeric characters or graphics (such as bar charts, graphics, plant layouts) that are either black, white, or a single color.

Contrary to a popular misconception, larger monitors do not necessarily have better picture resolution or the ability to increase the amount of intelligence available in the picture. All U.S. NTSC security monitors have 525 horizontal lines—regardless of their size or whether they are monochrome or color; therefore the vertical resolution is about the same regardless of the CRT monitor size. The horizontal resolution is determined by the system bandwidth. The only improvement to be made, assuming that the camera has good resolution, is to choose a monitor having resolution equal to or better than the camera or transmission link bandwidth. Chapter 7 gives more detailed characteristics of monochrome and color monitors used in the security industry.

2.9.1 Monochrome

By far the most popular monitor used in CCTV systems is the monochrome monitor, which comes in sizes ranging from viewfinder size, which has a 1-inch-diagonal CRT, up to a large display monitor, which has a 27-inch-diagonal (or larger) CRT. By far the most popular monochrome monitor size is the 9-inch diagonal, which optimizes video viewing for a person seated about 3 feet away. This is the typical

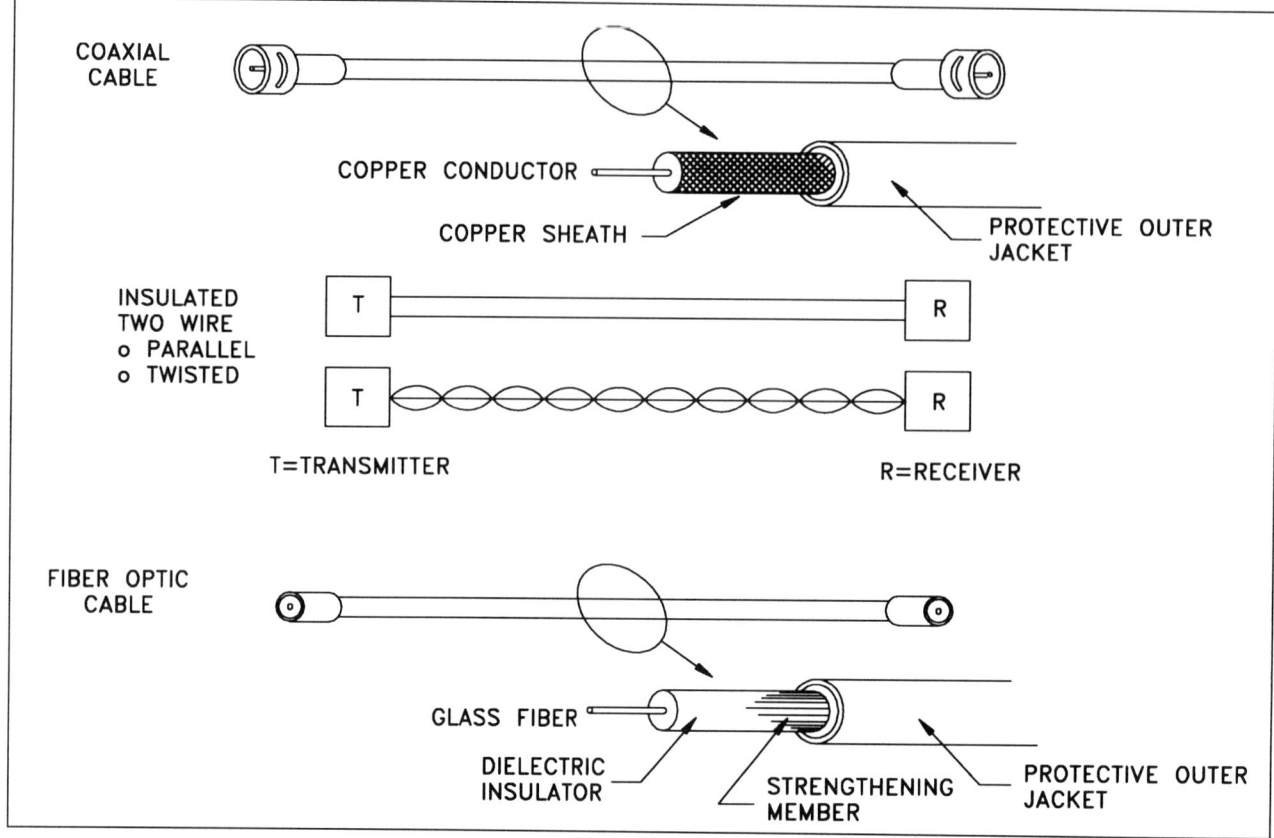

FIGURE 2-17 Hard-wired transmission means

distance in a security control room. A second reason for its popularity is that two of these monitors fit into the standard EIA 19-inch-wide rack-mount panel. Figure 2-21 shows this 9-inch monitor, a dual rack-mounted version, and a triple rack-mount version of a 5-inch-diagonal monitor.

Where space is at a premium, the triple rack-mounted monitor is popular, since three fit conveniently into the 19-inch EIA rack. To view the triple or 5-inch-diagonal monitor, the security guard should be about 1.5 feet away.

2.9.2 Color

Similar conditions hold true for color monitors, which range from 3- to 27-inch diagonal and have required viewing distances and capabilities similar to those of mono-chrome monitors. Since color monitors require three dif-ferent-colored dots to produce one pixel of information on the monitor, they have lower horizontal resolution than monochrome monitors.

2.9.3 Audio/Video

Many monitors have an audio channel built into them to produce audio and CCTV video information simultane-ously. Combination audio/video monitors with RF video signal inputs are also available.

2.10 RECORDERS

The television camera, transmission means, and monitor provide the remote eyes for the security guard in real time or slow scan, but as soon as the action or event is over, it disappears from the monitor screen forever. Unless it is recorded in some way. When a permanent record of the live video scene is required, a VCR or video magnetic or optical disk recorder is used (Figure 2-22).

2.10.1 Videocassette

Magnetic storage media are used universally to record the television image. The most popular system uses a VCR similar or identical to the home VCR. The VHS cassette format has become the standard, and the 8-mm Sony for-mat is also popular because of its smaller size. Super VHS and Hi-8, formats with higher resolution, are used in stra-tegic security applications.

VCRs can be subdivided into two classes: real-time and time-lapse. The commercial-grade real-time recorder is similar to that used in consumer electronics, but the re-corder hardware is more rugged and durable for the con-tinuous use required in security applications. The time-lapse recorder has significantly different mechanical and electrical features, permitting it to take snapshots of a

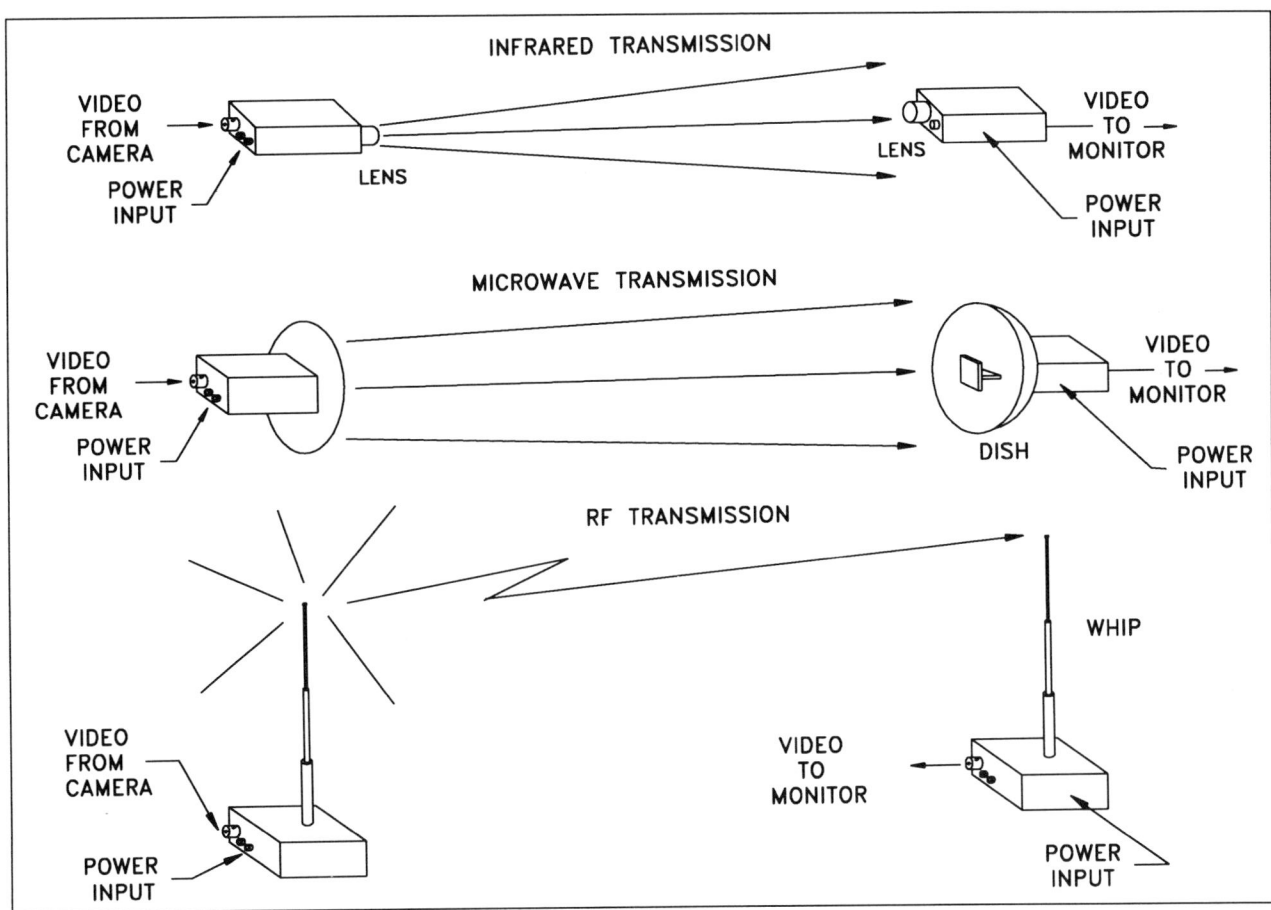

FIGURE 2-18 RF, microwave, and IR CCTV transmission links

scene at predetermined (user-selectable) intervals. It can also record in real time when activated by an alarm or other input command. Real-time recorders can record up to 6 hours in monochrome or color; time-lapse VCRs are available for recording time-lapse sequences up to 720 hours.

2.10.2 Magnetic and Optical Disk

A magnetic hard disk, similar to a microcomputer hard disk, can store many hundreds or thousands of images in analog or digital form. An optical disk has a much larger video image database capacity: many thousands or tens of thousands of images on a single optical disk. Both magnetic and optical disk systems have an advantage over VCRs with respect to storage and retrieval time of a particular video frame. Storage and retrieval times for video on magnetic hard disk are typically less than 1 second to less than $\frac{1}{20}$ of a second for monochrome, depending on the system, and 2 to 3 seconds down to $\frac{1}{5}$ of a second for color. VCRs on the other hand take many minutes to fast-forward or fast-rewind to seek a particular frame on the tape. Magnetic and optical disk video storage systems are available with removable disks, so that data can be transported to remote loca-

tions or stored in a vault for safekeeping. These cartridges are about the same size as VHS cassettes.

A mini magnetic disk storage system about the size of a 35-mm camera is available, which can store 50 fields (25 frames) of color or monochrome video on a 2 by 2-inch diskette. When played back, the diskettes permit random access of the video frames, with retrieval times of less than a second. Chapter 8 describes video recording equipment in detail.

2.11 HARD-COPY VIDEO PRINTERS

A hard-copy printout of a video image is often required as evidence in court, as a tool for apprehending a vandal or thief, or as a duplicate record of some document or person. The printout is produced by a hard-copy video printer, often a thermal printer that "burns" the video image onto coated paper. This technique, used by many hard-copy printer manufacturers, produces excellent-quality images in monochrome or color, many matching the camera or monitor resolutions. Figure 2-23 shows a monochrome thermal printer and a sample of the hard-copy image quality it produces.

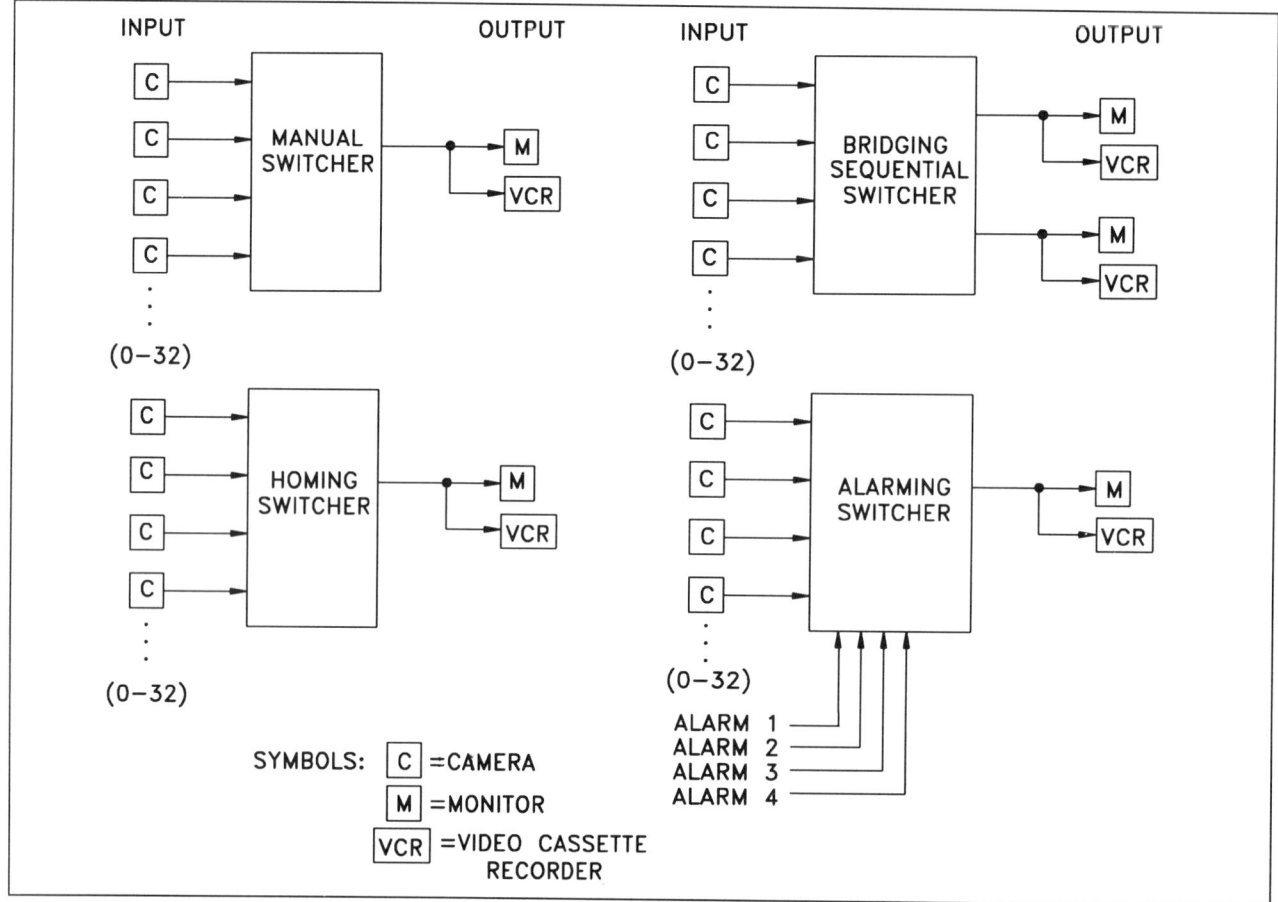

FIGURE 2-19 Basic switcher types

A standard hard-copy printout costs less than 10 cents. A color print costs less than a photographic print. During operation, the image displayed on the monitor or played back from the recorder is immediately memorized by the system and printed out in less than 10 seconds. This is particularly useful if an intrusion or unauthorized act has occurred and been observed by a security guard. An automatic alarm or a security guard can initiate printing the image of the alarm area or of the suspect, and the printout can then be given to another guard to take action. For courtroom uses, time, date, and any other information can be noted on the image. In a visitor-entry control application, a CCTV camera system with dual optics can be used with a video printer and time/date generator to create a temporary visitor badge (Figure 2-24). Split-image optics focus the visitor's face and personal information onto the camera. The composite image is then printed on the thermal video printer. Chapter 8 describes hard-copy video printer systems in detail.

2.12 ANCILLARY EQUIPMENT

Most CCTV security systems require additional accessories and equipment, including (1) camera housings, (2) camera pan/tilt mechanisms and mounts, (3) camera identifi-

ers, (4) video motion detectors, (5) image splitters/inserters, and (6) image combiners. These are described in more detail in Chapters 10 through 13. The two accessories most often used with the basic camera, monitor, and transmission link described previously are camera housings and pan/tilt mounts. Outdoor housings are used to protect the camera and lens from vandalism and the environment; indoor housings are used primarily to prevent vandalism and for aesthetic reasons. The pan/tilt mechanisms rotate and point the camera and lens via a remote control console.

2.12.1 Camera Housings

Indoor and outdoor CCTV housings protect cameras and lenses from dirt, dust, harmful chemicals, the environment, and vandalism. The most common housings are rectangular metal or plastic products, formed from indoor or outdoor plastic, painted steel, or stainless steel (Figure 2-25). Other shapes and types include corner-mount, ceiling-mount, and dome housings.

The rectangular-type housing is the most popular. It protects the camera from the environment, provides a window for the lens to view the scene, and is weatherproof and tamper resistant. The dome type is also popular, con-

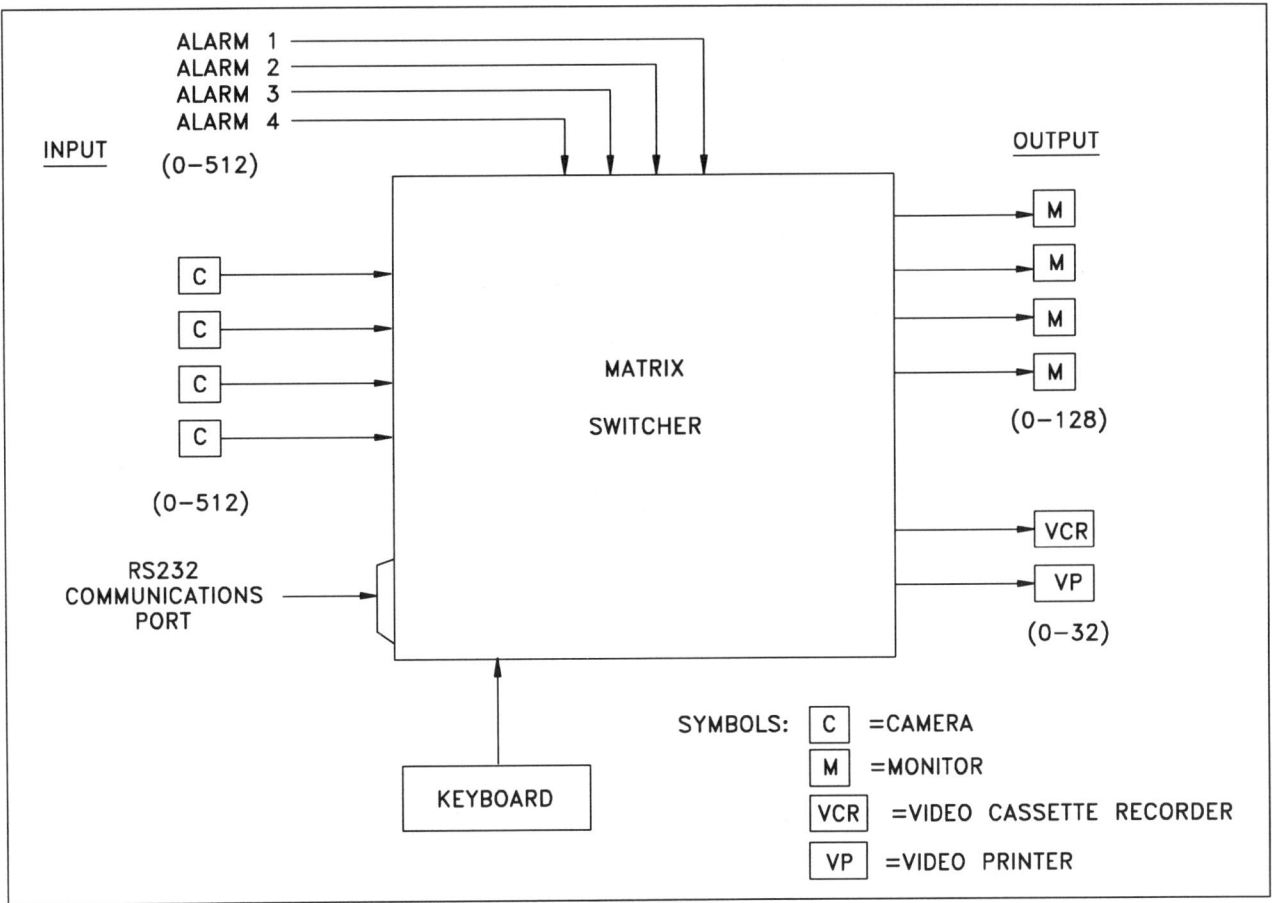

FIGURE 2-20 Microprocessor-controlled switcher and keyboard

sisting of a plastic hemispherical dome on the bottom half; the housing can be clear, tinted, or treated with a partially transmitting optical coating that allows the camera to see in any direction. In a freestanding application (e.g., on a pole, pedestal, or overhang), the top half of the housing consists of a protective cover and a means for attaching the dome to the structure. When the dome housing is mounted in a ceiling, a simpler housing cover is provided and mounted above the ceiling level to support the dome. There are many other specialty housings for mounting in or on elevators, ceilings, walls, tunnels, pedestals, hallways, and so forth. Chapter 11 describes camera housings and their specific applications in detail.

2.12.2 Pan/Tilt Mounts

To extend the angle of coverage of a CCTV lens/camera system, a pan/tilt mechanism is often used. Figure 2-26 shows two generic outdoor pan/tilt types: top-mounted and side-mounted camera.

The pan/tilt motorized mounting platform permits the camera and lens to rotate horizontally (pan) or vertically

(tilt) when it receives an electrical command from the remote monitor site. Thus the camera lens is not limited by its inherent FOV and can view a much larger area of a scene. A camera mounted on a pan/tilt platform is often provided with a zoom lens, which varies the FOV in the pointing direction of the camera/lens. The combination of the pan/tilt and zoom lens provides the widest angular coverage for television surveillance. But there is one disadvantage compared with the fixed camera installation. When the camera and lens are pointing in a particular direction via the pan/tilt platform, most of the other scene area the system is designed to cover is not being viewed. This dead area or dead time is unacceptable in many security applications, and therefore a careful consideration should be given to the adequacy of this wide-FOV pan/tilt system. Pan/tilt platforms range from small, indoor, lightweight units that only pan, up to large, outdoor, environmentally designed units carrying large cameras, zoom lenses, and large housings. Choosing the correct pan/tilt mechanism is important, since it generally requires more service and maintenance than any other part of the CCTV system. Chapter 10 describes several generic pan/tilt designs and their features.

FIGURE 2-21 Standard 5- and 9-inch monitors and EIA rack mounting: (a) single 9-inch monitor, (b) dual 9-inch monitor, (c) triple 5-inch monitor

FIGURE 2-22 VCR, hard disk, and floppy disk video storage media

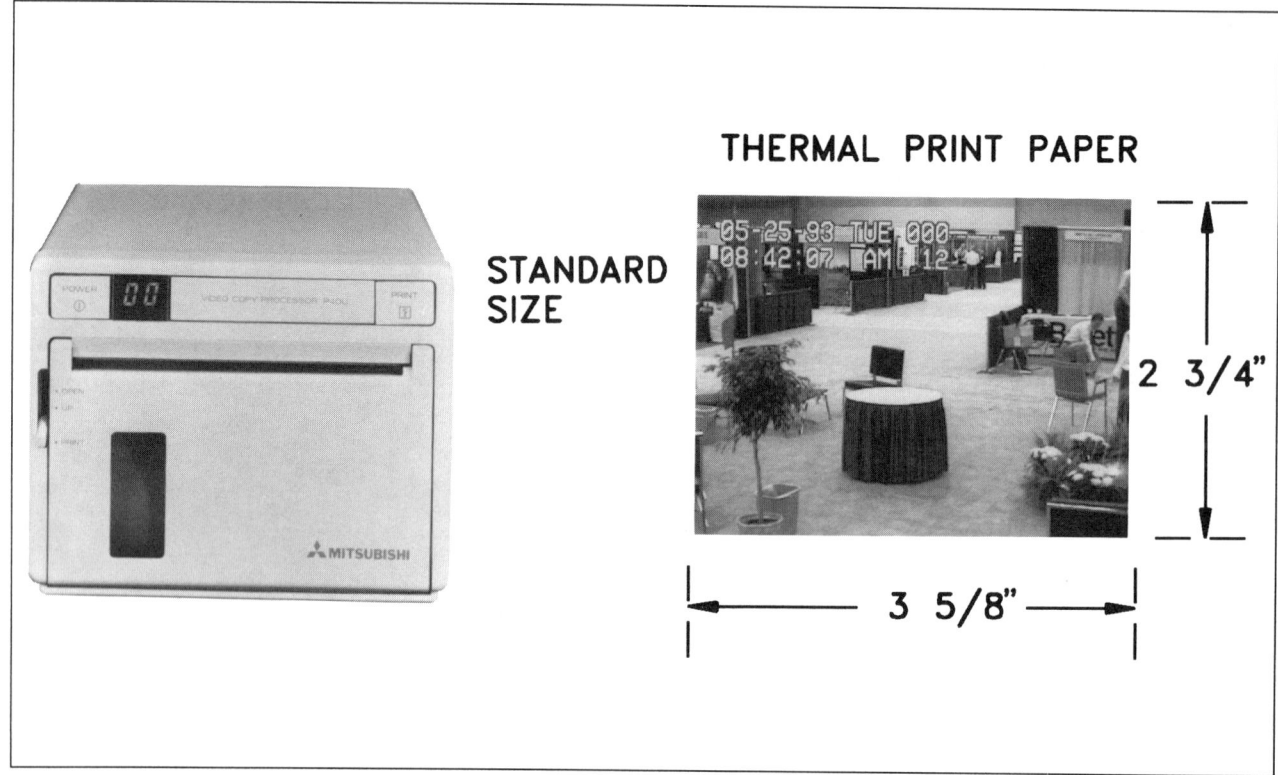

FIGURE 2-23 Thermal monochrome video printer and hard copy

2.12.3 Video Motion Detectors

Another important addition to a CCTV surveillance system is the video motion detector (VMD), which produces an alarm signal based on a change in the video scene. The VMD is connected to the camera video signal and, either by analog or digital means, stores the video frames, compares subsequent frames to the stored frames, and then determines whether the scene has changed. The VMD decides whether the change is significant and whether to call it an alarm to alert the guard or some equipment. Chapter 12 describes various VMDs, their capabilities, and their limitations.

2.12.4 Screen Splitter

The electronic or optical screen splitter takes a part of several camera scenes (two, three, or more), combines the scenes, and displays them on one monitor. The splitters do not compress the image. In an optical splitter, the image combining is implemented optically at the camera and requires no electronics. The electronic splitter/combiner

is located between the camera output and the monitor input. Chapter 13 describes these devices in detail.

2.12.5 Screen Compressor

A CCTV accessory that has become popular in recent years because of the proliferation of inexpensive digital solid-state storage devices is the electronic image compressor. This device, interposed between the camera and the monitor, accepts several camera inputs, memorizes the scenes from each camera, compresses them, and then displays more than one scene on a single video monitor. Equipment is available to provide 2, 4, 9, 16, and up to 32 separate video scenes on one single monitor. The most popular presentation is the quad screen (showing four pictures), which significantly improves camera viewing ability in multi-camera systems, decreases security guard fatigue, and requires three fewer monitors. Obviously there is a loss of resolution when more than one scene is presented on the monitor. The monitor resolution decreases as the number of scenes increases, that is, approximately one-quarter of the resolution is obtained on a quad

LAMP

FACE LENS

VISITOR

VIDEO PRINTER
WITH TIME/DATE

VISITOR BADGE
PRINTOUT

ACTUAL PRINT SIZE

FACE

IDENTIFICATION
CREDENTIAL

FIGURE 2-24 Videoscope and
sample visitor badge

FIGURE 2-25 Standard indoor/outdoor CCTV housings: (a) corner, (b) elevator corner, (c) ceiling, (d) outdoor environmental rectangular, (e) dome, (f) indoor rectangular

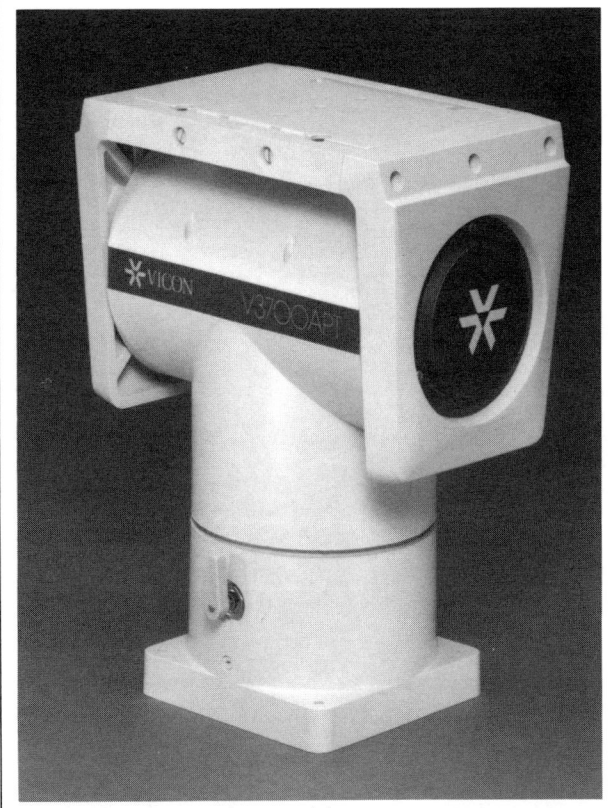

(a)

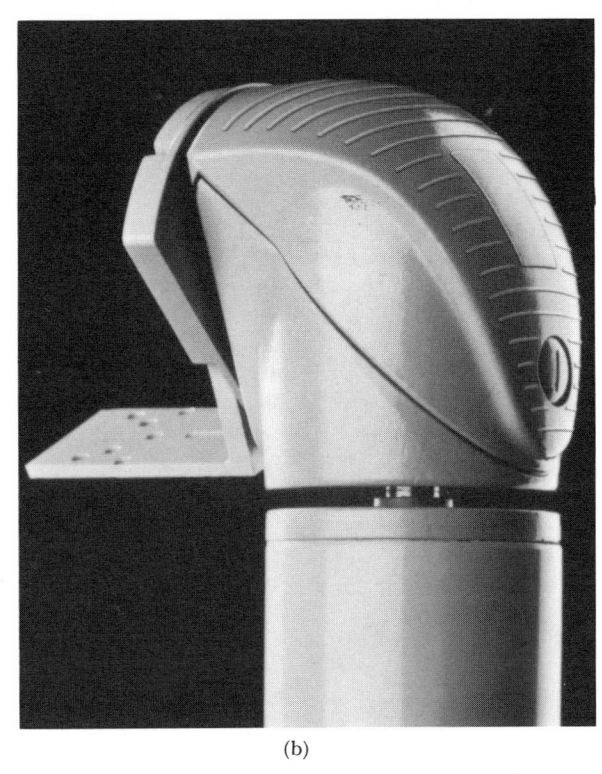

(b)

FIGURE 2-26   CCTV pan/tilt mechanisms: (a) top-mounted outdoor, (b) side-mounted outdoor

display (half in horizontal and half in vertical). Chapter 13 describes video screen compressors in detail.

2.12.6 Camera Identifier

When multiple cameras are used in a security system, some means must be provided to identify the camera. The system uses a camera identifier unit to electronically assign an alphanumeric code to each camera displayed on the monitor and/or recorded on a recorder.

2.12.7 Time/Date Generator

The time/date generator annotates the video picture with the time and date. This information is mandatory for any prosecution or courtroom procedure (see Chapter 13).

2.12.8 Character Generator

Alphanumeric and symbol character generators are available to annotate the video signal with English or foreign language instructions and/or status indicators useful to the security operator (see Chapter 13).

2.13 SUMMARY

CCTV serves as the remote eyes for management and the security force. It is a critical subsystem of any comprehensive security plan.

The CCTV camera operates very much like a person's eyes as they scan written material in order to read it. The eyes focus on the upper left corner of the page, move to the right, move quickly back to the left, and scan across the second line, quickly scan back across the third line, and so on down to the bottom of the page. But the camera skips every other line when it reads down and then comes back up the page and reads the lines it skipped. After the light image of the scene is converted to an electrical signal by the camera, the transmission means then transmits the information from the camera sensor to a receiver/monitor, which reconverts the electrical image back to a visual picture on the monitor.

Accessories available to augment and enhance the capabilities of the basic CCTV system include housings, switchers, pan/tilt mounts, annotation devices, magnetic and optical recorders, and hard-copy printout devices. When other alarm-input, intelligence-gathering, communication, and control devices are available, the video system interfaces with these equipments to provide an integrated security system.

PART 2

Chapter 3
Natural and Artificial Lighting

CONTENTS

3.1   OVERVIEW

Scene lighting affects the performance of any monochrome or color CCTV security system. Whether the application is indoor or outdoor, daytime or nighttime, the amount of available light and its color (wavelength) energy spectrum must be considered, evaluated, and compared with the sensitivity of the cameras to be used. In daytime applications some cameras must have protection from high light levels in the form of automatic-iris lenses and shutters. In nighttime applications the light level and characteristics of available and artificial light sources must be analyzed and matched to the camera's spectral and illumination sensitivities to ensure a good video picture. In applications where additional lighting can be installed, the available types of lamps—tungsten, tungsten-halogen, metal-arc, sodium, mercury, and others—must be compared. In applications where no additional lighting is permissible, the existing illumination level, color spectrum, and beam angle must be evaluated and matched to a CCTV lens/camera combination that will provide an adequate picture to gather the necessary scene intelligence.

An axiom in CCTV security applications is the more light, the better the picture. The quality of the monitor picture is affected by how much light is available and how well the sensor responds to the colors in the light source. This is particularly true when color cameras are used, since they need more light than monochrome cameras. The energy from light radiation is composed of a spectrum of colors, including "invisible light" produced by long-wavelength infrared (IR) and short-wavelength ultraviolet (UV) energy. Most monochrome CCTV cameras respond to visible and IR light; color cameras are made to respond to visible light only.

Although many consider lighting to be only a decorator's or architect's responsibility, the type and intensity is of paramount importance in any CCTV security system.

This chapter analyzes the available natural and artificial light sources and provides information to help in choosing an optimum light source or in determining whether existing light levels are adequate.

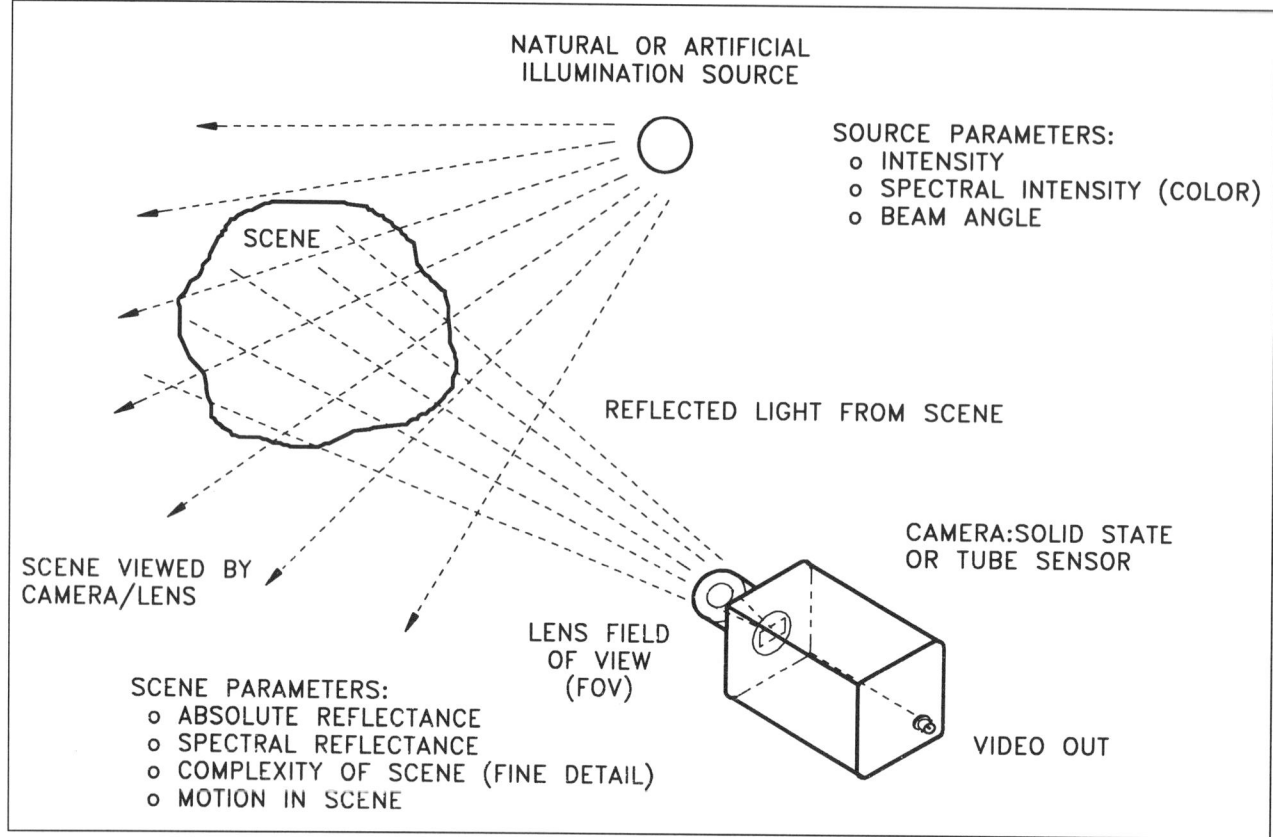

NATURAL OR ARTIFICIAL
ILLUMINATION SOURCE

SOURCE PARAMETERS:
o INTENSITY
o SPECTRAL INTENSITY (COLOR)
o BEAM ANGLE

SCENE

REFLECTED LIGHT FROM SCENE

CAMERA:SOLID STATE
OR TUBE SENSOR

SCENE VIEWED BY
CAMERA/LENS

LENS FIELD
OF VIEW
(FOV)

VIDEO OUT

SCENE PARAMETERS:
o ABSOLUTE REFLECTANCE
o SPECTRAL REFLECTANCE
o COMPLEXITY OF SCENE (FINE DETAIL)
o MOTION IN SCENE

FIGURE 3-1   CCTV camera, scene, and source illumination

3.2 CCTV LIGHTING CHARACTERISTICS

The illumination present in the scene determines the amount of light ultimately reaching the CCTV camera lens. It is therefore an important factor in the quality of the CCTV picture. The illumination can be from natural sources such as the sun, moon, or starlight or from artificial sources such as tungsten, mercury, fluorescent, sodium, metal-arc, or other lamps. Considerations about the *source* illuminating a scene include (1) source spectral characteristics, (2) beam angle over which the source radiates, (3) intensity of the source, (4) variations in that intensity, and (5) location of the CCTV camera relative to the source. Factors to be considered in the scene include (1) reflectance of objects in the scene, (2) complexity of the scene, (3) motion in the scene, and (4) degree of lighting contrast in the scene.

3.2.1 Scene Illumination

As stated previously, the CCTV camera image sensor responds to reflected light from the scene. To obtain a better understanding of scene and camera illumination, consider

Figure 3-1, which shows the illumination source, the scene to be viewed, and the CCTV camera and lens.

The radiation from the illuminating source reaches the television camera by first reflecting off the objects in the scene.

In planning a television system, it is necessary to know the kind of illumination and the intensity of light falling on a surface and how the illumination varies as a function of distance from the light source.

3.2.2 Light Output

The amount of light (lumens) produced by any light source is defined by a parameter called the candela (related to the light from one candle) (Figure 3-2).

One foot-candle (fc) of illumination is defined as the amount of light received from a 1-candela source at a distance of 1 foot. A light meter calibrated in foot-candles will measure 1 foot-candle at a distance of 1 foot from that source. As shown in Figure 3-2, the light falling on a 1-square-foot area at a distance of 2 feet is one-quarter foot-candle. This indicates that the light level varies inversely as the square of the distance between the source and observer. Doubling the distance from the source reduces the light

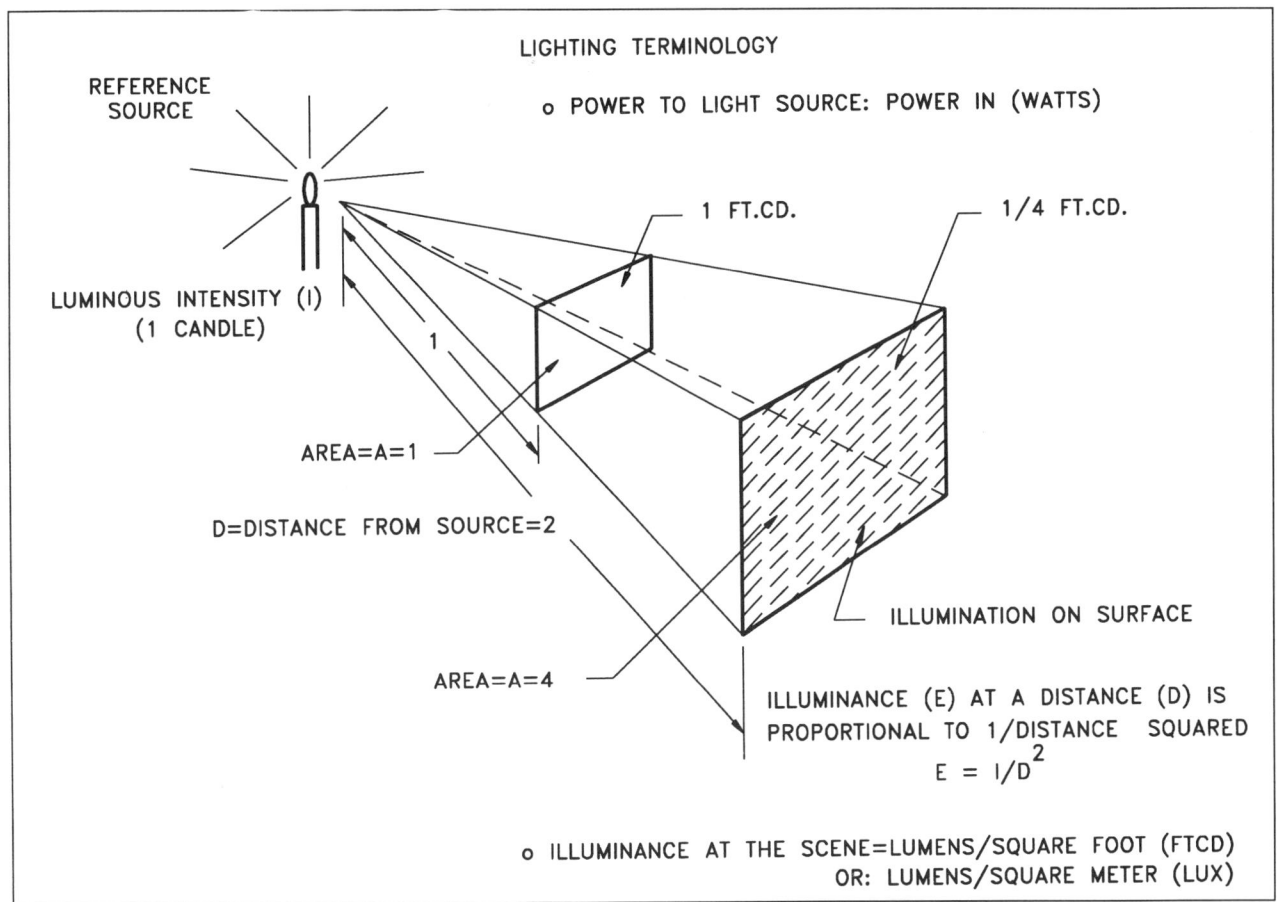

FIGURE 3-2 Illumination defined—the Inverse Square Law

level to one-quarter of its original level. Note that exactly four times the area is illuminated by the same amount of light—which explains why each quarter of the area receives only a quarter of the light.

3.2.3 Spectral Output

Since different CCTV camera types respond to different colors, it is important to know what type of light source is illuminating the surveillance area, as well as what type might have to be added to get the required television picture. Figure 3-3 shows the spectral light-output characteristics from standard tungsten, tungsten-halogen, and sodium artificial sources, as well as that from natural sunlight.

Superimposed on the figure is the spectral sensitivity of the human eye. Each source produces light at different wavelengths or colors. To obtain the maximum utility from any television camera, it must be sensitive to the light produced by the natural or artificial source. Sunlight, moonlight, and tungsten lamps produce energy in a range in which all TV cameras are sensitive. Silicon-tube and charge-coupled device (CCD) solid-state sensors are sensitive to visible and IR sources. (Caution: some CCD cameras have IR cut filters, which reduce this IR sensitivity.) Vidicon

and Newvicon tube cameras are more sensitive to the visible wavelengths produced by most natural and artificial light sources.

3.2.4 Beam Angle

Another characteristic important in determining the amount of light reaching a scene is the beam angle over which the source radiates.

Light sources are classified by their light-beam pattern: Do they emit a wide, medium, or narrow beam of light? The requirement for this parameter is determined by the field of view (FOV) of the camera lens used and the total scene to be viewed. It is best to match the camera lens FOV (including any pan and tilt motion) to the light-beam radiation pattern to obtain the best uniformity of illumination over the scene, and hence the best picture quality and light efficiency. Note: Most lighting manufacturers can tell you the coefficient of utilization (CU) for a specific fixture (luminaire). The CU expresses how much light the fixture lens directs to the desired location (e.g., CU = 75%).

Figure 3-4 shows the beam patterns of natural and artificial light sources. The natural sources are inherently wide,

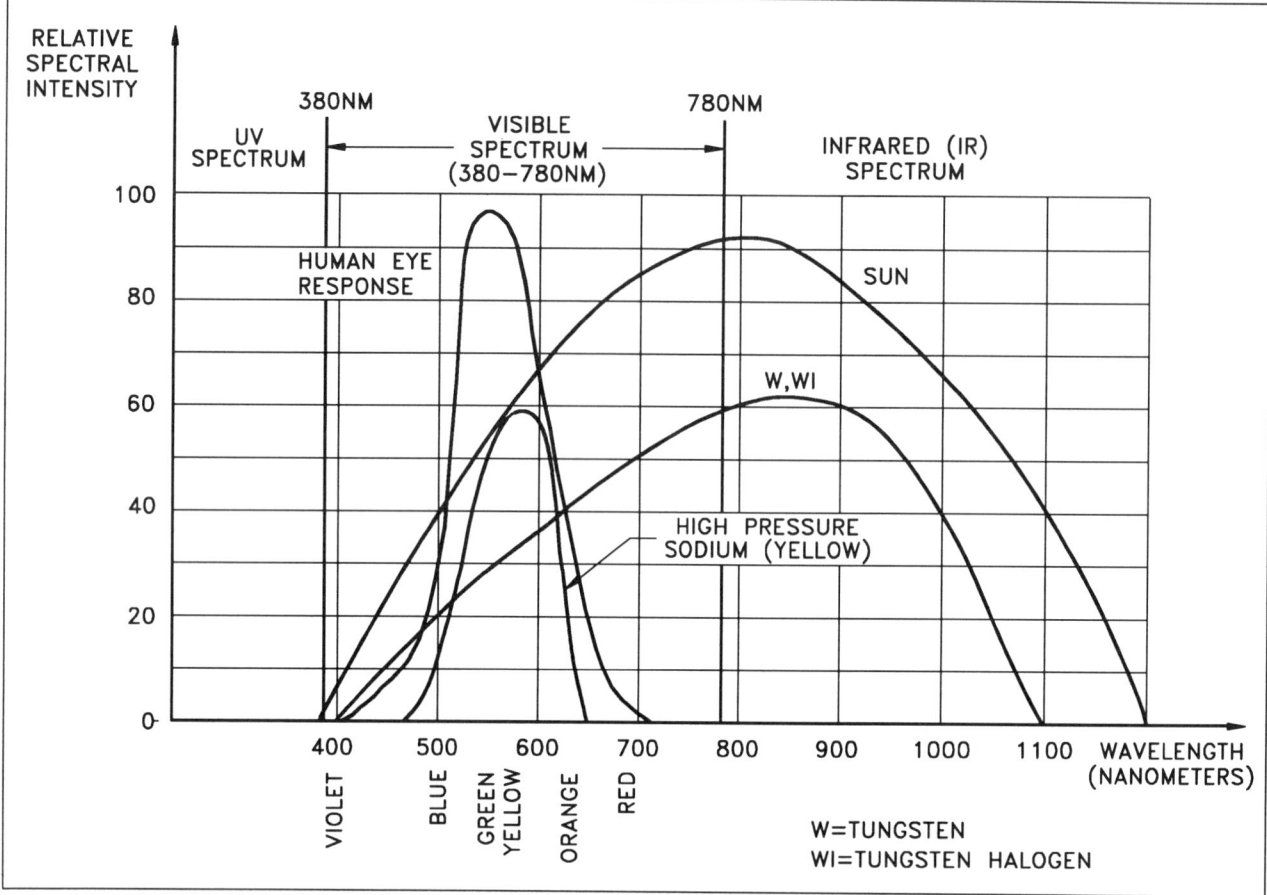

FIGURE 3-3 Light and IR output from common illumination sources

while the artificial sources are available in narrow-beam (a few degrees) to wide-beam (30 to 90 degrees) patterns.

The sun and moon, as well as some artificial light sources operating without a reflector, radiate over an entire scene. Artificial light sources and lamps almost always use lenses and reflectors and are designed or can sometimes be adjusted to produce narrow or wide-angle beams. If a large area is to be viewed, either a single wide-beam source or multiple sources must be located within the scene to illuminate it fully and uniformly. If a small scene at a long range is to be viewed, it is necessary to illuminate only that part of the scene to be viewed, resulting in a reduction in the total power needed from the source.

3.2.5 Natural Light Sources

There are two broad categories of light: natural and artificial. Natural light sources include the sun, moon, (reflected sunlight), and stars and contain most of the colors of the visible spectrum (blue to red), as shown in Figure 3-3. Sun- and moonlight contain IR radiation in addition to visible light spectra and are broadband light sources, that is, they contain all colors and wavelengths. Artificial sources can be broadband or narrowband, that is, containing only a lim-

ited number of colors. Monochrome cameras can use either narrow- or broadband sources, since they respond to the full spectrum of energy, and color rendition is not a concern. Broadband light sources containing most of the visible colors are necessary for good color rendition in color CCTV systems. (See the discussion of metal halides in Section 3.4.3.)

3.2.6 Outdoor Illumination

When television surveillance views an outdoor scene, the light source is natural or artificial, depending on the time of day. During the daytime, operating conditions will vary, depending on whether there is bright sun, clouds, overcast sky, or precipitation; thus the light's color or spectral energy, as well as its intensity, will vary. For nighttime operation the most widely used lamps are tungsten, tungsten-halogen, sodium, mercury, and high-intensity-discharge (HID) metal-arc types.

Before any camera system is chosen, the site should be surveyed to determine whether the area under surveillance will receive direct sunlight and whether the camera will be pointed toward the sun (to the south or west). Whenever possible, cameras should be pointed away from the sun

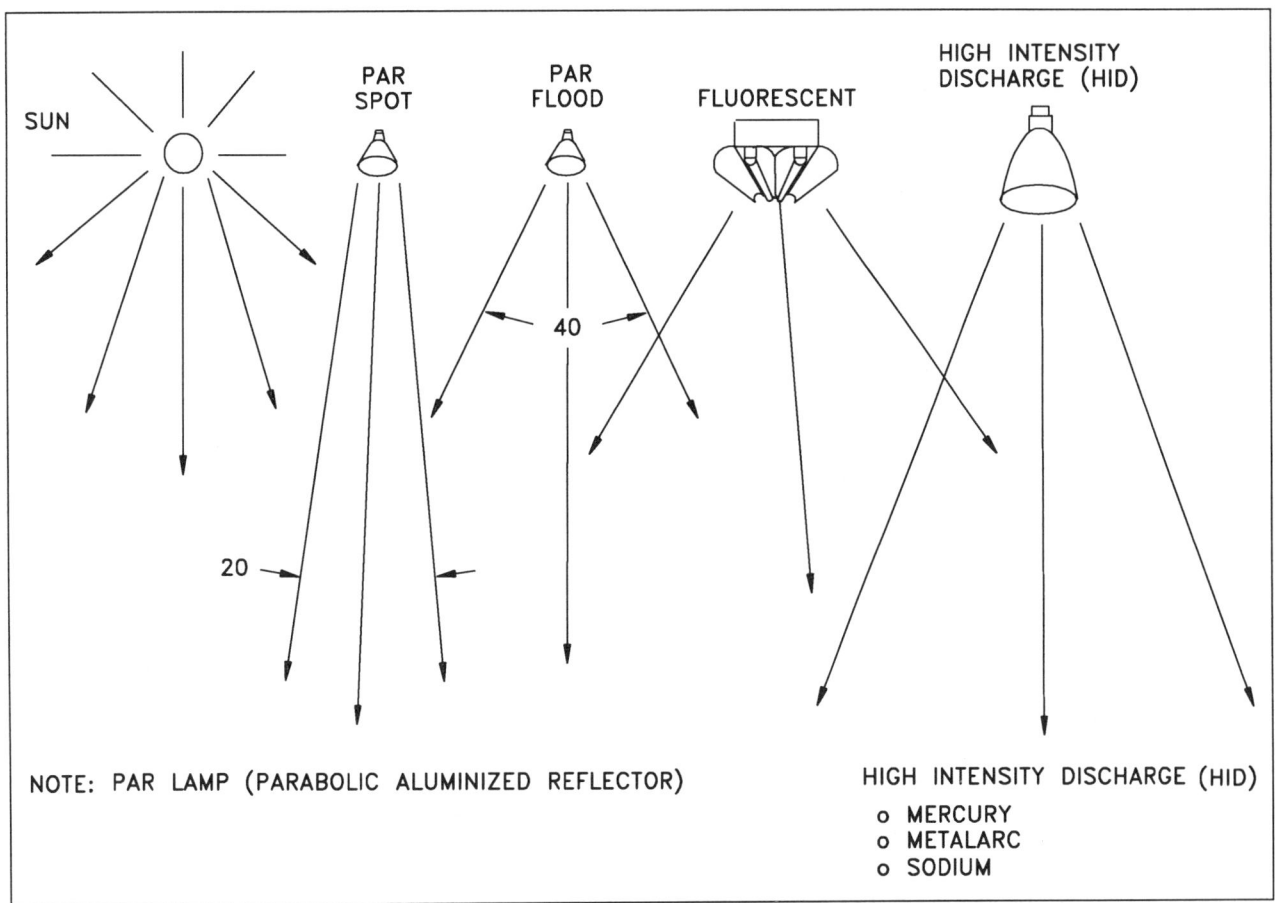

FIGURE 3-4 Beam patterns from common sources

to avoid additional glare and potential damage to the camera. Also, when the camera views a bright background or bright source, persons or objects near the camera may be hard to identify, since not much light illuminates them from the direction of the camera.

The CCTV camera for outdoor applications, where the light level and scene contrast range widely, will require some automatic light-level adjustment, usually with an automatic-iris lens or an electronic shutter in the camera. During nighttime or dusk operation, the camera system may see moonlight and/or starlight and reflected light from artificial illumination. The light level varies from a maximum of 10,000 fc (from natural bright sunlight) to a minimum of 1 fc (from artificial lamplight at night), giving a ratio of 10,000 to 1.

For indoor applications, tube and solid-state cameras (such as CCDs) usually have sufficient sensitivity and dynamic range to produce a good image and can operate with manual-iris lenses. In outdoor applications, the wide variation in light level necessitates the use of cameras such as silicon tube, CCD solid-state, SIT, ISIT, and ICCD, which all need some form of light compensation. Most outdoor cameras must have automatic-iris-control lenses or shuttered CCDs to adjust over the large light-level range encountered. Very often an expensive CCTV camera may cost

less than having to increase the lighting in a parking lot or exterior perimeter in order to obtain a satisfactory picture with a less expensive camera.

Every natural and artificial light source has a unique output spectrum of colors, which may or may not be advantageous for a particular camera. Monochrome CCTV systems for the most part cannot perceive the color distribution or spectrum of colors from different light sources. The picture quality of most monochrome systems depends solely on the total amount of energy emitted from the lamp—energy that the camera is sensitive to. When the lamp output spectrum of colors falls within the range of the sensor spectral sensitivity, then an optimum efficiency results.

The situation is more complex and critical for color CCTV systems. For the camera to be able to see *all* the colors in the visible spectrum, the light source must contain all the colors of the spectrum. To get a good color balance, the illumination source should match the sensor sensitivity. Most color cameras have what is called an automatic white-balance control, which automatically adjusts the camera electronics to obtain the correct color balance. The light source must still contain the colors in order for them to be seen on the monitor. Broadband light sources such as the sun, tungsten or tungsten-halogen, and xenon produce the best color pictures because they contain all the colors in the

Table 3-1 Light-Level Range from Natural Sources

LIGHTING CONDITION	LIGHT LEVEL	
	FTCD*	LUX**
UNOBSTRUCTED SUN	10,000	100,000
SUN WITH LIGHT CLOUD	7,000	70,000
SUN WITH HEAVY CLOUD	2,000	20,000
SUNRISE, SUNSET	50	500
TWILIGHT	.4	4
FULL MOON	.02	.2
QUARTER MOON	.002	.02
OVERCAST MOON	.0007	.007
CLEAR NIGHT SKY	.0001	.001
AVERAGE STARLIGHT	.00007	.0007
OVERCAST NIGHT SKY	.000005	.00005

*LUMENS PER SQUARE FOOT (FTCD)
** LUMENS PER SQUARE METER (LUX)
NOTE: 1 FTCD EQUALS APPROXIMATELY 10 LUX

spectrum. Narrowband light sources such as mercury-arc or sodium-vapor lamps do not produce a continuous spectrum of colors, so color is rendered poorly. A mercury lamp has little red light output, and therefore red objects appear nearly black when illuminated by a mercury arc. Likewise, a high-pressure sodium lamp contains large quantities of yellow, orange, and red light; a blue or blue/green object will look dark or gray or brown in its light. A low-pressure sodium lamp produces only yellow light and consequently is unsuitable for color CCTV.

Another aspect of artificial lighting is the consideration of the light-beam pattern from the lamp and the lens FOV. A wide-beam floodlamp will illuminate a large area with a fairly uniform intensity of light and therefore produce a more uniform picture. A narrow-beam light or spotlight will illuminate a small area; consequently, areas on the edges and beyond will be darker. A scene that is illuminated nonuniformly (i.e., with high contrast) will result in a nonuniform picture. For maximum efficiency, the camera lens FOV should match the lamp beam angle. If a lamp illuminates only a particular area of the scene, the camera-lens combination FOV should only be viewing the area illuminated by the lamp. This source beam angle problem does not exist for areas lighted by natural illumination such as the sun, which usually uniformly illuminates the entire scene.

All objects emit light when sufficiently hot. Changing the temperature of an object changes the intensity and color of the light emitted from it. For instance, iron glows dull red when first heated, then red-orange when it becomes hotter. In a steel mill, molten iron appears yellow-white because it is hotter than the red-orange of the lower-temperature iron. The tungsten filament of an incandescent lamp is hotter yet and emits nearly white light. Any object that is hot enough to glow is said to be incandescent: hence the

term for heated-filament bulbs. A meaningful parameter in color television is the color temperature or apparent color temperature of an object when heated to various temperatures.

A special radiating object emits radiation with 100% efficiency at all wavelengths when heated. Scientists call this object a blackbody radiator. The blackbody radiator emits energy in the ultraviolet, visible, and infrared spectrums following specific physical laws.

Tungsten lamps and the sun radiate energy like a blackbody because they radiate with a continuous spectrum, that is, they emit at all wavelengths (or colors). Other sources such as mercury, fluorescent, sodium, and metal-arc lamps do not emit a continuous spectrum but only produce narrow bands of colors: mercury produces a green-blue band, sodium produces a yellow-orange band.

3.3 NATURAL LIGHT

3.3.1 Sunlight

If the scene in Figure 3-1 is illuminated by sunlight, moonlight, or starlight, it will receive uniform illumination. If it is illuminated by several artificial sources, the lighting may vary considerably over the FOV of the camera and lens. Table 3-1 summarizes the overall light-level ranges, from direct sunlight to overcast starlight.

As mentioned, the measure of illumination is the foot-candle (fc), which is equivalent to lumens per square feet. For outdoor conditions, the camera system, which must operate over the full range from direct sunlight to nighttime conditions, must have an automatic light control means to compensate for this light-level change. Figure 3-5 summarizes the characteristics of natural sources, that is,

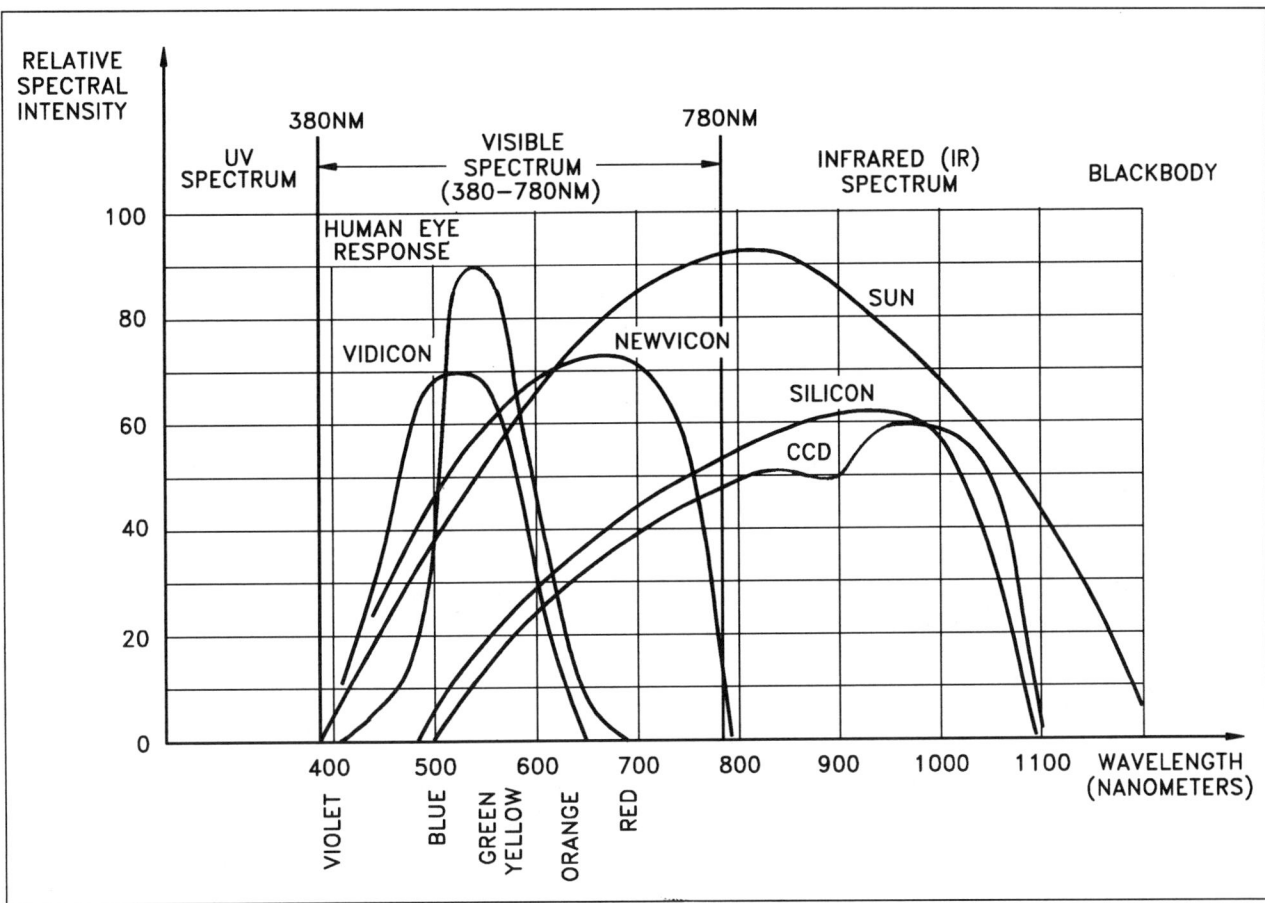

FIGURE 3-5 Spectral characteristics of natural sources and camera sensors

the sun, moon, and starlight, and how different camera types respond to them.

The sun illuminates outdoor scenes during the daylight hours. As an energy source, the sun is a continuum of all wavelengths and colors to which monochrome and color television cameras are sensitive. This continuum includes visible radiation in the blue, green, yellow, orange, red, and near-IR parts of the spectrum. All monochrome cameras are sensitive to the visible spectrum; some, particularly silicon-tube and CCD solid-state cameras, are sensitive to the visible and near-IR spectrum. Color cameras are sensitive to all the color wavelengths in the visible spectrum, as is the human eye. Color cameras are purposely designed to be insensitive to near-IR wavelengths.

During the first few hours of the morning and the last few hours of the evening, sunlight's spectrum is shifted toward the orange-red region, so things look predominantly orange and red. During the midday hours, when the sun is brightest and most intense, blues and greens are brightest and reflect the most balanced white light. For this reason a color camera must have an automatic white-balance control, which adjusts for color shift during the day so that the resulting picture represents the color in the scene reasonably well.

3.3.2 Moonlight and Starlight

After the sun has gone down in an environment without artificial lighting, the scene may be illuminated by the moon, the stars, or both. Since moonlight is reflected light from the sun, it contains most of the colors emitted from the sun. However, the low level of illumination reaching the earth prevents color cameras (and the human eye) from providing good color rendition.

3.4 ARTIFICIAL LIGHT

Artificial light sources consist of several types of lamps, which may be used in outdoor parking lots, storage facilities, fence lines, or in indoor environments for lighting rooms, hallways, work areas, elevator lobbies, and so on. Lamps come in two types: tungsten or tungsten-halogen lamps, which have solid filaments, and gaseous or arc lamps, which contain low- or high-pressure gas in an enclosed envelope. Arc lamps can be further classified into HID, low-pressure, and high-pressure short-arc types. HID lamps are used most extensively because of their high efficacy and long life. Low-pressure arc lamps include fluorescent and low-pressure sodium types, which are used in many indoor and outdoor installations. Long-arc xenon lamps are used in large outdoor sports arenas. High-pressure short-arc lamps find limited use in security applications that require a high-efficiency, well-directed narrow beam to illuminate a target at long distances (hundreds or thousands of feet). Such lamps include xenon, metal-halide, high-pressure sodium, and mercury. For covert security applications, some lamps are fitted with a visible-light-blocking filter so that only invisible IR radiation illuminates the scene.

A significant advance in tungsten lamp development is the use of a halogen element (iodine or bromine) in the lamp's quartz envelope, with the lamp operating in what is called the tungsten-halogen cycle. This operation increases a lamp's rated life significantly even though it operates at a high temperature and light output. Incandescent filament lamps are available with power ratings from a fraction of a watt to 10 kilowatts.

HID arc lamps comprise a broad class of light sources in which the arc discharge takes place between electrodes contained in a transparent or translucent bulb. The spectral radiation output and intensity are determined principally by the chemical compounds and gaseous elements that fill the bulb. The lamp is started using high-voltage ignition circuits. Some form of electrical ballasting is used to stabilize the arcs. In contrast, tungsten lamps operate directly from the power source. HID lamps take many forms. Compact short-arc lamps are only a few inches in size but emit high-intensity, high-lumen output radiation with a variety of spectral characteristics. In long-arc lamps, such as fluorescent, low-pressure sodium vapor, and xenon, their output spectral characteristics are determined by the gas in the arc or the tube-wall emitting material. The fluorescent lamp has a particular phosphor coating on the inside of the glass and bulb, which determines its spectral output. Power outputs available from arc-discharge lamps range from a few watts up to many tens of kilowatts.

The following sections describe some of the lamps in use today and their characteristics with respect to CCTV security applications.

3.4.1 Tungsten Lamps

The first practical artificial lighting took the form of an incandescent filament tungsten lamp. Introduced in 1907, these lamps used a tungsten mixture formed into a filament and produced an efficacy (ratio of light out to power in) of approximately 7 lumens per watt of visible light, a great increase over anything existing at the time. In 1913 ductile tungsten wire was fabricated, and coiled filaments were made possible. With this innovation, the luminous efficiency (efficacy) was increased to 20 lumens per watt.

Today the incandescent lamp is commonplace and is used in virtually every home, business, factory, school, hos-

pital, train station, airport, and so on. While its efficacy does not measure up to that of the arc lamp, the tungsten incandescent lamp nevertheless offers a low-cost installation for many applications. Since it is an incandescent source, it radiates all the colors in the visible spectrum as well as the near-IR spectrum, providing an excellent illumination source for monochrome and color CCTV. Its only two disadvantages when compared with arc lamps are (1) its relatively low efficiency, which makes it more expensive to operate, and (2) its relatively short operating life of several thousand hours.

Incandescent filament lamps are probably the most widely used sources of artificial illumination. The efficacy of the tungsten lamp increases with filament operating temperature and is nearly independent of other factors. However, lamp life expectancy decreases rapidly as lamp filament temperature increases. Maximum practical efficacy is about 35 lumens per watt in high-wattage lamps operated at approximately 3500 degrees Kelvin color temperature. This temperature approaches the melting point of tungsten. As a practical matter, a tungsten lamp cannot operate at this high temperature, since it will last only a few hours. At lower temperatures, life expectancy increases to several thousand hours, which is typical of incandescent lamps.

Incandescent lamp filaments are usually coiled to increase their efficiency. The coils are sometimes coiled again (coiled-coiled) to further increase the filament area and increase the luminance. Filament configurations are designed to optimize the radiation patterns for specific applications. Sometimes long and narrow filaments are used and mounted into cylindrical reflectors to produce a rectangular beam pattern. Others have small filaments so as to be incorporated into parabolic reflectors to produce a narrow collimated beam (such as a spotlight). Others have larger filament areas and are used to produce a wide-angle beam (such as a floodlight). Figure 3-6 shows several lamp configurations, and Figure 3-7 shows some standard lamp luminaires used in industrial, residential, and security applications.

The luminaire fixtures house the tungsten, HID, and low-pressure lamps described in the following sections. An incandescent lamp consists of a tungsten filament surrounded by an inert gas sealed inside a transparent or frosted-glass envelope. The purpose of the frosted glass is to increase the apparent size of the lamp, thereby decreasing its peak intensity and reducing glare and hotspots in the illuminated scene.

In conventional gas-filled, tungsten-filament incandescent lamps, tungsten molecules evaporate from the incandescent filament, flow to the relatively cool inner surface of the bulb wall (glass), and adhere to the glass, forming a thin film that gradually thickens during the life of the lamp and causes the bulb to darken. This molecular action reduces the lumen light output (resulting in a lumen loss) and

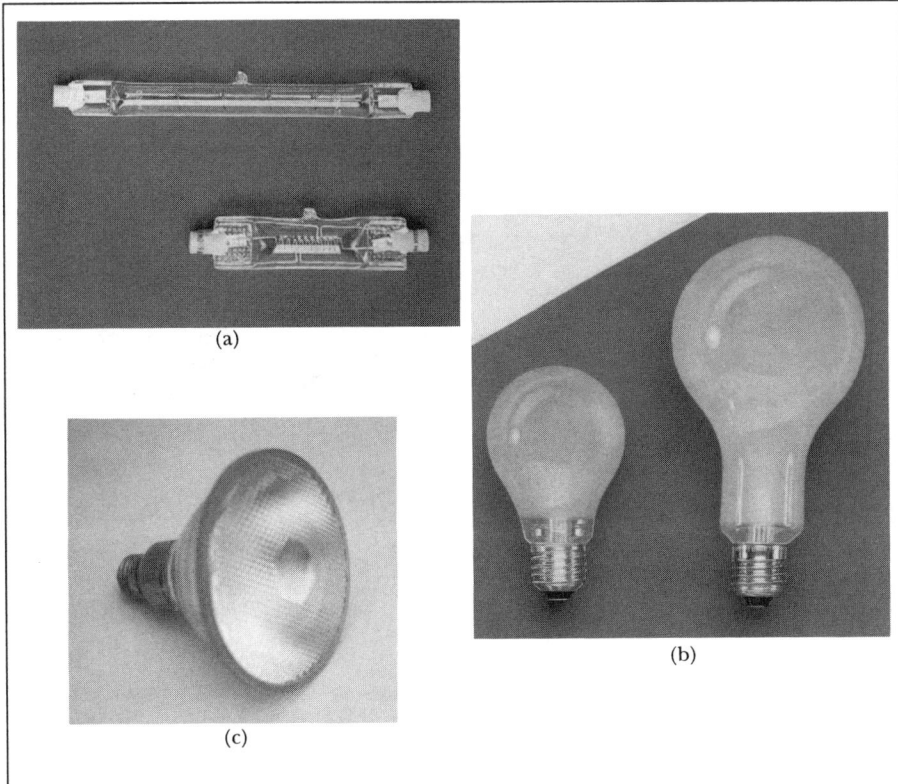

(a)

(b)

(c)

FIGURE 3-6 Generic tungsten and tungsten-halogen lamp configurations: (a) tungsten- halogen in quartz envelope, (b) tungsten filament, (c) tungsten-halogen in parabolic aluminized reflector

efficacy in two ways: First, evaporation of tungsten from the filament reduces the filament wire's diameter and increases its resistance, so that light output and color temperature increase. Second, the tungsten deposited on the bulb wall increases the opacity (reduces transmission of light through the glass) as it thickens. The blackened bulb wall absorbs an increased percentage of the light produced by the filament, thus reducing the light output. The tungsten-halogen lamp significantly improves on this deficiency in the incandescent lamp.

Figure 3-8 illustrates the relative amount of energy produced by tungsten-filament and halogen-quartz-tungsten lamps as compared with other arc-lamp types, including fluorescent, metal-arc, and sodium, in the visible and near-IR spectral range.

On an absolute basis, the energy produced by the tungsten lamp in the visible spectral region is significantly lower than that provided by other HID lamps. However, the total amount of energy produced by the tungsten lamp over the entire spectrum is comparable to that of the other lamps. Figure 3-8 shows the human eye response and spectral sensitivity of standard CCTV camera sensors.

3.4.2 Tungsten-Halogen Lamps

The discovery of the tungsten-halogen cycle significantly increased the operating life of the tungsten lamp. Tungsten-halogen lamps, like conventional incandescent lamps,

use a tungsten filament in a gas-filled light-transmitting envelope and emit light with a spectral distribution similar to that of a tungsten lamp. Unlike the standard incandescent lamp, the tungsten-halogen lamp contains a trace vapor of one of the halogen elements (iodine or bromine) along with the usual inert fill gas. Also, tungsten-halogen lamps operate at much higher gas pressure and bulb temperature than non-halogen incandescent lamps. The higher gas pressure retards the tungsten evaporation, allowing the filament to operate at a higher temperature, resulting in higher efficiencies than conventional incandescent lamps. To withstand these higher temperatures and pressures, the lamps use quartz bulbs or high-temperature "hard" glass. The earliest version of these lamps used fused quartz bulbs and iodine vapor and were called quartz iodine lamps. After it was found that other halogens could be used, the more generic *tungsten-halogen lamp* is now used.

The important result achieved with the addition of halogen was caused by the "halogen generative cycle," which maintains a nearly constant light output and color temperature throughout the life of the lamp and significantly extends the life of the lamp. The halogen chemical cycle permits the use of more compact bulbs than those of tungsten filament lamps of comparable ratings and permits increasing either lamp life or lumen output and color temperature to values significantly above those of conventional tungsten filament lamps.

Incandescent and xenon lamps are good illumination sources for IR CCTV applications when the light output is filtered with a covert filter (one that blocks or absorbs

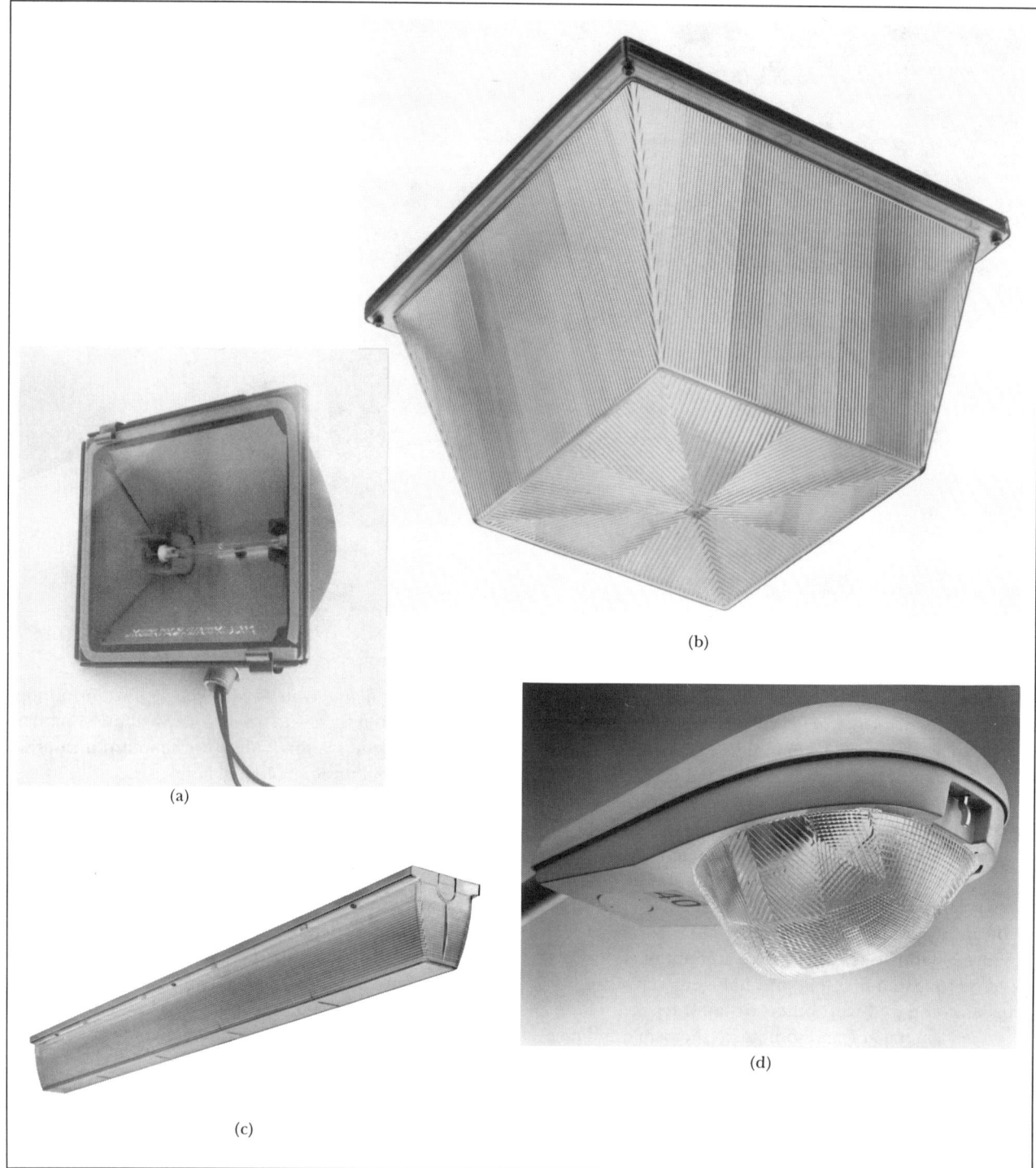

FIGURE 3-7 Standard lamp luminaires: (a) tungsten-halogen, (b) sodium, (c) fluorescent, (d) metal-arc

the transmission of visible radiation) and they transmit only the near-IR radiation. Figure 3-9 shows a significant portion of the emitted spectrum of the lamp radiation falling in the near-IR region that is invisible to the human eye but to which tube and solid-state silicon cameras are sensitive.

The reason for this is shown in Figure 3-9, which details the spectral characteristics of these lamps.

When an IR-transmitting/visible-blocking filter is placed in front of a tungsten-halogen lamp, only the IR energy illuminates the scene and reflects back to the CCTV camera lens. This combination produces an image on the television monitor from an illumination source that is invisible to the eye. This technology is commonly referred to as seeing in the dark: there is no visible radiation and yet a CCTV image is discernible. Solid-state CCD and silicon tube sensors are

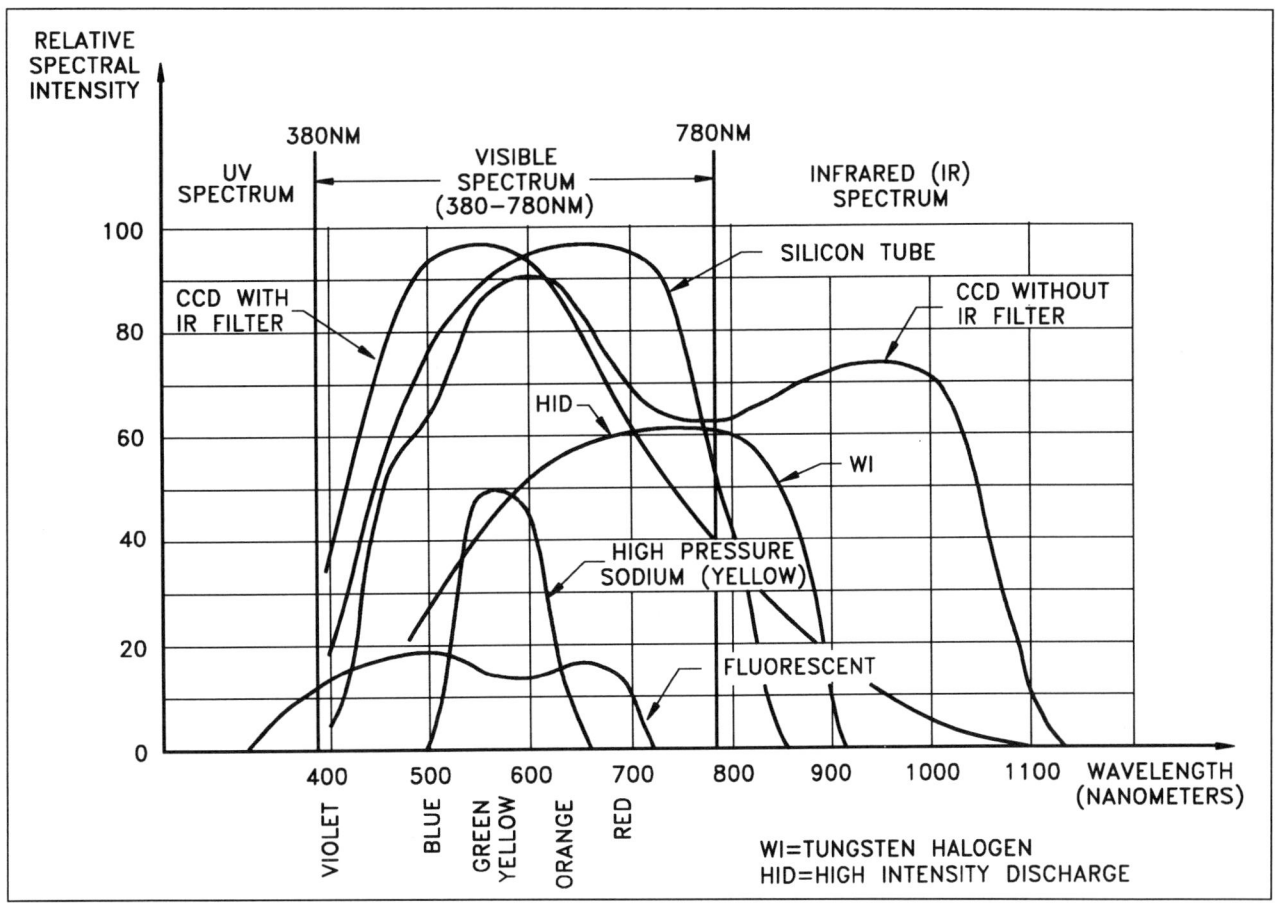

FIGURE 3-8 Spectral characteristics of lamps and camera sensors

responsive to this near-IR radiation. Since the IR region has no "color," color cameras are designed to be insensitive to the filtered IR energy. As can be seen in Figure 3-9, approximately 90% of the energy emitted by the tungsten-halogen lamp occurs in the IR region. However, only a fraction of this infrared light can be used by the silicon sensor, since the sensor is responsive only up to approximately 1100 nanometers. The remaining IR energy above 1100 nanometers does not help to produce the television image and manifests as heat, which does not contribute to the image. While the IR source is not visible to the human eye, it is detectable by silicon camera devices and other night vision devices (Chapter 15), which can be problematic in some covert applications.

3.4.3 High-Intensity-Discharge Lamps

A class of lamp in widespread use for general lighting and security applications is an enclosed arc type called a high-intensity-discharge lamp. There are three major types of HID lamps, each one having a relatively small arc tube

mounted inside a heat-conserving outer jacket and filled with an inert gas to prevent oxidation of the hot arc tube seals (Figure 3-10).

The three most popular HID lamps are (1) mercury in a quartz tube, (2) metal halide in a quartz tube, and (3) high-pressure sodium in a translucent aluminum-oxide tube. Each type differs in electrical input, light output, shape, and size. While incandescent lamps require no auxiliary equipment and operate directly from a suitable voltage, HID lamps—because they are discharge sources—require an electrical ballast and a high-voltage starting device. The high-voltage ignition provides the voltage necessary to start the lamp; once the lamp is started, the ballast operates it at the rated power (wattage) or current level. The ballast consumes power, which must be factored into calculations of system efficiency. HID lamps, unlike incandescent or fluorescent lamps, require several minutes to warm up before reaching full brightness. If turned off momentarily, they take several minutes before they can be turned on again (reignited).

The primary overriding advantages of HID lamps are their long life, provided they are operated at a minimum of several hours per start, and their high efficiency. Lamp

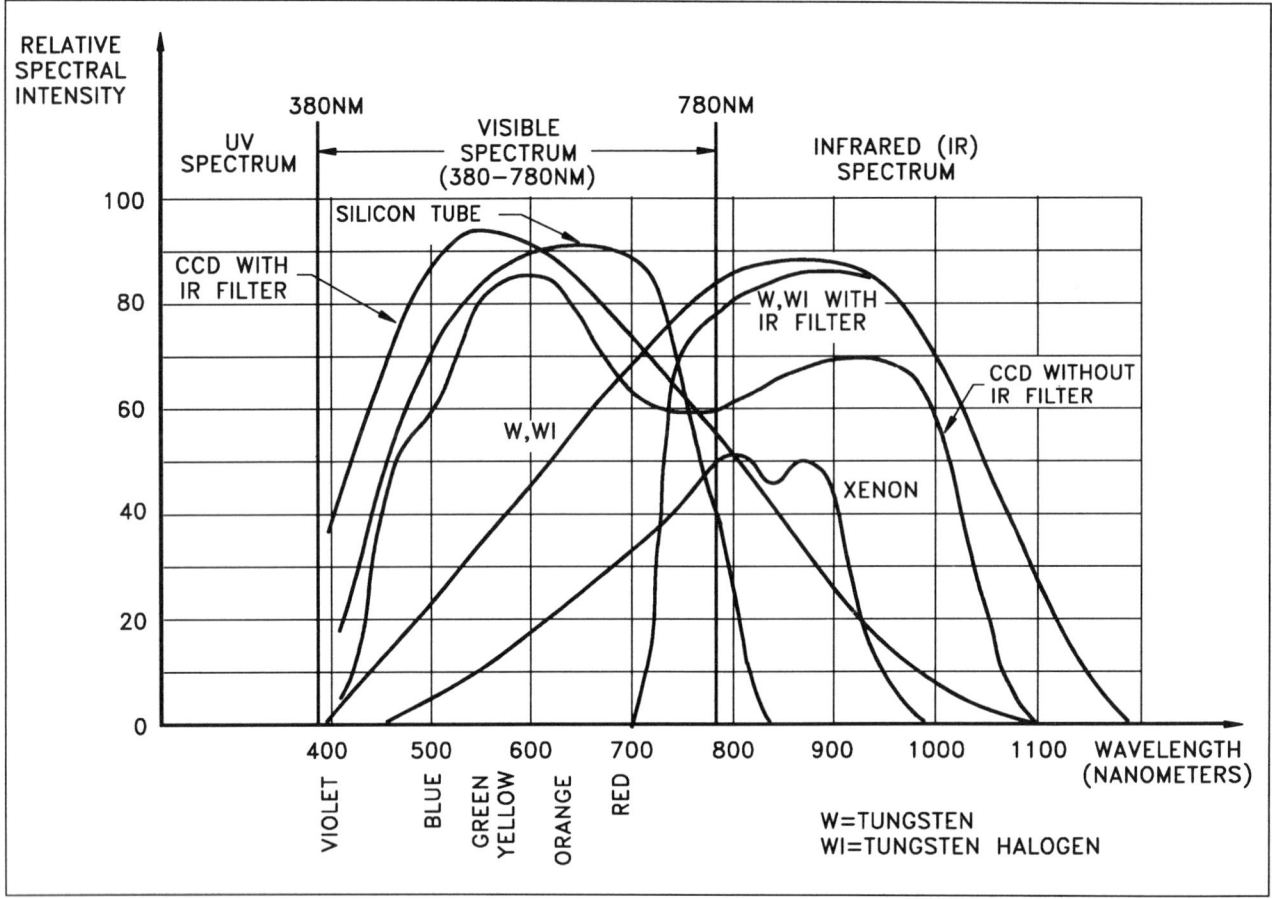

FIGURE 3-9   Filtered tungsten and xenon lamps vs. camera spectral sensitivity

lifetime is typically 16,000 to 24,000 hours, and light efficacy ranges from 60 to 140 lumens per watt. These lamps cannot be electrically dimmed without drastically affecting the starting warm-up luminous efficiency, color, and life.

HID lamps are the most widely used lamps for lighting industrial and commercial buildings, streets, sports fields, and so on. One disadvantage typical of short-arc lamps, as just mentioned, is their significant warm-up time—usually several minutes to ten minutes. If accidentally or intentionally turned off, these lamps cannot be restarted until they have cooled down sufficiently to reignite the arc, which may be 2 to 5 minutes, and then take an additional 5 minutes to return to full brightness. Dual-HID bulbs are now available, which include two identical HID lamp units, only one of which operates at a time. If the first lamp is extinguished momentarily, the cold lamp may be ignited immediately, eliminating the waiting time and allowing the first lamp to cool down.

The principle in all vapor-arc lighting systems is the same: an inert gas is contained within the tube to spark ignition. The inert gas carries current from one electrode to the other; the current develops heat and vaporizes the solid metal or metallic-oxide inside the tube; and then light is discharged from the vaporized substance, through the surface of the discharge tube, and into the area to be lighted.

Mercury HID lamps are available in sizes from 40 watts through 1500 watts. Spectral output is high in the blue region but extremely deficient in the red region; therefore, they should be used only in monochrome CCTV applications (Figure 3-11).

A second class of HID lamp is the metal-halide, which is filled with mercury-metallic iodides. These lamps are available with power ratings from 175 to 1500 watts. The addition of metallic salts to the mercury arc improves the efficacy and color by adding emission lines to the spectrum. With different metallic additives or different phosphor coatings on the outside of the lamp, the lamp color varies from an incandescent spectrum to a daylight spectrum. The color spectrum from the metal-halide lamp is significantly improved over the mercury lamp and can be used for monochrome or color CCTV applications.

The third class of HID lamp is the high-pressure sodium lamp. This lamp contains a special ceramic-arc tube material that withstands the chemical attack of sodium at high temperatures, thereby permitting high luminous efficiency and yielding a broader spectrum, compared with low-pres-

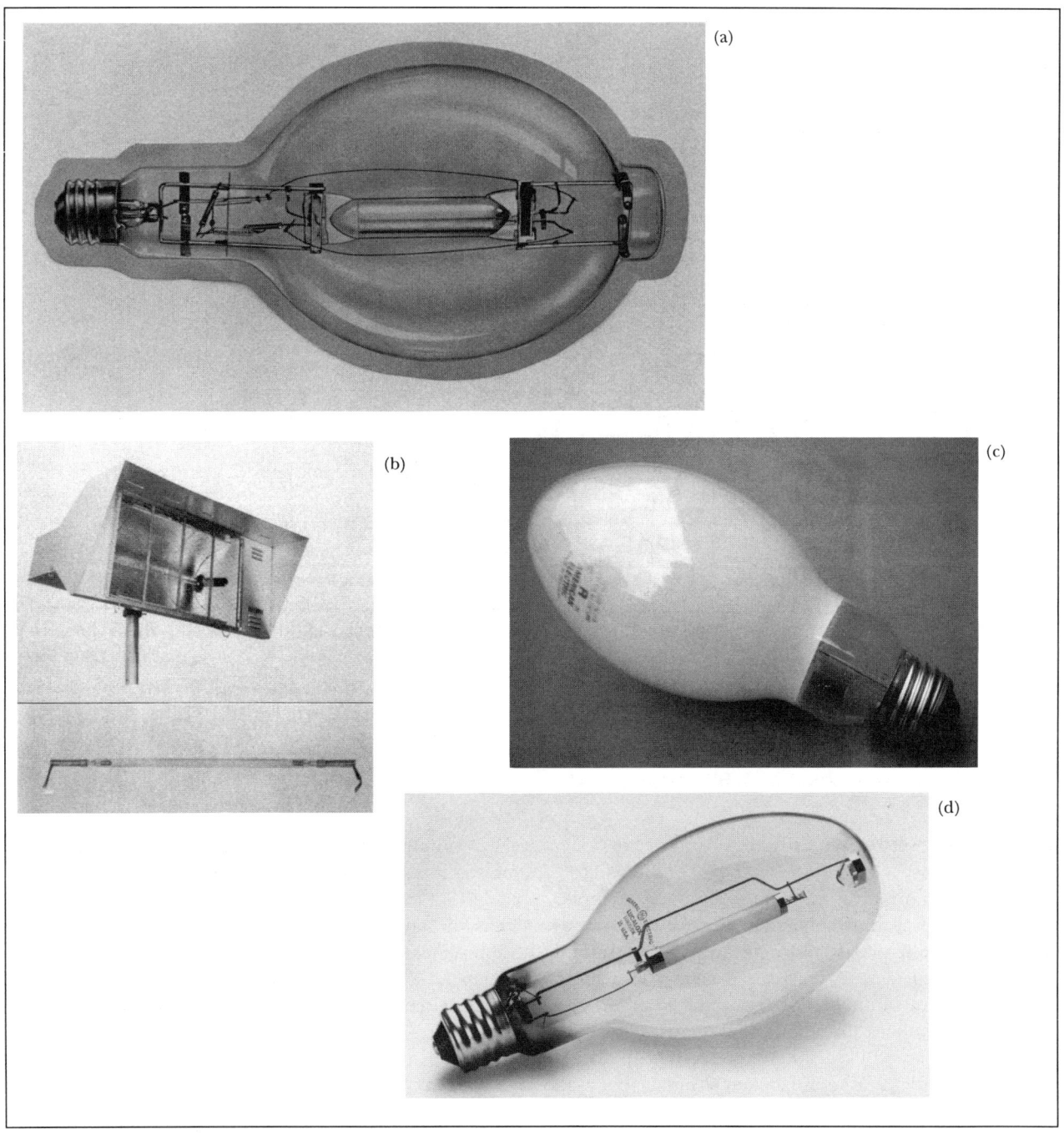

FIGURE 3-10 High-intensity-discharge lamps: (a) mercury, (b) xenon, (c) metal-arc, (d) sodium

sure sodium arcs. However, because the gas is only sodium, the spectral power distribution from the high-pressure sodium HID lamp is yellow-orange and has only a small amount of blue and green. For this reason the lamp is not the most suitable for good color CCTV security applications. The primary and significant advantage of the high-pressure sodium lamp over virtually all other lamps is its high luminous efficiency, approximately 60 to 140 lumens per watt. It also enjoys a long life, approximately 24,000

hours. The sodium lamp is an extremely good choice for monochrome surveillance applications.

HID lamps are filled to atmospheric pressure (when not operating) and rise to several atmospheres only when operating. This makes them significantly safer than short-arc lamps.

The choice of lamp is often determined by architectural criteria, but the CCTV designer should be aware of the color characteristics of each lamp to ensure their suitability for monochrome or color CCTV.

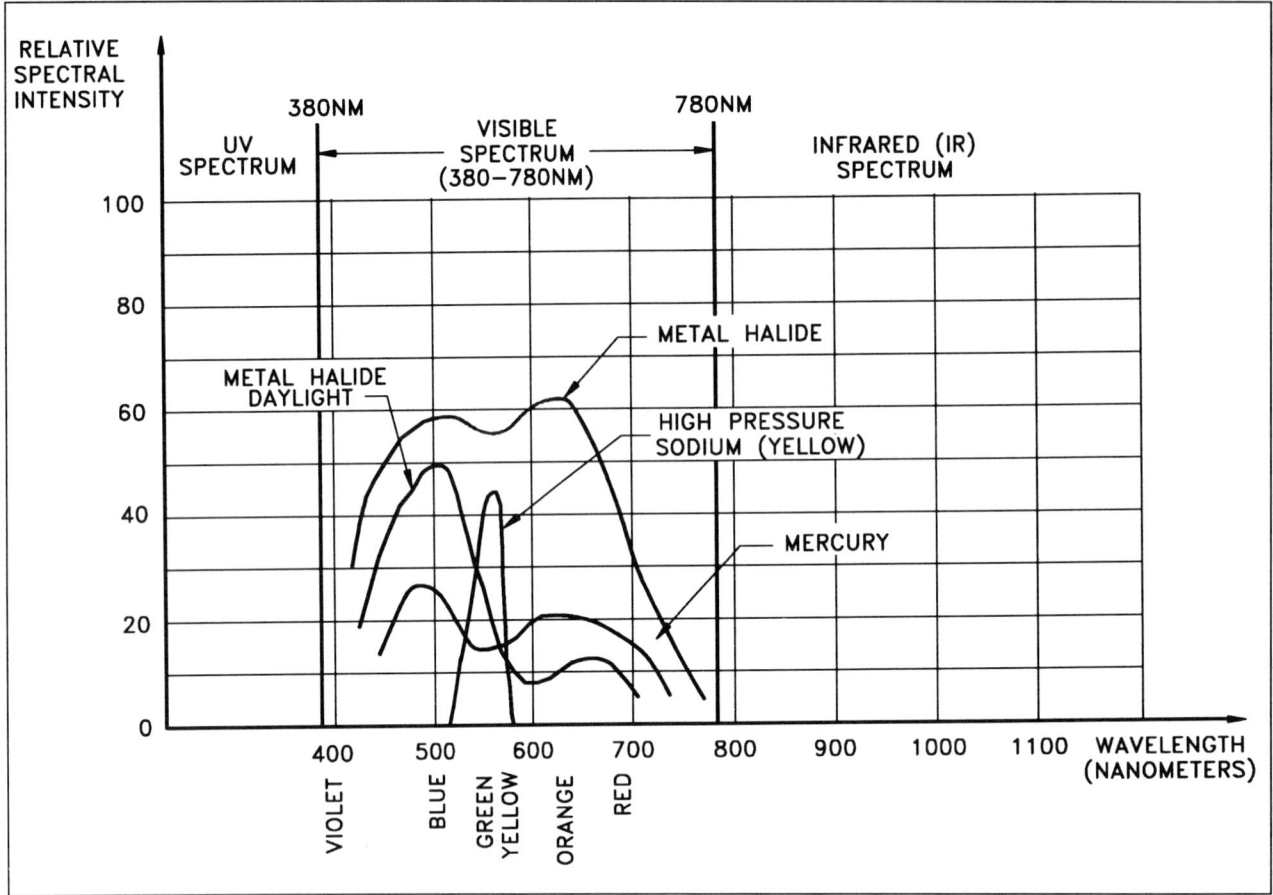

FIGURE 3-11   Spectral output from HID lamps

3.4.4 Low-Pressure Arc Lamps

Fluorescent and low-pressure sodium lamps are examples of low-pressure arc lamp illumination sources. These lamps have tubular bulb shapes and long arc lengths (several inches to several feet). A ballast is necessary to achieve proper operation, and a high-voltage pulse is required to ignite the arc and start the lamp.

The most common type is the fluorescent lamp, with a relatively high efficacy of approximately 60 lumens per watt. The large size of the arc tube (diameter as well as length) requires that it be placed in a large luminaire (reflector) to achieve a defined beam shape. For this reason, fluorescent lamps are used for large-area illumination and produce a fairly uniform pattern. The fluorescent lamp is sensitive to the surrounding air temperature, so it is primarily used for indoor lighting or in moderate temperatures. When installed outdoors in cold weather, a special low-temperature ballast must be used to ensure that the starting pulse is high enough to break over the arc and start the lamp.

The fluorescent lamp combines a low-pressure mercury arc with a phosphor coating on the interior of the bulb. The lamp arc produces ultraviolet radiation from the low-pressure mercury arc, which is converted into visible radiation by the phosphor coating on the inside wall of the outside tube. A variety of phosphor coatings is available to produce almost any color quality (Figure 3-12). Colors range from "cool white," which is the most popular variety, to daylight, blue white, and so on. The input power varies from 4 watts to approximately 200 watts; tube lengths vary from 6 inches to 56 inches (15 cm to 144 cm). Fluorescent lamps can be straight, circular, or U-shaped. Some special fluorescent lamps emit a continuous spectrum like an incandescent lamp; others emit a daylight spectrum.

A second class of low-pressure lamp is the sodium lamp, which outputs almost a single yellow color (i.e., nearly monochromatic). These lamps have ratings from 18 to 180 watts. The low-pressure sodium lamp has the highest efficacy output of any lamp type built to date, approximately 180 lumens per watt. While the efficacy is high, the lamp's monochromatic yellow limits it to security and roadway lighting applications that require only monochrome CCTV cameras. If used with color cameras, only yellow objects will appear yellow; all other objects will appear brown or black.

The low-pressure sodium light utilizes pure metal sodium with an inert-gas combination of neon-argon enclosed in a discharge tube about 28 inches long. The pressure in the tube is actually below atmospheric pressure, which causes the glass to collapse inward if it is ruptured—a good safety feature.

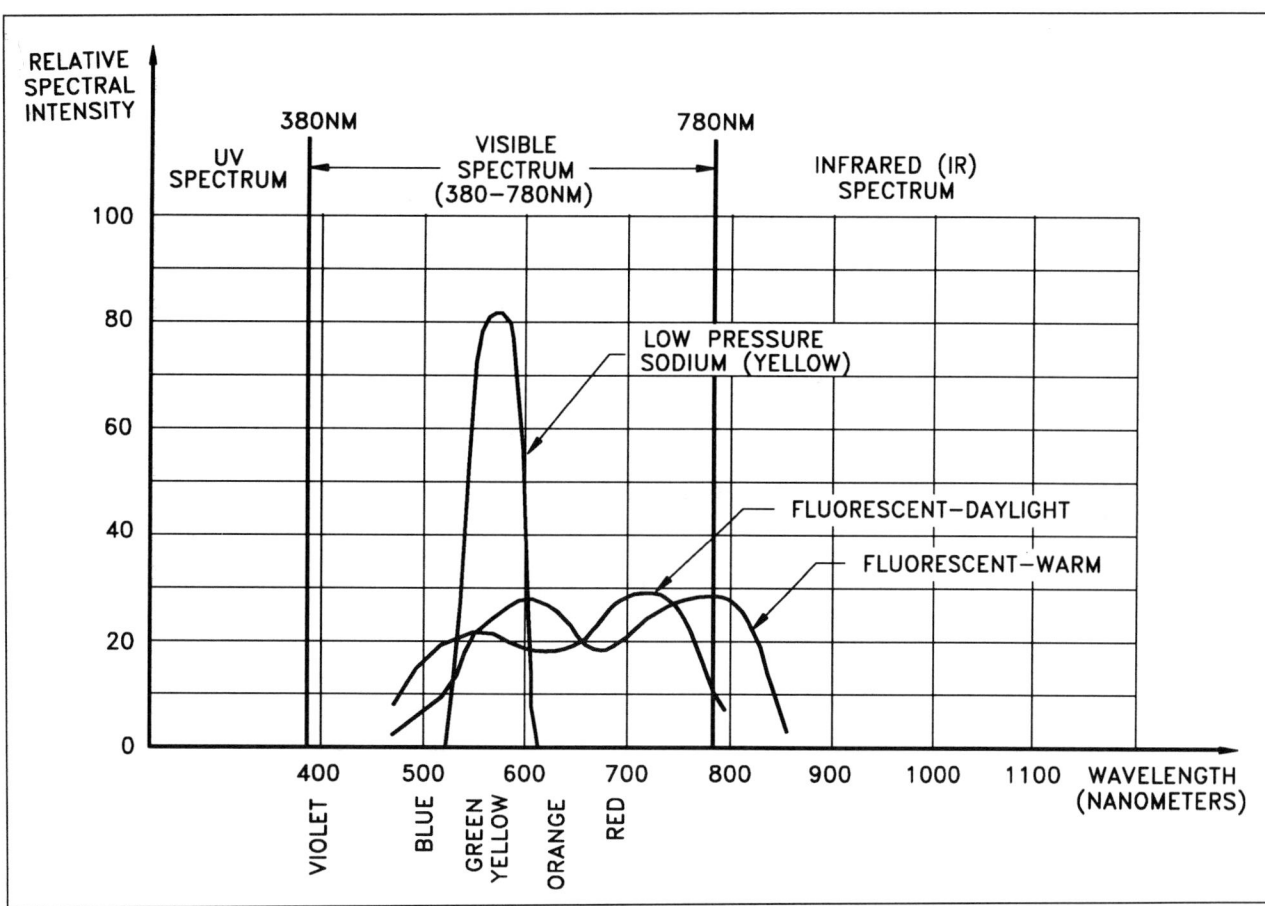

FIGURE 3-12 Light output from low-pressure arc lamps

A unique advantage of the low-pressure sodium amber light is its better "modeling" (showing of texture and shape) of any illuminated surface, for both the human eye and the CCTV camera. It provides more contrast, and since the monochrome CCTV camera responds to contrast, images under this light are clearer, according to some reports. The yellow output from the sodium lamp is close to the wavelength region at which the human eye has its peak visual response (560 nanometers).

Some security personnel and the police have identified low-pressure sodium as a uniquely advantageous off-hour lighting system for security, because the amber yellow color clearly tells people to keep out. This yellow security lighting also sends the psychological message that the premises are well guarded.

3.4.5 Compact Short-Arc Lamp

Enclosed arcs comprise a broad class of lamp in which the arc discharge takes place between two electrodes, usually tungsten, and is contained in a rugged transparent or frosted bulb. The spectrum radiated by these lamps is usually determined by the elements and chemical compounds inside.

One class of enclosed arc lamp is the compact short-arc lamp, whose arc is short compared with its electrode size, spacing, and bulb size and operates at relatively high currents and low voltages. Such lamps are available in power ratings ranging from less than 50 watts to more than 25 kilowatts. These lamps usually operate at low voltages (less than 100 volts), although they need a high-voltage pulse (several thousand volts) to start. The lamps generally operate on AC or DC power and require some form of current-regulating device (ballast) to maintain a uniform output radiation.

Several factors limit the useful life of compact lamps compared with HID lamps, especially the high current density required, which reduces electrode lifetime. Compact short-arc lamps generally have a life in the low thousands of hours and operate at internal pressures up to hundreds of atmospheres. Therefore they must be operated in protected enclosures and handled with care. The most common short-arc lamps are mercury, mercury-xenon, and xenon. Figure 3-13 shows the spectral output of mercury-xenon lamps. Short-arc xenon lamps are not common in security applications because of their high cost and short lifetime. However, they play an important role for IR sources used in covert surveillance.

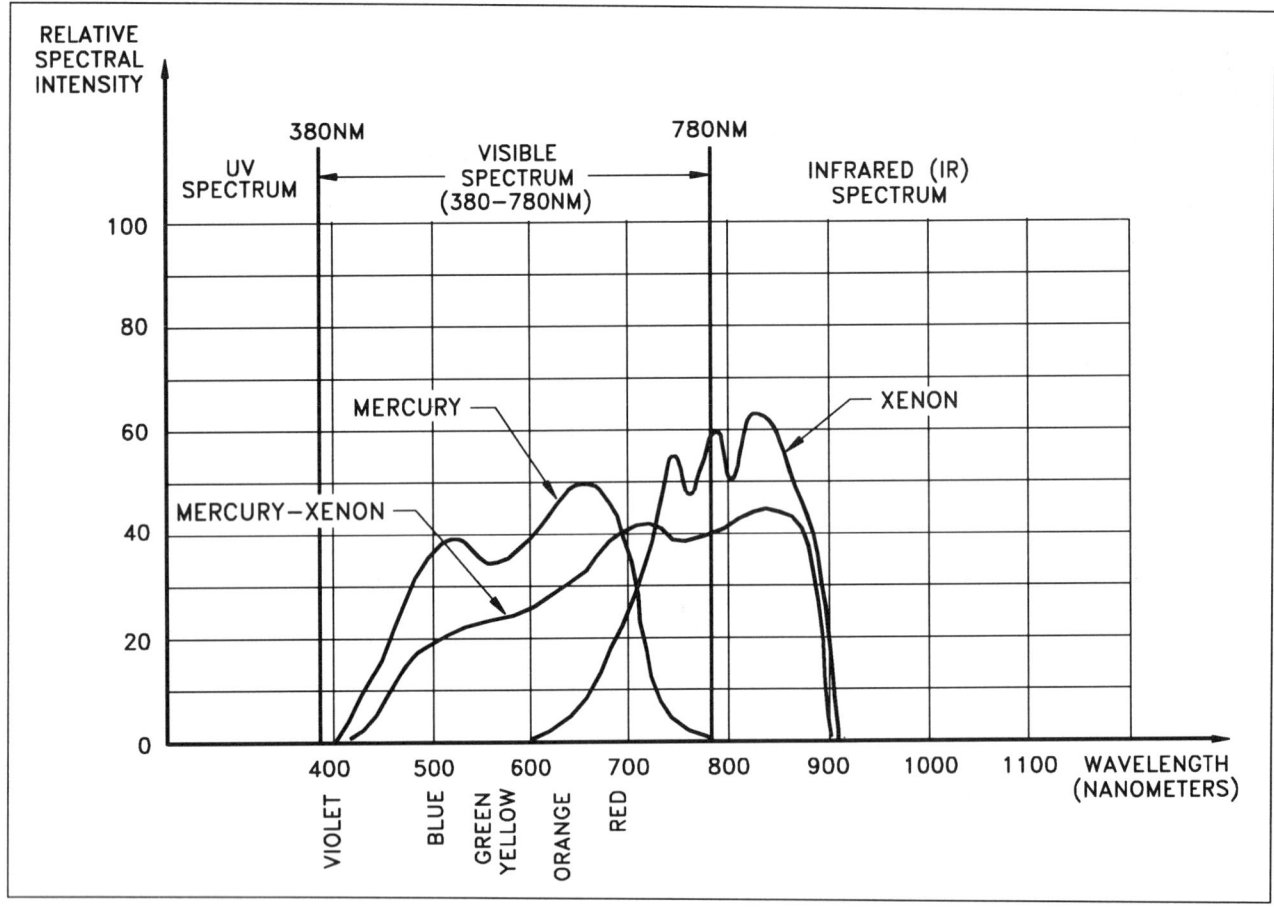

FIGURE 3-13 Spectral outputs of mercury-xenon lamps

Using a mercury arc lamp with monochrome CCTV cameras gives only fair results, since the light output is primarily in the blue region of the visible spectrum, to which silicon tube and solid-state cameras are insensitive. Although the vidicon tube camera has more sensitivity in the blue-green spectrum than do silicon or Newvicon tubes or CCD solid-state sensors, this camera is all but obsolete. So despite mercury lighting's good appearance to the human eye, typical solid-state and silicon tube cameras respond poorly to it.

The mercury-xenon lamp, which contains a small amount of mercury in addition to xenon gas, offers better color rendition than the mercury short-arc lamp. Immediately after lamp ignition, the output is essentially the same as the spectrum of a xenon lamp. As the mercury vaporizes over several minutes, the spectral output becomes that of mercury vapor, with light output in the blue, green, yellow, and orange portions of the spectrum. The xenon short-arc's luminous efficiency ranges from 20 to 53 lumens per watt over lamp wattage ranges of 200 to 7000 watts. In addition to the spectral output of the mercury-xenon lines, the lamp has a background continuum that improves color rendition.

The xenon short-arc lamp is filled to a pressure of several atmospheres. The color temperature of the ignited arc is approximately 6000 degrees K, which is almost identical to that of sunlight. However, since the xenon lamp output consists of specific colors as well as a continuum and some IR radiation, it does not produce the same color lighting as the sun (Figure 3-13). The greater percentage of the continuum radiation at all wavelengths closely matches the spectral radiation characteristics of sunlight. Thus, compared with all other short-arc lamps, the xenon lamp is the ideal artificial light choice for accurate color rendition. Moreover, the spectral output does not change with lamp life, so color rendition is good over the useful life of the lamp. Color output is virtually independent of operating temperature and pressure, thereby ensuring good color rendition under adverse operating conditions.

Xenon lamps are turned on with starting voltage pulses of 10 to 50 kilovolts (kV); typical lamps reach full output intensity within a few seconds after ignition. The luminous efficiency of the xenon lamp ranges from 15 to 50 lumens per watt over a wattage range of approximately 75 to 10,000 watts.

A characteristic unique to the compact short-arc lamp is the small size of the radiating source, usually a fraction of a millimeter to a few millimeters in diameter. Due to optical characteristics, one lamp in a suitable reflector can produce a very concentrated beam of light. Parabolic and spherical reflectors, among others, are used to provide optical control of the lamp output: the parabolic for search- or spotlights and the spherical for floodlights. Compact short-arc lamps are often mounted in a parabolic reflector to produce a highly collimated beam used to illuminate distant objects. This configuration also produces an excellent IR spotlight when an IR transmitting filter is mounted on the lamp. Even when not used for spotlighting, the small arc size of compact short-arc lamps allows the luminaire reflector to be significantly smaller than other lamp reflectors.

Mounting orientation can affect the performance of short-arc lamps. Most xenon lamps are designed for vertical or horizontal operation, but many mercury-xenon and mercury lamps must be operated vertically to prevent premature burnout.

3.4.6 Infrared Lighting

When conventional security lighting is not appropriate—for example, when the presence of a security system would attract unwanted attention, would alert intruders to surveillance systems, or would disturb neighbors—a covert IR lighting system is the answer.

There are two generic techniques for producing IR lighting. One method uses the IR energy from a thermal incandescent or xenon lamp. Such lamps are fitted with optical filters that block the visible radiation, so that only IR radiation is transmitted from the lamp housing to illuminate the scene. The second technique uses a nonthermal IR light-emitting diode (LED) or LED array to generate IR radiation through electronic recombination in a semiconductor device. Both techniques produce narrow or wide beams, resulting in excellent images when the scene is viewed with an IR-sensitive camera, such as a solid-state, silicon tube, or LLL-intensified camera.

3.4.6.1 Thermal Infrared Source

Xenon and incandescent lamps, which can illuminate a scene many hundreds of feet from the camera, produce sufficient IR radiation to be practical for a covert television system (Figure 3-9). Since thermal IR sources consume significant amounts of power and become hot, they may require a special heat sink or air cooling to operate continuously.

Figure 3-14 shows several tungsten and xenon thermal IR sources having built-in reflectors and IR transmitting (visual blocking) filters, which produce IR beams. An especially efficient configuration using a tungsten-halogen lamp as the radiating source and a unique filtering and cooling technique is shown in Figure 3-15.

The lamp system uses thin-film dichroic optical coatings (a light-beam splitter) and absorbing filters that direct the very-near IR rays toward the front of the lamp and out into the beam, while transmitting visible and long-IR radiation

to the back of the lamp, where it is absorbed by the housing. The housing acts as an efficient heat sink, effectively dissipating the heat. The system can operate continuously in hot environments without a cooling fan.

Figure 3-15 diagrammatically shows the functioning parts of a 500-watt IR illuminating source using a type PAR 56 encapsulated tungsten-halogen lamp. The PAR 56 lamp filament operates at a temperature of approximately 3000 degrees K and has an average rated life of 2000 to 4000 hours, depending on its operating temperature. The lamp's optical dichroic mirror coatings on the internal surfaces of the reflector and front cover lens are made of multiple layers of silicon dioxide and titanium dioxide. In addition to this interference filter, there is a "cold" mirror—a quartz-substrate shield—between the tungsten-halogen lamp and the coated cover lens to control direct visible-light output from the filament. The lamp optics have a visible absorbing filter between the lamp and the front lens, which transmits less than 0.1% of all wavelengths shorter than 730 nanometers. This includes the entire visible spectrum. The compound effect of this filtering ensures that only IR radiation leaves the front of the lamp and that visible and long-IR radiation (longer than is useful to the silicon television sensor) cannot leave the front of the lamp system. The lamp output is consequently totally invisible to the human eye.

The lamp system is available with different front lenses so different beam patterns can be used for various applications, covering wide scene illumination or long-range spotlighting. Table 3-2 summarizes the types of lenses available and the horizontal and vertical beam angles they produce. These beam angles vary from 12 degrees for a narrow beam (spotlight) to 68 degrees for a very wide beam (floodlamp).

3.4.6.2 Infrared-Emitting Diodes

Covert security systems are using a new generation of devices composed of an array of gallium arsenide (GaAs) semiconductor diodes that emit a narrow band of IR radiation and no visible light. These devices are very efficient—they typically convert 50% of electrical to optical (IR) radiation—operate just slightly above room temperature, and dissipate little heat, so they usually require little special cooling. The light is a PN-junction device that emits light (visible or IR) when electrically biased in a forward direction. Figure 3-16 shows an LED IR light source in diagrammatic form.

Electrical power enters the LED through two input wires and generates IR energy, which is directed toward the magnifying dome lens and emitted toward the scene. To adequately illuminate the entire CCTV scene requires an array of a hundred to several hundred diodes (mounted on a plane). The output power from each diode adds linearly, the sum total illuminating the target with sufficient IR energy to produce a good CCTV picture with solid-state or silicon tube cameras.

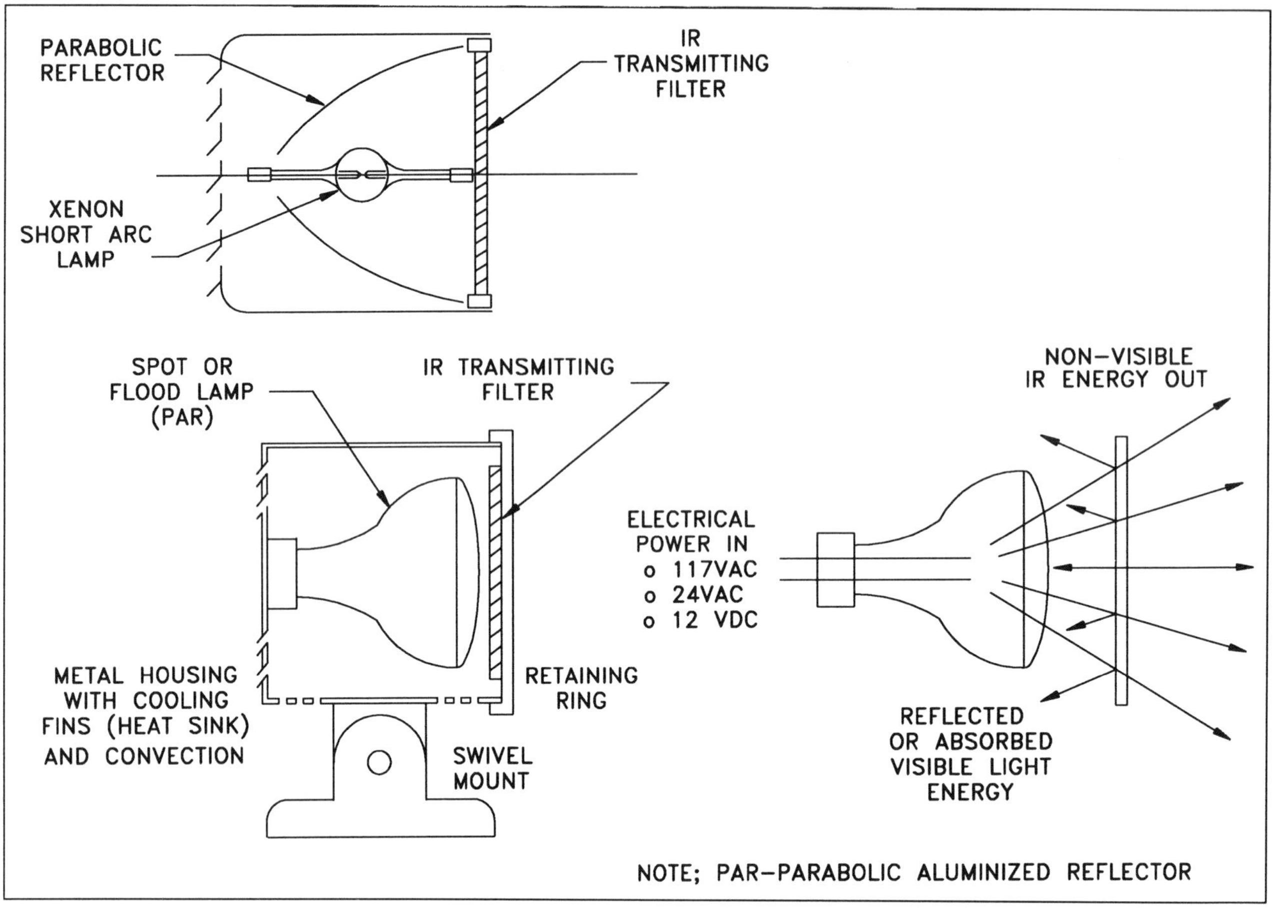

FIGURE 3-14 Thermal IR source configurations

Figure 3-16 shows a state-of-the-art GaAs array that produces a high-efficiency beam for covert applications. The IR light output from each diode adds up to produce enough radiation to illuminate the scene.

3.5 LIGHTING DESIGN CONSIDERATIONS

3.5.1 Lighting Costs

The cost of lighting an indoor or outdoor area depends on factors such as initial installation, maintenance, and operating costs (energy usage). The initial installation costs are lowest for incandescent lighting, followed by fluorescent lighting and then by HID lamps. All incandescent lamps can be connected directly to the electrical AC supply voltage, with no need for electrical ballasting or high-voltage starting circuits. All that is required is a suitably designed luminaire that directs the lamp illumination into the desired beam pattern. Some incandescent lamps are prefocused, with built-in luminaires to produce spot or flood beam coverage. Fluorescent lamps are installed in diffuse light reflectors and require only an igniter and simple ballast for starting and running. HID lamps require more

complex ballast networks, which are more expensive, larger and bulkier, consume electrical power, and add to installation and operating costs.

All lamps and lamp fixtures are designed for easy lamp replacement. Fluorescent and HID lamps, having ballast modules and high-voltage starting circuits, require additional maintenance, since they will fail sometime during the lifetime of the installation. Table 3-3 compares the common lamp types.

3.5.1.1 Operating Costs

To produce an effective CCTV security system, the security professional must be knowledgeable in the area of lighting. Energy efficiency is a prime consideration in any CCTV security system. Translated into dollars and cents, this relates to the number of lumens or the light output per kilowatt of energy input that additional lighting might cost or that could be saved if an LLL television camera system were installed.

Since the amount of lighting directly affects the quality and quantity of intelligence displayed on the CCTV security monitor, knowing the necessary type and amount of lighting is important. If the lighting already exists on the premises, the professional must determine quantitatively whether

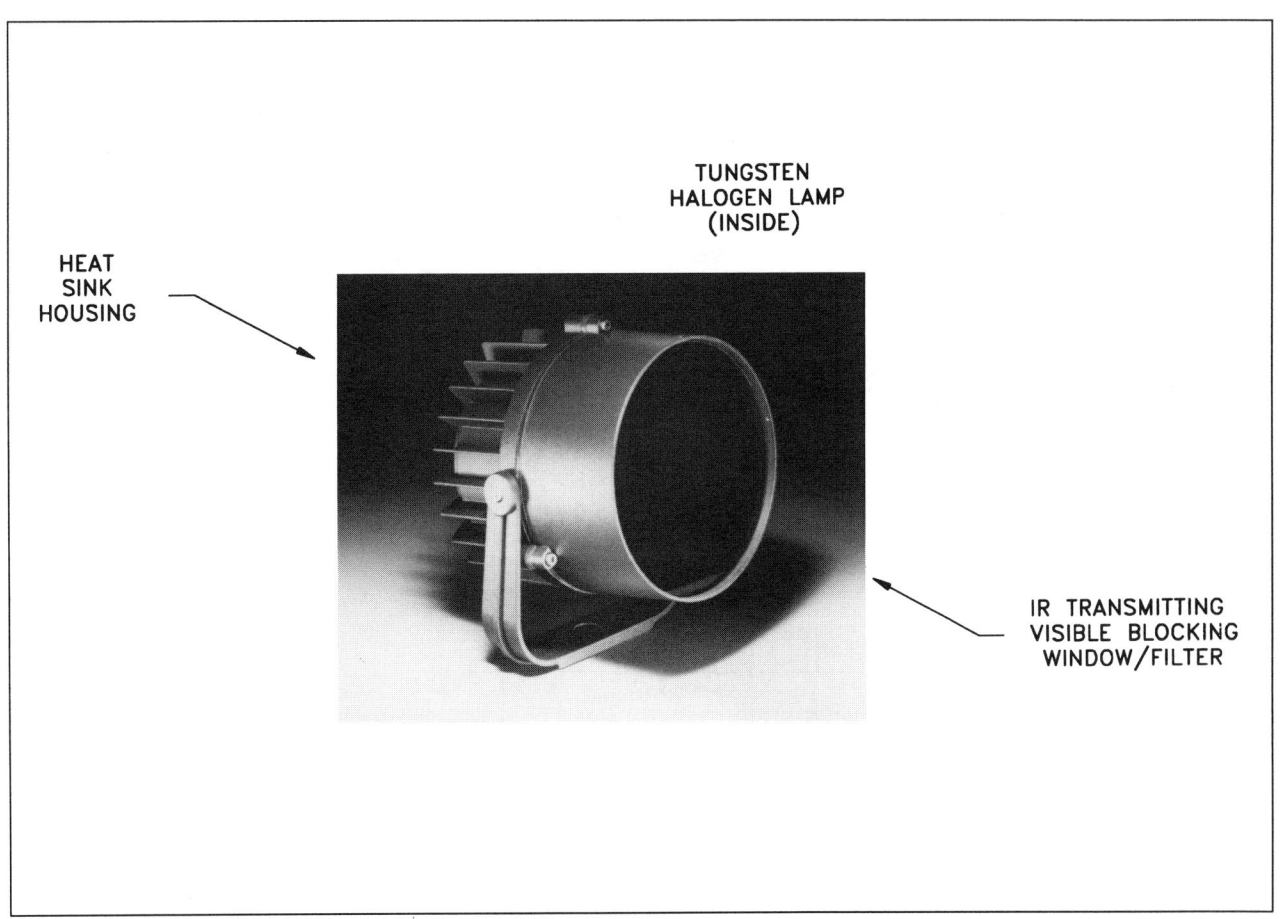

HEAT
SINK
HOUSING

TUNGSTEN
HALOGEN LAMP
(INSIDE)

IR TRANSMITTING
VISIBLE BLOCKING
WINDOW/FILTER

FIGURE 3-15 High-efficiency thermal IR lamp

the type of lighting is satisfactory and the amount of lighting is sufficient. The results of a site survey will determine whether more lighting must be added. Computer design programs are available to calculate the location and size of the lamps necessary to illuminate an area with a specified number of foot-candles. If adding lighting is an option, the

SOURCE	TYPE	INPUT POWER (WATTS) (VOLTAGE)	BEAM ANGLE (DEGREES)	MAXIMUM RANGE (FT)
WIDE FLOOD	FILTERED WI INCANDESCENT	100	60 HORIZ 60 VERT	30
SPOT	FILTERED WI INCANDESCENT	100	10 HORIZ 10 VERT	200
WIDE FLOOD	FILTERED WI INCANDESCENT	500	40 HORIZ 16 VERT	90
SPOT	FILTERED WI INCANDESCENT	500	12 HORIZ 8 VERT	450
FLOOD	FILTERED XENON ARC	400 (AC)	40	500
SPOT	FILTERED XENON ARC	400 (AC)	12	1500
FLOOD	LED	50 (12 VDC)	30	200
FLOOD	LED	8 (12 VDC)	40	70

WI=TUNGSTEN HALOGEN   LED=LIGHT EMITTING DIODE (880NM–DEEP RED GLOW, 950NM–INVISIBLE IR)
WI AND XENON THERMAL LAMPS USE VISUAL BLOCKING FILTERS

Table 3-2 Beam Angles for IR Lamps

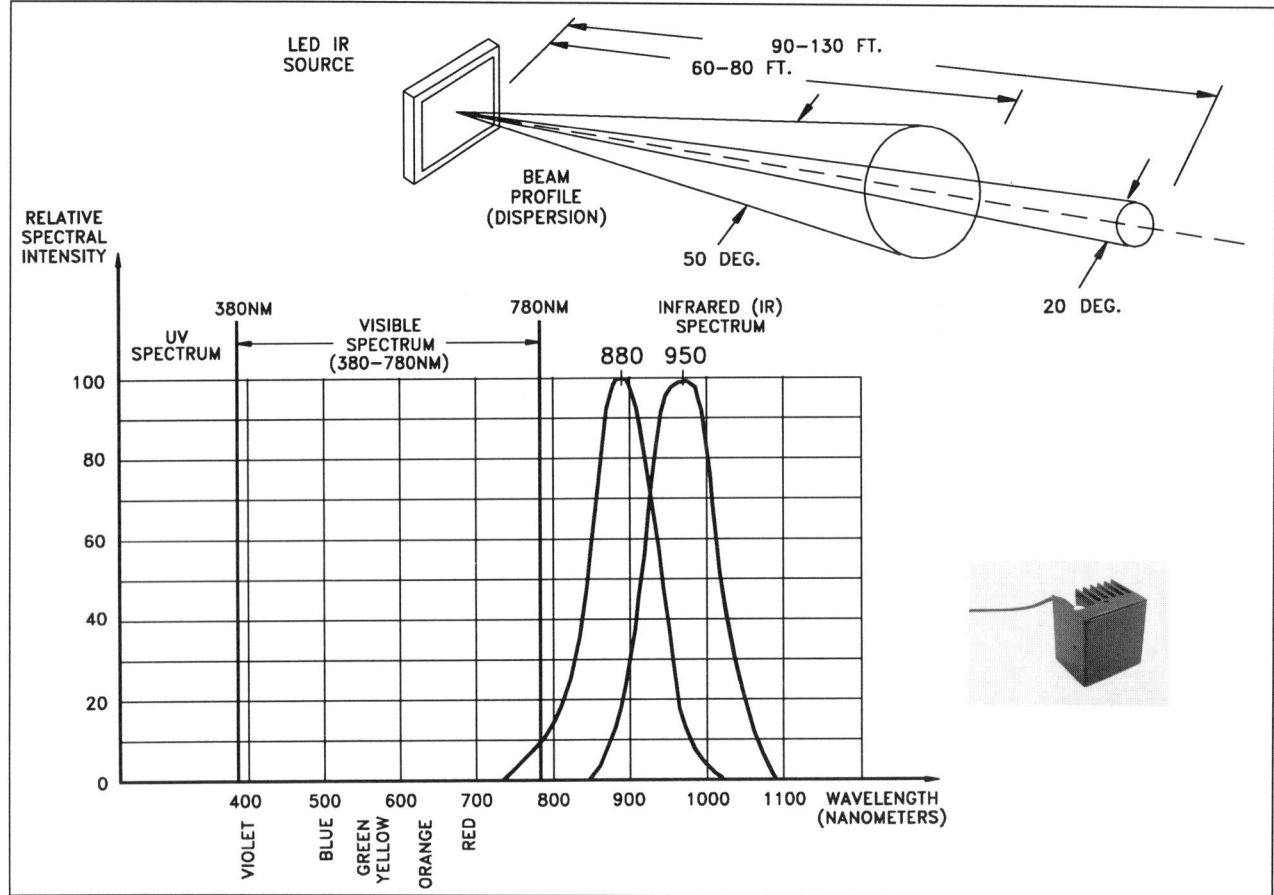

FIGURE 3-16 LED source output

analysis will compare that cost with the cost of installing more sensitive and expensive television cameras or lenses.

If the security system includes color television, the choice of lighting becomes even more critical. All color television cameras require a higher level of lighting than their monochrome counterparts. To produce a color television image having a signal-to-noise ratio or noise-free picture as good as a monochrome system, upwards of ten times more lighting is required. To obtain faithful color reproduction of facial tones, objects, and other articles in the scene, the light sources chosen or already installed must produce enough of these colors for the camera to detect and balance them. Since a large number of different generic lighting types are currently installed in industrial and public sites, the security professional must be knowledgeable in the spectral output of such lights.

Lighting-energy operating costs are related to lamp light output efficacy because operating each lamp type often costs more than the initial installation and maintenance costs put together. To appreciate the significant differences in operating costs for the different lamp types, Table 3-4 compares the average light output over the life of each lamp.

For the various models of incandescent, mercury vapor (HID), fluorescent, and high-pressure sodium lamps, lamp life in hours is compared with input power and operating cost, kilowatt-hours (kWh) used, based on 4000 hours of annual operation. The comparisons are made for lamps used in different applications, including dusk-to-dawn lighting, wall-mounted aerial lighting, and floodlighting. In each application, there is a significant saving in operational costs (energy costs) between the high-pressure sodium and fluorescent lamps as compared with the mercury vapor and standard incandescent lamps. Choosing the more efficient lamp over the less efficient one can result in savings of double or triple the operational costs, depending on the cost of electricity in a particular location.

3.5.1.2 Lamp Life

Lamp life plays a significant role in determining the cost-efficiency of different light sources. Actual lamp replacement costs and labor costs must be considered, as well as the additional risk of interrupted security due to unavailable lighting. Table 3-5 summarizes the average lamp life in hours for most lamp types in use today.

At the top of the list are the high- and low-pressure sodium lamps and the HID mercury vapor lamp, each providing approximately 24,000 hours of average lamp

TYPE	SPECTRAL OUTPUT	EFFICIENCY * LUMENS/WATT		LIFETIME (HOURS)	POWER RANGE (WATTS)	WARM–UP / RESTRIKE (MINUTES)
		INITIAL	MEAN			
MERCURY	BLUE–GREEN	32–63	25–43	16,000–24,000	50–1,000	5–7/3–6
HIGH PRESSURE SODIUM	YELLOW–WHITE	64–140	58–126	20,000–24,000	35–1,000	3–4/ 1
METAL ARC (METAL HALIDE) (MULTI–VAPOR)	GREEN–YELLOW	80–115	57–92	10,000–20,000	175–1,000	2–4/10–15
FLUORESCENT	WHITE	74–100	49–92	12,000–20,000	28–215	IMMEDIATE
INCANDESCENT: TUNGSTEN TUNGSTEN HALOGEN	YELLOW–WHITE YELLOW–WHITE	17–24	15–23	750–1,000 2,000	100–1,500	IMMEDIATE IMMEDIATE

* REFERRED TO AS EFFICACY IN LIGHTING (LUMENS/WATT)

Table 3-3 Comparison of Lamp Characteristics

TYPE	LIFETIME (HOURS)	INITIAL COST	OPERATING COST	TOTAL OWNING AND OPERATING COST
MERCURY	16,000–24,000	HIGH	MEDIUM	MEDIUM
HIGH PRESSURE SODIUM	20,000–24,000	HIGH	LOW	LOW
METAL ARC (METAL HALIDE) (MULTI–VAPOR)	10,000– 20,000	HIGH	LOW	LOW
FLUORESCENT	12,000– 20,000	MEDIUM	MEDIUM	MEDIUM
INCANDESCENT: TUNGSTEN TUNGSTEN HALOGEN	750–1,000 2,000	LOW	HIGH	HIGH

Table 3-4 Light Output vs. Lamp Type over Rated Life

life. Next, some fluorescent lamp types have a life of 10,000 hours. At the bottom of the list are the incandescent and quartz-halogen lamps, having rated lives of approximately 1000 to 2000 hours. If changing lamps is inconvenient or costly, high-pressure sodium lamps should be used in place of incandescent types. Using high-pressure sodium will save 12 trips to the site to replace a defective lamp, and having 12 fewer burned-out lamps will reduce the amount of time that security will be diminished.

Lamp designs require specifications of wattage, voltage, filament form, bulb type, base type, filling material (gas, vacuum, halogen), lumen output, color temperature, life,

TYPE	LIFETIME (HOURS)	POWER IN (WATTS)	LUMENS OUT (FTCD)
MERCURY	24,000	100 250 1000	4100 12100 57500
HIGH PRESSURE SODIUM	24,000	50 150 1000	4000 16000 140000
METAL ARC (METAL HALIDE) (MULTI–VAPOR)	7500 20000 3000	175 400 1500	14000 34000 155000
FLUORESCENT	18000 12000 10000	30 60 215	1950 5850 15000
INCANDESCENT: TUNGSTEN TUNGSTEN HALOGEN	2,000	250	4000

Table 3-5 Lamp Life vs. Lamp Type

TYPE	LOCATION	LIGHT LEVEL	
		FTCD	LUX
PARKING AREA	INDOOR	5–50	50–500
LOADING DOCKS	INDOOR	20	200
GARAGES–REPAIR	INDOOR	50–100	500–1,000
GARAGES–ACTIVE TRAFFIC	INDOOR	10–20	100–200
PRODUCTION/ASSEMBLY AREA ROUGH MACHINE SHOP/SIMPLE ASSY.	INDOOR	20–50	200–500
MEDIUM MACHINE SHOP/MODERATE DIFFICULT ASSY.	INDOOR	50–100	500–1,000
DIFFICULT MACHINE WORK/ASSY.	INDOOR	100–200	1,000–2,000
FINE BENCH/MACHINE WORK, ASSY.	INDOOR	200–500	2,000–5,000
STORAGE ROOMS/WAREHOUSES ACTIVE–LARGE/SMALL	INDOOR	15–30	150–300
INACTIVE	INDOOR	5	50
STORAGE YARDS	OUTDOOR	1–20	10–200
PARKING–OPEN (HIGH–MEDIUM ACTIVITY)	OUTDOOR	1–2	10–20
PARKING–COVERED (PARKING, PEDESTRIAN AREA)	OUTDOOR	5	50
PARKING ENTRANCES (DAY–NIGHT)	OUTDOOR	5–50	50–500

NOTE: 1 FTCD EQUALS APPROXIMATELY 10 LUX

Table 3-6 Recommended Light Levels for Typical Security Applications

and other special features. Color temperature, power input, and life ratings of a lamp are closely related and cannot be varied independently. For a given wattage, the lumen output and color temperature decrease as the life expectancy increases. Filament power (watts) is roughly proportional to the fourth power of filament temperature. So a lamp operated *below* its rated voltage has a longer life. A rule of thumb: Filament life is doubled for each 5% reduction in voltage. Conversely, filament life is halved for each 5% increase in voltage.

3.5.2 Security Lighting Levels

In addition to the lamp parameters and energy requirements, the size and shape of the luminaire, spacing between lamps, and height of the lamp above the surface illuminated must be considered. Although each CCTV application has special illumination requirements, primary responsibility for lighting is usually left to architects or illumination engineers. To provide adequate lighting in an industrial security or safety environment in building hallways, stairwells, outdoor perimeters, or parking lot facilities, different lighting designs are needed. Table 3-6 tabulates recommended light-level requirements for locations including parking lots, passenger platforms, building exteriors, and pedestrian walkways.

The CCTV designer or security director often has no option to increase or change installed lighting and must first determine whether the lighting is sufficient for the CCTV application and then make a judicious choice of CCTV camera to obtain a satisfactory picture. If lighting is not sufficient, the existing lighting can sometimes be aug-

mented by "fill-in" at selected locations to provide the extra illumination needed by the camera. Chapters 4, 5, and 15, which cover television lenses, cameras, and LLL cameras, respectively, offer some options for CCTV equipment when sufficient lighting is not available.

3.5.3 High-Security Lighting

Lighting plays a key role in maintaining high security in correctional facilities. Lighting hardware requires special fixtures to ensure survival under adverse conditions. High-security lamps and luminaires are designed specifically to prevent vandalism and are often manufactured using high-impact molded polycarbonate enclosures to withstand vandalism and punishing weather conditions without breakage or loss of lighting efficiency (Figure 3-17).

These luminaires are designed to house incandescent, HID, and other lamp types to provide the necessary light intensity and the full spectrum of color rendition required for monochrome and color television security systems. Most fixtures feature tamper-proof screws that prevent the luminaire from being opened by unauthorized personnel. For indoor applications, high-impact polycarbonate fluorescent lamps offer a good solution. The molded polycarbonate lenses have molded tabs that engage special slots in the steel-backed plate and prevent the luminaire from being opened, thereby minimizing exposure of the fluorescent lamps to vandalism. Applications include prison cells, juvenile-detention facilities, high-security hospital wards, parking garages, public housing hallways, stairwells, and underground tunnels.

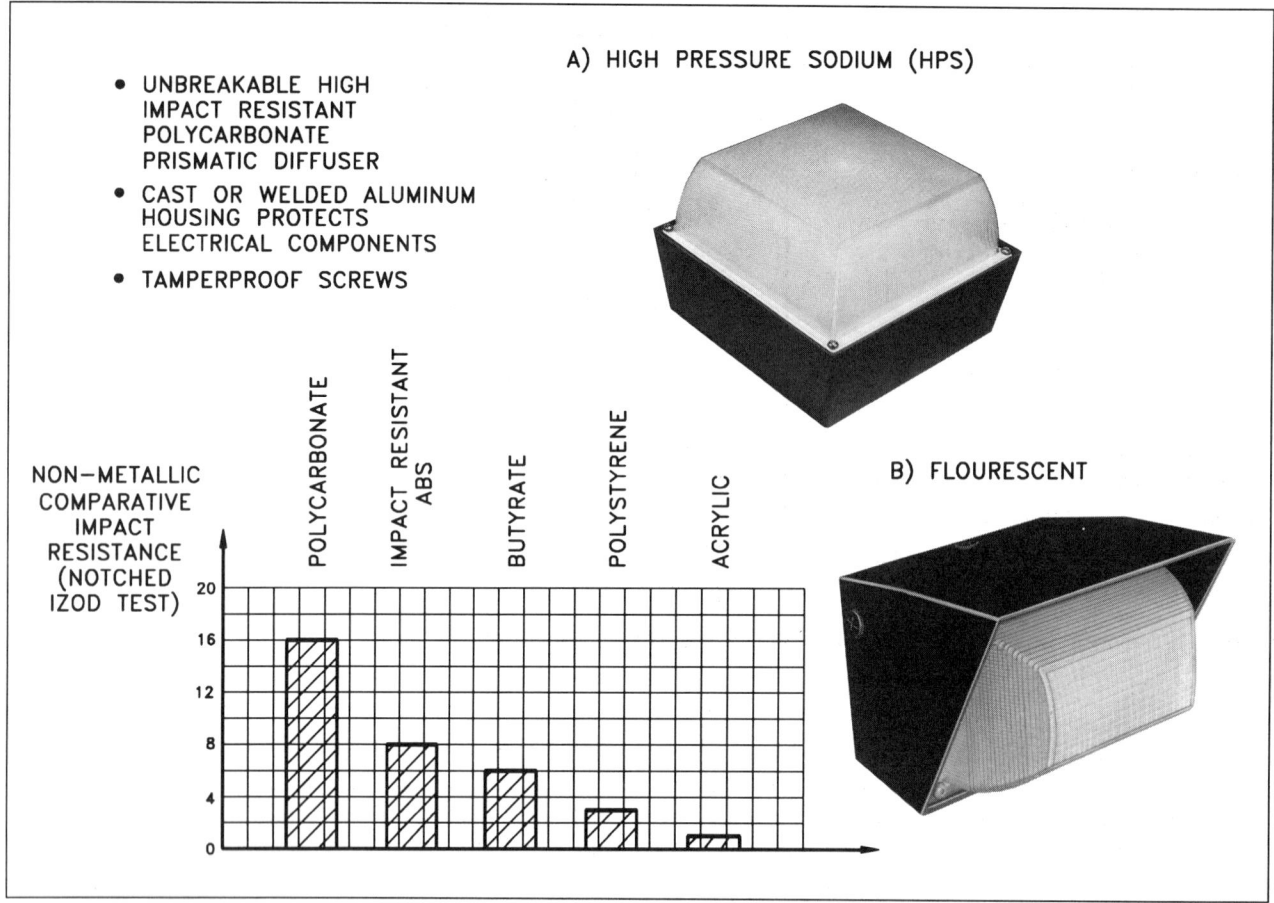

FIGURE 3-17 **High-security luminaires**

3.6 SUMMARY

The quality of the final CCTV picture and the intelligence it conveys are strongly influenced by the natural and/or artificial light sources illuminating the scene. For optimum results, an analysis of the source(s) parameters (spectrum, illumination level, beam pattern) must be made and matched to the spectral and sensitivity characteristics of the camera. Color systems require careful analysis and should be used only with natural illumination during daylight hours or with broad-spectrum color-balanced artificial illumination. If the illumination level is marginal, measure it with a light meter to quantify the actual light reaching the camera from the scene. If there is insufficient light for the standard CCTV camera, augment the lighting with additional fill-in sources or choose a more sensitive intensified LLL camera (Chapter 15). As with the human eye, lighting holds the key to clear sight.

Chapter 4

Lenses and Optics

CONTENTS

4.1 OVERVIEW

The function of the CCTV lens is to collect the reflected light from a scene and focus it onto the camera sensor. Choosing the proper lens is very important, since the lens determines the amount of light received by the camera sensor, the FOV, and the quality of the display. Understanding the characteristics of the lenses available and following a step-by-step design procedure simplifies the task and ensures a good design.

A CCTV lens functions like the human eye. Both collect light reflected from a scene or emitted by a luminous light source and focus the object scene onto some receptor—the retina or the camera sensor. The human eye has a fixed-focal-length (FFL) lens and variable iris, which compares to an FFL, automatic-iris CCTV lens. The eye has an iris diaphragm,

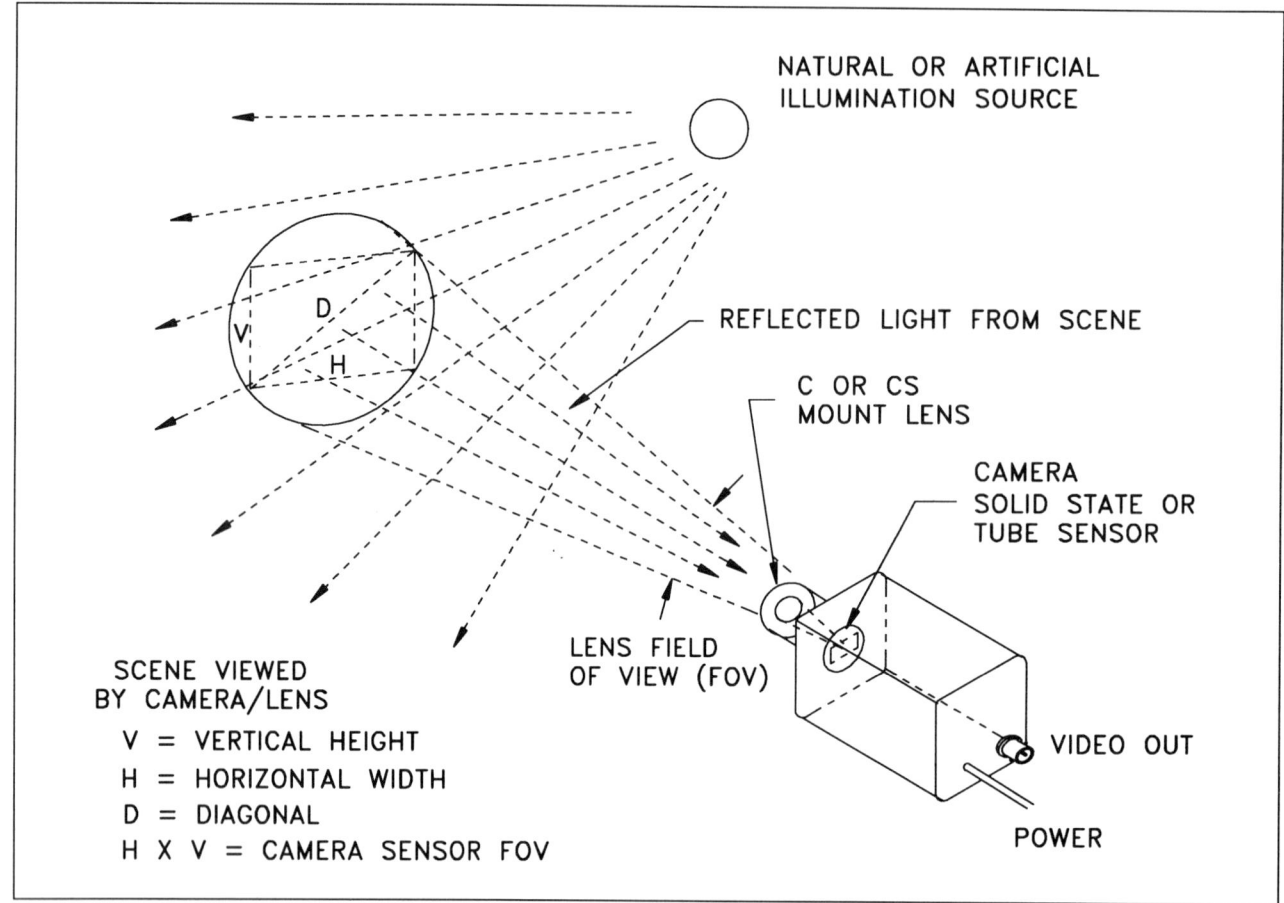

FIGURE 4-1 CCTV camera/lens, scene, and source illumination

which opens and closes just like an automatic-iris camera lens, to automatically adapt to the normal, large changes in light level. The iris—whether in the eye or in the camera—optimizes the light level reaching the receptor, thereby providing the best possible image. The iris in the eye is a muscle-controlled membrane; the automatic iris in a CCTV lens is a motorized device.

Of the many different kinds of lenses used in CCTV security applications, the most common is the FFL lens, which is found in the majority of installations. The FFL lens is available in wide, medium, and narrow FOVs. But for covering a wide scene and obtaining close-up views with the same camera, the variable-FOV zoom lens mounted on a pan/tilt platform is often used. For covert applications the pinhole lens is used, which has a small front diameter and can be hidden easily. There are many other specialty lenses, including split-image, fiber-optic, right-angle, and auto-focus.

4.2 LENS FUNCTIONS AND PROPERTIES

Lenses focus an image that the CCTV camera "sees" on its solid-state or tube sensor (Figure 4-1).

The human lens and a camera lens have many similarities: they both collect light and focus it onto a receptor

(Figure 4-2). But they have one important difference: the human lens has one fixed FL and the retina is one size, but the camera lens may have many different FLs and the sensor may have different sizes. Thus, the unaided human eye is limited to seeing a fixed and constant FOV, whereas the CCTV system can be modified to obtain a range of FOVs. To accommodate wide changes in light level, the eye has an automatic-iris diaphragm that opens and closes to optimize the light level reaching the retina, thereby providing the best image. The camera lens has an iris (some are manual, some are automatic) to regulate the light level reaching the sensor (Figure 4-3).

4.2.1 Focal Length and Field of View

CCTV lens magnification and FOV are based on the FL and retina size of the human eye. When a CCTV lens and sensor see the same basic picture, they are said to have the same FOV and magnification. In CCTV practice, a lens that has an FL and FOV similar to that of the human eye is referred to as a normal lens with a magnification of 1. The human eye's focal length—the distance from the center of the lens at the front of the eye to the retina in the back of the eye—is about 17 mm (0.67 in.) (Figure 4-2).

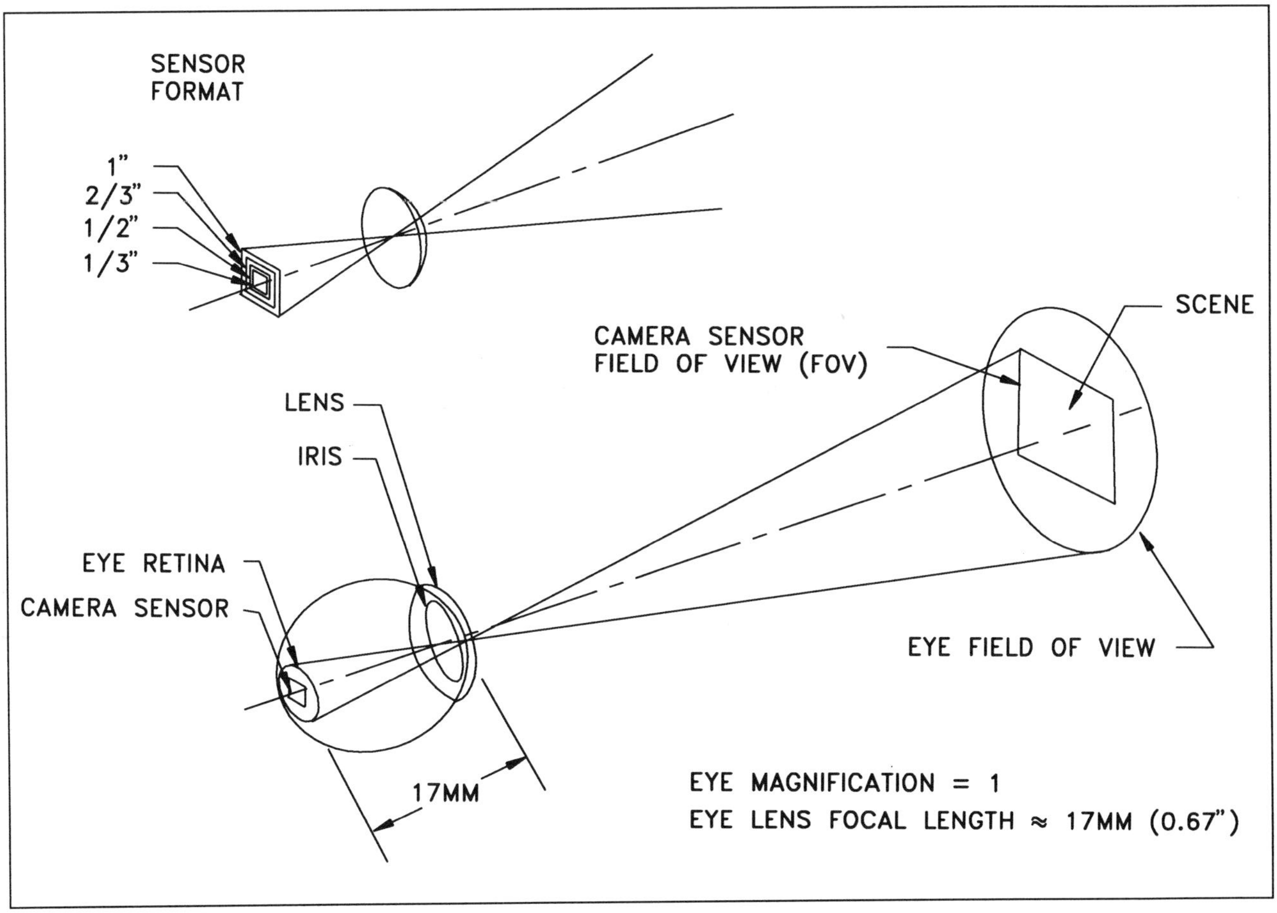

SENSOR
FORMAT

1"
2/3"
1/2"
1/3"

SCENE

CAMERA SENSOR
FIELD OF VIEW (FOV)

LENS

IRIS

EYE RETINA
CAMERA SENSOR

EYE FIELD OF VIEW

17MM

EYE MAGNIFICATION = 1
EYE LENS FOCAL LENGTH ≈ 17MM (0.67")

FIGURE 4-2   Comparing the human eye to a CCTV lens and camera sensor

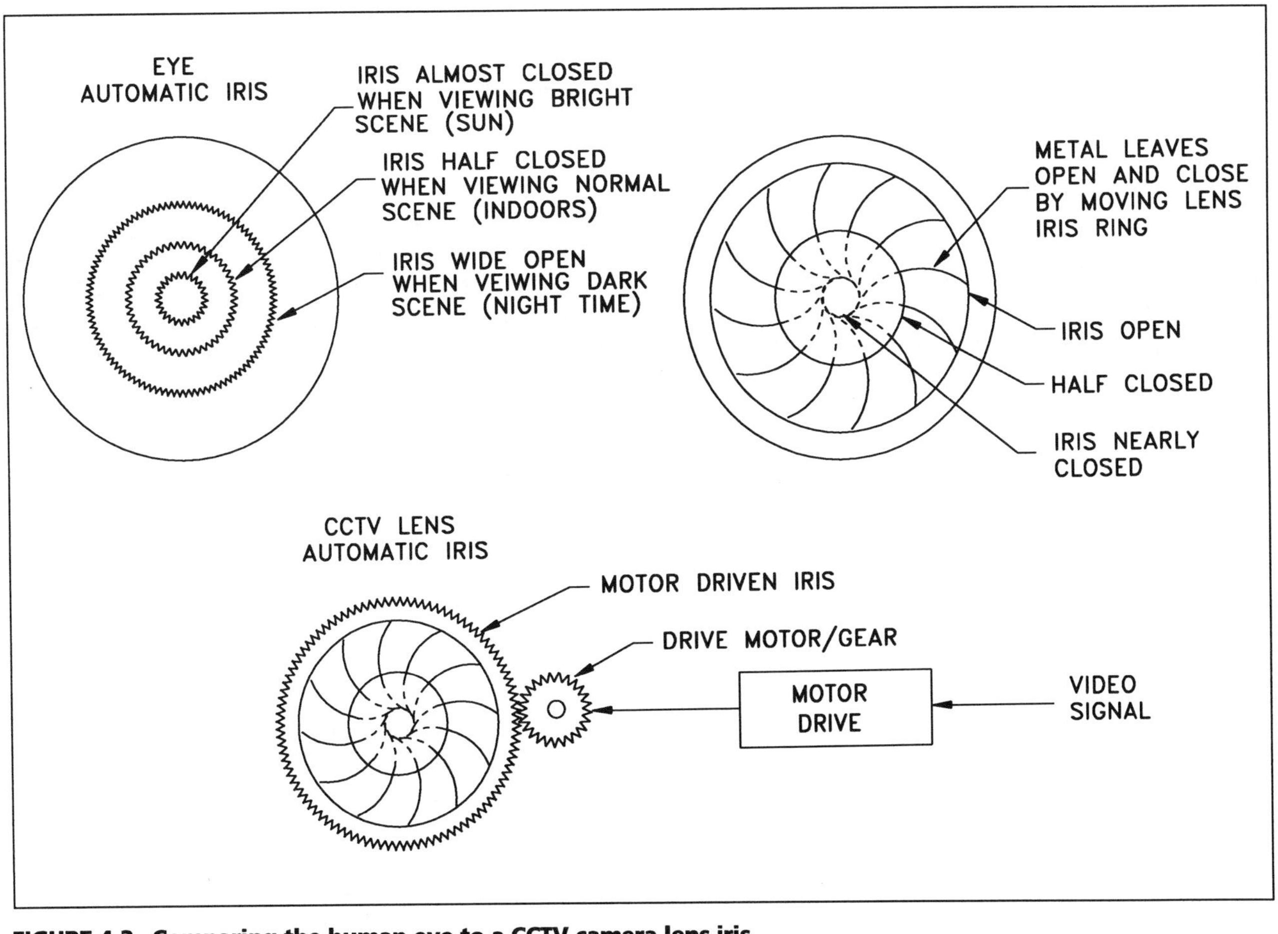

EYE
AUTOMATIC IRIS

IRIS ALMOST CLOSED
WHEN VIEWING BRIGHT
SCENE (SUN)

IRIS HALF CLOSED
WHEN VIEWING NORMAL
SCENE (INDOORS)

IRIS WIDE OPEN
WHEN VEIWING DARK
SCENE (NIGHT TIME)

METAL LEAVES
OPEN AND CLOSE
BY MOVING LENS
IRIS RING

IRIS OPEN

HALF CLOSED

IRIS NEARLY
CLOSED

CCTV LENS
AUTOMATIC IRIS

MOTOR DRIVEN IRIS

DRIVE MOTOR/GEAR

MOTOR
DRIVE

VIDEO
SIGNAL

FIGURE 4-3   Comparing the human eye to a CCTV camera lens iris

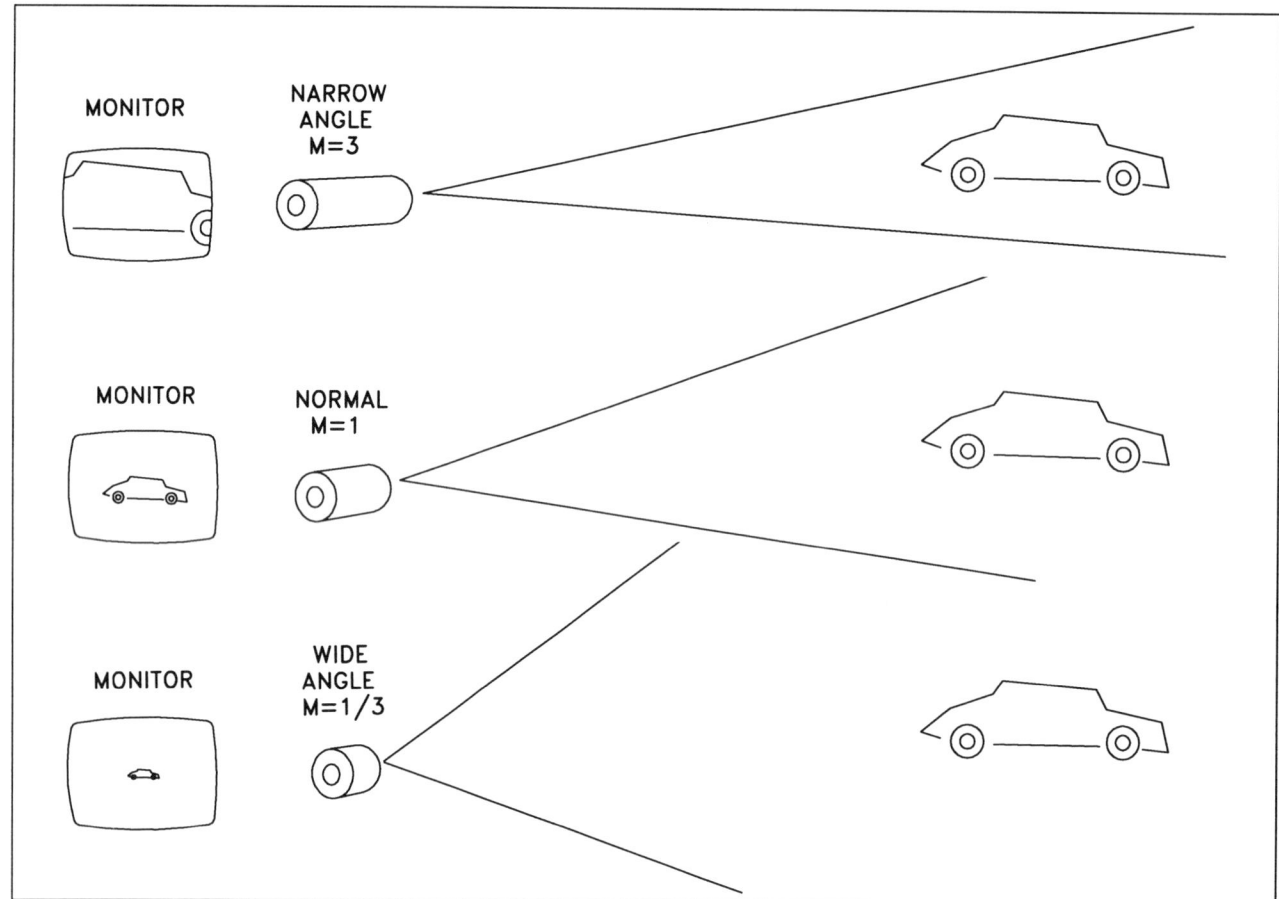

FIGURE 4-4   Lens FOV for magnifications of 1, 3, and ⅓

Most people see approximately the same FOV and magnification (M = 1). Specifically, the CCTV lens and camera format corresponding to the M = 1 condition is a 25-mm lens on a 1-inch format camera, a 16-mm lens on a ⅔-inch camera, a 12.5-mm lens on a ½-inch camera, or an 8-mm lens on a ⅓-inch camera. The 1-inch format designation was derived from the development of the vidicon television tube, which has a nominal tube diameter of 1 inch (25.4 mm) and an actual scanned area (active sensor size) approximately 16 mm in diameter. Figure 4-4 shows the FOV as seen with a lens having magnifications of 1, 3, and ⅓, respectively.

Lenses of shorter FL used with these sensors would be referred to as wide-angle lenses; lenses of longer FL are referred to as telephoto or narrow-angle lenses. Telephoto lenses used with CCTV cameras act like a telescope: they magnify the image viewed, narrow the FOV, and effectively bring the object of interest closer to the eye. While there is no device similar to the telescope for the wide-angle example, if there were, the device would broaden the FOV, allowing the eye to see a wider scene than is normal and at the same time cause the larger FOV to appear farther away from the eye. One can see this condition when looking through a telescope backwards. One similar device is the

automobile passenger side-view mirror, a concave mirror that causes the scene image to appear farther away, and therefore smaller (demagnified), than it actually is.

Just as your own eyes have a specific FOV—the scene you can see—so does the television camera. The camera FOV is determined by the simple geometry shown in Figure 4-5.

The scene has a width (W) and a height (H) and is a distance (D) away from the camera lens. Once the scene has been chosen, three factors determine the correct FL lens to use: (1) the size of the scene (H, W); (2) the distance between the scene and camera lens (D); and (3) the television image sensor size (⅓-, ½-, ⅔-, or 1-inch format).

Tables 4-1, 4-2, and 4-3 give scene-size values for the ½-, ⅔-, and ⅓-inch sensors, respectively, as a function of the distance from the camera to the object and FL. The tables include scene sizes for most available lenses ranging from 2.6 to 75 mm FL.

4.2.1.1 Field-of-View Calculations

There are many tables, graphs, nomographs, and linear and circular slide rules for determining the angles and sizes of a scene viewed at varying distances by a CCTV camera with a given sensor format and FL lens. One handy aid in

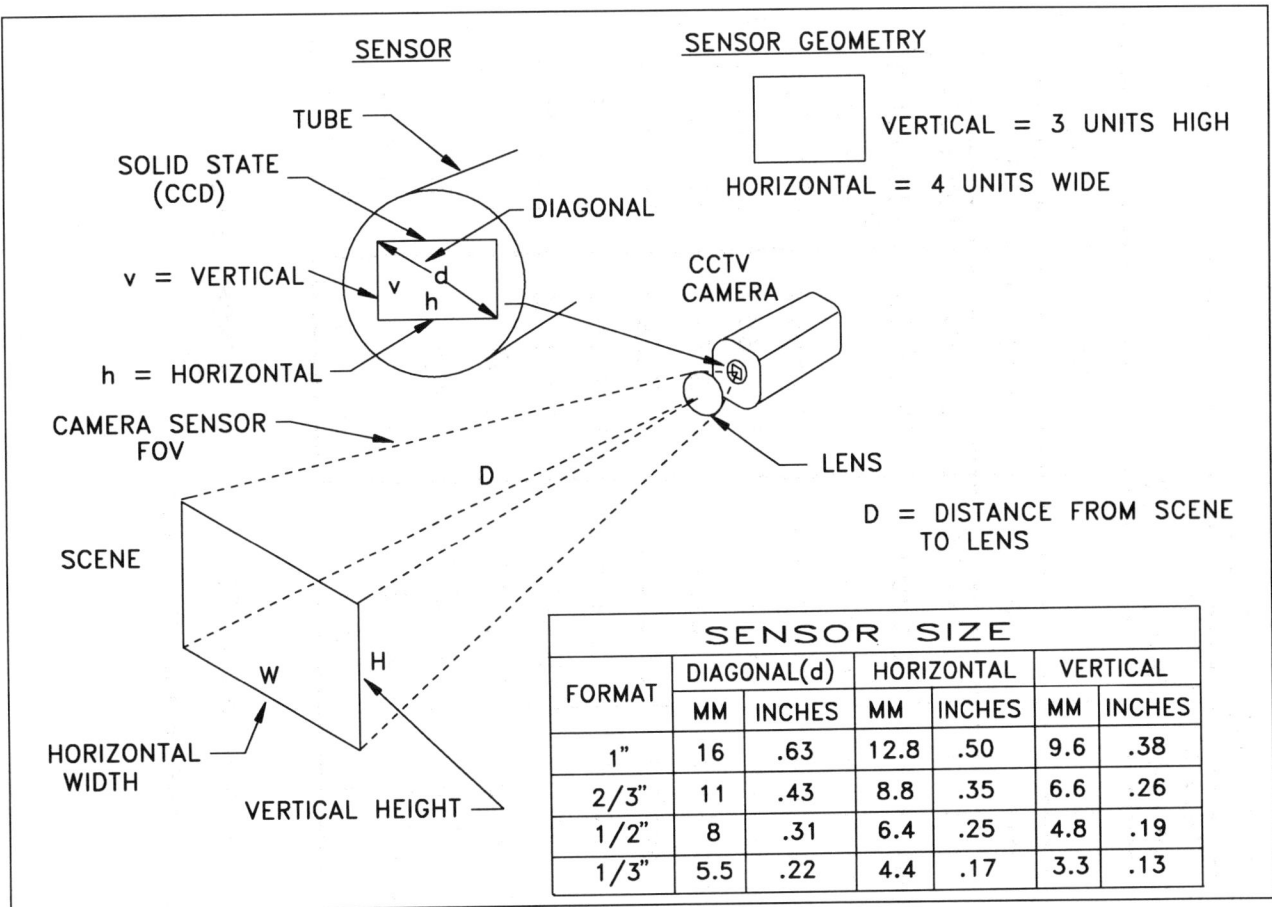

FIGURE 4-5 Camera/lens sensor geometry and formats

the form of transparent circular scales, called a Lens Finder Kit, eliminates the calculations required to choose a CCTV lens (see Section 4.2.5). Such kits are based on the simple geometry shown in Figure 4-6.

Since light travels in straight lines, the action of a lens can be drawn on paper and easily understood. Bear in mind that while commercial CCTV lenses are constructed from multiple lens elements, the single lens shown in Figure 4-6 for the purpose of calculation has the same effective FL as the CCTV lens. By simple geometry, the scene size viewed by the sensor is inversely proportional to the lens FL. Shown in Figure 4-6 is a camera sensor of horizontal width (h) and vertical height (v). For a ½-inch CCD sensor, this would correspond to h = 6.4 mm and v = 4.8 mm. The lens FL is the distance behind the lens at which the image of a distant object (scene) would focus. The figure shows the projected area of the sensor on the scene at some distance D from the lens. Using the eye analogy, the sensor and lens project a scene W wide × H high (the eye sees a circle). As with the human eye, the CCTV lens inverts the image, but the human brain and the electronics in the camera re-invert to provide an upright image.

Figure 4-6 shows how to measure or calculate the scene size (W × H) as detected by a rectangular CCTV sensor

format and lens with horizontal and vertical angular FOVs θ_H and θ_V, respectively.

To find the horizontal FOV θ_H, we use the geometry of similar triangles:

$$\frac{h}{W} = \frac{FL}{D}, \; W = \frac{h}{FL} \times D$$

(4-1)

The horizontal angular FOV θ_H is then derived as follows:

$$\frac{\tan \theta_H}{2} = \frac{h/2}{FL}$$

$$\frac{\theta_H}{2} = \tan^{-1} \frac{h}{2\,FL}$$

$$\theta_H = 2 \tan^{-1} \frac{h}{2\,FL}$$

(4-2)

For the vertical FOV, similar triangles gives

$$\frac{v}{H} = \frac{FL}{D}, \; H = \frac{v}{FL} \times D$$

(4-3)

1/2-INCH SENSOR FORMAT LENS GUIDE

CAMERA TO SCENE DISTANCE (D) IN FEET
WIDTH AND HEIGHT OF AREA (WxH) IN FEET

LENS FOCAL LENGTH (MM)	5 WxH	10 WxH	20 WxH	30 WxH	40 WxH	50 WxH	75 WxH
2.6	12.1x9.3	24.9x18.8	48.5x37.7	72.6x55.8	98.3x74.0	123.8x94.2	185.8x141.0
3.5	9.0x6.9	18.5x14.0	36.0x28.0	54.0x41.4	73.0X55.0	92.0X70.0	138.0X105.0
4.0	7.8x6.0	15.6x11.6	31.0x23.2	46.8x36.0	62.0x46.4	77.6x58.2	116.4x87.4
4.8	6.5x4.8	12.9x9.7	25.9x19.5	39.0x28.8	51.7x38.8	64.6x48.8	96.9x72.6
6.0	5.1x3.9	10.3x7.8	20.7x15.5	30.6x23.4	41.3x31.0	51.6x38.8	77.4x58.2
7.5	4.1x3.1	8.3x6.2	16.5x12.4	24.6x18.6	33.0x24.8	41.3x31.0	62.0x46.5
8.0	3.9x3.0	7.8x5.8	15.5x11.6	23.4x18.0	31.0x23.2	38.8x29.1	58.2x43.7
12.0	2.6x1.9	5.2x3.9	10.3x7.8	15.6x11.4	20.6x15.5	25.8x19.4	38.7x29.1
16.0	1.9x1.5	3.9x2.9	7.8x5.9	11.4x9.0	15.5x11.6	19.4x14.5	29.1x21.8
25.0	1.3x1.0	2.3x2.0	5.2x3.0	7.8x6.0	10.0x7.8	13.0x9.7	19.5x10.1
50.0	.62x.46	1.2x.93	2.5x1.9	3.7x2.8	4.9x3.7	6.2x4.7	9.3x7.1
75.0	.41x.31	.82x.62	1.7x1.2	2.5x1.9	3.3x2.5	4.1x3.1	6.2x4.7
150.0	.21x.16	.41x.31	.85x.60	1.3x1.0	1.7x1.3	2.1x1.6	3.1x1.9

NOTE: MANY MANUFACTURERS' 1/2-INCH LENSES WILL NOT WORK ON 2/3 OR 1-INCH SENSORS

Table 4-1   Scene-Sizes vs. FL and Camera-to-Scene Distance for 1/2-Inch Sensor

2/3 INCH SENSOR FORMAT LENS GUIDE

CAMERA TO SCENE DISTANCE (D) IN FEET
WIDTH AND HEIGHT OF AREA (WxH) IN FEET

LENS FOCAL LENGTH (MM)	5	10	20	30	40	50	75
	WxH	WxH	WxH	WxH	WxH	WxH	WxH
4.0	11.0x8.2	22.0x16.4	44.0x33.0	66.0x49.2	88.0x64.0	112.0x84.0	168.0x126.0
4.8	10.0x7.5	20.0x15.0	40.0x30.0	60.0x45.0	80.0x60.0	100.0x75.0	15.0x112.5
6.5	6.8x5.0	13.6x10.1	27.0x20.3	40.8x30.3	54.2x39.4	68.9x51.7	103.4x77.6
8.0	5.5x4.1	11.0x8.2	22.0x16.5	33.0x24.6	44.0x32.0	56.0x42.0	84.0x63.0
12.5	3.5x2.6	7.0x5.3	14.0x10.5	21.0x15.6	28.0x21.0	35.0x26.2	52.0x39.3
16.0	2.8x2.1	5.6x4.2	11.1x8.3	16.8x12.6	22.5x16.7	28.0x21.0	42.0x31.5
25.0	1.7x1.3	3.6x2.7	7.0x5.3	10.5x8.0	14.0x10.7	18.0x13.3	27.0x20.0
50.0	.84x.63	1.7x1.3	3.3x2.4	5.0x3.8	6.5x4.9	8.3x6.3	12.5x9.4
75.0	.58x.43	1.1x.98	2.2x1.6	3.5x2.6	4.3x3.2	5.5x4.1	8.3x6.2
100.0	.42x.32	.85x.65	1.7x1.2	2.5x1.9	3.3x2.5	4.2x3.2	6.3x4.2
150.0	.29x.22	.55x.49	1.1x.80	1.8x1.3	2.2x1.6	2.8x2.1	4.2x3.1

NOTE: NOT ALL MANUFACTURERS' 2/3–INCH LENSES WILL WORK ON 1–INCH SENSORS

Table 4-2   Scene Sizes vs. FL and Camera-to-Scene Distance for 2/3-Inch Sensor

1/3 INCH SENSOR FORMAT LENS GUIDE

LENS FOCAL LENGTH (MM)	CAMERA TO SCENE DISTANCE (D) IN FEET WIDTH AND HEIGHT OF AREA (WxH) IN FEET						
	5	10	20	30	40	50	75
	WxH	WxH	WxH	WxH	WxH	WxH	WxH
2.8	7.9x5.6	15.8x11.2	31.6x22.4	47.4x33.6	63.2x44.8	79.0x56.0	118.5x84.0
3.3	6.7x5.0	13.4x10.0	26.8x20.0	40.2x30.0	53.6x40.0	67.0x50.0	100.5x75.0
3.8	5.8x4.3	11.6x8.6	23.2x17.2	34.8x25.8	46.4x14.4	58.0x43.0	87.0x64.5
4.0	5.5X4.1	11.0X8.2	22.0x16.4	33.0x24.6	44.0x33.0	55.0x41	82.5x61.5
4.5	5.0x3.8	10.0x7.6	20.0x15.2	30.0x22.8	40.0x30.4	50.0x38	75.0x57.0
6.7	3.3x2.5	6.6x5.0	13.2x10.0	19.8x15.0	26.4x20.0	33.0x25	49.5x37.5
8.0	2.8X2.1	5.5x4.1	11.0x8.2	16.5x13.7	22.0x16.4	27.5x20.5	41.0x30.8
9.0	2.5x1.9	5.0x3.8	10.0x7.6	15.0x11.4	20.0x15.2	25.0x19.0	37.5x28.5

NOTE: MOST 1/3–INCH LENSES WILL NOT WORK ON 1/2, 2/3 OR 1–INCH SENSORS

Table 4-3   Scene Sizes vs. FL and Camera-to-Scene Distance for 1/3-Inch Sensor

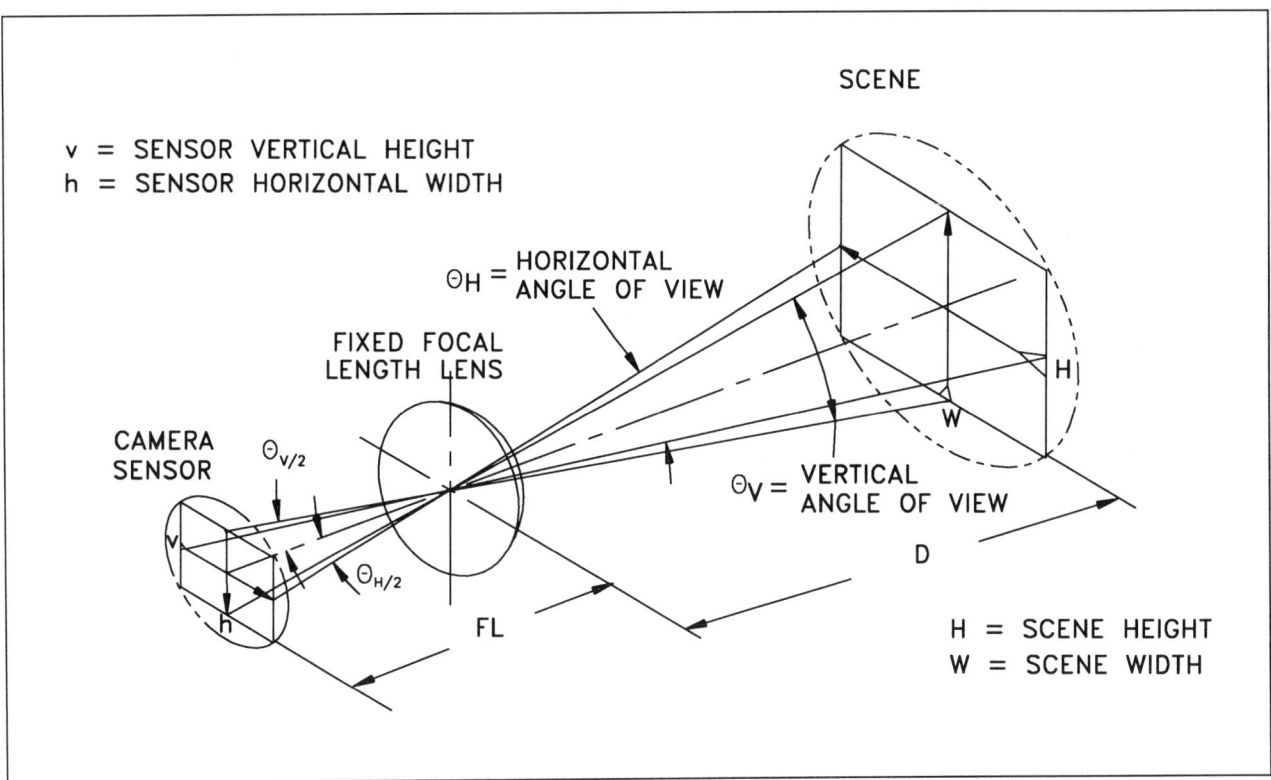

FIGURE 4-6 Sensor, lens, and scene geometry

The vertical angular FOV θ_V is then derived from the geometry:

$$\tan\frac{\theta_V}{2} = \frac{v/2}{FL}$$

$$\frac{\theta_V}{2} = \tan^{-1}\frac{v}{2\,FL}$$

$$\theta_V = 2\tan^{-1}\frac{v}{2\,FL}$$

(4-4)

Table 4-4 shows the angular FOV obtainable with some standard lenses from 2.6 to 75 mm in focal length. The values of angular FOV in Table 4-4 can be calculated from Equations 4-2 and 4-4.

4.2.1.2 Lens and Sensor Formats

FFL lenses must be used with either the image sensor size (format) for which they were designed or with a smaller sensor size. They cannot be used with larger sensor sizes because unacceptable image distortion and image darkening (vignetting) results at larger angles from the center of the FOV. Therefore, when a lens manufacturer lists a lens for a $\frac{2}{3}$-inch sensor format, it can be used on a $\frac{2}{3}$-, $\frac{1}{2}$-, or $\frac{1}{3}$-inch sensor but not on a 1-inch sensor without producing image vignetting. This problem of incorrect lens choice for a given format size occurs most often when a C-mount $\frac{1}{2}$-inch format lens is incorrectly used on a $\frac{2}{3}$- or 1-inch format camera. Since the lens

manufacturer does not "overdesign" the lens, that is, make glass lens element diameters larger than necessary, check the manufacturer's specifications for proper choice.

4.2.2 Magnification

CCTV magnification depends on two factors: (1) the lens FL and camera sensor format and (2) the monitor size. CCTV magnification is analogous to film magnification: the sensor is equivalent to the film negative, and the monitor is equivalent to the photo print.

4.2.2.1 Lens/Camera Sensor Magnification

The combination of lens focal length and camera sensor size defines the magnification at the camera location. For a specific camera, the sensor size is fixed. Therefore, no matter how large the image is at the sensor, the camera will see only as much of the image as will fit on the sensor.

CCTV lens magnification is measured relative to the eye (which is defined as a normal lens). The eye has approximately a 17-mm FL and a 25-mm FL lens on a 1-inch format. Therefore, the magnification of a 1-inch (16-mm format) sensor is

$$\underset{\text{(1- inch sensor)}}{\text{Magnification}} = \frac{\text{Lens Focal Length (mm)}}{25\ mm} = \frac{FL}{25\ mm}$$

(4-5)

CAMERA ANGULAR FIELD OF VIEW (FOV) (DEGREES)

LENS FOCAL LENGTH (MM)	1/3 INCH SENSOR		1/2 INCH SENSOR		2/3 INCH SENSOR	
	HORIZONTAL	VERTICAL	HORIZONTAL	VERTICAL	HORIZONTAL	VERTICAL
2.6	80.4	64.8	117.0	88.0	—	—
2.8	82	67	—	—	—	—
3.3	70	56	—	—	—	—
3.5	58.3	47.4	84.8	69.0	—	—
3.8	66.8	50.1	—	—	—	—
4.8	45.0	35.5	67.4	53.2	86.3	70.2
6.0	37.3	27.9	56.0	43.6	74.5	58.0
6.5	36.0	27	52.5	39.4	71.9	53.9
6.7	51.0	38.3	—	—	—	—
8.0	32.5	24.8	43.6	33.4	58.7	44.0
8.5	28.8	21.6	41.1	31.9	57.5	43.1
12.5	19.6	14.7	29.6	21.8	39.1	29.3
16.0	15.5	11.7	22.0	16.5	31.0	23.4
25.0	10.0	7.5	15.0	11.3	20.0	15.0
50.0	5.0	3.8	7.5	5.6	10.0	7.5
75.0	3.4	2.5	5.0	3.8	6.7	5.0

Table 4-4   Lens/Camera Angular FOV vs. Sensor Format and Lens Focal Length

For example, a 75-mm FL lens on a 1-inch format camera has the following magnification:

$$M_{(1-inch)} = \frac{FL}{25 \text{ mm}} = \frac{75 \text{ mm}}{25 \text{ mm}} = 3$$

For the ⅔-inch (11-mm format) sensor, the magnification is calculated as follows:

$$M_{(⅔-inch)} = \frac{\text{Lens Focal Length (mm)}}{16 \text{ mm}} = \frac{FL}{16 \text{ mm}}$$

(4-6)

For the ½-inch (8-mm format) sensor, the magnification is calculated as follows:

$$M_{(½-inch)} = \frac{\text{Lens Focal Length (mm)}}{12.5 \text{ mm}} = \frac{FL}{12.5 \text{ mm}}$$

(4-7)

For the ⅓-inch (5.5-mm format) sensor, the magnification is calculated as follows:

$$M_{(⅓-inch)} = \frac{\text{Lens Focal Length (mm)}}{8 \text{ mm}} = \frac{FL}{8 \text{ mm}}$$

(4-8)

The optical speed of a lens—how much light it collects and transmits to the camera sensor—is defined by a parameter called the f-number (f/#).

As the FL of a lens becomes longer, its optical aperture or diameter (d) must increase proportionally to keep the f-number the same. The f-number is related to the FL and the lens diameter (clear aperture) d by the following equation:

$$f/\# = \frac{FL}{d}$$

(4-9)

Long-FL lenses are larger (and costlier) than short-FL lenses, due to the cost of the larger optical elements.

4.2.2.2 Monitor Magnification

When the camera image is displayed on the CCTV monitor, a further magnification of the object scene takes place. The magnification is equivalent to the ratio of the monitor diagonal (d_m) to the sensor diagonal (d_s), or

$$M_{(monitor)} = \frac{d_m}{d_s}$$

(4-10)

For a 9-inch-diagonal monitor (d_m = 9 inches) and a ½-inch sensor format (d_s = 8 mm = 0.315 inch), Equation 4-10 computes as follows:

$$M = \frac{9}{0.315} = 28.57$$

Therefore, in the example, the scene from a 75-mm lens on a ½-inch format sensor displayed on a 9-inch monitor has an overall magnification of

$$M = 3 \times 28.57 = 85.71$$

Note that increasing the magnification by using a larger monitor does not increase the information in the scene; it only permits viewing from a greater distance.

4.2.3 Calculating the Scene Size

Equations 4-1 and 4-3 are used to calculate scene size. For example, calculate the horizontal and vertical scene size as seen by a ½-inch CCD sensor using a 12.5-mm FL lens at a distance D = 25 ft. A ½-inch sensor is 6.4 mm wide × 4.8 mm high.

From Equation 4-1 for horizontal scene width

$$\text{Scene width} = W = \frac{h}{FL} \times D$$

$$W = \frac{6.4 \text{ mm}}{12.5 \text{ mm}} \times 25 \text{ ft} = 12.8 \text{ ft}$$

For vertical scene height using Equation 4-1:

$$\text{Scene height} = H = \frac{v}{FL} \times D$$

$$H = \frac{4.8 \text{ mm}}{12.5 \text{ mm}} \times 25 \text{ ft} = 9.6 \text{ ft}$$

4.2.3.1 Converting One Format to Another

To obtain scene sizes (width and height) for a 1-inch sensor, multiply all the scene sizes in the ½-inch table (Table 4-1) by 2. For a ⅓-inch sensor, use Table 4-3 or divide all the scene sizes in the ⅔-inch table (Table 4-2) by 2.

Understanding Tables 4-1, 4-2, and 4-3 makes it easy to choose the right lens for the required FOV coverage. As an example, choose a lens for viewing all of a building 15 feet high by 20 feet long from a distance of 40 feet with a ½-inch format CCTV camera (Figure 4-7). From Table 4-1, a 12-mm FL lens will just do the job.

If a ⅔-inch format CCTV camera were used, a lens with an FL of 16 mm would be needed (from Table 4-2, a scene 16.7 feet high by 22.5 feet wide would be viewed).

If a ⅓-inch format CCTV camera were used, a lens with an FL of 8 mm would be used (from Table 4-3, a scene 16.7 feet high by 22.5 feet wide would be viewed).

4.2.4 Calculating Angular FOV

Equations 4-2 and 4-4 are used to calculate the angular FOV. Table 4-3 shows the angular FOV obtainable with some standard lenses from 2.6 to 75 mm focal length. For

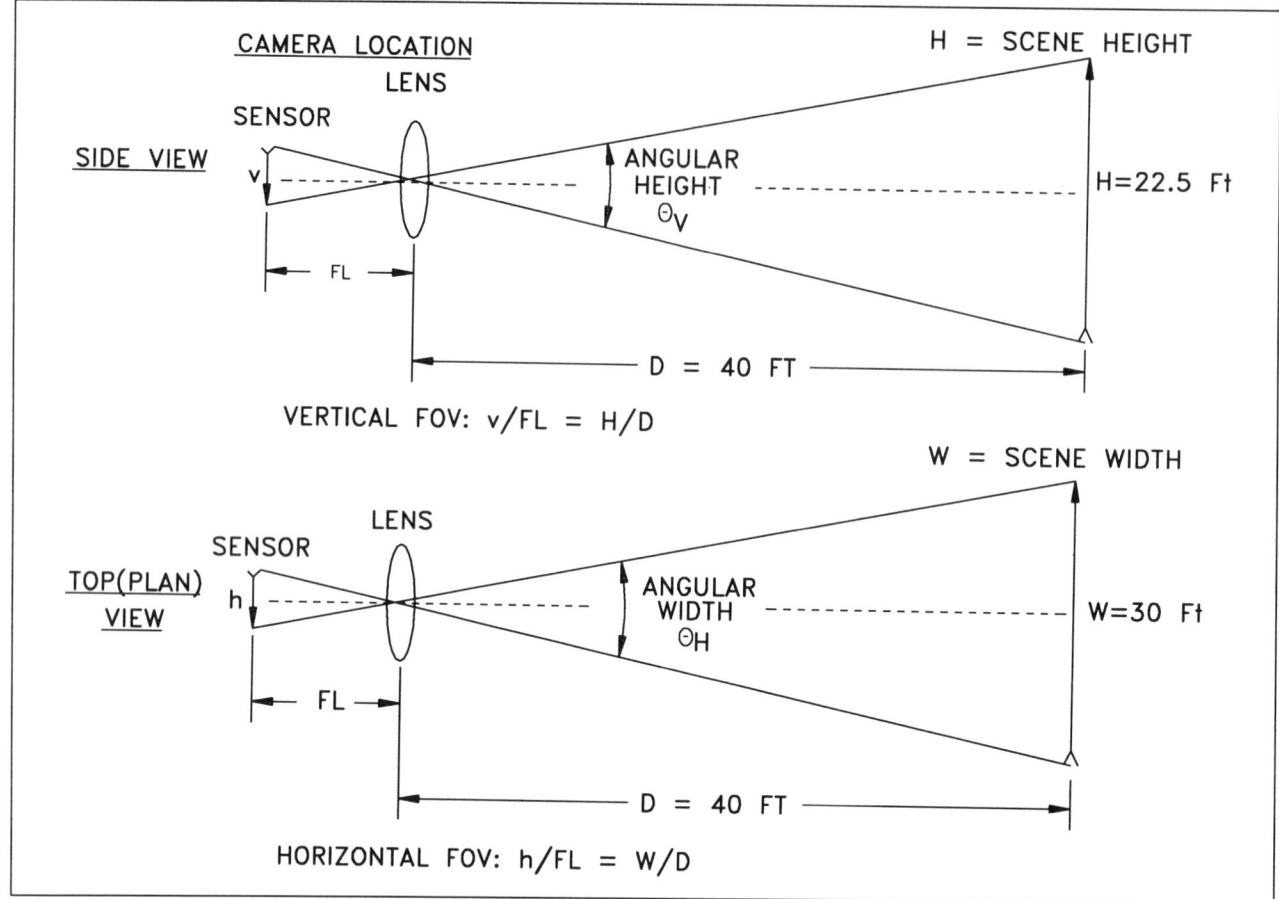

FIGURE 4-7 Calculating the focal length for viewing a building

the same example, calculate the horizontal and vertical angular FOVs θ_H and θ_V for a ½-inch CCD sensor using a 12.5-mm FL lens. The distance need not be supplied, since an angular measure is independent of distance.

From Equation 4-2, for horizontal angular FOV:

$$\tan \frac{\theta_H}{2} = \frac{h/2}{FL} = \frac{6.6 \text{ mm}/2}{12.5 \text{ mm}} = 0.264$$

$$\frac{\theta_H}{2} = 14.8 \text{ degrees}$$

$$\theta_H = 29.6 \text{ degrees}$$

From Equation 4-4 for vertical angular FOV:

$$\tan \frac{\theta_V}{2} = \frac{v/2}{FL} = \frac{4.8 \text{ mm}/2}{12.5 \text{ mm}} = 0.192$$

$$\frac{\theta_V}{2} = 10.9 \text{ degrees}$$

$$\theta_V = 21.8 \text{ degrees}$$

Table 4-4 summarizes angular FOV values for some of the popular lenses used on the ⅓-, ½-, and ⅔-inch sensors. To obtain the angular FOV for 1-inch sensors, *multiply* the

angles for the ½-inch sensor by 2. The angular FOV for ⅓-inch sensors can also be obtained by *dividing* the angles for the ⅔-inch sensor by 2.

4.2.5 Lens Finder Kit

Tables and slide rules for finding lens angular FOVs abound. Over the years many charts and devices have been available to simplify the task of choosing the best lens for a particular security application. Figure 4-8 shows how to quickly determine the correct lens for an application using the Lens Finder Kit.

There is a separate scale for each of the four camera-sensor sizes: ⅓-, ½-, ⅔-, and 1-inch (only the ½- and ⅔-inch are shown). Each scale shows the FL of standard lenses and the corresponding horizontal and vertical FOVs they let the camera see.

To use the kit, the plastic disk is placed on the facility plan drawing and the lens FL giving the desired camera FOV coverage is chosen. For example, a ½-inch format camera is to view a horizontal FOV (θ_H) in a front lobby 30 feet wide at a distance of 30 feet from the camera (Figure 4-9). What FL lens should be used?

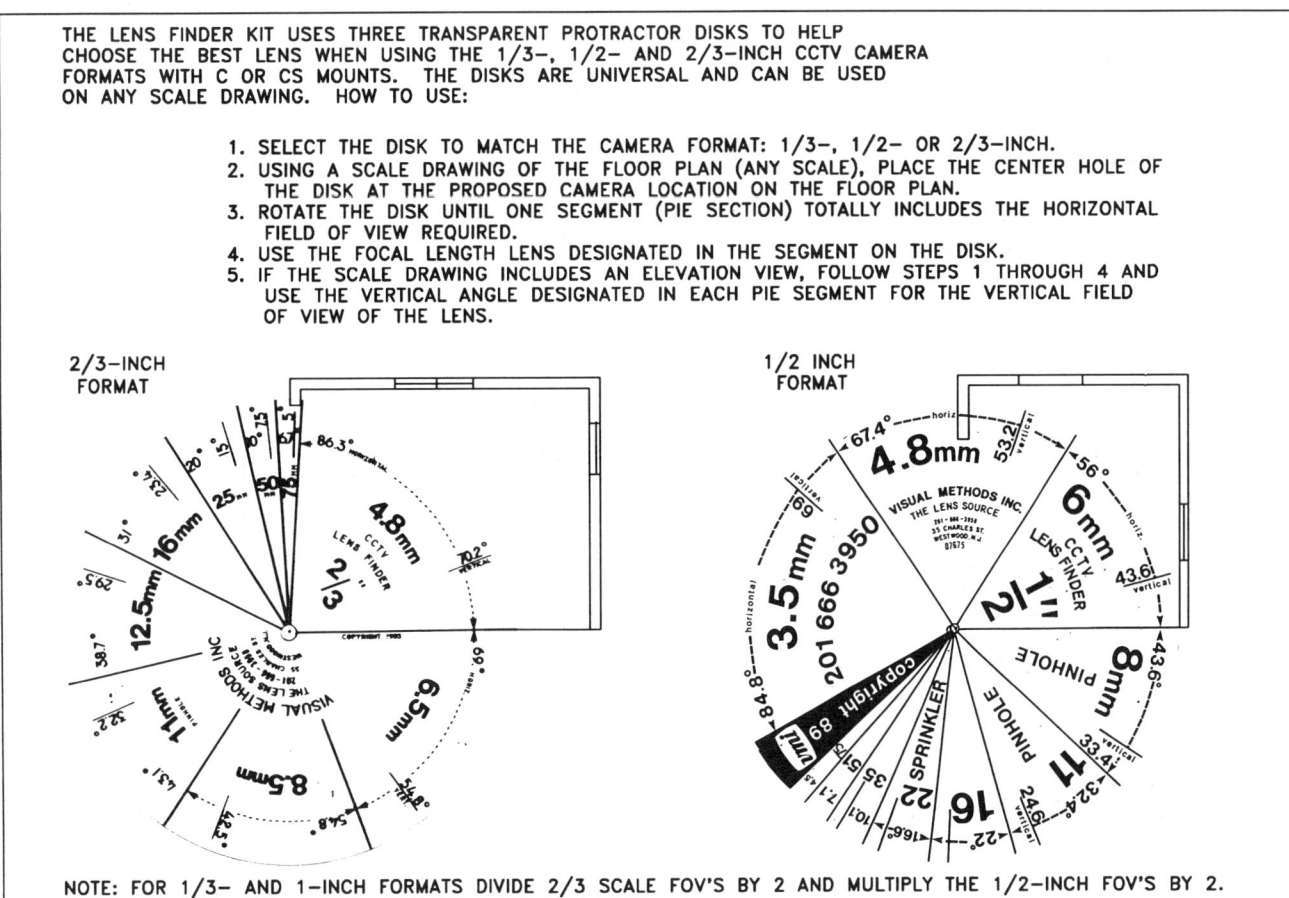

THE LENS FINDER KIT USES THREE TRANSPARENT PROTRACTOR DISKS TO HELP
CHOOSE THE BEST LENS WHEN USING THE 1/3-, 1/2- AND 2/3-INCH CCTV CAMERA
FORMATS WITH C OR CS MOUNTS. THE DISKS ARE UNIVERSAL AND CAN BE USED
ON ANY SCALE DRAWING. HOW TO USE:

1. SELECT THE DISK TO MATCH THE CAMERA FORMAT: 1/3-, 1/2- OR 2/3-INCH.
2. USING A SCALE DRAWING OF THE FLOOR PLAN (ANY SCALE), PLACE THE CENTER HOLE OF
   THE DISK AT THE PROPOSED CAMERA LOCATION ON THE FLOOR PLAN.
3. ROTATE THE DISK UNTIL ONE SEGMENT (PIE SECTION) TOTALLY INCLUDES THE HORIZONTAL
   FIELD OF VIEW REQUIRED.
4. USE THE FOCAL LENGTH LENS DESIGNATED IN THE SEGMENT ON THE DISK.
5. IF THE SCALE DRAWING INCLUDES AN ELEVATION VIEW, FOLLOW STEPS 1 THROUGH 4 AND
   USE THE VERTICAL ANGLE DESIGNATED IN EACH PIE SEGMENT FOR THE VERTICAL FIELD
   OF VIEW OF THE LENS.

NOTE: FOR 1/3- AND 1-INCH FORMATS DIVIDE 2/3 SCALE FOV'S BY 2 AND MULTIPLY THE 1/2-INCH FOV'S BY 2.

FIGURE 4-8 Choosing a lens with the Lens Finder Kit

To find the horizontal angular FOV θ_H, draw the following lines to scale on the plan: one line to a distance 30 feet from the camera to the center of the scene to be viewed, a line 30 feet long and perpendicular to the first line, and two lines from the camera location to the endpoints of the second 30-foot line. Place the $\frac{1}{2}$-inch Lens Finder Kit on the top view (plan) drawing with its center at the camera location and choose the FL closest to the horizontal angle required. A 6.0-mm FL lens is closest. This lens will see a horizontal scene width of 30.6 feet. Likewise for scene height: using the side-view (elevation) drawing, the horizontal scene height is 23.4 feet.

4.2.6 Optical Speed (f-number)

The optical speed or f-number (f/#) of a lens defines its light-gathering ability. The more light the lens can collect and transfer to the camera image sensor, the better the picture quality: that is, larger lenses can operate at lower light levels, and have increased depth of field. This light-gathering ability depends on the size of the CCTV optics: the larger the optics the more light that can be collected.

Most human eyes are the same size (approximately 7-mm lens diameter) and therefore have the same size lens. In television systems, however, the lens size (the diameter of the front lens) varies over a wide range; therefore, the optical speed of television lenses varies significantly. In fact, it varies as the square of the diameter of the lens. In practical terms, a lens having a diameter twice that of another will pass four times as much light through it. Like a garden hose, when the diameter is doubled, the flow is quadrupled (Figure 4-10).

The more light passing through a lens and reaching the CCTV sensor, the better the contrast and picture image quality. Lenses with low f-numbers, such as f/1.4 or f/1.6, pass more light than lenses with high f-numbers. The lens optical speed is related to the FL and diameter by the equation f/# = focal length/diameter. So the larger the FL given the same lens diameter, the "slower" the lens (less light reaches the sensor). A slow lens might have an f-number of f/4 or f/8. Most lenses have an iris ring usually marked with numbers such as 1.4, 2.0, 2.8, 4.0, 5.6, 8.0, 11, 16, 22, C, representing optical speed, f-numbers, or f-stops. The C indicates when the lens iris is closed and no light is transmitted.

The difference between each of the iris settings represents a difference of a factor of 2 in the light transmitted by the lens. Opening the lens from, say, f/2.0 to f/1.4 doubles the light transmitted. Only half the light is transmitted

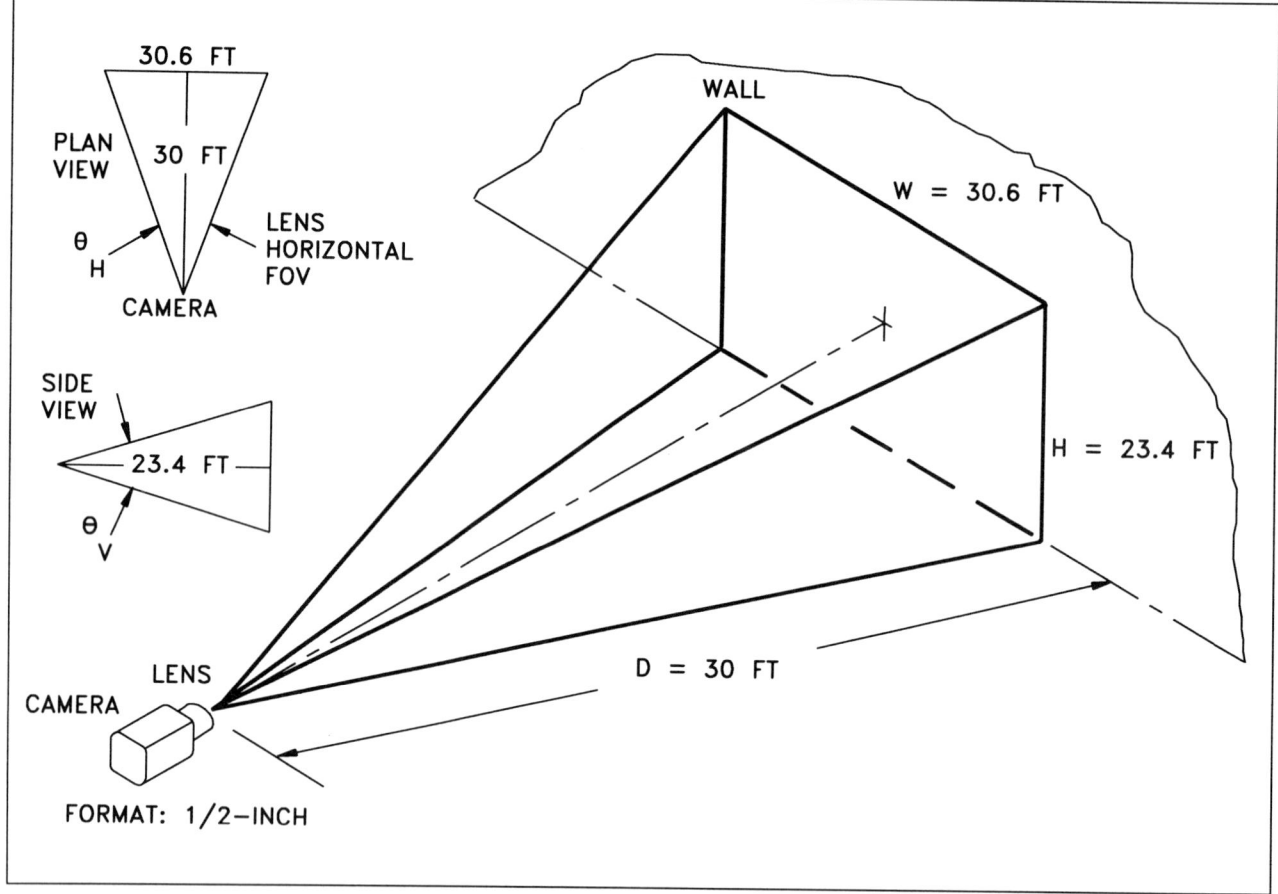

FIGURE 4-9   Determining lobby lens horizontal and vertical FOVs

when the iris opening is reduced from, say, f/5.6 to f/8. Changing the iris setting two f-numbers changes the light by a factor of 4 (or ¼), and so on. Covering the range from f/1.4 to f/22 spans a light-attenuation range of 256 to 1.

In general, faster lenses collect more light energy from the scene, are larger, and are more expensive. In calculating the overall cost of a television camera lens system, the more expensive, fast lens often overrides the higher cost incurred if a more sensitive camera were used or additional lighting were installed.

4.2.7 Depth of Field

The depth of field in an optical system is the distance that an object in the scene can be moved toward or away from the CCTV lens and still be in good focus. In other words, it is the *range* of distance from the camera lens in which objects in the scene remain in focus. Ideally this range would be very large: say, from a few feet from the lens to hundreds (or thousands) of feet, so that essentially *all* objects of interest in the CCTV scene would be in sharp focus. In practice this is not achieved because the depth of field is (1) inversely proportional to the focal length and (2) inversely proportional to the f-number. Medium to long FFL CCTV lenses operating at low f-numbers—say, f/1.2

to f/4.0—do not focus sharp images over their useful range of from 2 to 3 feet to hundreds of feet. When these lenses are used with their iris closed down to, say, f/8 to f/16, the depth of field increases significantly and objects are in sharp focus at almost all distances in the scene.

Short focal length lenses (2.7 to 5 mm) have a long depth of field. These lenses can produce sharp images from 1 foot to 50–100 feet even when operating at low f-numbers. On the other hand, long focal length lenses—say, 50–300 mm—have a short depth of field and can produce sharp images only over short distances and must be refocused when viewing objects at different distances in the scene.

4.2.8 Manual and Automatic Iris

The CCTV lens iris is manually or automatically adjusted to optimize the light level reaching the sensor (Figure 4-3). The manual-iris CCTV lens has movable metal "leaves" forming the iris. The amount of light entering the camera is determined by rotating an external iris ring, which opens and closes these leaves (Figure 4-11). Solid-state sensor (CCD, MOS) and silicon and Newvicon tube cameras can operate with manual-iris lenses over limited light-level changes but require automatic-iris lenses when used over their full light-level range, that is, from bright sunlight to

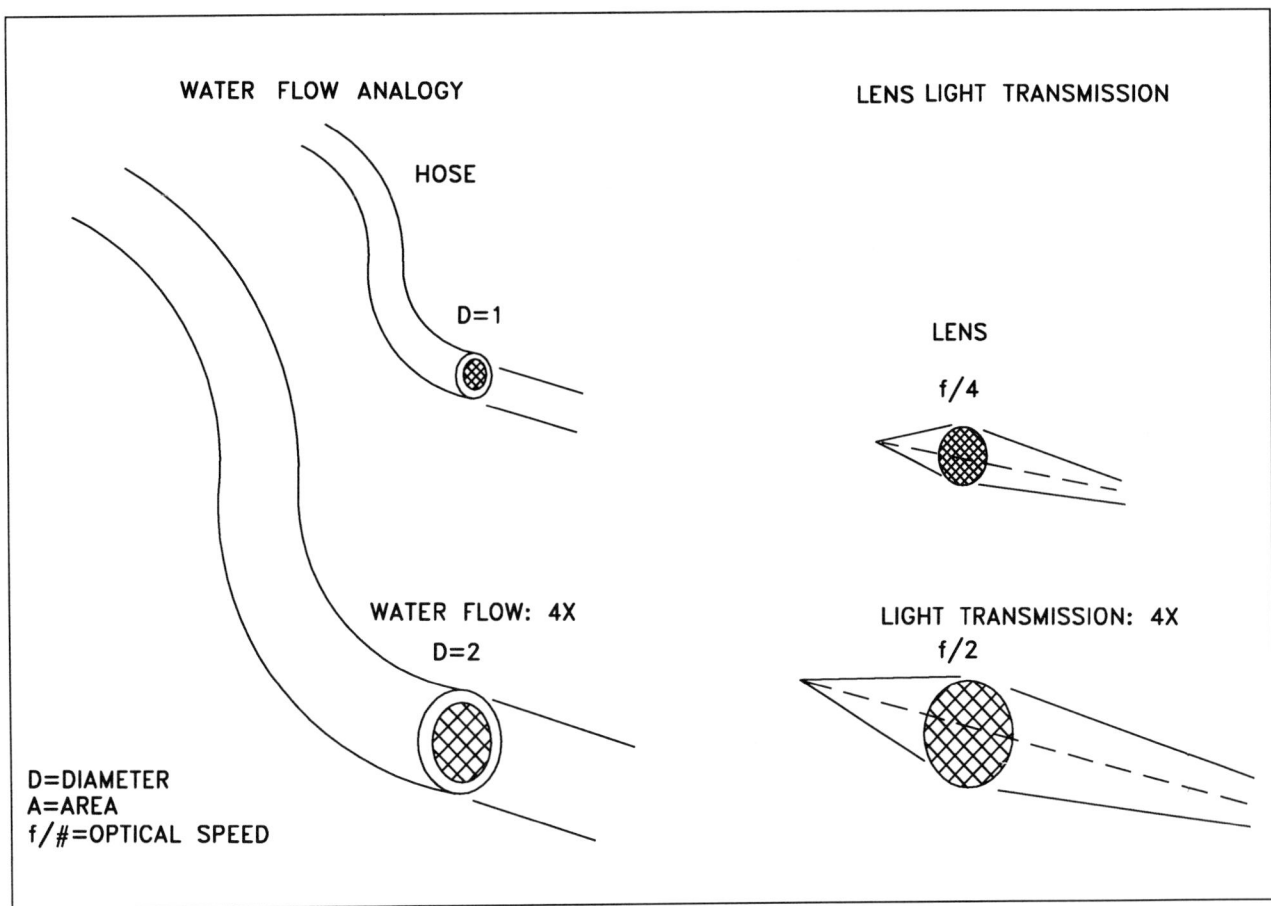

FIGURE 4-10 Light transmission through a lens

low-level nighttime lighting. Some solid-state cameras use electronic shuttering and do not require an automatic-iris lens. Vidicon tube cameras compensate fairly well over their useful range of operation without any change in the lens iris opening; they are not suited to automatic-iris operation unless the lens adjusts its aperture using light input sensed by a photocell in the lens.

Automatic-iris lenses should be used with cameras having fixed video gain in their system, which include silicon and Newvicon tubes and solid-state sensors. Automatic-iris lenses are more expensive than their manual counterparts, with the price ratio varying by about two or three to one.

4.2.8.1 Automatic-Iris Operation

Automatic-iris lenses have an electro-optical mechanism whereby the amount of light passing through the lens can be adjusted depending on the amount of light available from the scene and the sensitivity of the camera.

The intelligence used for adjusting the light passing through the lens is obtained from the camera video signal. The system works something like this: If a scene is too bright for the camera, the video signal will be strong (large in amplitude). This large signal will activate a motor or galvanometer that causes the lens iris circular opening to be-

come smaller in diameter, thereby reducing the amount of light reaching the camera. When the amount of light reaching the camera produces a predetermined signal level, the motor or galvanometer stops and maintains that level through the lens. Likewise if too little light reaches the camera, the video camera signal level is small, the automatic-iris motor or galvanometer opens up the iris diaphragm, allowing more light to reach the camera. In both the high- and low-light-level conditions, the automatic-iris mechanism produces the best contrast picture. Automatic-iris lenses are available to accomplish the full range of wide-angle to narrow-angle viewing. They are available with the automatic-iris compensation electronics either built into the lens housing or split (electronics in the camera and only the DC motor drive in the lens). Figure 4-12 shows some common automatic-iris lenses.

A feature available on some automatic-iris lenses, called average/peak response weighting, permits optimizing the picture still further based on lighting conditions. Certain scenes with high-contrast objects are better compensated for by setting the automatic-iris control to peak, so that the lens responds to the bright spots and highlights in the scene. Other low-contrast scenes or objects are better handled by setting the control to average. Figure 4-13 illustrates some actual scenes obtained when these adjustments are made.

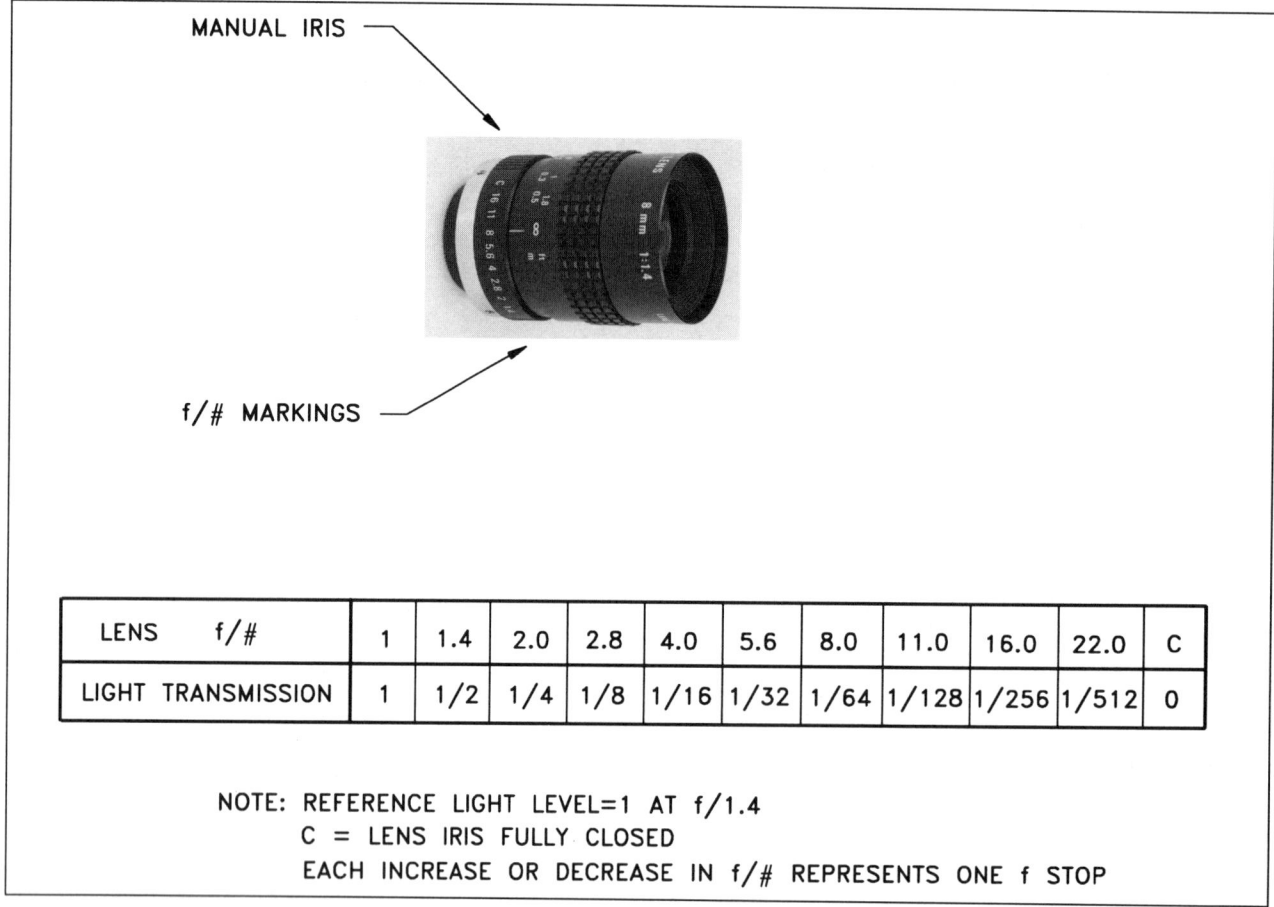

LENS f/#	1	1.4	2.0	2.8	4.0	5.6	8.0	11.0	16.0	22.0	C
LIGHT TRANSMISSION	1	1/2	1/4	1/8	1/16	1/32	1/64	1/128	1/256	1/512	0

NOTE: REFERENCE LIGHT LEVEL=1 AT f/1.4
C = LENS IRIS FULLY CLOSED
EACH INCREASE OR DECREASE IN f/# REPRESENTS ONE f STOP

FIGURE 4-11   Lens f-number vs. light transmission

4.3 STANDARD LENS TYPES

Television lenses come in many varieties, from the simple, small, and inexpensive "bottle cap," 16-mm, f/1.6 lenses to complex, large, expensive, motorized, automatic-iris zoom lenses. Each camera application has a specific scene to be viewed and specific intelligence to extract from the scene if it is to be useful in a security application. Simply monitoring a small front lobby or room to see if a person is present may require only a simple lens. On the other hand, trying to determine the activity of a person 200 feet away—perhaps to see if someone in a large showroom pockets an expensive diamond ring—requires a high-quality, long-FL lens and camera with high resolution. Covert pinhole lenses are often used for uncovering internal theft, shoplifting, or inappropriate actions or when covert surveillance is required. They can be concealed in inconspicuous locations, installed or moved on short notice, and can serve as remote, hidden eyes. The following sections describe common and special lens types used in CCTV surveillance applications.

4.3.1 Fixed Focal Length

The majority of lenses used in CCTV applications are referred to as FFL lenses. Most of these lenses are available with a manual iris, to adjust the amount of light passing through the lens and reaching the image sensor, and a manual focusing ring. The very short FL lenses (less than 6 mm) often lack a focusing ring or a manual iris. The 2.6-mm to 3.6-mm FL lenses are available only for 1/2-inch format cameras and show some image distortion in the picture. FFL lenses are the workhorses of the industry. Their attributes include low cost, ease of operation, and long life. Figure 4-14 shows some examples of standard FFL lenses.

Most FFL lenses are optically fast and range in speed from f/1.3 to f/1.8, providing sufficient light for most cameras to produce a good-quality picture. The manual-iris lenses are suitable for medium light-level variations when used with solid-state cameras and provide extended range when used with standard vidicon cameras, whose dynamic range is sufficiently high to cover wide light-level variations.

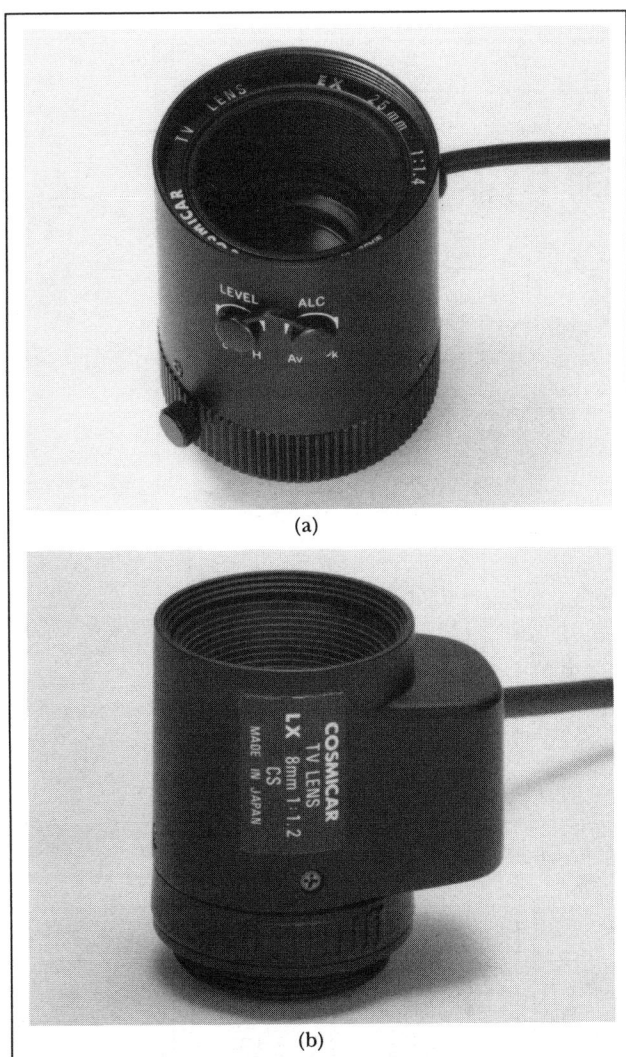

(a)

(b)

FIGURE 4-12   Automatic-iris FFL lenses: (a) DC motor in lens, (b) galvanometer drive

Most CCTV FFL lenses have a C or CS (1 inch × 32 threads per inch) mount. The C mount has been a standard in the CCTV industry for many years; the CS mount was introduced a few years ago to match the trend toward smaller camera sensor formats and their correspondingly smaller lens requirements. The C- and CS-mount lenses are used on ½- and ⅔-inch sensor format cameras, and the C-mount lens on ½-, ⅔-, and 1-inch tube or ½- and ⅔-inch solid-state sensor cameras. Commonly used FLs vary from 4.0 mm (wide angle) to 150 mm (telephoto), with longer-FL lenses (up to several thousand millimeters) available for special outdoor, long-range surveillance applications.

Lenses longer than approximately 150 mm FL are large and expensive. As the FL becomes longer, the diameter of the lens increases and costs escalate substantially. Most FFL

lenses are available in a motorized or automatic-iris form for use in remote-control applications. When used with low-light-level (LLL) tube, solid-state, or intensified cameras (Chapter 15), the iris or light-level control must be varied via an automatic iris and neutral density filters, depending on the scene illumination.

4.3.2 Wide-Angle Viewing

While the human eye has peripheral vision and can detect the presence and movement of objects over a wide angle (160 degrees), the eye sees a focused scene in only about the central 10 degrees of its FOV. No television camera has this unique eye characteristic, but a television system's FOV can be increased (or decreased) by replacing the lens with one having a shorter (or longer) FL. The eye cannot change its FOV.

To increase the FOV of a CCTV camera, a short-FL lens is used. The FOV obtained with wide-angle lenses can be calculated from Equations 4-5, 4-6, 4-7, and 4-8, or by using Table 4-3 or the Lens Finder Kit. For example, substituting an 8-mm FL, wide-angle lens for a 16-mm lens on a ⅔-inch format camera doubles the FOV. The magnification is reduced to one-half, and the camera sees "twice as much but half as well." By substituting a 4-mm FL lens for the 16-mm lens, the FOV quadruples. We see sixteen times as much scene area but one-fourth as well. Choosing different FL lenses brings trade-offs: Reducing the FL increases the FOV but reduces the magnification, thereby making objects in the scene smaller and less discernible (that is, decreasing resolution). Increasing the FL has the opposite effect.

A 4.8-mm FL lens is an example of a wide-angle lens; it has a 96-degree horizontal by 72-degree vertical FOV on a ⅔-inch sensor. A superwide FOV lens for a ½-inch sensor is the 2.6-mm FL lens, with an FOV approximately 117 degrees horizontal by 88 degrees vertical. Using a wide-angle lens reduces the resolution or ability to discern objects in a scene. Figure 4-15 shows a comparison of the FOV seen on ⅓-, ½-, and ⅔-inch format cameras with wide-angle, normal, and telephoto lenses.

4.3.3 Telephoto Viewing

When the lens FL increases above the standard 16 mm on a ⅔-inch camera, 12.5 mm on a ½-inch camera, or 8 mm on a ⅓-inch camera (the M = 1 condition), the FOV decreases and the magnification increases. Such a lens is called a telephoto or narrow-angle lens. The lens magnification is determined by Equations 4-5, 4-6, 4-7, and 4-8 for the four sensor sizes. (See also Table 4-4 and the Lens Finder Kit.)

Outdoor security applications often require viewing scenes hundreds or thousands of feet or even miles away

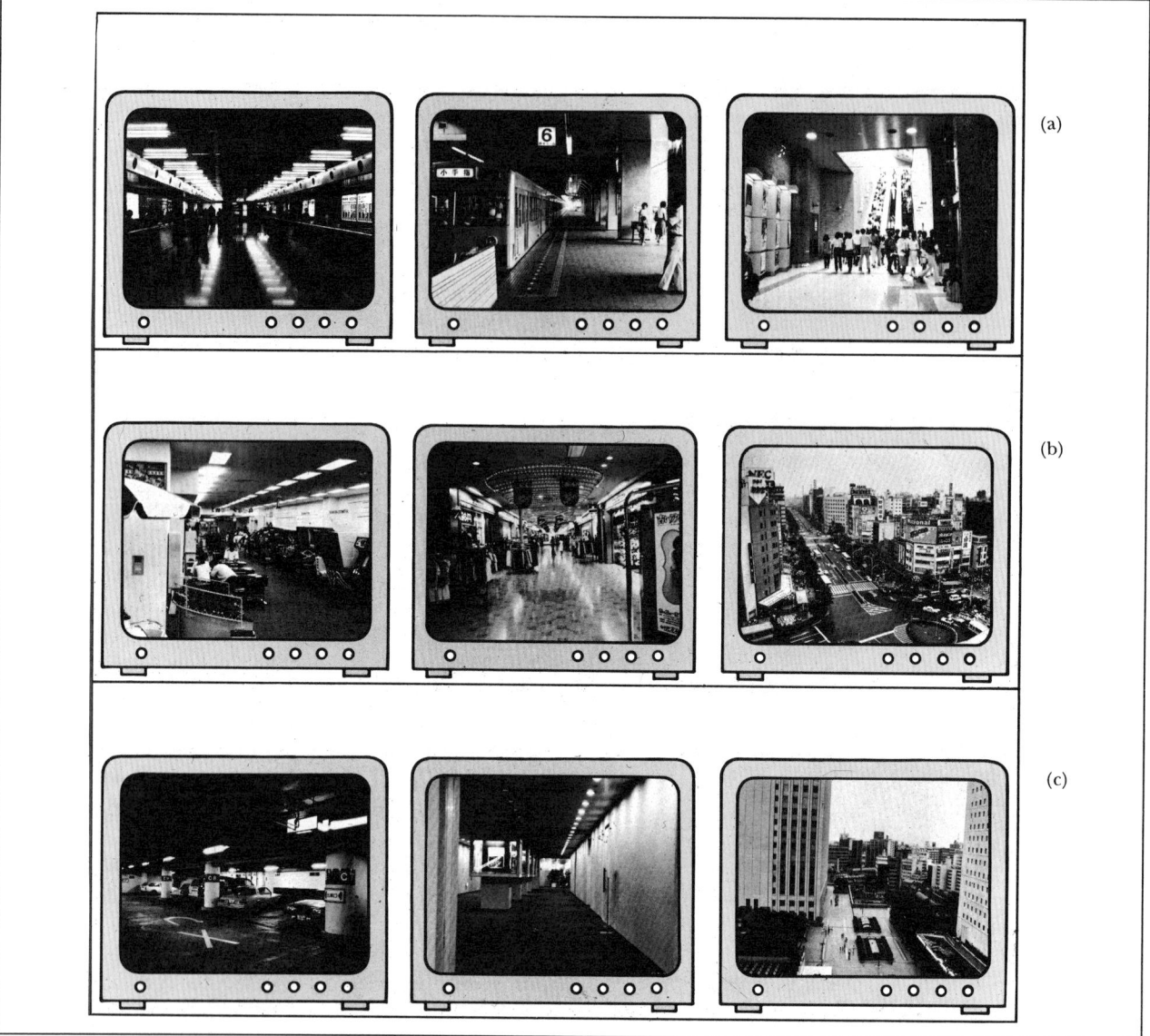

FIGURE 4-13 Automatic-iris enhanced video scenes: (a) high-contrast scenes optimized using peak-response weighting, (b) normal-contrast scenes optimized using medium-response weighting, (c) low-contrast scenes optimized using average-response weighting

from the lens and camera. To detect and/or identify objects, persons, or activity at these ranges requires long-FL lenses. Long-FL lenses between 150 and 1200 mm are usually used outdoors to view parking lots or other remote areas. These large lenses, which are used to view at long distances and/or under LLL conditions, require very stable mounts (or pan/tilt drives) to obtain good picture quality. The lenses must be large (4 to 12 inches in diameter) to collect enough light from the distant scene and produce usable f-numbers (f/1.5–f/8) for the CCTV camera to produce a good picture on the monitor.

For lenses with FLs from 2.6 mm up to several hundred millimeters, FFL lenses are refractive or glass type. Above approximately 300 mm FL, refractive glass lenses become large and expensive, and reflective mirror optics or mir-

ror/glass optics are used to achieve optically fast (low f-number) lenses with lower weight than refractive types. These long-FL telephoto lenses, called Cassegrain or catadioptric lenses, cost thousands of dollars. Figure 4-16 shows a 700-mm f/8.0 and a 170-mm f/1.6 lens used for long-range outdoor surveillance applications.

4.3.4 Zoom/Variable Focal Length

Zoom lenses are variable-FL lenses. The lens components in these assemblies are moved to change their relative physical positions, thereby varying the FL and angle of view through a specified range of magnifications.

FIGURE 4-14 Standard FFL lenses: (a) manual iris, (b) automatic iris

4.3.4.1 Zooming

Zooming is a lens adjustment that permits seeing detailed close-up shots of a subject (scene target) or a broad, overall view. Zoom lenses allow a smooth, continuous change in the angular FOV, so that angle of view can be made narrower or wider depending on the zoom setting. As a result, a scene can be made to appear close-up or far away, giving the impression of camera movement, even though the camera remains in a fixed position. Figure 4-17 shows the continuously variable nature of the zoom lens.

With a zoom lens, the FOV of the CCTV camera can be changed without replacing the lens. Several elements in these lenses are physically moved to vary the FL and thereby vary the angular FOV and magnification. Tables 4-1, 4-2, 4-3, and the Lens Finder Kit can be used to determine the FOV for any zoom lens. By adjusting its zoom ring setting, one can view narrow-, medium-, or wide-angle scenes. This allows a person to view a scene with a wide-angle perspective and then close in on one portion of the scene that is of particular interest.

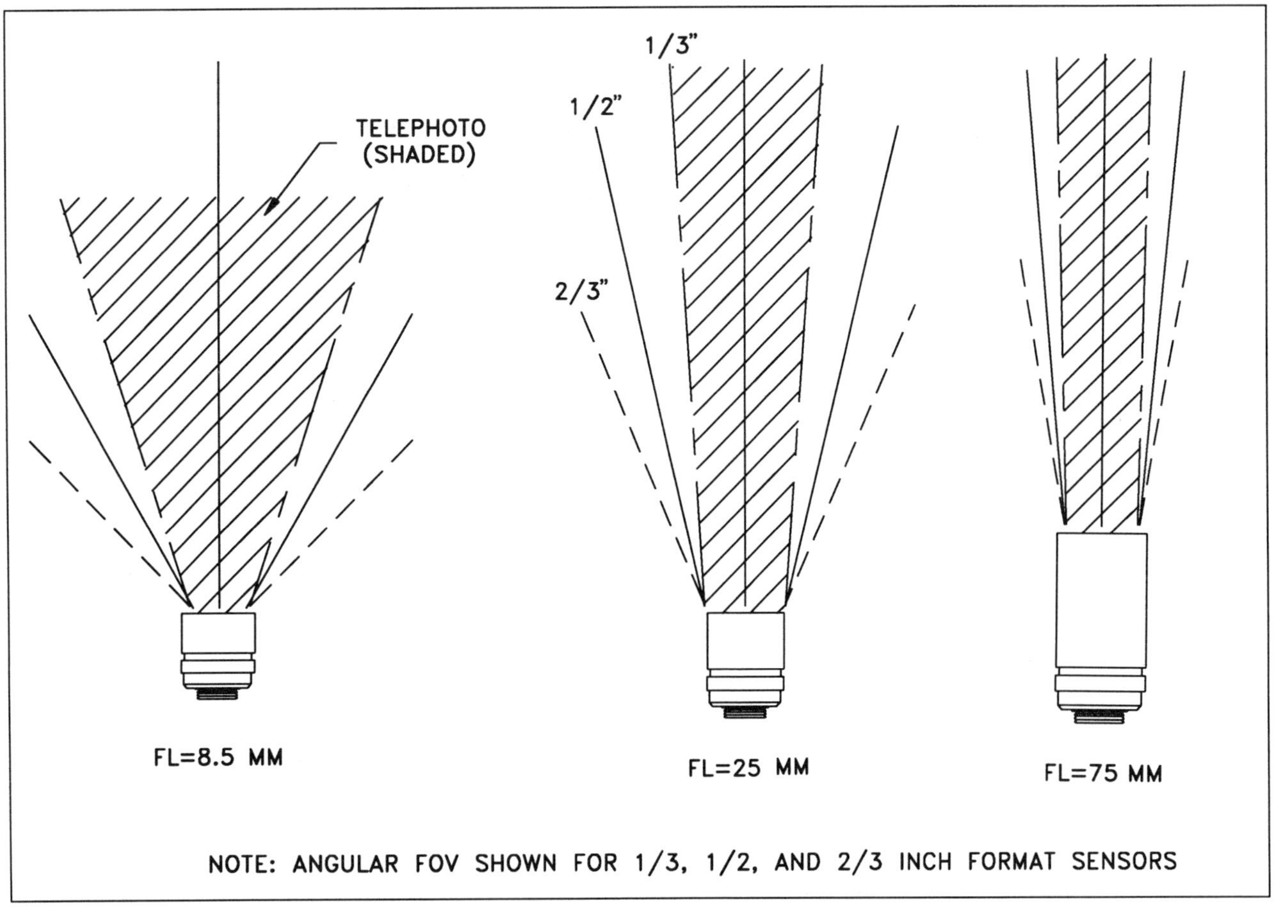

TELEPHOTO (SHADED)

1/3"

1/2"

2/3"

FL=8.5 MM

FL=25 MM

FL=75 MM

NOTE: ANGULAR FOV SHOWN FOR 1/3, 1/2, AND 2/3 INCH FORMAT SENSORS

FIGURE 4-15 Wide-angle, normal, and narrow-angle FFL lenses vs. format

FIGURE 4-16 Long-range, long-focal-length catadioptric lenses: (a) 700-mm FL, f/8, 8 inches long, 4⅜-inch diameter, 4 pounds; (b) 300-mm FL, f/1.8, t/2, 9 inches long, 9.5-inch diameter

(a)

(b)

In security surveillance applications, to make the zoom lens useful over a much wider FOV, the camera and lens are often mounted on a pan/tilt platform and controlled from a remote console. The pan/tilt positioning and variable angle, from wide to narrow to anywhere in between, provide a large dynamic FOV capability.

Before the invention of zoom optics, quick conversion to different FLs was achieved by mounting three different FFL lenses on a turret with a common lens mount in front of the CCTV camera sensor and rotating each lens into position, one at a time, in front of the sensor. The lenses usually had wide, medium, and short FLs to achieve different angular coverage. This turret lens was obviously not a variable-FL lens and had limited use.

4.3.4.2 Lens Operation

The zoom lens is a cleverly designed assembly of lens elements that can be moved to change the FL from a wide angle to a narrow angle (telephoto) while the image on the sensor remains in focus (Figure 4-18).

A variable-FL lens combines at least three groups of elements:

1. The front focusing objective group can be adjusted over a limited distance with an external focus ring to fine-focus the lens.
2. Between the front and rear groups is a movable zoom group, which is moved appreciably (front to back) using a separate external zoom ring. The zoom group

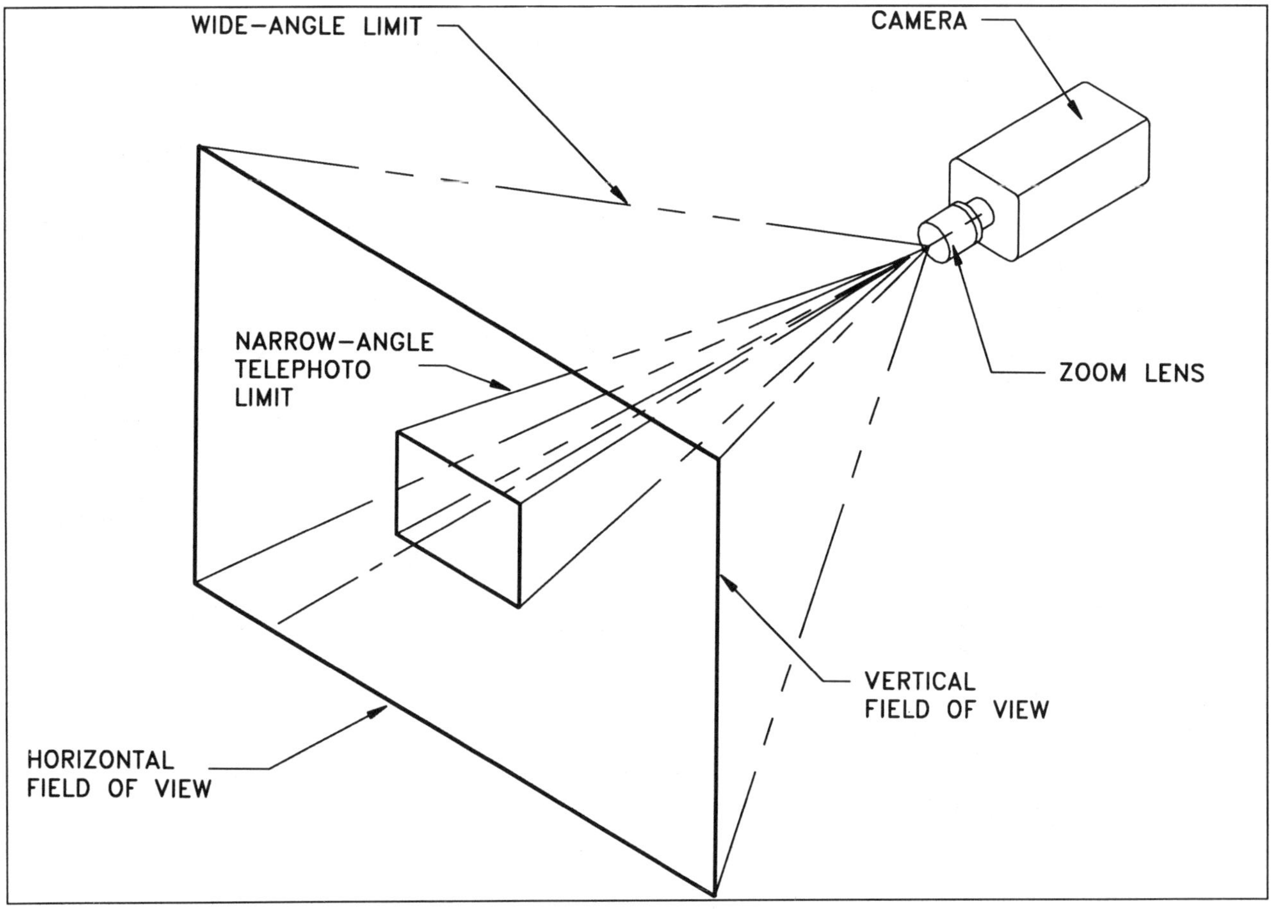

FIGURE 4-17 Zoom lens variable-focal-length function

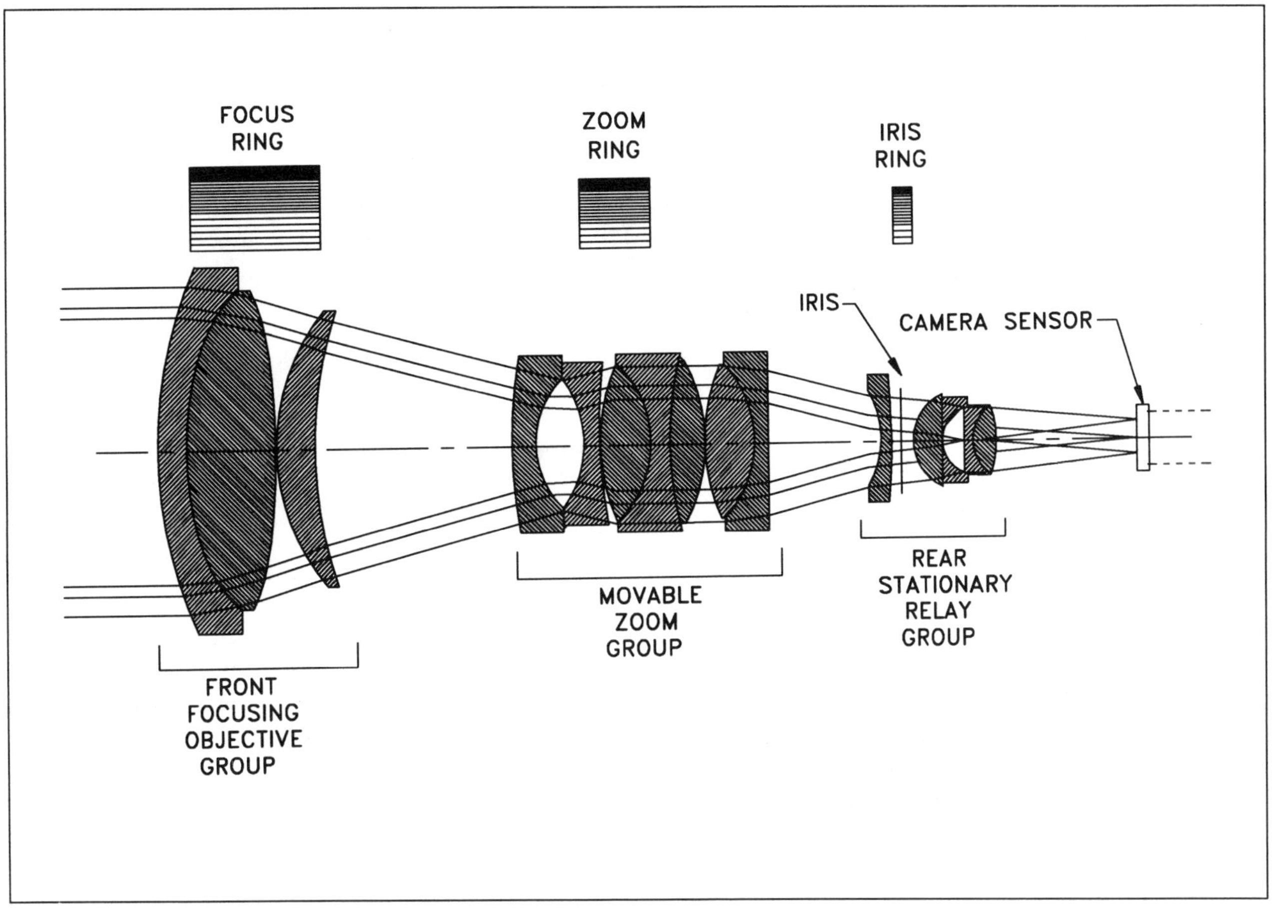

FIGURE 4-18 Zoom lens configuration

A)   MANUAL ZOOM LENS

FOCAL LENGTH : 9.5–152MM
OPTICAL SPEED: F/1.8 (9.5–114MM)
F/1.9 (152MM)

ZOOM RATIO: 9 TO 1

B) LONG RANGE MOTORIZED ZOOM LENS

FOCAL LENGTH: 13.5–600MM
OPTICAL SPEED: F/1.8 (13.5–350MM)
F/1.9 (600MM)

ZOOM RATIO: 44 TO 1

C)   COMPACT ZOOM LENS

FOCAL LENGTH: 8–48MM
OPTICAL SPEED: F/1.4
ZOOM RATIO: 6–1

FIGURE 4-19   Manual and motorized zoom lenses: (a) manual zoom lens—9.5–152-mm FL, f/1.8, 16:1 zoom ratio; (b) long-range motorized zoom lens; (c) compact zoom lens—8–48-mm FL, f/1.4, 6:1 zoom ratio

also contains corrective elements to optimize the image over the full zoom FL range. Connected to this group are other lenses that are moved a small amount to automatically adjust and keep the image on the sensor in sharp focus, thereby minimizing the external adjustment of the front focusing group.

3. At the camera end of the zoom lens is the rear stationary relay group, which determines the final image size when it is focused on the camera sensor.

Each group normally consists of several elements. When the zoom group is positioned correctly, it sees the image produced by the objective group and creates a new image from it. The rear relay group picks up the image from the zoom group and relays it to the camera sensor. In a well-designed zoom lens, a scene in focus at the wide-angle (short-FL) setting remains in focus at the narrow-angle (telephoto) setting and everywhere in between.

4.3.4.3 Optical Speed

Since the FL of a zoom lens is variable and its entrance aperture is fixed, its f-number is not fixed (see Equation 4-9). For this reason, zoom lens manufacturers often list

the f-number for the zoom lens at the wide and narrow FLs, with the f-number at the wide-angle setting being faster (more light throughput, lower f-number) than at the telephoto setting. For example, an 11–110-mm zoom lens may be listed as f/1.8 when set at 11-mm FL and f/4 when set at 110-mm FL. The f-number for any other FL in between the two settings listed is any number in between these two values.

4.3.4.4 Configurations

Many manufacturers produce a large variety of manual and motorized zoom lenses suitable for a wide variety of applications. Figure 4-19 shows three very different zoom lenses used for surveillance applications.

The manual zoom lens shown has a lens shade that minimizes off-axis stray light. The lens has a 9.5-to-152-mm FL (16-to-1 zoom ratio) and has an optical speed of f/1.8 from 9.5 mm to 114 mm and f/2.3 at 152 mm. The small zoom lens has 8–48-mm FL, has an optical speed of f/1.4, and is 2 inches in diameter by 3 inches long. The third lens shown has a large zoom ratio of 44:1. This expensive lens has an FL range of 13.5–600 mm and speed of f/1.8. It is 10 inches high by 11 inches wide by 21 inches long. The

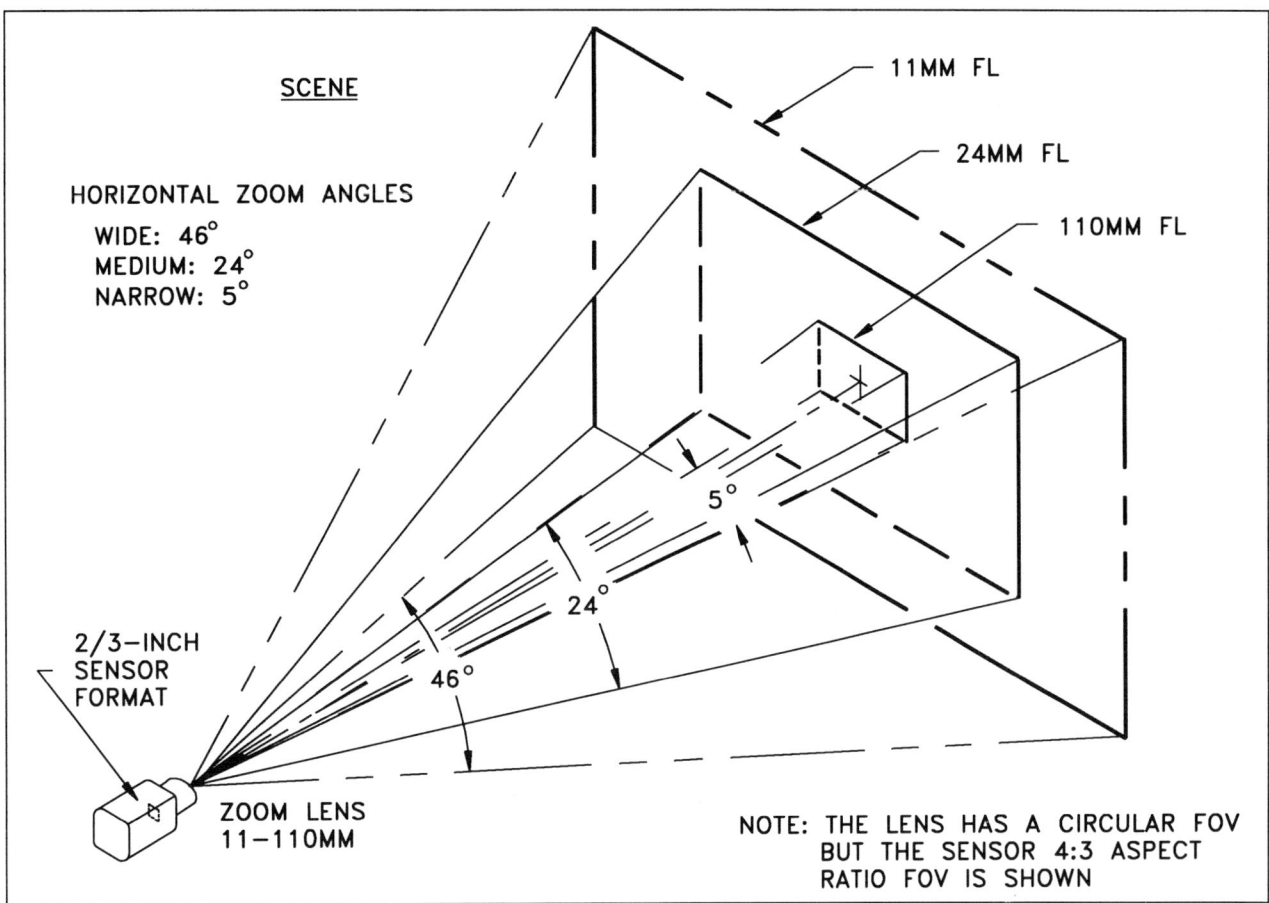

FIGURE 4-20   Zoom lens FOVs at different focal-length settings

13.5–600-mm lens has a built-in remotely controlled 2X magnifier to extend its FL to 27–1200 mm.

Figure 4-20 shows the FOVs obtained from an 11–110-mm FL zoom lens on a ²⁄₃-inch sensor camera at several zoom ring settings. Table 4-5 is a representative list of manual and motorized zoom lenses, from a small, light-weight, inexpensive 8–48-mm FL zoom lens to a large, expensive, 13.5–600-mm zoom lens used in high-risk security areas by industry, military, and government agencies.

Zoom lenses are available with magnification ratios from 6:1 to 44:1. Some have special features, including remotely controlled preset zoom and focus positions.

4.3.4.5 Manual or Motorized

The FL of a zoom lens is changed by moving an external zoom ring either manually or with an electric motor. When the zoom lens iris, focus, or zoom setting must be adjusted remotely, a motorized lens with a remote controller is used. Many zoom lenses have an automatic iris. The operator can control and change these parameter settings remotely using the toggle switch controls on the console. The motor and gear mechanisms effecting these changes are mounted on the exterior of the zoom lens.

4.3.4.6 Adding Pan/Tilt Mechanisms

Zoom lenses are often mounted on a pan/tilt mechanism so that the lens can be pointed in almost any direction (Figure 4-21). By varying the lens zoom control and moving the pan/tilt platform, a wide dynamic FOV is achieved. The pan/tilt and lens controller remotely adjusts pan, tilt, zoom, and focus. The lens usually has an automatic iris, but in some cases the operator can choose a manual- or automatic-iris setting on the lens or the controller. In surveillance applications, one shortcoming of a pan/tilt-mounted zoom lens is the occurrence of "dead zone" viewing areas, since the lens can't point in all directions at once.

4.3.4.7 Preset Zoom and Focus

In a computer-controlled surveillance system, a motorized zoom lens with an electronic preset function can be used. In a preset zoom lens, the zoom and focus ring positions are monitored electrically and memorized by the computer during system setup. They are then automatically repeated on command by the computer at a later time. In a pan/tilt surveillance application, this feature allows the computer to point the camera/lens combination according to a set of predetermined conditions: for example, (1) azimuth and

FOCAL LENGTH (MM)	ZOOM RATIO	FORMAT MAX (INCH)	f/# MAX	HORIZONTAL ANGULAR FOV (DEG)								DRIVE TYPE	MOUNT*
				1 INCH		2/3 INCH		1/2 INCH		1/3 INCH			
				WIDE	TELE	WIDE	TELE	WIDE	TELE	WIDE	TELE		
3.5–8.0	2.29X	1/3	1.8							68.3	30.8	MANUAL	CS
10–40	4X	1/2	1.8					35.2	8.8	24.2	6.1	MANUAL	C
7.5–75	10X	1/2	1.4					48.2	4.9	33.1	3.4	MOTORIZED	CS
8–48	6X	1/2	1.2					43.7	7.7	30.0	5.3	MOTORIZED	C
8–48	6X	1/2	1.4					42.8	7.7	29.4	5.3	MOTORIZED	CS
8.5–128	15X	2/3	1.7			54.7	4.0	38.3	2.8	26.3	1.9	MOTORIZED	C
10–140	14X	2/3	1.9			47.6	3.6	33.3	2.5			MOTORIZED	C
10-5–105	10X	2/3	1.4			45.9	4.9	33.9	3.5			MOTORIZED	C
11–110	10X	2/3	1.6			45.9	4.9	33.9	3.5			MOTORIZED	C
12.5–75	6X	1	1.2			38.8	6.7	28.7	4.9			MANUAL	C
13.5–600	44X	1	1.8	26.6	0.6	18.6	0.4	13.3	0.3			MOTORIZED	C
15–90	6X	1	1.8	46.1	8.4	32.2	5.5	24.1	4.1			MOTORIZED	C
15–180	12X	1	1.9	46.1	4.1	32.2	2.9	24.1	2.0			MOTORIZED	C
15–300	20X	1	2.8	47.4	2.5	33.2	1.8	23.7	1.3			MOTORIZED	C
16–160	10X	1	1.8	43.6	4.6	30.2	3.2	22.2	2.3			MOTORIZED	C
16–160	10X	1	1.8	43.6	4.6	30.2	3.2	22.2	2.3			MANUAL	C
17.5–105	6X	1	1.8	38.7	7.1	28.2	4.8	20.7	3.5			MANUAL	C

*ANY LENS WITH A C MOUNT CAN BE USED ON A CS MOUNT CAMERA

Table 4-5   Parameters of Commercially Available Zoom Lenses

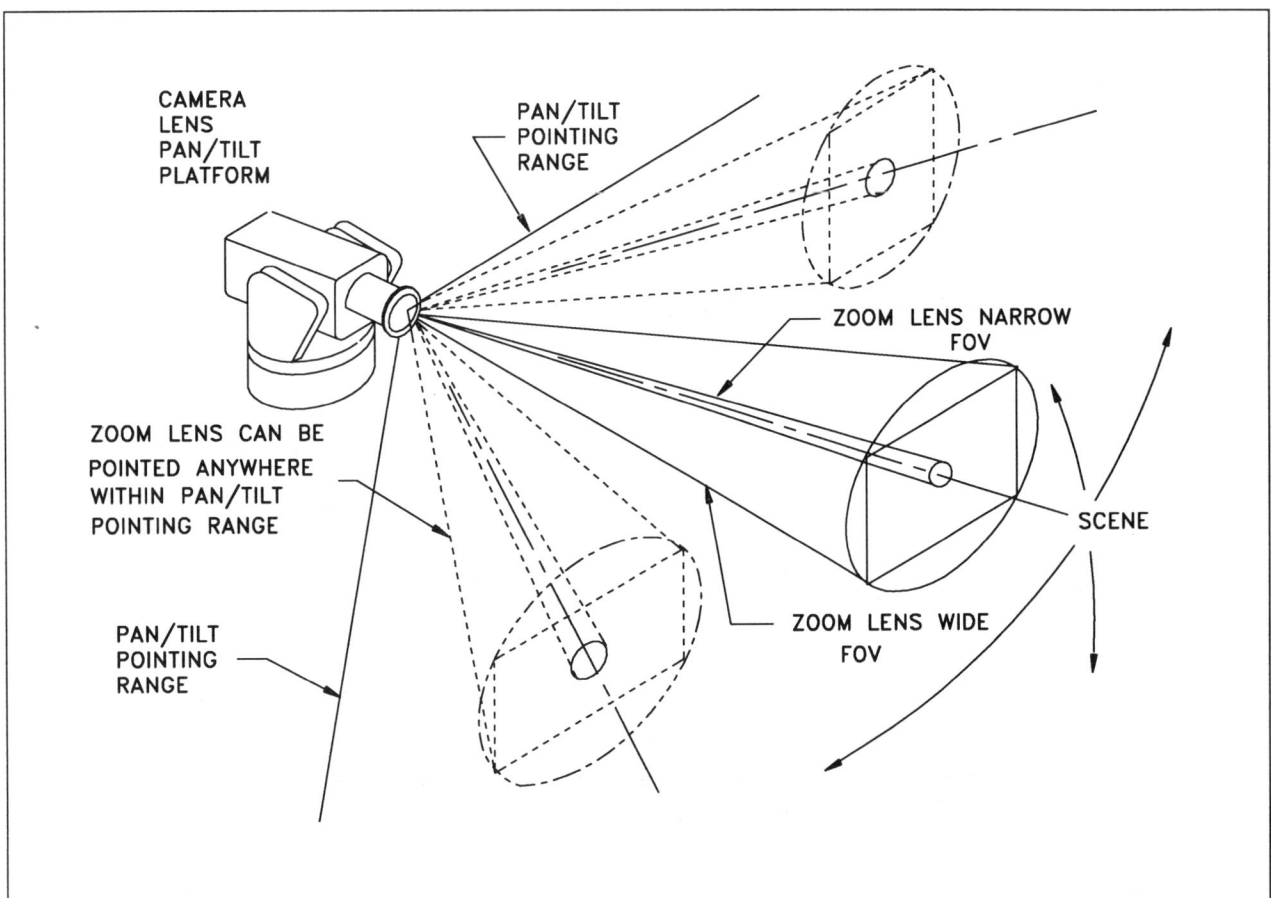

FIGURE 4-21 Dynamic FOV of pan/tilt-mounted zoom lens

elevation angle, (2) focused at a specific distance, and (3) iris set to a specific f-number opening. When a camera needs to turn in another direction in response to an alarm sensor, the preset feature eliminates the need for human response and significantly reduces the time it takes to acquire a new target.

4.3.4.8 Electrical Connections

The motorized zoom lens contains electronics, three motors, clutches, and end-of-travel limit switches to protect the lens and restrict the movement of the zoom, focus, and iris adjustment rings (Figure 4-22).

Since the electrical connections have not been standardized among manufacturers, the manufacturer's lens wiring diagram must be consulted for proper wiring. Figure 4-22 shows a typical wiring schematic for zoom, focus, and iris mechanisms. The zoom, focus, and iris motors are controlled with positive and negative DC voltages from the lens controller—using the polarity specified by the manufacturer. To provide commonality among manufacturers, the Closed Circuit Television Manufacturers Association group within the Electronic Industries Association has prepared guidelines to serve as an industry standard.

4.3.4.9 Initial Lens Focusing

To achieve the performance characteristics designed into a zoom lens, the lens must be properly focused onto the camera sensor during the initial installation. Since the lens operates over a wide range of focal lengths, it must be tested and checked to ensure that it is in focus at the wide-angle and telephoto settings. To perform a critical focusing of the zoom lens, the aperture (iris) must be wide open (set to the lowest f-number) for all back-focus adjustments of the camera sensor. This provides the conditions for a *minimum* depth of field, and the conditions to perform the most critical focusing. Therefore, adjustments must be performed in subdued lighting, or with optical filters in front of the lens, to reduce the light and allow the camera to operate within its optimum range. The following steps should be followed to focus the lens:

1. With the camera operating, view an object at least 75 feet away.
2. Make sure the lens iris is wide open, so that focusing is most critical.
3. Set the lens focus control to the extreme far position.

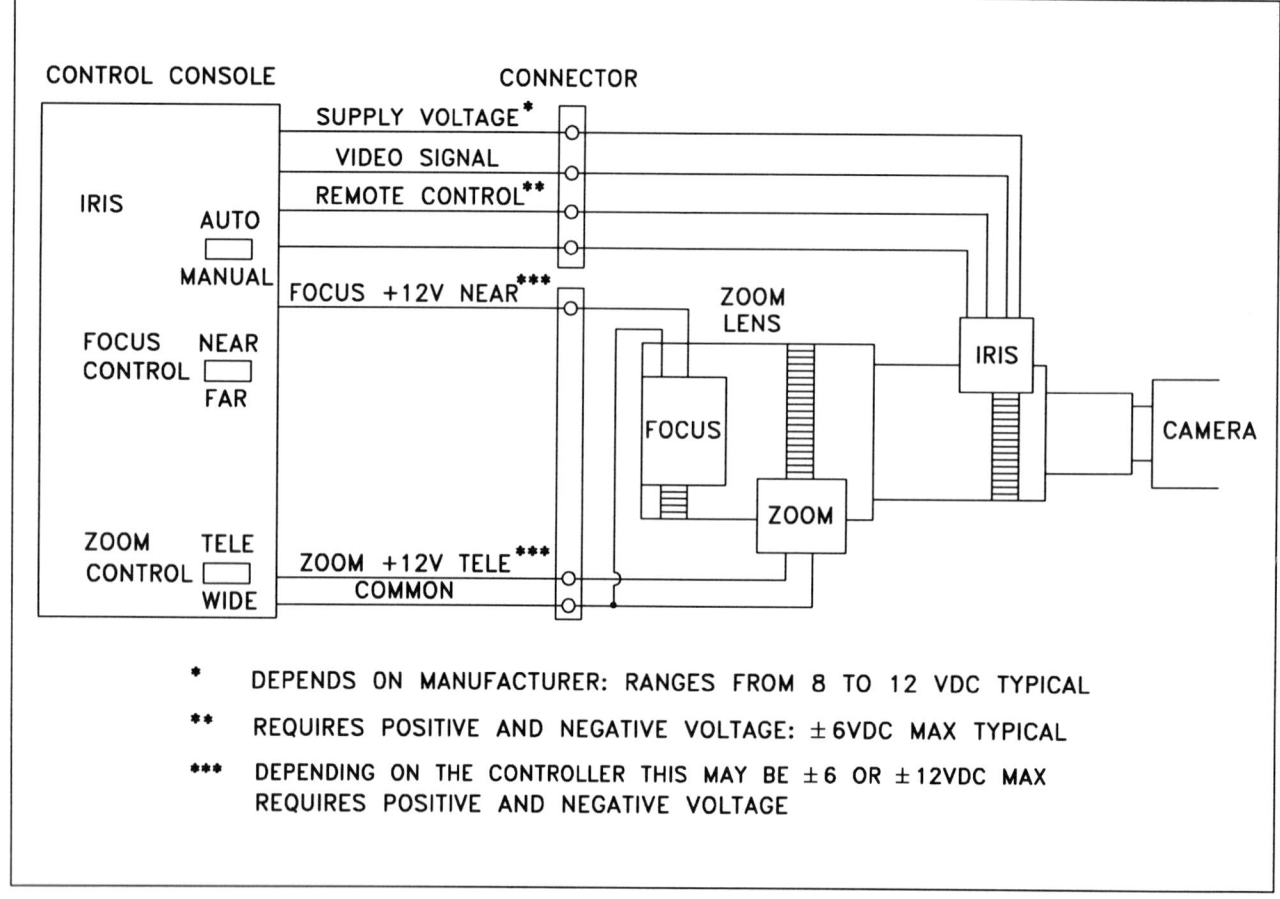

FIGURE 4-22 Motorized zoom lens electrical configuration

4. Adjust the lens zoom control to the extreme wide position (widest FOV).
5. Adjust the camera sensor position adjustment control to obtain the best focus.
6. Move the lens zoom to the extreme telephoto (smallest FOV) setting.
7. Adjust the lens focus control (on the controller) for the best picture.
8. Recheck the focus at the maximum FOV.

4.3.4.10 Auto-Focus Zoom Lens

Automatic-focusing capability is a very useful attribute in a zoom lens (or any lens). An automatic-focusing lens uses a ranging (distance-measuring) means to adjust the focus on the lens so that objects of interest remain in focus on the camera sensor even though they move toward or away from the lens (see Section 4.5.5). These lenses are particularly useful when a person (or vehicle, or other object) enters a camera's FOV and moves toward or away from the camera while the lens is zooming in or out or the camera is panning or tilting. To focus on the new object, the automatic-focus-

ing lens changes focus from the surrounding scene to the moving object, keeping the moving object in focus. Figure 4-23 is an example of an active IR focused, automatic-focus zoom lens having a 6 to 1 FL range.

The lens has an FL range of 11–66 mm (f/1.2 at the 11-mm FL setting) and is designed for ½- or ⅔-inch sensor formats. On a ⅔-inch format camera at the wide-angle setting, it covers a 43.6-degree horizontal by 33.4-degree vertical angular FOV. At the telephoto setting, it covers a 7.6-degree horizontal by 5.7-degree vertical FOV. The lens has automatic or manual iris control.

4.3.4.11 Pinhole Zoom Lens

Pinhole lenses with a small front lens element are commonplace in covert CCTV surveillance applications. Figure 4-24 shows a zoom pinhole lens with a 12.5–75-mm FL and a 6 to 1 zoom range.

Like other pinhole lenses, it is also available in a right-angle configuration. The large number of lens elements and the small entrance aperture (0.38 inches) make this an optically slow lens (high f-number). At the wide-angle setting it has an optical speed of f/8.0.

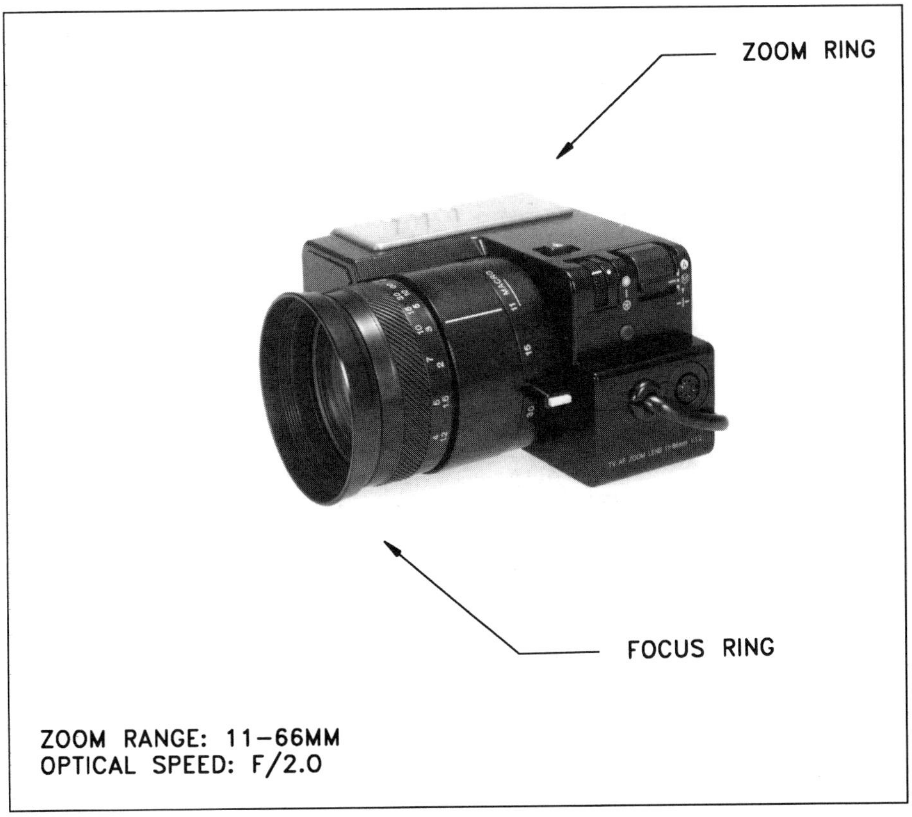

FIGURE 4-23   Auto-focus zoom lens

ZOOM RING

FOCUS RING

ZOOM RANGE: 11—66MM
OPTICAL SPEED: F/2.0

4.3.4.12 Checklist/Summary

The following questions should be asked when considering a zoom lens:

- Does the application require a manual or motorized zoom lens?

- What FOV is required? See Tables 4-1, 4-2, 4-3, 4-4, and the Lens Finder Kit.
- Is the scene lighting constant or widely varying? Is a manual and/or automatic iris required?
- What is the camera format ($\frac{1}{3}$-, $\frac{1}{2}$-, $\frac{2}{3}$-, or 1-inch)?

FIGURE 4-24   Pinhole zoom lens

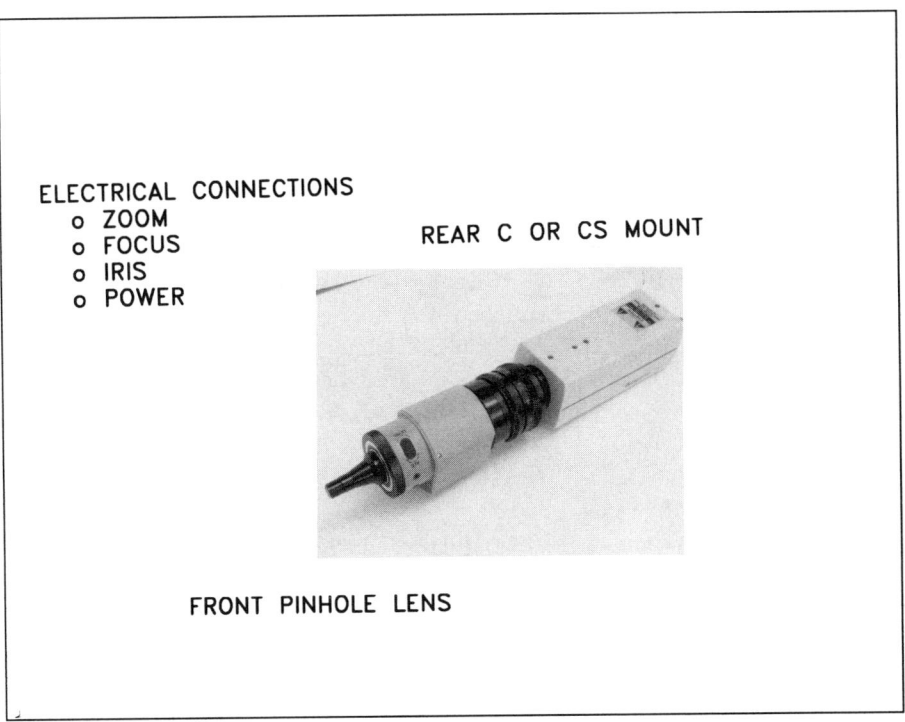

ELECTRICAL CONNECTIONS
   o ZOOM
   o FOCUS
   o IRIS
   o POWER

REAR C OR CS MOUNT

FRONT PINHOLE LENS

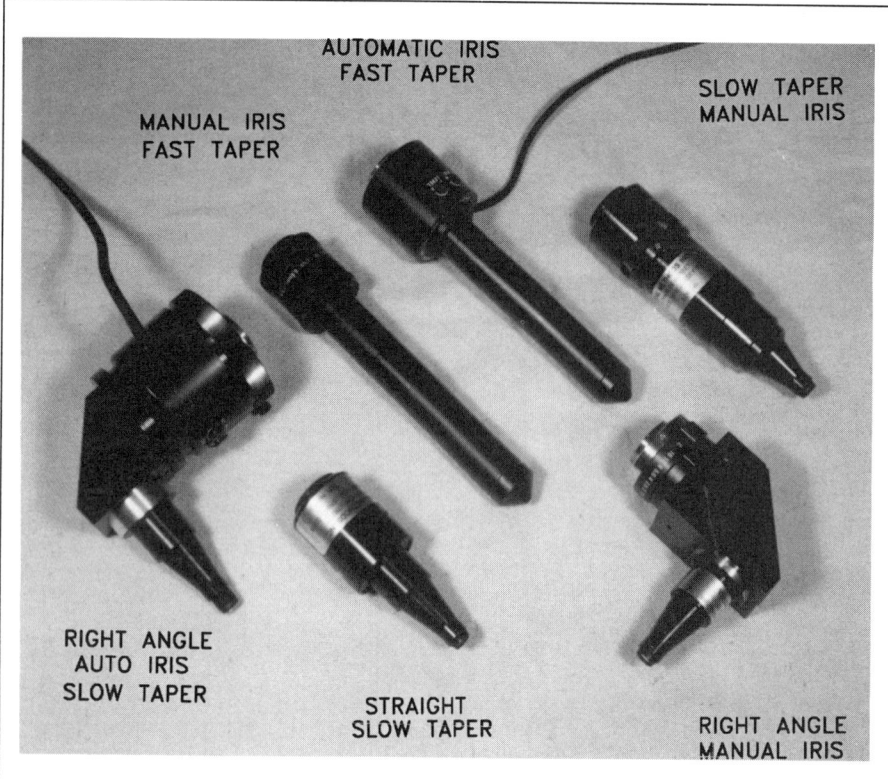

FIGURE 4-25   Straight and right-angle pinhole lenses

- What is the camera lens mount type—C, CS, bayonet, or other?

Zoom lenses play an important role in enhancing the viewing capability of a CCTV system. The advantage of the zoom lens over the FFL lens is its range of FLs, thereby accommodating a large number of FOVs with one lens. The increased complexity and precision required in the manufacture of zoom lenses makes them cost two to ten times as much as an FFL lens, but the added capability is often worth the extra cost and is unmatched by any FFL lens.

4.4 PINHOLE LENSES

A pinhole lens is a special security lens with a relatively small front diameter so that it can be hidden in a wall, ceiling, or some other object. Covert camera/pinhole lenses have been installed in emergency lighting fixtures, exit signs, ceiling-mounted lights, sprinkler heads, table lamps, and mannequins. Any object that can house the camera and pinhole lens, and can disguise or hide the front lens element, is a candidate for a covert installation. In practice the front lens is considerably larger than a pinhole, usually 0.06 to 0.38 inch in diameter, but nevertheless it can be successfully hidden from view. Variations of the pinhole lens include straight or right-angle, manual- or automatic-iris, narrow-taper or stubby-front shape, and mini-pinhole (Figure 4-25).

Whether to use the straight or right-angle pinhole lens depends upon the application. A detailed description and review of covert/pinhole lenses are presented in Chapter 14.

4.4.1 Generic

A feature that distinguishes two generic pinhole lens designs from each other is the shape and size of the front (Figure 4-26). The slow tapering design permits easier installation than the fast taper and also has a faster optical speed, since the larger front lens collects more light.

The optical speed (f-number) of the pinhole lens is important for the successful implementation of a covert camera system. The lower the f-number, the more light reaching the television camera and the better the television picture. An f/2.2 lens transmits 2.5 times more light than an f/3.5. The best theoretical f-number is equal to the FL divided by the entrance lens diameter (d). From Equation 4-9:

$$f/\# = \frac{FL}{d}$$

In practice the f-number obtained is worse than this because of various losses caused by imperfect lens transmission, in turn caused by reflection, absorption, and other lens imaging properties. For a pinhole lens, the light getting through the lens to the camera sensor is limited primarily by the diameter of the front lens or the mechanical

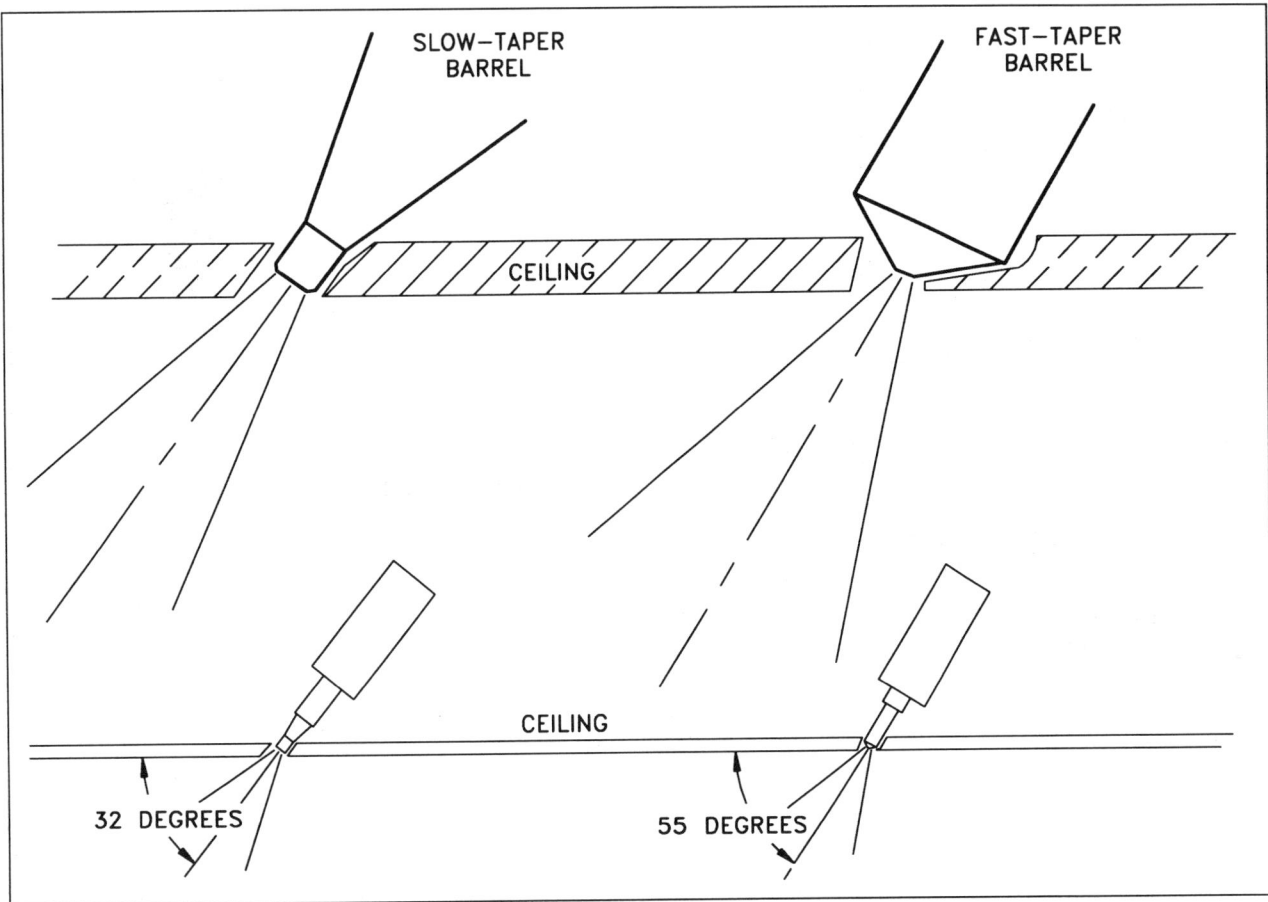

FIGURE 4-26   Generic pinhole lenses—slow and fast taper

opening through which it views. For this reason, the larger the lens entrance diameter, the more light gets through to the image sensor, resulting in a better picture quality, if all other conditions remain the same.

The amount of light collected and transmitted through the lens system (I) varies inversely as the square of the lens f-number:

$$I = \frac{K}{(f/\#)2}$$

For example, an f/2.0 lens transmits four times as much light as an f/4.0 lens. Small f-number differences make large differences in the television picture quality obtained. The f-number relationship is analogous to water flowing through a pipe. If the pipe diameter is doubled, four times as much water flows through it. Likewise, if the f-number is halved (that is, if the lens diameter is doubled), four times as much light will be transmitted through the lens.

4.4.2 Sprinkler Head

Of the large variety of covert lenses available for the security CCTV industry (pinhole, mini, fiber optic, and special covert configurations), one unique lens hides the pinhole lens

in a ceiling sprinkler fixture, making it extremely difficult for an observer standing at floor level to detect or identify the lens/camera. This unique device provides an extremely useful covert surveillance television system. Figure 4-27 shows the right-angle version of the sprinkler head lens.

This pinhole lens and camera combination is concealed in and above a ceiling, with only a modified sprinkler head in view below the ceiling. For investigative purposes, fixed pinhole lenses pointing in one specific direction are usually suitable. For looking in different directions, the sprinkler head and moving mirror are available in scanning (panning) configurations. (See Section 4.3.4.2 and Chapter 14.)

4.4.3 Mini-Pinhole

Another generic family of covert lenses is the mini-lens group (Figure 4-28). Because these lenses are very small, typically $3/8$-inch diameter by $1/2$-inch long, and optically fast, they can be used in places unsuitable for larger pinhole lenses. Mini-pinhole lenses have optical speeds of f/1.4 to f/1.8. An f/1.4 mini-pinhole lens transmits five times more light than an f/3.5 pinhole lens. Mini-pinhole lenses are available in FLs from 3.5 to 11 mm; when combined with a good CCD camera, they result in the fastest covert camera

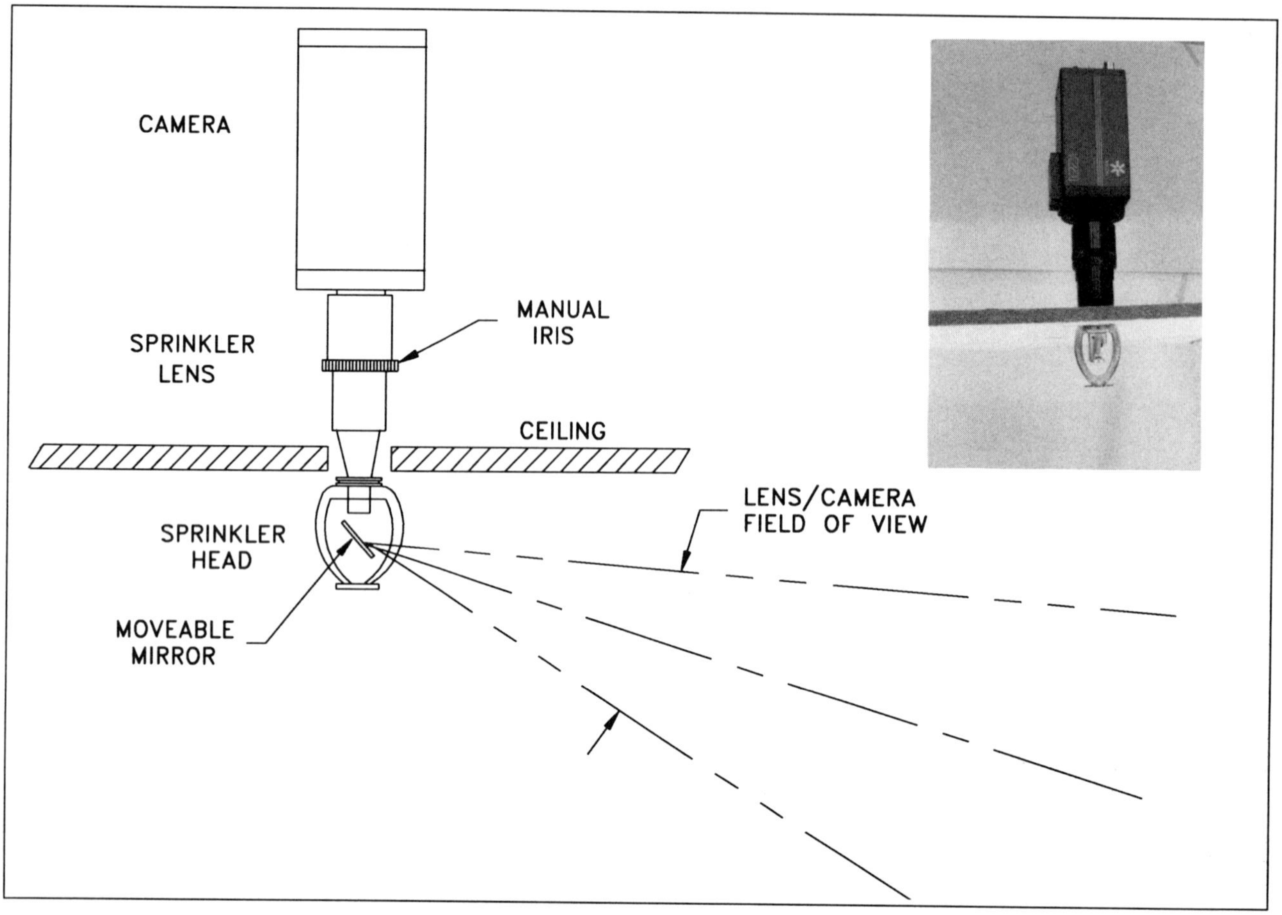

FIGURE 4-27   Sprinkler head pinhole assembly installation

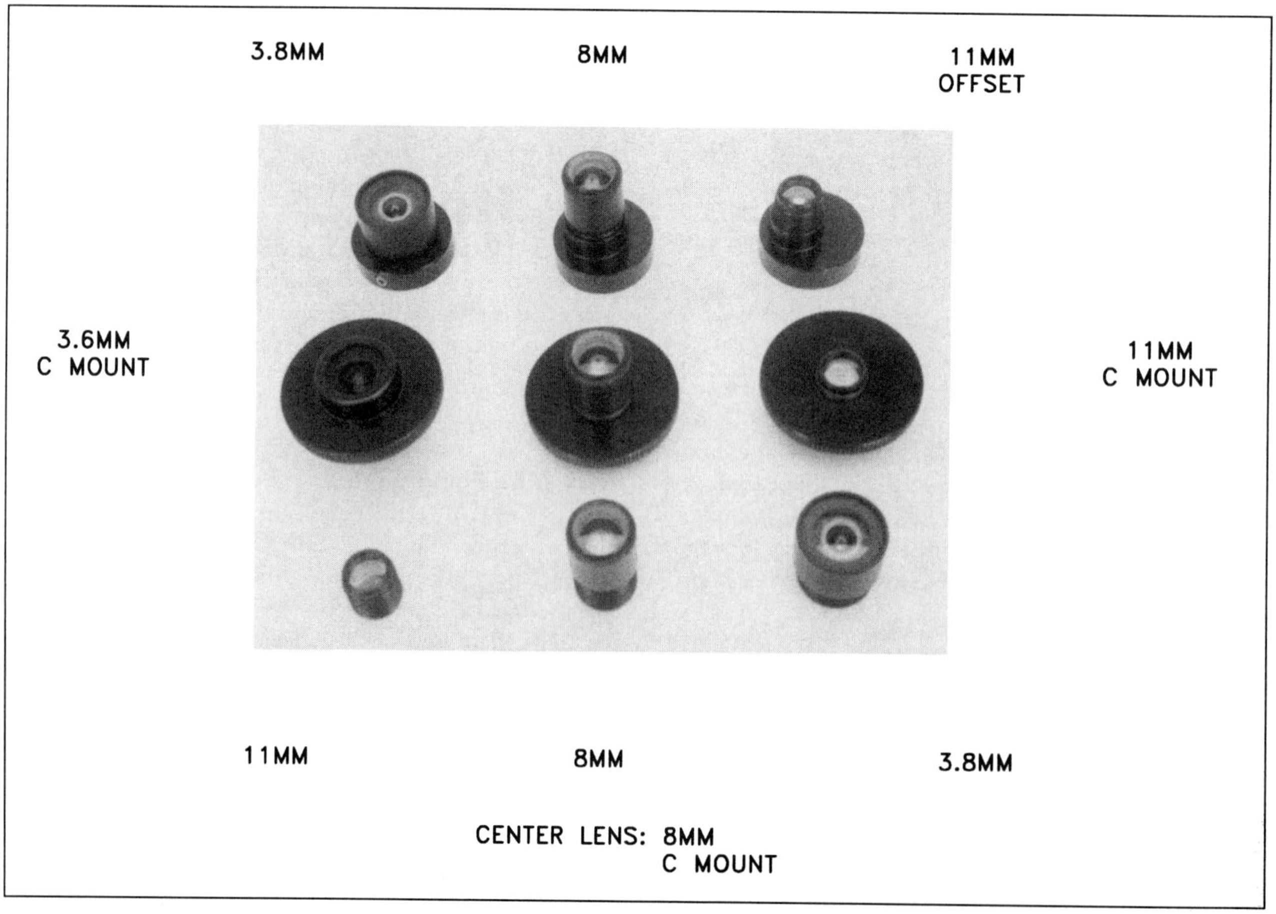

FIGURE 4-28   Mini-pinhole lenses

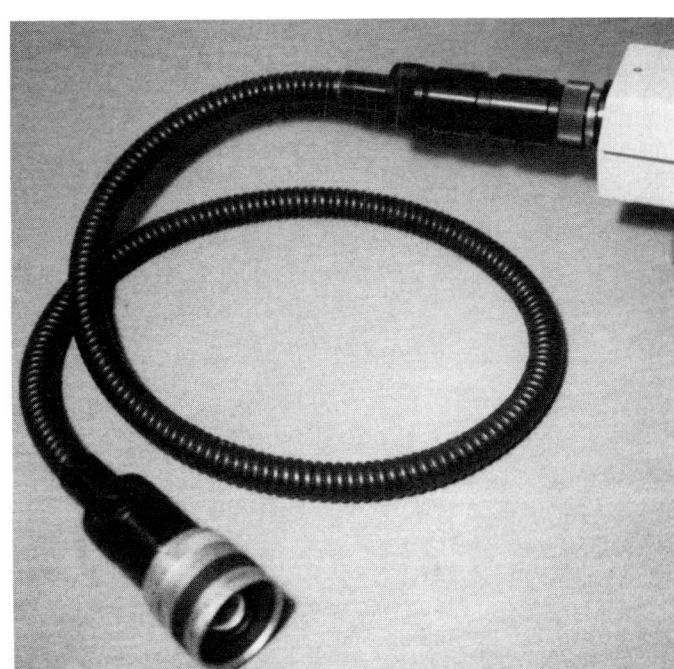

OBJECTIVE LENS: 8MM OR 11MM FL
FIBER TYPE: RIGID CONDUIT
RELAY LENS: M=1:1
MOUNT: C OR CS

OBJECTIVE LENS: ANY C OR CS MOUNT
FIBER TYPE: FLEXIBLE BUNDLE
RELAY LENS: M=1;1
MOUNT: C OR CS

FIGURE 4-29   Rigid and flexible fiber-optic lenses

available. A useful variation of the standard mini-lens is the off-axis mini-lens. This lens is mounted offset from the camera axis, which causes the camera to look off to one side, up, or down, depending on the direction chosen.

Chapter 13 describes pinhole and mini-lenses in detail.

4.5 SPECIAL LENSES

Special CCTV security lenses that deserve mention include fiber-optic and borescope, split-image, right-angle, relay, automatic-focus, stabilized, and long-range. These lenses are used when standard lenses are not suitable.

4.5.1 Fiber-Optic and Borescope

Difficult television security applications are sometimes solved through the use of coherent fiber-optic bundle lenses. Not to be confused with the single or multiple strands of fiber commonly used to transmit the television signal, the coherent fiber-optic lens has many thousands of individual glass fibers positioned adjacent to each other. It transmits a coherent image from an objective lens, over a

distance of several inches to several feet, where the image is then transferred again by means of a relay lens to the camera sensor. Since the picture quality obtained with fiber-optic lenses is not as good as that obtained with all glass lenses, such lenses should be used only when no other lens/camera system will solve the problem. Fiber-optic lenses are available in rigid or flexible configurations (Figure 4-29).

A high-resolution (450 TV lines) coherent fiber bundle consists of several hundred thousand glass fibers, which transfer a focused image from one end of the fiber to the other. *Coherent* means that each point in the image on the front end of the fiber bundle corresponds to a point at the rear end. To be properly called a fiber-optic lens, the fiber bundle must be preceded by an objective lens that focuses the scene onto the front end of the bundle and followed by a relay lens that focuses the image onto the sensor (Figure 4-30).

Fiber-optic lenses are typically used for viewing through thick walls or for installations where the camera must be several feet away from the front lens. For example, the camera may be on an accessible side of a wall and the lens on the inaccessible side. Thus the front lens can be moved substantially away from the camera sensor. Chapter 14

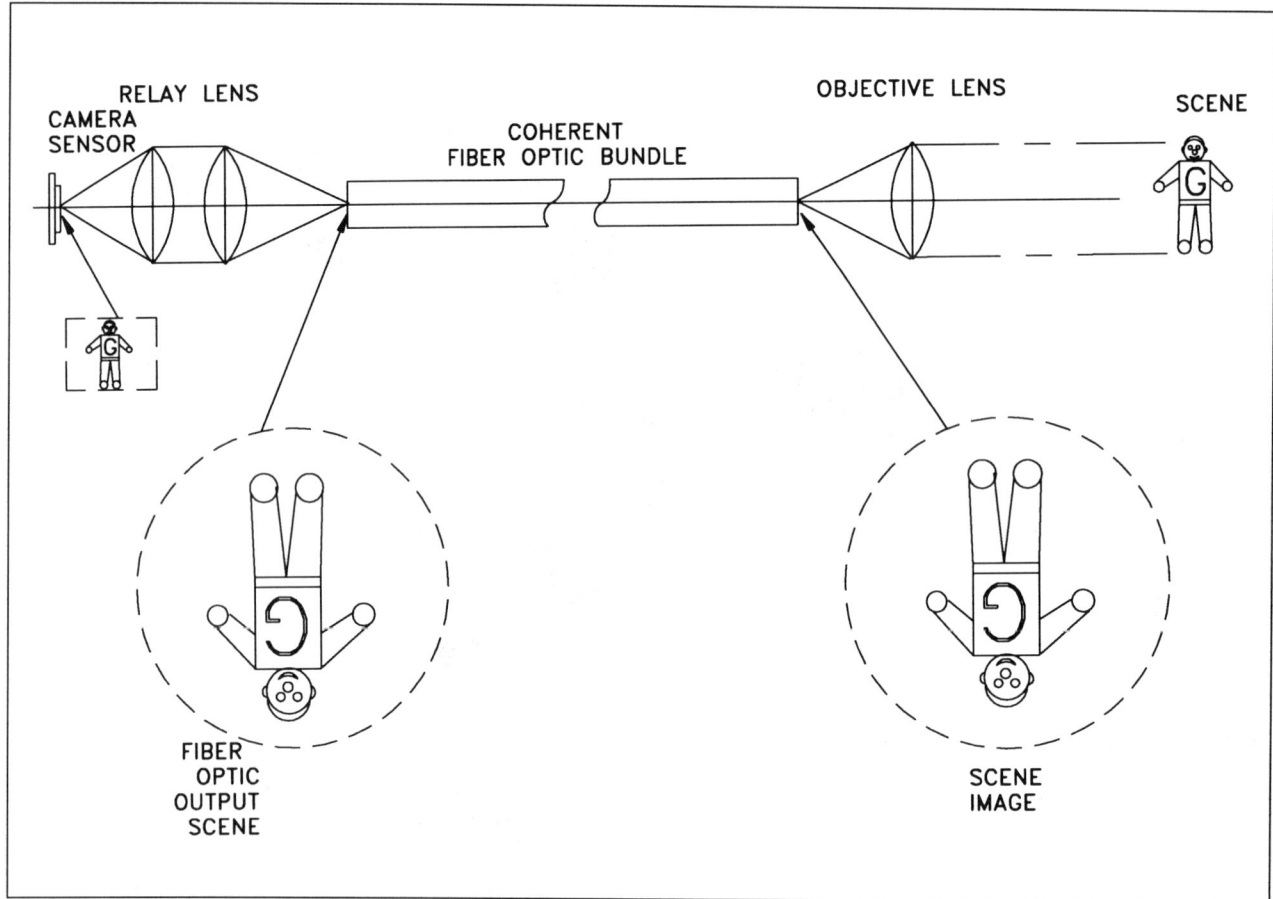

FIGURE 4-30 Fiber-optic lens configuration

shows how coherent fiber-optic lenses are used in covert security applications.

Another class of viewing optics for CCTV cameras is the borescope lens (Figure 4-31). The borescope lens consists of long rigid tubes (6 to 30 inches) from 0.04 to 0.5 inches in diameter with a single rod lens or multiple small lenses to transmit the image from a front lens to the rear lens and onto a camera sensor. The single rod lens uses a unique graded index (GRIN) glass rod, which refocuses the image along its length. Because of the small diameter of the rod and lenses, only a small amount of light is transmitted, resulting in slow f-numbers, typically f/11 to f/30. This slow speed limits borescope applications to well-illuminated environments and sensitive cameras. The image quality of the borescope lens is better than that of the fiber-optic lens, since it uses all glass lenses. Figure 4-31 shows two typical borescope lenses with tip diameters of 0.04 and 0.25 inches.

4.5.2 Split-Image Optics

A lens for imaging two independent scenes onto a single television camera is called an image-splitting or bifocal lens. This lens views two scenes with two separate lenses, with the same or different FLs (with the same or different magnifications), and combines them onto one camera sensor (Figure 4-32).

The split-image lens adapter accomplishes this with only one CCTV camera. The split-image lens has two C or CS lens ports (mounts), so that the two objective lenses view two scenes with the same camera. Depending on the orientation of the bifocal lens system on the camera, the image is split either vertically or horizontally. Any fixed-focus, pinhole, zoom, or other lens that mechanically fits onto the C- or CS-mount image splitter can be used. The adjustable mirror on the side lens permits looking in many directions. With the use of an additional external mirror, it can point at the same scene as the front lens. In this case, if the front lens is a wide-angle lens (6.5 mm FL) and the side lens is a narrow-angle lens (75 mm FL), a bifocal lens system results: one camera obtains simultaneously a wide-field and narrow-field coverage (Figure 4-32). The horizontal scene FOV covered by each lens is one-half of the total lens FOV. For example, with the 6.5-mm and 75-mm FL lenses, on a ⅔-inch camera and a vertical split (as shown), the 6.5-mm lens displays a 49.3 × 74-ft scene, and the 75-mm lens displays a 4.1 × 6.3-ft scene at a distance of 50 ft. The horizontal FOV of each lens has been reduced by one-half

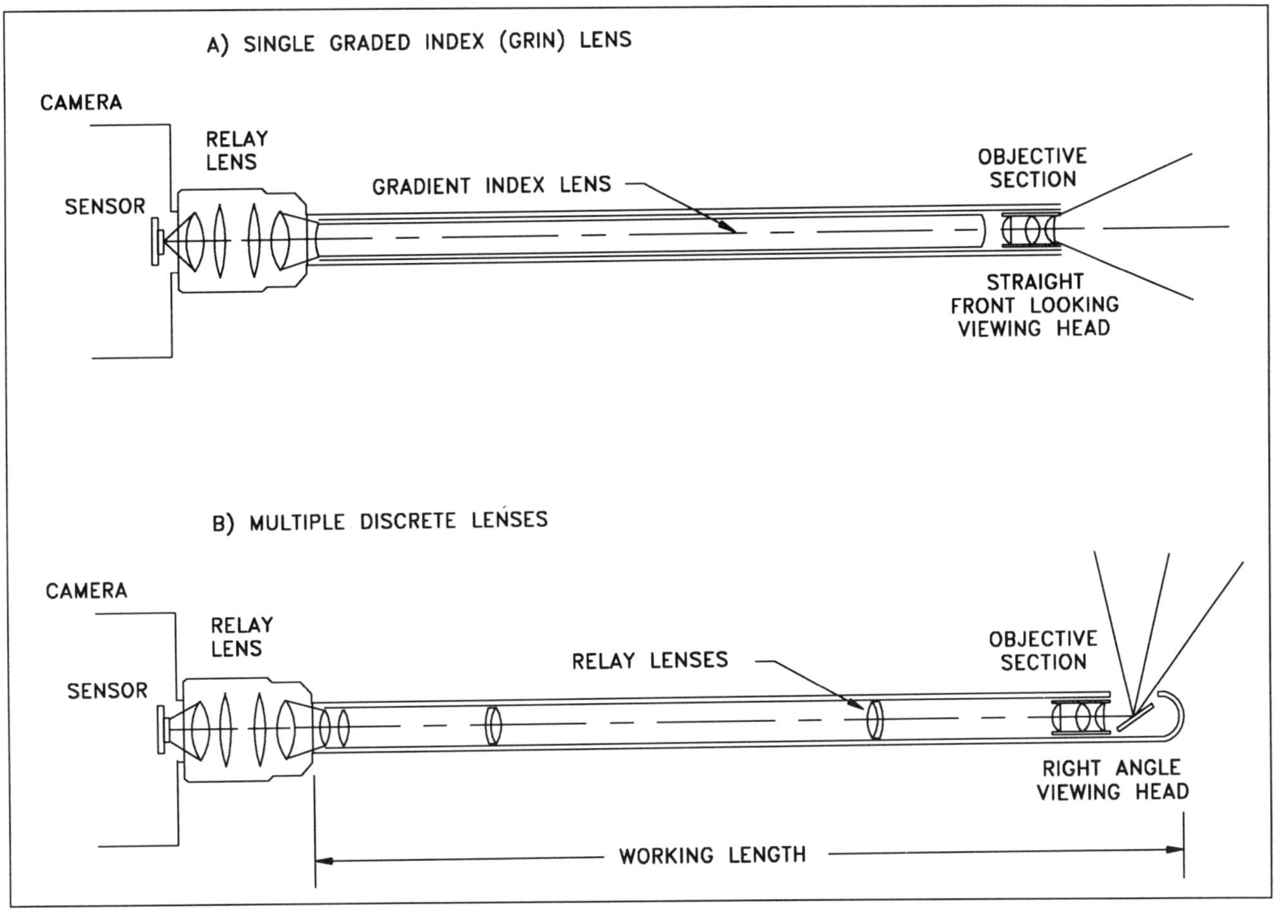

FIGURE 4-31   Borescope lens system

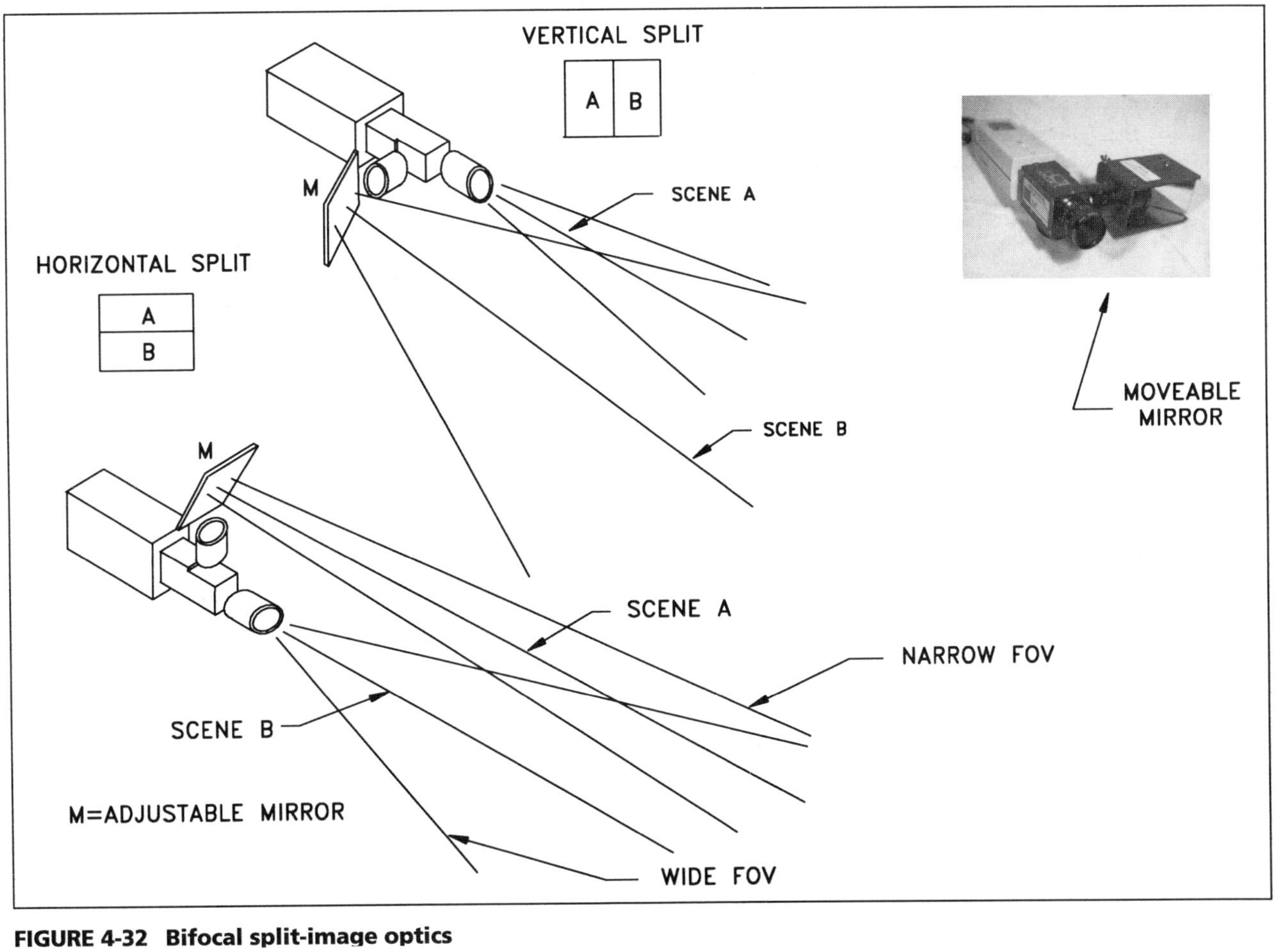

FIGURE 4-32   Bifocal split-image optics

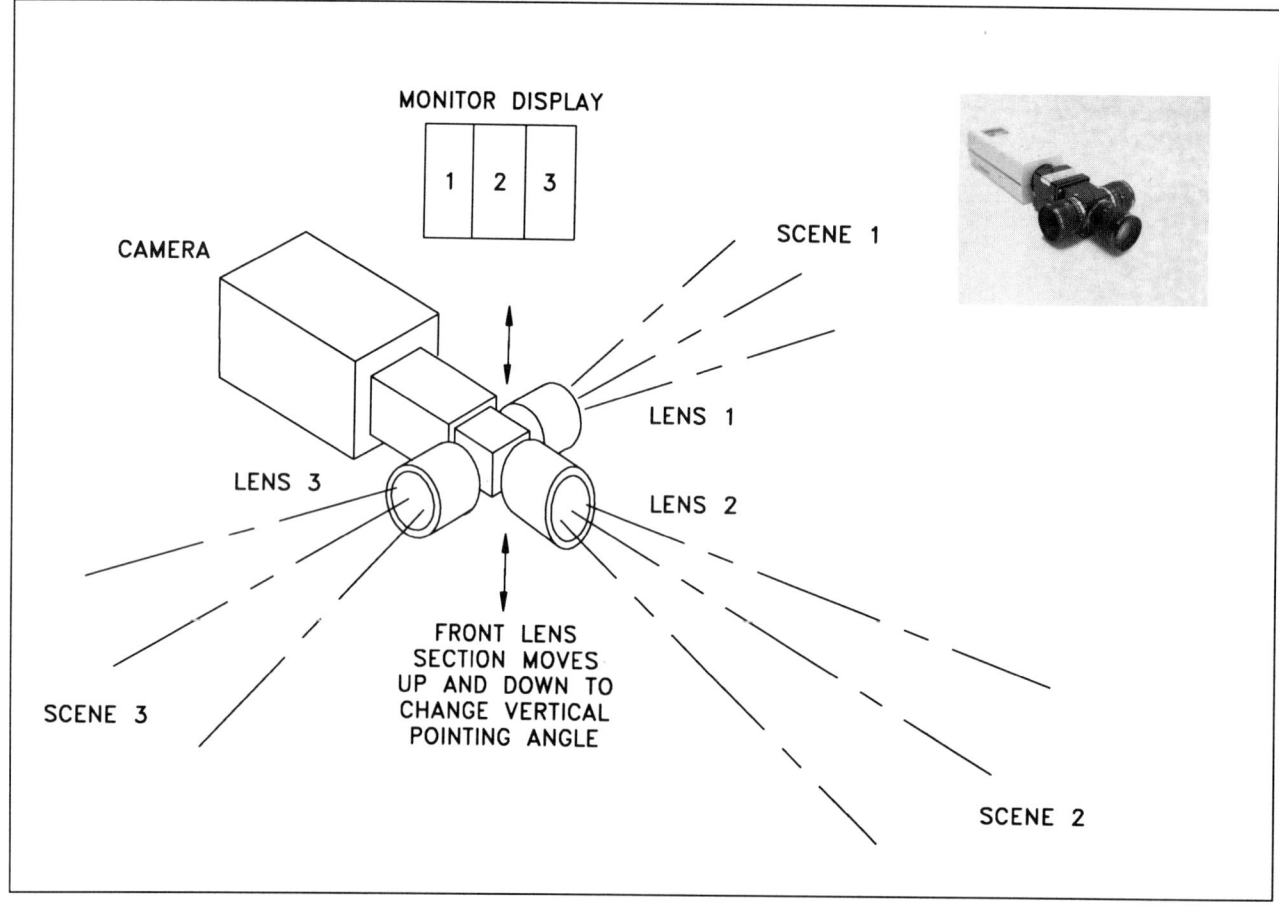

FIGURE 4-33 Tri-split lens views three scenes

of what each would see if the lens were mounted directly onto the camera (98.6 × 148 ft for the 6.5-mm lens and 8.2 × 12.6 ft for the 75-mm lens). By rotating the split-image lens 90 degrees about the camera optical axis, a horizontal split is obtained. The bifocal lens inverts the picture on the monitor, a condition that is corrected by inverting the camera.

A three-way optical image-splitting lens can view three scenes (Figure 4-33). The tri-split lens is useful primarily for viewing three hallways at one time, but it has found many other applications. The tri-split lens provides the ability to view three different scenes with the same or different magnifications with one camera on one monitor, without electronic splitters, thereby replacing two cameras, two electronic splitters, and two monitors. Each scene occupies one-third of the monitor screen. Adjustable optics on the lens permit changing the pointing elevation angle of the three front lenses so that they can look close-in for short hallway applications and all the way out (nearly horizontal) for long hallways. Like the bi-split lens, this lens inverts the monitor image, which is corrected by inverting the camera. Both the bi-split and tri-split lenses work on ⅓-, ½-, ⅔-, and 1-inch camera formats.

4.5.3 Right-Angle

The right-angle lens permits mounting a camera parallel to a wall or ceiling while the lens views the scene (Figure 4-34).

When space is limited behind a wall, as in an automatic teller machine, a ceiling, or an elevator cab, the right-angle lens is a solution. The right-angle optical system permits use of wide-angle lenses (2.6-mm, 110-degree FOV) looking at right angles to the camera axis. This cannot be accomplished by using a mirror and a wide-angle lens directly on the camera since the entire scene will not be reflected by the mirror to the lens on the camera. The edges of the scene will not appear on the monitor because of picture vignetting (Figure 4-35).

The right-angle adapter permits the use of any FL lens that will mechanically fit into its C or CS mount and works with ⅓-, ½-, ⅔-, or 1-inch camera formats.

4.5.4 Relay

The relay lens is used to transfer a scene image focused by any CCTV lens or coherent fiber-optic bundle onto the camera sensor (Figure 4-36).

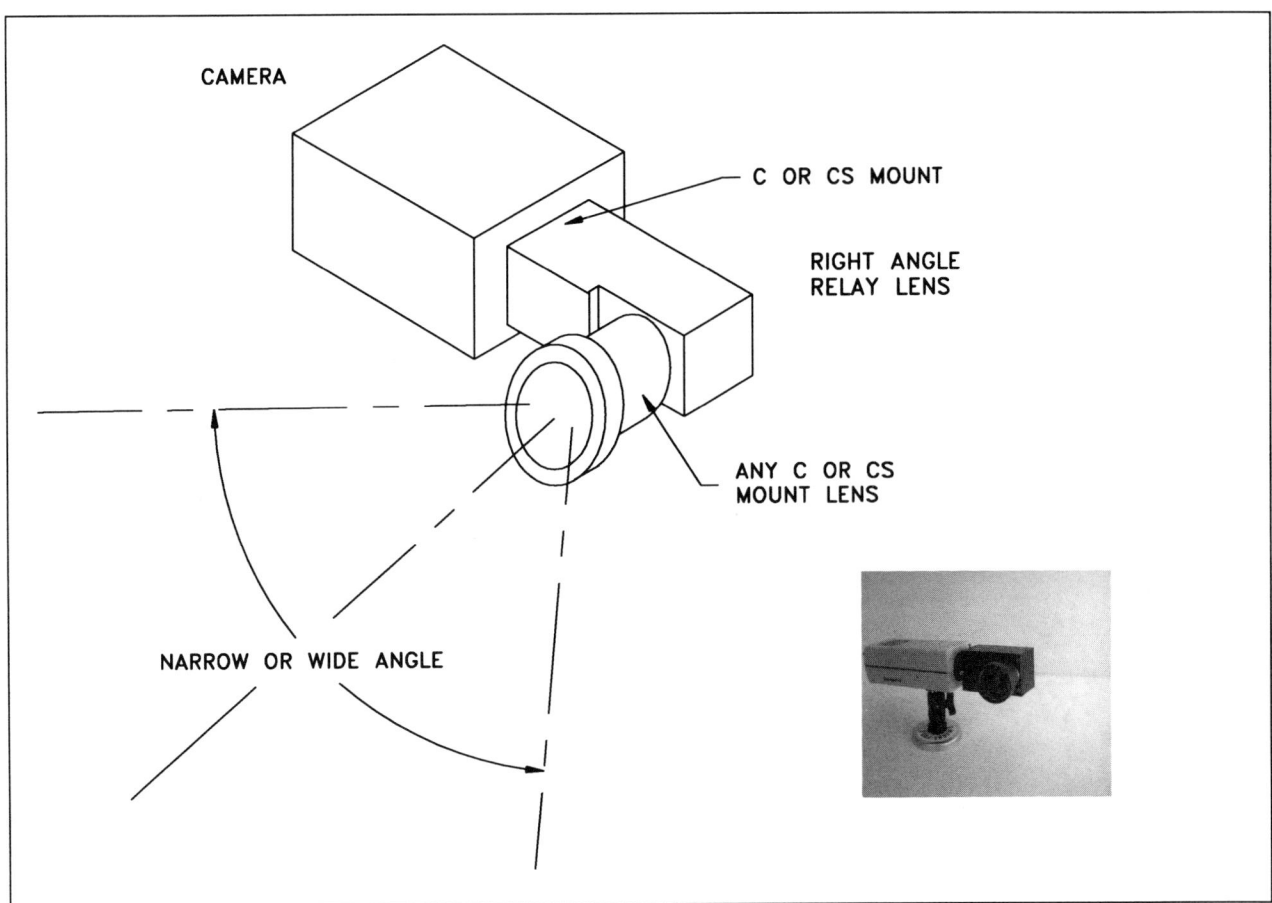

FIGURE 4-34 Right-angle lens

The lens must always be used with some other objective lens and does not produce an image in and of itself. When used with a fiber-optic lens, it re-images the scene located at the fiber bundle output end onto the sensor. When incorporated into split-image or right-angle optics, it re-images the "split" scene or right-angle scene onto the sensor. The relay lens can be used with a standard FFL, pinhole, zoom, or other lens as a lens extender with unity magnification, for the purpose of "moving" the sensor out in front of the camera.

4.5.5 Auto-Focus

Auto-focus lenses are not in common use in the security industry primarily because of the added cost over a standard manually focused lens. They were developed for the consumer VCR camcorder market, where zoom lenses are used almost exclusively. Section 4.3.4.4 describes an auto-focus zoom lens.

An automatic focusing lens uses a ranging (distance measuring) means to automatically adjust the focus on the lens so that objects of interest remain in focus on the camera sensor even though they move toward or away from the lens. These lenses are particularly useful when a person (or vehicle) enters a camera FOV and moves toward or away from the camera. The automatic focusing lens changes focus from the surrounding scene and focuses on the moving object (automatically), and keeps the moving object in focus.

Various types of automatic-focusing techniques are used, including (1) active IR ranging, (2) ultrasonic wave, and (3) solid-state triangulation. The active IR auto-focusing system uses the principle of triangulation. The lens has a light-emitting diode that emits a narrow beam of IR energy toward any subject in the central area of the zoom lens scene (Figure 4-37).

A receiver lens and dual (split) silicon cell sensor located on the opposite side of the lens optical axis and microprocessor electronics computes the subject-to-camera distance based on the CCD image sensor data obtained from the physical lens focusing ring position. The output from the microprocessor electronics adjusts a motor-driven focusing ring on the zoom lens to bring the centered target into sharp focus.

Auto-focus lenses do not work in all situations. If the subject is not reflective in the IR, if the subject is very "angular," so that very little of the IR beam is reflected back to the camera, or if the subject is beyond the operating range of the system, the system will not focus on the subject and remain focused on the general central scene area.

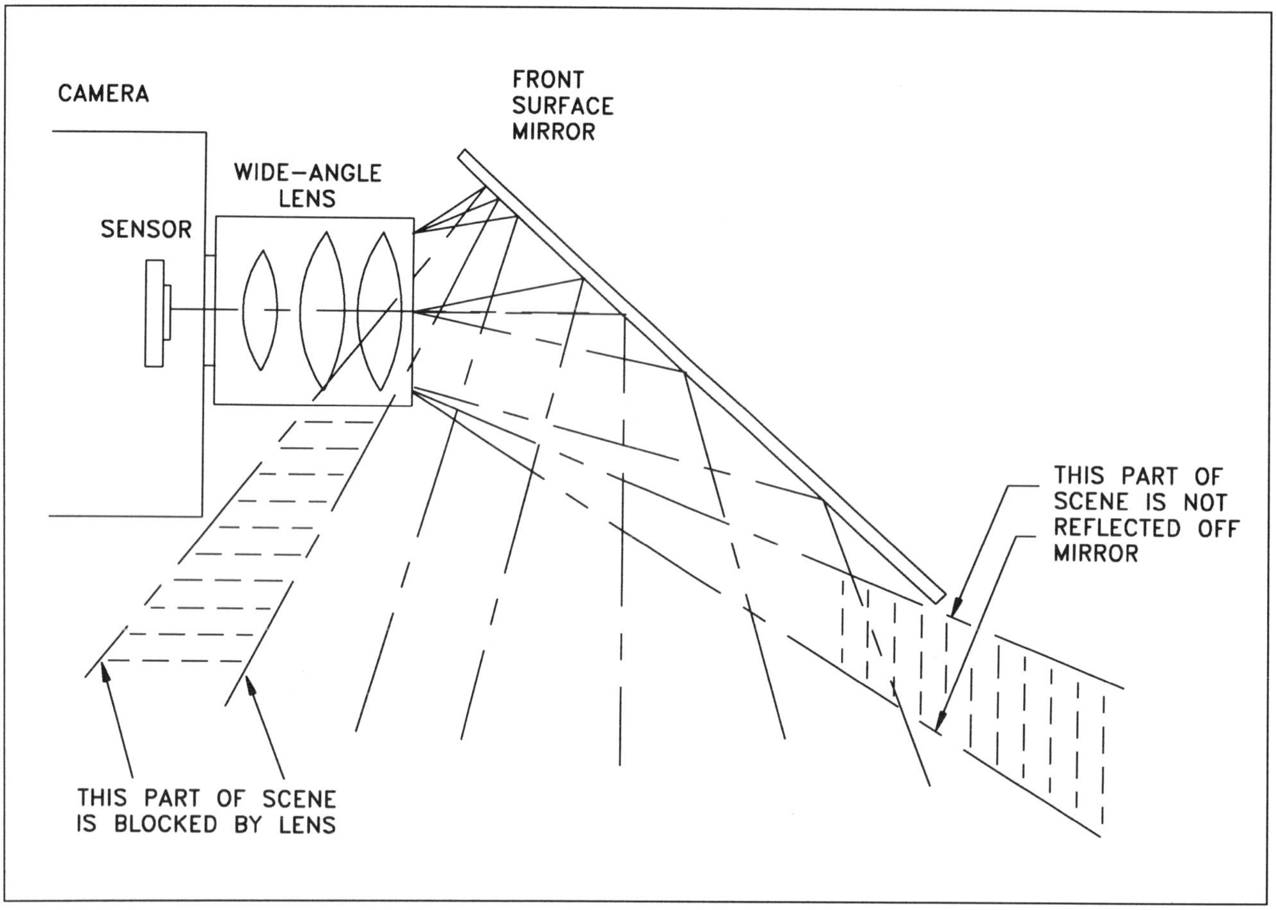

FIGURE 4-35 Picture vignetting from wide-angle lens and mirror

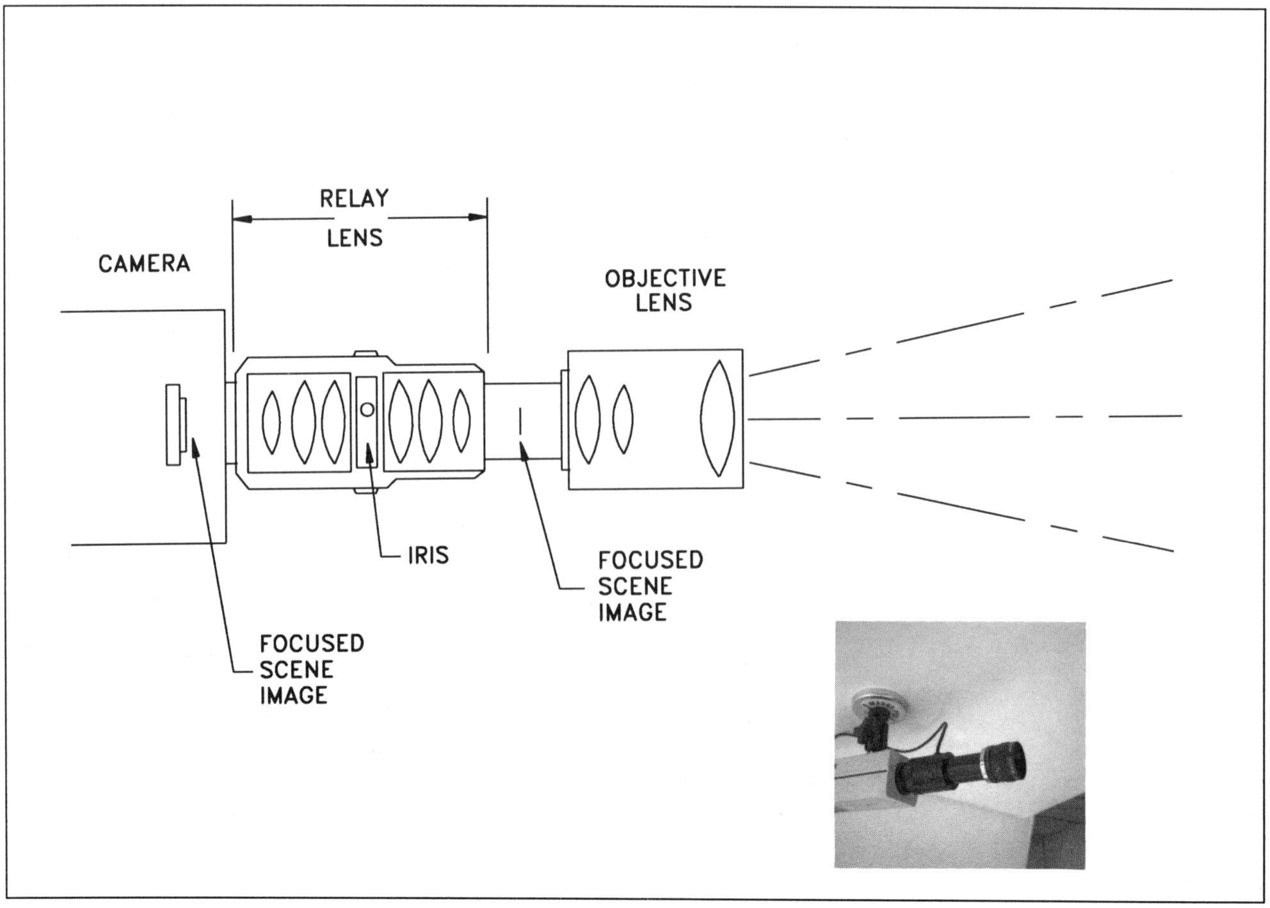

FIGURE 4-36 Relay lens adapter

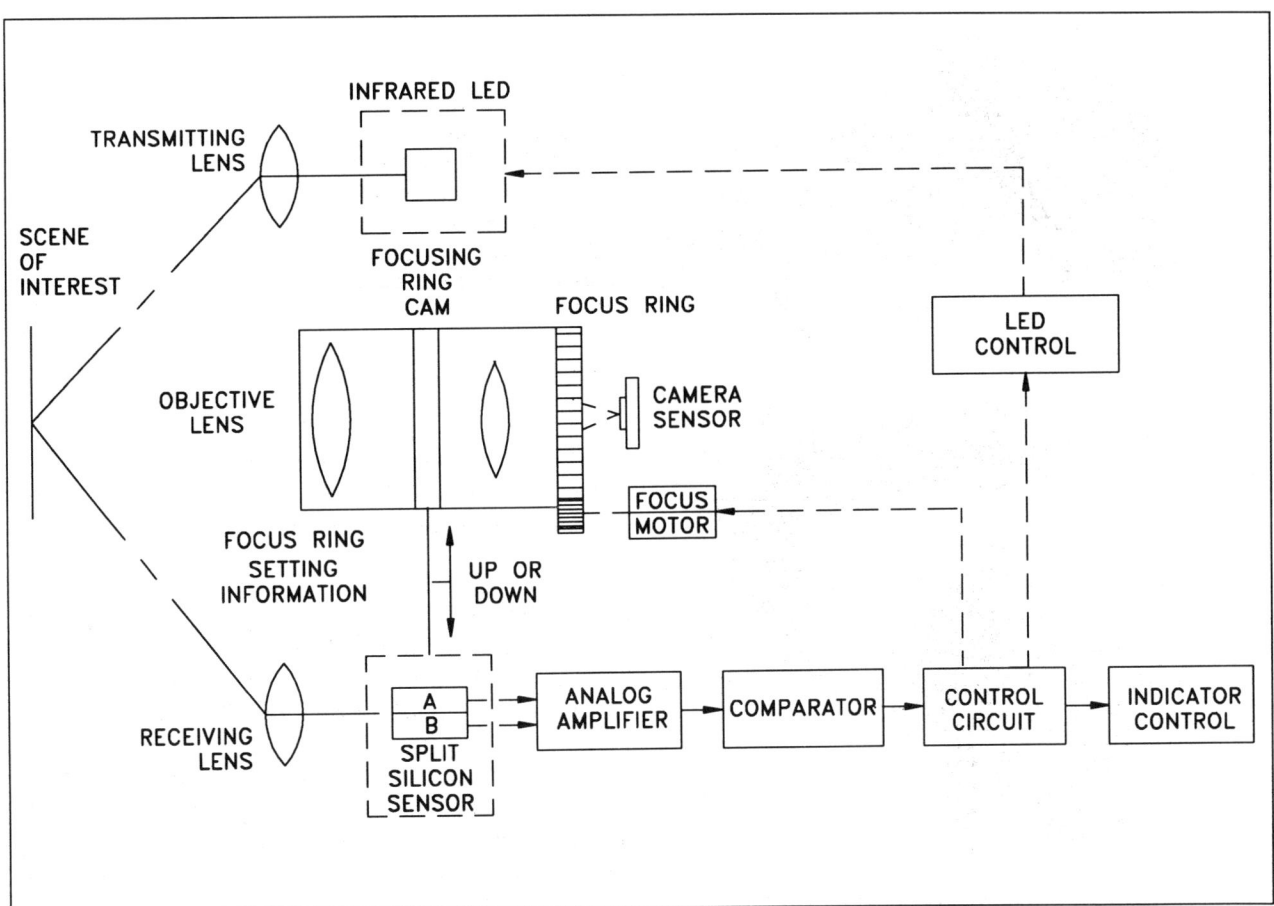

FIGURE 4-37   Block diagram of active infrared auto-focus zoom lens

4.5.6 Stabilized

In security applications where there is unwanted motion of the lens and camera with respect to the scene being viewed, a stabilized lens is used. Applications for these stabilized lenses include handheld cameras, ground vehicles, platforms, towers, airborne platforms, and shipboard use. Stabilized lenses can remove significant vibration in pan/tilt-mounted cameras that are buffeted by wind. This lens system has movable optical components that compensate for (move in the opposite direction to) the relative motion between the camera and scene.

Figure 4-38 shows two stabilized lenses and samples of pictures taken with and without the stabilization on.

4.6 COMMENTS—CHECKLIST AND QUESTIONS

- A standard objective lens inverts the picture image and the CCTV camera electronics re-invert the picture so that it is displayed right-side-up on the monitor.
- The 25-mm FL lens is considered the standard or reference lens for the 1-inch format sensor. This lens/camera combination is defined to have a magnification of M = 1

and is similar to the normal FOV of the human eye. The standard lens for a $^2/_3$-inch format sensor is 16 mm; for a $^1/_2$-inch sensor, 12.5 mm; and for a $^1/_3$-inch sensor, 8 mm. These configurations likewise produce a magnification of 1. They all have the same FOV and therefore see the same size angular FOV.

- A short-FL lens has a wide FOV (Table 4-1; 4.8 mm, $^1/_2$-inch sensor sees a 12.9 ft wide × 9.7 ft high scene at 10 ft).
- A long-FL lens has a narrow FOV (Table 4-1; 75 mm, $^1/_2$-inch sensor sees a 4.1 ft wide × 3.1 ft high scene at 50 ft).
- Using the 12.5-mm lens as a reference, with M = 1 on a $^1/_2$-inch format sensor, a 50-mm lens has a magnification of 4, and a 3.1-mm lens a magnification of $^1/_4$. Similar extensions can be made for the $^1/_3$-, $^2/_3$-, and 1-inch formats.
- What FOV is required? Consult Tables 4-1, 4-2, 4-3 and the Lens Finder Kit.
- Does the application require a manual or motorized zoom lens?
- Is the scene lighting constant or widely varying? Is a manual or automatic iris required?
- What is the camera format ($^1/_3$-, $^1/_2$-, $^2/_3$-, or 1-inch)?

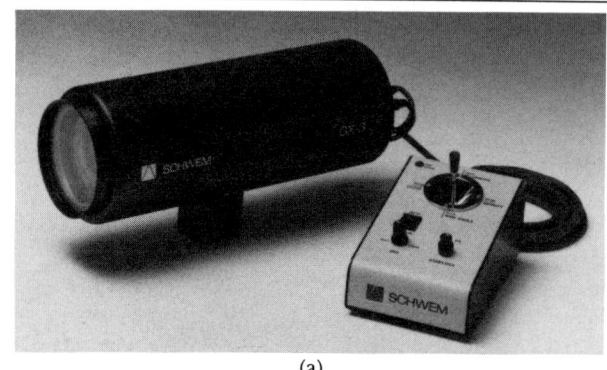

(a)

(b)

(c)

(d)

FIGURE 4-38 Stabilized lenses and results: (a) 6:1 zoom ratio, 12.5–75-mm FL, f/1.2, 4.5-inch diameter, 11 inches long, 6 pounds; (b) 5:1 zoom ratio, 60–300-mm FL, f/1.6, 4½ inches high by 13 inches long by 7¾ inches wide, 7.9 pounds, (c) unstabilized image, (d) stabilized image

• What type of camera lens mount—C, CS, bayonet, or other? (See Chapter 5, Section 5.7.)

4.7 SUMMARY

The task of choosing the right lens for a security application is an important part of designing a CCTV system. The large variety of types and variations in focal length make the proper choice a challenging one. The lens tables and Lens Finder Kit provide the tools needed to determine the FOV obtained with a specific FL lens and sensor size. To obtain good familiarity with the lens capabilities and trade-offs, however, several combinations should be tried and the results observed. Chapter 14, on covert CCTV systems, describes additional covert lenses.

Chapter 5
Cameras

CONTENTS

5.1 OVERVIEW

The CCTV camera's function is to convert the focused visual (or infrared) light image from the camera lens into a time-varying electrical video signal. The lens collects the reflected light from the scene and focuses it onto the camera image sensor. The camera processes the information from the sensor and sends it to a viewing monitor via coaxial cable or other transmission means. Figure 5-1 shows a simple CCTV camera/lens and monitor system.

The monochrome or color television camera in the security CCTV system analyzes the scene by scanning it in a series of either closely spaced lines (in a tube camera) or a horizontal and vertical array of pixels (in a solid-state camera). This technique generates and codes an electrical signal as a function of time, so that the scene can later be reconstructed on the monitor.

Unlike film cameras, human eyes, and LLL television image intensifiers, which see a *complete* picture one frame at a time, a television camera sees an image—point by point—until it has scanned the entire scene. In this respect the television scan is similar to the action of a typewriter: the type element starts at the left corner of the page and moves across to the right corner, completing a single line of type. The typewriter carriage then returns to the left side of the paper, moves down to the next line, and starts again.

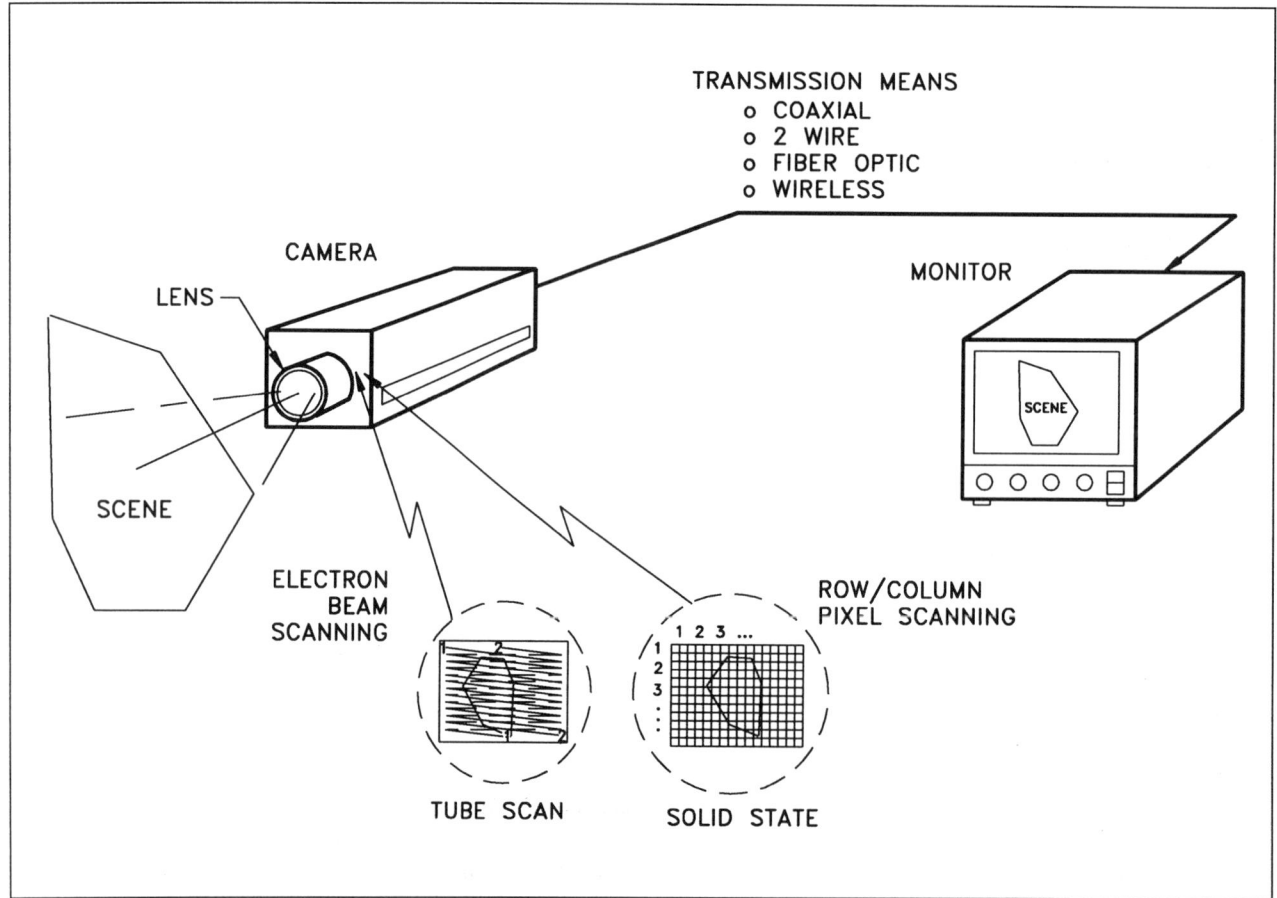

FIGURE 5-1   CCTV system with camera, lens, transmission means, and monitor

A CCTV scanner is like a typewriter that adds a second carriage return after each line, repeating the lines until it reaches the bottom of the page. This is how the scanner completes one field, or half the television picture. The scanner/typewriter then moves back up the page and begins typing on the second line at the left just below the first line. It continues this way, moving down two lines at a time, filling in the lines between the original lines, until the entire page is complete. In this way, the scanner completes the second field and produces one full video frame. This electronic process repeats (like putting in a new sheet of paper) for each frame.

Several different sensor types are available for CCTV security applications, including (1) vidicon, silicon diode, Newvicon, and Saticon tube types and (2) silicon CCD, MOS, and CID solid-state types. Although tube cameras have served well over the years, deficiencies in the standard vidicon, such as frequent tube replacement, image burn-in, and image lag, have been disadvantages. Now most tube cameras are being replaced by solid-state types. In LLL applications, tube and solid-state sensors are combined with SIT, ISIT, and ICCD intensifiers.

Color CCTV cameras have been available for security applications for many years, but their use has been severely restricted because of problems in manufacturing a reasonably priced color camera with stable color balance over its lifetime, or even for short periods of time. Once the technology for solid-state color cameras was developed for the consumer VCR and camcorder, color cameras became suitable for the security field. These cameras, consisting of a single solid-state sensor with an integral three-color filter and automatic white-balancing circuits, provide stable, long-lasting, and sensitive color cameras.

About ten years ago a new solid-state television sensor, the CCD, became a commercial reality. This new device is for all practical purposes replacing the television tube sensor and represents a significant advance in camera technology. The use of the solid-state sensor makes the camera 100% solid-state. (In the camera industry it is called a "chip" camera.) These solid-state cameras have several salient advantages over any and all tube cameras: long life, no aging, no image burn-in, low maintenance, good sensitivity, low power consumption, and small size.

Solid-state cameras use a silicon array of photosensor sites (pixels) to convert the input light image into an electronic video signal, which is then amplified and passed on to a monitor for display. Most solid-state sensors are

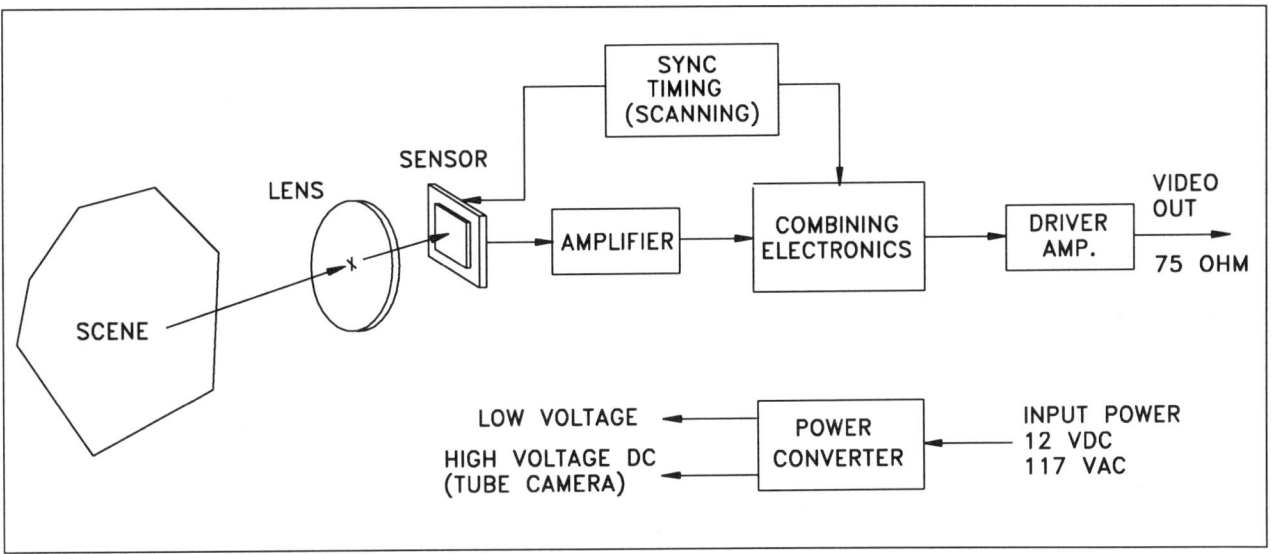

FIGURE 5-2   CCTV camera block diagram

charge-transfer devices (CTD), which come in three types, depending on manufacturing technology: (1) the CCD, (2) the charge-priming device (CPD), and (3) the charge-injection device (CID). A fourth sensor type is the MOS. By far the most popular devices used in security camera applications are the CCD, MOS, and CPD; the CID is reserved primarily for military and industrial applications.

Solid-state cameras are significantly smaller than their tube counterparts in both size and weight, and they consume less power. A packaged solid-state image sensor is typically $3/4$ inch by 1 inch by $1/4$ inch in size and its tube equivalent is $3/4$ inch in diameter by 5 inches long. Solid-state cameras consume approximately 2–5 watts, compared with 8–20 watts for tube sensors.

All security cameras have a lens mount in front of the sensor to mechanically couple an objective lens or optical system. The most popular mount is the C mount. The second most popular mount is called the CS mount and accounts for most of the remaining mounts. This recent new mounting configuration evolved primarily because of the use of smaller $1/2$-inch format cameras and correspondingly smaller objective lenses.

To produce a noise-free monochrome or color picture with sufficient resolution to identify objects of interest, the sensor must have sufficient sensitivity to respond to available natural daytime, nighttime, or artificial lighting. CCTV cameras have three general levels of sensitivity: (1) vidicon tube, (2) Newvicon and silicon tube, CCD, and MOS solid-state, and (3) LLL SIT, ISIT, and ICCD. The LLL cameras are described in Chapter 15. In subsequent sections, each parameter contributing to the function and operation of the security camera is described.

5.2 CAMERA FUNCTION

This section describes the functioning of the major parts of a CCTV camera—both tube and solid-state—and the video signal. For the tube sensor, magnetic or electrostatic deflectors scan the electron beam; for the solid-state sensor, electrical clocking circuits scan the pixel array. Figure 5-2 is a generalized block diagram of CCTV camera electronics.

The television camera consists of (1) an image sensor, (2) some form of electrical scanning system with synchronization, (3) timing electronics, (4) video amplifying and processing electronics, and (5) video signal synchronizing and combining electronics to produce a composite video output signal. To provide meaningful images when the scene varies in real time, scanning must be sufficiently fast—at least 30 frames per second—to capture and replay moving target scenes. The television camera must have suitable synchronizing signals so that a monitor or other recording equipment at the receiving location can be synchronized to produce a stable, flicker-free display or recording.

The following description of the television process is based on the tube camera, but it pertains to the solid-state camera as well.

The lens forms a focused image on the sensor. The electron beam moves across the picture image on the tube sensor in a process called linear (or raster) scanning. The television picture is formed by extracting the light level on the sensor by a succession of scanning lines that moves across the picture area. While the electron-beam motion across the target is very rapid, it is nevertheless linear and uniform. Since the brightness at each point in the target (sensor) varies as a function of the focused scene image,

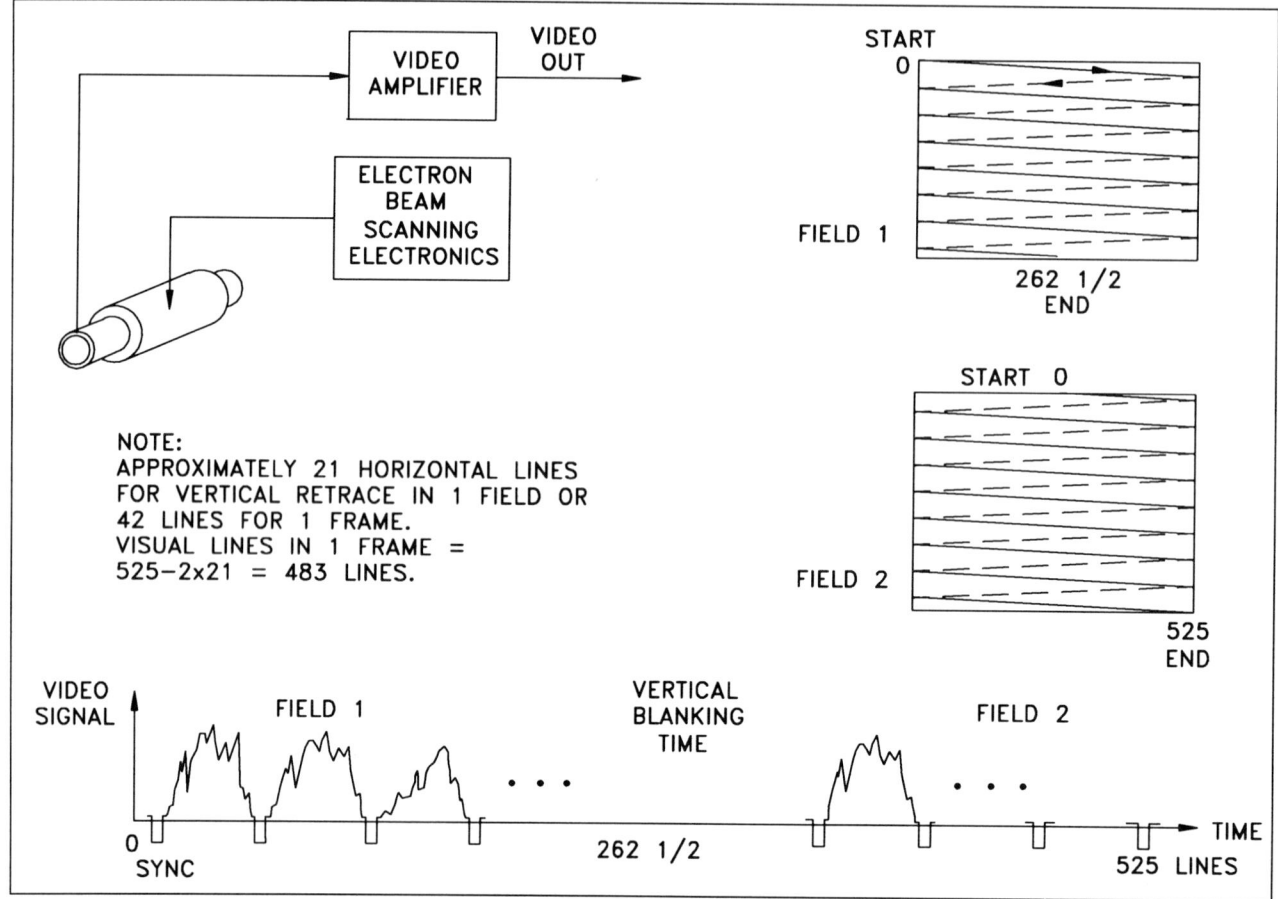

FIGURE 5-3   NTSC two-field, odd-line scanning process

the signal obtained is a representation of the scene intensity profile.

The entire frame is composed of two fields. In the U.S. NTSC system, based on the 60-Hz power line frequency and $\frac{1}{30}$ second per frame (30 frames per second), each frame contains 525 horizontal lines. In the European system, based on a 50-Hz power line frequency and $\frac{1}{25}$ second per frame, each frame has 625 horizontal lines.

Two methods of scanning are used: 2:1 interlace and random interlace. The interlace scanning technique is used to reduce the amount of flicker in the picture and improve motion display while maintaining the same television signal bandwidth. In both scanning methods, the electron beam scans the picture area twice, with each scan producing a field. In the NTSC system, each field contains $262\frac{1}{2}$ television lines. This scanning mode is called two-field, odd-line scanning (Figure 5-3).

Sixty fields are completed per second, and 30 frames are completed per second. With 525 line-periods per frame and 30 frames per second, there are 15,750 line-periods per second. In the standard NTSC system, the vertical blanking interval uses 21 lines per field, or a total of 42 lines per frame. Subtracting these 42 lines from the 525-line frame leaves 483 active lines per frame.

By convention, every camera's scanning beam and every receiver monitor's beam start from the upper left corner of the image and proceed horizontally across and slightly downward (for tube cameras) to the right of the sensor (or monitor screen). Each time the scanning beam reaches the right side of the image, it quickly returns to a point just below its starting point on the left side; this occurs during what is called the horizontal blanking interval. This process is continued and repeated until the scanned spot eventually reaches the bottom of the image, thereby completing one field. At this point the beam turns off again and returns to the top of the image; this time it is called the vertical blanking interval. For the second field (a full frame consists of two fields), the scan lines fall in between those of the first field. By this method the scan lines of the two fields are interlaced, which reduces image flicker and allows the signal to occupy the same transmission bandwidth. When the second field is completed and the scanning spot arrives at the lower right corner, it quickly returns to the upper left corner to repeat the entire process.

For the solid-state camera, in place of the moving electron beam in the tube, the light-induced charge in the individual pixels in the sensor must be clocked out of the sensor into the camera electronics (Figure 5-4).

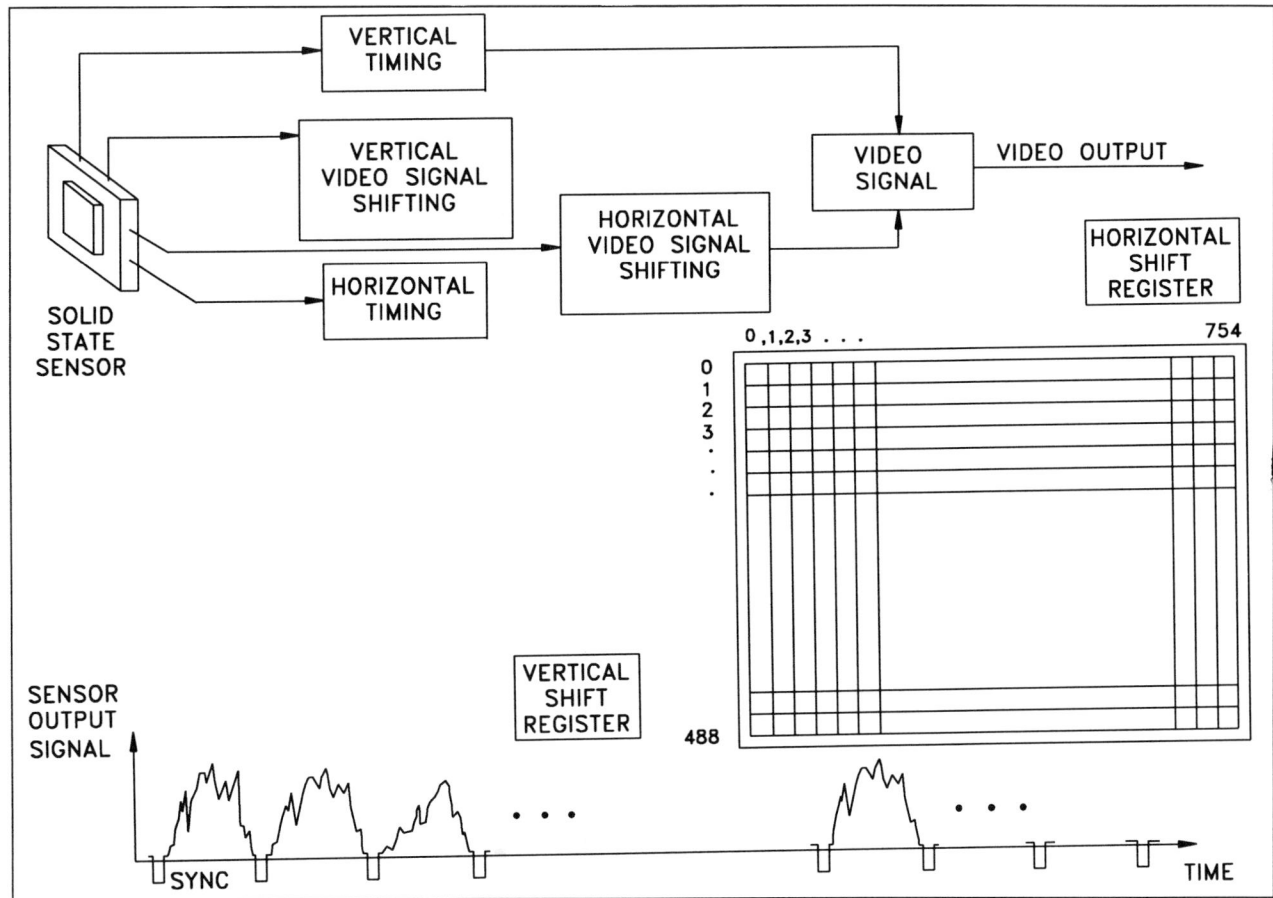

FIGURE 5-4 Solid-state camera scanning process

The time-varying video signal from the individual pixels clocked out in the horizontal rows and vertical columns likewise generates the two interlaced fields.

5.2.1 The Scanning Process

The camera sensor function is to convert a visual or IR light image into a temporary sensor image, which the camera scanning mechanism successively reads, point by point or line by line, to produce a time-varying signal representing the scene light intensity. In a color camera (Section 5.6), this function is accomplished threefold to convert the three primary colors—red, green, and blue—into an electrical signal. By scanning the target twice (remember the typewriter analogy), the sensor is scanned, starting at the top left side of the picture, and a signal representing the scene image is produced. First the odd lines are scanned, until one field of 262½ lines is completed. Then the beam returns to the top left of the sensor and scans the 262½ even-numbered lines, until a total picture frame of 525 lines is completed. Two separate fields of alternate lines are combined to make the complete picture frame every ⅟₃₀th of a second. This TV camera signal is then transmitted to the monitor, where it re-creates the picture in an inverse

fashion. This baseband video signal has a voltage level from 0 to 1 volt (1 volt peak to peak) and is contained in a 4 to 10-MHz bandwidth, depending on the system resolution.

5.2.2 The Television Signal

To better understand the television signal, look at the single horizontal line of the composite television signal shown in Figure 5-5.

The signal is divided into two basic parts: (1) the scene illumination intensity information and (2) the synchronizing pulses. Synchronization pulses with 0.1-microsecond rise and fall times contain frequency components up to 2.5 MHz. Other high frequencies are generated in the video signal when the electron beam scans image scene detail and encounters rapidly changing light-to-dark picture levels of small size. This frequency is represented by about 4.2 MHz and for good fidelity must be reproduced by the electronic circuits. These video signal high-frequency components during scene information represent rapid changes in the scene—either moving targets or very small objects. To produce a stable image on the monitor, the synchronizing pulses must be very sharp and the electronic bandwidth wide enough to accurately reproduce them.

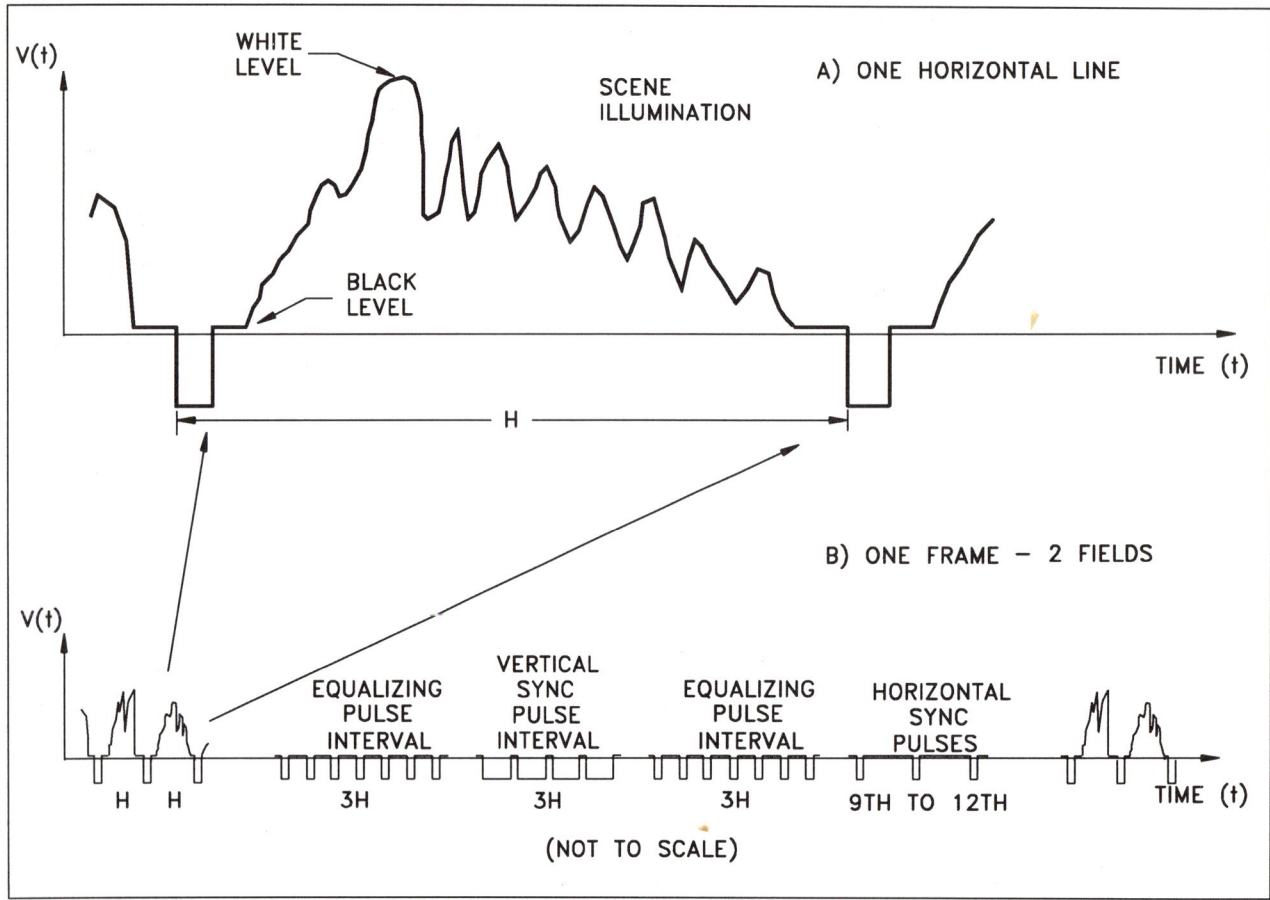

FIGURE 5-5 Monochrome NTSC CCTV video signal

5.3 BASIC SENSOR TYPES

Although cameras with tube-type sensors make up many of the present security installations, solid-state cameras will eventually replace all of them. Tube cameras use a scanning electron beam; solid-state cameras use moving electrical charges.

5.3.1 Tube Cameras

The tube camera video signal is extracted from the tube by means of an electron beam, which, in $1/30$th of a second, scans the entire target area exposed to the focused visible scene. This electron beam is emitted by an electron gun, which produces and accelerates electrons toward the sensor target photocathode, generating a signal proportional to the light intensity on the sensor.

Tube-camera electronics are solid-state except for the image-sensor tube. Like any vacuum tube, the average $2/3$-inch tube wears out in 1 to 2 years, but in LLL environments aging is accelerated. Also, the tube may have to be replaced early if it has a burned-in image—caused when the camera stares at the same scene for long periods—or if it views a bright light source.

Figure 5-6 illustrates the television camera tube technology and electron beam scanning-deflection technique necessary to produce a television camera signal.

The tube consists of a transparent window, a sensor target, and a scanning electron beam assembly. The electron beam is scanned across the sensor tube target area by means of electromagnetic coils positioned around the exterior of the tube, which deflect the beam horizontally and vertically. Depending on the camera design, the deflection means can be either magnetic, electrostatic, or a combination of both. The coils or plates deflect the electron beam across the target to produce the television picture.

Functionally, the camera lens focuses the scene image onto the target surface after it passes through the window. In the mechanism of electron beam scanning, the rear surface of the target (in the area being scanned) produces an electrical signal. Low-noise solid-state electronics then amplify this electrical signal (Figure 5-5) to about 1 volt and combine it with the synchronizing pulses used in scanning. This produces a composite video signal consisting of (1) an amplitude-modulated signal representing the instantaneous intensity of the light signal on the sensor and (2) synchronizing pulses.

The video signal is transmitted by cable or other means to the monitor, VCR, or printer, which uses the synchroniz-

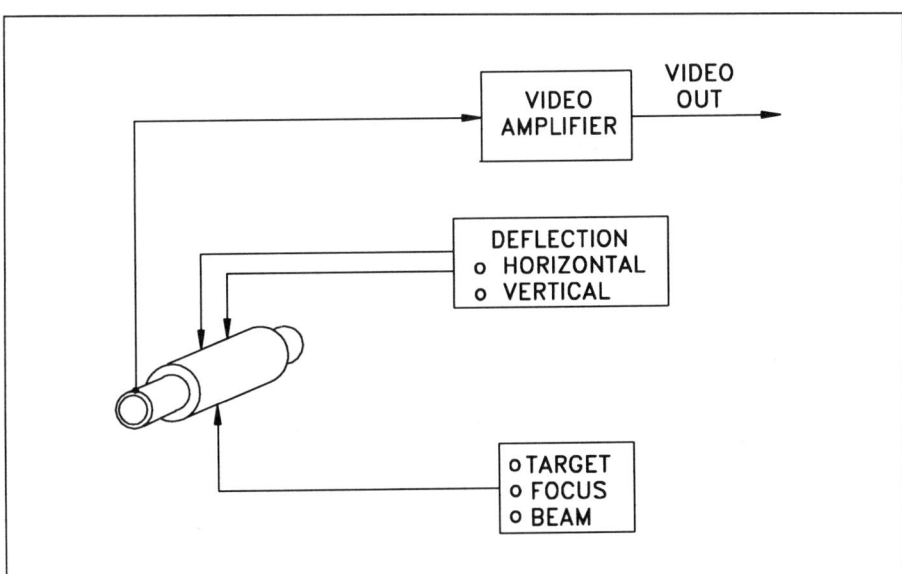

FIGURE 5-6 Tube-sensor-scanning and deflection-assembly diagram

ing signal to produce a stable display in synchrony with the original camera scanning signal. The monitor CRT uses the intensity-modulated video signal to display the time-varying light information in the scene synchronized with the original camera signal. The basic synchronizing frequency of 60 Hz is used because utility companies produce power at and maintain this frequency extremely accurately over long periods of time, and it provides an easy means for synchronizing signals throughout a network. Camera circuits convert the 60 Hz frequency to synchronizing pulses determined by the duration of a 60 Hz signal ($\frac{1}{60}$ second). (Color cameras use 59.94 Hz for their synchronization.) A complete video frame takes place every $\frac{1}{30}$th of a second (two fields every $\frac{1}{60}$th of a second) and is read out repeatedly to produce the television picture.

Most tube cameras can provide excellent resolution because the target is a homogeneous continuous surface. If the electron beam spot size is very small, high resolution—500 to 600 TV lines for $\frac{2}{3}$-inch cameras and 1000 TV lines are obtained when a 1-inch-diameter vidicon is used.

The silicon tube camera uses a similar scanning principle and physical construction, including deflection coils or plates, but has a special target called a silicon diode array. Prior to the innovation of the solid-state sensor, the silicon diode and Newvicon were the most popular sensitive near-IR tubes.

5.3.1.1 Vidicon

For many years the security industry has used the vidicon image tube camera. The most commonly used target material for the vidicon tube is antimony trisulphide, which works well for scenes illuminated with sunlight or artificial lighting. The tube is inexpensive, has excellent image resolution, and can be operated and controlled over wide variations in lighting levels from bright sunlight (10,000 fc) to relatively dim (1 fc) indoor scenes (10,000-to-1 variation

in light level). An important weakness is its susceptibility to image burn-in or damage when exposed to a bright light source such as the sun. Like all other vacuum tubes, the vidicon wears out and must be replaced periodically.

5.3.1.2 Newvicon and Silicon

For an application requiring a more sensitive camera, the silicon or Newvicon tube is a better choice. These more sensitive tube cameras operate down to dawn or dusk light conditions but require some form of automatic-iris lens mechanism when light levels vary by more than about 100 to 1. They are about 10 to 100 times more sensitive than vidicon tubes, depending on the spectral color content of the illumination. The silicon camera will not suffer image burn-in and can even be pointed directly at the sun without being damaged. The silicon diode tube has high sensitivity in the red region of the visible spectrum and the near-IR spectrum and can "see in the dark" when the scene is illuminated with an IR source. The resolution of the silicon diode tube is not as good as that of the vidicon or Newvicon. The Newvicon tube maintains the resolution of the vidicon and has reduced image burn-in susceptibility, but it can be damaged by high-intensity lights or the sun.

5.3.2 Solid-State Types

The solid-state sensor (CCD, MOS, and so on) CCTV camera performs like the tube camera but has a significantly different sensor and scanning system. No electron beam scans the sensor; an array of pixels replaces the sensor tube. The typical sensor has hundreds of pixels in the horizontal and vertical directions, equivalent to several hundred thousand pixels over the entire sensor area. A pixel, the smallest sensing element on the sensor, converts light energy into an electrical charge and a signal. Thousands of pixels placed side

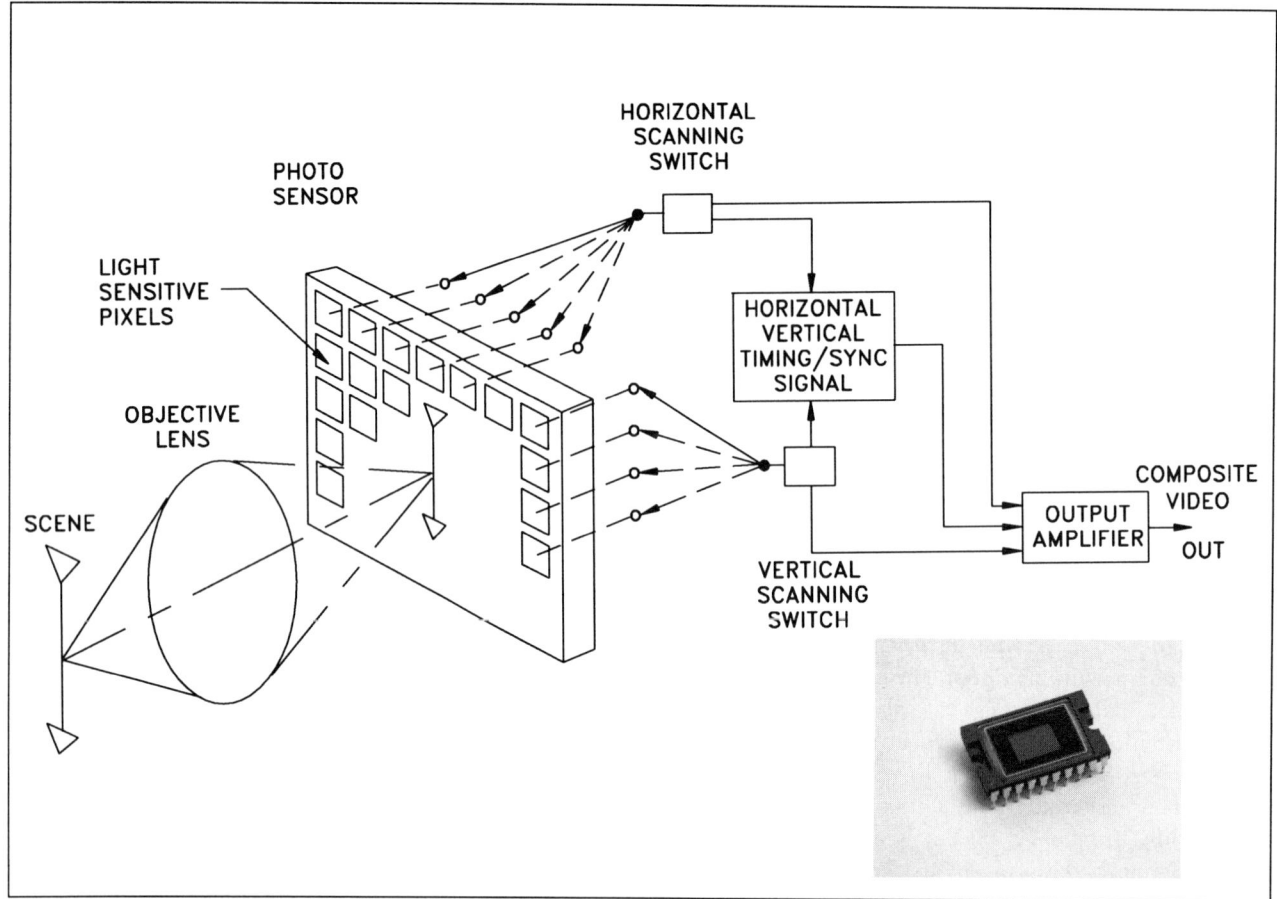

FIGURE 5-7 Schematic representation of solid-state sensor scanning and timing diagram

by side on the sensor respond to the light level of the scene and produce an electrical signal. Arranged in a checkerboard pattern, sensor pixels come in a certain number of rows and columns, as specified by the manufacturer.

Solid-state image sensors are available in several types, but all fall into two basic categories: CTD and MOS. The generic (CTD) class can further be divided into CCD, CPD, and CID. Of these three types, the CCD is by far the most popular. The CID has specialized applications generally found not in the security area but in industrial and military television systems. Its unique feature is its ability for random scanning. The MOS sensor also enjoys widespread use in the security field.

5.3.2.1 CCD

Solid-state CCD sensors are a family of image-sensing silicon semiconductor components invented at Bell Telephone Laboratories. The CCD imagers used in security CCTV are small, rugged, and low in power consumption (Figure 5-7).

By placing pixels in a line and stacking multiple lines, an area array detector is created. As the CCTV lens focuses a single point of the scene on each pixel, the incident light on each pixel generates an electron charge "packet" whose intensity is proportional to the incident light. Each charge packet corresponds to a pixel. Each row of pixels represents

one line of horizontal video information. If the pattern of incident radiation is a focused light image from the optical lens system, then the charge packets created in the pixel array are a faithful reproduction of that image.

In the process called charge coupling, the electrical charges are collectively transferred from each CCD pixel to an adjacent storage element by use of external synchronizing or clocking voltages. In solid-state sensors the image scene is moved out of the silicon sensor via timed clocking pulses that in effect push out the signal, line by line, at a precisely determined clocked time. The amount of charge in any individual pixel varies widely, depending on the light intensity, and represents a single point of the intelligence in the picture. To produce the equivalent of scanning (as in the tube-sensor device), a periodic waveform called a clock voltage is applied to the CCD sensor, causing the discrete charge packets in each pixel to move out for processing and transmission. The area image sensor used in CCTV has both vertical and horizontal transfer clocking signals, as well as storage registers, to deliver an entire field of video information once during each integration period, $\frac{1}{30}$th of a second in the NTSC system.

All CCD image sensors consume low power and operate at low voltages. They are not damaged by intense light but suffer some saturation and blooming under intense illumi-

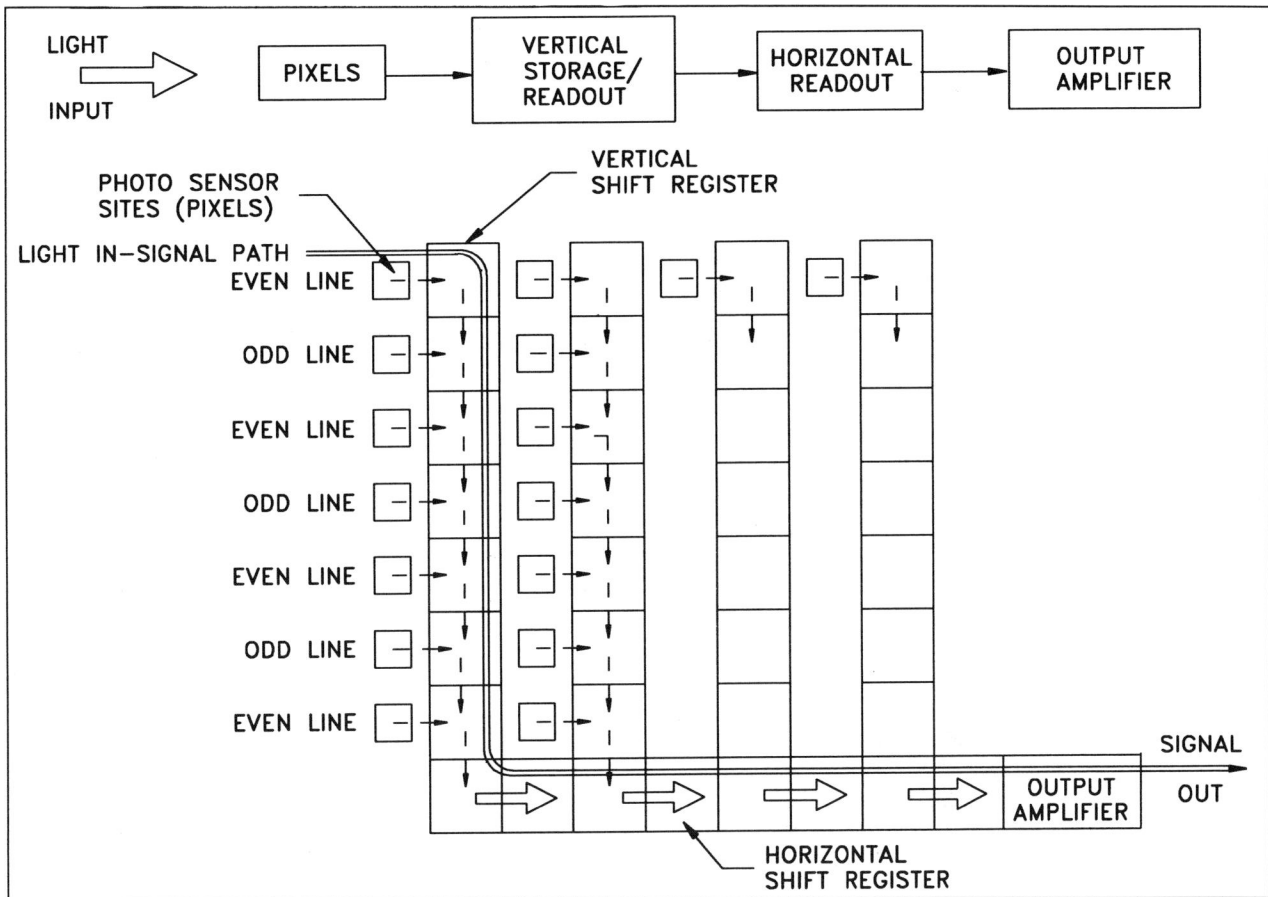

FIGURE 5-8   Interline-transfer CCD sensor layout

nation. More recent devices contain anti-blooming geometry and exposure control (electronic shuttering). Typical device parameters for a CCD available today are 488 by 380 pixels (horizontal by vertical) in formats of ⅓-, ½-, and ⅔-inch, in a 4 by 3 aspect ratio. Typical dynamic ranges are 1000 to 1 without shuttering, and 100,000 to 1 with electronic shuttering.

Compared with the MOS imaging device, the CCD provides higher sensitivity. Under fluorescent lighting, the interline transfer (ILT) CCD does not produce flicker in the picture, as do the frame transfer (FT) storage CCD and MOS-type sensors.

CCD sensors have 3 to 5 times greater packing density than the next most dense MOS large-scale integrated circuit. This is primarily because the basic CCD storage element requires no electrical contacts.

5.3.2.2 MOS

The MOS-type sensor exhibits high picture quality but has a lower sensitivity than the CCD. In the MOS device, the electric signals are read out directly through an array of MOS transistor switches, rather than line by line as in the CCD sensor.

5.3.2.3 CID

Another generic solid-state camera is the CID, invented at the General Electric Company. Unlike all other solid-state devices, the camera can address or scan any pixel in a random sequence rather than in the row/column sequence used in the other sensors. Though this capability has not been utilized in the security field, industrial and military applications take advantage of it when scan sequences or patterns other than the standard NTSC sequence are required. When scanned in the normal NTSC pattern, the CID is sensitive and has attributes similar to those of other solid-state cameras. The CID is less sensitive to overload from bright lights in the scene, which produce a black vertical line in overloaded CCDs.

5.3.2.4 Interline and Frame Transfer

There are several different CCD sensor pixel architectures used by different manufacturers. The two most common CCD types are the ILT and FT types.

Figure 5-8 shows the pixel organization and readout technique for the ILT CCD image sensor.

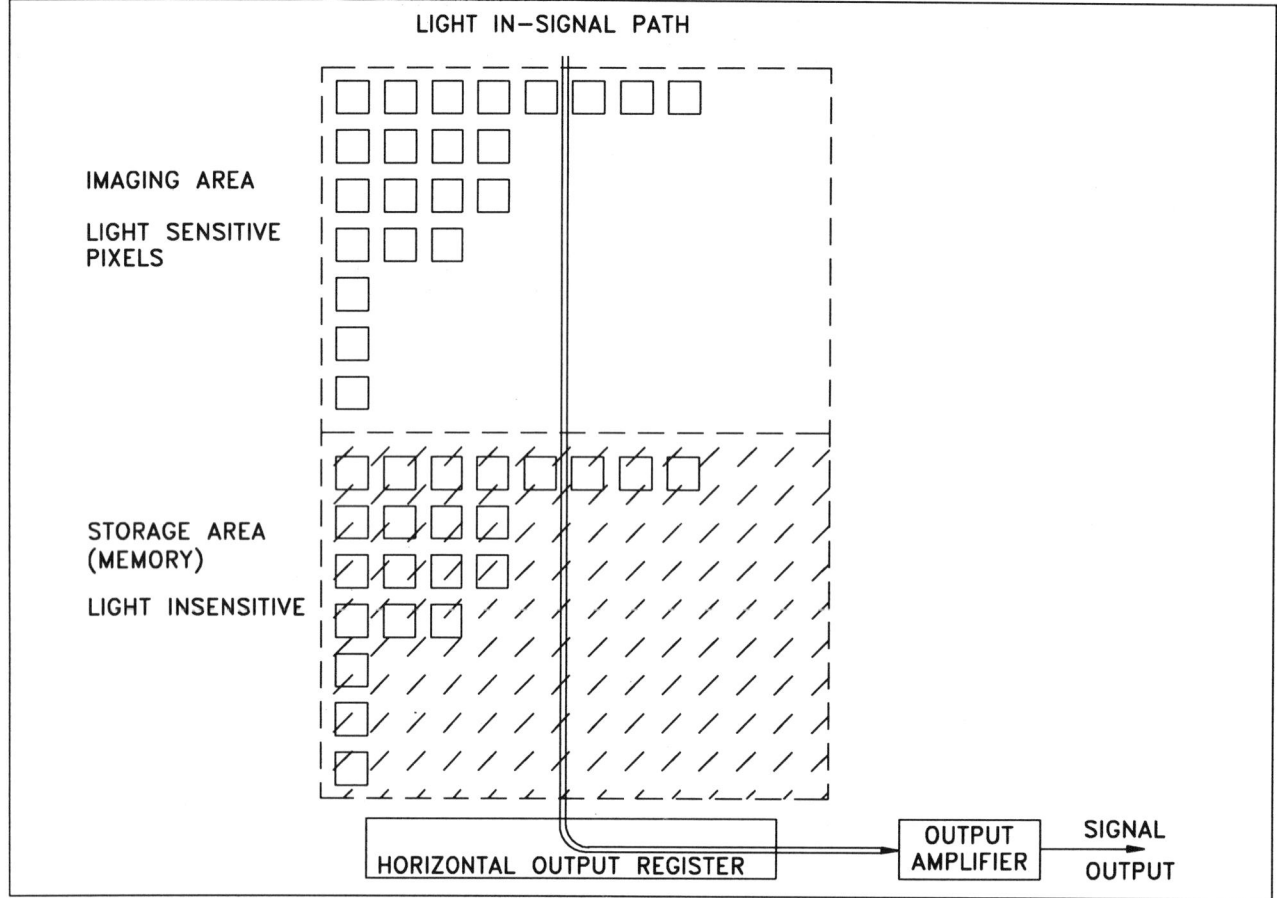

FIGURE 5-9   Frame-transfer CCD sensor layout

The pixel organization in the ILT sensor has precisely aligned photosensors and vertical shift registers arrayed interlinearly, and a horizontal shift register linked with the vertical shift registers as shown. The photosensor sites sense light variations, which generate electronic charges proportional to the light intensity. The charges are passed into the vertical shift registers simultaneously and then transferred to the horizontal shift register successively until they reach the sensor output amplifier. The camera electronics further amplify and process the signal. Each pixel and line of information in the ILT device is transferred out of the sensor array line by line, eventually clocking out all 525 lines, thereby scanning the entire sensor and producing a frame of video. This sequence is repeated to produce a continuous television signal.

In the FT CCD, the entire 525 lines are transferred out simultaneously and stored temporarily in a silicon buffer located in an adjacent silicon array (Figure 5-9).

5.3.2.5 Smear and Infrared Cut Filter

Most CCD image sensors have wide spectral ranges and are usually useful over the entire visible range and into the near-IR spectral region (above 800 to 900 nanometers). However, most solid-state cameras exhibit an image smear when the illumination source contains visible and IR radia-

tion, with the IR radiation causing most of the smearing. Figure 5-10 shows the spectral response of the CCD sensor.

The loss in resolution is due to the action of the IR image photons, which generate electrons that penetrate deeper in the silicon: those electrons get lost, and poor resolution results. For this reason CCD sensors are often supplied with an IR blocking filter, which reduces sensitivity but increases resolution. More recently developed sensors have eliminated most of this problem.

When the CCD camera is pointed toward a strong light source or bright object, the sensor is overloaded due to the high sensitivity of the CCD imager in the near-IR region, and a bright light band is produced on the monitor display above and below the object. This phenomenon, called image smear, is caused by the extra energy from the light source extending out into other pixel sites and in the vertical sites above and below the illuminated pixel. When the longer-wavelength IR light enters the CCD imager, it passes through the photosensor and generates a charge underneath, causing the smear phenomenon. To prevent the smear, an IR filter is interposed between the sensor and the lens, thereby reducing the IR energy received by the sensor. The use of the IR filter alters the spectral response of the CCD imager (Figure 5-10). The two curves represent the sensor with and without the IR filter in place. When IR

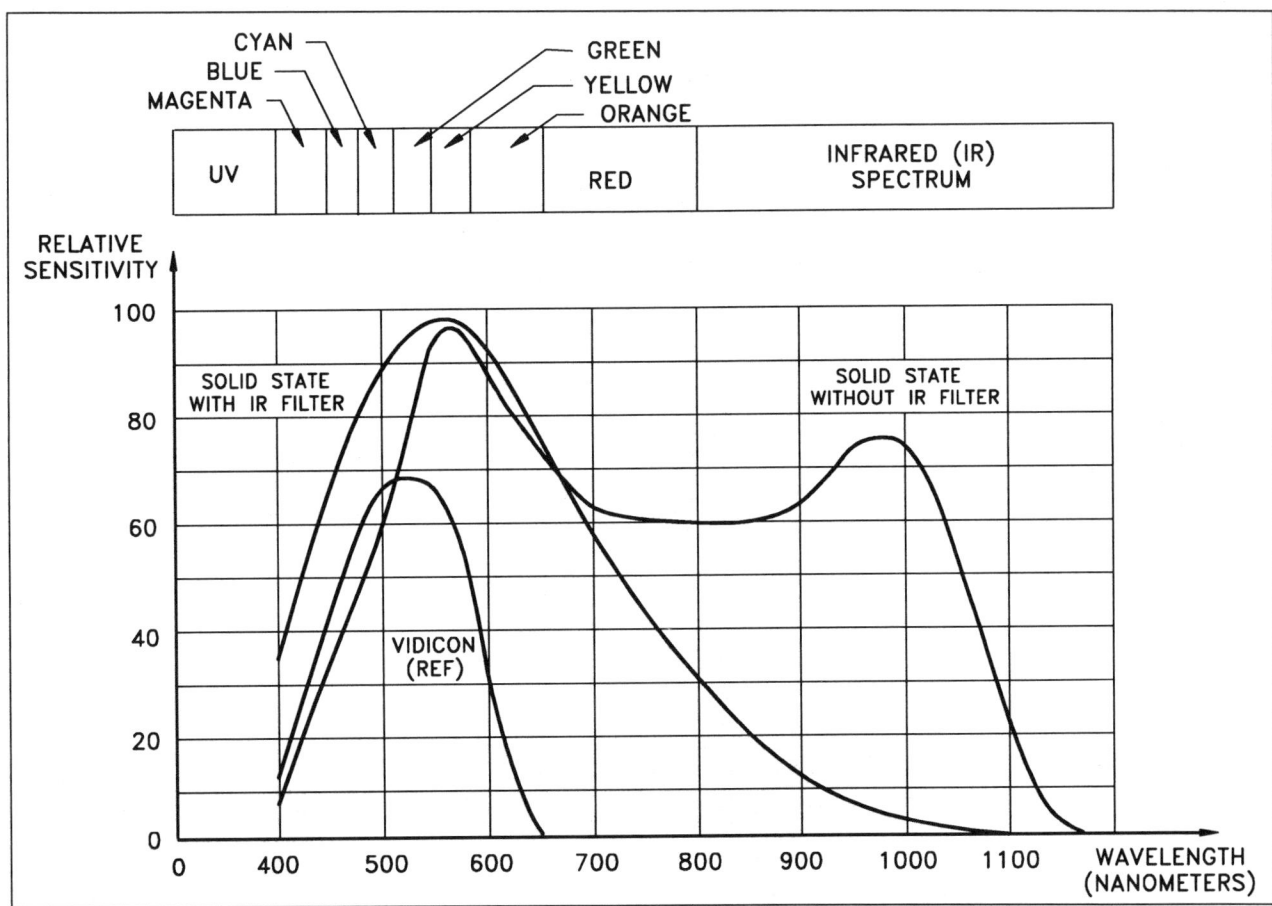

FIGURE 5-10   CCD spectral response and IR blocking filter transmission

overload is not expected but there is IR radiation in the illuminating source, it is advantageous to use the sensor without the IR cut filter in place, thereby increasing signal strength and picture quality. If the illuminating source contains a bright spot of IR radiation, such as from sunlight or a car headlight, the IR cut filter should be used to prevent sensor overload.

5.3.2.6 Geometric Accuracy

One of the significant advantages of charge-coupled image sensors over vacuum-tube sensors is the precise geometric location of the pixels with respect to one another. In a camera tube the video is "read" from a photosensitive material by a scanning electron beam. The exact position of the beam is never precisely known because of some uncertainty in the sweep circuits resulting from random electrical noise, variations in power-supply voltage, or other variations. In a CCD sensor, the locations of the individual photosensor pixel sites are known exactly, since they are determined during the manufacture of the sensor. Such geometric accuracy is especially important for proper alignment in color cameras to maintain color purity. As with the tube-sensor cameras, solid-state sensors are read out in two separate fields—first all the even-numbered pixels and then

all the odd-numbered pixels in each column, rather than the odd and even elements in each column serially.

5.3.3 Low-Light-Level SIT, ISIT, and ICCD

The most sensitive CCTV cameras are the LLL SIT, ISIT, and ICCD. These cameras share many of the characteristics of the tube types described earlier but include means to respond to much lower light levels. The SIT, ISIT, and ICCD cameras are used to view scenes illuminated by natural moonlight, starlight, or some other very low light level artificial illumination.

These cameras use image intensifiers coupled to imaging tubes or solid-state sensors and are capable of viewing scenes hundreds to thousands of feet from the camera under nighttime conditions. The method uses a light intensifier tube or microchannel plate (MCP) intensifier to amplify the available light up to 50,000 times. The SIT tube camera combines a silicon target vidicon structure with a stage of light amplification ahead of it, coupled electronically to the silicon diode target (Figure 5-11).

The ICCD camera (Figure 5-11a) is an LLL camera class whose sensitivity approaches that of the best SIT cameras and eliminates the blurring characteristics of the SIT under

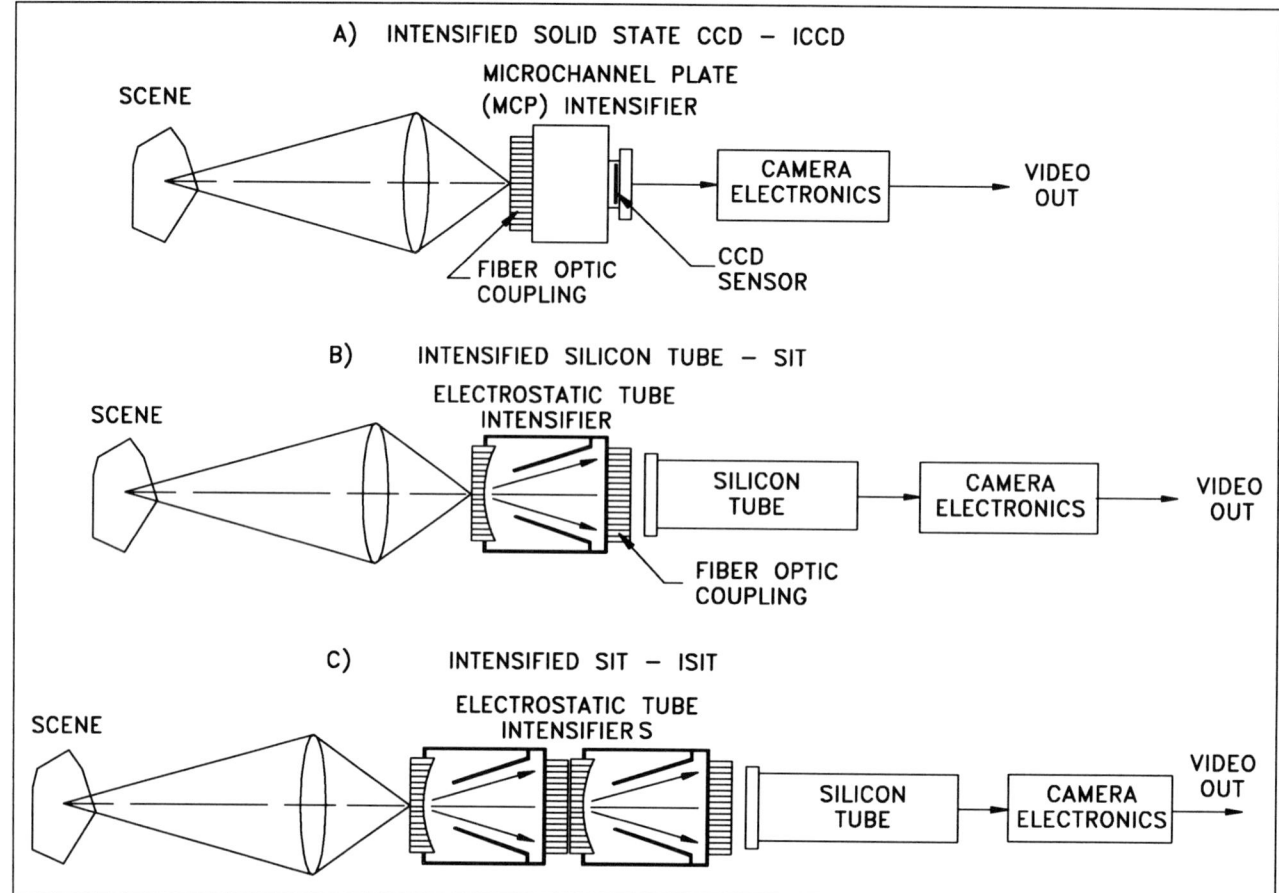

FIGURE 5-11   Intensified solid-state and silicon tube cameras

very low light level conditions. The ICCD camera combines an intensifier (tube or MCP) with a CCD image sensor to provide a sensitivity similar to that of a SIT camera.

The SIT and ISIT cameras (Figure 5-11b) have an image intensifier located between the camera lens and a silicon tube to amplify the incoming light before it reaches the tube.

Complete SIT tube and ICCD camera systems have sufficient sensitivity and automatic light compensation to be used in surveillance applications from full sunlight to quarter moonlight conditions. They are provided with automatic light-level mechanisms having a 100 million to 1 light-level range and built-in tube protection to prevent tube sensor burnout or overload when viewing bright scenes.

For viewing the lowest light levels, a fully compensated ISIT camera provides the wide dynamic range from full sunlight to starlight conditions, with a 4 billion to 1 automatic light-level range control. Though large, these cameras have been used in many LLL applications. The ISIT camera uses a SIT tube with an additional light amplification stage and is the lowest light-level camera available today. Some resolution is lost in the ISIT camera compared

with the SIT camera, but if ultra-LLL performance is required (without supplemental lighting), ISIT is the only choice.

For dawn and dusk outdoor illumination, the best CCD cameras can barely produce a usable video signal. SIT and ICCD cameras can operate under the light of a quarter moon with 0.001 fc. The ISIT camera can produce an image from only 0.0001 fc, which is the light available from stars on a moonless night. SIT, ISIT, and ICCD offer a 100 to 1000 times improvement in sensitivity over the best CCD cameras because they intensify light, whereas the tube and CCD cameras only detect it. A detailed description of these special LLL cameras is given in Chapter 15.

5.4 RESOLUTION

CCTV resolution is a critical measure of the television picture quality: the higher the resolution, the higher the level of information. CCTV resolution is measured by the number of horizontal and vertical TV lines that can be discerned in the monitor picture.

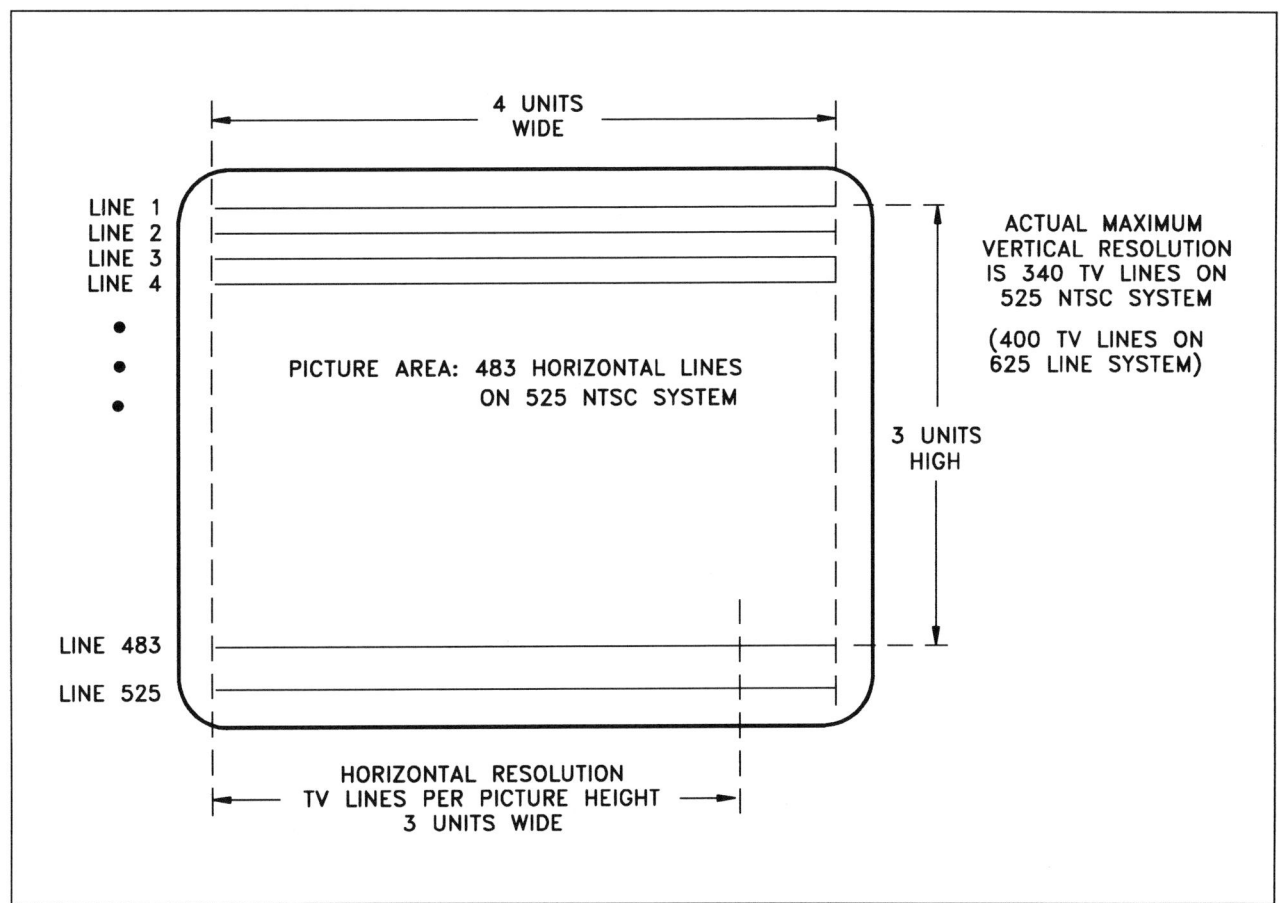

FIGURE 5-12   CCTV vertical and horizontal resolution

5.4.1 Vertical

Vertical resolution is derived from the 504 effective scanning lines in the 525-line NTSC television system (Figure 5-12).

For example, the camera will dissect a vertical line appearing in the scene into 483 separate sections. Since each scanning line has a discrete width, some of the scene detail between the lines will be lost. As a general rule (called the Kell factor), approximately 30% of any scene is lost. Therefore, the standard 525-line NTSC television system produces 340 TV lines of resolution (483 effective lines × 0.7). In any standard 525-line CCTV system, the maximum achievable vertical resolution is approximately 350 TV lines. In a 625-line system, the maximum achievable vertical resolution is 408 TV lines.

5.4.2 Horizontal

The NTSC standard provides a full video frame composed of 525 lines, with 483 lines for the image and two vertical blanking intervals composed of 21 retrace lines each. The TV industry also adopted a viewing format with a width-to-height ratio of 4 to 3 and specifies horizontal resolution in TV lines per picture height. The horizontal resolution on the monitor tube depends on how fast the electron beam can change its intensity as it traces the image on a horizontal line. This in turn is determined by the maximum speed or frequency response (bandwidth) of the television electronics and television signal. In other words, while the vertical resolution is determined solely by the number of lines chosen—and thus not variable under the U.S. standard of 525 lines—the horizontal resolution depends on the electrical performance of the individual camera, transmission, and monitor system. Most standard cameras with a 6-MHz bandwidth produce a horizontal resolution in excess of 450 TV lines. System horizontal resolution will be limited to approximately 80 lines per megahertz of bandwidth. The traditional method of testing and presenting CCTV resolution test results is to use the Electronic Industries Association (EIA) resolution target (Figure 5-13).

The minimum-spaced discernible black-and-white transition boundaries in the two wedge areas are the vertical (horizontal wedge) and horizontal (vertical wedge) limiting resolution values. Various industries using electronic imaging devices have specified resolution criteria dependent on the particular discipline involved. In the CCTV industry, the concept of TV lines is defined as the resolution parameter. The solid-state-imaging industry has adopted pixels as its

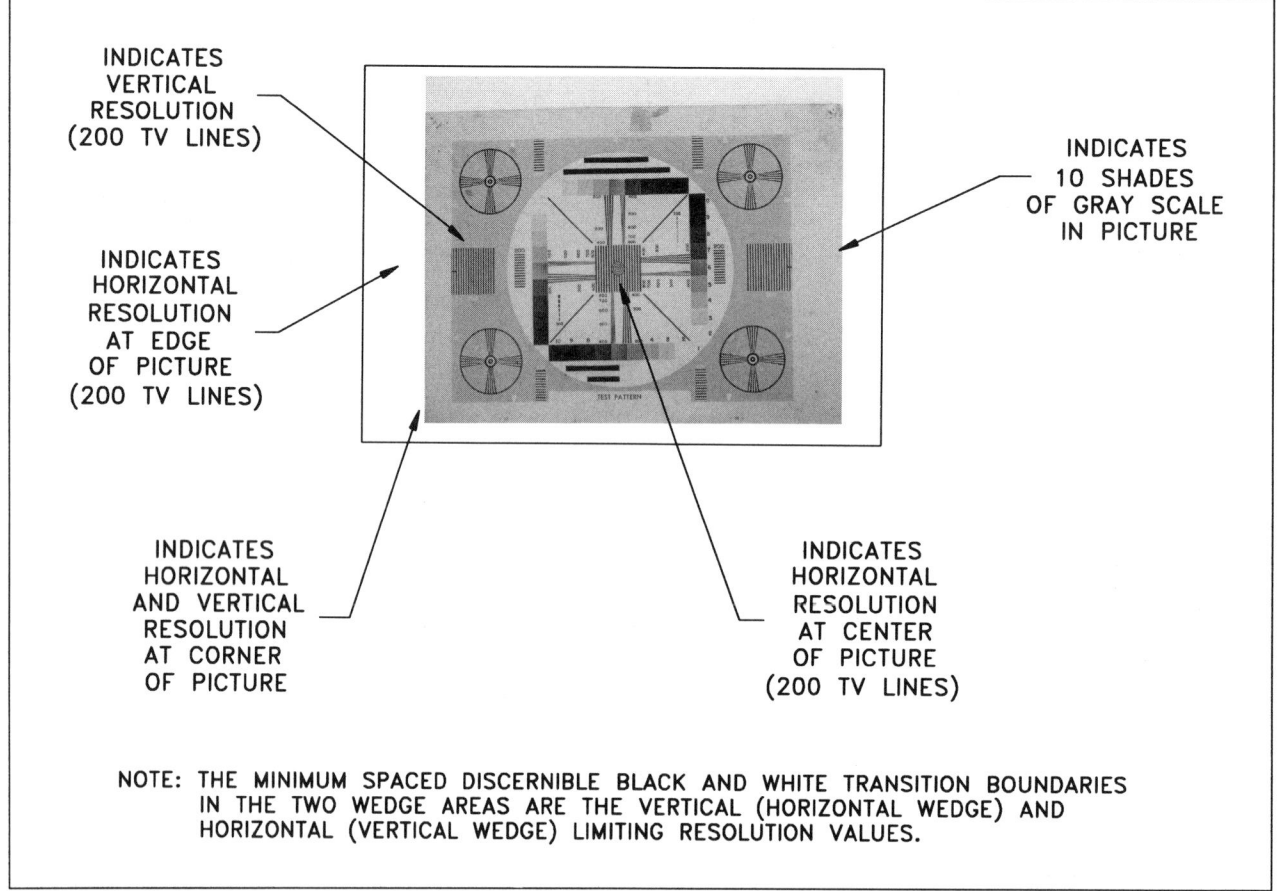

INDICATES
VERTICAL
RESOLUTION
(200 TV LINES)

INDICATES
HORIZONTAL
RESOLUTION
AT EDGE
OF PICTURE
(200 TV LINES)

INDICATES
10 SHADES
OF GRAY SCALE
IN PICTURE

INDICATES
HORIZONTAL
AND VERTICAL
RESOLUTION
AT CORNER
OF PICTURE

INDICATES
HORIZONTAL
RESOLUTION
AT CENTER
OF PICTURE
(200 TV LINES)

NOTE: THE MINIMUM SPACED DISCERNIBLE BLACK AND WHITE TRANSITION BOUNDARIES
IN THE TWO WEDGE AREAS ARE THE VERTICAL (HORIZONTAL WEDGE) AND
HORIZONTAL (VERTICAL WEDGE) LIMITING RESOLUTION VALUES.

FIGURE 5-13   EIA resolution target

parameter. To obtain TV-line resolution equivalent when the number of pixels are specified, multiply the number of pixels by 0.75. In photography, line pairs or cycles per millimeter is the resolving power notation.

While all these parameters are useful, they tend to be confusing. For the purposes of CCTV security applications, the use of parameters besides TV line is minimized. These other parameters are defined as follows:

- One cycle is equivalent to one line pair.
- One line pair is equivalent to two TV lines.
- One TV line is equivalent to 0.75 pixels.

One cycle is equivalent to one black-and-white transition and represents the minimum sampling information needed to resolve the elemental areas of the scene image.

A figure of merit for solid-state CCTV cameras is the total number of pixels reproduced in a picture area. A typical value is 200,000 pixels for a good 525-line CCTV system.

A parameter deserving mention that is used in lens, camera, and image-intensifier literature is the modulation transfer function (MTF). This concept was introduced to assist in predicting the overall system performance when cascading several devices such as the lens, camera, transmission medium, and monitor in one system. The MTF provides a figure of merit for a part of the system (such as the

camera or monitor) acting alone or when the parts are combined with other elements of the system. It is used particularly in the evaluation of LLL devices (Chapter 15).

One comparison made between solid-state and tube cameras is resolution, that is, how much detail appears in the picture. The resolution for a good tube security camera is 500 to 600 TV lines. Solid-state data sheets often quote the number of pixels instead of resolution. However, unless the number of pixels is converted into equivalent TV lines, resolution is not known. To approximate the horizontal resolution from the horizontal pixel count, multiply the number of horizontal pixels by 0.75. Only recently have solid-state sensors been available that can match the resolution of the average-to-better tube cameras. Improved horizontal resolution in solid-state cameras has been achieved by increasing the number of pixels and other electronic processing means. Table 5-1 summarizes the state of the art in solid-state sensors and gives information on the horizontal and vertical pixels available for representative ⅓-, ½-, and ⅔-inch format types.

When monochrome solid-state sensor cameras were first introduced, the sensors had a maximum horizontal resolution of approximately 200 TV lines per picture height. These early low-resolution sensors had 288 horizontal by 394 vertical pixels. Present-day sensors have horizontal resolu-

FORMAT	TYPE	PIXELS		TOTAL	RESOLUTION (TVL)		COMMENTS
		HORIZONTAL	VERTICAL		HORIZONTAL	VERTICAL	
1/3	MONOCHROME	510	492	250,920	380	350	
	MONOCHROME	512	582	297,984	380	350	
1/2	COLOR	682	492	355,544	430	350	REMOTE HEAD
	MONOCHROME	682	492	335,544	500		REMOTE HEAD
	COLOR	756	484	365,904	570	→	
	MONOCHROME	768	494	379,392	460		
	MONOCHROME	510	492	250,920	380		SMALL PC BOARD
	MONOCHROME	512	492	250,920	380	350	SMALL PC BOARD
2/3	MONOCHROME	768	493	378,624	570	350	HIGH SENSITIVITY
	MONOCHROME	754	484	273,460	565	350	HIGH SENSITIVITY

* HORIZONTAL RESOLUTION APPROXIMATELY 0.75 x NUMBER OF HORIZONTAL PIXELS
** VERTICAL RESOLUTION IS LIMITED BY 525 NTSC AND 625 CCIR HORIZONTAL LINE SCAN RATE

Table 5-1 Resolution of Representative Solid-State Cameras

tions of 500 to 600 TV lines per picture height. Medium-resolution camera sensors have 510 (H) × 492 (V) pixels, and high-resolution cameras have 739 (H) × 484 (V) pixels.

Improvements in the resolution of solid-state sensors to match the best tube sensors have resulted from various approaches: increased pixel density, image-shift enhancement, zigzag pixel format, and photoconductive overlay. The most successful increase in resolution has come from increased pixel density. These strides in decreasing the pixel size have resulted from the techniques used to manufacture very large scale integrated (VLSI) devices for computers. Image sensors are also VLSI devices. In the manufacture of such devices, silicon wafer yield increases approximately as the cube of the device area decreases. Thus there has been a substantial motivation to improve the resolution while decreasing the sensor size. The majority of solid-state sensors in use today have a $\frac{1}{2}$- or $\frac{2}{3}$-inch image format, but there is a rapid movement away from $\frac{2}{3}$-inch to the $\frac{1}{2}$- and $\frac{1}{3}$-inch formats.

Several other techniques are used to improve resolution. In one camera configuration, image-shift enhancement results in a doubling of the ILT CCD imager horizontal resolution by shifting the visual image in front of the CCD sensor by one-half pixel. This technique simultaneously reduces aliasing, which causes a foldback of the high-frequency signal components, resulting in "herringbone" or jagged edges. Often seen when viewing plaid patterns, screens, and so on with medium- to low-resolution solid-state cameras, aliasing reduces resolution and causes considerable loss in picture intelligence.

Another technique used to improve the horizontal resolution without increasing the pixel count is offsetting each row of pixels by one-half pixel, effecting a zigzag of the pixel rows. This arrangement, in conjunction with corresponding clocking, allows simultaneous readout of two horizontal rows and nearly doubles the horizontal resolution compared with conventional detectors with identical pixel counts. One sensor with 384 horizontal pixels per line provides approximately 280 TV lines per picture height in non-zigzag mode; but using this technique, the sensor demonstrates excellent horizontal resolution at over 500 TV lines per picture height.

Further illustrating the high-resolution capability of the ILT structure, one manufacturer has developed an 800 (H) × 490 (V) image sensor in a $\frac{1}{2}$-inch format exhibiting a horizontal resolution of 560 TV lines per picture height.

5.4.3 Static versus Dynamic

So far static resolution has been described, that is, the resolution achieved when a camera views a stationary scene. But when a camera views a moving target—a person walking through the scene, a car passing by—or scans a scene, we define a new parameter, called dynamic resolution. Under either the moving-target or scanning condition, extracting intelligence from the scene depends on resolving, detecting, and identifying fine detail. Solid-state cameras inherently have higher dynamic resolution than tube cameras. When scanning a scene or viewing a moving target, solid-state sensors are far superior to the tube type, especially under LLL conditions. Many tube cameras provide adequate dynamic resolution if the light level is medium to high (50 fc and above). The solid-state camera has the ability to resolve rapid movement without degradation in resolution under almost all suitable lighting conditions.

When high resolution is required while viewing very fast moving targets, solid-state cameras with an electronic shutter are used to capture the action. Many solid-state cameras have a variable-shutter-speed function, with common shutter speeds of $\frac{1}{60}$, $\frac{1}{1000}$, $\frac{1}{2000}$. This shuttering technique is equivalent to using a fast shutter speed on a film camera. The ability to shutter solid-state cameras results in advantages similar to those obtained in photography: the moving object or fast-scan panning that would normally produce a blurred image can now produce a sharp one. This technique's only disadvantage is that since a decreased amount of light enters the camera, the scene lighting must be adequate for the system to work successfully.

5.5 SENSOR FORMAT

There are four existing image format sizes for solid-state and tube sensors: $\frac{1}{3}$-, $\frac{1}{2}$-, $\frac{2}{3}$-, and 1-inch. All sensor formats have a horizontal-by-vertical geometry of 4 by 3, as defined in the EIA and NTSC standards. The most common formats used in the security industry are the $\frac{1}{3}$-, $\frac{1}{2}$-, and $\frac{2}{3}$-inch sizes. For a given lens, the $\frac{1}{3}$-inch format sensor sees the smallest scene image and the 1-inch sees the largest, with the $\frac{1}{2}$- and $\frac{2}{3}$-inch format cameras seeing proportionally in between. Specifically, the ratio of scenes is $\frac{1}{3}$:$\frac{1}{2}$:$\frac{2}{3}$:1. The $\frac{2}{3}$- and $\frac{1}{2}$-inch solid-state formats are presently the most popular, and the $\frac{1}{3}$-inch format is finding increased use. The $\frac{2}{3}$- and 1-inch vidicons and other tube cameras, while popular years ago and still in extensive use today, are being sold only for replacement or for high-resolution applications. The SIT and ISIT tube cameras using the 1-inch tube to provide LLL capabilities are likewise being replaced by their solid-state counterpart, the ICCD. As a basis for comparison with film formats, Figure 5-14 shows the four CCTV image formats, the Super 8 film camera, the 16-mm semiprofessional film camera, and the 35-mm film camera still in use for surveillance and bank holdup cameras. Table 5-2 lists the CCTV image format sizes available.

5.5.1 Tube Sensor

The physical target area in tube cameras is circular and corresponds to the diagonal of the lens image circle. The active target is the inscribed 4 × 3 aspect ratio rectangular area scanned by the electron beam. The target area in the

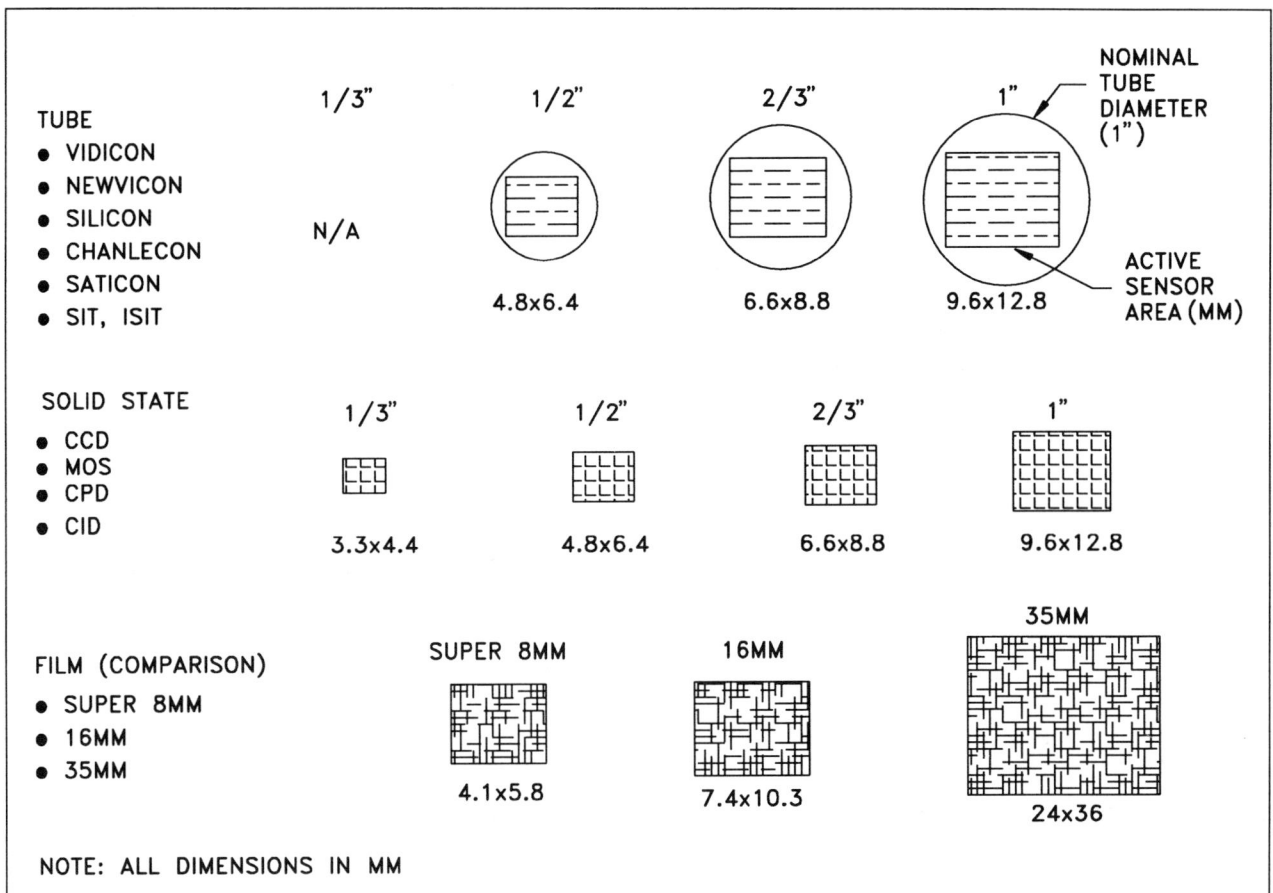

FIGURE 5-14   Standard sensor format sizes

solid-state sensor is the full sensor 4×3 format array, since each pixel is used in the image.

Tube-camera sensors are available in ½-, ⅔-, and 1-inch formats. The larger tube sizes are more sensitive than their smaller counterparts and have higher resolution. The camera sensor format is important since it determines the lens format size with which it must operate, and along with the lens focal length (FL), sets the CCTV system field of view.

As a general rule, the larger the tube size, the larger the diameter of the lens glass size, which translates to increased weight and cost. Any lens designed for a larger format can be used on a smaller camera. The opposite is not true: a lens designed for a ½-inch format will not work properly on a ⅔- or 1-inch format camera and will produce vignetting.

5.5.2 Solid-State Sensor

CCTV solid-state camera sensors are available in ⅓-, ½-, and ⅔-inch formats. The sensor arrays are rectangular in shape and have the sizes listed in Table 5-2 and shown in

Figure 5-14. For solid-state sensors, larger does not mean better. Significant progress has been made in producing exceptionally high quality ⅓- and ½-inch format sensors that rival the sensitivity of some of the large tube and solid-state sensors.

5.6 COLOR CAMERAS

The television broadcast industry has used single-tube and three-color-tube cameras extensively. However, until recently the security industry has not used tube color cameras to any appreciable degree, for two reasons: (1) relatively high cost for an inferior picture and (2) low equipment stability and reliability. The inferior picture quality and low stability of such tube cameras, as well as their low resolution and sensitivity and their color drift (that is, the need for periodic color adjustment) also kept consumer demand low. But the development of the superior solid-state CCD color camera for the VCR home consumer market accelerated the use of color CCTV in the security industry.

SENSOR SIZE						
FORMAT	DIAGONAL(d)		HORIZONTAL		VERTICAL	
	MM	INCHES	MM	INCHES	MM	INCHES
1"	16	.63	12.8	.50	9.6	.38
2/3"	11	.43	8.8	.35	6.6	.26
1/2"	8	.31	6.4	.25	4.8	.19
1/3"	5.5	.22	4.4	.17	3.3	.13
1/4" *	4	.16	3.2	.13	2.4	.10
1/6" *	2.75	.11	2.2	.09	1.7	.07

* LATEST SENSOR FORMATS

Table 5-2   CCTV Camera Sensor Formats

5.6.1 Solid-State

Solid-state color cameras are available in ⅓-, ½-, and ⅔-inch formats with CCD and MOS sensors. Most used in security applications have single-chip sensors with three-color stripe filters integral with the image sensor. Typical color sensitivities for these cameras are from 0.7 to 1.4 fc (7 to 15 lux) for full video, which is less sensitive than their monochrome counterpart. The resolution of most color cameras ranges from 330 to 400 TV lines, corresponding to standard VHS and 8-mm VCR capabilities. Cameras with higher resolutions of 420 to 460 TV lines are available for use with the higher-resolution S-VHS and Hi-8 recorders. Color cameras with a ⅔-inch format producing resolution of 250 to 350 TV lines require on the order of 780 (H) by 490 (V) pixels. This pixel array represents 380,000 picture elements, which result in a good color image.

The color tube and early versions of the CCD cameras had external white-balance sensors and circuits to compensate for color changes. However, most new designs incorporate automatic white-balance compensation as an integral part of the camera. This auto-white-balance is required so that when the camera is initially turned on, it properly balances its color circuits to a white background, which in turn is determined by the type of illumination at the scene. The camera constantly checks the white-balance circuitry and makes any minor compensation for variations in the illumination color temperature (that is, the spectrum of colors in the viewed scene).

5.6.2 Single- versus Three-Sensor

There are presently two techniques to produce a color video signal: single-sensor and three-sensor with prism.

The first and most common technique uses a single color sensor (Figure 5-15). This camera has a complex color-imaging sensor containing three integral optical filters that produce the three primary colors—red (R), green (G), and blue (B)—which are sufficient to reproduce all the colors in the visible spectrum. The three color filters divide the total number of pixels on the sensor by three, so that each filter type covers one third of the pixels. The sensor is followed by video electronics and clocking signals synchronized to produce the composite video color signal.

Since this camera has only one sensor, the light from the lens must be split into thirds, thereby decreasing the overall camera sensitivity by three. Since each resolution element is composed of three colors, the resolution likewise is reduced by a factor of three. However, because of its relatively low cost, this single-sensor camera is more widely used than the three-sensor prism type.

Figure 5-16 shows the second technique used to produce a color video signal: a beam-splitting prism is interposed between the lens and three solid-state sensors.

The function of the prism is to split the full visible spectrum into the three primary colors, R, G, and B. Each individual sensor has its own video electronics and clocking signals synchronized together to eventually produce three separate signals proportional to the RGB color content in

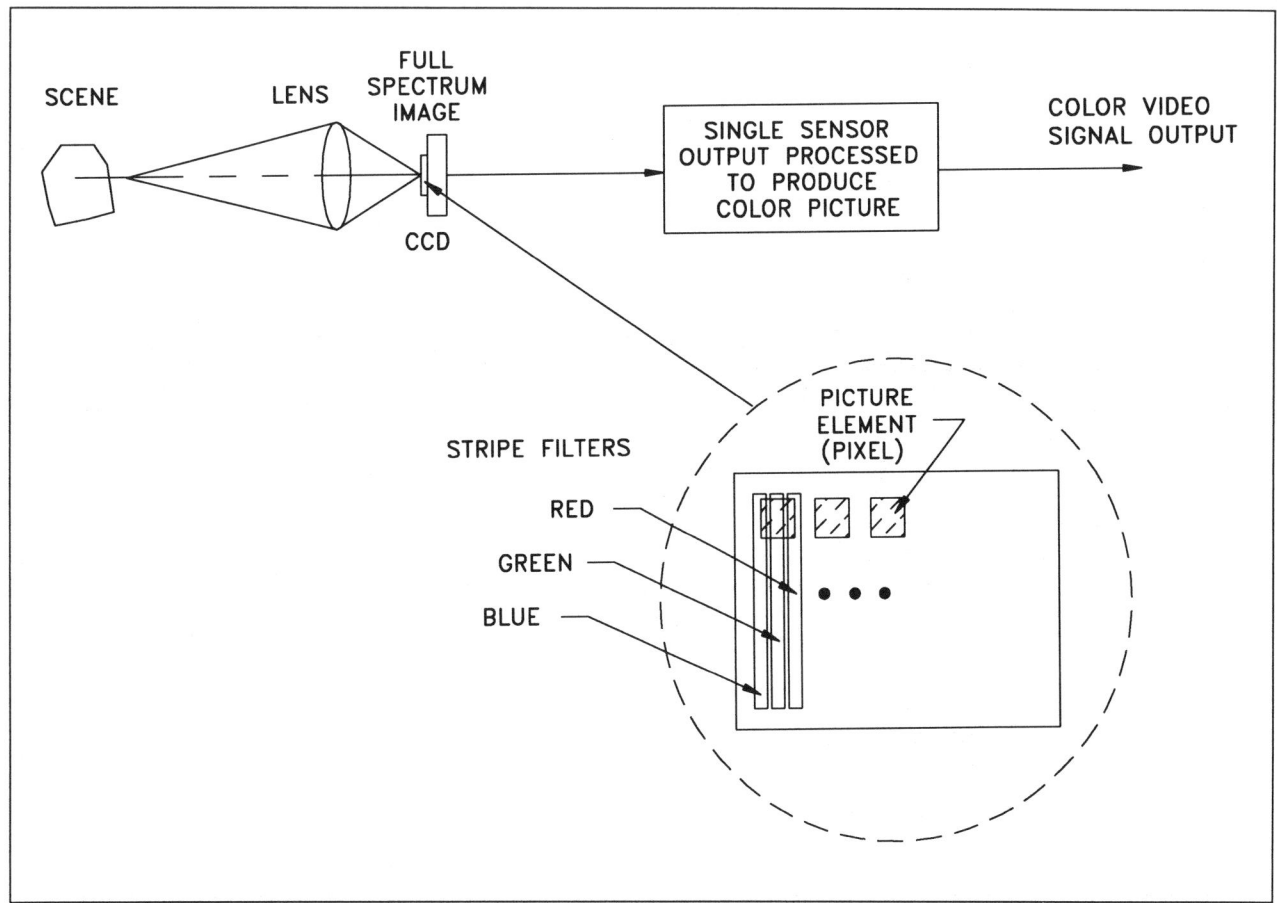

FIGURE 5-15 Single-sensor color camera

the original scene. The camera provides a composite video output signal that is an algebraic combination of these three signals in a specific ratio, modulated according to the NTSC format representing the scene intensity, that is, the monochrome picture, which is later used by the color (or monochrome) monitor displaying the picture. This prism and three-sensor technique is inherently more costly than a single-sensor camera, but it results in superior color fidelity and higher resolution and sensitivity.

One excellent color camera using three CCDs to detect the color picture uses a 2.25-inch-diameter bayonet-type lens mount to accommodate the three-sensors and prism. After the prism separates the incoming white light into the three primary colors, the output signals from each sensor are mixed electronically to produce the composite color picture. When the displayed results of this camera are compared with those of a camera that processes the signals in a single CCD sensor, there are three times the number of pixels to sense each individual color. The result is a picture that has almost three times higher resolution and sensitivity and a picture closer to the true colors in the scene. Resolution is 550 (H) by 350 TV lines (V). This camera is well suited for the higher-resolution S-VHS and Hi-8 VCRs now available for higher-resolution require-

ments (see Chapter 8). S-VHS and Hi-8 recorders use the Y-signal (luminance) and C-signal (chrominance) that appear as separate outputs on the camera. The camera is also compatible with conventional composite video signal systems. Because the camera uses three CCDs, the minimum subject illumination is only 1 fc, or the light from a single candle. Such a state-of-the-art security color camera, because it operates at low light levels and with high resolution, can be used for identifying personnel or objects in surveillance or access control applications.

5.6.3 The Color Signal

The color camera encodes more than just the light-intensity information. It must encode the spectral distribution of the scene illumination into the RGB color components.

The color video signal is far more complex than its monochrome counterpart, and the timing accuracy, linearity, and frequency response of the electronic circuits are more critical to achieve high-quality color pictures. The color video signal contains seven components necessary to extract the color and intensity information from the picture scene and later reproduce it on a color monitor:

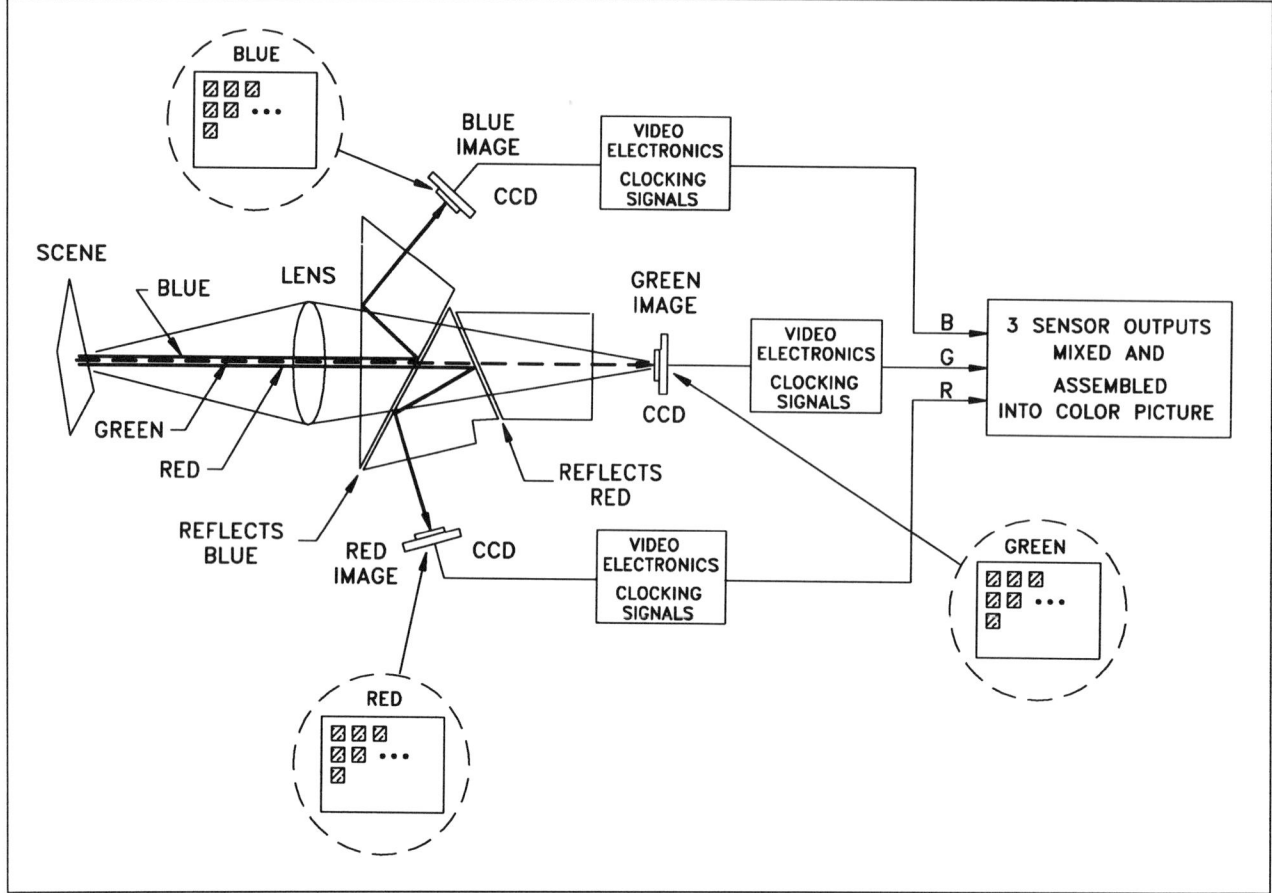

FIGURE 5-16 Three-sensor/three-prism color camera

1. horizontal line synchronizing pulses
2. color synchronization (burst signal)
3. setup (black) level
4. luminance (gray-scale) level
5. color hue (tint)
6. color saturation (vividness)
7. field synchronizing pulses

Figure 5-17 shows the television waveform with some of these components.

Horizontal line synchronization pulses. The first component of the composite video signal has three parts: (1) the front porch, which isolates the synchronization pulses from the active picture information of the previous line; (2) the back porch, which isolates the synchronization pulses from the active picture information of the next scanned line; and (3) the horizontal line sync pulse, which synchronizes the receiver to the camera.

Color synchronization (burst signal). The second component, the color synchronization (burst), is a short burst of color information used as a phase synchronization for the color information in the color portion of each horizontal line. The front porch, synchronization pulse, color burst, and back porch make up the horizontal blanking interval. This color burst signal, occurring during the back-porch interval of the video signal, serves as a color synchronization signal for the chrominance signal.

Setup. The third component of the color television waveform is the setup or black level, representing the video signal amplitude under zero light conditions.

Luminance. The fourth component is the black-and-white picture detail information. Changes and shifts of light as well as the average light level are part of this information.

Color, hue, and saturation. The fifth and sixth components are the color, hue, and color saturation information. This information is combined with the black-and-white picture detail portion of the waveform to produce a color image.

Field synchronization pulse. The field synchronization pulses maintain vertical synchronization and proper interlace.

These seven components form the composite waveform for a color video signal.

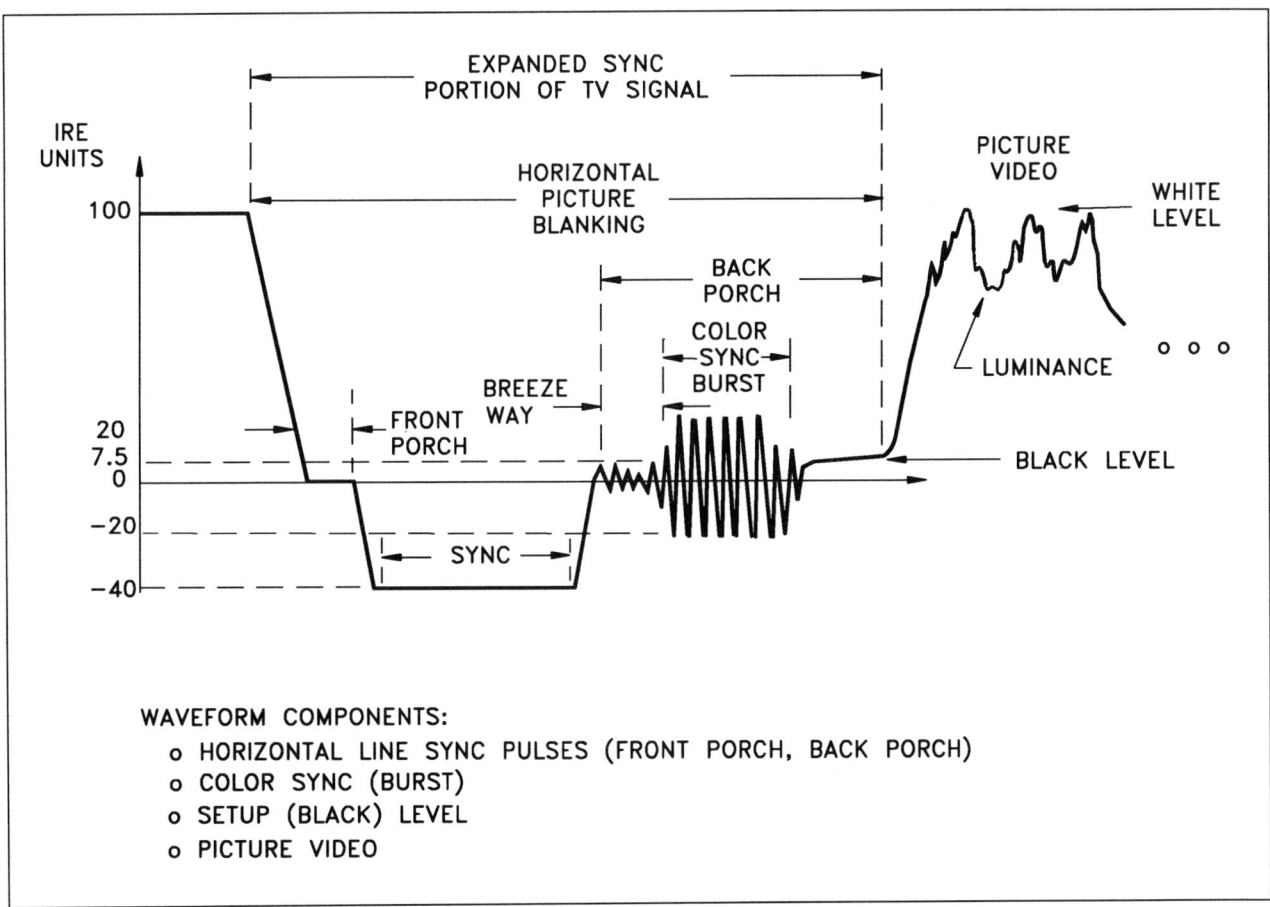

FIGURE 5-17   Color signal components

5.7 CAMERA LENS MOUNT

Several lens-to-camera mounts are standard in the CCTV industry. They are not mechanically or optically interchangeable, so the lens mount must match the camera mount. The two widely used C and CS mounts, 10 and 12 mm mini-lens mounts, and the infrequently used bayonet mounts are described in the following sections.

5.7.1 C and CS Mounts

For many years, all 1-, ⅔-, and ½-inch cameras used an industry-standard mount called the C mount to couple the lens to the camera. Figure 5-18 shows the mechanical details of the C and CS mounts.

The C-mount camera has a 1-inch-diameter hole with 32 threads per inch (TPI). The lens has a matching thread (1-32 TPI) that screws into the camera thread. The distance between the lens rear mounting surface and the image sensor for the C mount is 0.69 inches (17.526 mm).

With the introduction of smaller-format cameras and lenses, it became possible and desirable to reduce the size of the lens and the distance between the lens and the sensor. A mount adopted by the industry for ⅓- and ½-

inch-sensor-format cameras is the CS mount, which matches the C mount in diameter and thread. However, the distance between the lens rear mounting surface and the image sensor for the CS mount is 0.492 inches (12.5 mm), 0.2 inches (5 mm) shorter than for the C mount. Since the lens is 5 mm closer to the sensor, the lens can be made smaller in diameter. A C-mount lens can be used on a CS-mount camera if a 5-mm spacer is interposed between the lens and the camera and if the lens format covers the camera format size. The advantage of the CS-mount system is that the lens can be smaller, lighter, and less expensive than its C-mount counterpart. The CS-mount camera is completely compatible with the common C-mount lens when a 5-mm spacer ring is inserted between the C-mount lens and the CS-mount camera. The opposite is not true: a CS-mount lens *will not* work on a C-mount camera. Table 5-3 summarizes the present lens mount parameters.

5.7.2 Mini-Lens Mounts

The proliferation of small mini-lenses (as described in Chapter 4) and small CCD cameras (as discussed in this chapter) has led to widespread use of nonstandard lens/camera mounts. These nonstandard mounts having

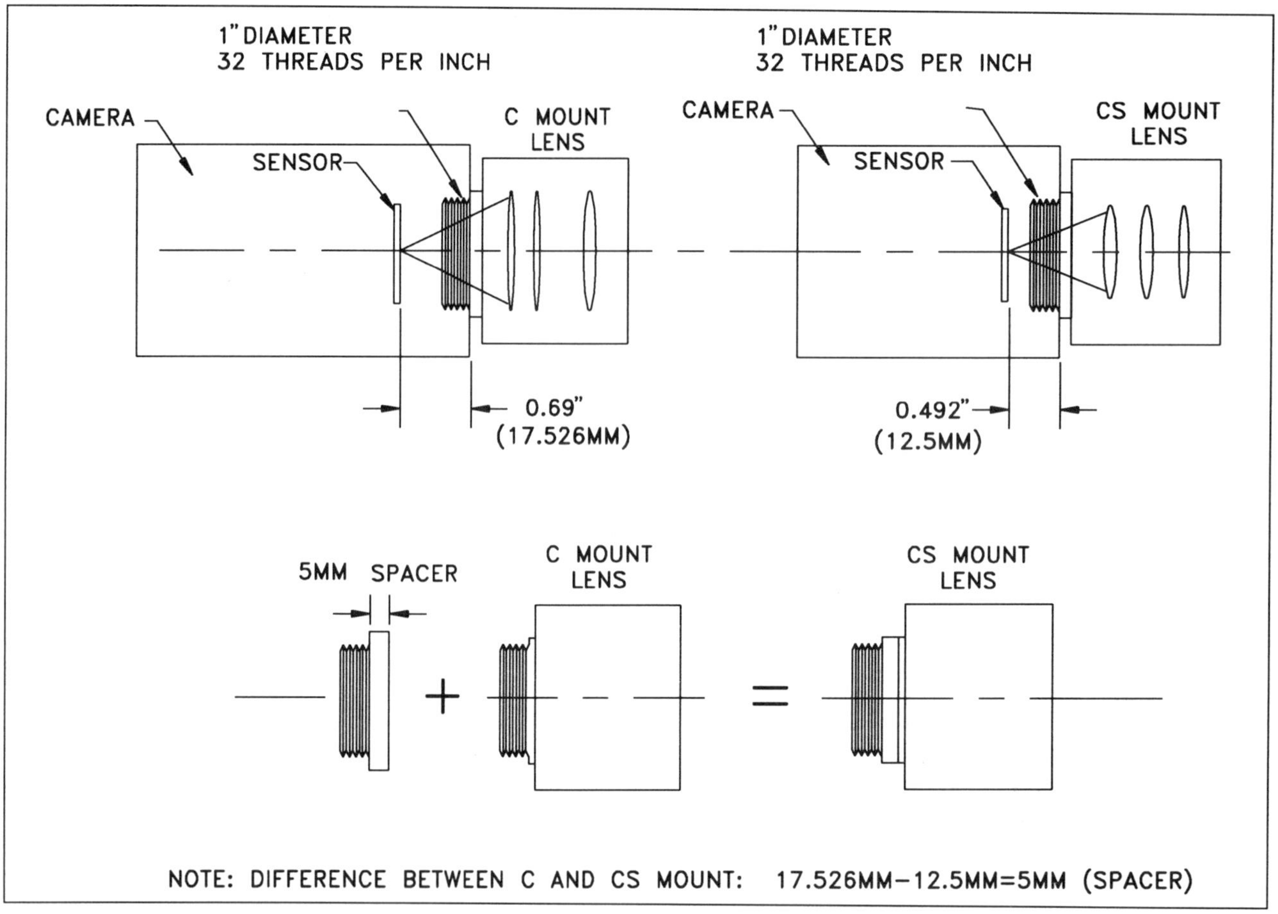

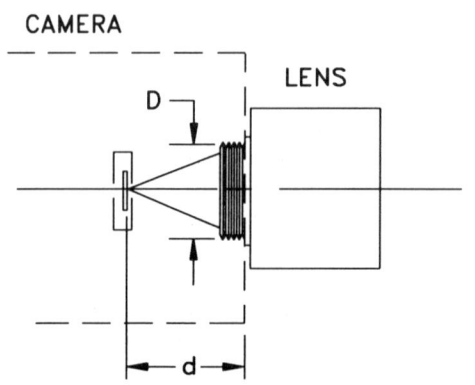

1" DIAMETER
32 THREADS PER INCH

CAMERA

SENSOR

C MOUNT
LENS

0.69"
(17.526MM)

1" DIAMETER
32 THREADS PER INCH

CAMERA

SENSOR

CS MOUNT
LENS

0.492"
(12.5MM)

5MM SPACER   C MOUNT
LENS

+

CS MOUNT
LENS

=

NOTE: DIFFERENCE BETWEEN C AND CS MOUNT: 17.526MM−12.5MM=5MM (SPACER)

FIGURE 5-18 Camera/lens mount characteristics

Table 5-3 Standard Camera/Lens Mount Parameters

MOUNT TYPE	MOUNTING SURFACE TO SENSOR DISTANCE (d)		MOUNT TYPE
	INCH	MM	DIAMETER (D)
C	.069	17.526	THREAD: 1−INCH DIA. 32 TPI
CS	0.492	12.5	THREAD: 1−INCH DIA. 32 TPI

TPI−THREADS PER INCH

CAMERA

LENS

D

d

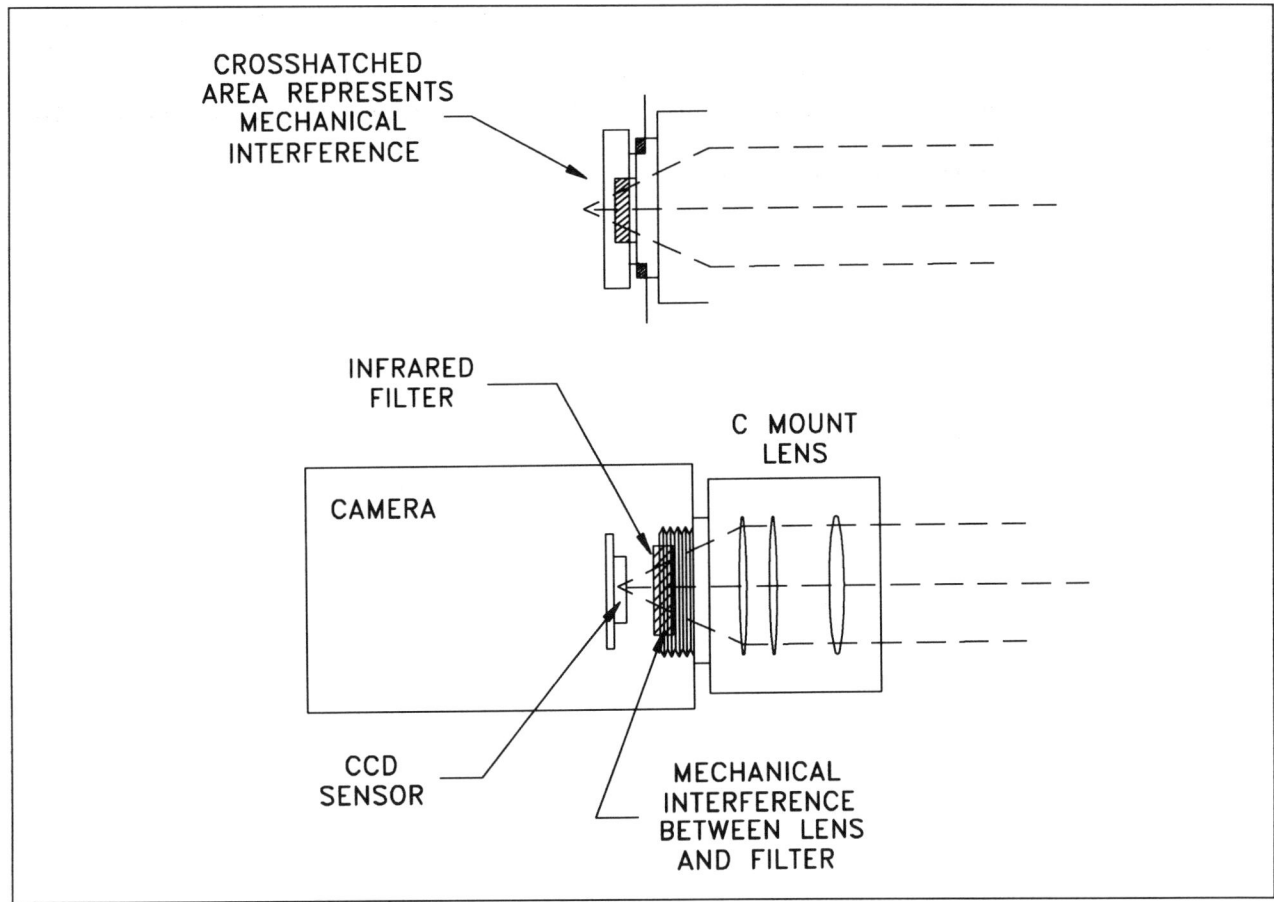

FIGURE 5-19   Lens/mount interferences

metric size threads are supplied by the camera and lens manufacturers in many diameter sizes and thread pitches. Two widely used sizes are the 10-mm and 12-mm diameter with 0.5-mm pitch. Variants of these include several with coarse pitches of 1.5 to 2.0 mm. Eventually some new standards will evolve.

5.7.3 Bayonet Mount Parameters

The large 2.25-inch-diameter bayonet mount is used primarily in the industrial, broadcast, and military fields with three-sensor color cameras. It is only in limited use in the security field.

5.7.4 Lens/Mount Interferences

Figure 5-19 illustrates a potential problem with some lenses when used with CCD or solid-state cameras.

Some of the shorter-FL lenses (such as 2.6, 3.5, and 4.8 mm) have a snout that protrudes behind the C or CS mount and can interfere with the filter or window used with the solid-state sensor. This mechanical interference prevents the lens from fully seating in the mount, thereby causing the image to be out of focus. Most lens and camera manufacturers are aware of the problem and for the most part have designed lenses and cameras that are compatible. However, since the potential problem exists, the security designer should be aware of it.

5.8 SUMMARY

The single most significant advance in CCTV technology was the development of the solid-state camera image sensor. These sensors offer such a compelling advantage over vacuum-tube image sensors because of solid-state reliability, inherent long life, low cost when manufactured in quantity, low-voltage operation, low power dissipation, geometric reproducibility, no image lag, and visible and near-IR response. Chip cameras have provided the reliability and stability needed in monochrome and color security systems.

The availability of solid-state color cameras has made a significant impact on the security CCTV industry. Color cameras provide enhanced television surveillance because of their increased ability to display objects and persons. The lighting available in most security applications is sufficient for most color cameras to have satisfactory sensitivity and resolution. CCD color cameras have excellent color rendition, maintain color balance, and need no color rebalancing when light level or lighting color temperature varies.

ICCD cameras are solid-state cameras coupled to tube or microchannel plate intensifiers. They are replacing SIT tube intensifiers for LLL applications (Chapter 15). The smaller size and lower power consumption make ICCDs a superior choice over SIT cameras.

Chapter 6
Video Signal Transmission

CONTENTS

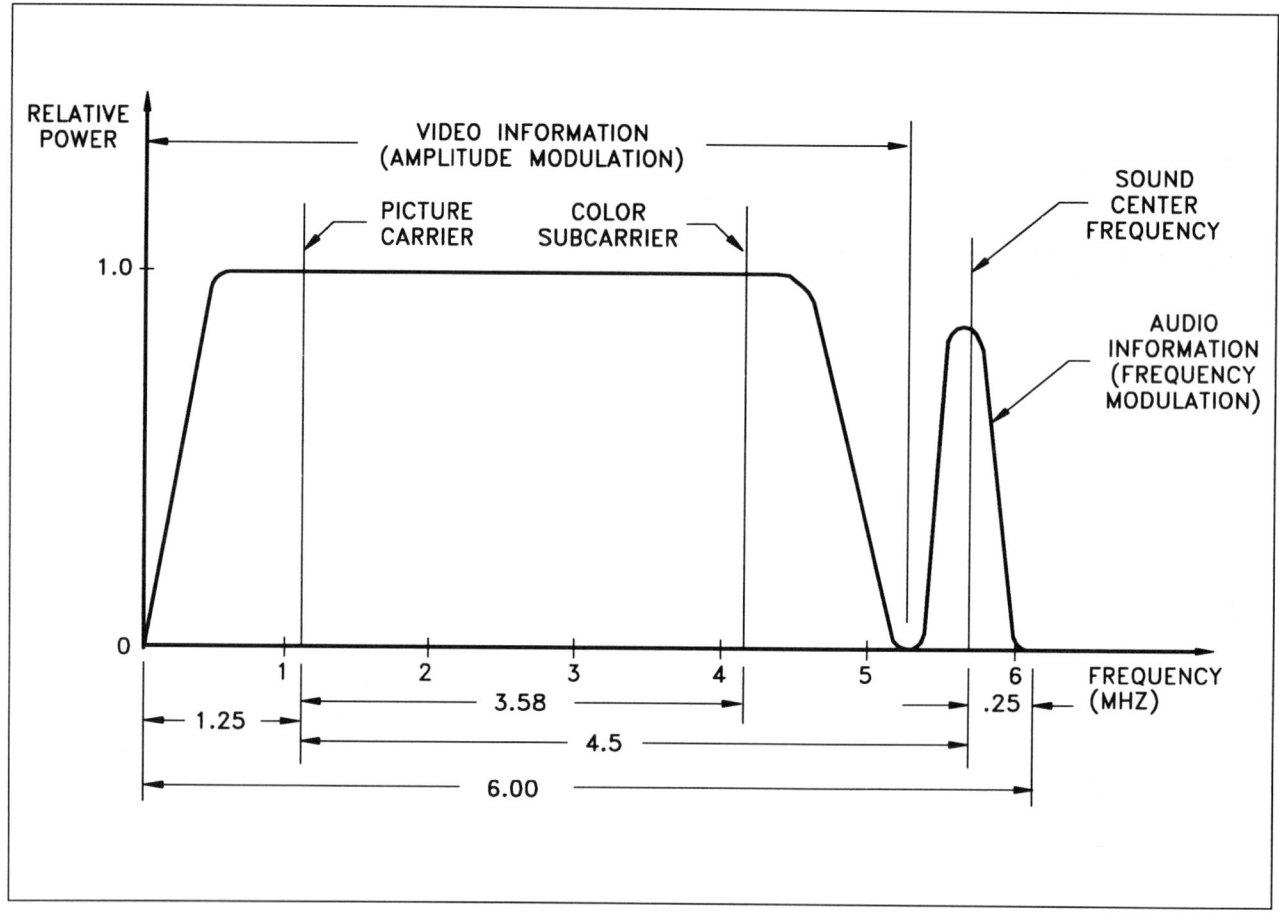

FIGURE 6-1 Single-channel CCTV bandwidth requirements

6.1 OVERVIEW

In its most common form, the CCTV video signal is transmitted at baseband frequencies over coaxial cable. This chapter identifies the problems associated with transmitting the CCTV signal from the camera site to the remote monitoring location; the chapter also discusses copper-wire, fiber-optic, and through-the-air wireless transmission techniques.

Electrical-wire techniques include coaxial cable and two-wire, which use direct modulation in real time; and slow-scan, which operates in non–real time. Coaxial cable is specially constructed for transmitting all video frequencies with minimum distortion or attenuation. Two-wire systems using standard conductors (intercom wire, etc.) use special transmitters and receivers that preferentially boost (amplify) the high video frequencies to compensate for their loss over the wire length. As the name implies, slow-scan reduces the speed of video picture transmission a rate at which individual snapshots (frames) of video are transmitted.

Correct video signal transmission is one of the most important aspects of a CCTV system. Each CCTV channel is allocated approximately 6 MHz in bandwidth; however, for security and safety applications, monochrome picture transmission needs only 4.2 MHz in bandwidth. Figure 6-1 shows the single-channel television bandwidth requirements for monochrome and color systems.

Based on information from other chapters, it is not difficult to specify a good lens, camera, monitor, and VCR to produce a high-quality picture. However, if transmission means does not deliver an adequate signal from the camera to the monitor, VCR, or printer, an unsatisfactory picture will result. The final picture is only as good as the weakest link in the system—usually the transmission means. Achieving good signal transmission requires the system specifier and installer to know the correct transmission type, use high-quality materials, and practice professional installation techniques. The installing dealer often ends up choosing the transmission means, even though he or she may lack expertise in all the necessary disciplines and lack knowledge of the equipment required for an optimum system. The result: a weakest-link transmission means that degrades the specifications for camera, lens, monitoring, and recording equipment.

Fiber-optic technology has been the most important advance in CCTV signal transmission in many decades. The use of fiber-optic cable has significantly improved the picture quality of the transmitted video signal and provided a more secure, reliable, and cost-effective transmission link. Not only is the technology available, but so is the installation capability of most security installers. The advantages of fiber optics over electrical coaxial-cable or two-wire systems include

1. high bandwidth (resulting in higher resolution or simultaneous transmission of multiple television signals),
2. no electrical interference to or from other electrical equipment or sources,
3. strong resistance to tapping (providing a secure link), and
4. no environmental degradation (unharmed by corrosion, moisture, and other conditions).

Fiber optics offers a new technology to transmit high-bandwidth, high-quality, multiplexed television pictures, audio, and control signals over a single fiber.

Wireless transmission techniques use radio frequencies (RF) in the very high frequency (VHF) and ultra high frequency (UHF) bands, as well as microwave frequencies in the S and X bands. Low-power microwave and RF systems can transmit up to several miles with excellent picture quality, but most require an FCC license for operation. One microwave system has a built-in intrusion-detection function: if a person walks between the microwave transmitter and receiver, an alarm is registered. Cableless video transmission using infrared (IR) atmospheric propagation is also discussed. IR laser transmission requires no FCC approval but has a shorter range—generally several thousand feet or less in good visibility, and a few hundred feet in poor visibility.

In bidirectional communications, control signals are sent in the opposite direction to the television signal picture information, and audio is sent in both directions. Wireless systems permit independent placement of the CCTV camera in locations that might be inaccessible for coaxial or other cables.

An elegant and powerful technique for transmitting CCTV pictures anywhere on earth is slow-scan. In slow-scan, a real-time video signal is electronically converted to a non–real-time signal and transmitted over any audio communications channel (3000-Hz bandwidth). Unlike conventional video transmission, in which a real-time signal changes every 1/30th of a second, slow-scan transmission sends a single snapshot of a scene over a period of 1 to 72 seconds (that is, in non–real time), depending on the resolution specified. The immediate effect of this snapshot or sampling technique is the loss of real-time motion in the scene. This effect is similar to that of opening your eyes once every second or once every minute (or somewhere in between). However, such slow-motion pictures are adequate for many applications. The slow-scan system is often used in conjunction with an alarm system, such as a video motion detector. After the system alarm has been activated, the slow-scan equipment begins sending pictures, once every few seconds at low resolution—perhaps 200 TV lines—or every 32 seconds at high resolution—perhaps 500 TV lines.

A basic understanding of the capabilities of the aforementioned techniques, as well as the advantages and disadvantages of different transmission means, is essential to optimize the final CCTV picture and avoid costly retrofits. Understanding the transmission requirements when choosing the transmission means and hardware is important because it constitutes the most labor-intensive portion of the CCTV installation. Specifying, installing, and testing the television signal and communication cables for intrabuilding and interbuilding wiring represents the major labor cost in the television installation. If the incorrect cable is specified and must be removed and replaced with another type, serious cost increases result. In the worst situation, where cables are routed in underground outdoor conduits, it is imperative to use the correct size and type of cable so as to avoid retrenching or replacing cables.

6.2 COAXIAL CABLE

Coaxial cable is used widely for short to medium distances (several hundred to several thousand feet) because its electrical characteristics best match those required to transmit the full-signal bandwidth from the camera to the monitor. The CCTV signal is composed of slowly varying (low-frequency) and rapidly varying (high-frequency) components. Most wires of any type can transmit the low frequencies (20 Hz to a few thousand Hz); for example, practically any wire can carry a telephone conversation. It takes the special coaxial-cable configuration to transmit the full spectrum of frequencies from 20 Hz to 6 MHz without attenuation, as required for high-quality CCTV pictures.

There are basically four types of coaxial cable for use in video transmission systems:

1. 75-ohm unbalanced indoor cable,
2. 75-ohm unbalanced outdoor cable,
3. 124-ohm twin-axial balanced indoor cable, and
4. 124-ohm twin-axial balanced outdoor cable.

The cable construction for the coaxial and twin-axial types is shown in Figure 6-2.

The choice of a particular coaxial cable depends on the environment in which it will be used and the electrical characteristics required. By far the most common coaxial cables are the RG59/U and the RG11/U, having a 75-ohm

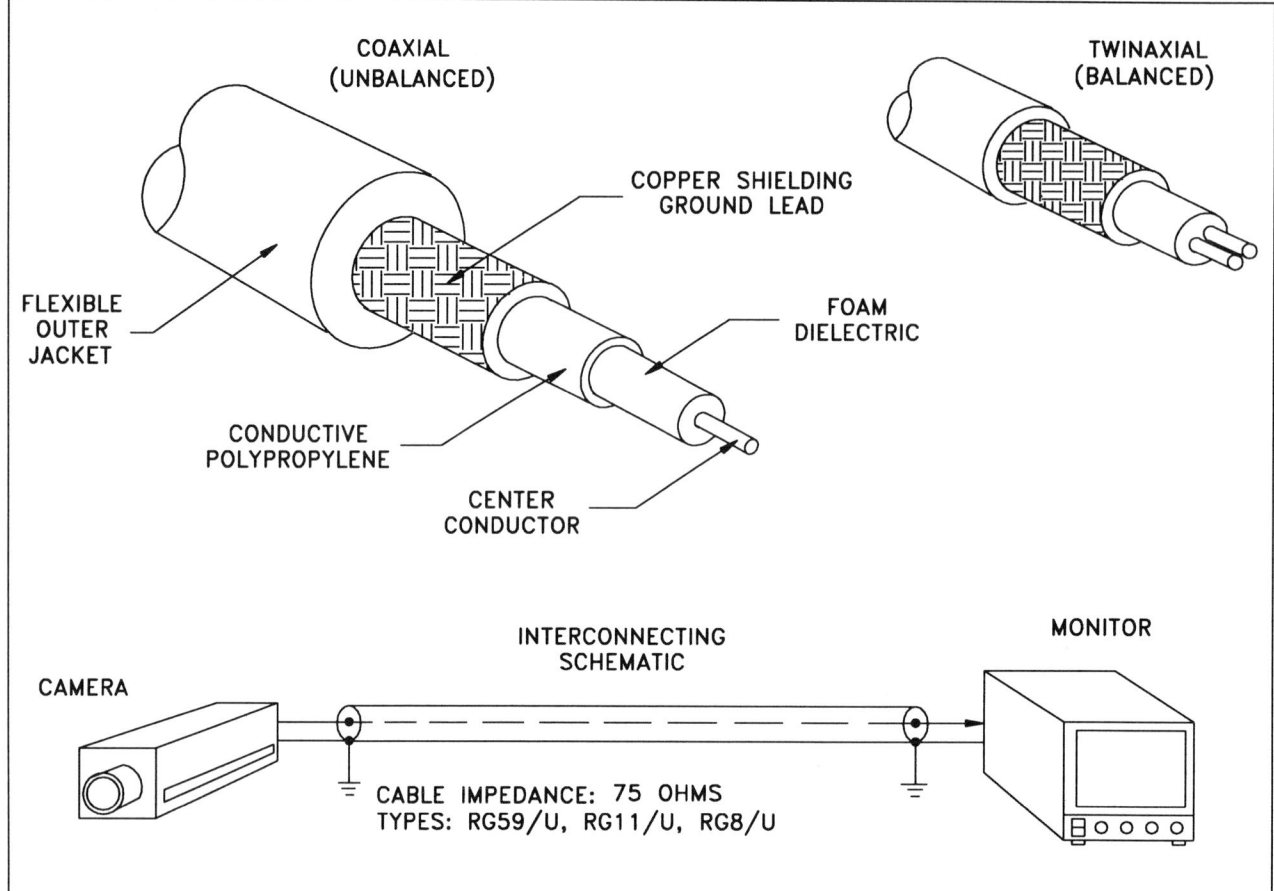

FIGURE 6-2 Coaxial-cable construction

impedance. For short camera-to-monitor distances (a few hundred feet), preassembled or field-terminated lengths of RG59/U coaxial cable with BNC connectors at each end are used. The BNC connector is a rugged video and RF connector in common use for many decades and the connector of choice for all baseband video connections. Short preassembled lengths of 5, 10, 25, 50, and 100 feet, with BNC-type connectors attached, are available. Long cable runs (several hundred feet and longer) are assembled in the field, made up of a single length of coaxial cable with a connector at each end. For most interior CCTV installations, RG59/U (0.25-inch diameter) or RG11/U (0.5-inch diameter) 75-ohm unbalanced coaxial cable is used. When using the larger diameter RG11/U cable, a larger UHF-type connector is needed. When long cable runs of several thousand feet or more are required, particularly between several buildings or if electrical interference is present, the balanced 124-ohm coaxial cable or fiber-optic cable should be considered. When the camera and monitoring equipments are in two different buildings, and likely at different ground potentials, an unwanted signal may be impressed on the video signal, which shows up as an interference (wide bars on the video screen) and makes the picture

unacceptable. A two-wire balanced or fiber-optic cable eliminates this condition.

Television camera manufacturers generally specify the maximum distance between camera and monitor over which their equipment will operate when interconnected with a specific type of cable. Table 6-1 summarizes the transmission properties of coaxial and twin-axial cables when used to transmit the CCTV signal.

In applications with cameras and monitors separated by several thousand feet, video amplifiers are required. Located at the camera output and/or somewhere along the coaxial-cable run, they permit increasing the camera-to-monitor distance to 3400 feet for RG59/U cable and to 6500 feet for RG11/U cable.

The increased use of color television in security applications requires the accurate transmission of the CCTV signal with minimum distortion by the transmitting cable. High-quality coaxial-cable and fiber-optic (see Section 6.4) installations satisfy these requirements.

While a coaxial cable is the most suitable hard-wire cable to transmit the video signal, video information transmitted through coaxial cable over long distances is attenuated differently depending on its signal frequencies. Figure 6-3

COAXIAL TYPE	MAXIMUM RECOMMENDED CABLE LENGTH (D)				CONDUCTOR (GAUGE)	NOMINAL DC RESISTANCE (OHMS/1000FT)
	CABLE ONLY		CABLE WITH AMPLIFIER			
	FEET	METERS	FEET	METERS		
RG59/U	750	230	3,400	1,035	22 SOLID COPPER	10.5
RG59 MINI	200	61	800	250	20 SOLID COPPER	41.0
RG6/U	1,500	455	4,800	1,465	18 SOLID COPPER	6.5
RG11/U	1,800	550	6,500	1,980	14 SOLID COPPER	1.24

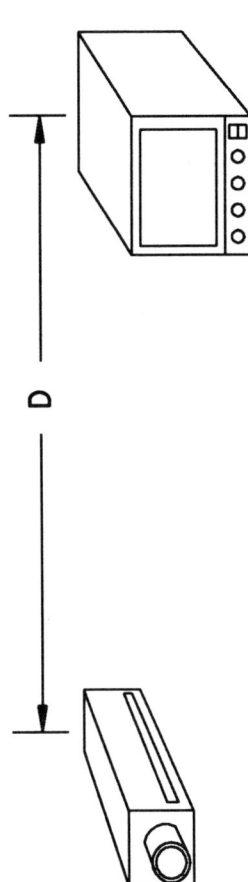

NOTES: 1. IMPEDANCE FOR ALL CABLES = 75 OHMS

Table 6-1 Coaxial-Cable Run Capabilities

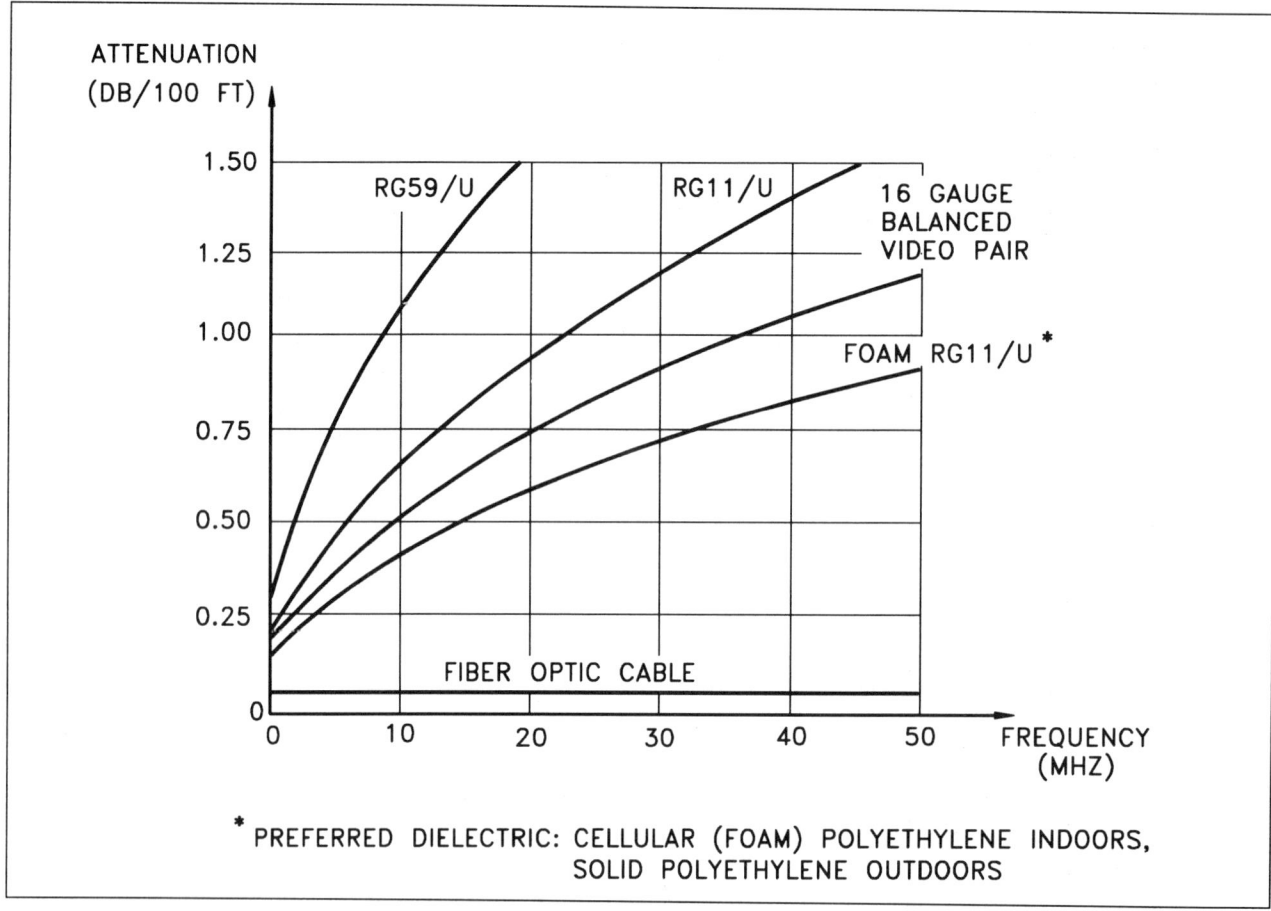

FIGURE 6-3 Coaxial-cable signal attenuation versus frequency

illustrates the attenuation as a function of distance and frequency as exhibited by standard coaxial cables.

The attenuation of a 10-MHz signal is approximately three times greater than that of a 1-MHz signal when using a high-quality RG11/U cable. In CCTV transmission, a 3000-foot cable run would attenuate the 5-MHz portion of the video signal (representing the high-resolution part of the picture) to approximately one-fourth of its original level at the camera; a 1-MHz signal would be attenuated to only half of its original level. At frequencies below 500 kHz, the attenuation is generally negligible for these distances. This variation in attenuation as a function of frequency has an adverse effect on picture resolution and color quality. The signal deterioration appears on monitors in the form of less definition and contrast and poor color rendition. For example, severe high-frequency attenuation of a signal depicting a white picket fence against a dark background would cause the pickets to merge into a solid, smearing mass, resulting in less intelligence in the picture.

The most commonly used standard coaxial is RG59/U, which also has the highest signal attenuation. For a 6-MHz bandwidth, the attenuation is approximately 1 dB per 100 feet, representing a signal loss of 11%. A 1000-foot run would have a 10-db loss—that is, only 31.6% of the video signal would reach the monitor end.

In a process called vidiplexing, special CCTV cameras transmit both the camera power and the video signal over a single coaxial cable (RG59/U or RG11/U). This single-cable camera reduces installation costs, eliminates power wiring, and is ideal for hard-to-reach locations, temporary installations, or camera sites where power is unavailable.

6.2.1 Unbalanced Single-Conductor Cable

The most widely used coaxial cable for CCTV security transmission and distribution systems is the unbalanced coaxial cable, represented by the RG59/U or RG11/U configurations. This cable has a single conductor with a characteristic impedance of 75 ohms, and the video signal is applied between the center conductor and a coaxial braided or foil shield (Figure 6-2).

Single-conductor coaxial cables are manufactured with different impedances, but video transmission uses only the 75-ohm impedance, as specified in EIA standards. Other cables that may look like the 75-ohm cable have a different electrical impedance and will not produce an acceptable television picture when used at distances of 25 or 50 feet or more.

The RG59/U and RG11/U cables are available from many manufacturers in a variety of configurations. The primary difference in construction is the amount and type of shielding and the insulator (dielectric) used to isolate the center conductor from the outer shield. The most common shields are standard single copper braid, double braid, or aluminum foil. Aluminium-foil type should not be used for any CCTV application. Common dielectrics are foam, solid plastic, and air, the latter having a spiral insulator to keep the center conductor from touching the outer braid. The cable is called unbalanced because the signal current path travels in the forward direction from the camera to the monitor on the center conductor and from the monitor back to the camera again on the shield, which produces a voltage difference (potential) across the outer shield. This current (and voltage) has the effect of unbalancing the electrical circuit.

For short cable runs (a few hundred feet), many of the deleterious effects of the coaxial cable—such as signal attenuation, hum bars on the picture, deterioration of image resolution and contrast—are not observed. However, as the distance between the camera and monitor increases to 1000–3000 feet, all these effects come into play. In particular, high-frequency attenuation sometimes requires equalizing equipment in order to restore resolution and contrast.

Video coaxial cables are designed to transmit maximum signal power from the camera output impedance (75 ohms) to the receiver monitor or recorder input impedance (75 ohms) with a minimum signal loss. If the cable characteristic impedance is not 75 ohms, excessive signal loss and signal reflection from the receiving end will occur and cause a deteriorated picture.

The cable impedance is determined by the conductor and shield resistance of the core dielectric material, shield construction, conductor diameter, and distance between the conductor and the shield. As a guide, the center conductor resistance for an RG59/U should be approximately 15 ohms per 1000 feet, and for an RG11/U cable, approximately 2.6 ohms per 1000 feet. Table 6-2 summarizes some of the characteristics of the RG59/U and RG11/U coaxial cables.

6.2.1.1 Connectors

Coaxial cables are terminated with several types of connectors: the PL-259, used with the RG11/U cable, and the BNC, used with the RG59/U cable. Older systems use the PL-259 connector on the RG59/U cable, but this practice has been discontinued in favor of the smaller BNC connector. The F type is an RF connector used in cable TV systems. Figure 6-4 illustrates these connectors.

The BNC has become the connector of choice in the CCTV industry because it provides a reliable connection with less signal loss, has a fast and positive twist-on action, and has a small size, so that many connectors can be installed on a chassis when required. There are essentially three types of BNC connectors available: (1) solder, (2) crimp-on, and (3) screw-on.

The most durable and reliable connector is the solder type, but it is also the most difficult to assemble. The solder type is used when the connector is installed at the point of manufacture or in a suitably equipped electrical shop. It is not usually installed or repaired in the field.

The crimp-on and screw-on types are the most commonly used in the field, during installation and repair of a system. Either type can be successfully assembled with few tools in most locations. The crimp-on type uses a sleeve, which is attached to the cable end after the braid and insulation have been properly cut back; it is crimped onto the outer braid and the center conductor with a special crimping plier. When properly installed, this cable termination provides a reliable connection.

To assemble the screw-on type, the braid and insulation are cut back and the connector slid over the end of the cable and then screwed on. This too is a fairly reliable type of connection, but it is not as durable as the crimp-on type, since it can be inadvertently unscrewed from the end of the cable. The screw-on termination is less reliable if the cable must be taken on or off many times. The screw-on connector does provide a means for easily removing the connector and installing a new one if necessary. Both the crimp-on and screw-on types are popular field techniques for installing BNC connectors onto RG59/U coaxial cables.

6.2.1.2 Amplifiers

When the distance between the camera and monitor exceeds the recommended length for the RG59/U and RG11/U cables, it is necessary to insert a video amplifier to boost the video signal level. The video amplifier is inserted at the camera location or somewhere along the coaxial cable run between the camera and monitor location (Figure 6-5).

The disadvantage of locating the video amplifier somewhere along the coaxial cable is that since the amplifier requires a source of AC (or DC) power, the power source must be available at its location. Table 6-1 compares the cable-length runs with and without a video amplifier. Note that the distance transmitted can be increased more than fourfold with one of these amplifiers.

When the output from the camera must be distributed to various monitors or separate buildings and locations, a distribution amplifier is used (Figure 6-5). This amplifier transmits and distributes monochrome and color video signals to multiple locations. In a quad unit, a single video input to the amplifier results in four identical, isolated video outputs capable of driving four 75-ohm RG59/U or RG11/U cables.

The distribution amplifier is in effect a power amplifier, boosting the power from the single camera output so that multiple 75-ohm loads can be driven.

A potential problem with an unbalanced coaxial cable is that the video signal is applied across the single inner conductor and the outer shield, thereby impressing a small

ATTENUATION (DB) @ 5–10 MHZ

CABLE TYPE	100 FT	200 FT	300 FT	400 FT	500 FT	1,000 FT	1,500 FT	2,000 FT
RG59/U	1.0/1.8[+]	2.0	3.0	4.0	5.0	10.0	15.0	20.0
RG59 MINI	1.3	2.6	3.9	5.2	6.5	13.0	19.5	26.0
RG6/U	.8/1.43	1.6	2.4	3.2	4.0	8.0	12.0	16.0
RG11/U	.51/.91	1.02	1.53	2.04	2.55	5.1	7.66	10.2
8281[*]	.6/2.0	1.2	1.8	2.4	3.0	6.0	9.0	12.0
2422/UL1384	3.96	7.9	11.9	18.8	19.8	39.6	59.4	79.2
2546[**]	1.82	3.6	5.5	7.3	9.1	18.2	27.3	36.4
2947[**][++]	1.85	3.7	5.6	7.4	9.3	18.6	27.6	37.2
RG179B/U	2.0/6.5	4.0	6.0	8.0	10.0	20.0	30.0	40.0

[*] MOGAMI INC
[**] SAXTON INC
[+] @ 50 MHZ
[++] DUAL COAX

NOTE: IMPEDANCE FOR ALL CABLES = 75 OHMS

DB LOSS	1	2	3	4.5	6	8	10.5	14	20
% SIGNAL REMAINING	90	80	70	60	50	40	30	20	10

Table 6-2 Coaxial-Cable Attenuation vs. Length

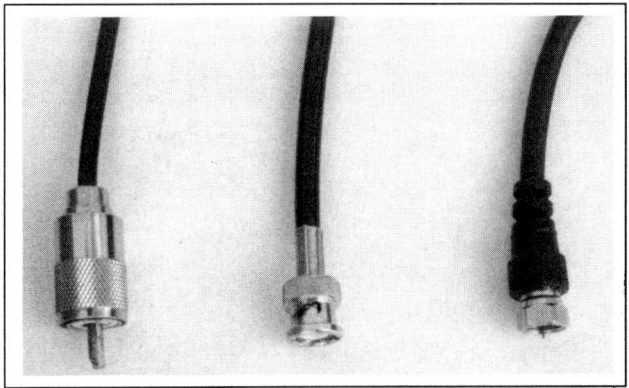

FIGURE 6-4   BNC, UHF, and F coaxial-cable connectors

voltage (hum voltage) on the signal. This hum voltage can be eliminated in many cases by using an isolation amplifier, a balanced coaxial cable, or fiber optics.

6.2.2 Balanced Two-Conductor Cable

Less familiar to the CCTV industry than the unbalanced cables just described, balanced coaxial cables consist of a pair of inner conductors, each insulated from the other and

having identical diameters. The two conductors are surrounded by additional insulation, a coaxial-type shield, and an outer insulating protective sheath. Balanced coaxial cables have been used for many years by telephone companies and others for transmitting video information and other high-frequency data and offer many advantages for long cable runs. The balanced cable has a characteristic impedance of 124 ohms and is often called a video pair. Because of the two center conductors and additional insulation, the outside diameter is larger (typically 0.5 inch) than that of the unbalanced RG59/U cable. Likewise, its cost, weight, and volume are higher than those of the unbalanced cable. Since the polarity on balanced cables must be maintained, the connector types are usually polarized (keyed). If nonpolarized connectors are used, a male-female type is used to insure proper connection. Figure 6-6 shows the construction and configuration of a balanced-coaxial-cable system.

The primary purposes for using balanced cable are to increase transmission range and to eliminate the picture degradation found in some unbalanced applications. Unwanted hum bars (dark bars on the television picture) are introduced in unbalanced coaxial transmission systems when there is a difference in voltage between the two ends of the coaxial cable (see Section 6.2.5). This can often occur

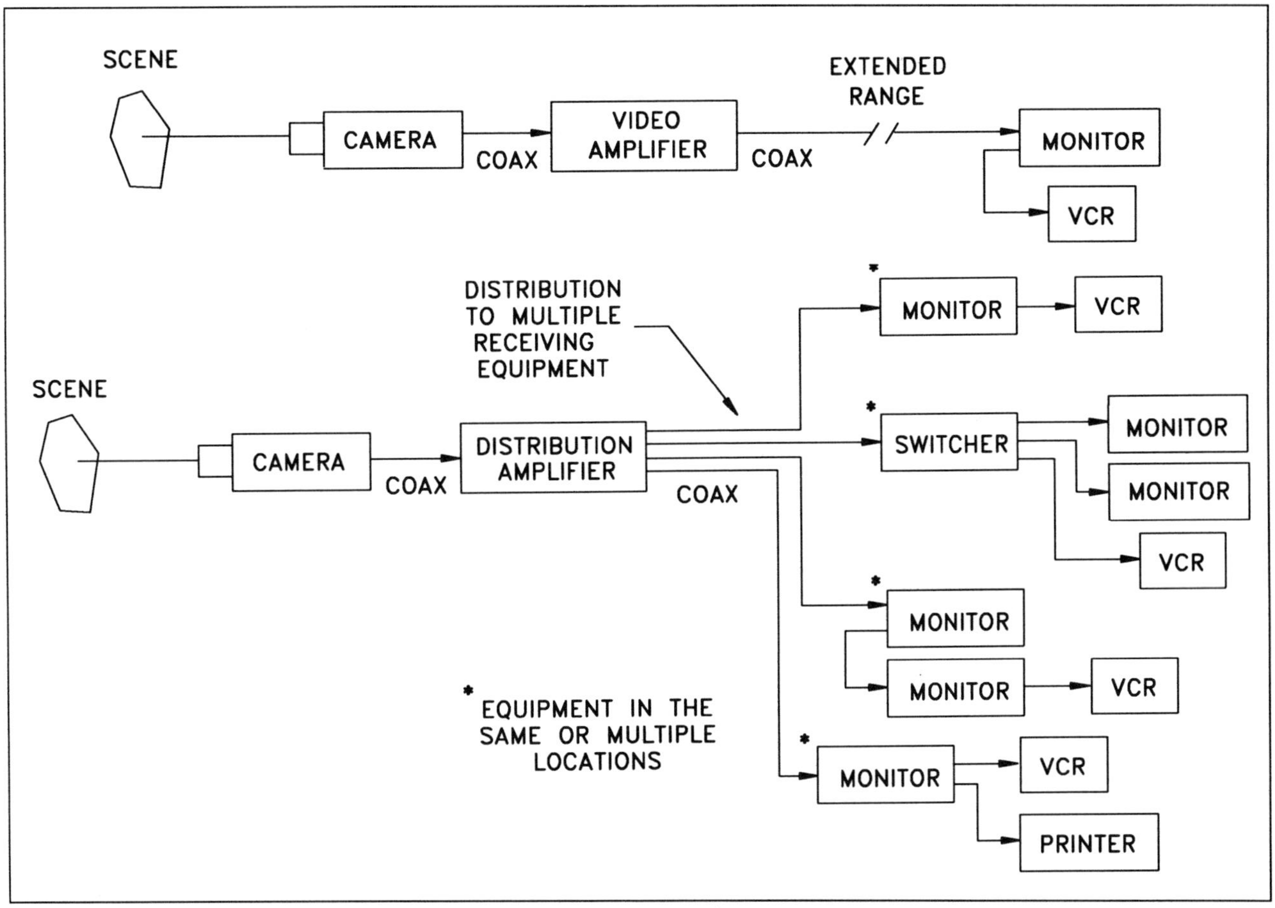

FIGURE 6-5   Video signal amplifiers to extend range and/or distribution

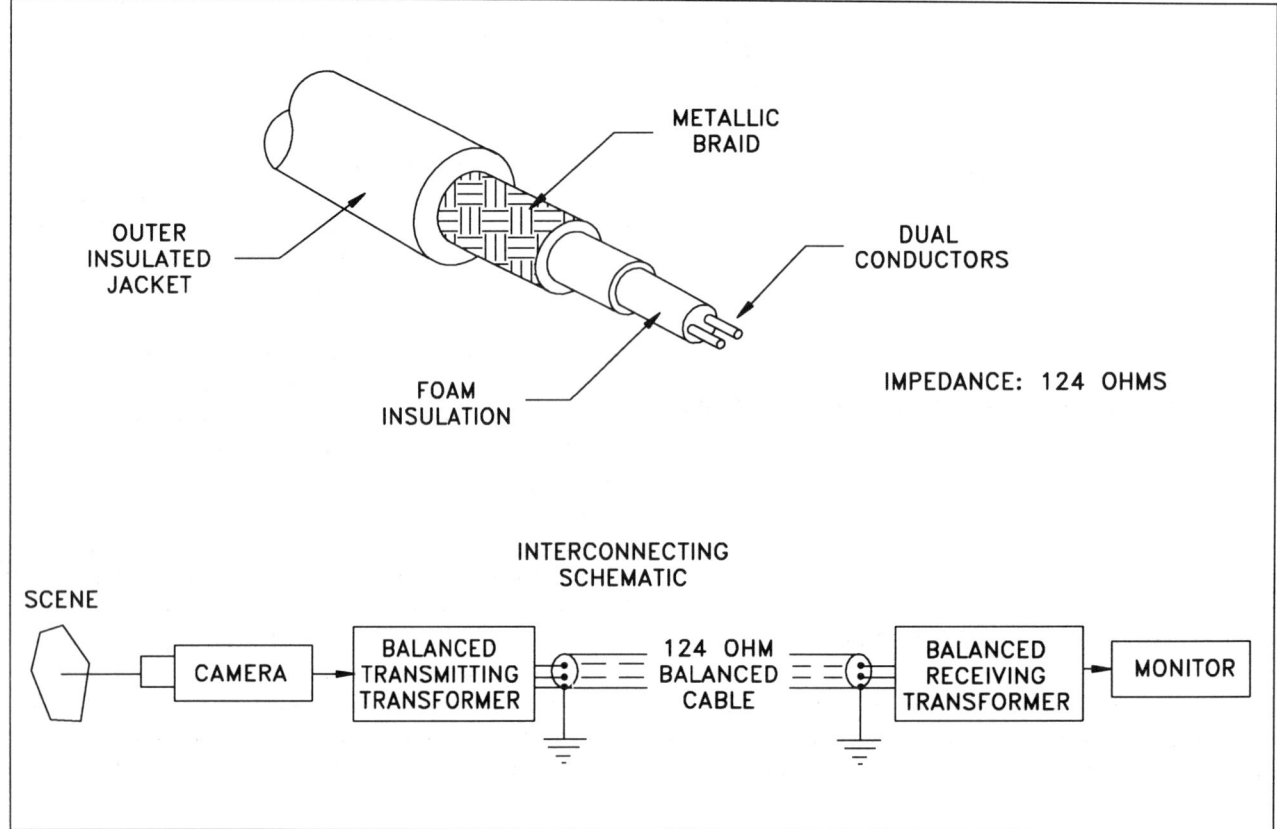

FIGURE 6-6   Balanced-coaxial-cable construction and interconnection

when two ends of a long cable run are terminated in different buildings, or when electrical power is derived from different power sources—in different buildings or even within the same building.

Since the signal path and the hum current path through the shield of an unbalanced cable are common and result in the hum problem, a logical solution is to provide a separate path for each. This is accomplished by applying the signal between each center conductor of two parallel unbalanced cables (Figure 6-6). The shields of the two cables carry the ground currents while the two conductors carry the transmitted signal. This technique has been used for many years in the communications industry to reduce or eliminate hum. Since the transmitted video signal travels on the inner conductors, any noise or induced AC hum is applied equally to each conductor. At the termination of the run (for example, the monitor), the disturbances are cancelled while the signal is directed to the load unattenuated. This technique in effect removes the unwanted hum and noise signals.

While the balanced transmission line offers many advantages over the unbalanced line, it has not been in widespread use in the CCTV security industry. The primary reason is the need for transformers at the camera-sending and monitor-receiving ends, as well as the need for two-conductor coaxial cable. All three hardware items require additional cost as compared with the unbalanced single-

conductor coaxial cable. The use of fiber optics, described in Section 6.4, solves all these problems.

6.2.3 Indoor Cable

Indoor coaxial cable is smaller in diameter than outdoor cable, uses a braided shield, and is much more flexible than twin-axial. This allows it to be formed around corners, hidden under moldings, and so on. However, since the outside diameter of the indoor cable is smaller, to maintain the correct electrical impedance the inner conductor must be proportionally smaller in diameter. This decrease in the cable conductor diameter causes a corresponding increase in the cable signal attenuation, especially at high frequencies. Therefore, the RG59/U indoor cable cannot be used for long lengths. The inner conductor of the RG59/U is smaller than that in the RG11/U cable and is more susceptible to damage. Since the impedance of any coaxial cable is directly related to the spacing between the inner conductor and the shield, any change in this distance—caused by tight bends, kinking, indentations, or other factors—will change the cable impedance, possibly resulting in picture degradation. Since indoor cabling and connectors need no protection from water, solder, crimp-on, or screw-on connectors can be used.

6.2.4 Outdoor Cable

Television signal transmission in outdoor applications places additional requirements on the coaxial cable. Environmental factors such as precipitation, temperature changes, humidity, and corrosion are present for both aboveground and buried installations. Other aboveground considerations include wind loading, rodent damage, and electrical storm interference. For direct burial applications, ground shifts, water damage, and rodent damage are potential problems. Most outdoor coaxial cabling is ½ inch in diameter or larger, since the insulation qualities in outside protective sheathing must be superior to those of indoor cables. Because outdoor cables are larger in diameter, their electrical qualities are considerably improved over small-diameter indoor RG59/U cables. Outdoor cables have large inner-conductor diameters (approximately 16 gauge), resulting in much less signal loss than the smaller center conductor used in indoor RG59/U cables (approximately 18 gauge). Outdoor cables are designed and constructed to take much more physical abuse than the indoor RG59/U cable. However, most outdoor cables are not very flexible; care must be taken with extremely sharp bends. As a rule of thumb, outdoor cabling should always be used for cable runs of more than 1000 feet, regardless of the environment.

Outdoor video cable may be buried, run along the ground, or suspended on utility poles. The exact method should be determined by the length of the cable run, the economics of the installation, and the particular environment. Environment is an important consideration. The attenuation characteristics of coaxial cable vary as its ambient temperature changes, due to the physical expansion and contraction of the diameter and length of the cable. At a frequency of 4.5 MHz, every 1-degree Fahrenheit change in the temperature of the cable causes a 0.1% change in its attenuation, with the impedance increasing as the temperature rises and decreasing as it falls.

In locations with severe weather, such as electrical storms or high winds, it is prudent to locate the coaxial cable underground, either direct-buried or enclosed in a conduit. This method isolates the cable from the severe environment, improving the life of the cable and reducing signal loss caused by the environment. In locations having rodent or ground-shift problems, enclosing the cable in a separate conduit will protect it. For short cable runs between buildings (less than 600 to 700 feet) and where the conduit is waterproof, indoor RG59/U cable is suitable.

While there are about 25 different types of RG59/U and about 10 different types of RG11/U cable, only a few are suitable for CCTV systems. For optimum performance, choose a cable that has 95% copper shield or more and a copper or copper-clad center conductor. The copper-clad center conductor has a core of steel and copper cladding, has higher tensile strength, and is more suitable for pulling through conduit over long cable runs. While cable with 65% copper shield is available, 95% shielding or more should be used to reduce and prevent outside electromagnetic interference (EMI) signals from penetrating the shield, causing spurious noise on the video signal. The 95% shielding eliminates most of the outside EMI and therefore reduces the chance that signals will reach the center conductor and appear as noise on the monitor. A coaxial cable with 95% shield and a copper center conductor will have a loop resistance of approximately 16 to 57 ohms per 1000 feet, depending on the type of video cable.

6.2.5 Potential Problems

Various potential problems should be considered when utilizing coaxial cables for signal transmission in indoor or outdoor applications. Indoor application problems include (1) different ground potentials at the ends of the coaxial cable or at different television equipment locations in a building and (2) coaxial cable running through or near other electrical power distribution equipment or machinery, producing high electromagnetic fields that can cause interference and noise in the television signal. In outdoor applications, the two problems just mentioned must be considered, in addition to environmental conditions caused by lightning storms or other high-voltage noise generators, such as transformers on power lines, electrical substations, automobile/truck electrical noise, or other EMI.

6.2.5.1 Ground Loops and Hum Bars

The ground loop is by far the most troublesome and noticeable video cabling problem (Figure 6-7). It is most easily detected before connecting, by measuring the electrical voltage between the coaxial-cable shield and the chassis to which it is being connected. If the voltage difference is 1 or 2 volts or more, there is a potential for a hum problem. As a precaution, it is generally good practice to measure the voltage difference before connecting the cable and chassis for a system with a long run or between any two electrical supplies, to prevent any damage to the equipment from voltage differences between the two grounds.

Probably more than half of the large multiple-camera systems installed contain some amount of distortion in the video picture. This is caused by random or periodic noise or, if more severe, by hum bars. The hum bar appears as a horizontal distortion across the monitor at two locations, one-third and two-thirds of the way down the picture. If the camera is synchronized or power-line-locked, the bar will be stationary on the screen. If the camera is not line-locked, the distortion or bar will roll slowly through the picture, recurring over and over. Sometimes the hum bars are accompanied by sharp tearing regions across the monitor or erratic horizontal pulling at the edge of the screen (Figure 6-7). This phenomenon is caused by the effect of

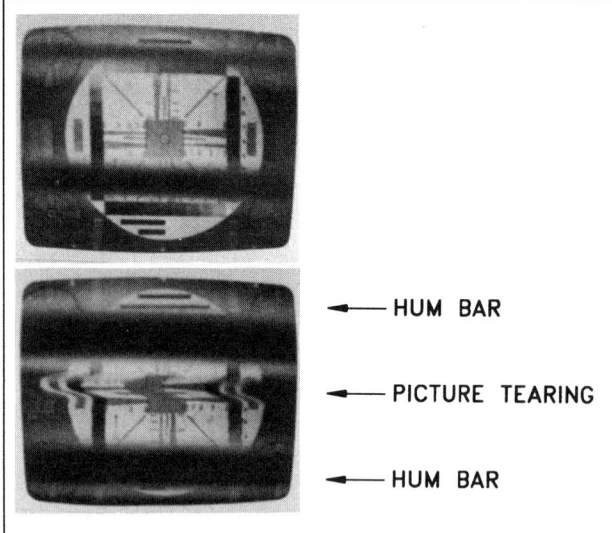

FIGURE 6-7   Hum bars caused by ground loops

the high voltages on the horizontal synchronization signal. Other symptoms include uncontrolled vertical rolling of the scene on the screen when there are very high voltages present in the ground loop.

Interference caused by external sources or voltage differences can to some extent be predicted prior to installation. The hum bar and potential difference between two electrical systems usually cannot be determined until the actual installation. The system designer should try to anticipate the problem and, along with the user, be prepared to devote additional equipment and time to solve it. The problem is not related to equipment at the camera or monitor end or to the cable installed; it is strictly an effect of the particular environment encountered, be it EMI interference or difference in potential between the main power sources at each location. The ground problem can occur at any remote location, and it can be eliminated inexpensively with the installation of an isolation amplifier. Another solution, described in Section 6.4, is the use of fiber-optic transmission means, which eliminates electrical connections entirely.

One totally unacceptable solution is the removal of the third wire on a three-prong electrical plug, which is used to ground the equipment chassis to earth ground. Not only is such removal a violation of local electrical codes and Underwriters Laboratory (UL) recommendations, it is a hazardous procedure. If the earth ground is removed from the chassis, a voltage can appear on the camera, monitor or other equipment chassis, producing a "hot" chassis that, if touched, can shock any person with 60 to 70 volts.

When video cables bridge two power distribution systems, ground loops occur. Consider the situation (Figure 6-8) in which the CCTV camera receives AC power from power source A, while some distance away or in a different building the CCTV monitor receives power from distribution system B.

The camera chassis is at 0 volts (connected to electrical ground) with reference to its AC power input A. The monitor chassis is also at 0 volts with respect to its AC distribution system B. However, the level of the electrical ground in one distribution system may be higher (lower) than that of the ground in the other system; hence a voltage potential can exist between the two chassis. When a video cable is connected between the two distribution system grounds, the cable shield connects the two chassis and an alternating current flows in the shield between the units. This extraneous voltage (causing a ground-loop current to flow) produces the unwanted hum bars in the video image on the monitor.

The second way in which hum bars can be produced on a television monitor is when two equipment chassis are mechanically connected, such as when a camera is mounted on a pan/tilt unit. If the camera receives power from one distribution system and the chassis of the pan/tilt unit is grounded to another system with a different level, a ground loop and hum bars may result. The size and extent of the horizontal bars depend on the severity of the ground potential difference.

6.2.5.2 Electrical Interference

In the case of EMI, a facility site survey should be made of the electromagnetic radiation present from any electrically noisy power distribution equipment. The cables should then be routed around such equipment so that there is no interference with the television signal.

When a site survey indicates that the coaxial cable must run through an area containing large electrical interfering signals (EMI) caused by large machinery, high-voltage power lines, refrigeration units, microwaves, truck ignition, radio or television stations, fluorescent lamps, two-way radios, motor-generator sets, or other sources, a better shielded cable, called a triaxial cable, or fiber optics may be the answer. The triaxial cable has a center conductor, an insulator, a shield, a second insulator, a second shield, and the normal outer polyethylene or other covering to protect it from the environment. The double shielding significantly reduces the amount of outside EMI radiation that gets to the center conductor.

The number of horizontal bars on the monitor can indicate where the source of the problem is. If the monitor has six dark bars, multiplying 6 by 60 equals 360, which is close to a 400-cycle frequency. This interference could be caused by an auxiliary motor-generator set often found in large factory machines operating at this frequency. To correct the problem, the cable could be rerouted away from the noise source, replaced with a balanced coaxial or triaxial cable, or for 100% elimination of the problem, upgraded to fiber-optic cable.

If lightning and electrical storms are anticipated and signal loss is unacceptable, outdoor cables must be buried underground and proper high-voltage-suppression cir-

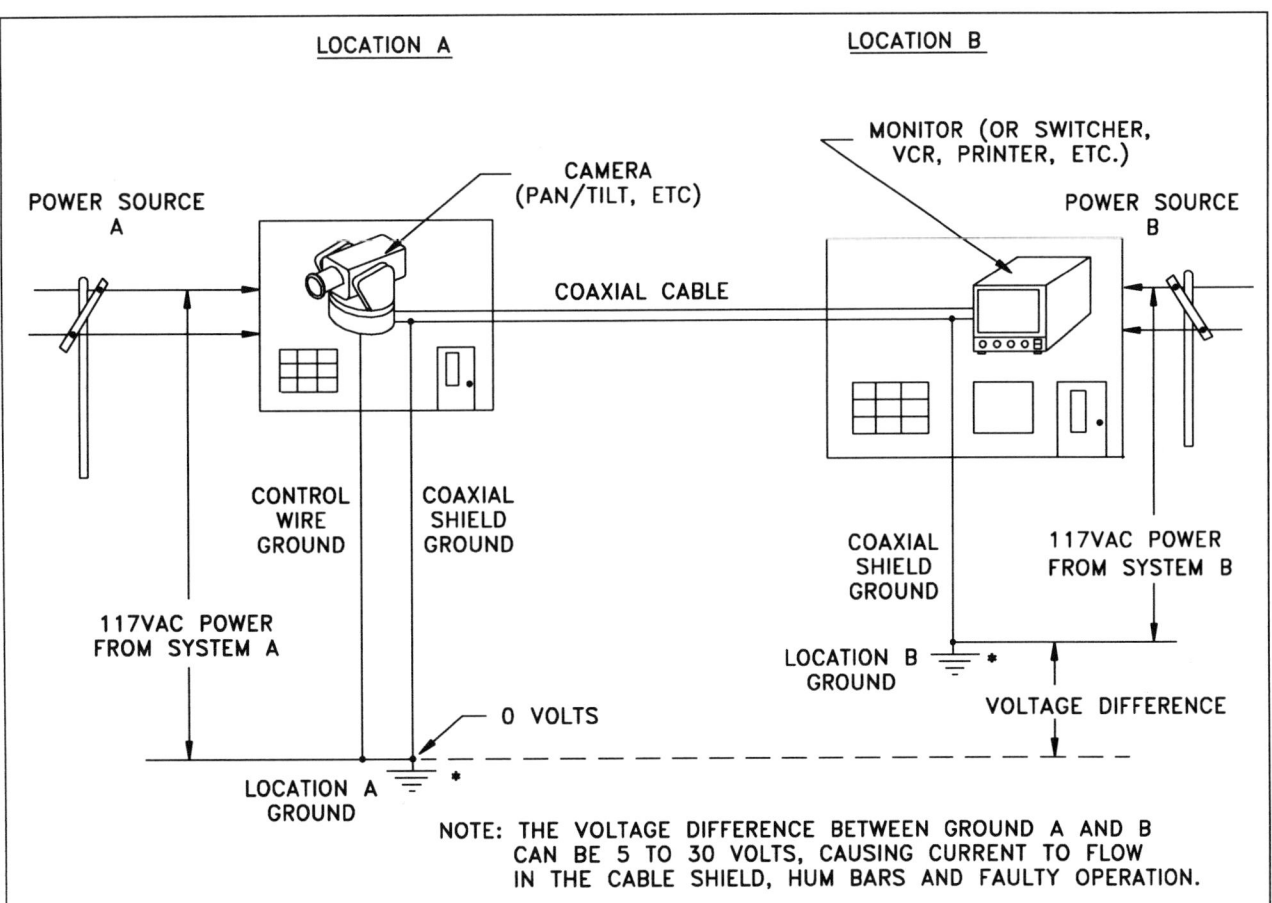

LOCATION A

LOCATION B

POWER SOURCE
A

CAMERA
(PAN/TILT, ETC)

MONITOR (OR SWITCHER,
VCR, PRINTER, ETC.)

POWER SOURCE
B

COAXIAL CABLE

CONTROL
WIRE
GROUND

COAXIAL
SHIELD
GROUND

COAXIAL
SHIELD
GROUND

117VAC POWER
FROM SYSTEM B

117VAC POWER
FROM SYSTEM A

LOCATION B
GROUND

VOLTAGE DIFFERENCE

0 VOLTS

LOCATION A
GROUND

NOTE: THE VOLTAGE DIFFERENCE BETWEEN GROUND A AND B
CAN BE 5 TO 30 VOLTS, CAUSING CURRENT TO FLOW
IN THE CABLE SHIELD, HUM BARS AND FAULTY OPERATION.

FIGURE 6-8 Two-source AC power distribution system

cuitry must be installed at each end of the cable run and on the input power to the television equipment.

In new installations with long cable runs (several thousand feet to several miles) or where different ground voltages exist, a fiber-optic link is the better solution—although balanced systems and isolation amplifiers can often solve the problem.

6.2.6 Aluminum Cable

Although coaxial cable with aluminum shielding provides 100% shielding, it is generally reserved for RF cable television (CATV) and master television (MATV) signals used for home video cable reception. This aluminum-shield type should *not* be used for CCTV for two reasons: (1) it has higher resistance and (2) it distorts horizontal synchronization pulses.

The added resistance—approximately seven times more than that of a 95% copper or copper-clad shield—increases the video cable loop resistance, causing a reduction in the video signal transmitted along the cable. The higher loop resistance means a smaller video signal reaches the monitoring site, producing less contrast and an inferior picture. Always use a good-grade 95% copper braid RG59/U cable

to transmit the video signal up to 1000 feet and an RG11/U to transmit up to 2000 feet.

Distortion of the horizontal synchronization pulse causes picture tearing on the monitor, depicting straight-edged objects with raggedy edges.

6.2.7 Plenum Cable

Another category of coaxial cable is designed to be used in the plenum space in large buildings. This plenum cable has a flame-resistant outside covering and very low smoke emission. The cable can be used in air-duct air-conditioning returns and does not require a metal conduit for added protection. The cable, designated as "plenum rated," is approved by the National Electrical Code and UL.

6.3 TWO-WIRE CABLE

It would be convenient, cheap, and simple if the CCTV signal could be transmitted over an existing two-wire system. However, the previous section described why coaxial cable is required for optimum transmission of high video frequencies. A standard two-wire telephone electrical system lacks the necessary characteristics to transmit all of the high-frequency information required for a good-resolution monochrome or color picture.

A technique does exist for transmitting television pictures over a dedicated pair of wires, which can be run parallel or as a twisted pair but not run through a telephone switching station. The system uses a small transmitter and receiver—one at each end of the pair of wires—and transmits the picture over maximum distances of 3000 to 10,000 feet. Picture resolution of approximately 350 TV lines is obtained with this non–coaxial-cable system. A second, more complex method called slow-scan transmits video snapshots—not real-time images—over a voice-grade communications link through a telephone switching station.

6.3.1 Real-Time Transmission

As described in Section 6.2, coaxial cable is needed to transmit high-quality television pictures several hundred or several thousand feet. However, some transmitters and receivers will transmit an acceptable video picture over a standard two-wire path using twisted or untwisted wires. This two-wire path can be a telephone system, intercom system, or any other two wires that have *continuous* conductive paths from the camera to the monitor location. By means of high-frequency emphasis in the transmitter or receiver, or both, the normal attenuation produced by the two-wire system is compensated for by excessive amplification of the high frequencies.

The system will not work if there is no conductive (resistive copper) path for the two wires. The signal path cannot have electrical switching circuits between the camera and monitor location. Two-wire systems sometimes require adjustments in the transmitter and/or receiver to optimize the compensation, that is, to produce a transmission as close as possible to that which could be achieved with a coaxial cable. In two-wire systems the picture at the monitor is slightly delayed, which is a problem only if other cameras are switched into the system. A second consideration is the loss of some picture detail or resolution, since the highest frequencies in the CCTV picture are not fully transmitted over the long distance. The resolution obtained is approximately 350 TV lines. While 350 TV line resolution does not equal the 550 TV line camera and monitor resolution, it can very often supply a satisfactory picture for surveillance and access control applications.

The two-wire system costs five to seven hundred dollars, since an additional transmitter and receiver are required. This cost may be small compared with the cost of installing a new coaxial cable from the camera to the monitor location. If a medium-resolution picture is acceptable and a two-wire system already exists, this is a cost-effective technique. Figure 6-9 illustrates the block diagram and connections for a two-wire transmitter and receiver pair.

Internal adjustments equalize frequency transmission over the full bandwidth of the television signal. The transmitter unit converts the camera signal impedance (75 ohms) to match the wire impedance and provides the frequency compensation required. The receiver unit reconstructs the signal and transmits it the short distance to the television monitor via coaxial cable. The receiver unit has frequency compensation networks to optimize the video picture on the monitor. Both the transmitter and receiver are powered by either 12-volt DC or 117- or 24-volt AC.

The transmission path is a dedicated two-wire twisted cable pair having no bridging circuitry or any other connections. Mechanical splices and connectors are permissible. If wires owned by the telephone company are used, a telephone system interface unit may be necessary to meet all Bell System picture-phone interface specifications.

The following examples are for two-wire transmission of a video picture with 350 TV line resolution: If the two wires are 16 American Wire Gauge (AWG) and have interwire capacitance of 30 picofarads per foot, a transmission distance of 9660 feet can be obtained. For the same conditions with a 20 AWG wire, approximately 3960 feet can be obtained. When the system is used at shorter distances, higher resolution results. The system can be operated over wire ranges from 14 to 24 AWG.

A system is available to transmit color television pictures in a manner similar to that of the monochrome system just described. This transmitter/receiver pair transmits the color video image through a twisted-pair, two-conductor cable over distances of 3000 feet using a 24-AWG or larger wire.

When it is necessary to transmit the video picture only several hundred feet, a two-wire AC power line system may be used. This is accomplished by converting the camera video signal into a frequency-modulated (FM) signal, superimposing it on the AC power line voltage, and transmitting it along with this voltage. At the receiving end, an FM demodulator extracts the video signal from the power cable and converts it to a standard CCTV signal for presentation on the monitor. One significant limitation imposed by this method is that both transmitter and receiver operate on the same phase (power line circuit). While this setup finds limited applications in most security systems, it offers a unique solution to some difficult transmission problems.

A variation of this system—called real-time duplex television—permits transmitting CCTV pictures with one-half the resolution of a normal picture simultaneously in both directions. Two-way voice communication and control signals for switching cameras, adjusting lenses, and performing other functions can be added (Figure 6-10).

6.3.2 Non–Real-Time Slow-Scan Transmission

The coaxial-cable transmission techniques described in the previous sections and fiber-optic cable (Section 6.4) account for the majority of transmission means from the

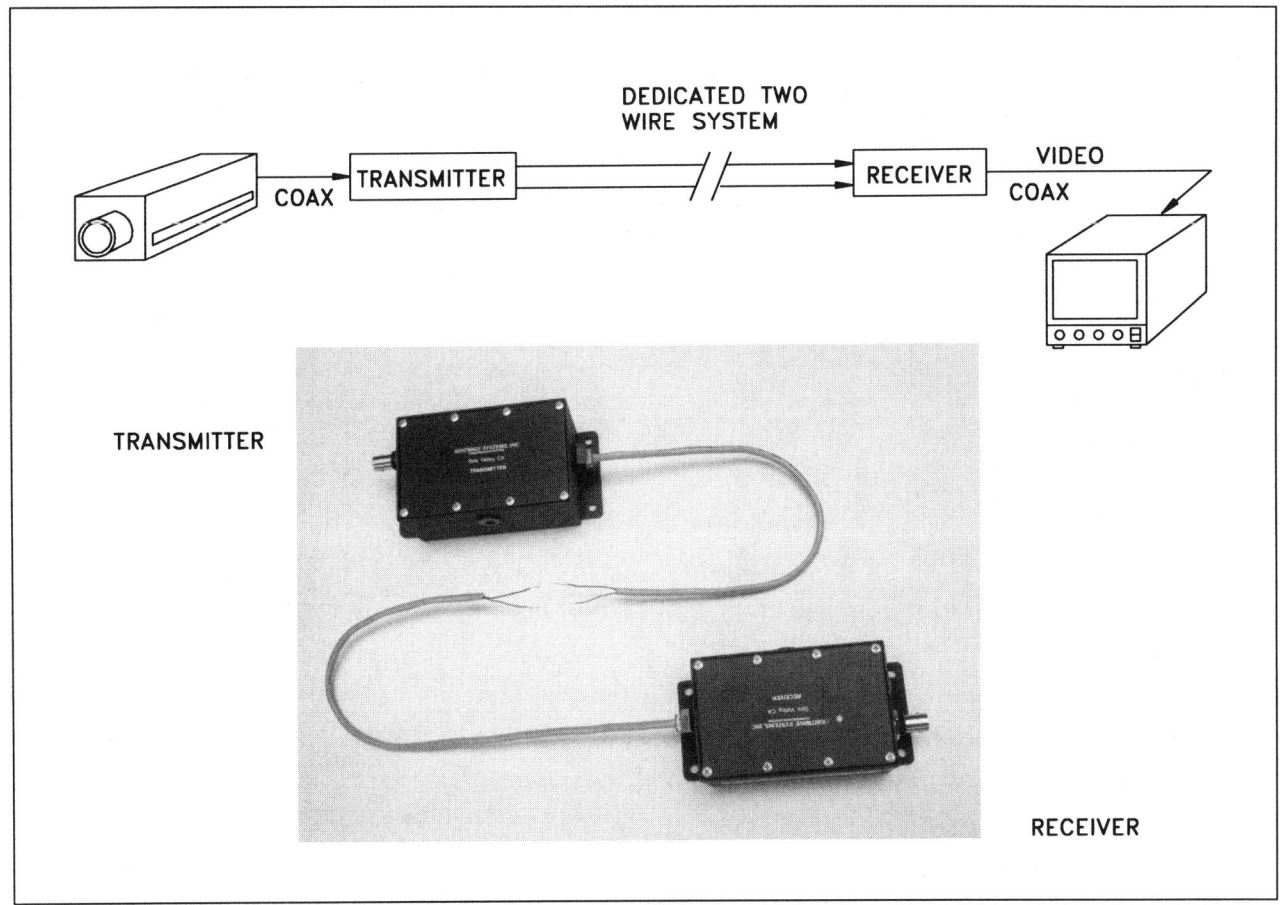

FIGURE 6-9   Real-time two-wire video transmission system

remote camera site to the monitoring equipment. There are, however, many instances when the television picture must be transmitted over very long distances—tens, hundreds, or thousands of miles, or across continents—for which coaxial cables are not a practical solution. Two-wire, coaxial, or fiber-optic cables for real-time transmission are also not practical in a metropolitan area where a television picture must be transmitted from one building through congested city streets to another building out of sight of the first. This section describes a non–real-time two-wire technology called slow-scan, which permits the transmission of a CCTV picture from one location to any other location in the world, providing a two-wire or wireless voice-grade link (telephone line) is available.

The wireless transmission systems described in Section 6.3.1 all result in real-time television transmission. A scheme for transmitting the television picture over large distances, even anywhere in the world, uses slow-scan television transmission (Figure 6-11). This non–real-time technique involves storing one television picture frame (snapshot) and sending it slowly over a telephone or other audio-grade network anywhere within a country or to another country. The received picture is reconstructed at the remote receiver to produce a continuously displayed televi-

sion snapshot. Each snapshot takes anywhere from several to 72 seconds to transmit, with a resulting picture having from low to high resolution, depending on the speed of transmission. A time-lapse effect is achieved, and every scene frame is transmitted spaced from several to 72 seconds apart.

Since slow-scan uses ordinary voice-grade phone lines, the picture is transmitted anywhere for the price of a phone call. The transmitted picture is displayed on a monitor, recorded on videotape, printed out on a hard-copy video printer, or recorded on audiotape (the signal has an audio bandwidth from 300 to 3000 Hz and can be stored on audiotape). Since slow-scan uses the worldwide switched telephone network as the transmission medium, it overcomes possible right-of-way problems with short-range direct connections. Slow-scan also sidesteps the siting and licensing problems associated with microwave and RF transmission.

6.3.2.1 Principle of Operation

The slow-scan system consists of two fairly complex electronic storage and conversion units: a transmitter and a receiver. The transmitter converts the real-time CCTV signal,

CAMERA
COAX
TRANSMITTER:
• VIDEO
• AUDIO
• ALARM
RECEIVER:
• AUDIO
• CONTROL
DEDICATED TWO WIRE SYSTEM
• SHIELDED 2 WIRE
• TWISTED PAIR
• TELEPHONE
RECEIVER:
• VIDEO
• AUDIO
• ALARM
TRANSMITTER:
• AUDIO
• CONTROL
VIDEO
COAX
MICROPHONE
SPEAKER
COMMAND FUNCTIONS
ALARM INPUTS
MICROPHONE
SPEAKER
ALARM OUTPUT DEVICES

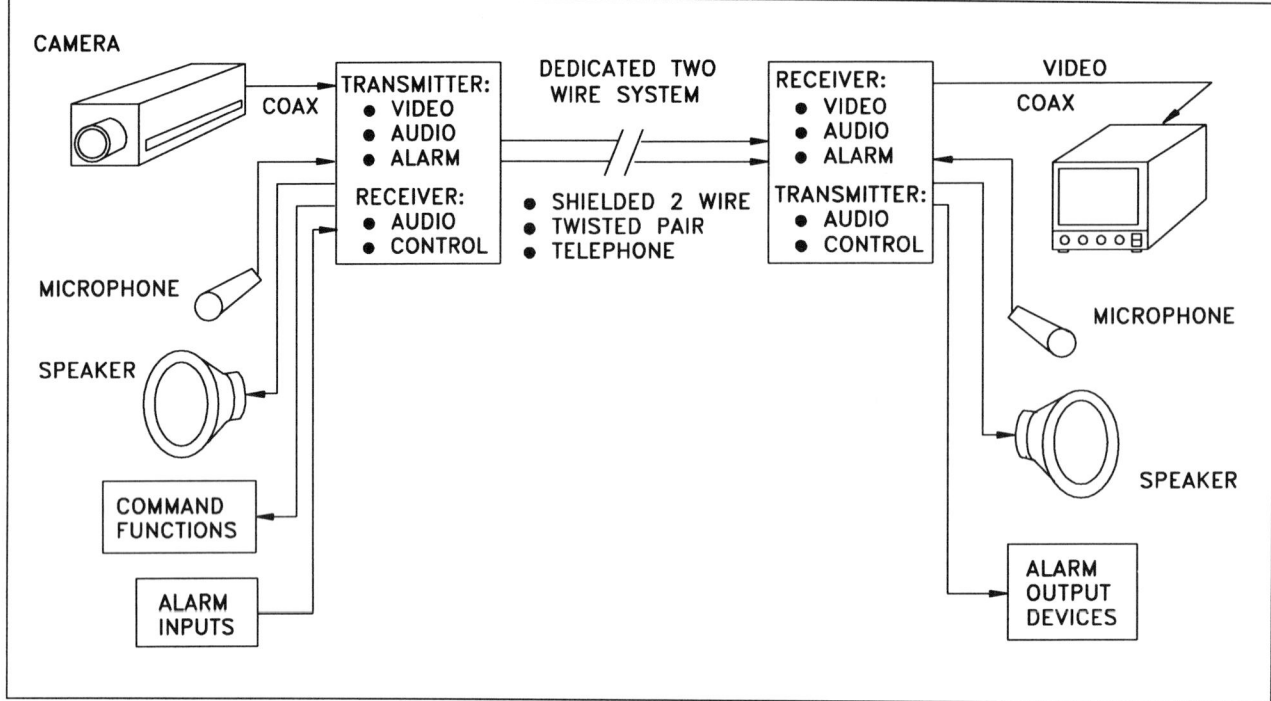

FIGURE 6-10 Real-time transmission system with video, audio, alarms, and controls

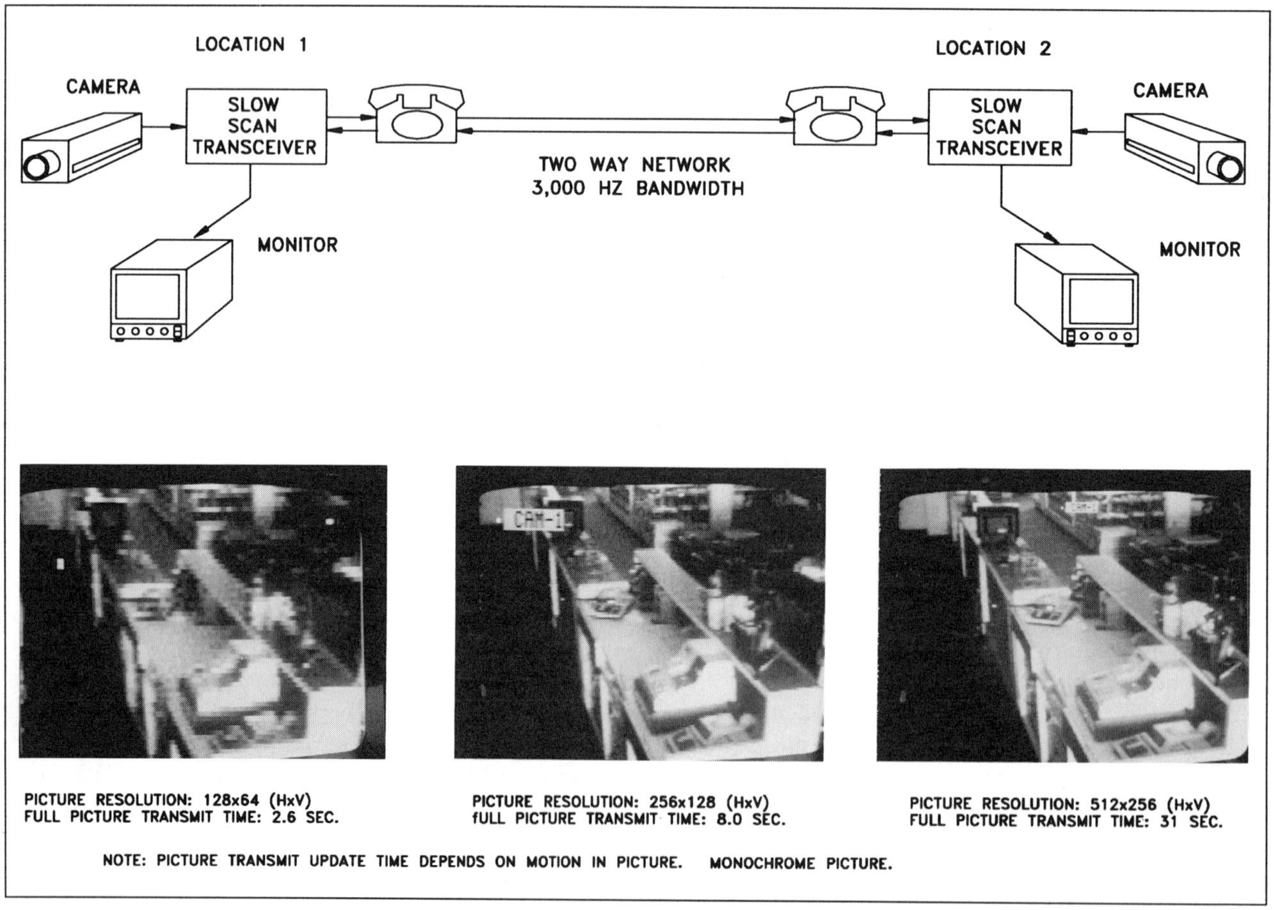

LOCATION 1
LOCATION 2
CAMERA
SLOW SCAN TRANSCEIVER
TWO WAY NETWORK
3,000 HZ BANDWIDTH
SLOW SCAN TRANSCEIVER
CAMERA
MONITOR
MONITOR
CAM-1
PICTURE RESOLUTION: 128x64 (HxV)
FULL PICTURE TRANSMIT TIME: 2.6 SEC.
PICTURE RESOLUTION: 256x128 (HxV)
FULL PICTURE TRANSMIT TIME: 8.0 SEC.
PICTURE RESOLUTION: 512x256 (HxV)
FULL PICTURE TRANSMIT TIME: 31 SEC.
NOTE: PICTURE TRANSMIT UPDATE TIME DEPENDS ON MOTION IN PICTURE. MONOCHROME PICTURE.

FIGURE 6-11 Slow-scan television transmission and transmitted pictures over telephone lines

which can change every $\frac{1}{30}$th of a second, into a stored video image—a snapshot of one of the fields or frames composing the video signal. A frame takes $\frac{1}{30}$th of a second to memorize, and a field $\frac{1}{60}$th of a second. The transmitter electronics frame-grab (digitize) one of the pictures and stores it in temporary memory. Depending on the system resolution, the video image is digitized into a 128×128, 256×256, or a 512×512 (or a combination thereof) pixel array. The gray scale of each pixel in the array is memorized, with anywhere from 64 to 256 different levels, so that the combination of the pixel locations and their gray-scale levels makes up the recorded snapshot information.

Once the field or frame is stored in the transmitter, the transmitter begins, on command from the receiver, to transmit the information for each pixel (that is, the pixel's location and its gray-scale level), at a rate set by the receiver control. Because transmission occurs over the telephone line, the rate must be within its voice bandwidth of 300 to 3000 Hz. One by one the pixel information is transmitted to the receiver, just as any voice, modulated voice, or digital information would be sent.

The receiver electronics accept the audio-bandwidth signal information with the pixel information, stores it in active memory, and creates the entire frame as originally present in the transmitting unit. While it is storing the frame, the receiver unit sends the information to the monitor via coaxial cable to display and "paint" the incoming signal on the monitor. This process can take from several to 72 seconds, depending on the field (frame) transmit time selected. After the receiver has received all information for the single frame, the transmitter snaps another picture and repeats the process.

Through this operation, specific frames are serially captured, sent down the telephone line, and reconstructed by the slow-scan system. Once the receiver has stored the digital picture information, if the transmitter is turned off or the video scene image does not change, the receiver continues to display the last video frame continuously (30 frames/second) as a still image. The image stored in the receiver or transmitter changes when the system is commanded, manually or automatically, to take a new snapshot. Figure 6-12 illustrates the salient difference between real-time television transmission and non–real-time.

In Figure 6-12a, successive frames of the real-time picture are shown, with time progressing downward. The first frame occurs at time zero, the second $\frac{1}{30}$th of a second later, and the third $\frac{2}{30}$th of a second later. These three frames represent three sequential frames in a realtime television signal. In a real-time transmission system, these three frames are transmitted serially, or one after the other, so that they appear on the monitor in the exact time relationship as they occurred in the camera: frame 1, then frame 2, then frame 3. This sequence produces a scene that is continually changing on the monitor and follows the motion in the camera scene, and it looks like the normal, smoothly changing television picture.

Figure 6-12b shows the relationship of non–real-time or slow-scan television transmission. At the camera site the first frame starts at time zero, the second frame at $\frac{1}{30}$th of a second, and the third frame at $\frac{2}{30}$th of a second (the same as for real-time). Before these frames are transmitted over the audio-grade transmission link, the signal is processed at the camera site in a transmitter processor. The processor captures frame 1 from the camera, that is, it memorizes (digitizes) the CCTV picture. The processor then slowly (at 2 seconds per frame, as shown in the figure) transmits the television frame element by element, line by line, until the receiver processor located at the monitor site has accepted all 525 lines in that frame.

The significant difference between real-time and slow-scan transmission is the time it takes to transmit the picture. In the real-time case, it is $\frac{1}{30}$th of a second, the *real time* of the frame. In the case of the slow-scan, it may take 2, 4, 8, 32, up to 72 seconds to transmit that single frame to the monitor site. The price paid for the slow-scan method is that all of the camera frames cannot be transmitted. The next frame that can be transmitted is at least several seconds later than the initial frame transmitted. Obviously, the motion that occurs in the scene between time zero and several seconds later will not be displayed. One or two seconds later, the next frame is captured by the slow-scan transmitter processor and again slowly sent to the receiver. This procedure continues indefinitely.

Some systems incorporate special compression techniques and modulation algorithms so that only the changing parts of a scene (i.e., the movements) are transmitted. Another technique first transmits areas of high scene activity with high resolution and then areas of lower priority. These techniques increase the transmission of the intelligence in the scene.

The reason for sacrificing a real-time picture is that the slow-scan system transmits a picture over virtually any audio communications link. By increasing transmission time of the frame from $\frac{1}{30}$th of a second to several seconds, the choice of cable or transmission path changes significantly. For slow-scan it is possible to send the full video image on a twisted-pair or telephone line or any communications channel having a bandwidth equivalent to audio frequencies, that is, up to only 3000 Hz (instead of a bandwidth up to 4.2 MHz, as needed in real-time transmission). So all existing satellite links, mobile telephones, and other connections can be used. A variation of this equipment for an alarm application can store multiple frames at the transmitting site, so if the information to be transmitted is an alarm, this alarm video image can be stored for a few seconds (every $\frac{1}{30}$ second) and then slowly transmitted to the remote monitoring site frame by frame, thereby transmitting all of the alarm frames.

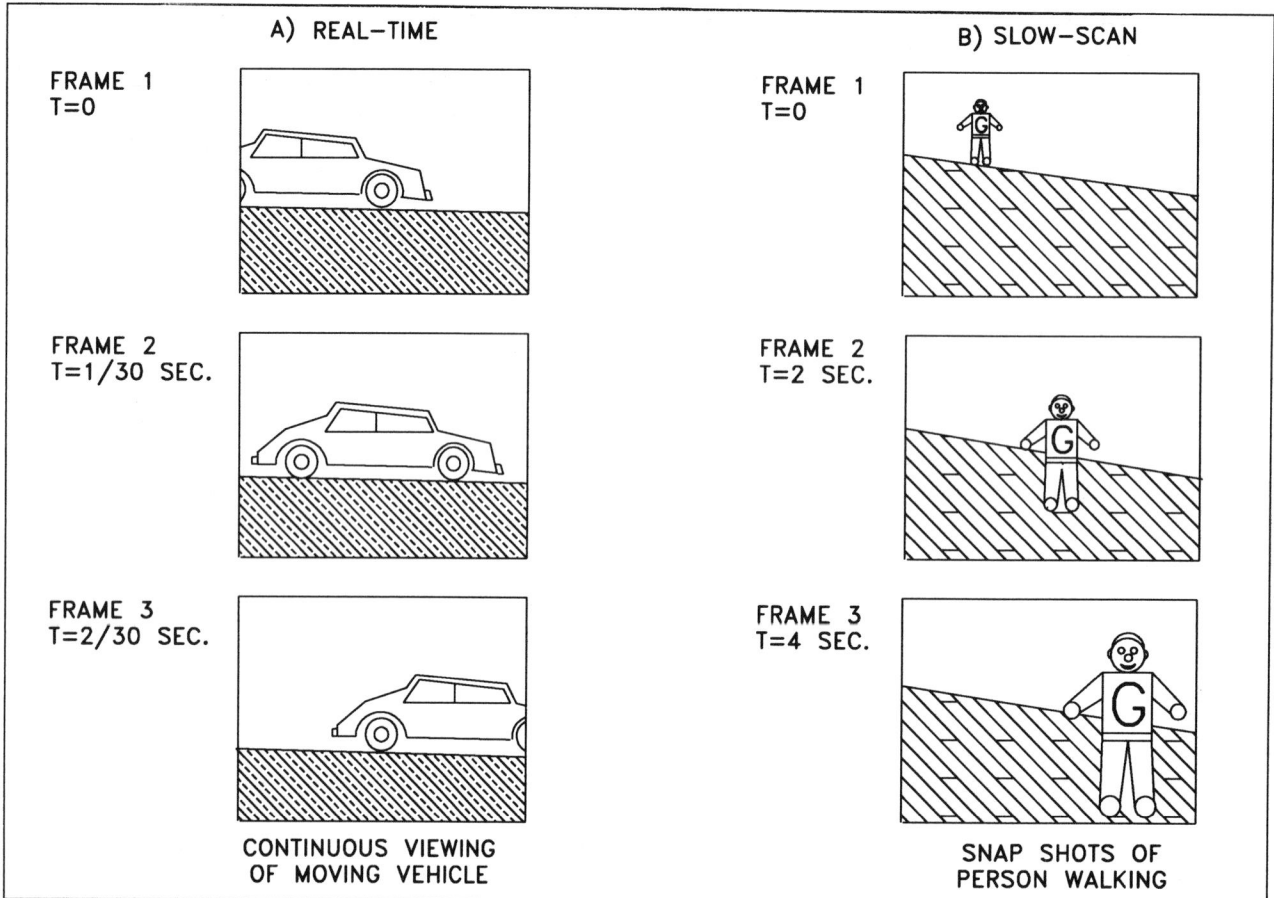

FIGURE 6-12 Real-time television transmission versus non–real-time (slow-scan)

6.3.2.2 Hardware

Slow-scan or phone-line TV hardware consists of a transmitter unit at the camera location and a receiver unit at the monitoring location. In some systems the slow-scan transmitter incorporates a built-in switcher to transmit video from multiple cameras sequentially to the monitoring site. Likewise, image-splitting optics or electronic splitters can take pictures from various cameras, combine them, and send them over a single slow-scan link to the monitoring site. Slow-scan equipment is available for monochrome and color CCTV with variable scan rates. Figure 6-13 shows a slow-scan system with selectable resolution that can send a monochrome or color video picture anywhere in the world.

6.3.2.3 System Use

Slow-scan systems are easy to install. They are composed of two telephone interface modules: the transmitter sends the modulated video signal (FM voice-grade transmission) and the receiver demodulates the signal and converts it to a video signal, which is then displayed on the monitor. Figure 6-14 shows the block diagram for interconnecting a duplex slow-scan system capable of sending one to four (quad) video images, as selected from the remote end.

The camera output is connected to the transmitter input, and the transmitter output is connected to the two-wire phone line system. The receiver input is connected to the other end of the telephone line, and the receiver output is connected to the monitor. The image on the monitor appears as a still, freeze-frame image: it does not change until the camera sends the next picture. The operator can trip front-panel switches on the receiver for different transmission times, thereby changing the resolution. For example, if an alarm is triggered, a fast, low-resolution picture is sent in 1 to 8 seconds. If the scene deserves further scrutiny with higher resolution, the operator selects a higher resolution and longer transmission time, say 32 seconds, to allow better identification of the cause of the alarm. This equipment opens a new dimension to remote monitoring via CCTV.

Controls on the receiver module include scan rate to change resolution (measured in TV lines or pixels), and transmit modes. For instance, the equipment can transmit a single full-screen image from one camera or a quad display from four cameras at once. The slow-scan system can be interfaced with a printer to provide a hard-copy printout of the freeze-frame image. Likewise, it can be connected to a video recorder, to record exactly what the monitor dis-

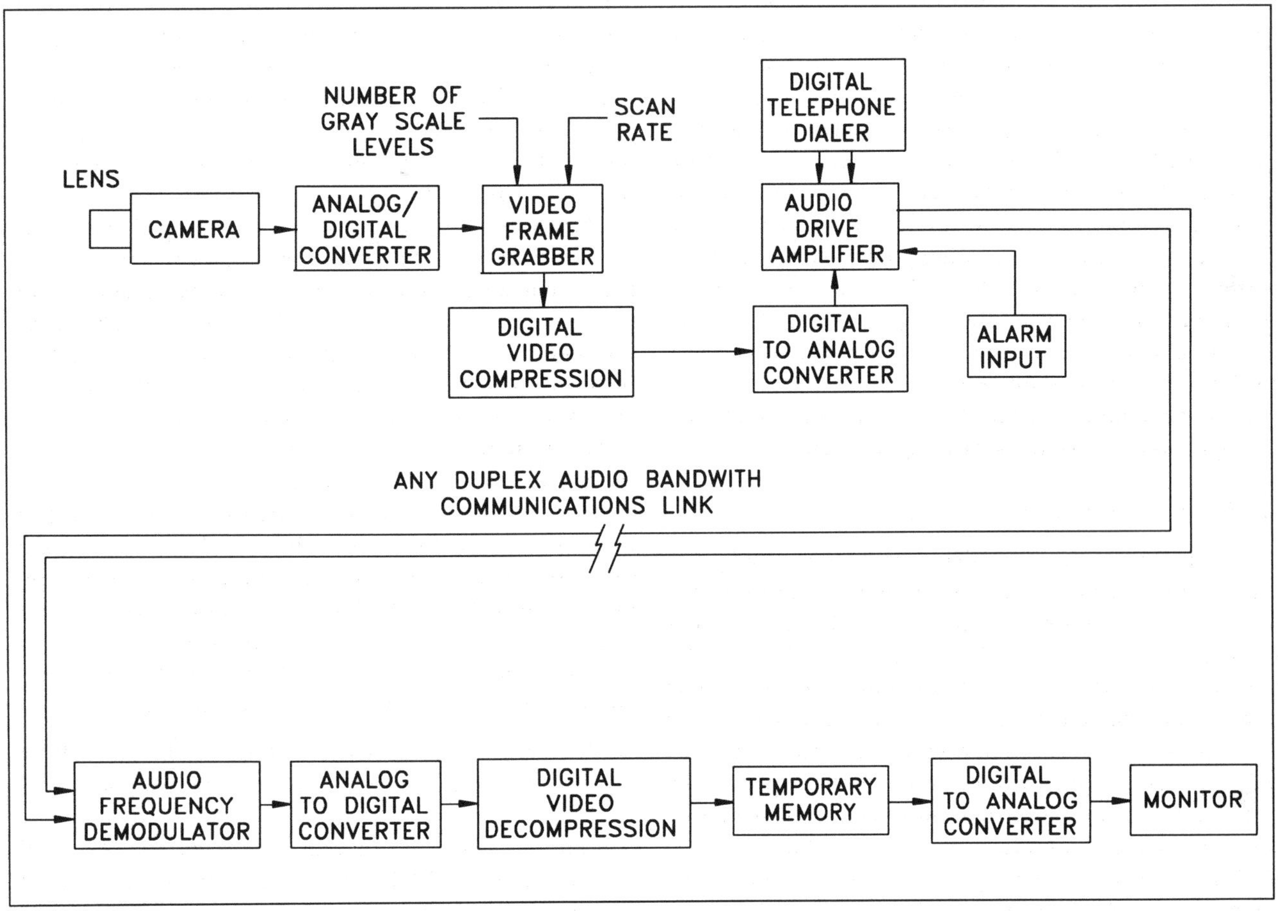
NUMBER OF GRAY SCALE LEVELS
SCAN RATE
DIGITAL TELEPHONE DIALER
LENS
CAMERA
ANALOG/ DIGITAL CONVERTER
VIDEO FRAME GRABBER
AUDIO DRIVE AMPLIFIER
DIGITAL VIDEO COMPRESSION
DIGITAL TO ANALOG CONVERTER
ALARM INPUT
ANY DUPLEX AUDIO BANDWITH COMMUNICATIONS LINK
AUDIO FREQUENCY DEMODULATOR
ANALOG TO DIGITAL CONVERTER
DIGITAL VIDEO DECOMPRESSION
TEMPORARY MEMORY
DIGITAL TO ANALOG CONVERTER
MONITOR

FIGURE 6-13 Slow-scan system block diagram—simplex (one-way)

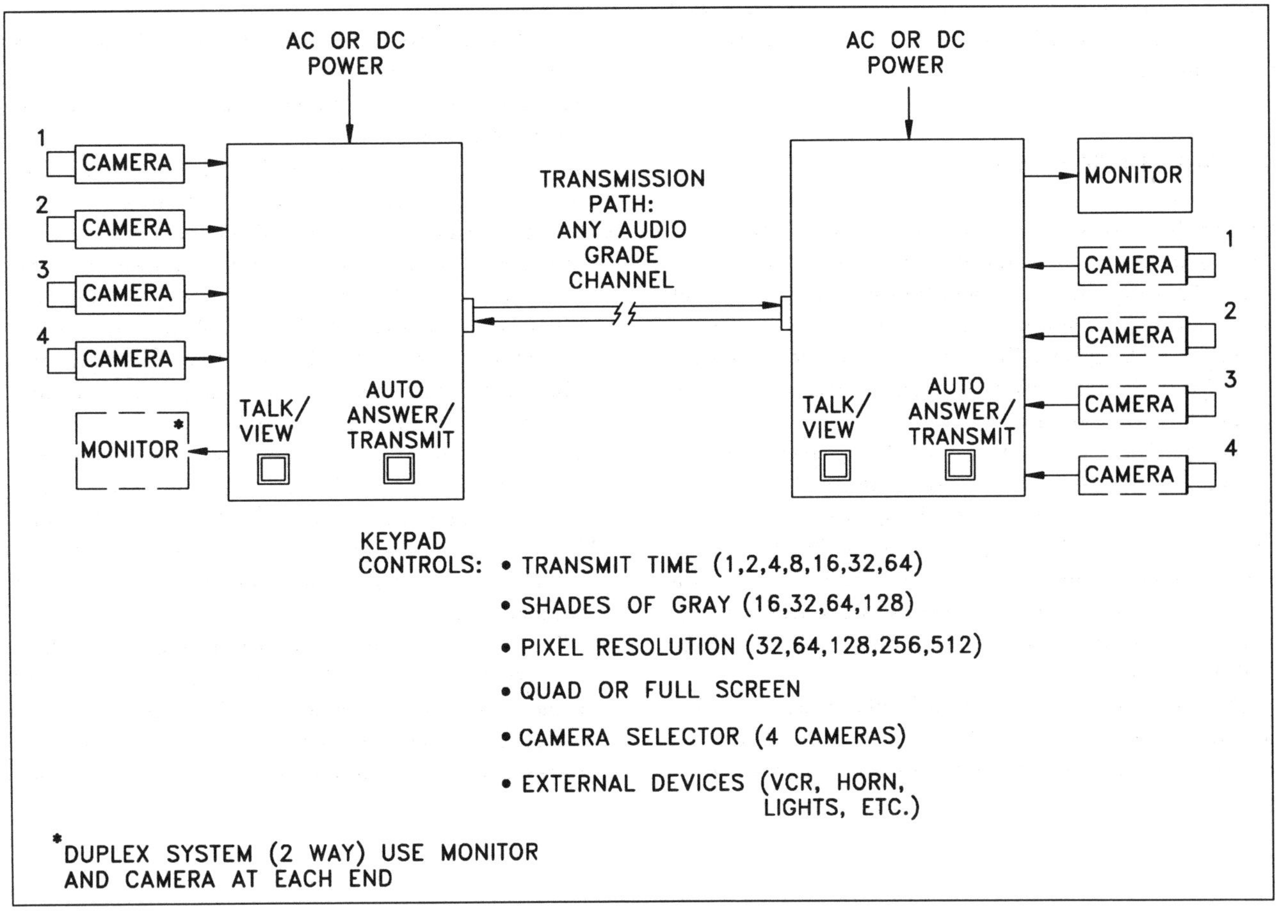
AC OR DC POWER
AC OR DC POWER
1 CAMERA
2 CAMERA
3 CAMERA
4 CAMERA
MONITOR*
TALK/ VIEW
AUTO ANSWER/ TRANSMIT
TRANSMISSION PATH: ANY AUDIO GRADE CHANNEL
MONITOR
CAMERA 1
CAMERA 2
CAMERA 3
CAMERA 4
TALK/ VIEW
AUTO ANSWER/ TRANSMIT
KEYPAD CONTROLS:
• TRANSMIT TIME (1,2,4,8,16,32,64)
• SHADES OF GRAY (16,32,64,128)
• PIXEL RESOLUTION (32,64,128,256,512)
• QUAD OR FULL SCREEN
• CAMERA SELECTOR (4 CAMERAS)
• EXTERNAL DEVICES (VCR, HORN, LIGHTS, ETC.)
*DUPLEX SYSTEM (2 WAY) USE MONITOR AND CAMERA AT EACH END

FIGURE 6-14 Slow-scan interconnecting diagram and controls

plays, or connected to an audio recorder, to record the received audio transmission, providing a very low cost video storage medium. Later the audiotape can be played back into the receiver to display the freeze-frame video images sequentially, one at a time, on a video monitor. This feature is particularly useful if continuous freeze-frame pictures are transmitted, no activity or alarm is occurring, but a permanent record is desired. Storing these images on audiotape is far less expensive than recording them on video. Transporting the audiotapes is also easier, and they provide a degree of security because the video images on the audiotape are in effect "scrambled": only the slow-scan receiver hardware can "unscramble" them.

Table 6-3 summarizes the characteristics of several slow-scan systems available for security applications.

6.3.2.4 Resolution versus Transmit Time

Video picture resolution is determined primarily by the transmit time. A short transmission time of 1 to 8 seconds provides low resolution—typically 128×128 pixels. A long transmission time of 32 to 72 seconds provides high-resolution pictures—512×512 pixels. A second factor affected by transmission time is the gray-scale level transmitted. If only a few gray-scale levels are transmitted, say two to four, as with a paper photocopy, the receiver obtains poor gray-scale rendition and less picture information. Such a result is associated with a short transmission time of 1 to 8 seconds. High levels of gray scale, up to 256 levels, require more information and longer transmission times, which are obtained from the 32-to-72-second transmissions. Therefore, one must compromise between the scan time, the resolution, and the gray scale, based on the particular security application at hand.

6.4 FIBER-OPTIC CABLE

One of the most significant advances in communications and signal transmission has been the innovation of fiber optics. However, the concept of transmitting video signals over fiber optics is not new. In the early 1970s, manufacturers began making glass fibers that were sufficiently low-loss to transmit light signals over practical distances of hundreds or a few thousand feet.

Why use fiber-optic transmission when coaxial cables can provide adequate CCTV signal transmission? Today's high-performance CCTV systems require greater reliability and more "throughput," that is, getting more signals from the camera end to the monitor end, over greater distances, and in harsher environments. The fiber-optic transmission system preserves the quality of the video signal and provides a high level of security.

The information-carrying capacity of a transmission line, whether electrical or optical, increases as the carrier frequency increases. The carrier for fiber-optic signals is light, which has frequencies several orders of magnitude (1000 times) greater than radio frequencies. Likewise, the higher the carrier frequency the larger the bandwidth that can be modulated onto the cable. Some transmitters and receivers permit multiplexing multiple television signals, control signals, and duplex audio onto the same fiber optic because of its wide bandwidth.

The clarity of the picture transmitted using fiber optics is now limited only by the camera, environment, and monitoring equipment. Fiber-optic systems can transmit signals from a camera to a monitor over great distances—typically several miles—with virtually no distortion or loss in picture resolution or detail. Figure 6-15 shows the block diagram of the hardware required for a fiber-optic transmission system.

The system uses an electrical-to-optical signal converter/transmitter, a fiber cable for sending the light signal from the camera to the monitor, and a light-to-electrical signal receiver/converter to transform the signal back to a baseband CCTV signal required by the monitor. At both camera and monitor ends, the standard RG59/U coaxial cable is used to connect the camera and monitor to the system.

A glass-fiber-optic-based video link offers distinct advantages over copper-wire or coaxial-cable transmission means:

1. The system transmits information with greater fidelity and clarity over longer distances than a wire or coaxial cable.
2. The fiber is totally immune to all types of electrical interference—EMI or lightning—and will not conduct electricity. It can come in contact with high-voltage electrical equipment or power lines without a problem.
3. The fiber does not create ground loops and deleterious hum bars or picture tearing.
4. The fiber can be maintained while the transmitting or receiving equipment is still energized.
5. The fiber can be used where electrical codes and common sense prohibit the use of copper wires.
6. The cable will not corrode and the glass fiber is unaffected by most chemicals. The direct-burial type of cable can be laid in most kinds of soil or exposed to most corrosive atmospheres inside chemical plants or outdoors.
7. Since there is no electrical connection of any type, the fiber poses no fire hazard to any equipment or facility in even the most flammable atmosphere.
8. The fiber is virtually unaffected by atmospheric conditions, so the cable can be mounted aboveground and on telephone poles. When properly applied, the cable is stronger than standard electrical wire or coaxial cable and will therefore withstand far more stress from wind and ice loading.
9. Whether made up of single or multiple fibers, fiber-optic cable is much smaller and lighter than a coaxial cable of similar information-carrying capacity. So it is easier to handle and install and uses less conduit or duct space. An optical cable weighs 8 pounds per 3300

| RESOLUTION (PIXELS) * | | GRAY SCALE | | RESOLUTION (TV LINES) | | PERCENT PICTURE CHANGE | SCAN TRANSMIT TIME (SEC.) * |
HORIZONTAL	VERTICAL	SHADES	BITS	HORIZONTAL	VERTICAL		
128	128	16	4	90	90	10	0.2
	→	16	4	→	→	100	3.0
		32	5			10	0.3
		32	5			100	4.0
→		64	6			10	0.4
	128	64	6	90	90	100	5.0
240	200	32	5	168	140		12.0
256	256	16	4	180	180	10	0.5
		16	4			100	8.0
		32	5			10	1.0
→	→	32	5	→	→	100	10.0
		64	6			10	1.8
** 256	256	64	6	180	180	100	18.0
** 512	480	64	6	358	336	100	4.0 ***

* PIXEL=PICTURE ELEMENT, 1 PIXEL EQUIVALENT TO APPROXIMATELY 0.7 TV LINES

** INDUSTRY NORM TRANSMITTED OVER 9,600 BAUD MODEM

*** 1–4 SECONDS WITH DYNAMIC COMPRESSION

Table 6-3 Slow-Scan System Parameters

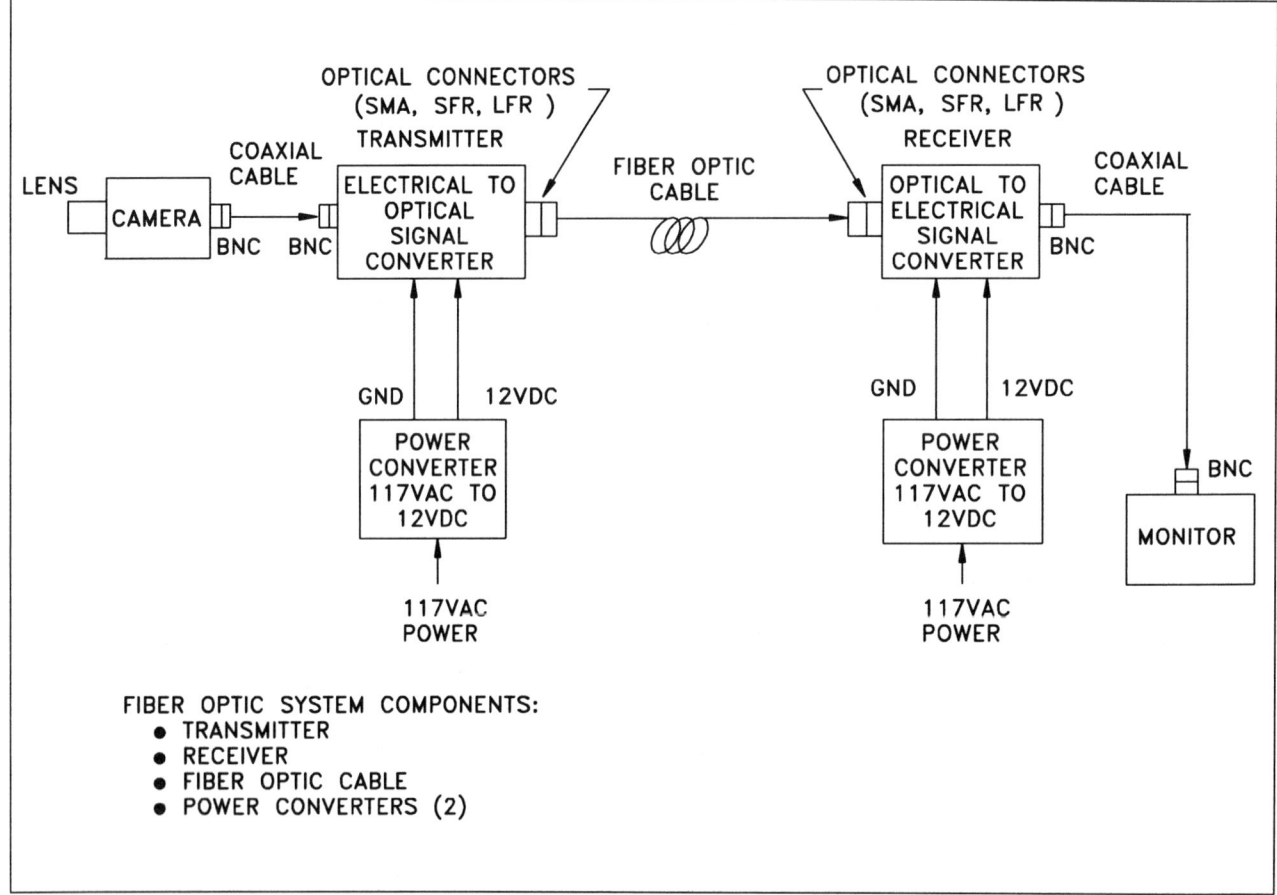

FIGURE 6-15 Fiber-optic transmission system

feet and has an overall diameter of 0.156 inches; a coaxial cable weighs 330 pounds per 3300 feet and is approximately 0.25 inches in diameter.

10. It transmits the video signal more efficiently (i.e., with lower attenuation) and since it needs no repeater (amplifier), it is more reliable and easier to maintain.

11. It is a more secure transmission medium, since it is not only hard to tap but an attempted tap is easily detected.

The economics of using a fiber-optic system is complex. Users evaluating fiber optics should consider the costs beyond those for the components themselves. The small size, light weight, and flexibility of fiber optics often present offsetting cost advantages. The prevention of unanticipated problems such as those just listed can easily offset any increased hardware costs of fiber optic systems.

With such rapid advances, the security system designer should consider fiber optics the optimum means to transmit high-quality television signals from high-resolution monochrome or color cameras to a receiver (monitor, switcher, recorder, printer, and so on) without degradation. This section reviews the attributes of fiber optic systems, their design requirements, and their applications.

6.4.1 Background

While the transmission of optical signals in fibers was investigated in the 1920s and 1930s, it was not until the 1950s that VanHeel, Hopkins, and Kapany developed the flexible fiber-scope, now widely used in medicine. During this period Kapany invented the practical glass-coated (clad) glass fiber and coined the term *fiber optics.*

Clad fiber was actively investigated in the 1960s by K. C. Kao and G. A. Hockham, researchers at Standard Telecommunications Laboratories in England, who proposed that this type of waveguide could form the basis of a new transmission system. In 1967 typical attenuations of the fiber measured more than 1000 dB per kilometer (which were impractical for transmission purposes), and researchers focused on reducing these losses. Figure 6-16 shows a comparison of fiber-optic transmission versus other electrical transmission means.

Three years later, investigators Kapron, Keck, and Maurer at Corning Glass Works announced reduction of losses to less than 20 dB per kilometer in fibers hundreds of meters long. In 1972 Corning announced a 4-dB-per-kilometer cable, and in 1973 Corning broke this record with a

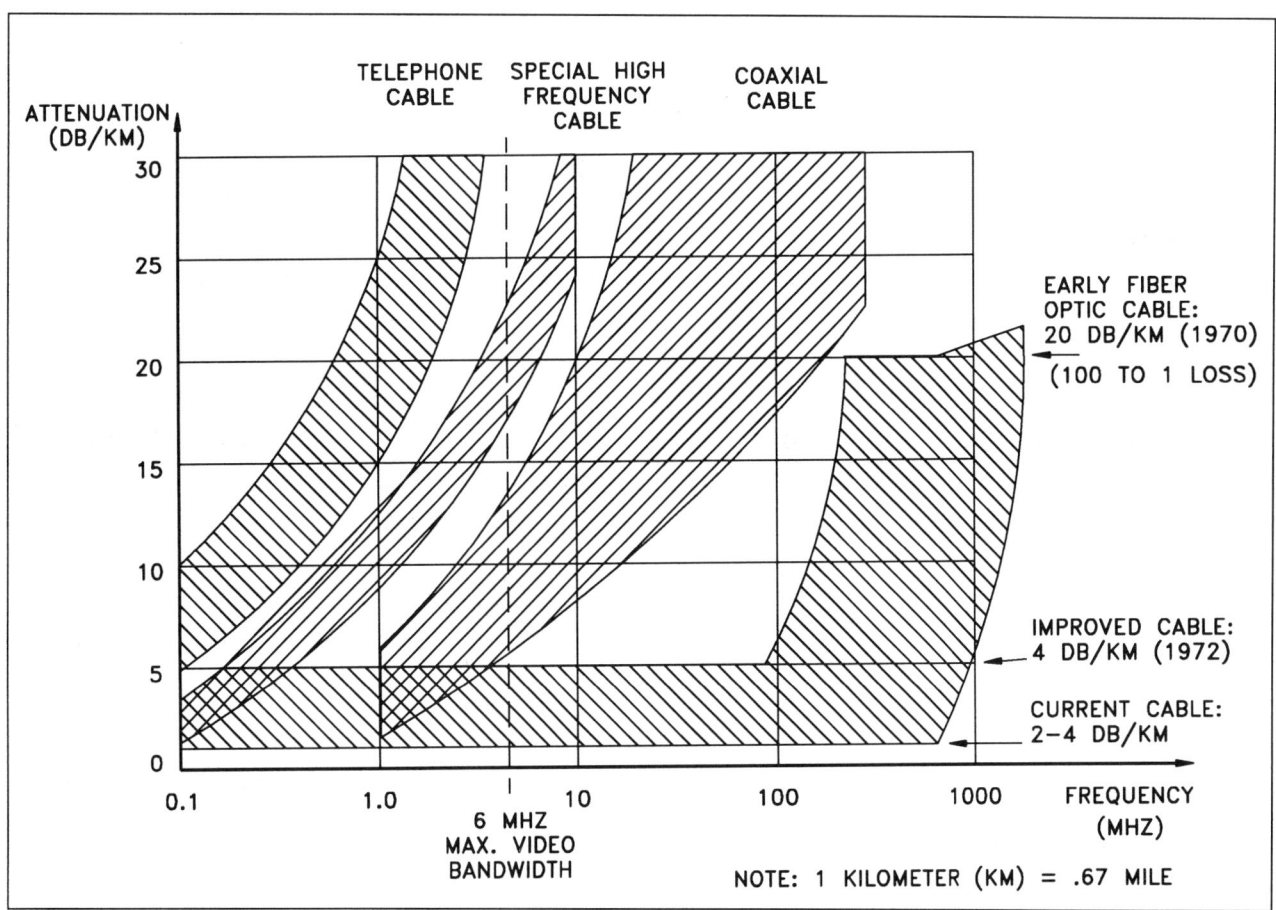

FIGURE 6-16   Attenuation versus frequency for fiber-optic cable and other media

2-dB-per-kilometer cable. This low-loss achievement made a revolution in transmission of wide-bandwidth, long-distance communications inevitable: it was only a matter of time before electrical and coaxial cables would be replaced by fiber-optic cables for communication systems, including CCTV.

6.4.2 Fiber-Optic Theory

In addition to the fiber-optic cable, which replaces the coaxial or two-wire cable in electrical transmission systems, the fiber-optic system uses a transmitter at the camera end and a receiver at the monitor end (Figure 6-15).

The following sections describe these three components. By far the most critical is the fiber-optic cable, since it must transmit the light signal over a long distance without attenuating the signal appreciably or distorting it (changing its shape or attenuation at high frequencies). As shown in Figure 6-15, the signal from the camera is sent to the transmitter via standard coaxial cable. At the receiver end, the output from the receiver is sent via standard coaxial cable to the monitor or recording system.

The optical transmitter at the camera end converts the electrical CCTV analog signal into a corresponding optical signal. The output from the transmitter is an optical signal generated by either a light-emitting diode (LED) or an injection laser diode (ILD) emitting IR light.

The fiber-optic cable consists of one or more glass fibers, each acting as a waveguide or conduit for one video optical signal. The glass fibers are enclosed in a protective outer jacket whose construction depends on the application.

The fiber-optic receiver collects the light from the end of the fiber-optic cable and converts the optical signal back into an electrical signal having the same waveform and characteristics as the original CCTV signal at the camera. This CCTV signal is then sent to the monitor.

The only variation in this block diagram for a single camera is the inclusion of a connection, splice, or repeater that may be required if the cable run is very long (many miles). The connector physically joins the output end of one cable to the input end of a second cable. The splice reconnects two fiber ends so as to make them continuous. The repeater amplifies the light signal to make it large enough to provide a good signal at the receiver end.

How does the fiber-optic transmission system differ from the electrical cable systems described in the previous

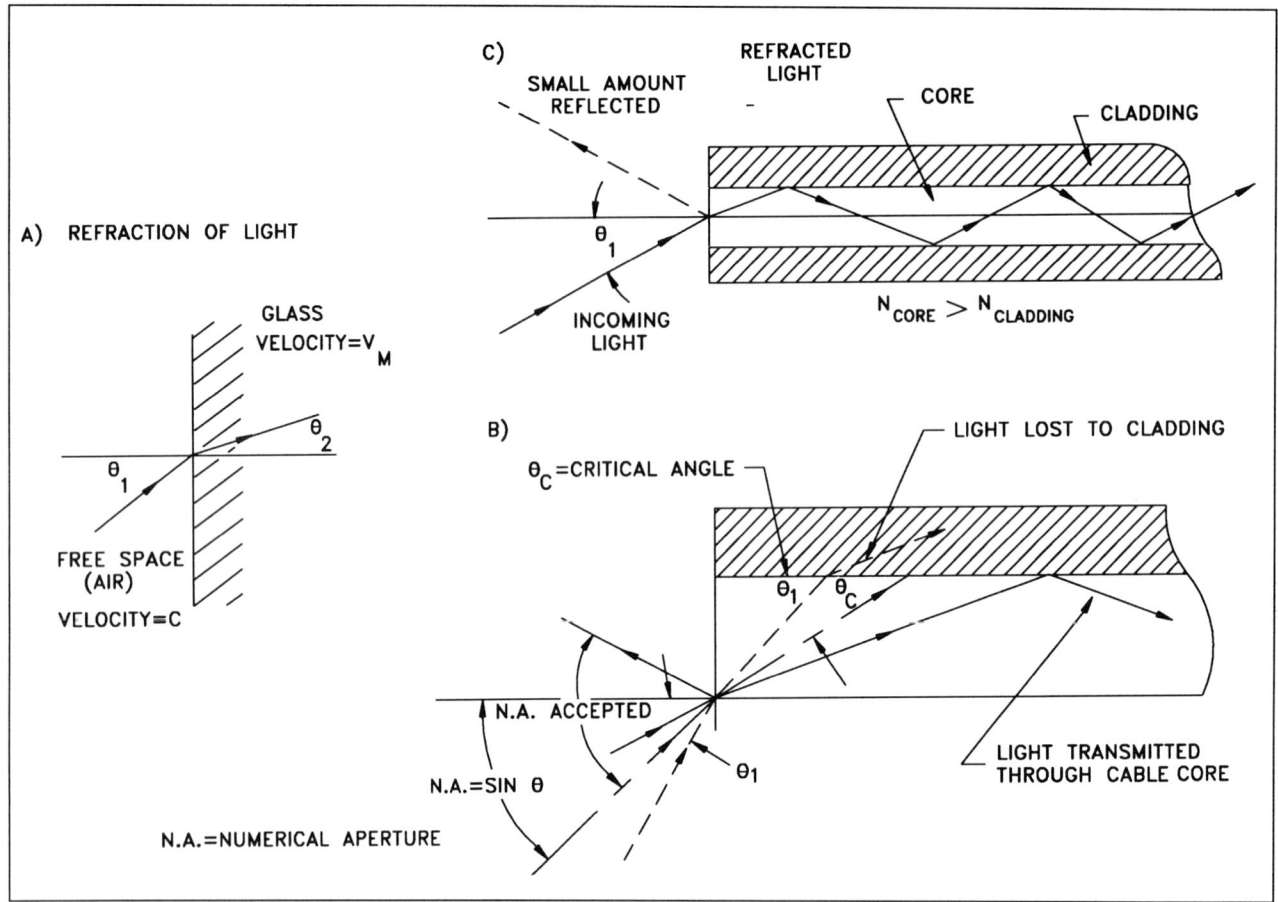

FIGURE 6-17   Light reflection/transmission in fiber optics

sections? From the block diagram (Figure 6-15), it is apparent that two new hardware components are required: a transmitter and receiver (Sections 6.4.3 and 6.4.4). The transmitter provides an amplitude- or frequency-modulated representation of the CCTV signal at a near-IR wavelength, which the fiber optic transmits, and at a level sufficient to produce a high-quality picture at the receiver end. The receiver collects whatever light energy is available at the output of the fiber-optic cable and converts it efficiently, with all the information from the CCTV signal retained, into an electrical signal that is identical in shape and amplitude to the camera output signal.

The fiber-optic cable efficiently transmits the modulated light signal from the camera end over a long distance to the monitor, while maintaining the signal's shape and amplitude. Fiber-optic cable characteristics are totally different from those of coaxial cable or two-wire transmission systems.

Before discussing the construction of the fiber-optic cable, we will briefly describe the transmitting light. In any optical material, light travels at a velocity (V_m) characteristic of the material, which is lower than the velocity of light (C) in free space or air (Figure 6-17a).

The ratio (fraction) of the velocity in the material compared with free space defines the refractive index (n) of the material:

$$n = \frac{C}{V_m}$$

When light traveling in a medium of a particular refractive index strikes another material of a lower refractive index, the light is bent toward the interface of the two materials (Figure 6-17b).

If the angle of incidence is increased, a point is reached where the bent light will travel along the interface of the two materials. This is known as the critical angle (θ_C). Light at any angle greater than the critical angle is totally reflected from the interface and follows a zigzag transmission path (Figure 6-17b, c). This zigzag transmission path is exactly what occurs in a fiber-optic cable: the light entering one end of the cable zigzags through the medium and eventually exits at the far end at approximately the same angle.

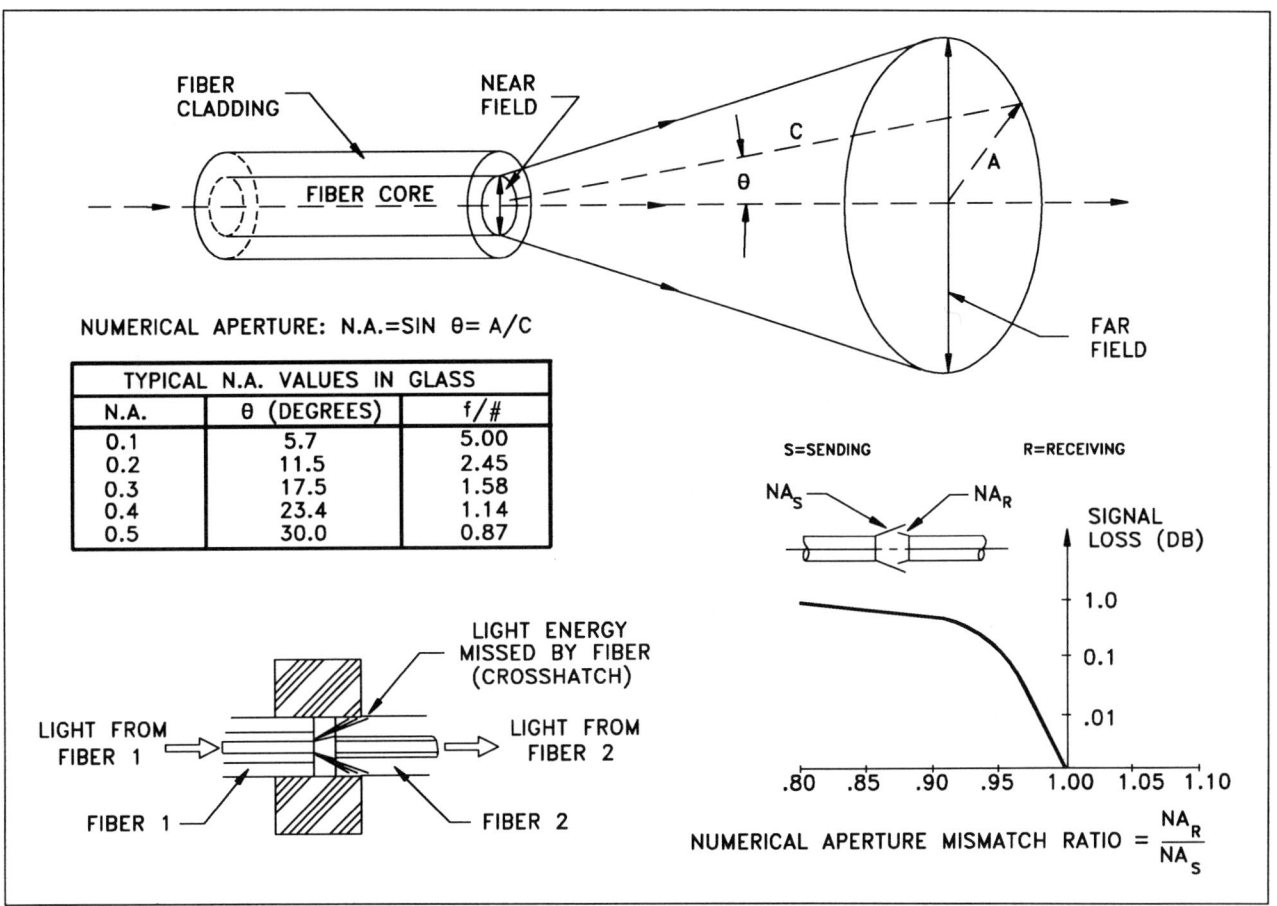

NUMERICAL APERTURE: N.A.=SIN θ= A/C

TYPICAL N.A. VALUES IN GLASS

N.A.	θ (DEGREES)	f/#
0.1	5.7	5.00
0.2	11.5	2.45
0.3	17.5	1.58
0.4	23.4	1.14
0.5	30.0	0.87

$$\text{NUMERICAL APERTURE MISMATCH RATIO} = \frac{NA_R}{NA_S}$$

FIGURE 6-18   Fiber-optic numerical aperture

As shown in Figure 6-17c, some incoming light is reflected from the fiber-optic end and never enters the fiber.

In practice, an optical fiber consists of a core, a cladding, and a protective coating. The core material has a higher index of refraction than the cladding material and therefore the light, as just described, is confined to the core. This core material can be plastic or glass, but glass provides a far superior performance (lower attenuation and greater bandwidth) and therefore is more widespread for long-distance applications.

One parameter often encountered in the literature is the numerical aperture (NA) of a fiber optic, a parameter that indicates the angle of acceptance of light into a fiber—or simply the ease with which the fiber accepts light. The NA is an important fiber parameter that must be considered when determining the signal-loss budget of a fiber-optic system. To visualize the concept, picture a bottle with a funnel (Figure 6-18). The larger the funnel angle, the easier it is to pour liquid into the bottle. The same concept holds for the fiber. The wider the acceptance angle, the higher the NA, the larger the amount of light that can be funneled into it from the transmitter. The larger or higher an optical fiber NA, the easier it is to launch light into the fiber, which correlates to higher coupling efficiency. Since fiber-optic systems are often coupled to LEDs, which are the light generators at the transmitter, and since LEDs have a less-concentrated, diffuse output beam than an ILD, fiber optics with high NAs allow more collection of the LED output power.

In order for the light from the transmitter to follow the zigzag path of internally reflected rays, the angles of reflection must exceed the critical angle. These reflection angles are associated with "waveguide modes." Depending on the size (diameter) of the fiber-optic core, one or more modes are transmitted down the fiber. The characteristics and properties of these different cables carrying single-mode and multimode fibers are discussed in the next section.

Like radio waves, light is electromagnetic energy. The frequencies of light used in fiber-optic video, voice, and data transmission are approximately 3.6×10^{14}, which is several orders of magnitude higher than the highest radio waves. Wavelength (the reciprocal of frequency) is a more common way of describing light waves. Visible light with wavelengths from about 400 nanometers (nm) for deep violet to 750 nm for deep red covers only a small portion of the electromagnetic spectrum (see Chapter 3). Fiber-optic video transmission uses the near-IR region, extending from approximately 750 to 1500 nm, since glass fibers

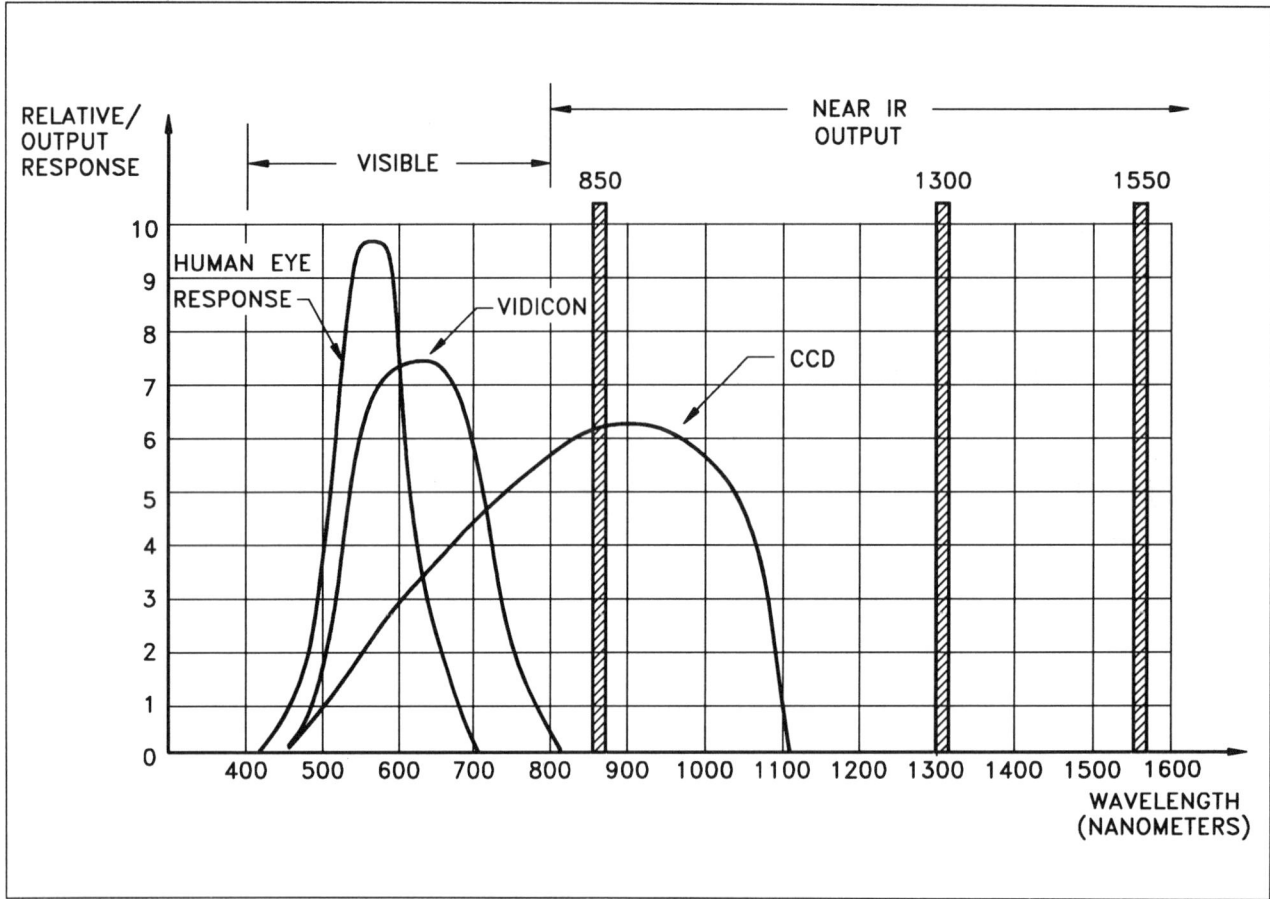

FIGURE 6-19   Fiber-optic transmission wavelengths

propagate light at these wavelengths most efficiently and efficient detectors (silicon and germanium) are available to detect such light.

6.4.3 Fiber-Optic Cable

The most significant part of the fiber-optic signal transmission system is the glass fiber itself, a thin strand of very pure glass approximately the diameter of a human hair. It transmits specific light frequencies with extremely high efficiency. Most fiber-optic systems operate at IR wavelengths (frequencies) of 850, 1300, or 1550 nm. Figure 6-19 shows where these near-IR light frequencies are located with respect to the visible light spectrum.

Most short (i.e., several miles long) fiber-optic security systems operate at a wavelength of 850 nm rather than 1300 or 1550 nm, because 850-nm LED emitters are more readily available and less expensive than their 1300-nm or 1550-nm counterparts. Likewise, IR detectors are more sensitive at 850 nm. LED and ILD radiation at the 1300 and 1550 nm wavelengths is transmitted along the fiber-optic cables more efficiently than at the 850-nm frequency; they are used for much longer run cables (hundreds of miles).

6.4.3.1 Fiber Types

Three types of fibers are used in security systems: (1) multimode step-index (rarely), (2) graded-index, and (3) single-mode (monomode). These three types are defined by the index of refraction (n) profile of the fiber and the cross section of the fiber core. All three types have different properties and are used in different applications.

6.4.3.1.1 Multimode Step-Index Fiber Figure 6-20a illustrates the physical characteristics of the multimode step-index fiber. The fiber consists of a center core of index n = 1.47 and outer cladding of index n = 2. Light rays enter the core and are reflected a multiple number of times down the core and exit at the far end. Since this fiber propagates many modes, it is called multimode step-index. The multimode step-index is usually 50, 100, or even 200 microns (0.002, 0.004, or 0.008 inches) in diameter. The fiber core itself is clad with a thin layer of glass having a sharply different index of refraction. Light travels down the fiber, constantly being reflected back and forth from the interface between the two layers of glass. Light that enters the fiber at a sharp angle is reflected at a sharp angle from the interface and is reflected back and forth many more times, thus traveling more slowly through the fiber than light that

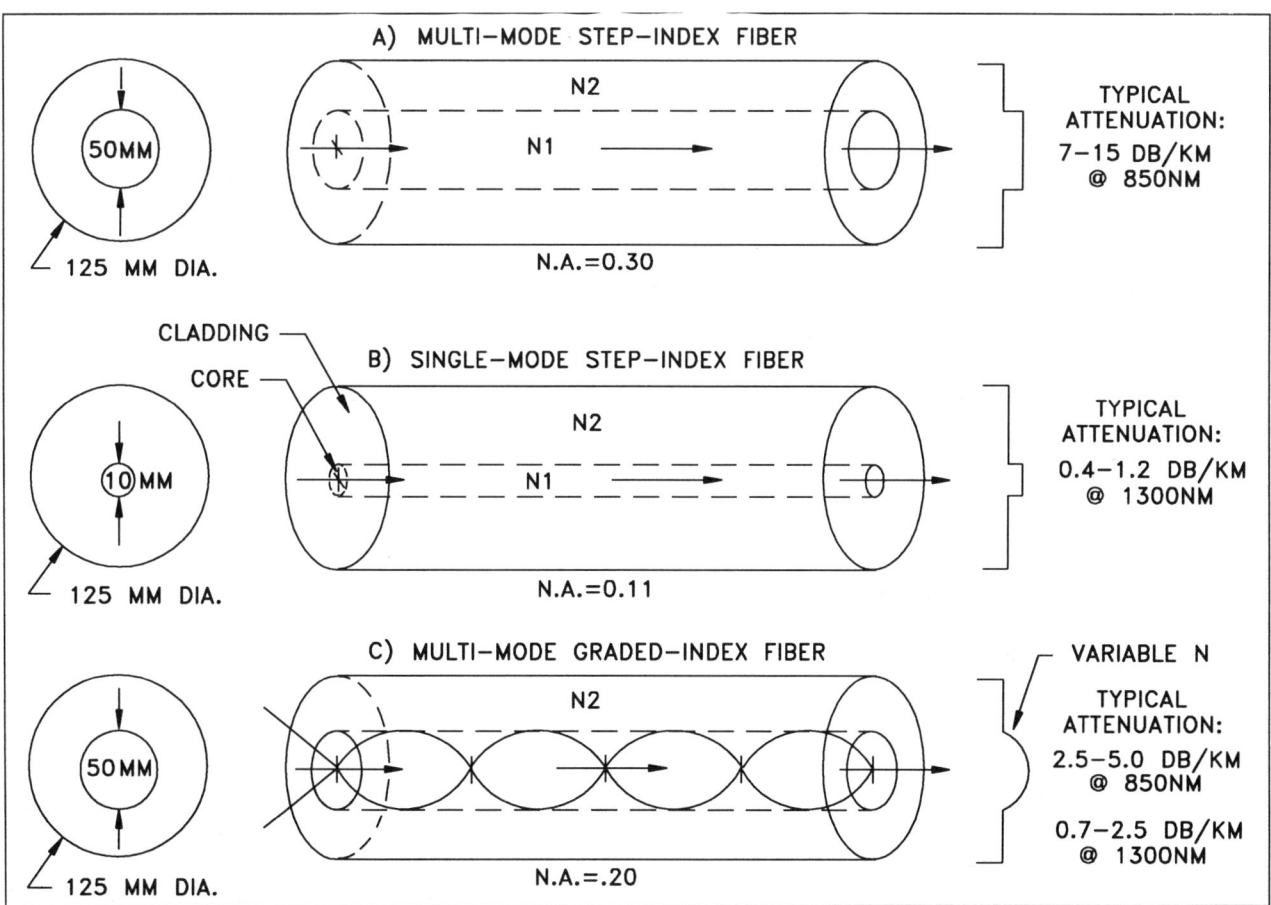

FIGURE 6-20   Multimode and single-mode fiber-optic cable

enters at a shallow angle. The difference in the arrival time at the end of the fiber limits the bandwidth of the step-index fiber, so that most such fibers provide good signal transmission up to a 20-MHz signal for about 1 kilometer. This limitation is more than adequate for many CCTV applications.

6.4.3.1.2 Single-Mode Fiber The single-mode-fiber cable is a step-index fiber cable in which the diameter of the fiber core is much smaller than that of the step-index multimode cable (Figure 6-20b). The single-mode fiber is only 5 to 10 microns (0.0002 to 0.0004 inches) in diameter. In the multimode fiber, the multimode light propagates in many modes involving many repeated reflections. In the single-mode cable with the small core, only a few reflections of light occur. Therefore the fundamental mode or single mode travels down the length of the fiber-optic cable without multiple reflections at the core/clad boundary. By decreasing the fiber cable diameter, transmission is significantly improved and increased with respect to bandwidth and distance. This increase seems attractive on the surface, but it generally does not translate into a good choice for many CCTV applications. The small diameter (10 microns) that must be coupled to the transmitter source and the receiver detector provides an extremely difficult connector

problem. Thus most security installations do not utilize the single-mode fiber. Such fiber should be considered only when very long transmission runs (many miles) are required.

6.4.3.1.3 Multimode Graded-Index Fiber The multimode graded-index fiber is the workhorse of the CCTV security industry (Figure 6-20c). Its low power attenuation—less than 3 dB (50% loss) per kilometer at 850 nm makes it well suited for short and long cable runs. Most fibers are available in 50-micron-diameter core with 125-micron total fiber diameter (exclusive of outside protective sheathing). Graded-index fiber sizes are designated by their core/cladding diameter ratio, thus the 50/125 fiber has 50-micron-diameter core and a 125-micron cladding. The typical graded-index fiber has a bandwidth of approximately 1000 MHz per km and is one of the least expensive fiber types available. The 50/125 fiber provides high efficiency when used with a high-quality LED transmitter or, if very long distances or very wide bandwidths are required, an ILD source. Table 6-4 lists some of the common cable sizes available.

For the graded-index fiber, the index of refraction (n) of the core is highest at the center and gradually decreases as the distance from the center increases (Figure 6-20c). Light in this type of core travels by refraction: the light rays are continually

FIBER TYPE	DIAMETER * (MICRONS)			TYPICAL CABLE PARAMETERS					
				SINGLE FIBER		2 FIBER		4 FIBER	
	CORE	CLADDING	BUFFERING	OD** (MM)	WEIGHT (KG/KM)	OD** (MM)	WEIGHT (KG/KM)	OD** (MM)	WEIGHT (KG/KM)
10/125	10	125	250	3.0	6.5	3.5	6.8	7.1	50
50/125+	50	125	250	2.6	6.5	3.4x6	22	8	55
62.5/125	62.5	125	250	3.0	6.4	3.0x6.1	18	9.4	65.5
100/140	100	140	250	2.6	6.5	3.4	22	7.1	50

*OD = OUTSIDE DIAMETER (1 MICRON=.00004 INCH)

**CABLE OUTSIDE DIAMETER

+MOST WIDELY USED IN SECURITY APPLICATIONS

1MM=1000 MICRONS

1 KG/KM = 0.671 LB/1000FT

Table 6-4   Standard Fiber-Optic-Cable Sizes

bent toward the center of the fiber-optic axis. In this manner the light rays traveling in the center of the core have a lower velocity due to the high index of refraction, and the rays at the outer limits travel much faster. This effect causes all the light to traverse the length of the fiber in nearly the same time and greatly reduces the difference in arrival time of light from different modes, thereby increasing the fiber bandwidth carrying capability.

The graded-index fiber satisfies long-haul, wide-bandwidth security system requirements that cannot be met by the multimode step-index fiber. At the same time, because of its larger diameter, the graded-index fiber does not impose the connector problems of the single-mode cable.

6.4.3.2 Fiber-Cable Construction

A fiber-optic cable consists of several components. The optical fiber is generally surrounded by a tube of plastic that is substantially larger than the fiber itself. Over this tube is a layer of Kevlar reinforcement material. The entire assembly is then coated with an outer jacket, typically polyvinyl chloride (PVC). This construction is generally accepted for use indoors or where cable is easily pulled through dry conduit.

There are two main approaches to giving primary protection to a fiber: the tight buffer and the loose tube (Figure 6-21). The tight buffer uses a dielectric (insulator) material such as PVC or polyurethane applied tightly to the fiber. For medium- and high-loss fibers (step-index type), such cable-induced attenuation is small compared with overall attenuation. The tight buffer offers the advantages of smaller bend radii and better crush resistances than loose-tube cabling. These advantages make tightly buffered fibers useful in applications of short runs where sharp bends are

encountered or where cables may be laid under carpeted walking surfaces.

Microbends caused by tight buffers are eliminated by placing the fiber within a hard plastic tube that has an inside diameter several times larger than the diameter of the fiber. Such loose-tube cabling isolates the fiber from the rest of the cable, allowing the cabling to be twisted, pulled, and otherwise stressed with little effect on the fiber. Fibers for long-distance applications typically use a loose tube, since decoupling of the fiber from the cable allows the cable to be pulled long lengths during installation. The tubes may be filled with jelly to protect against moisture, which can condense and freeze, damaging the fiber.

Fiber-optic cable can be further classified into indoor and outdoor types. They differ in the jacket surrounding the fiber and the protective sheath that gives it sufficient tensile strength to be pulled through a conduit or overhead duct or strung on poles. Single indoor cables (Figure 6-22) consist of the clad fiber-optic cable surrounded by a Kevlar reinforcement sheath, wrapped in a polyurethane jacket for protection from abrasion and the environment. The outdoor cable has additional protective sheathing for additional environmental protection.

Plenum cables are available for indoor applications that require specific smoke- and flame-retardant characteristics, do not require the use of a metal conduit, and are classified by UL or other codes.

When higher tensile strength is needed, additional strands of Kevlar are added outside the polyethylene jacket and another polyethylene jacket provided over these Kevlar reinforcement elements. Some indoor cables utilize a stranded-steel central-strength member or nonmetallic Kevlar. Kevlar is preferred in installations located in explosive areas or areas of high electromagnetic interference,

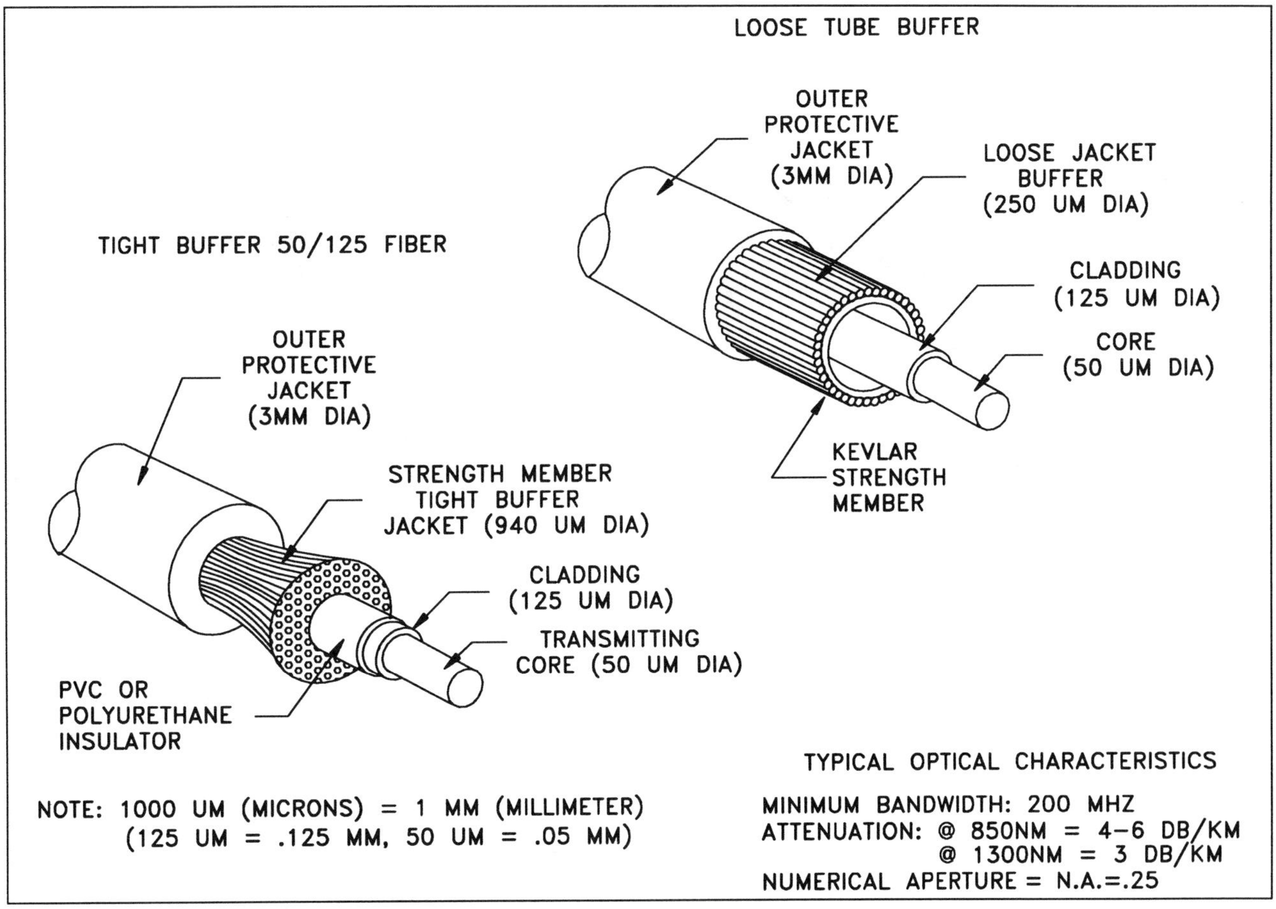

LOOSE TUBE BUFFER
OUTER PROTECTIVE JACKET (3MM DIA)
LOOSE JACKET BUFFER (250 UM DIA)
CLADDING (125 UM DIA)
CORE (50 UM DIA)
KEVLAR STRENGTH MEMBER
TIGHT BUFFER 50/125 FIBER
OUTER PROTECTIVE JACKET (3MM DIA)
STRENGTH MEMBER TIGHT BUFFER JACKET (940 UM DIA)
CLADDING (125 UM DIA)
TRANSMITTING CORE (50 UM DIA)
PVC OR POLYURETHANE INSULATOR
NOTE: 1000 UM (MICRONS) = 1 MM (MILLIMETER)
(125 UM = .125 MM, 50 UM = .05 MM)
TYPICAL OPTICAL CHARACTERISTICS
MINIMUM BANDWIDTH: 200 MHZ
ATTENUATION: @ 850NM = 4–6 DB/KM
@ 1300NM = 3 DB/KM
NUMERICAL APERTURE = N.A.=.25

FIGURE 6-21 Tight-buffer and loose-tube single-fiber-optic-cable construction

where nonconducting strength members are desirable. Kevlar rather than steel is used predominantly in fiber-optic cables. The mechanical properties of cables typically found on data sheets include crush resistance, impact resistance, bend radius, and strength.

An outdoor cable or one that will be subjected to heavy stress—in long-cable-run pulls in a conduit or aerial application—uses dual Kevlar/polyethylene layers as just described. The polyethylene coating also retards the deleterious effects of sunlight and weather.

INDOOR
SINGLE
DUPLEX
PVC Outer Jacket
Braided Kevlar Strength Member
Loose Tube
Buffered Optical Fiber
OUTDOOR
KEVLAR REINFORCEMENT
TIGHT BUFFER
LOW SMOKE AND FLAME SPREAD FLUOROPOLYMER JACKET
OPTICAL FIBER

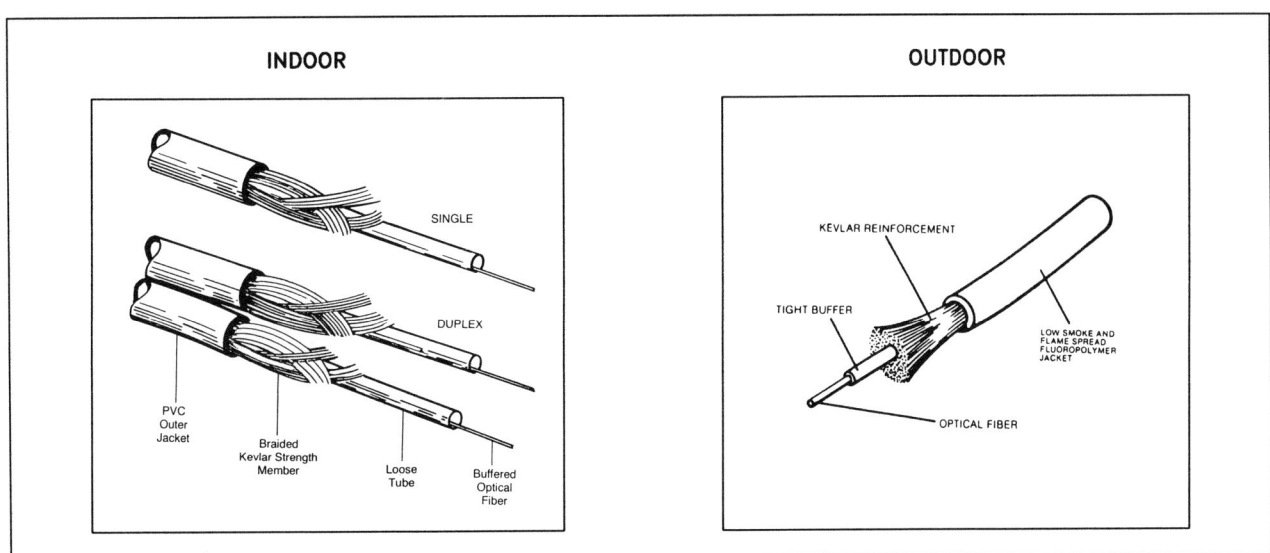

FIGURE 6-22 Indoor and outdoor fiber-optic-cable construction

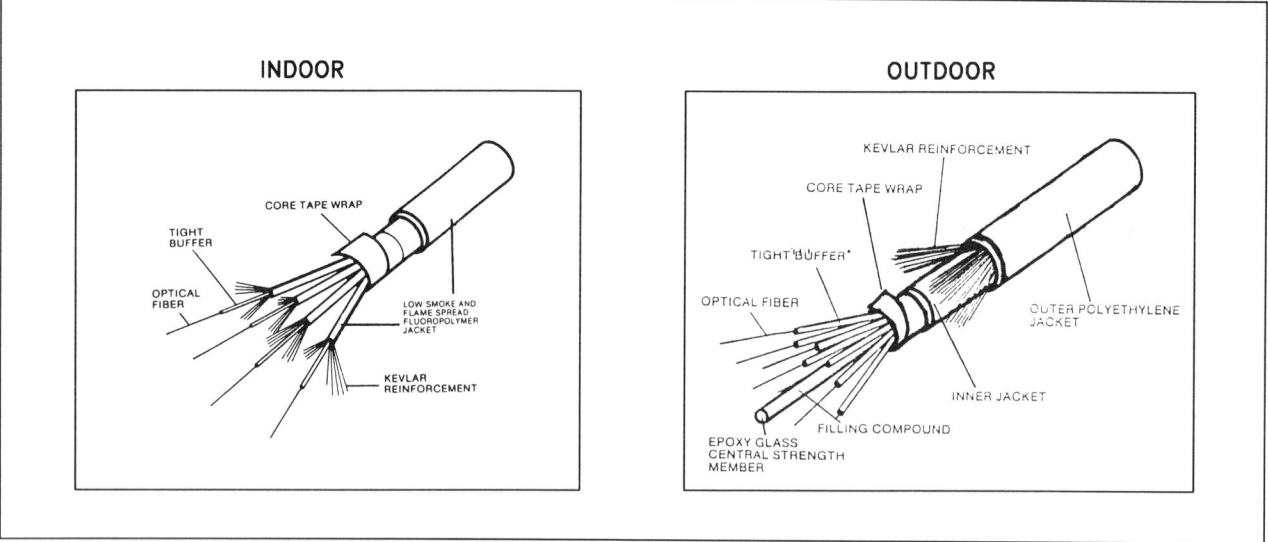

FIGURE 6-23   Multi-conductor fiber-optic cable

When two fibers are required, two single cable structures may be paired in Siamese fashion (side by side) with a jacket surrounding around them.

If additional fiber-optic runs are required, multifiber cables (having four, six, eight, or ten fibers) with similar properties are used (Figure 6-23). The fibers are enclosed in a single or multiple buffer tube array around a tensile-strength member composed of Kevlar and then surrounded with an outer jacket of Kevlar.

6.4.3.3 Fiber-Cable Sizes

Multimode graded-index fiber is available in several primary core sizes: 50/125, 62.5/125, 85/125, and 100/140. Table 6-4 summarizes the properties of five different fiber-cable types, indicating the sizes, numerical aperture, attenuation, and bandwidth. The first number (50) refers to the core outside diameter size, the second (125) to the glass fiber outside diameter (the sizes exclude reinforcement or sheathing). The fiber size is expressed in microns, which is equivalent to one one-thousandth of a millimeter (1/1000 mm). By comparison, the diameter of a human hair is about 0.002 inches or 50 microns.

Each size has advantages for particular applications, and all four are EIA standards. The most popular and least expensive multimode fiber is the 50/125, used extensively in security CCTV. Compared with the other three larger fibers, its geometric tolerances are more critical because of its smaller core size. It has the lowest NA of any multimode fiber, which allows the highest collection of light and highest bandwidth. Because 50/125 has been used for many years, established installers are experienced and comfortable working with it. Many connector types are available for terminating the 50/125 cable. The 62.5/125 cable is an alternative to the 50/125 fiber.

Unlike 50/125 and 62.5/125, which were developed for telephone networks, the 85/125 was developed specifically for computer or digital local networks where short distances are required. The slightly larger 85-micron size permits easier connector specifications and LED source requirements.

The 100/140 multimode fiber was developed in response to computer manufacturers, who wanted an LED-compatible, short-wavelength, optical-fiber data link that could handle higher data rates than coaxial cable. While this fiber was developed for the computer market it is excellent for short-haul CCTV security applications. It is least sensitive to fiber-optic geometry variations and connector tolerances, which generally means lower losses at joint connections. This is particularly important in industrial environments where the cable may be disconnected and connected many times. The only disadvantage of 140-micron-outside-diameter is that it is nonstandard, so available connectors are fewer and more expensive than those for the 125-micron size.

6.4.4 Connectors and Fiber Termination

This section describes fiber-optic connectors, techniques for finishing the fiber ends when terminated with connectors, and potential connector problems. For very long cable runs, joining and fusing the actual glass-fiber core and cladding is done by a technique called splicing. Splicing joins the two lengths of cable by fusing the two fibers (locally melting the glass) and physically joining them in a permanent connection (Section 6.4.5).

Coaxial cables require connectors at their ends to interface with the connectors on cameras, switchers, monitors, and so on. Likewise, fiber-optic cables require connectors to couple the optically transmitted signal from the transmitter into the fiber-optic cable, and at the other end to couple the light output from the fiber into the receiver. The term *connector* refers to the terminating devices that are used to connect the cable to the transmitter or receiver, or to join two lengths of cable together. If the fiber-optic cable run is very long or must go through various barriers (e.g., walls), the total run is often fabricated from sections of fiber-optic cable and each end joined with connectors. This is equivalent to an interline coaxial connector.

A large variety of optical connectors is available for terminating cables. Most are based on butt coupling of cut and polished fibers to allow direct transmission of optical power from one fiber core to the other. Such a connection is made using two mating connectors, precisely centering the two fibers into the connector ferrules and fixing them in place with epoxy. The ferrule and fiber surfaces at the ends of both cables are then ground and polished to produce a clean optical surface. The two most common types are the cylindrical and cone ferrule connectors.

6.4.4.1 Factors Affecting Coupling Efficiency

The efficiency of light transfer from the end of one fiber-optic cable to the following cable or device is a function of six different parameters: (1) fiber-core lateral or axial misalignment, (2) angular core misalignment, (3) fiber end separation, (4) fiber distortion, (5) fiber end finish, and (6) Fresnel reflections. Of these loss mechanisms, distortion loss and the effects of fiber end finish can be minimized by using proper techniques when the fibers are prepared for termination. A chipped or scratched fiber end will scatter much of the light signal power, but proper grinding and polishing minimize these effects in epoxy/polish-type connectors.

Lateral misalignment of fiber cores causes the largest amount of light loss, as shown in Figure 6-24a. An evaluation of the overlap area of laterally misaligned step-index fibers indicates that a total misalignment of 10% of a core diameter yields a loss of greater than 0.5 dB. This means that a fiber core of 0.002 inches (50 microns) must be placed within 0.0001 inches of the center of its connector for a worst-case lateral misalignment loss of 0.5 dB. While this dimension is small, the connection is accomplished in the field.

Present connector designs maintain angular alignment well below one degree (Figure 6-24b), which adds only another 0.1 dB (2.3%) of loss for most fibers.

Fiber-end-separation loss depends on the NA of the fiber. Since the optical light power emanating from a transmitting fiber is in the form of a cone, the amount of light coupled into the receiving fiber or device will decrease as the fibers are moved apart from each other (Figure 6-24c). A separation distance of 10% of the core diameter using a fiber with an NA of 0.2 can add another 0.1 dB of loss.

Fresnel losses usually add another 0.3 to 0.4 dB when the connection does not use an index-matching fluid (Figure 6-24d).

The summation of all of these different losses often adds up to 0.5 to 1.0 dB for ST type (higher for SMA 1906) terminations and connections (Table 6-5).

6.4.4.2 Cylindrical Ferrule Connector

In the cylindrical ferrule design, the two connectors are joined and the two ferrules are brought into contact inside precisely guiding cylindrical sleeves. Figure 6-25 shows the geometry of this type of connection.

Lateral offset in cylindrical ferrule connectors is usually the largest loss contributor. In a 50-micron graded-index fiber, 0.5 dB (12%) loss results from a 5-micron offset. A loss of 0.5 dB can also result from a 35-micron gap between the ends of the fibers, or from a 2.5-degree tilted fiber surface. Commercial connectors of this type reach 0.5 to 1 dB (12–26%) optical loss for ST type and higher for SMA 1906. Optical-index-matching fluids in the gap further reduce the loss.

6.4.4.3 Cone Ferrule Connector

The cone ferrule termination technique centers the fiber in one connector and insures concentricity with the mating fiber in the other connector using a cone-shaped plug instead of a cylindrical ferrule. The key to the cone connector design is the placement of the fiber-centering hole (in the cone) in relationship to the true center, which exists when the connector is mated to its other half. A fiber (within acceptable tolerances) is inserted into the ferrule, adhesive is added, and the ferrule is compressed to fit the fiber size while the adhesive sets. The fiber faces and ferrule are polished to an optical finish and the ferrule (with fiber) is placed into the final alignment housing. Most low-loss fiber-optic connections are made utilizing the cone-shaped plug technique.

The two most popular cone-shaped designs are the small-fiber resilient (SFR) bonded connector and the SMA, a redesigned coaxial connector style (Figure 6-26). Both use the cone-shaped ferrule, which provides a reliable, low-cost, easily assembled termination in the field. Both connectors can terminate fibers with diameters of 125-micron cladding.

The technique eliminates almost all fiber and connector error tolerance buildup that normally causes light losses. It makes use of a resilient material for the ferrule, metal for the construction of the retainer assembly, and a rugged metallic connector for termination. The fiber alignment is repeatable after many connects and disconnects due to the tight interference fit of the cone-shaped ferrule into the cone-shaped mating half. This cone design also forms a sealed interface for a fiber-to-fiber or fiber-to-active-device junction, such as fiber cable to transmitter or fiber cable to receiver. Tolerances in the fiber diameter are absorbed by the resiliency of the plastic ferrule. This connector offers a

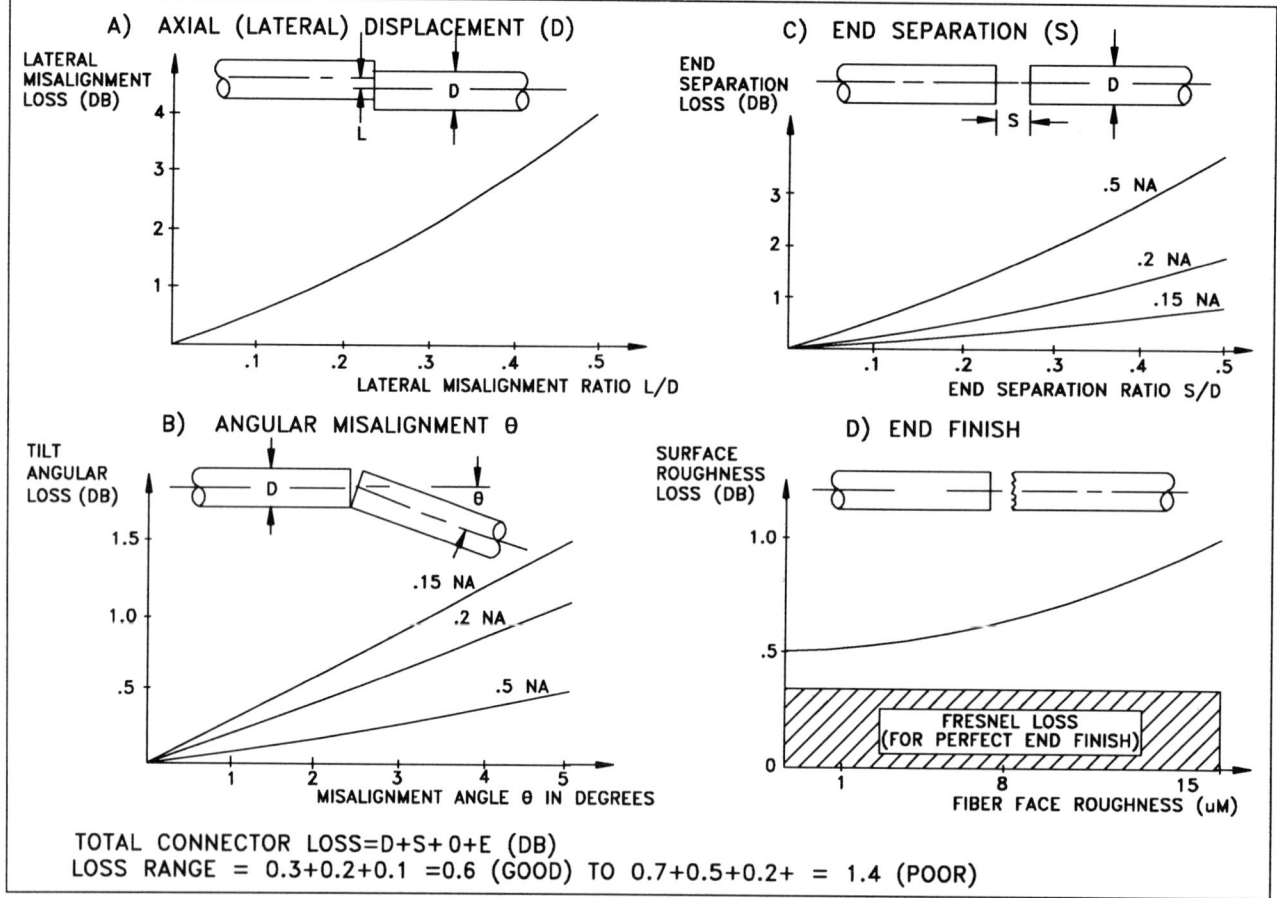

FIGURE 6-24   Factors affecting fiber-optic coupling efficiency

maximum signal loss of 1.0 dB (26%) and provides repeatable coupling and uncoupling with little increase in light loss.

The popular SMA-style connector is compatible with many other SMA-manufacturer-type connectors and terminates the 125-micron fibers. Internal ferrules ensure axial fiber alignment to within 0.1 degree. The SMA connector has a corrosion-resistant metal body and is available in an environmentally sealed version.

6.4.4.4 Single-Mode Connections

When terminating and connecting single-mode fibers approximately 10 microns in diameter, alignment and other problems are magnified by almost a factor of ten over multimode assemblies. Thus single-fiber types and attendant connectors are not as common in the CCTV security field, although the situation is rapidly changing. The single-mode fiber-optic transmission system scheme should be avoided unless an expert is available for its design and implementation.

6.4.5 Fiber-Optic Termination Kits

An efficient fiber-optic-cable transmission system relies on a high-quality termination of the cable core and cladding. This step requires use of perhaps unfamiliar but easy tech-niques with which the installer must be acquainted. Fiber-terminating kits are available from most fiber-cable, connector, and accessory manufacturers. Figure 6-27 shows a complete kit, including all grinding and polishing compounds, alignment jigs, tools, and instructions.

Manufacturers can provide descriptions of the various techniques for terminating the ends of fiber-optic cables, including cable preparation, grinding, polishing, testing, and so forth.

6.4.6 Splicing Fibers

Although not often required, splicing of single- and multimode fibers is sometimes necessary. In systems with long fiber-optic-cable runs (longer than 2 km), it is necessary to splice cable sections together rather than connect them by terminating the fiber ends and using connectors. A splice made between two fiber-optic cables can provide a connection with only one-tenth the optical loss as that obtained when a connector is inserted between fibers. Good fusion splices made with an electric arc produce losses as low as 0.05 to 0.1 dB (1.2 to 2.3% loss). While making a splice via a fusing technique is more difficult and requires more

CABLE LOSS TYPE	TYPICAL LOSS		COMMENTS
	(DB)	(%)	
AXIAL–LATERAL DISPLACEMENT (10%)	0.55	12.0	MOST CRITICAL FACTOR
ANGULAR MISALIGNMENT (2 DEGREES)	0.30	6.7	FUNCTION OF NUMERICAL APERTURE
END SEPARATION (AIR GAP)	0.32	7.0	ESSENTIALLY ELIMINATED USING INDEX MATCHING FLUID
END FINISH: A)ROUGHNESS (1 MICRON) B)NON PERPENDICULAR	0.50 0.25	11.0 5.6	INCLUDES FRESNEL LOSS (.35 DB) LOSS NOT COMMONLY FOUND
CORE SIZE MISMATCH: 1% DIAMETER TOLERANCE ±5% DIAMETER TOLERANCE	0.17 0.83	4.0 18.0	LOSS OCCURS ONLY WHEN LARGER CORE COUPLES INTO SMALLER CORE
NUMERICAL APERTURE (NA) DIFFERENCE OF ±0.02 (2%)	1.66	31.6	CRITICAL FACTOR WHEN NA_S IS LARGER THAN NA_R

S=SENDING FIBER
R=RECEIVING FIBER

NOTE: DB = DECIBELS = 10 LOG $\frac{POWER_S}{POWER_R}$

Table 6-5 Fiber-Optic-Connector Coupling Losses

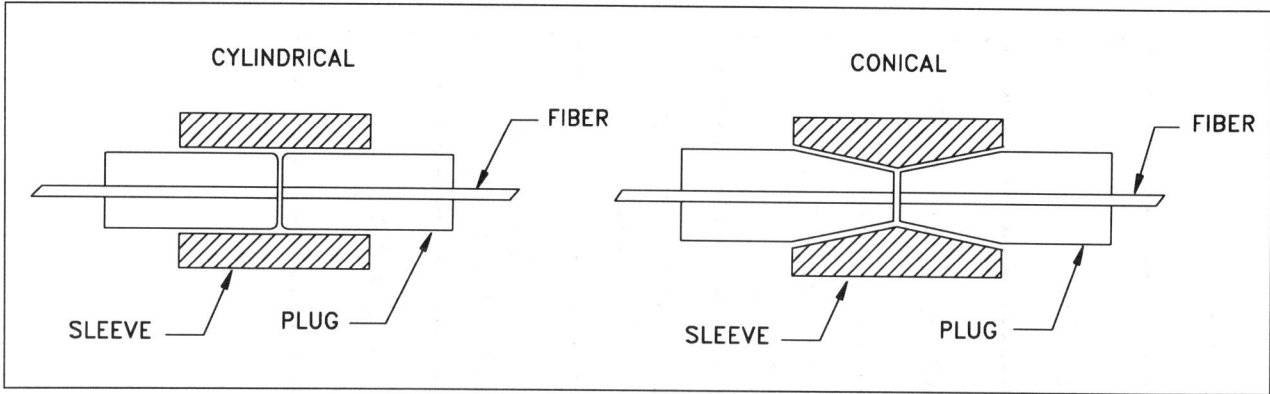

FIGURE 6-25 Cylindrical and conical butt-coupled fiber-optic ferrule connectors

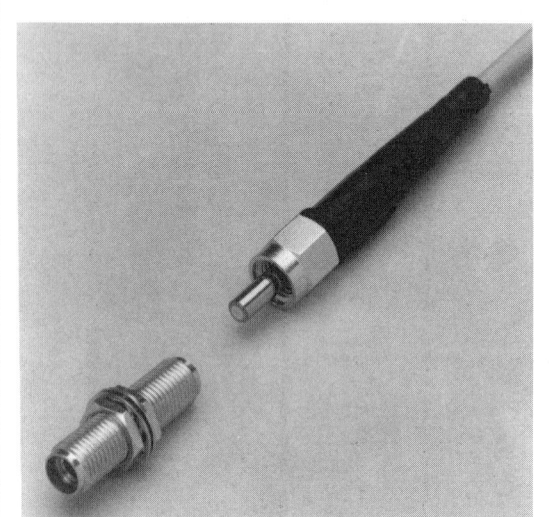

TYPE: SMA
THREADED—SCREW ON

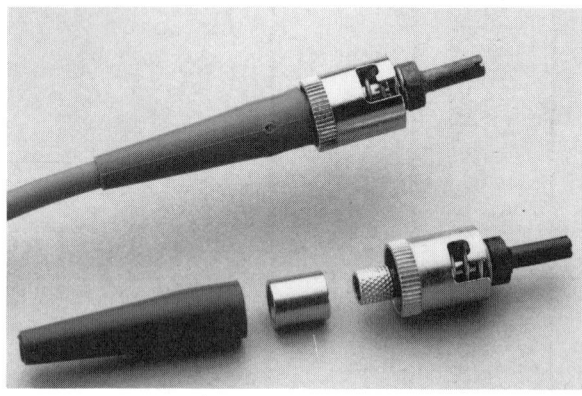

TYPE: ST(SFR) *
POLARIZED AND SPRING LOADED
QUARTER TURN BAYONET LOCK

* SMALL FIBER RESILIENT

FIGURE 6-26   SMA and SFR connectors

equipment and skill than terminating the end of a fiber with a connector, it is worth the cost if it eliminates the use of an in-line amplifier. The splice can also be used to repair a damaged cable, eliminating the need to add connector terminations, which would decrease the light signal level.

Although terminating or splicing multimode (either step-index or graded-index) fibers is relatively straightforward and accomplished in the shop or in the field by experienced personnel, making terminations on the single-mode (10-micron-core) type of fiber is quite difficult. It is used only in very long fiber-optic runs and accomplished by experienced installers.

6.4.7 Fiber-Optic Transmitter

Referring to Figure 6-15, the fiber-optic transmitter is the electro-optic transducer between the camera television electrical signal output and the light signal input to the fiber-optic cable. The function of the transmitter is to efficiently and *accurately* convert the electrical video signal into an optical signal and couple it into the fiber optic. The transmitter electronics convert the amplitude-modulated CCTV signal through the LED or ILD into an AM or FM light signal, which faithfully represents the CCTV signal. The transmitter consists of an LED for normal security applications or an ILD when a very long range transmission is required. LEDs are used for most CCTV security applications. Figure 6-28 illustrates the block diagram for the transmitter unit.

6.4.7.1 Light-Emitting Diode

The LED light source is a semiconductor device made of gallium arsenide (GaAs) or a related semiconductor compound with the ability to linearly convert an electrical video signal to an optical signal, thereby making it very suited for CCTV transmission applications at distances up to several

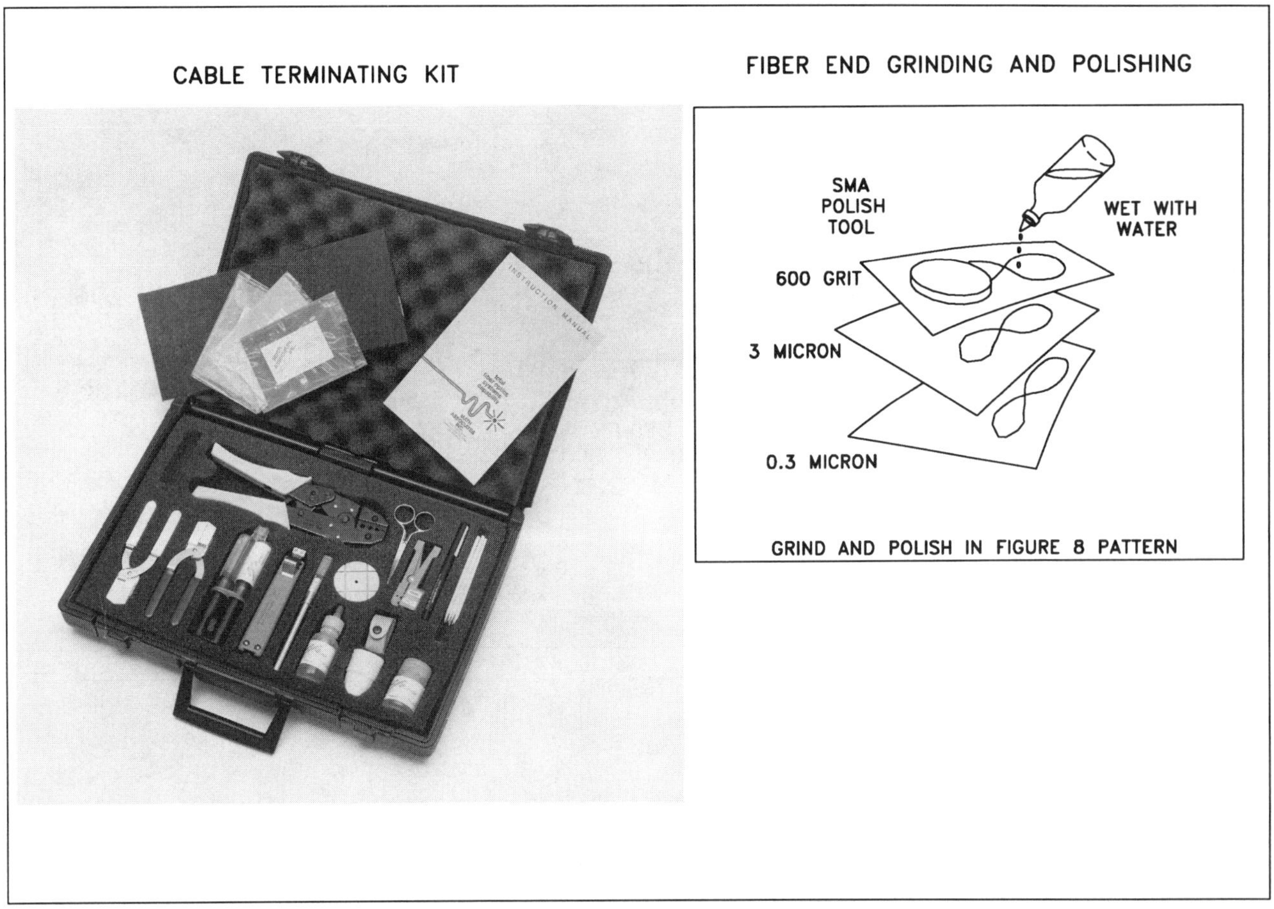

CABLE TERMINATING KIT

FIBER END GRINDING AND POLISHING

FIGURE 6-27   Fiber-optic termination kit

miles and at frequencies up to 80 MHz. The LED is a diode junction that spontaneously emits nearly monochromatic (single wavelength or color) radiation into a narrow light beam when current is passed through it. While the ILD is more powerful, the LED is more reliable, less expensive, and easier to use. The ILD is used with single-mode, long-haul, wide-bandwidth fiber-optic applications.

The LED's main requirements as a light source are (1) to have a fast operating speed to meet the bandwidth requirements of the video signal, (2) to provide enough optical power to provide the receiver with a signal-to-noise (S/N) ratio suitable for a good television picture, and (3) to produce a wavelength that takes advantage of the low-loss propagation characteristics of the fiber.

6.4.7.2 Modulation

For CCTV applications, the electrical video signal from the camera is amplitude- or frequency-modulated and converted to current or carrier variations in the LED. The LED optical output power varies directly proportionally to the electrical input signal for AM (it's constant for FM).

850-nm LEDs are best suited for video analog applications, since they can be amplitude-modulated, that is, the electrical video signal can be converted to a light output signal as a nearly linear function of the LED drive current. This produces a very faithful transformation of the electrical video information to the light information that is transmitted along the fiber-optic cable.

6.4.7.3 Cone Angle

The parameters that constitute a good light source for injecting light into a fiber-optic cable are those that produce as intense a light output into as small a cone diameter as possible. Another factor affecting the light-transmission efficiency is the cone angle of the LED output that can be accepted by and launched down the fiber-optic cable. Figure 6-29 illustrates the LED-coupling problem.

The entire output beam from the LED (illustrated by the cone of light) is not intercepted or collected by the fiber-optic core. This unintercepted illumination loss can be a problem when the light-emitting surface is separated from the end of the fiber core. Most LEDs have a lens at the

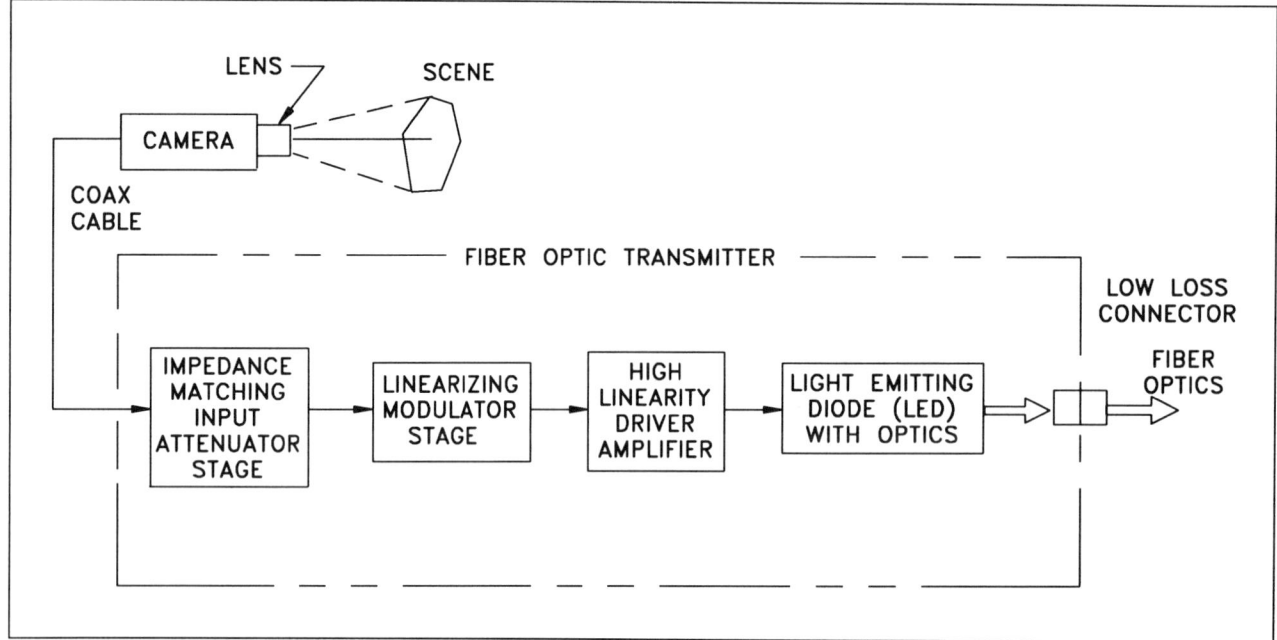

FIGURE 6-28 Block diagram of LED fiber-optic transmitter

surface of the LED package to collect the light from the emitting source and concentrate it onto the core of the fiber.

6.4.7.4 Wavelength

An important characteristic of the transmitter output is the wavelength of the emitted light. This should be compatible with the fiber's minimum-attenuation wavelength, which is 850 nm (in the IR region) for most CCTV fiber-optic cable. The wavelength of light emitted by an LED depends on the semiconductor material composition. Pure GaAs diodes emit maximum power at a wavelength of 940 nm (near-IR), which is undesirable because most glass fibers have a high attenuation at that wavelength. Adding aluminum to GaAs to produce a GaAlAs diode yields a maximum power output at a wavelength between 800 and 900 nm, with the exact wavelength determined by the percentage of aluminum. In most transmitters today, the emitting wavelength is 850 nm, which matches the maximum transmission capability of the glass fiber.

Alternative transmitting wavelengths are 1060, 1300, and 1550 nm, which are regions where glass fibers exhibit a lower attenuation and dispersion than at 850 nm. These wavelengths are produced by combining the element indium with gallium arsenide (InGaAs) and are used in some long-distance transmission applications.

6.4.8 Fiber-Optic Receiver

The term *receiver* at the output end of the fiber-optic cable refers to both a light-detecting transducer and its related electronics, which provides any necessary signal conditioning to restore the signal to its original shape at the input, as well as additional signal amplification. To interface the receiver with the optical fiber, the proper match between light source, fiber-optic cable, and light detector is required. In the AM transmission system, the optical power input at the fiber is modulated so the photodetector operating in the photocurrent mode must provide good linearity, speed, and stability. The photodiode produces no electrical gain and is therefore followed by circuits that amplify electrical voltage and power to drive the coaxial cable. Figure 6-30 illustrates the block diagram for the receiver unit.

As light exits from the receiver end of an optical fiber, it spreads out with a divergence approximately equal to the acceptance cone angle at the transmitter end of the fiber. Photodiodes are packaged with lenses on their housings so that the lens collects this output energy and focuses it down onto the photodiode-sensitive area.

The most common fiber-optic receiver uses a photodiode to convert the incident light from the fiber into electrical energy. After the light energy is converted into an

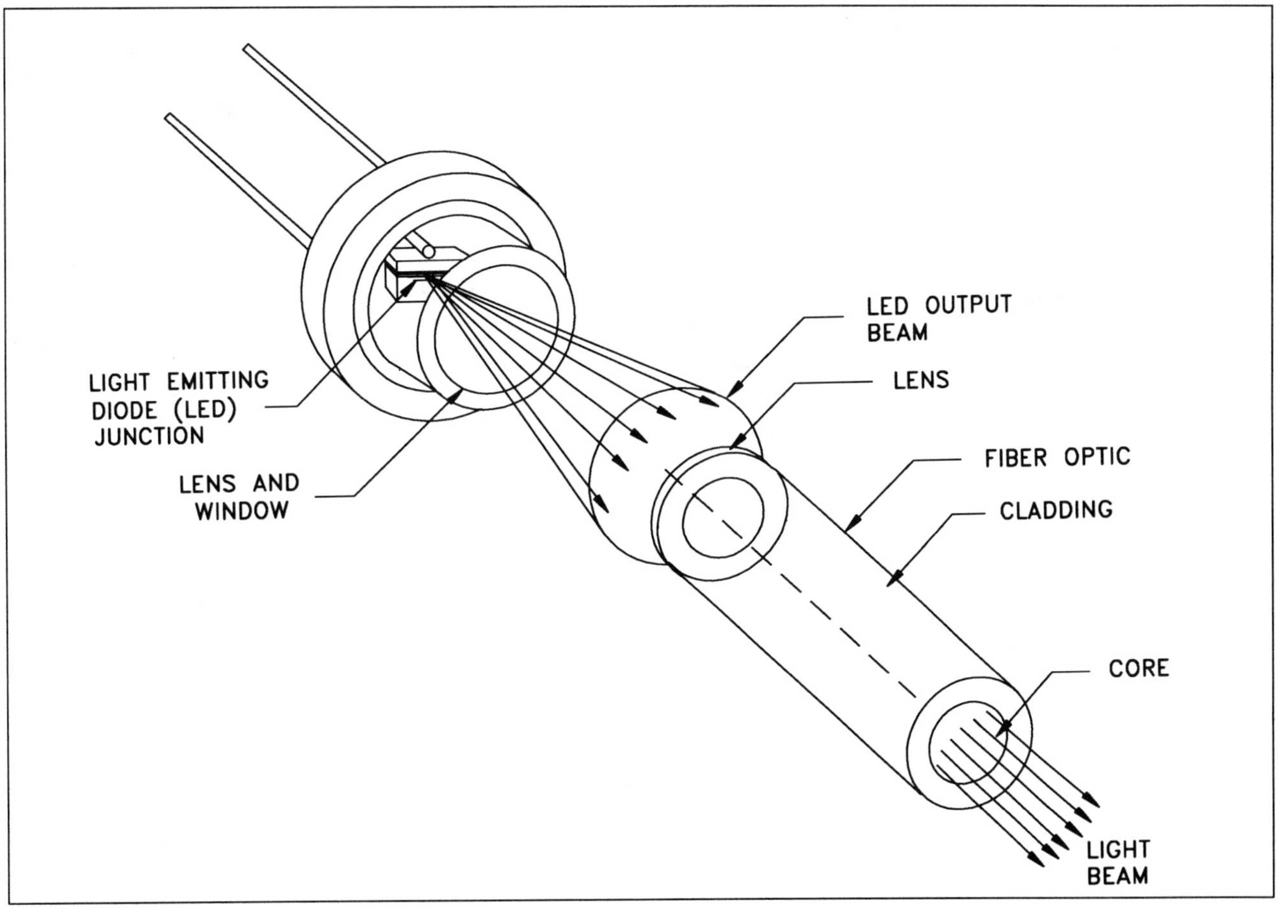

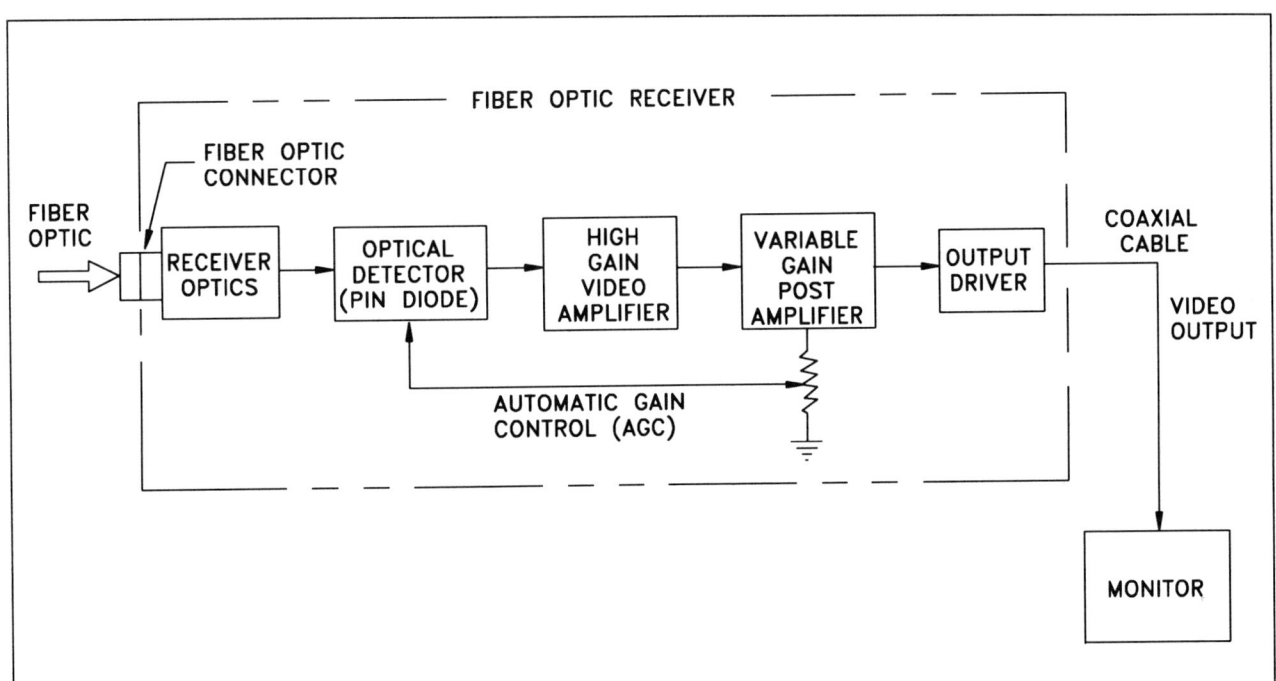

FIGURE 6-29   LED light-beam-output cone angle

FIGURE 6-30   Block diagram of fiber-optic receiver

electrical signal by the photodiode, it is linearly amplified and conditioned to be suitable for transmission over standard coaxial cable (RG59/U or RG11/U) to a monitor or recorder.

6.4.9 Multisignal Single-Fiber Transmission

In the previous sections on fiber-optic transmission cable, the cable's wide signal bandwidth capability was not discussed to any great extent. Transmitting a single video signal on a single fiber easily fits within the bandwidth capability of most fiber-optic cables. Modulators and demodulators (special transmitters and receiver-transceivers) permit transmission of bidirectional video, audio, and control signals over a single optical-fiber cable. Using the full-duplex capabilities of the system, the transceiver at the camera transmits video and audio signals from the camera location to the monitor location while simultaneously receiving audio, control, or camera genlock signals from the transceiver at the monitor location. All transmissions occur via the same single optical-fiber cable.

One system providing this performance uses a 50/125-micron cable operating at the 850-nm wavelength with a transmission range of 1 mile. In another system using time multiplexing, if only control information must be transmitted, the system can control normal lens and pan/tilt functions over a single fiber-optic cable at distances up to 3 miles. The transmitter and receiver contain all circuitry and controls for the transmission of pan/tilt, zoom, focus, and iris information, as well as contact-closure channels to perform other functions.

6.4.10 Fiber-Optic Advantages

Why go through all the complexity and extra expense of converting the electrical video signal to a light signal and then back again? Fiber optics offers several very important features that no electrical cabling system offers, including: (1) complete electrical isolation; (2) complete noise immunity to RF interference (RFI), electromagnetic interference (EMI), and electromagnetic pulse (EMP); (3) transmission security (i.e., the cable is hard to tap); (4) no spark or fire hazard or short-circuit possibility; (5) absence of crosstalk; and (6) no RFI/EMI radiation. Table 6-6 compares the features of coaxial and fiber-optic transmission.

6.4.10.1 Electrical Isolation

The complete electrical isolation of the transmitting section, that is, the camera, lens controller, pan/tilt, and related equipment, from the receiving section, that is, the monitor, recorder, printer, switching network, and so on, is very important for interbuilding and intrabuilding loca-

tions when a different electrical power source is used for each location. Using fiber-optic transmission prevents all possibility of ground loops and ground voltage differences that could require the redesign of a coaxial-cable-based system.

6.4.10.2 RFI, EMI, and EMP Immunity

When a transmission path runs through a building or outdoors past other electrical equipment, the site survey usually cannot uncover all possible contingencies of existing RFI/EMI noise. This is also true of EMP and lightning strikes. Therefore, using fiber optics in the initial design prevents any problems caused by such noise sources.

6.4.10.3 Transmission Security

Since the fiber optic has no electrical noise to leak and no visible light, there is excellent, inherent transmission security. Methods for compromising the fiber-optic cable are difficult, and the intrusion is usually detected. To tap a fiber-optic cable, the bare fiber in the cable must be isolated from its sheath without breaking it. This will probably end the tapping attempt. If the bare fiber is successfully isolated, an optical tap must be made, the simplest of which is achieved by bending the fiber into a small radius and extracting some of the light. If a measurable amount of power is tapped (which is necessary for a useful tap), the tap can be detected by monitoring the power at the system receiver. In contrast, tapping a coaxial cable is easy to do and hard to detect.

6.4.10.4 No Fire Hazards

Since no electricity is involved in any part of the fiber-optic cable, there is no chance or opportunity for sparks or electrical short circuits, and hence no fire hazards. Short circuits and other hazards encountered in electrical wiring systems can start fires or cause explosions. When a light-carrying fiber is severed, there is no spark, and a fiber cannot short-circuit in the electrical meaning of the term.

6.4.10.5 Absence of Crosstalk

Because the transmission medium is light, there is no crosstalk between any of the fiber-optic cables. Therefore there is no degradation due to the close proximity of cables in the same bundle, as there can be when multiple channels are encased in the same electrical cable.

6.4.10.6 No RFI or EMI Radiation

Fibers do not radiate energy. They generate no interference to other systems. Therefore, the fiber-optic cable will not emit any measurable EMI/RFI radiation, and other cabling in the vicinity will suffer no noise degradation. There are no FCC requirements for fiber-optic transmission.

DESIGNATION	CABLE TYPE		ATTENUATION @ 5–10 MHZ			OUTSIDE DIAMETER (INCHES)	WEIGHT (LBS.) PER 100 FT
			DB/100FT.	DB/1000FT.	DB/KM		
RG59	COAXIAL		1.0	10.0	32.8	.242	3.5–4.0
RG59 MINI	COAXIAL		1.3	13.0	42.6	.135	1.5
RG6	COAXIAL		.8	8.0	26.2	.272	7.9
RG11	COAXIAL		.51	5.1	16.7	.405	9–11
2422/UL1384	MINI–COAX		3.96	39.6	129.9	.079	0.9
2546	MINI–COAX		1.82	18.2	59.7	.13	1.4
2947	DUAL MINI–COAX		1.85	18.5	60.7	.118x2	2.5
RG179B/U	MINI–COAX		2.0	20.0	65.6	.089	1.0
10/125	FIBER OPTIC	850NM*	—	—	—	.036	—
		1300NM	.01–.02	.1–.2	.4–.8		
50/125	FIBER OPTIC	850NM	.12–.21	1.2–2.1	4–7.0	.12	.50–1.0
		1300NM	.09–.18	.9–1.8	3–6.0		
140/200	FIBER OPTIC	850NM	.08–.18	.8–1.8	2.5–6.0	.244	—
		1300NM	.02–.14	.2–1.4	.8–4.5		

*TRANSMISSION WAVELENGTH (NANOMETER–NM)

1 KILOMETER (KM) = 3280 FEET (FT)
1 MILE (MI) = 1.609 KILOMETERS (KM)
1 POUND (LB) = .454 KILOGRAMS (KG)

Table 6-6 Comparison of Fiber-Optic and Coaxial-Cable Transmission

6.4.11 Fiber-Optic-Transmission Checklist

The following questions should be asked when considering the use or design of a new fiber-optic transmission system.

1. What are the lengths of cable runs? If over 500 feet, fiber optic should be considered. In screen rooms or tempest areas, runs as short as 10 feet sometimes require fiber-optic cable.

2. What size core/clad-diameter fiber should be used? The most common diameter is 50/125 microns, but 62.5 is becoming more popular.

3. What wavelength should be used—850, 1060, 1300, or 1550 nm? The most common wavelengths are 850 and 1300 nm.

4. How many fibers are necessary for transmitting video, audio, and controls? Should single- or multi-fiber cable be used? The entire CCTV security system should be designed and the number of required one-way (simplex) and two-way (duplex) transmission paths listed.

5. Are the cable runs going to be indoors or outdoors? Separate the indoor and outdoor requirements and list the types of fiber required.

6. If outdoors, will the fiber be strung on poles, surface-mounted on the ground, undergo direct burial, or pass through a conduit? Choose cable according to manufacturers' recommendations.

7. If indoors, will it be in a conduit, cable tray or trough, plenum, or ceiling? Choose cable according to manufacturers' recommendations.

8. What temperature range will the fiber-optic cable experience? Most cable types will be suitable for most indoor environments. For outdoor use, the cable chosen must operate over the full range of hot and cold temperatures expected and must withstand ice and wind loading.

9. Are there any special considerations such as water, ice, chemicals? See manufacturers' specifications for extreme environmental hazards.

10. Are there special safety codes, such as plenum use or heavy abrasion? Fiber-optic cable is available with plenum-grade or special abrasion-resistant construction.

11. Should spare cables be included? Each design is different, but it is prudent to include one or more spare fiber-optic cables to account for cable failure or future system growth. The number of spares also depends on how easy or difficult it is to replace a cable or add to existing cables. For example, is the cable direct-buried or in a conduit?

6.5 WIRELESS TRANSMISSION

Most CCTV security systems transmit video, audio, and control signals via coaxial cable, two-wire, or fiber-optic transmission means. These techniques are cost-effective, reliable, and provide an excellent solution for transmission.

However, there are applications and circumstances that require wireless transmission of CCTV and other signals.

The CCTV signal can be transmitted from the camera to the monitor through the atmosphere, without having to connect the two with hard wire or fiber. The most familiar technique is the transmission of commercial television signals from some distant transmitter tower to consumer TV sets, broadcast through the atmosphere on very high frequency (VHF) and ultra high frequency (UHF) RF channels.

Commercial broadcasting is of course rigidly controlled by the FCC, whose regulations dictate its precise usage. Microwave transmission is also controlled by the FCC. Rules are set forth and licenses are required for certain frequencies and applications, which limits usage to specific purposes. The U.S. government currently exercises strict control over transmission of wireless video via RF and microwave. Up until recently, RF and microwave atmospheric video transmission links were limited to governmental agencies (federal, state, and local) that could obtain the necessary licenses. Now some low-power RF transmitters and receivers suitable for short links (less than a mile) are available for use without an FCC license, and some microwave links are licensable by private users after a frequency check is made.

Some examples of wireless TV transmission described in the following sections include microwave (ground-to-ground station, satellite), RF over VHF or UHF channels, and light-wave transmission using IR beams. The hardware cost of RF, microwave, and light-wave systems is considerably higher than any of the copper-wire or fiber-optic systems; such systems should be used only when absolutely necessary. When their use avoids expensive cable installations (such as across roadways), or in temporary or covert applications, they become cost-effective.

The results obtainable with hard-wired copper-wire or fiber-optic video transmission are usually predictable, with the exception of interference that might occur due to the copper wire cables running near electromagnetic radiating equipment. The results obtained with wireless transmission are generally not as predictable, because of the variable nature of the atmospheric path and materials through which RF, microwave, or light signals must travel, as well as the specific transmitting and propagating characteristics of the particular wavelength or frequency of transmission. Each of the three wireless transmitting regimes acts differently because of the wide diversity in frequencies at which they transmit.

The RF link constitutes the lowest carrier frequency (Figure 6-31) and therefore penetrates many visually opaque materials, goes around corners, and does not require a line-of-sight path (that is, a receiver in sight of the transmitter) when transmitting from one location to another. The radio frequencies are, however, susceptible to attenuation and reflection by metallic objects, ground terrain, or large buildings or structures, and hence they sometimes produce unpredictable results.

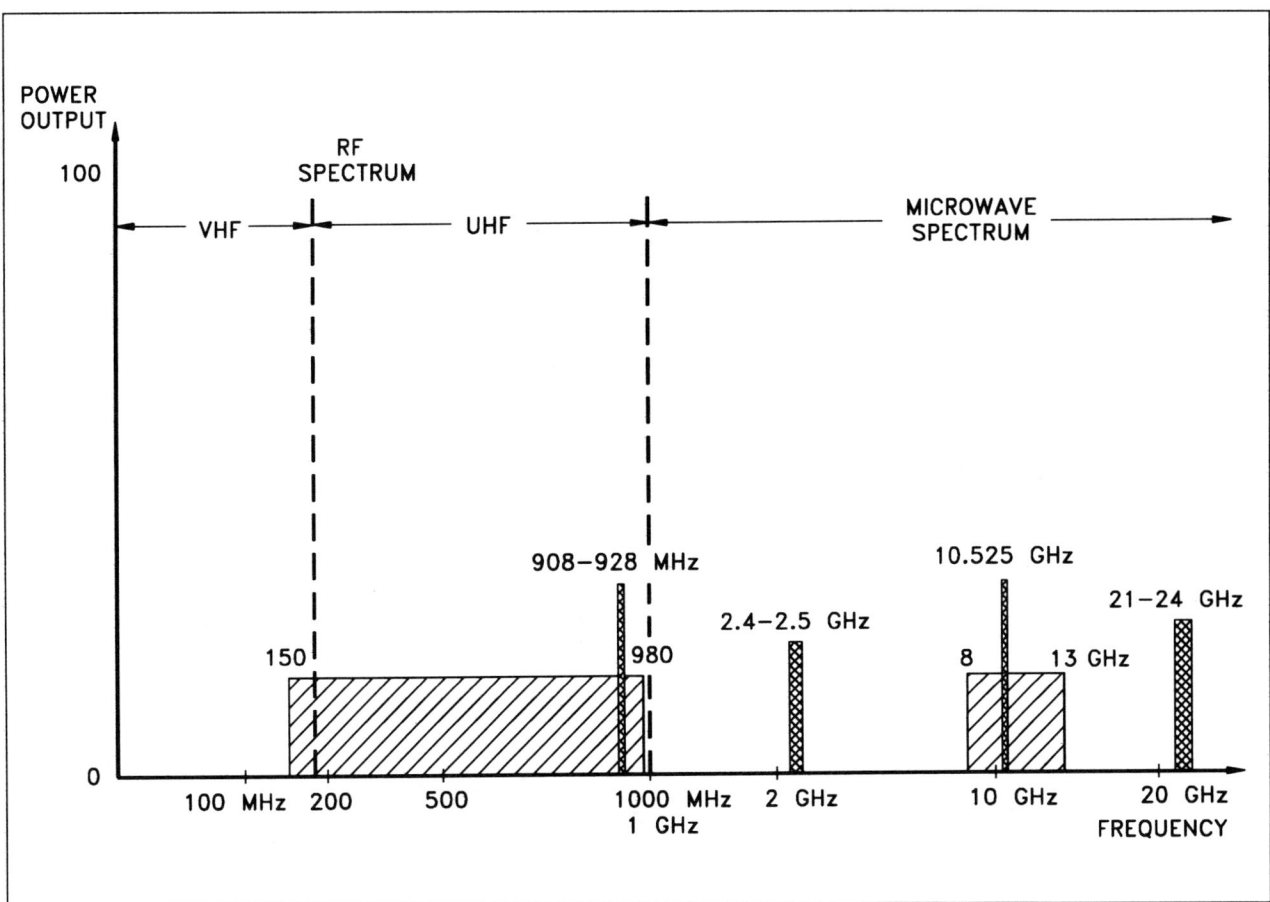

FIGURE 6-31 Wireless television transmission frequencies

The microwave frequencies require an unobstructed line of sight; any metallic or wet objects in the transmission path cause severe attenuation and reflection, often rendering a system useless. However, metallic poles or flat surfaces can sometimes be used to reflect the microwave energy, allowing the beam to turn a corner. Reflection of this type does reduce the energy reaching the receiver and the effective range of the system. Some microwave frequencies penetrate dry nonmetallic structures such as wood or drywall walls and floors, so that non–line-of-sight transmission is possible.

The frequency range most severely attenuated by the atmosphere and blocked completely by any opaque object is a light-wave signal in the near-IR region, which is strongly attenuated by heavy fog or precipitation, often severely reducing its effective range as compared with clear-line-of-sight, clear-weather conditions. As would be expected, the IR-wavelength system requires a clear line of sight with no opaque obstructions whatsoever between the transmitter and receiver. The IR beam can be reflected off one or more mirrors to go around corners. The IR systems' advantages over RF and microwave links are security (since it is hard to tap a narrow light beam) and high bandwidth (able to carry multiple channels).

6.5.1 Microwave Transmission

Microwave systems applicable to television transmission use have been allocated frequencies in bands from 1 to 40 gigahertz (GHz) (see Table 6-7).

Microwave frequencies, which approach light-wave frequencies, are usually transmitted and received by parabolically shaped reflector antennas or metallic horns. These reflectors must be contoured accurately so as to establish a precise beam pattern. Microwave television propagation characteristics are such that transmission can be achieved only if a line of sight exists between the transmitting and receiving antenna—unless metal reflectors and beam benders can compensate. Even when a line of sight exists, there can be signal fading, caused primarily by changes in atmospheric conditions between the transmitter and receiver, a problem that must be taken into account in the design. This fading can result at any frequency but in general is more severe at higher microwave frequencies (20 to 40 GHz).

6.5.1.1 Terrestrial Equipment

For terrestrial use, several manufacturers provide reliable microwave transmission equipment suitable for transmitting

BAND	FREQUENCY RANGE (GHZ) *	COMMERCIAL EQUIPMENT	
		FREQUENCY (GHZ)	COMMENTS
975	.75–1.12	1.7–1–9	VIDEO TRANSMITTER
L	1.12–1.7	2.3–2.5	VIDEO TRANSMITTER
S	2.6–3.95	2.45	CONSUMER MICROWAVE OVEN
G	3.95–5.85	—	
C	4.9–7.05	—	
J	5.85–8.2	8.4–8.6	
H	7.05–10.0	—	
X	8.2–12.4	10.4–10.6	REFERRED TO AS 3 CM BAND
M	10.0–15.0	10.525	VIDEO, AUDIO, INTRUSION, 1 KM RANGE
P	12.4–18.0	10.35–10.8	VIDEO TRANSMITTER
N	15.0–22.0	21.2–23.2	VIDEO–UP TO 3 MI RANGE
K	18.0–26.5	24.125	VIDEO, VOICE, INTRUSION, 1500 FT RANGE
R	26.5–40.0	—	

* GIGAHERTZ (GHZ) = 1000 MHZ

Table 6-7 Microwave Television Transmission Frequencies

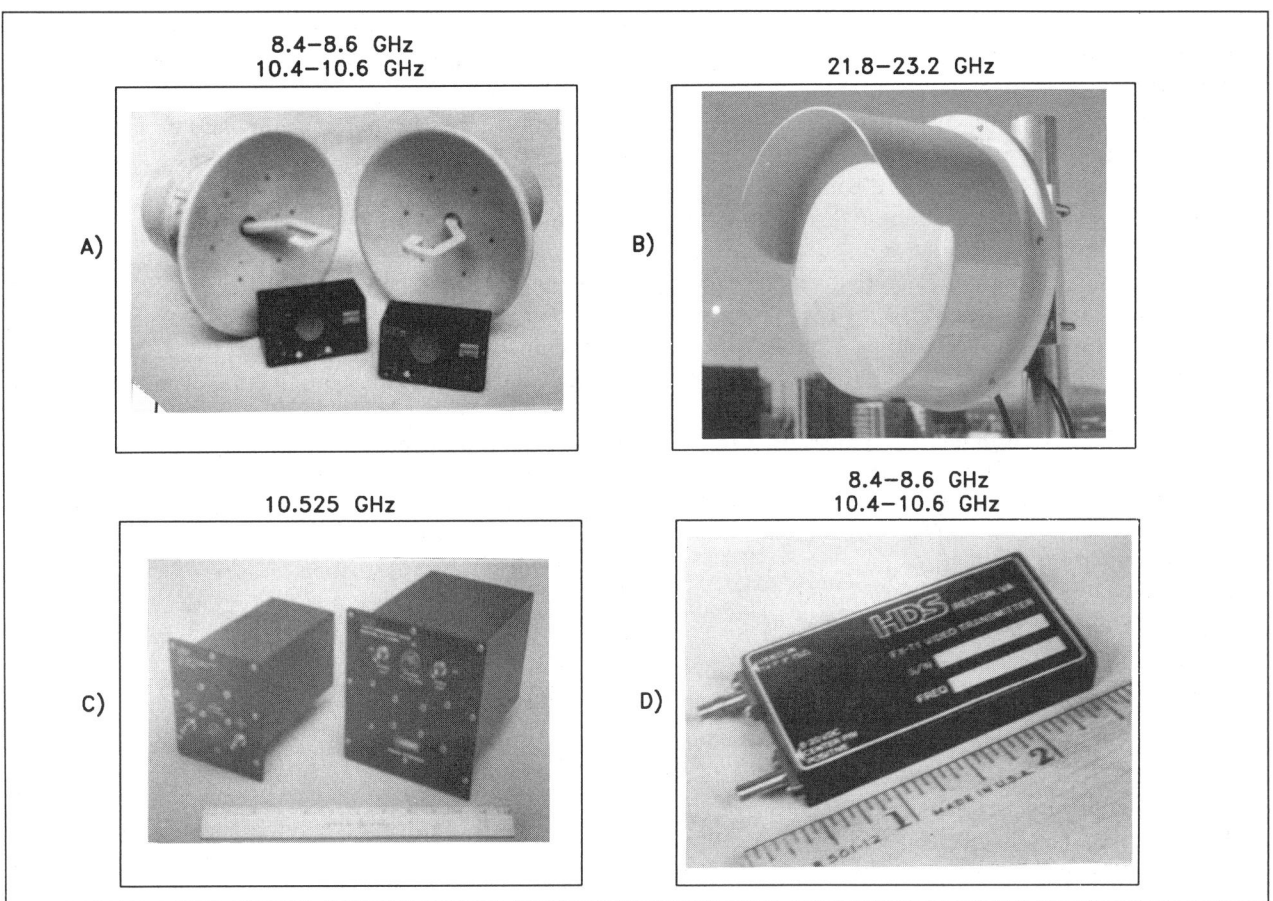

FIGURE 6-32 Monochrome/color microwave transmission systems

video, audio, and control signals over distances from several hundred feet to 10 to 20 miles in line-of-sight conditions.

One system transmits a single NTSC video channel and two 20-kHz audio channels over a distance of 1 mile. A high-gain directional antenna is available to extend the system operating range to several miles. Figure 6-32a shows the transmitter and receiver units. This system operates at a carrier frequency of 2450 to 2483.5 MHz with a power output of 1 watt. The transmitter and receiver operate from 11 to 16 volts DC derived from batteries, an AC-to-DC power supply, or 12-volt DC vehicle power. A small power supply that converts 117 volts AC to 12 volts DC permits operation from an AC source. The microwave transmitter utilizes an omnidirectional antenna. A high-gain, low-noise receiver collects the microwave transmitter signal with an omnidirectional or directional antenna. The system has a selectable video bandwidth from 4.2 MHz for enhanced sensitivity or 8 MHz for high resolution and has a single or dual audio subcarrier channel for audio communications between the two sites. It transmits monochrome or color video with excellent quality.

The 2450–2483.5-MHz band is available for a variety of industrial applications and requires an FCC license for operation. The system operates indoors or outdoors, uses frequency modulation, and provides immunity from vehi-

cles, power lines, and other AM-type noise sources. Another benefit in operating at a microwave frequency is that it prevents casual interception by most of the population, thereby making the transmission semicovert. The microwave frequency utilized has the ability to penetrate dry walls and ceilings and reflect off metal surfaces.

Another microwave system designed for outdoor use with a longer range (up to 3 miles) is shown in Figure 6-32b. This system uses directional antennas pointed toward each other to provide the necessary signal at the receiver from the transmitter.

The system is weatherproof, pedestal mounted, and designed for permanent installation. It transmits an excellent full-color or monochrome picture over an FM carrier in a frequency range of 21.8 to 23.2 GHz with a video bandwidth of 10 MHz. In addition to the video channel, the system is capable of providing up to three voice or data (control) channels. The data channels may be used to control pan/tilt, zoom, focus, and iris at the camera location. As with the previous system, FCC licensing is required and can be obtained for government and industrial users, providing an authorized interference survey is made to verify that no interference will result in other equipment.

A third system in widespread use in government and industrial applications operates at 10.525 GHz (X band)

and transmits an excellent FM color or monochrome video signal over a distance of 3500 feet while also operating as a video intrusion-detection system (Figure 6-32c). When the transmitter and receiver are located to intercept movement by a person, vehicle, or other object passing through the beam, the beam interception causes the microwave detection system to register an alarm. Prior to the alarm, a video camera located at the distant end transmits a picture of the intruder. After the intrusion is cleared, the video camera can continue transmitting the video picture. This system operates from 11 to 16 volts DC, and an AC-to-DC adapter permits 117 or 24 volts AC operation. The system is designed for small-size, lightweight, all-weather operation. Output power of the transmitter is 15 milliwatts. A built-in horn antenna with a gain of 16 dB provides a narrow 20-degree beamwidth, resulting in good signal level over ranges of the order of 1 kilometer. The receiver receives the transmitted signal through a built-in 16-dB horn antenna. The received signal is amplified and demodulated, and the detected video output is at the standard 1-volt level to drive a monitor.

There are other variations of this transmitter/receiver system:

1. Operation in any frequency band from 8.5 to 12.4 GHz with output powers up to 100 milliwatts.
2. Operation as a command-and-control unit providing a 14-channel Touch-Tone system for transmitting the control signal information. An encoder is used at the transmitter and a decoder at the receiver to remotely control functions such as power on/off, lens focus, zoom, iris, and camera motion.
3. An audio channel to provide simplex (one-way) or duplex (two-way) communications.
4. Further expansion to provide the user with the ability to sequence through and transmit the video outputs from two to eight surveillance cameras. In this system, the receiver and control units are located at the monitor site and the transmitter and sequencer units are located with the CCTV cameras. The outputs of up to eight cameras are fed to the sequencer unit via eight coaxial cables. The operator at the receiver end controls the sequencing of the eight cameras and has the option to manually advance through the eight cameras, have the eight cameras sequence automatically, or change the camera dwell time from 3 to 33 seconds.

Figure 6-32d shows examples of very small wide-beam, short-range microwave transmitters used for covert or portable video systems.

6.5.1.2 Satellite Equipment

Microwave transmission of video signals can be accomplished via satellite. Such systems are in extensive use for earth-to-satellite-to-earth communications, in which one ground-based antenna transmits to an orbiting synchro-nous satellite repeater, which relays the microwave signal at a shifted frequency to one or more receivers on earth (Figure 6-33).

While this type of communication and transmission for video security applications is not widespread, it is definitely a promising technique. The satellites used for transmission are in a synchronous orbit at an altitude of 22,300 miles and appear stationary with respect to the earth.

Satellites are placed in a synchronous or stationary orbit to permit communications from any two points in the continental United States by a single "up" and a single "down" transmission link. Consequently, a characteristic of domestic satellite video communications is that the transmission cost is independent of terrestrial distance. It takes 0.15 seconds for a microwave signal traveling at the speed of light to make a one-way journey to or from the satellite. Therefore, there is a 0.3-second delay between transmission and reception of the video carrier, independent of ground distance. This delay is not usually a problem for transmission of video security signals; however, it must be kept in mind when synchronization of different incoming video signals is required.

Two antenna types are used for satellite transmission systems, single beam and multiple beam. A single-beam antenna reflects signals from a satellite to a single feed horn. In order to target a second satellite, the whole body of the antenna with the feed horn fixed in position is moved. The most familiar single-beam antenna, the prime-focus parabola type, theoretically concentrates all incoming signals directed parallel to its axis to a single point. Any signals traveling from a direction other than that of the satellite target are reflected away from the focal point. Multiple-beam antennas use a common reflector surface with one movable feed horn or multiple feed horns so that more than one satellite can be simultaneously detected without moving the antenna.

The signal level reaching the feed horn depends on the size and shape of the antenna. The quality of an antenna is determined by how well it concentrates the radiation intercepted from a target satellite to a single point and by how well it ignores noise and unwanted signals coming from sources other than the target satellite. Three interrelated concepts—gain, beam width, and noise temperature—describe how well an antenna performs. Antenna gain is a measure of how many thousands of times a satellite signal is concentrated by the time it reaches the focus of the antenna. For example, a typical well-built 10-foot-diameter prime-focused antenna dish can have a gain of 40 dB, which is a factor of 10,000 power gain, which means that the signal is concentrated 10,000 times higher at the focal point than anywhere on the antenna. This gain is primarily dependent on three factors:

Dish size. As the size of a dish increases, more radiation from space is intercepted. Thus if the area of an antenna is doubled, the gain is doubled.

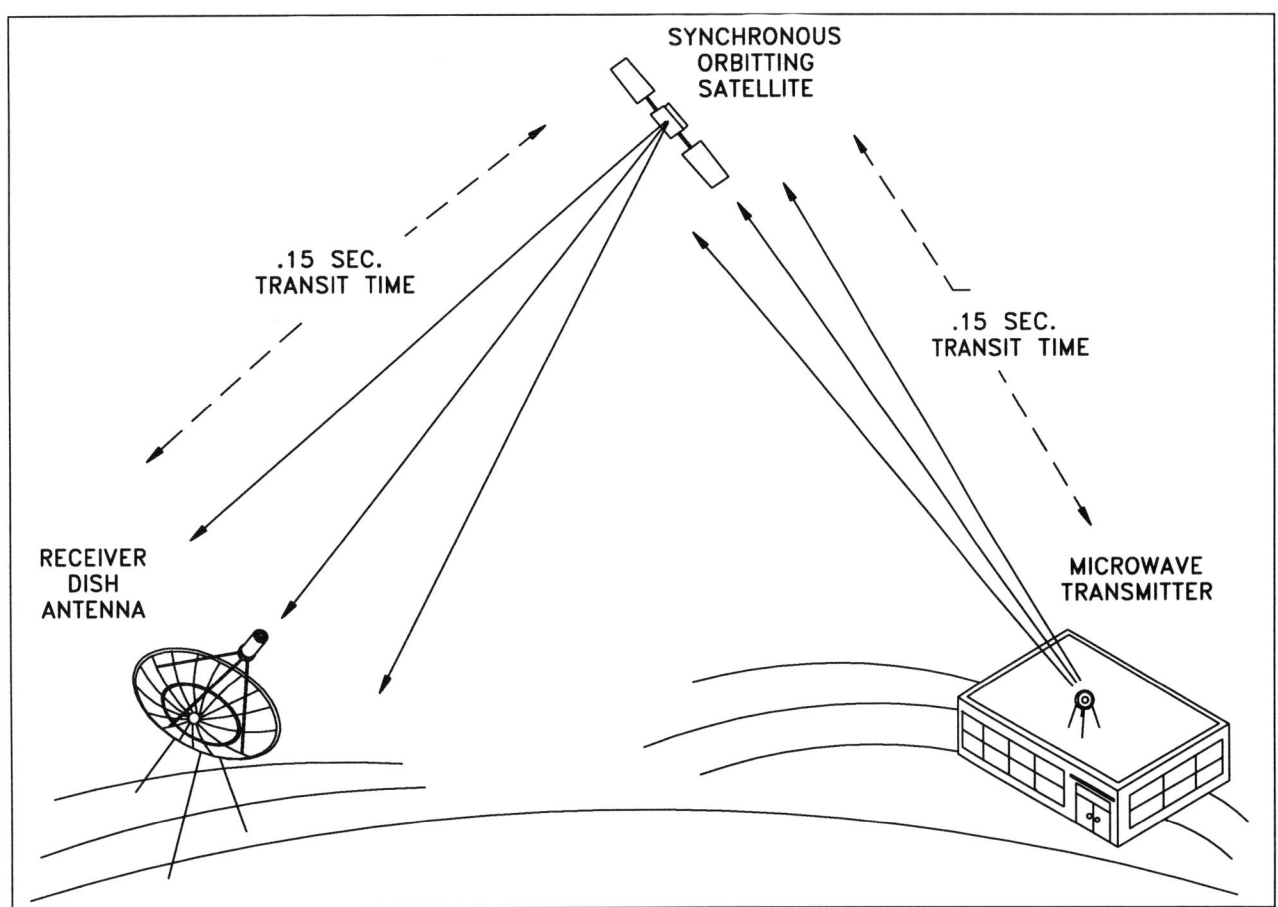

FIGURE 6-33 Satellite television transmission system

Frequency. Gain increases with increasing frequency because higher-frequency microwaves, being closer to the frequency of light, behave a little more like light. Thus they do not spread out like waves in water but can be focused more easily into straight lines like beams of light. Since the gain of a microwave antenna is proportional to the square of the frequency, a signal with twice the frequency is concentrated by an antenna with four times the gain. As an example, if the gain is 10,000 when a signal of 5 GHz is received, then it will have a gain of 40,000 at 10 GHz.

Surface accuracy. Gain is further determined by how accurately the surface of an antenna is machined or formed to exactly a parabolic or other selected shape, and how well the shape is maintained under wind loading, temperature changes, or other environmental conditions.

Antenna beam width is a measure of how well the dish can "see" into a very narrow region of space. This is an important property because satellites separated by 4 degrees or less are very close together when viewed from earth. A good antenna will see only a narrow beam width and will be able to pick out a satellite with this 4-degree angular spacing. A poor-quality dish will see too much

extraneous noise: it will receive less signal energy from the satellite of interest and pick up unwanted energy.

Large dish-shaped antennas are seen outside motels, commercial and government buildings, private homes, and many other places. Most dish antennas focus on one earth-orbiting satellite at a time and concentrate the faint signals into a device called a feed horn. The feed horn directs the microwave signal into a low-noise amplifier (LNA), which amplifies the weak signal and eventually transmits it by cable to the monitoring location.

Figure 6-34 shows a block diagram of a satellite receiver system. Signals concentrated by an antenna are captured by the feed horn. This device channels the signal to an LNA. Feed horns, also known as waveguides, are metal, hollow pipes of circular, rectangular, or other cross-sectional shapes that transmit the microwaves. Like a fiber-optic cable that transmits light along its path, the feed horn receives the signal from the antenna and directs it to the LNA, which takes this small signal and amplifies it by approximately 100,000 times while keeping the noise level as low as possible.

The LNA is the first active electronic component in the receiving system that acts on the video signal. The LNA is

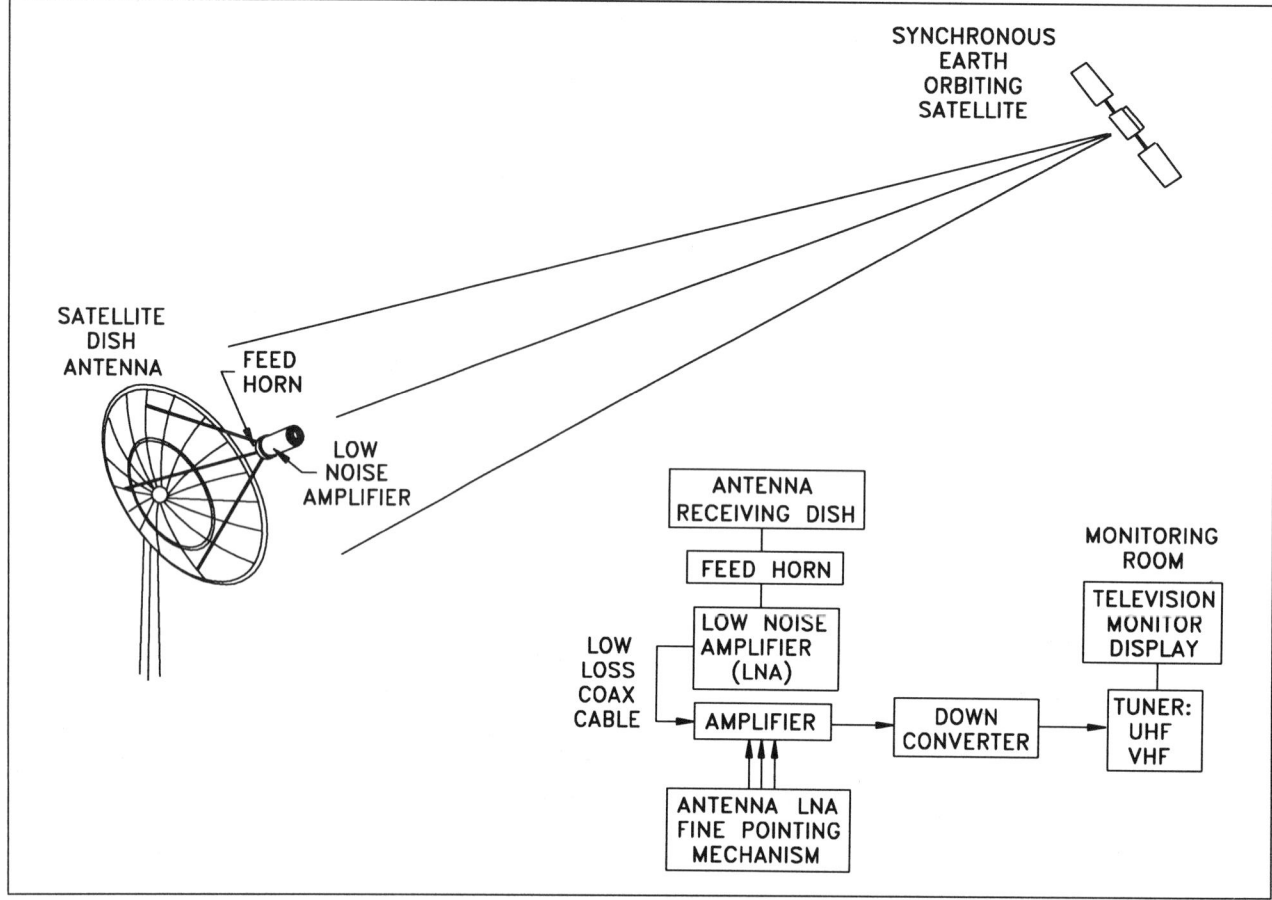

FIGURE 6-34   Satellite television receiver system

analogous to the audio preamplifier in that it provides the first critical preamplification. Its noise characteristics generally determine the quality of the final video image seen on the monitor.

The microwave signal from the LNA is transmitted via coaxial cable to a down converter, which converts the satellite microwave signal to a lower frequency. Since the signal level is still very low, a special low-loss coaxial cable must be used and the signal run must be as short as possible. Increasing cable run decreases signal level, thereby decreasing the final S/N ratio. The down-converted microwave signal is eventually converted to VHF or UHF and displayed on a television receiver.

Today most satellite receivers generate a baseband signal containing all the video and audio information as well as the necessary synchronization pulses to create a television picture that can be fed directly into a television monitor. Unlike a monitor, however, a conventional television set needs its incoming signal to be modulated, because the TV set has built-in demodulating circuits. So a standard home TV set requires another device besides the receiver to process most direct satellite signals. Such a go-between is the VCR, which, like a TV monitor, processes baseband

signals. So the satellite video receiver feeds its signals directly to the VCR, whose built-in modulator allows the signal to be properly interpreted by a television set.

6.5.1.3 Interference Sources

Transmission interference occurs when unwanted signals are received along with the desired satellite signal. Of the several types of interference, perhaps the most common and irritating is caused by reception of nearby microwave signals using the same or adjacent frequency band. Microwaves reflecting off buildings or even passing cars are responsible for the interference. Very often, just moving the microwave antenna several feet can significantly reduce the interfering signal levels.

Other interference includes stray signals from adjacent satellites, or from uplink or downlink interference. Finally, a predictable form of interference is caused by the sun. Twice a year the sun lines up directly behind each satellite for periods of approximately ten minutes per day for two or three days. Since the sun is a source of massive amounts of radio noise, no transmissions can be received from satellites during these sun outage times. This unavoidable type of interference can be expected during the normal course of operation of an earth satellite station.

6.5.2 Radio Frequency Transmission

Radio frequency is a wireless video transmission means used primarily by government agencies, although it is finding increasing use in commercial security applications as a result of the removal of some government restrictions. Video transmitters and receivers transmit a monochrome or color video signal over distances of several hundred feet to several miles using small, portable, battery-operated equipment emitting 150 to 980 MHz, covering the full VHF and UHF frequency spectrums. Table 6-8 summarizes the channel frequencies available.

While RF transmission provides significant advantages when a wired system is not possible, there are FCC restrictions limiting the use of many such transmitters to government applications. Only low-power transmitters are available for commercial applications. Any RF systems used outside the United States require the approval of the foreign government.

6.5.2.1 Transmission Path Considerations

The video signal transmission by RF means can follow either commercial broadcasting standards, in which the visual signal is an AM signal, or noncommercial standards, which generally use an FM signal. In the commercial standard, the audio frequency on the carrier is FM. In both systems the video input signal ranges from a few hertz to 4.5 MHz. For the low-powered transmitter/receiver systems used in security applications, FM modulation has provided far superior performance (increased range and lack of interference) and is the preferred method. The range obtained with an FM RF transmitter is from three to four times that of the AM type.

If the transmitter were to use standard AM commercial broadcast video standards, operating on one of the designated VHF or UHF channels, any consumer-type receiver could receive and display video. This potential is obviously a disadvantage for covert security surveillance. For this reason and to avoid general nuisance, the FCC prohibits the use of RF transmitters operating on the commercial VHF and UHF channels. In the case of FM video transmission, many consumer receivers, though not designed to receive such signals, do indeed display FM signals with some degree of picture quality, because of nonlinear and sporadic operation of various receiver circuits. Likewise, the FCC does not permit the commercial use of FM or other modulation techniques in the commercial VHF and UHF channels.

Very low power RF transmission in the 902–928-MHz range has now been approved for general security applications without an FCC license.

6.5.2.2 Equipment

Several manufacturers produce wireless video RF links (Figure 6-35). For many years manufacturers have had the technical ability to manufacture low-, medium-, and high-power RF video links for commercial broadcasting and government applications. These links were not available to any industrial or nongovernment users because of FCC restrictions. In the past few years, however, many types of RF links have become available to all users. This equipment operates on FCC-assigned frequencies with specific transmitter output power levels. These general-purpose RF links operate in the 902–928-MHz range at output field strengths of from 50 to 250 milliwatts per meter at 3 meters.

Figure 6-35 illustrates three RF video transmission systems. A low-power 100-milliwatt AM transmitter and receiver having a range of 100 feet and operating at 920 MHz is shown in Figure 6-35a. A higher-power FM system capable of different factory-set carrier frequencies and having a power output (switchable) from 1 watt to 8 watts and operating anywhere in the VHF and UHF range is available to government agencies (Figure 6-35b). Figure 6-35c shows an FM transmitter with 50-milliwatt output, operating anywhere in the 902–928-MHz range.

A short dipole stub antenna (whip) can be mounted directly onto the chassis-mounted BNC (or N-type) output connector on the transmitters. In place of the whip antenna, a coaxial cable can feed a remotely located antenna having a better vantage point for transmission of the video signal. This antenna-feed coaxial cable must have a 50-ohm video impedance rather than the usual 75-ohm video impedance, since these RF circuits are designed with a 50-ohm input and output impedance. Using a 75-ohm cable at the transmitter output or receiver input will always seriously degrade the performance of the system even if the cable is short (1 or 2 feet). RG58/U and RG/8 coaxial cables have a 50-ohm impedance.

Another transmitter available is a multiple standard (US—NTSC; foreign—CCIR) with a switch-selectable frequency (using plug-in crystals to determine frequency) having a separate input for video camera and VCR, and a wired remote-control input to select VCR or camera and/or frequency. The unit provides excellent transmission of monochrome or color television with an RF output power of 3 watts. This system uses standard commercial modulation with an AM video signal and FM audio. The video bandwidths for the three different standards used worldwide are as follows: U.S.—4.5 MHz; CCIR—5.0 MHz; and U.K.—5.5 MHz. The audio subcarrier frequencies are as follows: U.S.—4.5 MHz with 25-kHz peak deviation; CCIR—5.5 MHz, 50-kHz; U.K.—6.0 MHz, 50 kHz.

Figure 6-36 shows a graph illustrating the approximate distance between transmitters and receiver antennas (range) versus power for good video transmission.

The range values are for smooth and obstacle-free terrain applications using a dipole antenna at the receiver and a

BAND	CHANNELS	FREQUENCY RANGE (MHz)	COMMERCIAL EQUIPMENT	USAGE RESTRICTIONS
VHF–LOWBAND	2–6	54–88	LOW–MEDIUM POWER, RANGE UP TO SEVERAL MILES	GOVERNMENT, LAW ENFORCE–MENT ONLY.
FM RADIO	—	88–108	—	—
VHF–HIGHBAND	7–13	174–216	LOW–MEDIUM POWER, RANGE UP TO SEVERAL MILES	GOVERNMENT, LAW ENFORCE–MENT ONLY.
SECURITY	—	350–950	SINGLE CHANNEL TRANS–MITTER/RECEIVER	GOVERNMENT, LAW ENFORCE–MENT ONLY.
UHF	14–83	470–890	LOW–MEDIUM POWER, RANGE UP TO SEVERAL MILES	GOVERNMENT, LAW ENFORCE–MENT ONLY.
SECURITY	—	908–928	LOW POWER, NO FCC LICENSE REQUIRED	NO RESTRICTIONS, NO FCC LICENSE REQUIRED

Table 6-8   RF Video Transmission Frequencies

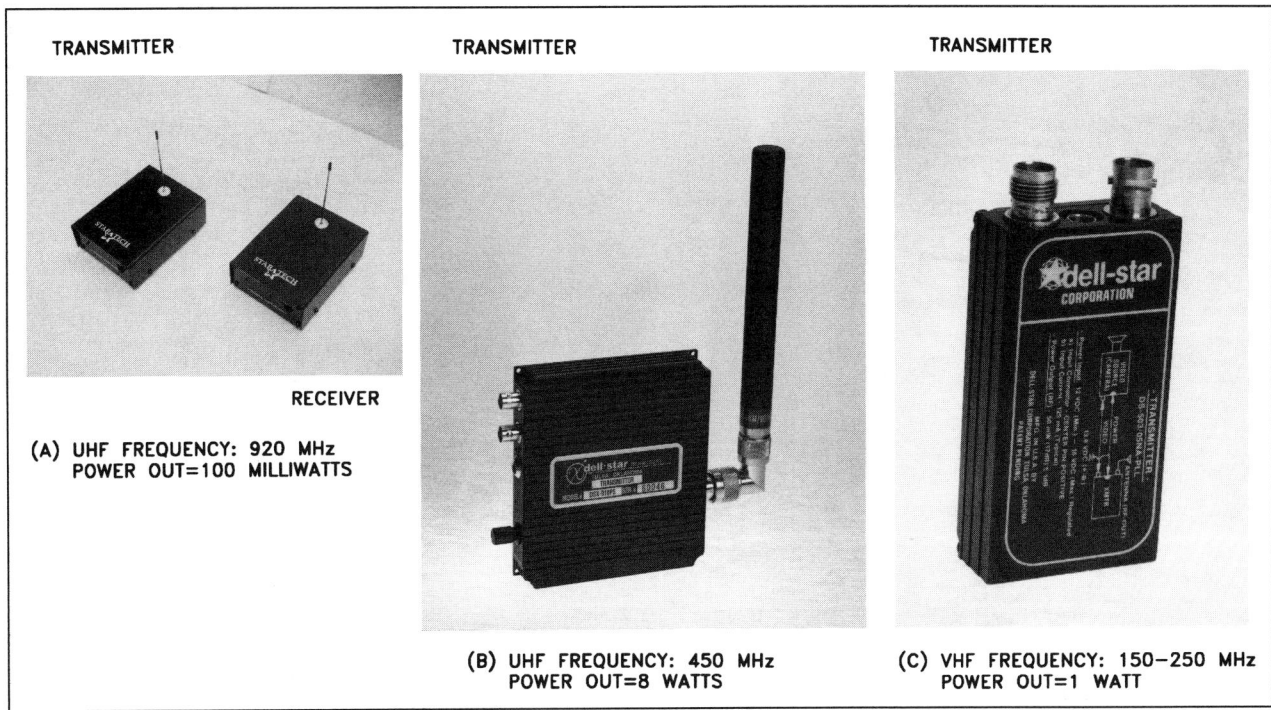

TRANSMITTER

RECEIVER

(A) UHF FREQUENCY: 920 MHz
    POWER OUT=100 MILLIWATTS

TRANSMITTER

(B) UHF FREQUENCY: 450 MHz
    POWER OUT=8 WATTS

TRANSMITTER

(C) VHF FREQUENCY: 150–250 MHz
    POWER OUT=1 WATT

FIGURE 6-35   RF video transmitters

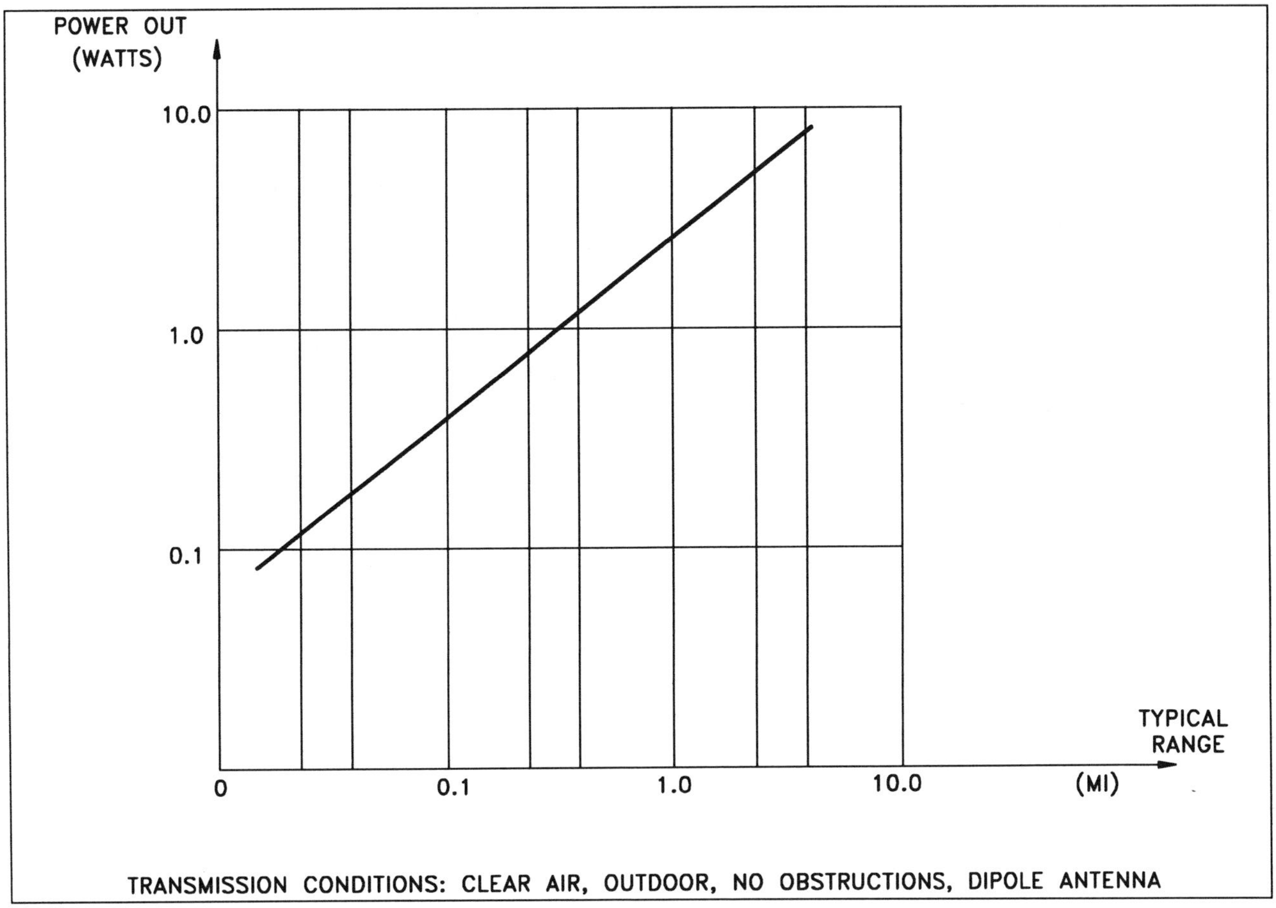

POWER OUT
(WATTS)

10.0

1.0

0.1

0   0.1   1.0   10.0   (MI)

TYPICAL
RANGE

TRANSMISSION CONDITIONS: CLEAR AIR, OUTDOOR, NO OBSTRUCTIONS, DIPOLE ANTENNA

FIGURE 6-36   Transmitter RF power out vs. transmission range

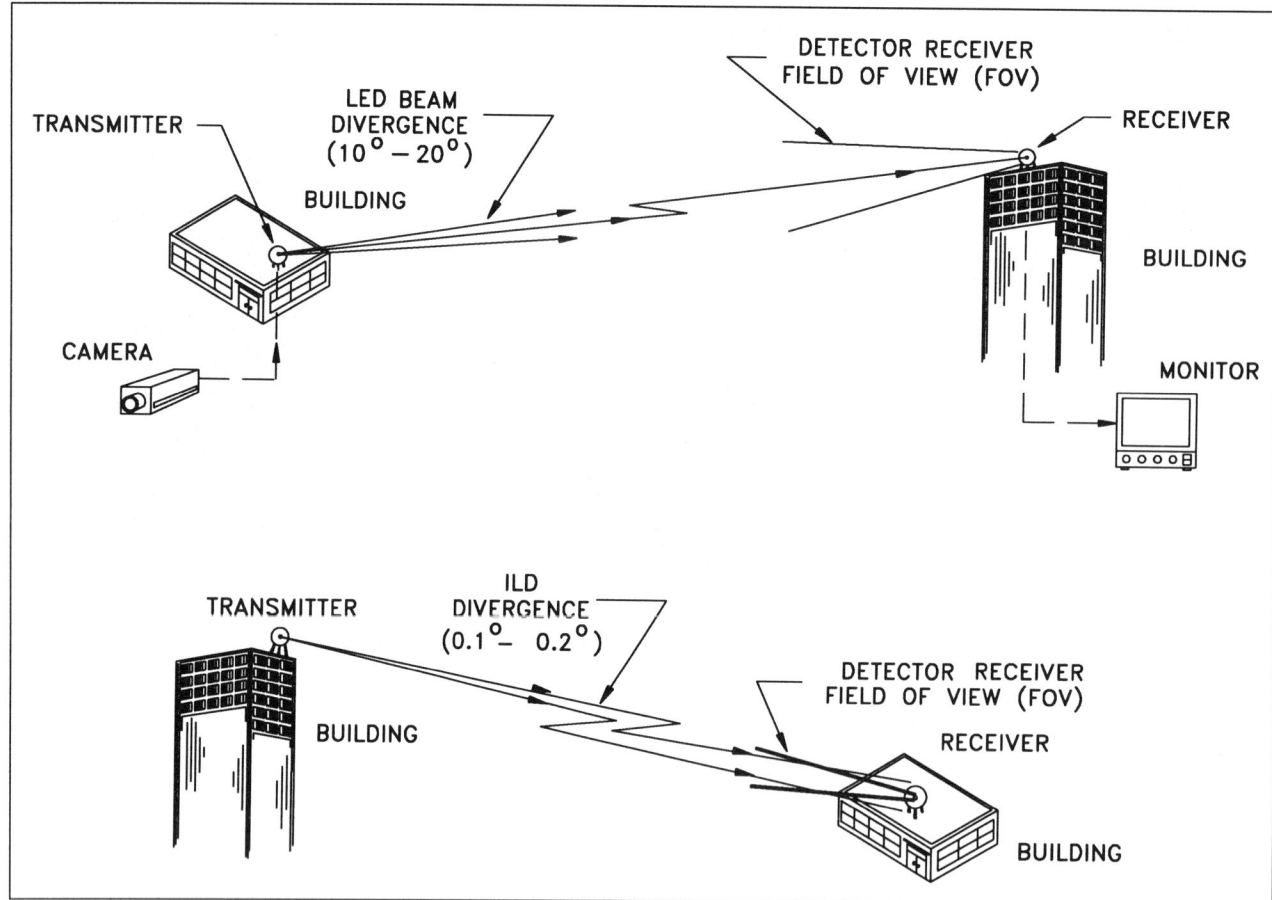

FIGURE 6-37   IR atmospheric video transmission system

quarter-wave whip antenna on the transmitter. The numbers obtained should be used as a guide only. Actual installation and experience with specific equipment on-site will determine the actual quality of the video image received.

6.5.3 Infrared Atmospheric Transmission

A technique for transmitting a video signal by wireless means uses propagation of an IR beam of light through the atmosphere (Figure 6-37).

The light beam is generated by either an LED or ILD in the transmitter. The receiver in the optical communication link uses a silicon-diode IR detector, amplifier, and output circuitry to drive the 75-ohm coaxial cable and monitor. Primarily the transmitter-to-receiver distance and the link's security requirements determine which type of diode is used. Short-range transmissions of up to several hundred feet are accomplished using the LED. To obtain good results for longer ranges, up to several miles under clear atmospheric conditions, the ILD must be used.

The LED system costs less and has a wider beam, 10 to 20 degrees wide, making it relatively simple to align the

transmitter and receiver. The beam width of a typical ILD transmitter is 0.1 or 0.2 degrees, making it more difficult to align and requiring that the mounting structure be very stable in order to maintain alignment. To ensure a good, stable signal strength at the receiver, the ILD transmitter and receiver must be securely mounted on the building structure. Additionally, the building structure must not sway, creep, vibrate, or produce appreciable twist due to uneven thermal heating (sun loading).

LED and ILD systems can transmit the IR beam through most transparent window glazing; however, glazing with high tin content severely decrease signal transmission, thereby producing poor video quality. The suitability of the window can be determined only by testing the system. Since many applications require the IR beam to pass through windowpanes across a city street or between two buildings, tests should be performed prior to designing and installing such a system.

The primary advantages of the ILD system are long-range (under clear atmospheric conditions) and secure video, audio, and control signal transmission. ILD atmospheric links are hard to tap because the tapping device—a laser receiver—must be positioned into the laser beam, which is hard to accomplish undetected.

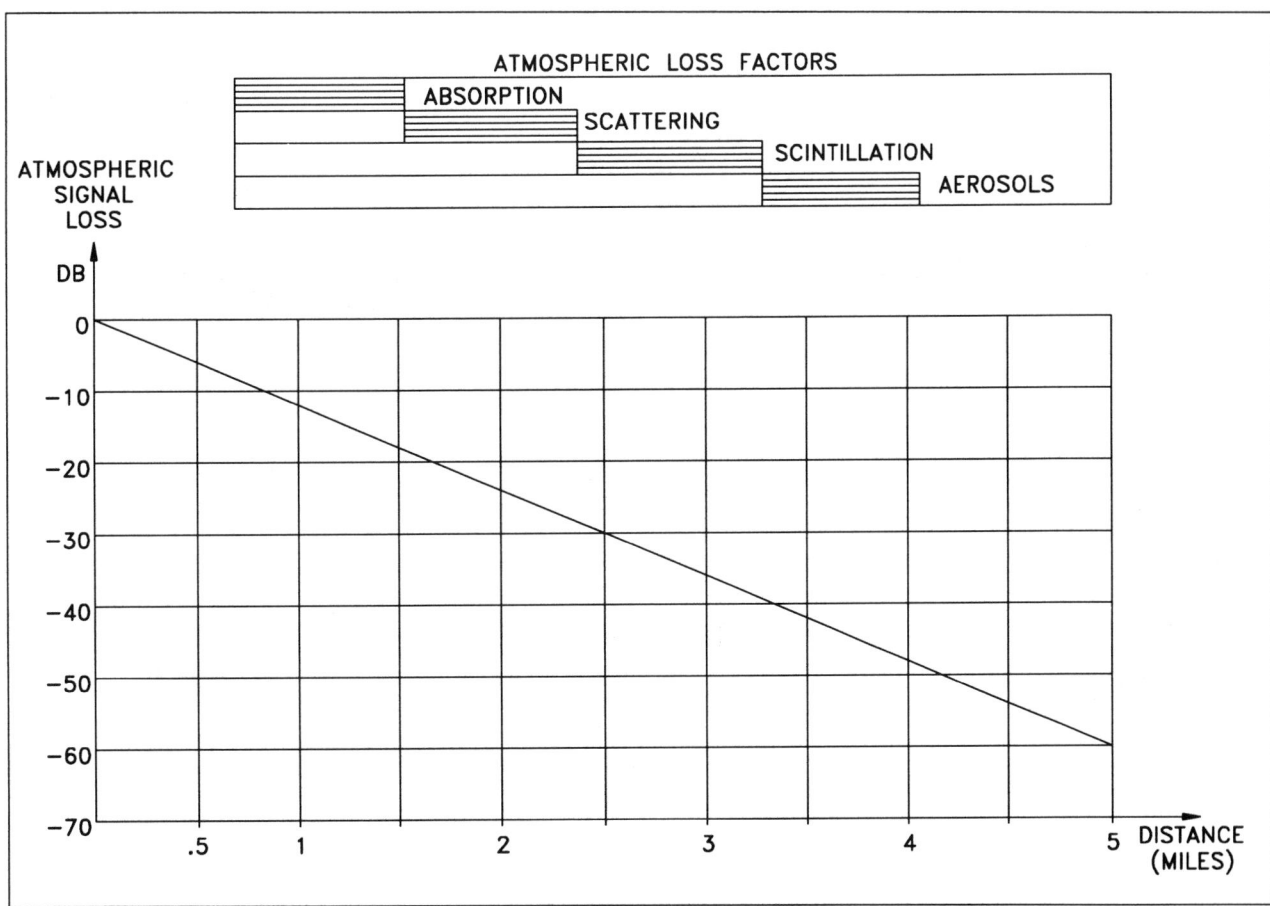

FIGURE 6-38 Atmospheric absorption factors and visibility

6.5.3.1 Transmission Path Considerations

Several transmission parameters must be considered in any atmospheric transmission link. LED and ILD atmospheric transmission systems suffer video signal transmission losses caused by atmosphere path absorption. Molecular absorption is always present when a light beam travels through a gas (air). At certain wavelengths of light, the absorption in the air is so great as to make that wavelength useless for communications purposes. Wavelength ranges in which the attenuation by absorption is tolerable are called atmospheric windows. These windows have been extensively tabulated in the literature. All LED and ILD systems operate in these atmospheric windows.

Another cause of light signal absorption is particles such as dust and aerosols, which are always present in the atmosphere to some degree. These particles may reach very high concentrations in a geographical area near a body of water. In these locations, improved performance can be achieved by locating the link as high above the ground as possible.

Fog is a third factor causing severe absorption of the IR signal. In fog-prone areas, local weather conditions must be considered when specifying an atmospheric link, since the presence of fog will greatly influence link downtime. Figure 6-38 shows the predicted communication range versus visibility for a practical GaAs LED or ILD atmospheric communications system.

In addition to signal loss, the atmosphere contributes signal noise, since it exhibits some degree of turbulence. Turbulence causes a refractive index variation in the signal path (similar to the heat waves seen when there is solar heating in air—the mirage effect) and its subsequent wind-aided turbulent mixing. The net effect of this turbulence is to move or bend the IR beam in an unpredictable direction, so that the transmitted radiation does not reach the remote receiver. To compensate for this turbulence, the transmitter beam is made wide enough so that it is highly unlikely that the beam will miss the receiver. This wider beam, however, results in lower beam intensity, so the received signal on average will be less than from a narrower beam.

6.5.3.2 Equipment

The transmitter and receiver used in atmospheric IR transmission systems are very similar to those used in the fiber-optic-cable transmission system (Section 6.4). The primary differences are in the type of LED (or ILD) in the transmit-

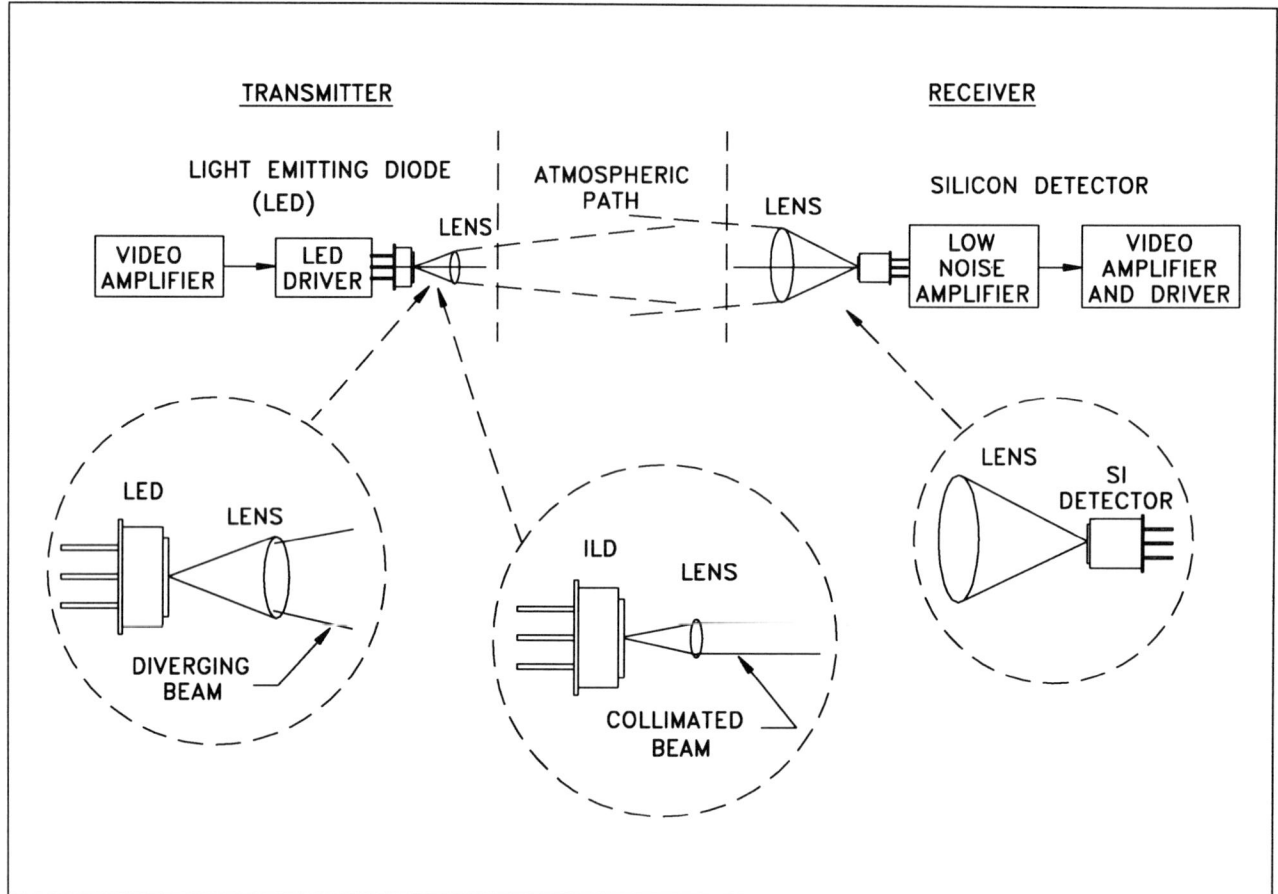

FIGURE 6-39   Block diagram of IR video transmitter and receiver

ter, and the optics in both the transmitter and the receiver (Figure 6-39).

The optics in the transmitter must couple the maximum amount of light from the emitter into the lens and atmosphere, that is, to produce the specified beam divergence depending on LED or ILD usage. The receiver optics are made as large as practical (several inches in diameter) to maximize transmitter beam collection, thereby achieving the highest possible signal-to-noise ratio. An example of an atmospheric IR link is shown in Figure 6-40.

The system has a range of approximately 3000 feet and operates at 12 volts DC. For outdoor applications, the transmitter is mounted in an environmental housing with a thermostatically controlled heater and fan, as well as a window washer and wiper.

6.6 SIGNAL MULTIPLEXING, SCRAMBLING, AND CABLE TV

It is sometimes desirable or necessary to combine several video signals onto one communications channel and transmit them from the camera location to the monitor location. This technique is called multiplexing. Some systems allow multiplexing video, control, and audio signals onto one channel. When it comes to protecting the integrity of the information on a signal, high-level security applications sometimes require the scrambling of video signals. The video scrambler is a privacy device that alters a television camera output signal to reduce the recognizability of the transmitted signal when displayed on a standard monitor/receiver. The descrambler device restores the signal to permit retrieval of the original video information.

6.6.1 Audio and Control Signal Multiplexing

Signal multiplexing has been used to combine audio and control functions (up to 16 functions) in time-division or frequency-division multiplexing. One system uses the telephone Touch-Tone system, which is standard throughout the world. With this system, an encoder generates a number code corresponding to the given switch (digit) closure. Each switch closure produces a dual Touch-Tone signal, which is uniquely defined and recognized by the remote receiver station. All that is needed for transmitting the signal is a twisted-pair or telephone-grade line. With such a system, audio and all of the conceivable camera functions (pan, tilt, zoom focus, on/off, sequencing, and others) can be controlled with a single cable pair or single transmission

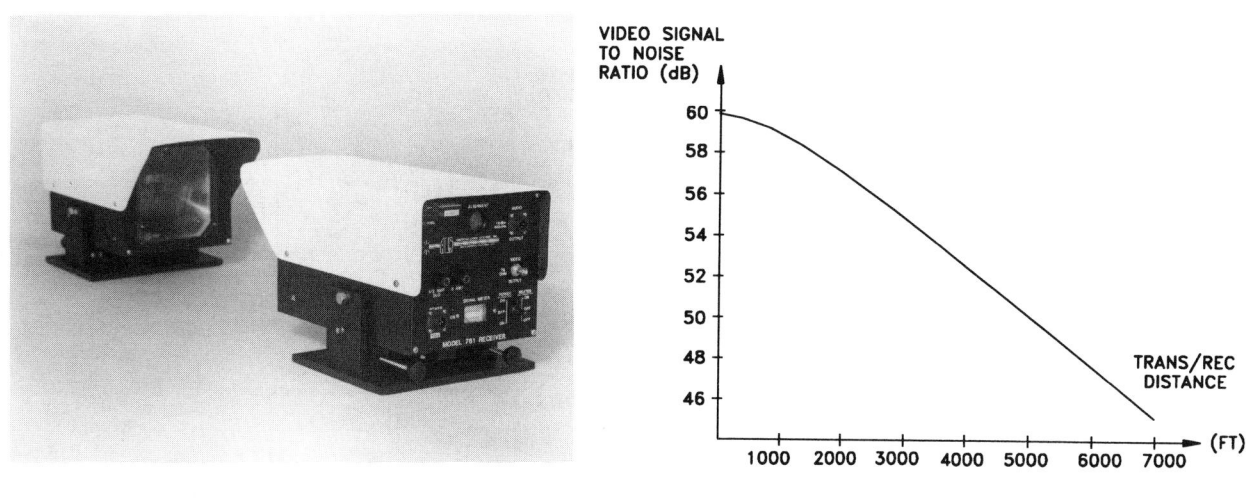

VIDEO STANDARD: NTSC, PAL, SECAM (525 TV LINES, 60 HZ OR 625 TV LINES, 50 HZ)
MONOCHROME, COLOR

RANGE: 1 MILE +
LICENSE REQUIREMENT: NONE
TYPE: SIMPLEX (ONE DIRECTION)
SIGNAL BANDWIDTH: 5.5 MHZ ±1dB, 7 MHZ ±3dB

TRANSMITTER: LIGHT EMITTING DIODE (860–900NM)
PEAK POWER OUTPUT: 30MW
BEAM DIVERGENCE: 3 MILLIRADIANS
POWER: AC, 115/220V, 50/60HZ, 25VA
DC, 12VDC, 12 WATTS

RECEIVER: SILICON AVALANCHE DETECTOR
FIELD OF VIEW: 3.75 MILLIRADIANS
POWER: AC, 115/220V, 50/60HZ, 25VA
DC, 12VDC, 12 WATTS

FIGURE 6-40 IR video transmitter and receiver hardware

channel. This concept offers a powerful means for controlling remote equipment with an existing transmission path.

It is sometimes advantageous to combine several video and/or audio and control signals onto one transmission channel. This is true when a limited number of cables are available or when transmission is wireless. If cables are already in place or a wireless system is required, the hardware to multiplex the various functions onto one channel is cost-effective. Multiplexing of video signals is used in many CATV installations, whereby several VHF and/or UHF video channels are simultaneously transmitted over a single coaxial cable or microwave link. In CCTV systems, modulators and demodulators are available to transmit the video control signals on the same coaxial cable used to transmit the video signal.

6.6.2 Signal Scrambling

The key element in any video scrambling system is to modify one or more basic CCTV signal parameters to prevent an ordinary television receiver or monitor from being able to receive a recognizable picture. The challenges in scrambling-system design are to make the signal secure without

degrading the picture quality when it is reconstructed, to minimize the increase in bandwidth or storage requirements for the scrambled signal, and to make the system cost-effective.

There are basically two classes of scrambling techniques. The first modifies the signal with a fixed algorithm, that is, some periodic change in the signal. These systems are comparatively simple and inexpensive to build and are common in CATV pay television, as well as in some security applications. The signals can easily be descrambled once the scrambling code or technique has been discovered. It is relatively straightforward to devise and manufacture a descrambling unit to recover the CCTV signal.

One of the earliest techniques for modifying the standard television signal is called video inversion, in which the polarity of the video signal is inverted so that a black-on-white picture appears white-on-black (Figure 6-41). While this technique destroys some of the intelligence in the picture, the content is still recognizable. Some scrambling systems employ a dynamic video-inversion technique: a parameter such as the polarity is inverted every few lines or fields in a pseudo-random fashion to make the image even more unintelligible. Another early technique was to suppress the vertical and/or horizontal synchronization pulses

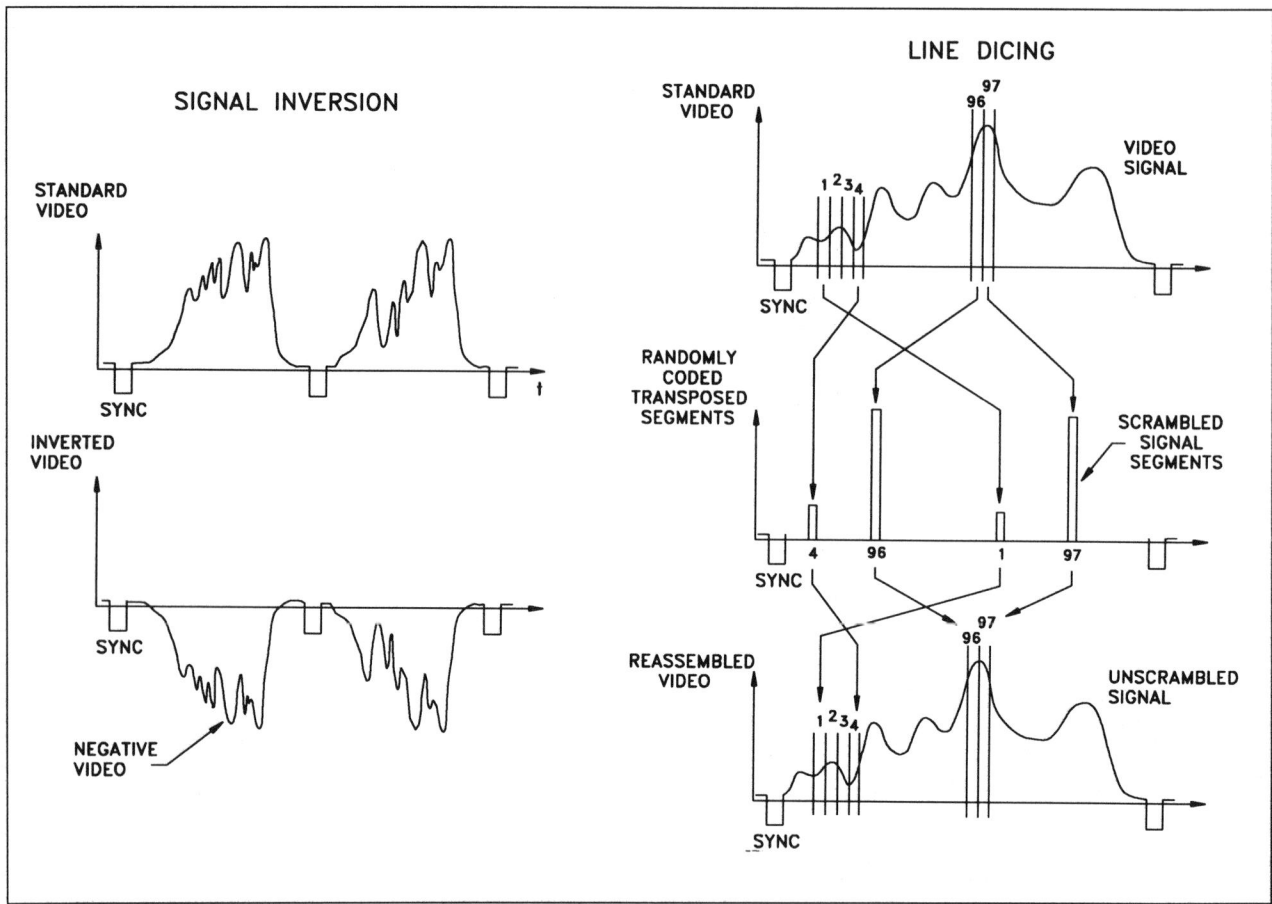

FIGURE 6-41 Video scrambling techniques

to cause the picture to roll or tear on the television monitor. Likewise, this technique produced some intelligence loss, but some television receivers could still lock on to the picture or a descrambler could be built to re-insert the missing pulses and synchronize the picture, making it intelligible again.

A second class of scrambling systems using much more sophisticated techniques modifies the signal with an algorithm that continually changes in some unpredictable or pseudo-random fashion. These more complex dynamic scrambler systems require some communication channel between the transmitter and the receiver in order to provide the descrambling information to the receiver unit, which reconstructs the missing signal. This descrambling information is communicated either by some signal transmitted along with the television image or by some separate means, such as a different channel in the link. The decoding signal can be sent by telephone or even by mail, on a cassette recording.

In a much more secure technique known as line dicing, each line of the television image is cut into segments that are then transmitted in random order, thereby displacing the different segments horizontally into new locations (Figure 6-41). A picture so constructed on a standard receiver has no intelligence whatsoever. Related to line dicing is a

technique known as line shuffling, in which the scan lines of a television signal are sent not in the normal top-to-bottom image format but in a pseudo-random or unpredictable sequence.

In the most sophisticated dynamic scrambling systems, utilized for direct-broadcast satellites and multichannel applications, the video and audio signals are scrambled in a way that cannot be decoded even by the equipment manufacturer without information from the signal operator. For example, the audio signal can be digitized and then transmitted in the vertical blanking interval, the horizontal blanking interval, or on a separate subcarrier of the television signal.

One very secure system employs encoders and decoders to provide two levels of concealment and six levels of security to foil attempts (by computer or otherwise) to decode the video encryption.

It is often necessary to scramble the audio signal in addition to the video signal, using techniques such as frequency hopping, adapted from military technology. Similar to line dicing, this technique breaks up the audio signal into many different bits coming from four or five different audio channels, and by jumping from one to another in a pseudo random fashion, scrambles the audio signal. The descrambler is equipped to tune to the different audio channels in synchrony with the transmitting signal, thereby recovering the audio information. The descrambling information transmitted to the receiver can be encrypted using a form of the data encryption adopted by the National Bureau of Standards in Washington.

The system can scramble still monochrome or color surveillance pictures. In this system, horizontal lines are completely concealed in the video image signal. The equipment is compatible with VHS, Beta, 8-mm, and U-Matic VCR machines. A six-level encryption algorithm, separate video and audio encryption algorithms, decoder ID, password, day/night window, and tamper-proof design are provided. This equipment uses a proprietary encryption algorithm that randomly changes frame by frame, resulting in virtually an infinite number of encryption combinations.

Some applications of encrypted video are uses in federal prison systems, where physical movement of prisoners to courts of law is expensive and sometimes dangerous. The ability to fully encrypt both video transmission and videocassettes makes videoconferencing possible among prisoners, lawyers, district attorneys, and others.

Witness-protection programs especially benefit from this added level of security. Encrypting video depositions and arraignments can prevent their release to the public and the media before a trial, preventing mistrials.

In military applications, sensitive video information on personnel, security plans, new weapons, transport vehicles, planes, electronics, and so on can be scrambled and delivered via tape, videoconference, or base cable TV systems.

While there are many other applications for encryption equipment, its relative high cost permits its use in only those applications that require the most secure systems. As with other technologies, a tradeoff for very high level scrambling security equipment is high cost. Generally, the more complicated and secure the scrambling system is, the more costly the equipment both at the transmitter and receiver.

6.6.3 Cable TV

Cable television (CATV) systems distribute multiple video channels in the VHF or UHF bands, using a coaxial or fiber-optic cable or an RF or microwave link. Consumer-based CATV employs this modulation-demodulation scheme, using a coaxial or fiber-optic cable. At the receiver end the signal is demodulated and the multiple camera signals are separated and presented on multiple monitors or switched one at a time (Figure 6-42).

The multiplexing technique is often used when video information from a large number of cameras must be transmitted to a large number of receivers in a network. These systems require careful analysis and choice of the modulator, coaxial-cable type, and demodulator. Table 6-9 summarizes the VHF and UHF television frequencies used in these CATV RF transmission systems.

In CATV distribution systems, the equipment accepts baseband (composite video) and audio channels and linearly modulates them to any user-selected RF carrier in the UHF (470-to-770-MHz) spectrum. The modulated signal is then passed through an isolating combiner, where it is merged (multiplexed) with the other signals. The combined signal is then transmitted over a communications channel and separated at the receiver end into individual video and audio information channels.

Cable costs are significantly reduced by modulating multiple channels on a single cable. Since the transmission is done at RF frequencies, design and installation is far more critical as compared with baseband CCTV. High-quality CATV systems are now installed with fiber-optic cable for medium to long distances, and fiber-optic or coaxial cable for distribution within a building.

6.7 CHECKLIST

The following checklists for coaxial and fiber-optic cable transmission systems show some items that should be considered when designing and installing a CCTV security project.

6.7.1 Coaxial Cable

1. When using coaxial cable, always terminate all unused inputs and unused outputs in their respective impedances.
2. When calculating coaxial-cable attenuation, always figure the attenuation at the highest frequency to be used; that is, when working with a 6-MHz bandwidth, refer to the cable losses at 6 MHz.
3. In long cable runs do not use an excessive number of connectors, since each causes additional attenuation. Avoid splicing coaxial cables without the use of proper connectors, since incorrect splices cause higher attenuation and can cause severe reflection of the signal and thus distortion.
4. For outdoor applications, be sure that all connectors are waterproof and weatherproof; many are not, so consult the manufacturer.
5. Try to anticipate ground loop problems if unbalanced-coaxial-cable video runs between two power sources are used. Use fiber optics to avoid the problem.
6. Using a balanced coaxial cable (or fiber-optic cable) is usually worth the increased cost in long transmission systems. When connecting long cable runs between several buildings or power sources, measure the voltage before attempting to mate the cable connectors. Be careful, since the voltage between the cable and the connected equipment may be of sufficient potential to be harmful to you.

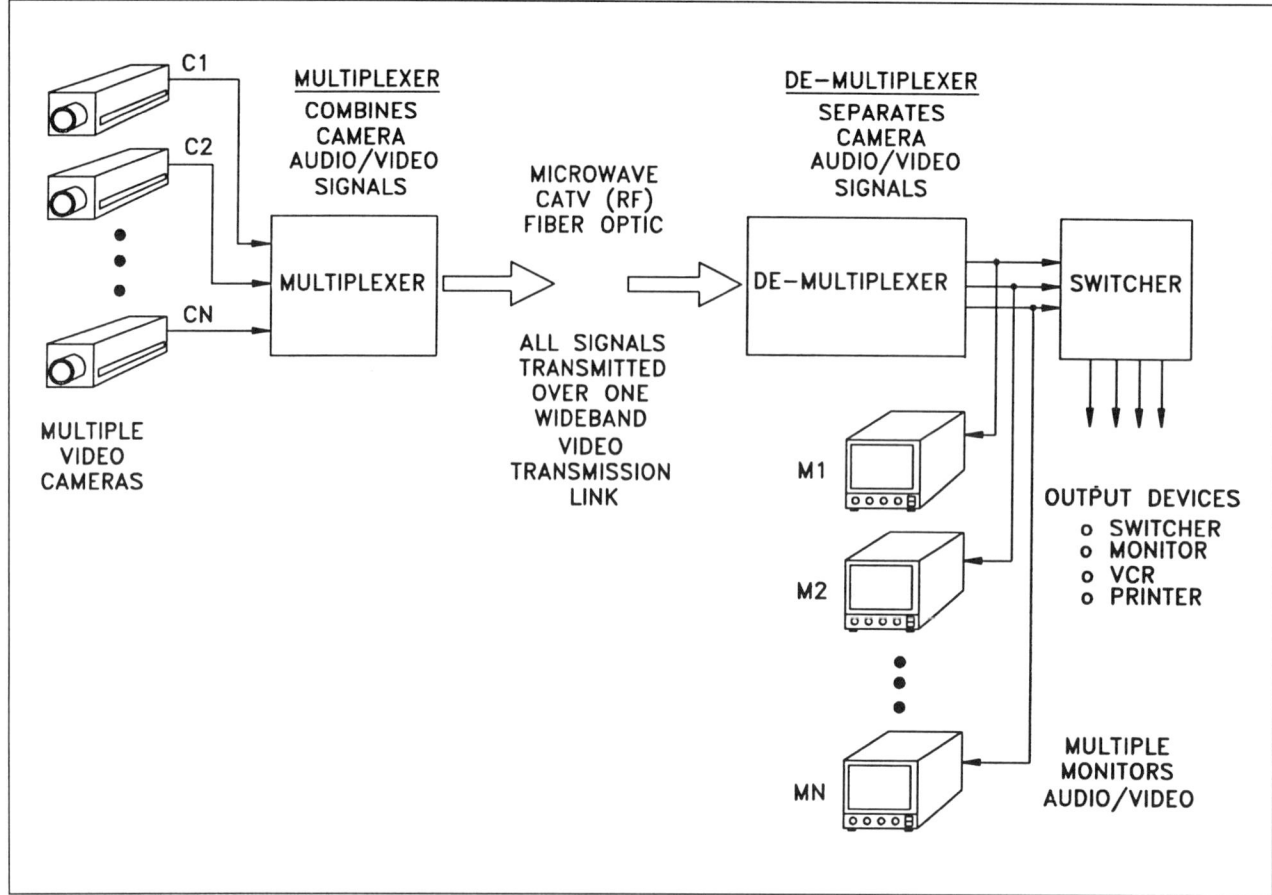

FIGURE 6-42   Multiplexed video transmission system

BAND	CHANNEL DESIGNATION	FREQUENCY RANGE VISUAL CARRIER (MHZ)
LOW–BAND VHF	2 → 6	55.25 → 83.25
MID–BAND VHF	A–2, A–2, A → I	109.25 → 169.25
HIGH–BAND VHF	7 → 13	175.25 → 211.25
SUPERBAND	J → W	217.25 → 295.25
HYPERBAND	AA → RR	301.25 → 403.25

EACH CHANNEL 6 MHZ BANDWIDTH

Table 6-9   Allocated CATV RF Transmission Frequencies

7. Do not run cable lines adjacent to high-power RF sources such as power lines, heavy electrical equipment, other RFI sources, or electromagnetic sources. Good earth ground is essential when working with long transmission lines. Be sure that there is adequate grounding, and that the ground wire is eventually connected to a waterpipe ground.

6.7.2 Fiber-Optic Cable

1. Consider the use of fiber optics when the distance between camera and monitor is more than a few hundred feet (depending on the environment), if it is a color system, if the camera and monitor are in different buildings or powered by different AC power sources.
2. If the cable is outdoors and above ground, use fiber optics to avoid atmospheric disturbance from lightning.
3. If the cable run is through a hazardous chemical or electrical area, use fiber optics.
4. Use fiber optics when a high-security link is required.

6.8 SUMMARY

Video signal transmission is a key component in any CCTV installation. Success requires a good understanding of transmission systems.

Most systems use coaxial cable, but fiber-optic cable is gaining acceptance because of its better picture quality (particularly with color) and lower risk factor with respect to ground loops and electrical interference. In special situations where coaxial or fiber-optic cable is inappropriate, other wired or wireless means are used, such as RF, microwave, or light-wave transmission. For very long range applications, non–real-time slow-scan systems are appropriate.

Many security system designers consider cabling to be less important than choosing the camera and lens and other monitoring equipment in a CCTV application. Often they attempt to cut costs on cabling equipment and installation time, since they often make up a large fraction of the total system cost. Such equipment is not visible and can seem like an unimportant accessory. However, such cost-cutting can drastically weaken the overall system performance and picture quality.

Chapter 7
Monitors and Terminals

CONTENTS

7.1 OVERVIEW

The ultimate detector in any CCTV imaging system is the human eye. While image intensifiers, low-light-level cameras, and high-resolution cameras may be coupled via switchers and other equipment to a television monitor, the final image presented relies on the human eye for interpretation. Security designers and users must keep in mind that the human operator will make a final analysis of any monitor scene at any security console. The video images presented on the monitors must have high enough resolution and quality to permit the security guard to assess the situation at hand.

The average adult watches about 4 hours of television a day and can sit through documentaries, comedies, or news programs and obtain the "intelligence" presented on the screen. The viewer can lie down, go to the refrigerator during a commercial, or turn off the television if things get too boring. On the other hand, the average CCTV security console operator sees a boring picture most of the time. The guard may watch a deserted hallway, an empty parking lot, a back door, a front lobby, or an elevator cab, all from his or her own chair for 6 or 8 hours a day.

Ultimately, the security operator's ability to act as an alert observer determines the effectiveness of the CCTV system. To ensure success, the security director should analyze the security guard's duties along with the security console hardware. Some criteria include:

1. the number and location of monitors;
2. the time required to effectively observe each monitor, depending on the activity at each location;
3. the amount of scene detail needed in each monitor;
4. the types of security the guard must perform and the events he or she must observe in each scene;
5. the guard's knowledge of the actual scene, that is, what the area looks like in person.

These and other parameters are analyzed in the following sections.

A second part of the chapter analyzes the monitoring hardware used for CCTV security systems. This hardware consists of a variety of monochrome and color monitors with cathode ray tubes (CRTs) varying in size from 5 inches (diagonal) to 23 inches, a size that might be found in the security console center. The monitor size depends in part on how many monitors are required for the console, how much room is available, and how many guards will view the monitors. Also, this chapter considers the problem of how many cameras will be viewed sequentially or simultaneously on a single monitor, including discussion of monitors displaying 1, 2, 4, 9, 16, and 32 individual camera pictures on a single monitor.

A very useful console monitor accessory is the touch screen. The touch screen allows an operator to input a command to the security system by touching defined locations on the monitor. The touch-screen system is particularly useful when guard personnel must react quickly and decisively and when suitable graphics can appear on the monitor screen to effect an action. At present the technique is not in widespread use, but as system complexity and knowledge of its availability increase, more security systems will incorporate it.

7.2 HUMAN ENGINEERING

Security directors and managers must recognize that one very important factor in the successful operation of any CCTV security system is the performance of the security guards monitoring it. Progressive security companies have investigated guard performance and implemented fatigue-reducing procedures to alleviate problems associated with monitoring multiscreen consoles, such as backache, stress, headache, and eyestrain. The primary cause of guard fatigue and boredom is constant staring at monitors and their associated glare.

One technique for reducing operator boredom and fatigue is to have a manager or someone else enter the console area at periodic intervals to speak with the guard, ask of any concerns, praise his or her accomplishments, and in general break the monotony. Managers have addressed these problems by rotating the guard staff and assigning them shorter periods of time at the console monitors. Security personnel are often required to perform auxiliary duties that can serve to relieve boredom and improve performance. However, such duties should not detract from the attention needed to respond properly to the many real and false alarms that occur during a security guard shift.

Other technical solutions to improve guard performance include using sequential CCTV switchers or multiple camera scenes on a single monitor to reduce the number of monitors; using audio and visual alert signals; and creating ergonomic console designs to reduce fatigue, eyestrain, and boredom. Sensor alarms, video motion detectors, or simple switchers to call up and display a monitor scene that needs attention reduce the burden of the guard and allow him or her to operate more efficiently. An important design criterion in any security system is the optimum design of the security console. If in-house design experience is not available, outside designers should be consulted to accomplish this crucial task.

Other factors affecting efficiency and quality of security guard performance include environmental factors such as room temperature, lighting, and noise. Temperatures that are set too high produce fatigue. Background noises should be held to a minimum, as they are distracting and lead to lower performance; however, a complete absence of noise will induce sleepiness. Lights mounted overhead or behind the guard can cause glare on monitor screens, as can daylight. To significantly reduce this problem, manufacturers produce glare-reducing monitor screen overlays (Section 7.11).

For maximum interpretation of the television scene, console operators should be able to relate to a scene on the monitor through actual physical objects at the scene location. Thus, operators should be required to visit the physical area and become familiar with it. They should stand at the camera location, observe the scene, and become acquainted with objects in it, note the camera's blind spots, and understand the surveillance requirements. Operators should be tested regularly to determine whether they can recognize individuals, objects, and other events and activities on the monitor screen.

For tolerable monitor image distortion, the maximum vertical viewing angle between observer line of sight and monitor is approximately 30 degrees (Figure 7-1). The maximum horizontal viewing angle in both directions is approximately 45 degrees. The optimum distance of the operator from the monitor depends on the monitor size. Table 7-1 summarizes the suggested maximum and minimum viewing distances for different-size monitors.

A semicircular monitor console is optimum for viewing multiple monitors. An observer's station should not consist of more than approximately 15 monitors. Figure 7-2 illustrates two console configurations for a single-guard operation. Note that the monitors are slanted from top to bottom. The semicircular configuration permits optimum viewing of the monitors with minimum distortion and minimum head rotation to observe all monitors.

Figure 7-3 illustrates a fairly complex monitor security console, in which two guards monitor approximately 25 monitors. Note that the monitor banks are arranged in a near-semicircular configuration and that the top- and lower-bank monitors are directed in toward the center, so that the guard viewing the monitor bank at central eye level will be looking nearly perpendicularly at each of the monitors, thereby producing minimum distortion.

There are some hardware options for making the console monitoring area more efficient and conducive to optimum guard performance. These hardware factors include

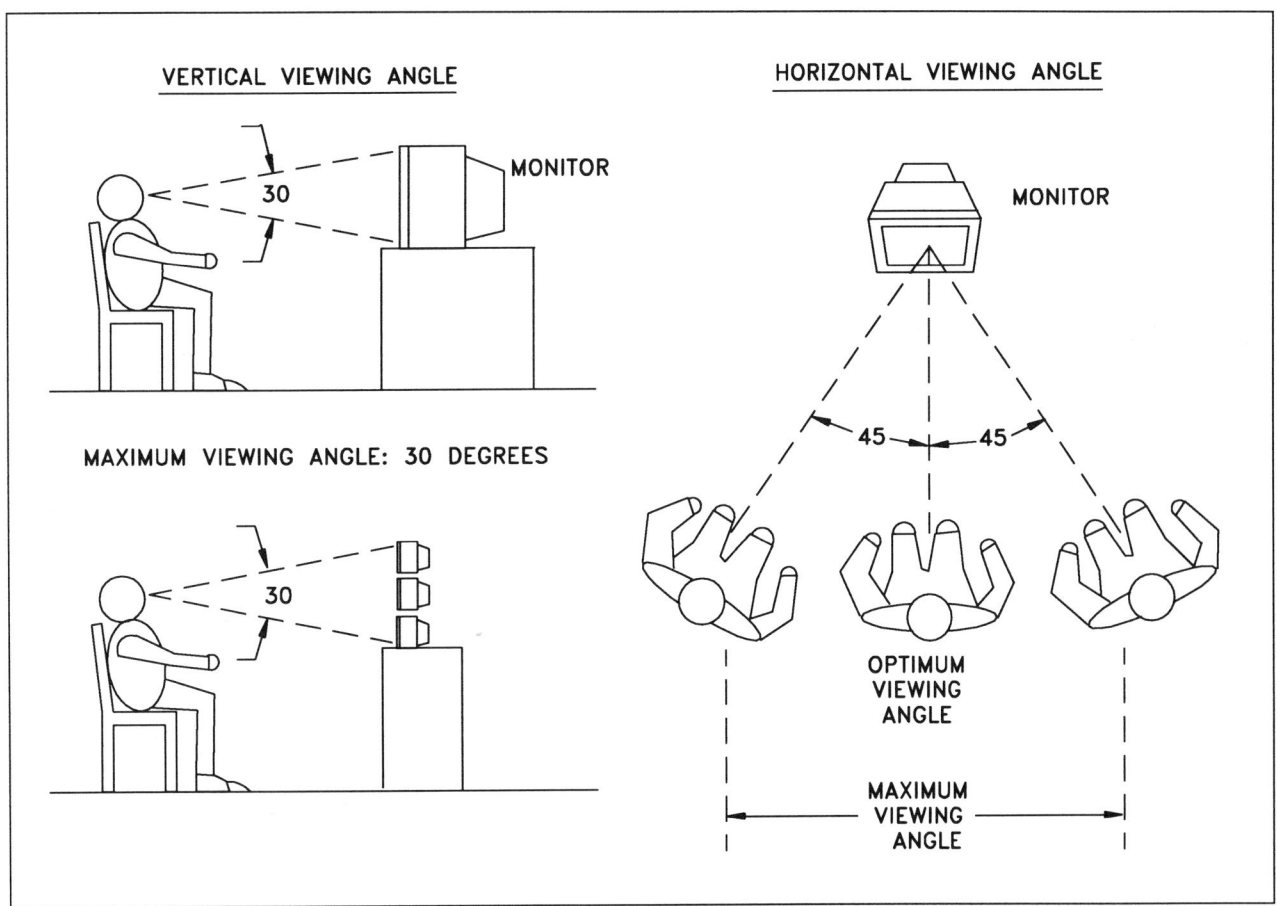

FIGURE 7-1 CCTV monitor viewing angles

MONITOR SIZE (DIAGONAL–INCHES)	VIEWER TO SCREEN* DISTANCE (INCHES)
5	12–24
7	16–35
9	20–42
10	22–48
12	24–60
13	26–65
17	30–75
19	35–88
20	36–96

Table 7-1 Viewing Distance vs. Monitor Size

sequential switchers, pan/tilt cameras, motion sensors, two-way intercom, VCRs, and video printers to carry out some of the functions guards might normally perform.

The use of sequential switchers, in which many cameras are switched through a single monitor, reduces the number of monitors a guard must view. If a sequential switcher is used, it is recommended that a monitor view no more than 12 different cameras, to minimize guard fatigue. The maximum number depends on the activity in each scene. For more active applications, such as motion in the scene or some form of activity, 4 or 8 monitors might be the maximum recommended. Too many cameras with too many monitors increases stress, with a commensurate decrease in the effectiveness of the guard. Experience has shown that 15 monitors is considered maximum for any operator, with the main viewing screen having a 17-to-21-inch diagonal and the other monitors a 9-inch diagonal. The use of split screens with multiple scenes on one monitor to reduce guard fatigue is described later.

Monitors used for security operations have screen sizes (diagonal measurement) of 5, 7, 9, 14, 17, 19, 21, or 23 inches. By far the most popular size is the 9-inch tube. There are several reasons for this:

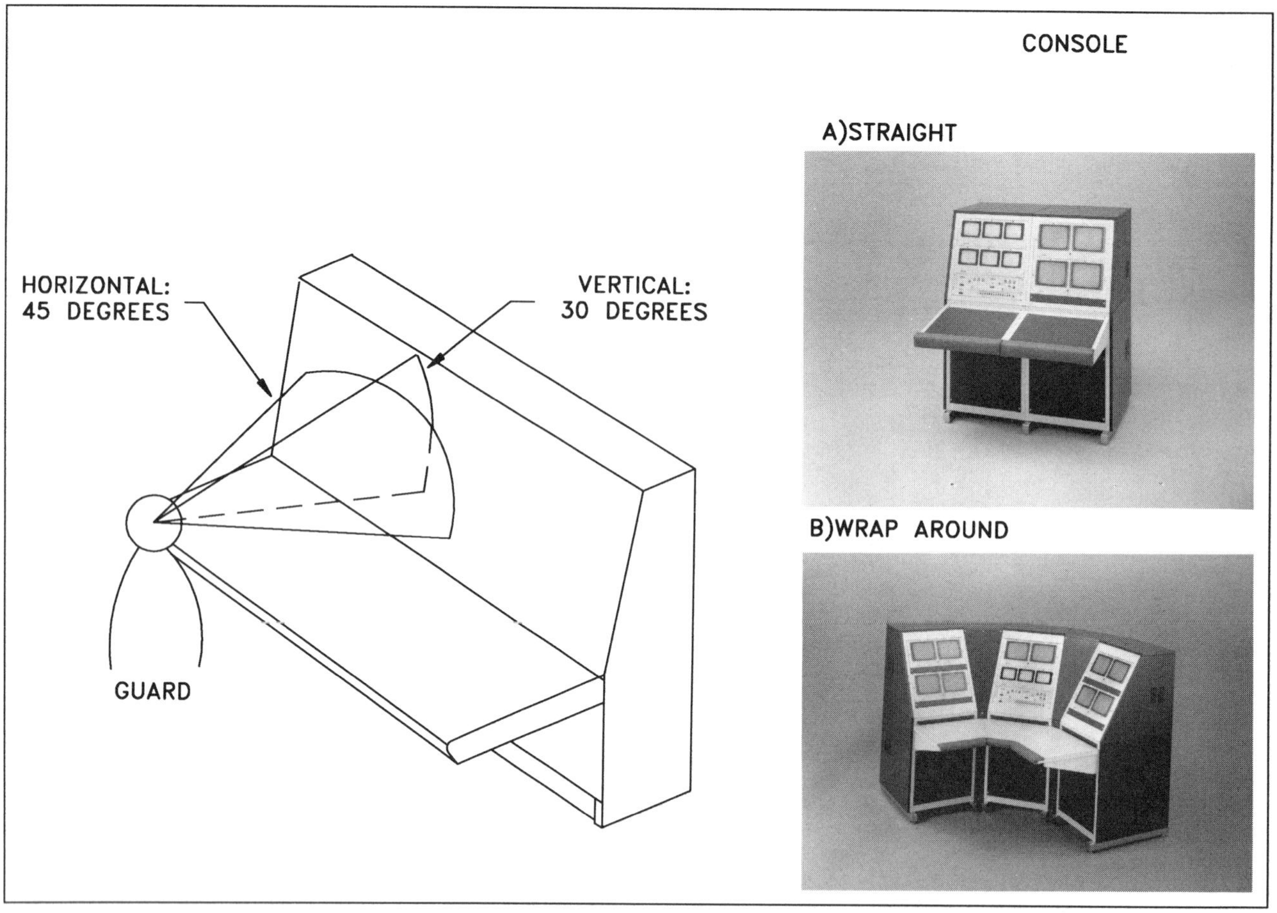

CONSOLE

A)STRAIGHT

HORIZONTAL:
45 DEGREES

VERTICAL:
30 DEGREES

GUARD

B)WRAP AROUND

FIGURE 7-2   Single-guard security console configuration

WRAP AROUND CONSOLE

SLOPED
MONITOR
RACKS

HORIZONTAL:
45 DEGREES

VERTICAL:
30 DEGREES

GUARD   GUARD

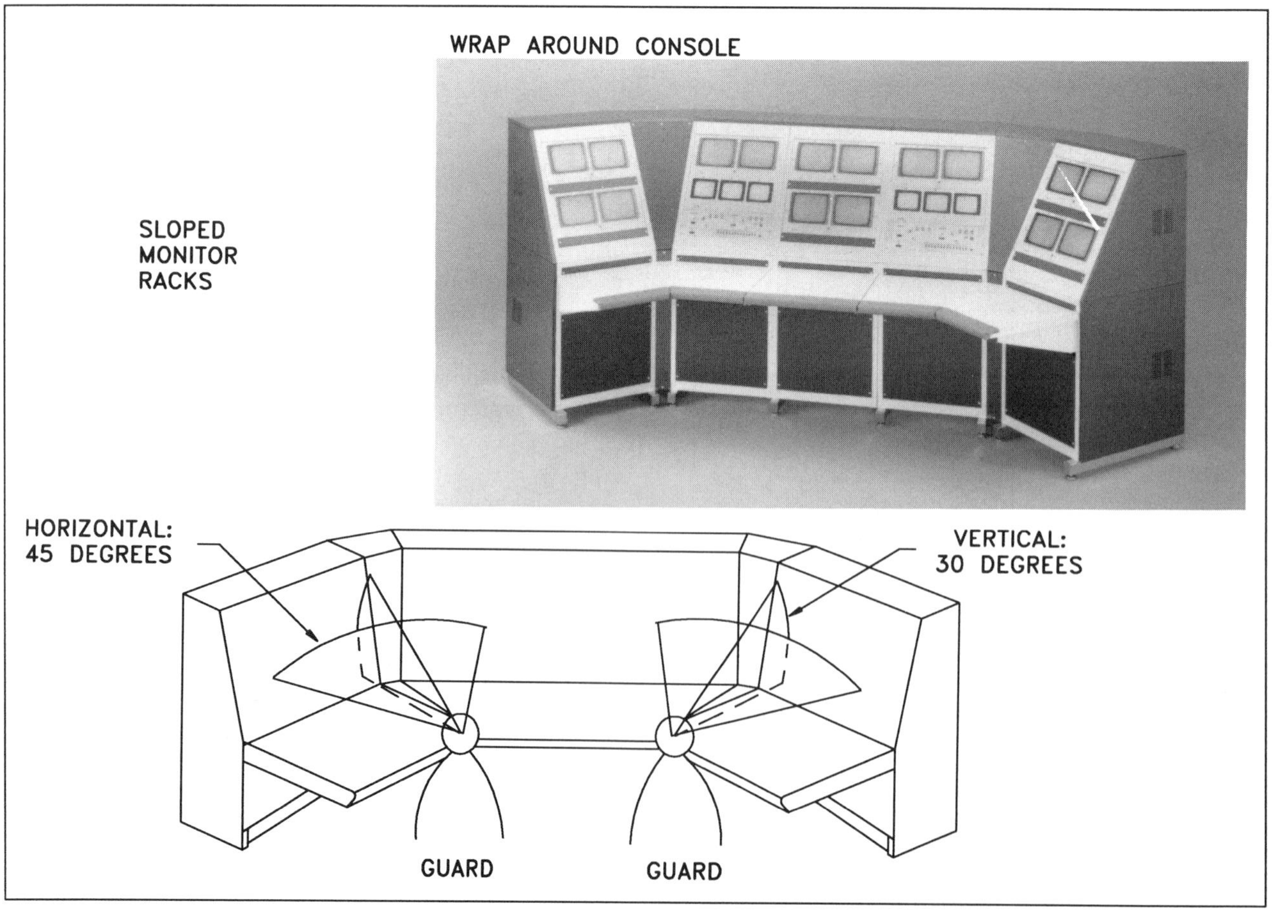

FIGURE 7-3   Two-guard security console configuration

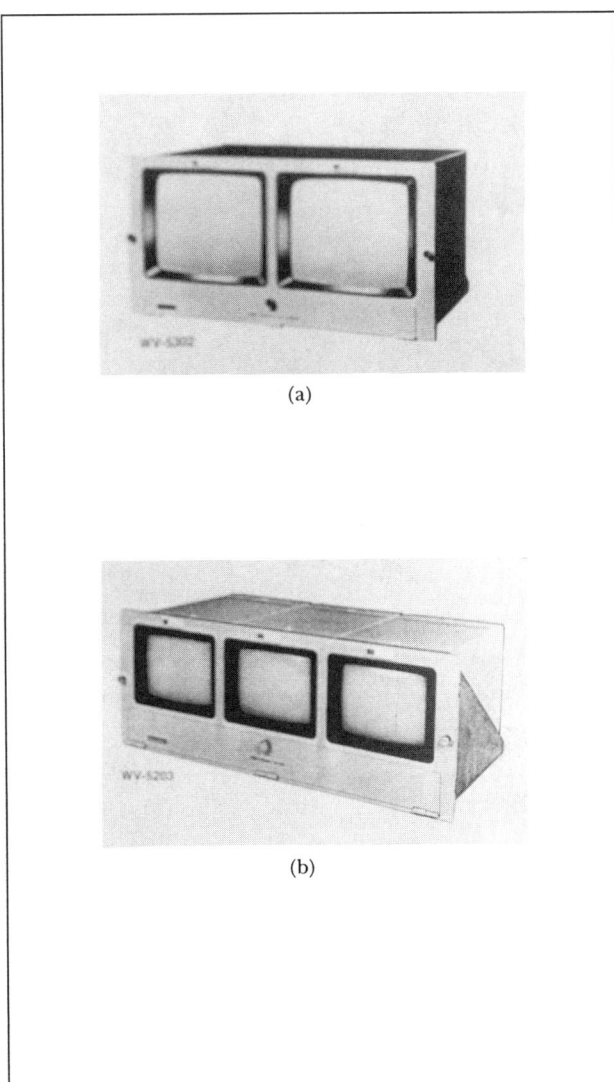

(a)

(b)

FIGURE 7-4 Dual and triple rack-mounted monitors: (a) dual monitors—9-inch diagonal, (b) triple monitors—5-inch diagonal

1. The 9-inch diagonal screen provides the highest resolution from the monitor.
2. Using a 9-inch monitor, the optimum viewing distance of the security operator from the monitor is approximately 3 feet, a convenient distance found in many console rooms.
3. The 9-inch monitor size is such that two 9-inch monitors fit side by side in the standard EIA 19-inch rack configuration. This is an important consideration, since space is generally at a premium in the console room and optimum integration and placement of the monitors is important.

In applications where more monitors are required and closer viewing is possible or resolution can be sacrificed, a triple-monitor arrangement in a 19-inch rack is available.

The monitors are 5-inch diagonal and mounted three across. Figure 7-4 illustrates the dual and triple rack-mounting arrangement.

The size of monitor (5 to 23 inches diagonal) chosen depends in part on how many security operators will be in the monitoring console room. The question of how many cameras to sequence through one monitor is an important one. If each camera is displayed on its own monitor, the guard can view all cameras at any time without switching. However, he cannot realistically view all of the monitors simultaneously. When there are few cameras and perhaps up to 10 monitors, the guard can view them all fairly well simultaneously. When there are perhaps 20 or 30 monitors, some of them must be combined (split-screen) and switched through a single monitor to reduce the number of monitors in the console. With a guard seated 2 to 3 feet from the nearest monitor, the optimum screen size is 9 inches diagonal.

One disadvantage of switching cameras through a monitor is that the guard may miss an event on a single monitor because it was showing the view from another camera. One way to overcome this problem is to install an alarming device (video motion detector, switch, infrared detector) that automatically switches the monitor to the camera in the area of activity. Most systems fall in between the extremes of many monitors each showing a single camera view and one monitor showing many camera views. When there are several critical scenes to be viewed, each area should be equipped with a dedicated monitor (a camera directly connected to a monitor) without any switching, to ensure that the guard always has 100% viewing of that camera scene.

Another technique to view all scenes all the time is to use a split-screen monitor presentation, in which 2, 4, 9, 16, or 32 scenes are simultaneously displayed on a single monitor. While this technique does provide multiple scenes on a single monitor, it decreases the resolution of each picture proportionally, thereby making identification of action, person, or object more difficult. If a guard needs to see the picture with better resolution, he or she can switch from split-screen presentation to single-screen presentation of the camera of interest.

The horizontal resolution of a 9-inch monochrome monitor is approximately 700 TV lines; for a 9-inch color monitor, approximately 330 TV lines. The horizontal resolution of a typical 5-inch monochrome monitor is 500 TV lines or higher; for a 5-inch color monitor, approximately 350 TV lines. Vertical resolution is about 350 TV lines on both types.

Most monitors are available for 117-volt, 60-Hz (or 230-volt, 50-Hz) AC operation; some are available for 12-volt DC operation. Video signal connections are made via BNC connectors on the rear, which are terminated in either 75 ohms or a high impedance (100,000 ohms) via a switch. If only one monitor is used, the switch on the rear of the monitor is set to the 75-ohm or low-impedance position for

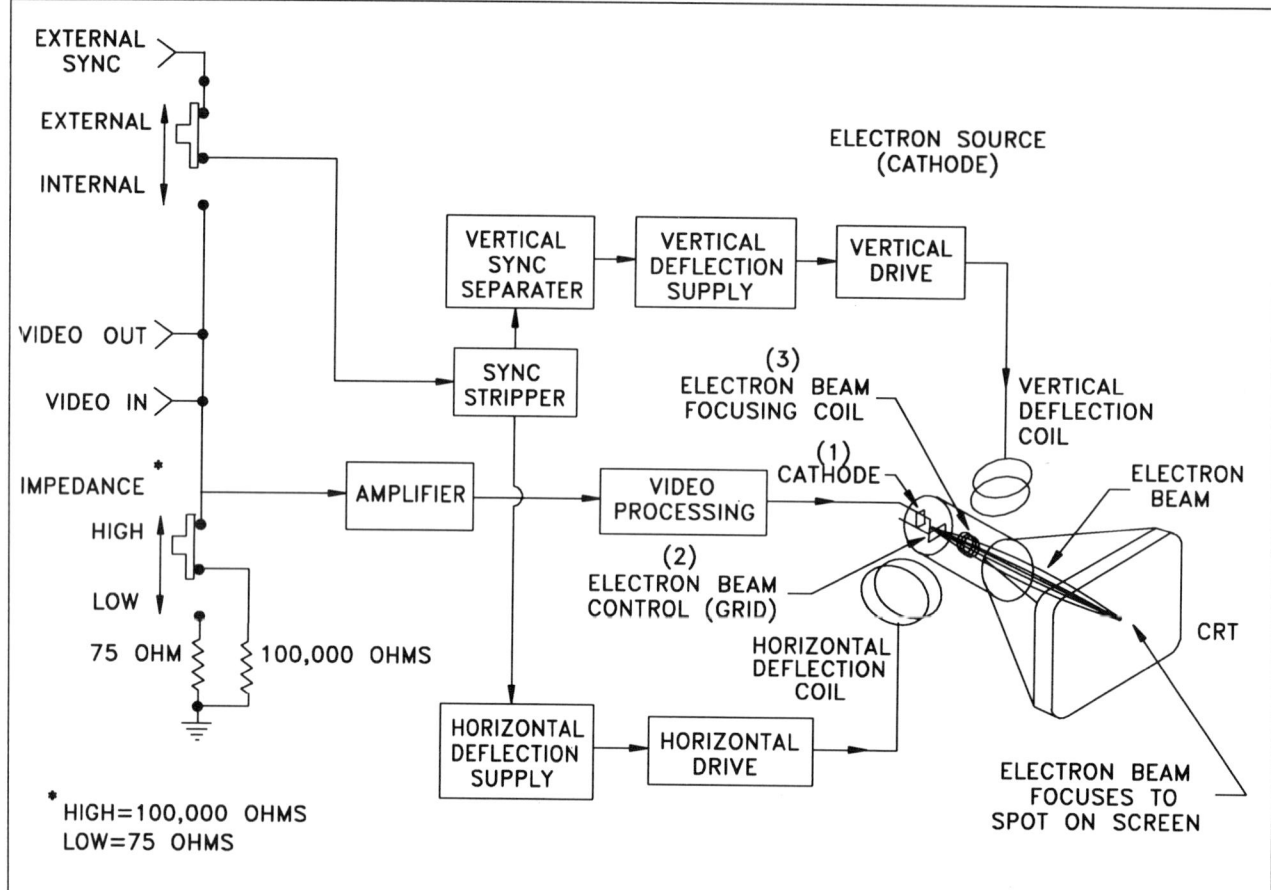

FIGURE 7-5 Monochrome monitor block diagram

best results. If multiple monitors are used, all but the last monitor in the series is set to the high-impedance position. The last monitor is set to the low-impedance position. All cameras and monitors have a 4 × 3 geometric display format, that is, the horizontal-to-vertical size is 4 to 3, respectively. If a VCR or video printer is connected, all the monitors are set to high impedance and the VCR or printer set to low impedance. (VCR and printer manufacturers usually set the impedance to 75 ohms at the factory.)

Once the single-camera-lens system is understood, much of the design of multiple camera systems has been accomplished. One choice remains:

1. Should each camera be displayed on an individual monitor?
2. Or should several camera scenes be displayed on one monitor?
3. Or should the picture from each individual camera be switched to a single monitor via an electronic switcher.

If the scene activity, that is, the number of people passing into or out of an area, is relatively high, all cameras should be displayed on separate monitors. For installations with infrequent activity or casual surveillance, a manual, automatic, or other switcher or combiner (split-screen) should be used.

Since each installation requires a different number of monitors and has different monitoring criteria depending on the type of surveillance required, each installation becomes a custom design. The final layout and installation of a system should be a collaboration between the security department, management, outside consultants, and any others having security experience.

7.3 MONOCHROME MONITOR

In its simplest form, the monochrome monitor consists of these parts (Figure 7-5):

1. an input-terminating circuit,
2. a video amplifier and driver,
3. a sync stripper,
4. vertical-deflection circuitry,
5. horizontal-deflection circuitry, and
6. a CRT.

Shown in the block diagram are the video-processing circuits to remove the scene information from the video signal and present it to the CRT. The sync-stripper circuit and vertical- and horizontal-deflection electronics drive the vertical and horizontal coils on the CRT to produce a raster scan. The video signal presented to the monitor is a negative sync type, with the scene signal amplitude modulated as the positive portion of the signal (see Figure 5-4). Via frequency-selective circuits, the horizontal and vertical synchronization pulses are separated and passed on to the horizontal and vertical drive circuits. The scanning signals producing the horizontal and vertical deflection of the electron beam on the CRT are like those used in the camera from which the signal has been obtained.

The CRT consists of (1) a cathode (source) that emits electrons to "paint" the picture, (2) a grid (valve) that controls the flow of the electrons as they pass through it, and (3) an electron beam that passes by another set of electrodes that focus the beam down to a spot, and (4) a phosphor-coated screen that produces the visible picture.

When the focused beam passes through the field of the tube's deflection yoke (coils), it is deflected by the yoke to strike the appropriate spot of the tube's phosphor screen. By varying the voltage on the horizontal and vertical coils, the electron beam and spot are made to move across the CRT with the familiar raster pattern. The screen then emits light with an intensity proportional to the beam intensity, resulting in the television image on the monitor. The CRT monitor accomplishes all this using relatively simple and inexpensive components. In this way the scene received by the camera is reconstructed at the monitor.

The CCTV monitor accepts the normal CCTV baseband signal (20 Hz–6.0 MHz) and displays the image on the CRT phosphor. The monitor circuitry is much the same as in an ordinary household television receiver but lacks the electronic means to receive the VHF or UHF broadcast signals and electronic detecting components. All sizes of monochrome and color monitors used for CCTV surveillance accept the standard 525-line NTSC input signal. The video signal is input to the monitor via a BNC connector terminated by one of two input impedances: 75 ohm, which matches the coaxial-cable impedance; or high impedance (generally 10,000 to 100,000 ohms), which does not match the coaxial-cable impedance and is generally used only when the monitor will be terminated by some other equipment such as a looping monitor, a VCR, or some other device with a 75-ohm impedance (Figure 7-6).

If two or more monitors receive the same CCTV signal from the same source, only one of the monitors—the last one in line—should be set to the 75-ohm position. (If a VCR and not a second monitor is used, the VCR automatically terminates the coaxial cable with a 75-ohm resistor.)

The operator controls available on most monochrome monitors are power on/off, contrast, brightness, horizontal hold, and vertical hold. Three other controls sometimes available via screwdriver adjust or in the rear of the monitor are horizontal size, vertical size, and focus.

As shown in the block diagram in Figure 7-6, the monitor has two BNC input connectors in parallel. When only one monitor is used, the impedance switch is moved to the 75-ohm, low-impedance position, which terminates the co-

axial cable and optimizes the signal received by the monitor. If more than one monitor or auxiliary equipment is used, the terminating switch is left in the high-impedance position, opening the connection to the 75-ohm resistor so that the final termination is determined by a second monitor or some other equipment, such as a VCR. Some monitors contain an external synchronization input connector so that the monitor may be synchronized from a central, or external, source.

7.3.1 Technology

With the widespread use of computer displays in many security departments and the availability of various technologies for displaying data and video images (such as liquid crystal displays), it may be surprising that the CRT display is still used in virtually all security console environments. The continuing success of the CRT is based on an extremely simple concept with a relatively simple structure having the electronic circuitry support of semiconductor technology. While the CRT still utilizes vacuum-tube technology, its combination with semiconductor technology provides the most cost-effective solution to displaying a video image, be it monochrome or color.

Over the years, the CRT monitor has become less expensive, higher in quality and longer in lifetime. These monitors have electronics that provide for optimum analog and digital displays of video frame storage pictures and microcomputer outputs. The CRT bends an electron beam at extremely high speed with exactly timed movement and gating to compose a dense, complex picture. Electron beams can be deflected quickly so that pictures on the screen can be refreshed without noticeable flicker. The CRT's energetic electron beam strikes and excites the phosphor screen, which has a high luminous efficiency—another advantage over competing display technologies. Other techniques are limited to specific applications, such as portable equipment that uses liquid crystal, electro-luminescent, and plasma displays. The only disadvantage of the CRT is its relative size, particularly its depth. However, if there is sufficient space behind the monitor screen (which there generally is) in the console room environment, there really is no disadvantage to the CRT monitor size.

7.3.2 CRT Costs

CRT displays cost far less than any competing technologies because of their simple construction and long history of high-volume production. Over the years, the associated electronics that support the CRT display have experienced cost reductions due to high volume. In particular, the current expanding use of personal computer systems has significantly increased the volume of tube and monitor manufacture, thereby reducing cost.

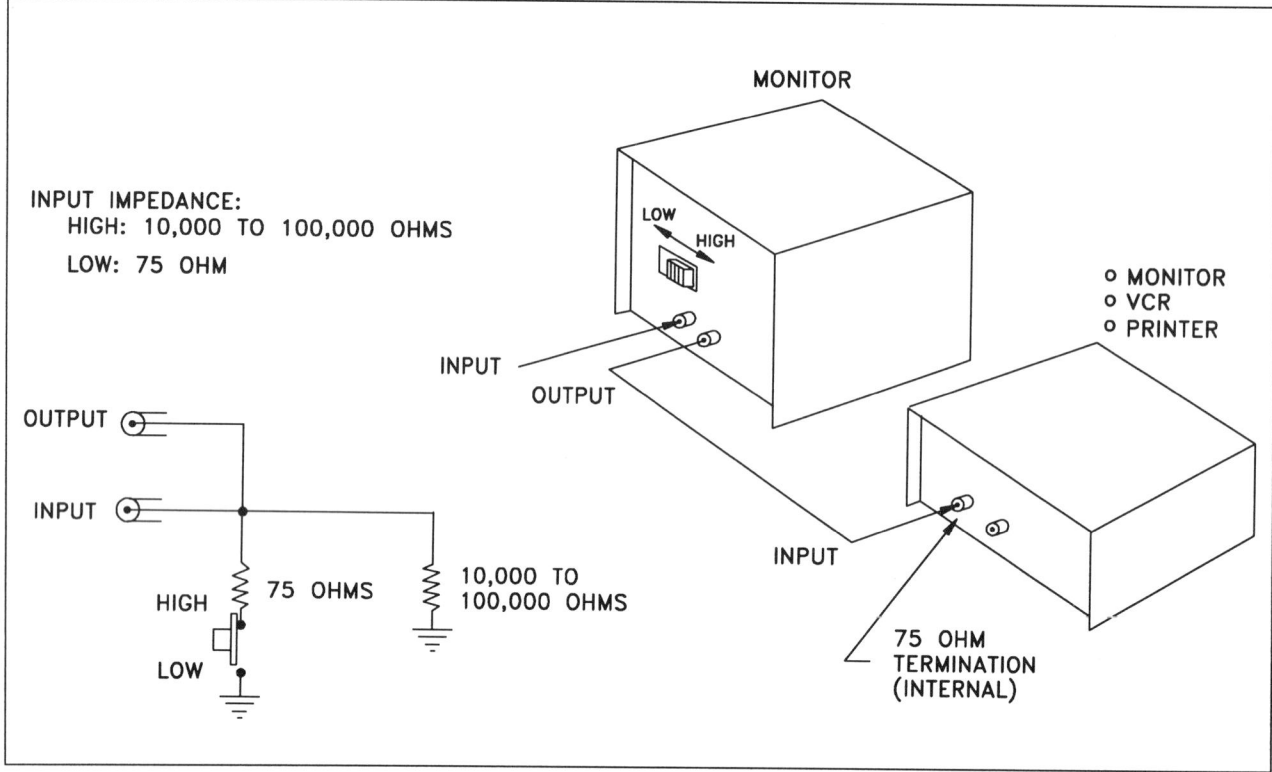

FIGURE 7-6   CCTV monitor terminations and connections

There are essentially three components composing the CRT: the electron gun, the glass envelope, and the phosphor screen, on which the image appears. While the color CRT is a bit more complex, the basic construction is similar, and high-volume production keeps prices down. The monochrome CRT tube is relatively easy to manufacture, since the screen consists of a uniform coating of a single material, the phosphor. It is far easier to apply this coating than to make a flat-panel screen (LCD, for example), with its array of discrete pixels or cells and attendant connections. The yield obtained with CRTs is also significantly higher than other techniques, since the human eye is far less sensitive to variations in phosphor flaws than it is to pixel or cell failures. As with the vidicon tubes compared with the CCD imagers, the homogeneous and continuous phosphor layer has very high resolution, since it is continuous, as contrasted to the high-density flat-panel cells (pixels). Consequently, the resolution of a monochrome CRT is limited only by the electron beam diameter and the electronic video bandwidth (determining how fast the beam can turn on and off). This gives the CRT a cost advantage that increases as the resolution increases as compared with a flat screen having individual pixels. For a nominal 500-line video CRT display, only a single coating of phosphor and one electron gun is required, which is considerably easier to produce than a flat-panel display, which must achieve a very high yield on as many as 250,000 cells.

7.3.3   CRT Lifetime

The lifetime of standard CRTs is legendary, especially under adverse operating conditions: consider the abuse that standard consumer TVs receive. The CCTV monitors may be cycled on and off and adjusted over a wide range of brightness and contrast beyond their design limits and still operate satisfactorily for many years.

7.3.4   Spot Size

The light beam spot size is the diameter of the focused electron beam, which ultimately determines the resolution and quality of the picture. This size should be as small as possible to achieve high resolution. Typical spot sizes range from about 0.1 to 1.0 mm; the spot size is smallest at the center of the CRT and largest at the corners. Deflection to the edges elongates the spot and decreases resolution.

7.3.5   Phosphors

A CRT's phosphor glows for a time determined by the phosphor material. The phosphor must be matched to the refresh rate. The use of the P4 (white) phosphor has been widespread as the standard monochrome television monitor phosphor. It is capable of achieving good focus and small spot size. Its low cost and ready availability contribute to its continued popularity in monitors. P4 is a medium- to medium-short-persistence phosphor. The "glow" activated by the electron beam fades away fairly rapidly, leaving no cursor trail or temporary "ghost" scene when the monitor is turned off. The P4 phosphor is moderately resistant to phosphor burn, a term used to describe the permanent dark pattern of the often repeated portions of the video image on the CRT tube face. The susceptibility to burn is somewhat proportional to persistence, with longer-persistence phosphors more liable to burn.

7.3.6 Interlacing and Flicker

Interlacing is one of the most common and cost-effective methods used to achieve increased resolution at conventional 60-Hz horizontal scan rates. One critical design consideration in interlaced operation is that a long-persistence phosphor such as P39 must be used; P4 phosphor is not suitable for interlaced operation. (The European equivalent of P4 is W.) The glow of short-to-medium-persistence phosphors (P4) begins to fade before it can be refreshed. At the standard U.S. noninterlaced 60-Hz refresh rate, this presents no problem: the viewer's eye retains the image long enough to make any fading imperceptible.

In an interlaced monitor, the beam skips every other row of phosphor as it moves down the CRT face in successive horizontal scans. Only half the image is refreshed in a vertical sweep cycle, so the frame-refresh rate is effectively 30 Hz. The eye cannot retain the image long enough to prevent pronounced flicker in the display if a short-to-medium-persistence phosphor is being used. The glow of the phosphor therefore must persist long enough to compensate for the slower refresh rate.

The "flicker threshold" of the human eye is about 50 Hz, with a short-to-medium-persistence phosphor. Monitor manufacturers designing for European-standard 50-Hz operation therefore pay particular attention to the phosphor used.

Although it has become acceptable to call 60-Hz noninterlaced displays "flicker free," a large percentage of the population can see flicker at 60 Hz in peripheral vision when P4-type phosphors are used. A 19-inch CRT viewed at 27 inches covers more than the central cone vision, and therefore most people see some flicker. While this situation is not ideal, it cannot be overcome, because of the inherent 60-Hz power-line frequency.

Interlaced scanning is the method used in most CCTV systems, and since television scene content consists primarily of large white areas, no objectionable flicker is apparent. In computer alphanumeric/graphic displays, all display data consist of small bright or dark elements. Consequently, an annoying flicker results when alphanumeric/graphic data are displayed using interlaced scanning, unless a

longer-persistence phosphor is used. Therefore, the phosphor type used in the video monitor is different from the general requirement for computer terminal monitors.

7.3.7 Brightness

The luminance (brightness) of the display picture is proportional to the electron beam power, while the resolution depends on the diameter of the beam. Both of these properties are determined by the electron gun. Very high resolution monitors are available having a resolution of 3000 lines—close to the ergonomic limit of the human eye. While present security systems do not take advantage of this high resolution, some systems display 1000-line horizontal resolution.

7.3.8 Beam Deflection

Most CRT monitors use magnetic deflection. Figure 7-5 illustrates the placement of vertical and horizontal deflection coils at the neck of the CRT. When current flows through the coils, a horizontal magnetic field is produced across the neck, and the amount of vertical deflection of the electron beam depends on the strength of the magnetic field, and hence the current through the coil. The direction of the beam deflection—up or down while passing through the horizontal coil—depends on the polarity of the field. Likewise for the vertical deflection coil: the electron beam is deflected depending on the strength of the magnetic field, which in turn is dependent on the vertical deflection current. The combination of both the horizontal and vertical coils simultaneously causes the raster scan on the CRT. As with the scanning in a tube or solid-state television camera, the television monitor has an aspect ratio of 4 to 3 (the diagonal is 5 units). The size of the tube is measured from one corner of the screen to the opposite corner and referred to as the diagonal. Therefore, the tube on a 9-inch monitor has a useful viewing diagonal of 9 inches.

7.3.9 Standards

In the NTSC television format, there are about 495 horizontal lines (any number of lines between 482 and 495 may be transmitted at the discretion of the television station). To produce satisfactory horizontal picture definition—that is, a gray scale and sufficient number of gradations from dark and light per line—a bandwidth of 4.2 MHz is required.

CCTV monitors generally conform to EIA specifications EIA-170, RS-330, RS-375, RS-420, and most often to UL specification 1410. The analog circuitry is usually capable of reproducing a minimum of 10 discernible shades of gray, as described in the RS-375 and RS-420 specification.

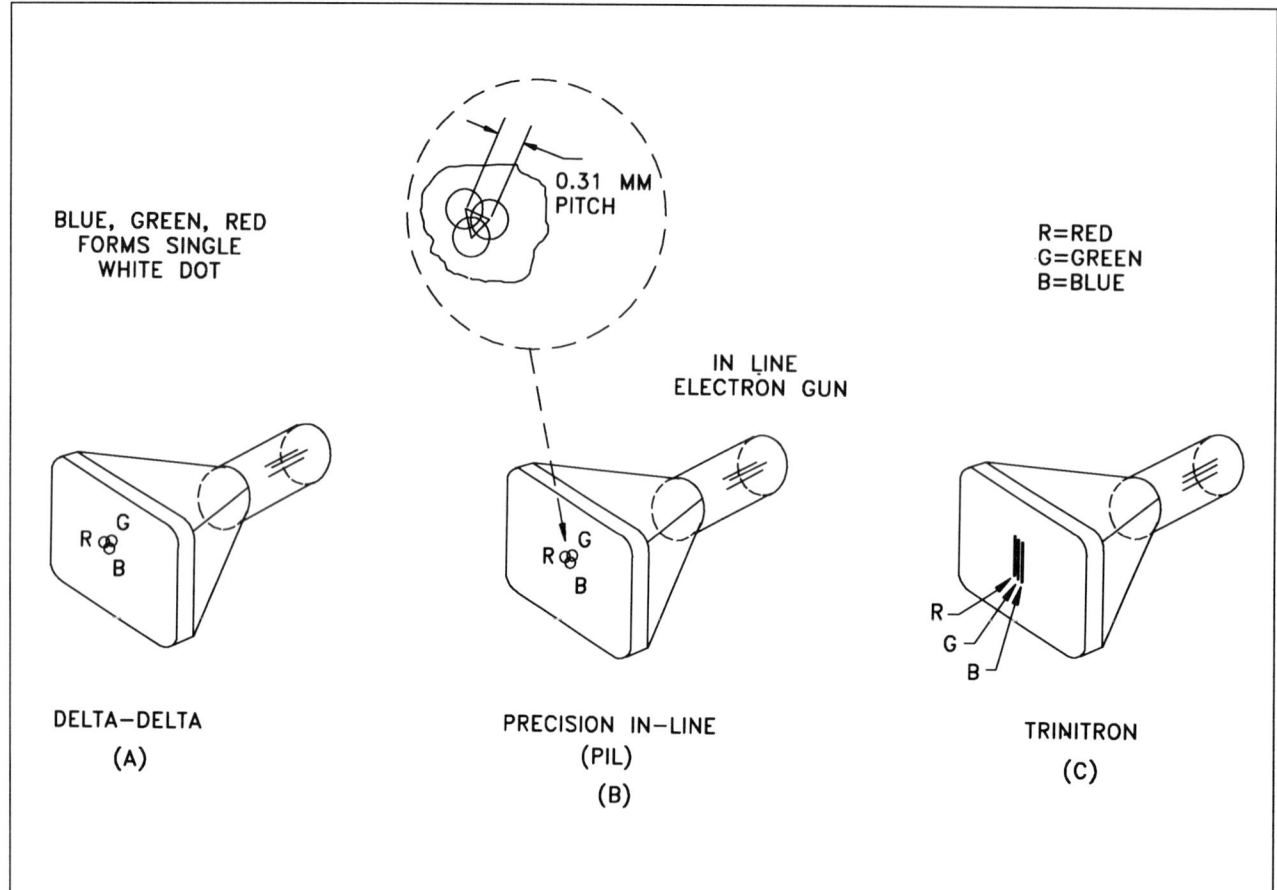

FIGURE 7-7 Color monitor technology

7.4 COLOR MONITOR

Until recently, the major CRT technology used in color monitors employed three electron guns (one for each primary color) arranged in a triangle (Figure 7-7a). This was called the delta-delta system. A device called a shadow mask aligned each electron gun output so that the beam fell on the proper phosphor dot. This provided the highest resolution possible but required the guns to be aligned manually by a technician, as well as expensive convergent-control circuitry.

Today the most widely used CRT color technique is the precision in-line (PIL) tube, which overcomes most of these difficulties (Figure 7-7b). The PIL tube uses the shadow mask found in its predecessors, but the electron guns are in a single line. The shadow mask is a thin steel screen in the CRT containing fine holes that concentrate the electron beam. The spacing between the holes is termed the dot pitch or dot-trio spacing and determines the tube's resolution capability. The highest-resolution production PIL tube has approximately a 0.31-mm pitch. The PIL tube is preconverged by the manufacturer so there is no adjustment in the field. There is a slight decrease in resolution for the PIL as compared with the original delta-delta, but

this is a small sacrifice considering that no field adjustment is required.

A third CRT color tube scanning technique is called the Trinitron (a trademark of Sony Corporation) and consists of alternate red, green, and blue (R, G, B) vertical stripes (Figure 7-7c). The composite video input signal in the color monitor contains the information for the correct proportion of R, G, B signals to produce the desired color. It also contains the vertical and horizontal synchronization input needed to steer the three video signals to the correct color guns. Composite video color monitors decode the signal and provide the proper level to generate the desired output from the three electron guns.

7.5 VIDEO MONITOR VERSUS COMPUTER TERMINAL

The significant difference between the video monitor and graphic or computer terminal display is that the video monitor displays a continuous scene image while the computer displays alphanumeric characters or graphics. Generally, flicker will show up sooner on the alphanumeric display, since the information to be displayed has fine detail and generally a light and dark character.

7.6 SPLIT-SCREEN MONITOR PRESENTATION

A significant reduction in the number of monitors required in a security console room is obtained through the use of image-combining equipment. While the monitors for these displays are the same as those for single-image displays, the image-combining devices permit displaying multiple camera scenes on one monitor. Chapter 13 describes hardware to accomplish this function. The hardware takes the form of image-combining optics or electronic combining circuits to produce multiple images—from 2 to 32 images on one monitor screen.

7.7 AUDIO/VIDEO

As with home television cameras and receivers, some monitors are equipped with audio amplifiers and speakers so that video and audio from the camera location are displayed on and heard from the monitor. The video input impedance is 75 ohms and the audio is 600 ohms.

7.8 SCREEN SIZE VERSUS RESOLUTION

The monitor screen size chosen for any particular application depends primarily on the subject-to-monitor distance. For one-on-one or field-test applications, small screens ranging in size from a few inches up to 9 inches are recommended.

In security console environments, the optimum screen size for most applications is a 9-inch diagonal. Often if the installation is medium to large there are several larger monitors included—17 through 23 inch—since there may be other personnel in the environment who will be viewing these larger monitors to obtain an overview of the security operation. The 9-inch monitor usually provides the highest resolution.

7.9 FLAT SCREEN

A unique flat-screen CRT monitor system has been developed by Sony Corporation and is used in many portable and small-screen monitoring applications (Figure 7-8). The flat CRT monitor is excellent for a test monitor to align video cameras and lenses in the field. The system uses a new flat-screen technology that provides high resolution, high brightness, and low power consumption in a display assembly less than 2 inches thick. Up to 600-line resolution is available in the 4-inch-diagonal model. A 2-inch-diagonal

model is also available. Warm-up time is fast: less than 2 seconds after it is switched on. Operation can be on either DC or AC, making it a completely portable monitor.

As shown in the side view of the deflection tube, the electron gun is at the base, projecting the electron beam up toward the screen on the rear of the tube, which contains the phosphor coating. The electron beam energizes the phosphor, directing its light forward toward the observer. An advantage of this is that the phosphor is irradiated by the electron beam from the front, producing a very efficient electrical-to-light output design. The bright picture is a result of this front-reflection screen. The 2-inch diagonal system has a horizontal resolution of approximately 280 TV lines. While this represents a low resolution, it nevertheless can serve as a useful monitor in field service and maintenance applications. Another use for the flat screen is in portable surveillance equipment, which must consume low power, be small, and be lightweight.

7.10 MULTISTANDARD, MULTISYNC

Multistandard, multisync, and multivoltage television monitor-receiver combinations are available that operate on the U.S. NTSC (525 TV lines) and the European CCIR (625 TV lines) standards. Color systems operate on the NTSC, PAL, and Secam formats. These monitors are not in high production since they are more expensive than single-standard types. The multisync monitors are used primarily in computer displays where the computer program has a scan rate different from the 60-Hz (or 50-Hz) rate and the monitor must synchronize to that other scan rate. These systems are used with many new graphic programs that display personnel data records and photo identification, security graphics, alarm scenes, and so on. Multivoltage monitors operate from 90 to 270 volts AC, 50–60 Hz for worldwide use.

7.11 ANTI-GLARE SCREENS

A common problem associated with television monitor viewing is the glare coming from the screen when ambient lighting located above, behind, or to the side of the monitor reflects off the front surface of the monitor. This glare reduces the picture contrast and produces unwanted reflections. Monitors are often used in environments that have medium-to-bright lighting. In well-designed security console rooms where the designer has taken monitor glare into consideration at the outset, glare will not significantly reduce screen intelligibility or viewer fatigue.

For best results, face the monitor in the direction of a darkened area of the room. Keep lights away from the direction in which the monitor is pointing, which is either behind the person looking at the monitor or from the

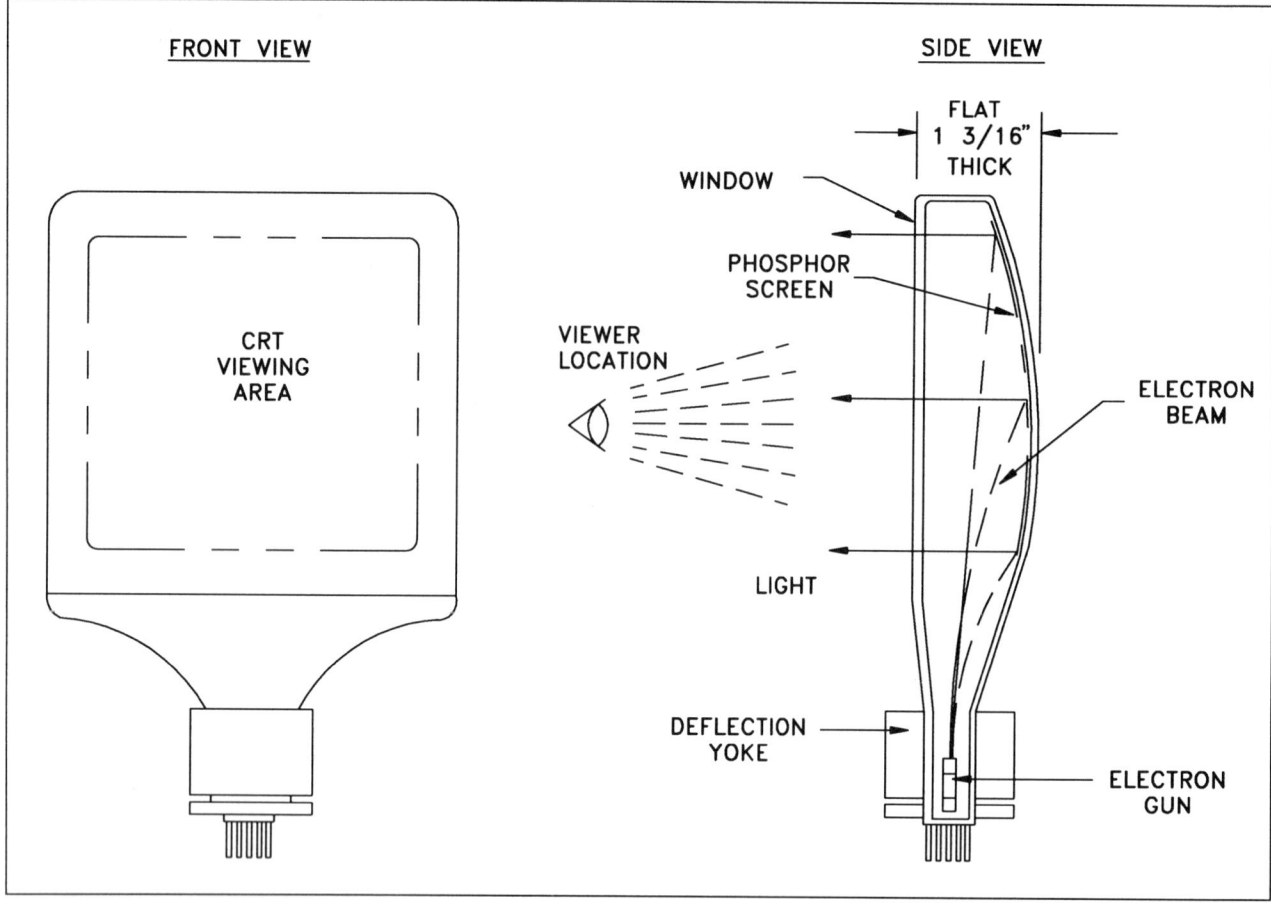

FIGURE 7-8 Flat-screen CRT monitor

ceiling above. If there are windows to the outside of the building where bright sunlight may come through, point the monitors away from the outside windows and toward the inside walls. When this cannot be accomplished and annoying glare would produce fatigue and reduce security, various anti-glare filters can be applied to the front of a monitor to reduce the glare and increase the contrast of the picture on the monitor. With a well-designed anti-glare screen and proper installation, glare and reflection levels can be reduced significantly. Figure 7-9 is an unretouched photograph showing the contrast enhancement (glare reduction) provided by one of these filters.

The anti-glare treatment on the filter front surface reduces the annoying reflections from overhead and outdoor ambient lighting. The benefits of the glare elimination are improved readability of the screen and a reduction of user error and fatigue.

These optical filters are manufactured from polycarbonate or acrylic plastic materials and are suitable for indoor applications. Polycarbonate filters used in outdoor applications can withstand a wider temperature range and are therefore more suitable. The filters come in a range of colors, with the most common being neutral density (gray),

green, yellow, or blue. The colored filters are used on graphic and data display computer terminals, whereas the neutral-density types are used for monochrome and color monitors. While the use of a color filter on a monochrome monitor might appear a little unusual, it can nevertheless increase the contrast and ability to see the video image on the screen. In the case of color displays, the lighter gray filters should be used for glare reduction.

Glare-reducing filters are chemical and abrasion resistant, and easily mounted and cleaned. To clean the filter, a nonabrasive cloth or paper wipe should be used with a mild window-cleaning fluid, mild detergent, ammonia and water, or propanol alcohol.

7.12 VIDEO GENLOCK, GRAPHICS, TEXT

There are CCTV security applications in which it is desirable to combine the CCTV image with computer text and/or graphics. To accomplish this, the video and computer display signals must be synchronized, combined, and sent to a monitor. Equipment is available to perform this integration (see Chapter 13, Section 13.3).

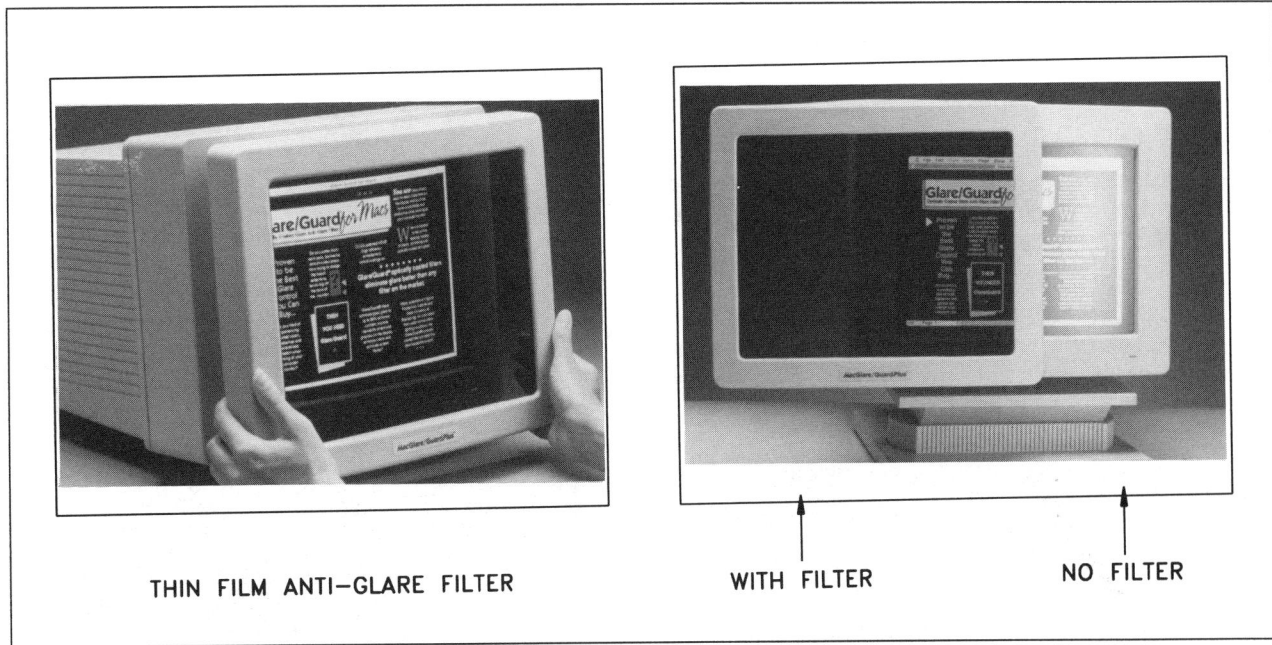

THIN FILM ANTI—GLARE FILTER   WITH FILTER   NO FILTER

FIGURE 7-9 Anti-glare CRT monitor screens

7.13 INTERACTIVE TOUCH SCREEN

Many monitors in security applications display the outputs from graphics and/or an alphanumeric database generated by a computer. Some advanced systems operate with computer software and hardware that permits interacting with the screen display by touching the screen at specific locations and causing specific actions to occur. The devices are called touch-screen templates and are mounted on the front of the monitor. The touch screen permits the operator to activate a program or hardware change by touching a specific location on the screen.

Touch-screen interaction between the guard and the hardware and CCTV system has obvious advantages. It frees the guard from a keyboard and provides a faster input command. Also, the guard does not have to memorize keyboard commands and type the correct keys. There is also less chance for error with the touch-screen input, since the guard can point to a particular word, symbol, or location on the screen with better accuracy and reliability.

There are several different types of touch screens available, using different principles of operation. One technique uses an LED and photo-transistor detector to create an invisible grid of IR light in front of the monitor. When the IR beam is interrupted by a finger or other stylus, a signal is returned to the computer electronics to perform a predetermined action. The space within the frame attached to the front of the monitor forms the touch-active area, and a microprocessor calculates where the person has touched

the screen. Figure 7-10a shows such a touch screen installed on a monitor.

Since there is no film or plastic material placed in front of the monitor, there is no change or reduction in optical clarity of the displayed picture.

A second type of touch-screen panel consists of a transparent, conductive polyester sheet over a rigid acrylic backplane; both are affixed to the front of the display to form a transparent switch matrix (Figure 7-10b). The switch matrix assembly has 120 separate switch locations that can be labeled with words or symbols on the underlying display, or a scene can be divided into 120 separate locations and interacted with by the operator. Individual touch cells may be grouped together to form larger touch keys via programming commands in the software. This type of touch screen has the disadvantage that not all the light can pass through the touch screen (toward the operator) and therefore the picture contrast is reduced. Typical light transmission for this type of design is 65 to 75%.

A third type of interactive touch-screen accessory consists of an optically clear Mylar-polyester membrane that is curved around the monitor's front glass screen and consists of an assembly of electrically conductive elements that are transparent to the user. When the conductive surface of the Mylar is pressed against the conductive surface of the glass by the security operator, a voltage change on the screen results. This change in voltage is detected by the monitor electronics, which communicate with the security system to indicate that the person has touched the screen at a particular location.

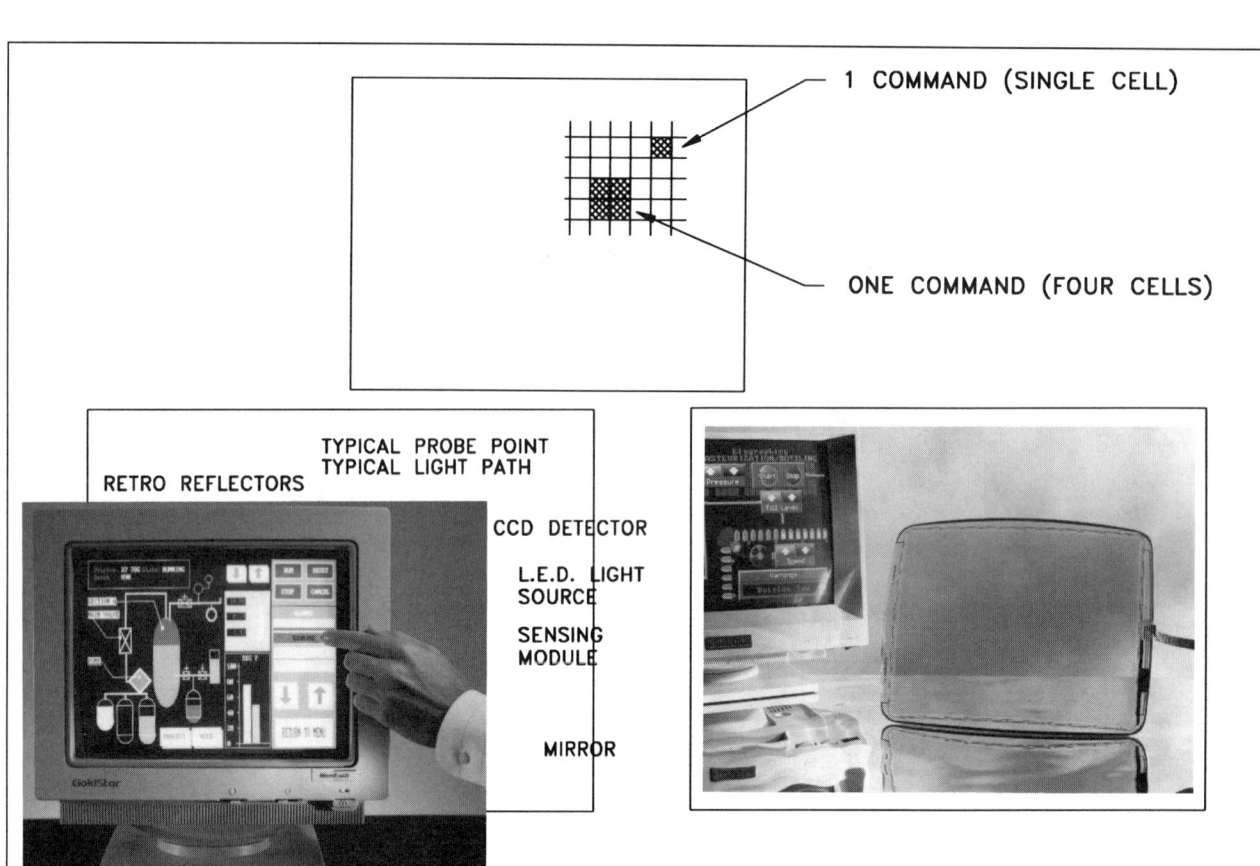

1 COMMAND (SINGLE CELL)

ONE COMMAND (FOUR CELLS)

TYPICAL PROBE POINT
TYPICAL LIGHT PATH

RETRO REFLECTORS

CCD DETECTOR

L.E.D. LIGHT
SOURCE

SENSING
MODULE

MIRROR

LED – REFLECTORS – PHOTODETECTOR ARRAY   CONDUCTIVE POLYESTER

FIGURE 7-10   Monitor touch screen

LCD MONITOR

FLAT 4" CRT MONITOR

FIGURE 7-11   Flat-screen receiver/monitor

7.14 RECEIVER/MONITOR—VIEWFINDER

Various manufacturers produce small lightweight television monitors or receiver/monitors, which accept baseband CCTV signal inputs and/or VHF/UHF commercial RF channels and are powered by 6, 9, or 12 volts DC (Figure 7-11).

These television monitors are particularly useful in portable and mobile surveillance applications for servicing and law enforcement. Often the portable surveillance camera will be transmitting the video signal (perhaps also audio) via an RF or UHF video transmitter operating at or near one of the commercial channels. The small receiver-monitors with 1.5- to 5-inch-diagonal CRT or LCD displays can receive and display the transmitted video signal and have an output to provide the baseband CCTV signal for a VCR or hard-copy video printer at the receiver site. These devices usually have medium resolution (250 to 400 lines), which is often sufficient to provide useful security information.

7.15 SUMMARY

CCTV console design requires careful consideration from the human engineering aspect (ergonomics) as well as hardware capability. The number and placement of monitors with respect to the guard(s) must be analyzed carefully in the early stages of design. A list of guard functions and duties should be generated and reviewed with management and security personnel at the facility and critiqued by a professional consultant. The best CCTV system will be ineffective if proper attention is not paid to the human engineering aspects of the console room design.

Chapter 8

VCRs, Frame-Storage Devices, and Hard-Copy Printers

CONTENTS

8.1 OVERVIEW

Early real-time video recording systems used reel-to-reel tape media, which required changing the tape reels, were unreliable, and were prone to tape damage. The arrival of the real-time color videocassette recorder (VCR) permitted loading and unloading of the videotape without user contact with the tape. The VCR was convenient, safe, and reliable. While designed and marketed for the home consumer, this medium's basic philosophy and technical characteristics provided a powerful recording means for security and industrial applications. The addition of specialized functions, including alarm activation and time-lapse recording, further enhanced its usefulness.

Recording and preserving the CCTV picture has become increasingly important, since many security systems run unattended for long periods of time, and recorded video scenes are often used to apprehend and prosecute thieves and violators. The video recorder converts the incoming video camera signal into a recorded form on magnetic tape. Later, during playback, it reconstructs the video signal to a form suitable for displaying on a CCTV monitor or printing on a hard-copy video printer.

A camera image played back on a monitor from a good quality videotape recording may be almost as valuable as a live observation. The CCTV image on a monitor is fleeting, and unless it is recorded, it is of use only once—at the time of the occurrence.

A second class of VCR is called time-lapse. The time-lapse VCR records single pictures at closely spaced time intervals longer than the real-time $1/30$-second frame time. The value of such convertibility (from real-time to time-lapse and vice versa) is that routine events of no special significance can be recorded in a time-lapse mode, thus conserving tape, while permitting the real-time recording of significant security events such as a break-in when they occur. When a security event of significance occurs, the VCR switches to real-time recording, since it is important to record the video scene with high resolution and in real time. These real-time events are the ones that the security guard would consider important and normally view on a command monitor.

The primary reason for using time-lapse recording with security CCTV is to permit the use of a single VCR tape cassette for as long a period as possible without having to change the cassette. Although economical because fewer cassettes are used, it is operationally important to record for as long as possible without having to change the tape when a recorder must be left unattended. The time-lapse recorder permits many more hours of recorded events to be accumulated on a single cassette than would be possible in real-time recording, and avoids running out of tape when no one is available to replace it.

An important but often overlooked consideration when using a VCR is safeguarding the tape during a break-in. Obviously, if the tape is stolen or destroyed by the intruder, the incriminating video evidence is lost. The VCR machine should be located where it cannot be tampered with, destroyed, or stolen. It should be either hidden or stored in a secure room or vault. Protective steel cabinets and lock-boxes are also available.

Some VCR time-lapse machines can be interfaced with a personal computer to allow remote control of the VCR. In other computer-interface applications, a business can electronically link the recorder to a cash register and other point-of-sale terminals, thereby allowing the cash transaction to appear directly with the scene as viewed by the camera at the point-of-sale terminal.

8.2 VIDEOCASSETTE RECORDER

Reel-to-reel videotape recorders (VTRs) were brought to the security field to provide real-time and time-lapse recording of the television picture. The recording system required the operator to manually thread the tape from the tape reel through the recorder onto an empty take-up reel—similar to threading an 8-mm or 16-mm film projector. The ease of use of videocassettes and VCRs has resulted in wide acceptance of this recording medium. Most VCRs use the Victor Home System (VHS) videocassette as the recording medium. A second choice was the Sony $\frac{1}{2}$-inch Beta format, which is no longer in use. The newer Sony 8-mm format is gaining in popularity because of its smaller, compact size. Another format, the VHS-C, is a compact version of the VHS format but has found limited use in the security market.

Present real-time VCR systems record 2, 4, or 6 hours of continuous monochrome or color video, with more than 300 lines of resolution, on one VHS or 8-mm cassette. Time-lapse recorders have total recording times up to 960 hours. Most time-lapse recorders are provided with an alarming mode, which switches the recorder to real-time recording when an alarm condition occurs, and the alarm signal (usually a sensitive switch closure) triggers the recorder alarm input. Every television scene that can be displayed on a monitor is capable of being recorded on a VCR as a permanent record for later use. When necessary, the time and date are superimposed on the video scene and the audio signal recorded on an adjacent audio track. The videotapes can be erased and reused many times, so that tape cost is relatively low.

With respect to picture "quality," the weakest link in the CCTV security system is usually the VCR, which falls short when recording and playing back a high-resolution television picture. In comparison to lenses, cameras, transmission means, and monitors, video recorders in the security field have not kept pace. A VCR operating in either real-time or time-lapse mode with resolution matching that of good cameras and monitors would be considerably more expensive. The VCR is a complex piece of machinery consisting of delicate electromechanical, electrooptical, and electronic components, which all must be manufactured and made to operate in concert to high precision. Two relatively new VCR formats, called S-VHS and Hi-8, have increased the resolution and picture quality noticeably while still maintaining the basic VHS and 8-mm physical cassette formats. Horizontal resolution of 400 TV lines is achieved with a new recording format on high-resolution ferric-oxide magnetic tape. To obtain picture quality matching the best cameras and monitors, VCRs are available from the broadcast industry that have high resolution and excellent picture quality, but the cost for this equipment is prohibitive in the normal commercial security budget.

Present VCRs permit capturing as much as 40 days' worth of CCTV information on one 2-hour cassette when in the time-lapse recording mode. Time-lapse recording makes maximum use of the space available on the videocassette by turning the video heads on and off at a preselected rate, slowing the tape, and thereby slowing the recording rate. Instead of recording at a normal rate of 60 fields per second (30 frames), a time-lapse VCR records one picture every fraction of a second or every few seconds. Any moderately sophisticated CCTV security system installed today incorporates one or more time-lapse video recorders.

8.2.1 VCR Formats

VCR systems are available in 8-mm-wide, $\frac{1}{2}$-inch-wide, and $\frac{3}{4}$-inch-wide magnetic tape formats. While the $\frac{3}{4}$-inch tape format produces a picture with higher resolution and sig-

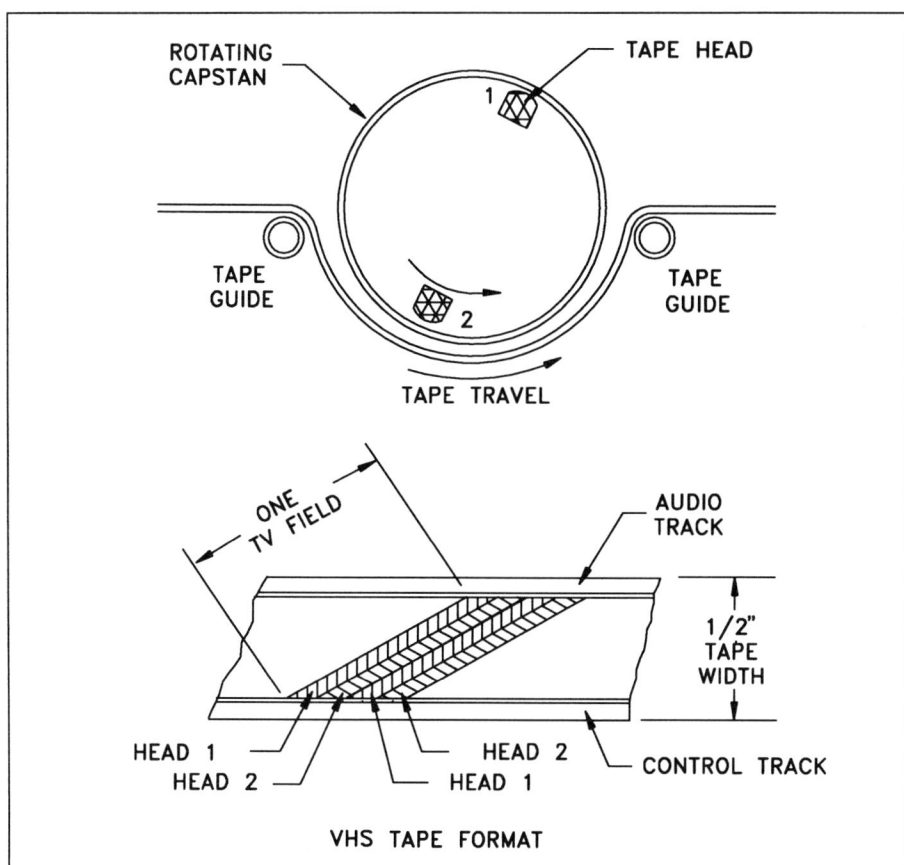

FIGURE 8-1 VHS video-cassette recorder geometry and format

nal-to-noise ratio than the narrower 8-mm (approximately ¼ inch) and ½-inch formats, the ½-inch format has become the standard of choice for consumer home-viewing systems as well as security systems. When the ½-inch systems initially came to market, two formats were available with the ½-inch tape width: Beta and VHS. As a matter of history, although the Sony Beta format provided superior-quality video recording and playback, marketing demands made the VHS system all but universal in the ½-inch format. Within the past few years, new and more compact tape formats and recorders have been developed. One introduced by Sony Corporation uses an 8-mm-wide tape cartridge. A second compact version of the VHS format, called VHS-C, was introduced by the JVC Company (a subsidiary of Victor Home System). While most video recorders for security applications still use the standard VHS cassette format, future systems will use the more compact Sony 8-mm cassette.

8.2.1.1 VHS Format

Standard continuous recording times for VHS tapes in real-time recording modes are 2, 4, and 6 hours. When these cassettes are used in time-lapse recorders where a single field (or frame) or a selected sequence of fields is

recorded, up to 960 hours can be recorded on a single 2-hour VHS cassette. The VHS cassette recorder is a relatively small and lightweight machine and provides up to 6 hours of real-time recording and playing time. This is accomplished through state-of-the-art high-density recording media and a simple, reliable loading mechanism. VHS VCR equipment is available in recorder and/or playback, as camera-recorder (camcorder) configurations. A brief description of the operation of the VHS recorder follows (Figure 8-1).

Videotape recorders record the video scene on magnetic tape using the same laws of physics as audiotape recorders—in fact, a narrow audio track (occupying a small width of the tape) along one edge of the videotape operates in exactly the same manner as conventional audiotape recorders, but at a slower speed. The challenging aspect of recording a video picture on a magnetic tape as compared with an audio signal is that the standard U.S. NTSC video signal has a wide bandwidth and includes frequencies above 4 MHz (4 million cycles per second) and below 30 Hz, as compared with an audio signal with frequencies between 20 and 20,000 Hz.

To record the high video frequencies, the tape must slide over the recording head at a speed of approximately 6 meters per second or faster. Reel-to-reel recorders could

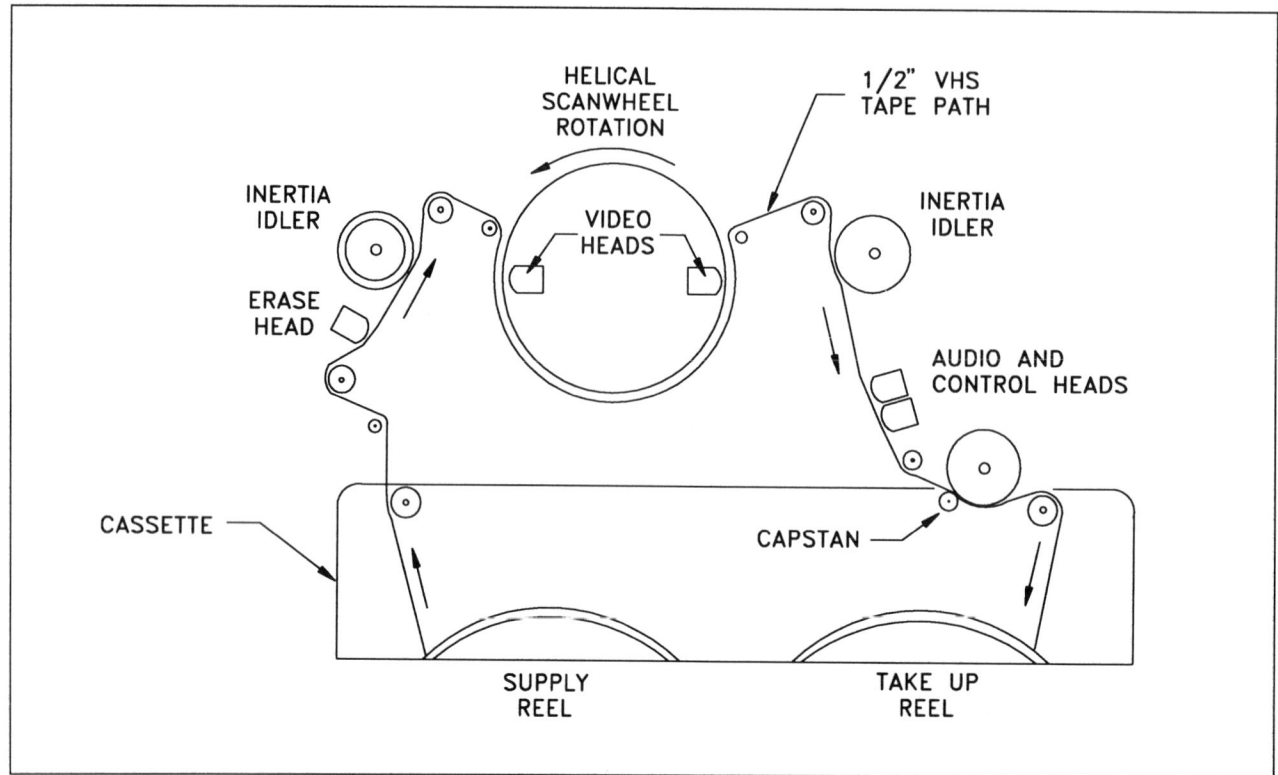

FIGURE 8-2 VHS recorder technology

never reach a reliable 6-meter-per-second tape speed. Consequently, all present VCRs have a helical-scan design, in which the magnetic tape is pulled slowly around a drum of rapidly moving tape heads (Figure 8-2).

The tape spirals (wraps) around a revolving drum, generally for exactly half a turn. The video recording heads are attached to the edge of a rotating disc inside the drum. This design allows the tape head, traversing a circular path with the same radius as the drum, to scan diagonal tracks on the tape at the high speed required to record the video signal. Tape speed need only be fast enough to provide the desired track pitch, which is an order of magnitude (one-tenth) slower than the linear recording-head tape speed.

The audio is recorded along one edge of the tape as a single (mono) or a dual (stereo) channel. The audio recording is completely conventional. Often a separate erase head is mounted just before the audio head to facilitate dubbing a new audio signal alongside the video signal.

Along the other edge of the tape is the control track, normally a 30-Hz square-wave signal (NTSC system), which synchronizes the video frames on the tape with the monitor during playback. A full-track erase head on the drum is available for erasing any prerecorded material on the tape. Video signals are recorded as tape passes over the drum, and soon after it leaves the drum the audio and control track signals are recorded onto it simultaneously. The tape

speed is 33.4 mm per second and the head drum diameter is 62 mm, making it the smallest two-headed VCR with a $\frac{1}{2}$-inch format.

8.2.1.2 8-mm Sony Format

The most recent addition to videocassette recording equipment is the 8-mm format by the Sony Corporation (Figure 8-3). This system uses a small cassette cartridge approximately the size of an audiocassette and an 8-mm wide (approximately $\frac{1}{4}$-inch) magnetic tape. An obvious advantage of this type of system is its small size as compared with either the VHS or Beta tape formats having the $\frac{1}{2}$-inch format.

Even though the format and cassette are significantly smaller, the equipment maintains image quality and system capability similar to that of the larger formats. Present cassette running times are $\frac{1}{2}$ hour, 1 hour, and 2 hours. The equipment available for using these cassettes is a camcorder, a recorder-player combination with separate camera, and a complete camera-recorder-player combination. In one form the total VCR weighs 2.4 pounds (1.1 kilograms) and operates on 117 volts AC power or 12 volts DC batteries (Figure 8-4a).

The recorder's overall size is 7 inches by 7 inches by 3 inches high. The recorder accepts a standard video signal

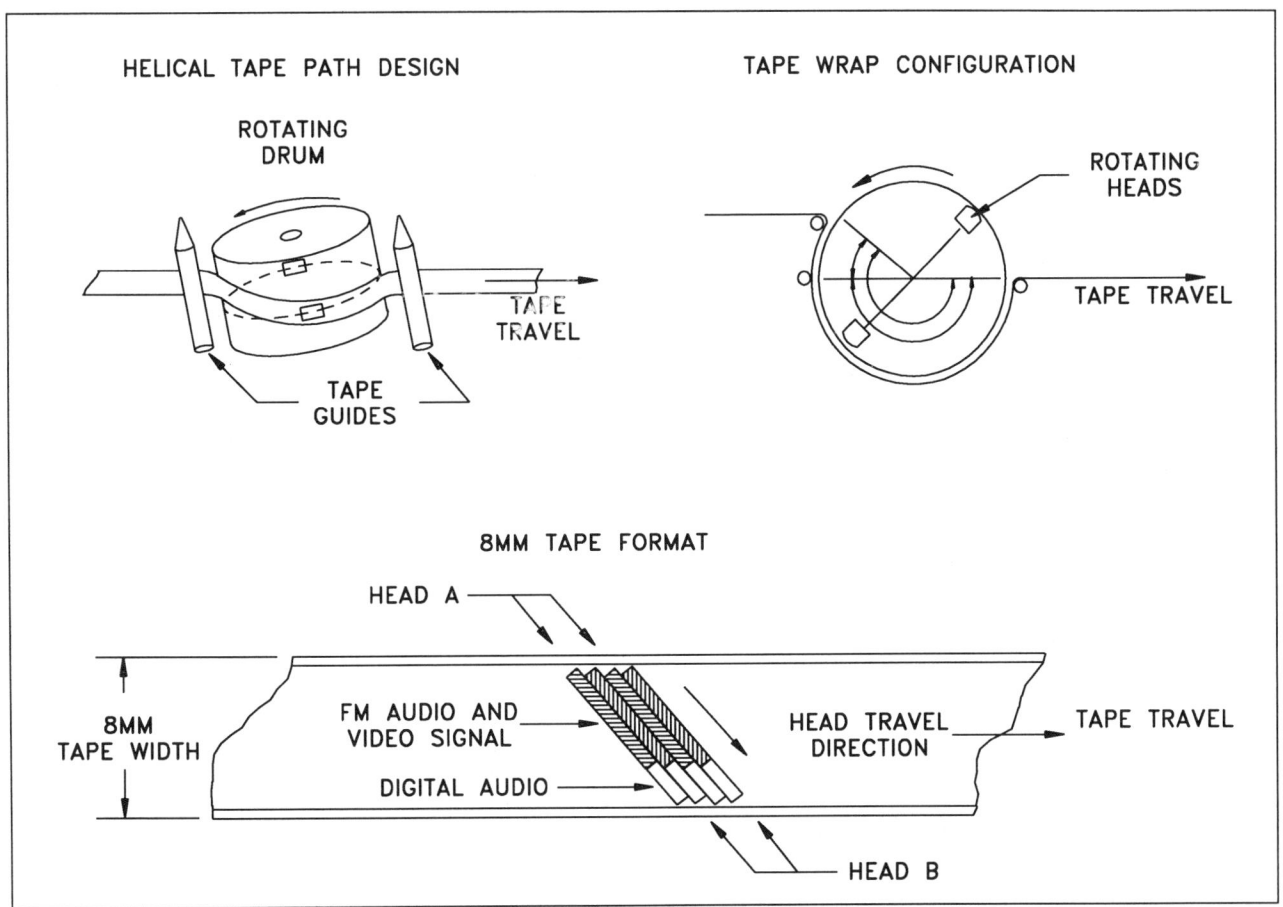

FIGURE 8-3 Sony 8-mm videocassette recorder and format

from any monochrome or color video camera and plays it back on either baseband video (30 Hz to 4 MHz) or on a standard TV receiver through a small RF modulator on VHF channel 3 or 4.

In the camcorder version, the system weighs 2.2 pounds (1 kilogram) and contains a lens, camera, and recorder in a compact housing 2.5 inches wide by 4.5 inches high by 7 inches (Figure 8-4b). This camcorder records 30, 60, or 120 minutes on the 8-mm cassette. The configuration is particularly suitable for covert applications requiring a small, lightweight system. The camera CCD image sensor has 250,000 pixels, providing high resolution, and automatic-iris and white-balance circuits producing a high-fidelity color picture.

A monochrome time-lapse version of the 8-mm-format VCR provides still-frame video recording and jitter-free stop motion. It provides an impressive 380 horizontal by 350 vertical TV lines of monochrome resolution, and because it is a true still-frame recorder, the playback picture quality is very high regardless of the recording speed. The time-lapse recording speed can be varied from 12 to 384 hours on a single 90-minute, 8-mm videocassette. The recorder permits viewing of a single frame or field in the still mode, with the ability to switch between odd and even fields. The machine contains variable-speed playback, providing ease

of use and versatility. The 8-mm video format U-loading (unique tape path around heads and capstan drive) system permits high speed bidirectional search and frame-by-frame or field-by-field shuttling in both forward or reverse direction, while maintaining picture clarity.

As with VHS machines, many 8-mm VCRs are available with built-in time/date generators and alarm recording mode for unattended operation over extended periods of time.

8.2.1.3 VHS-C Format

The VHS-C tape format makes use of a small tape cartridge slightly larger than the Sony 8-mm and the VHS electronics and encoding scheme (Figure 8-5).

When introduced, it was expected to be the future small VCR format for portable equipment; however, the Sony 8-mm format has become more popular. The VHS-C cartridge is played back on a standard VHS machine with a VHS-C-to-VHS cartridge adapter.

8.2.1.4 Super-VHS and Hi-8 Formats

The resolution obtained with standard color VHS and 8-mm VCRs, whether operating in real-time or time-lapse mode, is 230–240 TV lines, which is not sufficient for many

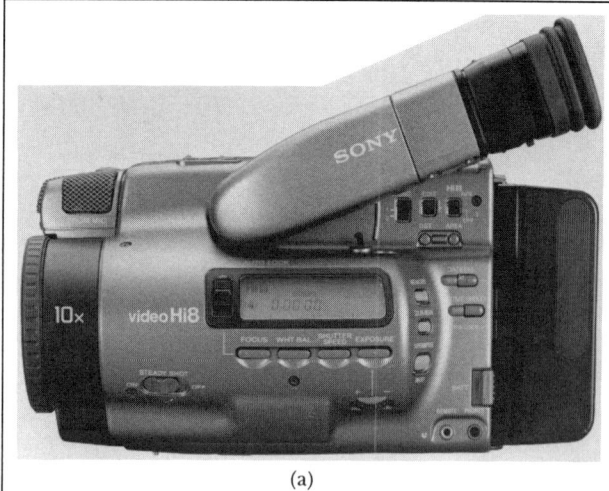

(a)

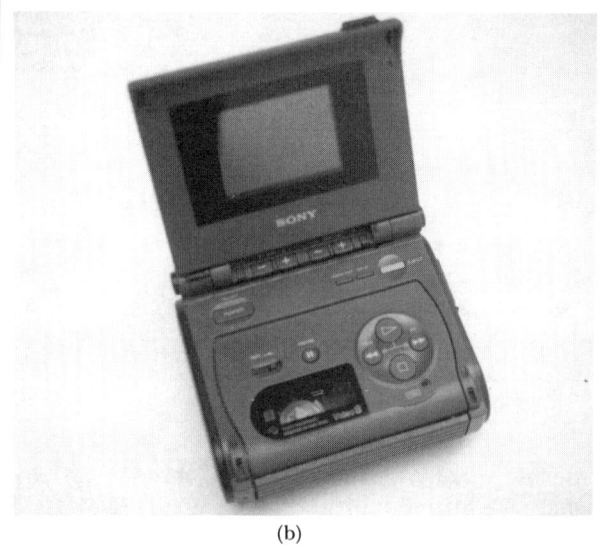

(b)

FIGURE 8-4   8-mm Sony camcorder (a) and VCR (b) with LCD monitor

security applications. Monochrome time-lapse recorders provide 350-TV-line resolution. The new color S-VHS and Hi-8 format real-time and time-lapse recorders increase the horizontal resolution to more than 400 TV lines, suitable for facial identification and other critical security applications.

As with the standard VHS and 8-mm systems, there is no compatibility between the S-VHS and Hi-8 formats. There is some compatibility between VHS and S-VHS, and between 8-mm and Hi-8 VCRs. This is important for users who cannot make a switch to the higher-resolution formats all at once and do not want to make existing equipment obsolete.

To obtain the higher resolution, the S-VHS and Hi-8 systems use a wider-luminance (Y) bandwidth and frequency deviation than the standard formats. In the standard VHS system, the luminance information is in a 3.4-to-4.4-MHz bandwidth. For the S-VHS system, it is in-

creased to 5.4 to 7.0 MHz. In the 8-mm system, it is 4.2 to 5.4 MHz for the standard and 5.0 to 7.0 MHz for the Hi-8 system. Table 8-1 compares the VHS, S-VHS, 8-mm, and Hi-8 recording parameters.

To increase the purity of the picture color, the chrominance signal is separated out from the luminance to prevent cross-color and dot interference. This reduces unwanted artifacts such as picture ghosts, multiple outlines of images, and images moving through the picture. The separate chrominance (C) and luminance information are sent to the S-VHS or Hi-8 high-resolution monitor via a Y/C cable equipped with a four-pin "S" connector.

There is a misconception that the monitor must be equipped with an "S" connector to see the 400 TV lines of horizontal resolution. This is not the case. A high-resolution monitor equipped with direct video input can still give the same 400-TV-line resolution, but because the chrominance and luminance signals are not separated, chroma noise is returned to the conventional VHS (8-mm) levels. Some important differences between and features of the standard VHS and 8-mm, and the S-VHS and Hi-8 formats are as follows:

- S-VHS and Hi-8 recordings cannot be played back on conventional machines.
- S-VHS and Hi-8 videocassettes require high coercivity, fine-grain cobalt–ferric-oxide and metal tapes to record the high-frequency, high-bandwidth signals.
- All S-VHS and Hi-8 recorders can record and play back in standard mode. The cassettes have a special sensing notch that automatically triggers the VCR to switch to the correct mode.

8.2.1.5 Magnetic Tape Types

Developments in video recording in recent years have focused on VCR and camcorder size, recording formats, and playing times, all of which play key roles in the recording and playing back of high-quality video images and sound. However, the magnetic tape also plays a critical role in determining the final quality of the video picture. Videotape manufacturers have been working on ways to enhance the video picture and sound quality. Improving materials and applying them to a tape base in production have resulted in significant progress in picture quality and maintaining "clean" pictures (low signal dropout and noise) over long periods of time and after many tape replays. For security applications, it is important to choose a high-quality tape with matched characteristics for the VCR equipment and format used.

Most videotape, when grouped by size, falls into three categories: 8-mm, VHS, and VHS-C (Table 8-2). Further distinctions among videotape types are made on the basis of the kinds of magnetic particle used. Three common types are cobalt–ferric-oxide, metal-particle or metal-evaporated, and cobalt-magnetite. The development of powerful new magnetic substances has led to the creation of a variety

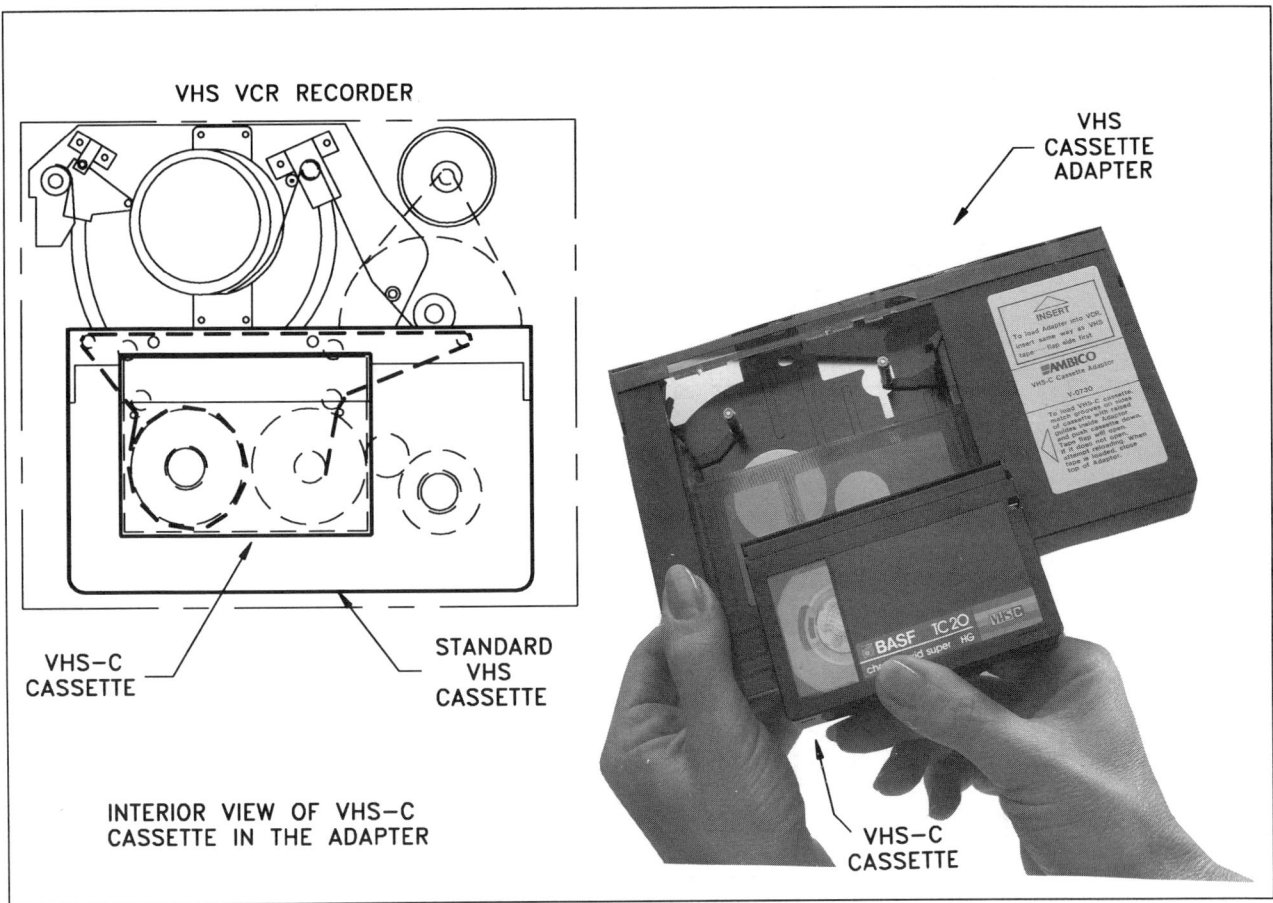

FIGURE 8-5 VHS-C videocassette recorder

of high-performance tapes and the consequent high-resolution S-VHS and Hi-8 VCR formats.

Cobalt–ferric-oxide videotapes are manufactured by combining minute, needle-shaped cobalt–ferric-oxide crystals, each about 0.2 micron long, with a bonding material or binder. The oxide particles and binder are applied uniformly to the tape base. Before the coated tape dries, it is passed through a uniform magnetic-energy field, causing

VCR TAPE FORMAT	LUMINANCE (Y) BANDWIDTH ( ) (MEGAHERTZ)	CHROMINANCE (C) CENTER FREQUENCY (KILOHERTZ)	RESOLUTION (TV LINES)
VHS	3.4 – 4.4 (1.0)	629	330 MONOCHROME 240 COLOR
VHS–C	3.4 – 4.4 (1.0)	629	400–440 MONOCHROME 400–440 COLOR
S–VHS	5.4 – 7.0 (1.6)	629	400–440 MONOCHROME 400–440 COLOR
8MM			300
Hi8	5.0 – 7.0 (2.0)		400

NOTE: STANDARD CCTV BANDWIDTH IS 0 TO 4.2 MEGAHERTZ
COLOR SIGNAL: CENTERED AT 3.58 MHZ
COLOR INFORMATION ON VCR CENTERED AT 629 KILOHERTZ

Table 8-1 VHS, S-VHS, 8-mm, and Hi-8 Parameter Comparison

TAPE FORMAT	MAXIMUM RESOLUTION (TV LINES)	TAPE WIDTH MM (IN)	CASSETTE SIZE LxWxH (MM)*	PLAYING TIME (HRS)		
				STANDARD (SP)	LONG (LP)	EXTENDED (EP)
VHS–C	240	12.7 (1/2)	188x104x25	0.33	0.66	1.0
VHS: T–60	240	12.7 (1/2)	188x104x25	1	2	3
T–120	240	12.7 (1/2)	188x104x25	2	4	6
S–VHS	400	12.7 (1/2)	188x104x25	2	4	6
**8MM P6–60	240	8.0 (0.31)	95x62.5X15	1	2	–
P6–120	240	8.0 (0.31)	95x62.5X15	2	4	–
**Hi8 P6–60	400	8.0 (0.31)	95x62.5X15	1	2	–
P6–120	400	8.0 (0.31)	95x62.5X15	2	4	–

*1 INCH = 25.4 MM
**AVAILABLE FORMATS: P6–15, P6–30, P6–90, P6–120

Table 8-2 Videocassette Recorder Tape Formats

the needle-shaped particles to align in the same direction. After drying, a calendering treatment produces an ultra-smooth, mirror-like finish on the tape surface. A back coating is then applied to the base film and it is slit into 0.5-inch or 8-mm widths and wound onto tape cassettes.

By 1976, when the VHS format became available, most tapes used cobalt–ferric oxide, but its magnetic strength (coercivity) was insufficient to obtain a high-quality video image and the tapes suffered from magnetic instability. In the 1970s, tape manufacturers successfully developed epitaxial growth technologies that produced tapes able to significantly improve the ability to record video. As there are different techniques for applying the magnetic particles to the base, each manufacturer uses a different name for this process: Hitachi Maxell calls it epitaxial, TDK calls it Avilyn, and Fuji Photo Film calls it Beridox. Present standard VHS tapes use one of these similar technologies.

In the specifications for videotape, two characteristics often listed are coercive force and residual magnetic flux density. Coercive force or coercivity refers to the tape's ability to maintain its magnetic quality: tapes with a higher coercivity retain their magnetism over longer periods of time and permit superior recording at higher frequencies. Residual magnetic flux is related to signal output: the higher it is, the higher the signal output. VHS tapes have a coercive force of 700 oersted (Oe) and a residual magnetic flux density of 1500 Gauss. By comparison, S-VHS tapes typically have 900 Oe and 1700 Gauss.

To attain high resolution and clear reproduction of image details, it is important to provide perfect contact between the video head and the tape's magnetic surface. To this end, tape manufacturers mix ultra-fine-grain magnetic particles, apply them uniformly, and smooth them by calendering. The most recent videotapes use cobalt–ferric-oxide particles 0.17 micron in size for VHS and 0.14 micron for S-VHS. To make use of these high densities, it is imperative that the tape and VCR recording head be kept clean.

Cobalt–ferric-oxide tapes for VHS camcorders and VCRs are available in four grades: Pro, Hi-Fi, HG (high-grade), and Standard. Since tape is available in many different grades, for different equipment types, consult the equipment manufacturer for proper usage.

The most recent innovation in VHS is a double-coating technology developed by Fuji Photo Film. This technology produces a 0.3-micron lower layer that has excellent low-frequency-range characteristics and a 0.4-micron upper layer with ultra-fine-grain magnetic particles that offer superior medium- and high-frequency recording characteristics. This double-coating method provides low-noise and high-image-quality recording. These new tapes are referred to as Super-VHS Pro and Super VHS and extend the frequency response to produce improved signal-to-noise and resolution.

Another tape material having superior properties is called magnetite. This material is naturally black and therefore does not require black light-shielding materials in the magnetic coating (used by photosensors to detect the end of the tape). More important, standard videotape collects dust and other foreign particles, which stick to the tape and cause dropouts (momentary losses of picture). In contrast, the high electric conductivity of magnetite prevents electrification of the magnetic layer, so that there is no "static cling," and virtually no dust adheres to the tape. This factor, plus the carbon back-coating, serves to greatly reduce dropouts.

In metal tapes, the manufacturer uses pure iron particles as the magnetic substance, resulting in a 1500-Oe coercive force and a residual magnetic flux density of 2500 Gauss. Like ferric-oxide tape, metal tape comes in two versions: with particles 0.2 micron long for standard 8-mm recording and 0.15 micron for Hi-8 systems. Since pure iron oxidizes easily, the metal tapes are coated with a thin layer of ceramic film, which also increases the durability of the tape. In metal tape production, an 80-meter length of tape 13 microns thick is wound onto an 8-mm cassette, providing 90 minutes of play. By reducing the thickness to 10 microns, 120 minutes are obtained. The high density of recording 0.7-micron wavelength means the tape must be protected from dust and finger contact, so the tape cassette has a special design that sandwiches the tape in the lid portion of the cassette shelf. The 0.49-micron wavelength of Hi-8 recording is even shorter. Manufacturers offer two types of tape: evaporated-metal and new particle orientation technology. The evaporated-metal tape uses no binder. Instead, a perfect metal alloy magnetic layer is formed on a tape base by vacuum evaporation. This results in a drastic enhancement of the density of the magnetic material. Table 8-3 summarizes available video magnetic tape characteristics.

Equipment using the VHS, S-VHS, 8-mm, and Hi-8 formats can provide many different modes of operation, including 2-hour standard play (SP), 4-hour long play (LP), and 6-hour extended play (EP) in standard VHS formats and 24, 48, 72, 120, 180, 240, 480, 600, 720, and 960 hour and one-shot single field in time-lapse. In the one-shot mode, the user can select the number of fields being recorded and automatically set the recorder to record that number of fields at various intervals.

8.3 TIME-LAPSE VIDEOCASSETTE RECORDER

The time-lapse recorder is a real-time VCR that pauses to record a single CCTV field (or frame) every fraction of a second or number of seconds, based on a predetermined time interval (Figure 8-6).

Standard VCRs record the video scene in real time: the fields or frames displayed by the camera are sequentially recorded on the tape and then played back in real time, slow-motion, or a frame at a time. In time-lapse mode, the VCR records only selected fields (or frames) a fraction of the time. Time-lapse recorders have the ability to record both in real time and in a variety of time-lapse ratios, which

TAPE TYPE	MATERIAL	COERCIVE FORCE OERSTAD (Oe)	RESIDUAL MAGNETIC FLUX DENSITY* (GAUSS)	MAGNETIC PARTICLE SIZE (MICRONS)
VHS, VHS-C	COBALT-FERRIC OXIDE COBALT-MAGNETITE	700 -	1,500 1,800	0.17 0.14
S-VHS	COBALT-FERRITE OXIDE	900	1,700	0.14
8MM-ME 8MM-MP	METAL EVAPORATED (ME) METAL PARTICLE (MP)	1,500 1,500	2,500 2,500	0.2 0.15
Hi8-MP-DC**	METAL PARTICLE (MP)	1,600	3,400	-

* RETENTIVITY (Br)
** DOUBLE COATED—IMPROVES FREQUENCY RESPONSE—PICTURE QUALITY

Table 8-3 Video Magnetic Tape Properties

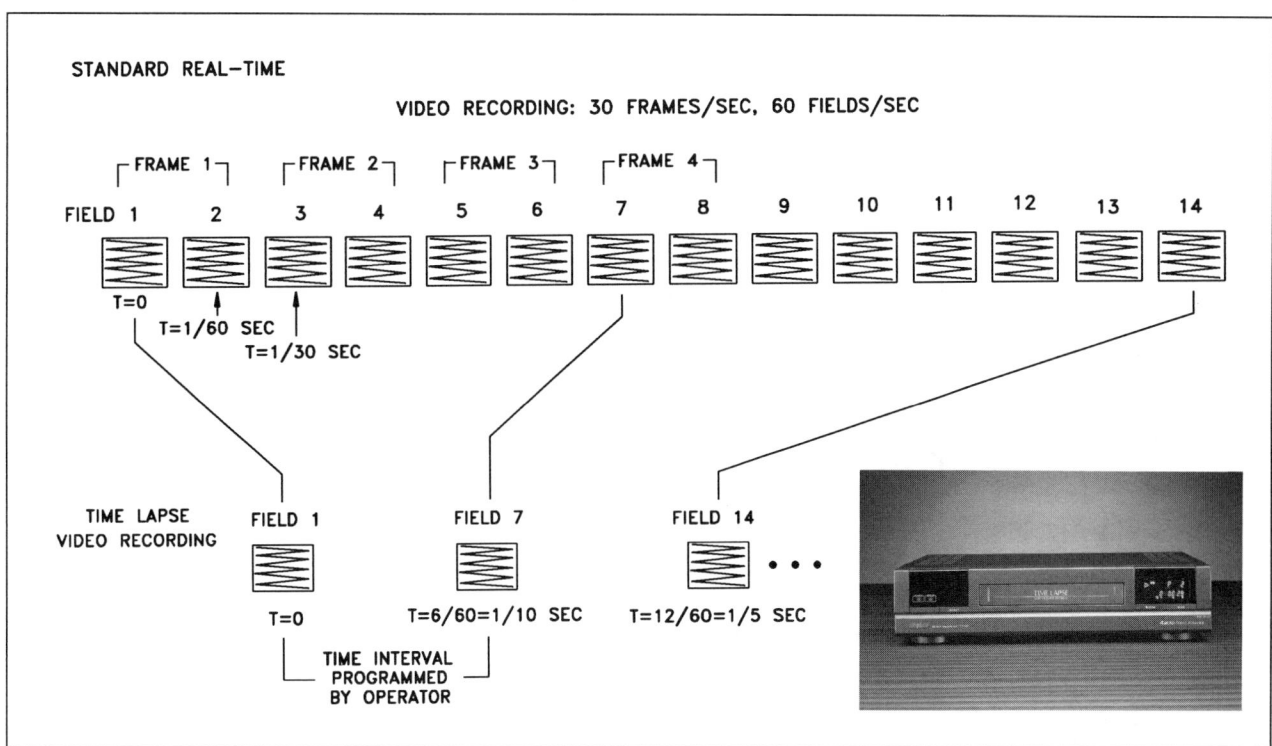

FIGURE 8-6 Time-lapse videocassette recorder

are operator-selected either manually or automatically. The automatic switchover from time-lapse mode to real-time mode is triggered by some auxiliary input to the VCR, such as a signal from an alarm device or a video motion detector. In this mode, the recorder records real-time for a predetermined length of time after the alarm is received, and then returns to the time-lapse recording mode until another alarm is received.

Time-lapse video recording consists of selecting specific image fields or frames to be recorded at a slower rate than they are being generated by the camera. The CCTV camera generates 30 frames (60 fields) per second. One TV frame consists of the interlaced combination of all the even-numbered lines in one field and all the odd-numbered lines in the second field. Each field is essentially a complete picture of the scene but viewed with only half the vertical resolution ($262\frac{1}{2}$ horizontal lines). Therefore, by selecting individual fields—as most time-lapse VCRs do—and recording them at a rate slower than 60 per second, the time-lapse VCR records less resolution than available from the camera.

When the tape is played back later at the same speed at which it was recorded and viewed on the monitor, the pictures on the monitor will appear as a series of animated still scenes.

Table 8-4 presents a comparison of time-lapse modes as a function of time-lapse ratio, total recording period, recording interval fields, or frames-per-second recorded. From Table 8-4, it is apparent that the larger the time-lapse ratio, the fewer pictures recorded over any period of time.

For example, for a time-lapse ratio of 6:1, the recorder captures 10 pictures (fields) per second, whereas in real time (or 1:1) it captures 60. Although the recorder is only recording individual fields spaced out in time, if nothing significant is occurring during these times, no information is lost.

The choice of the particular time-lapse ratio for an application depends on various factors, which may include the following:

1. the length of time during which the VCR will record on a 2-, 4-, or 6-hour videocassette
2. the type and number of significant alarm events that are likely to occur, and the duration of the alarms
3. the time period that must elapse before the cassette can be replaced or reused
4. the time-lapse ratios available on the VCR (not all manufacturers provide the same values of time-lapse ratio)

To record as much information as possible, it is advisable to select as low a time-lapse ratio as is consistent with the requirement. Only marginal economic savings can be obtained by sparing use of tape, since the cassette tape is relatively inexpensive. By careful analysis of the operating conditions and requirements, it is possible to record events without sacrificing important information—and at substantially less tape cost than real-time recording.

In the time-lapse recording mode, the videotape speed is much slower than the real-time speed, since the video

TOTAL RECORDING PERIOD*		TIME-LAPSE RATIO	RECORDING/PLAYBACK SPEED (RECORDING INTERVAL)		RECORDING/PLAYBACK (PICTURES/SECOND)	
HOURS	DAYS		1 FIELD PER ___ SEC.	1 FRAME PER ___ SEC.	FIELDS	FRAMES
2**	.083	1:1	0.017	0.034	60	30
12	.50	6:1	0.1	0.2	10	5
24	1	12:1	0.2	0.4	5	2.5
48	2	24:1	0.4	0.8	2.5	1.25
72	3	36:1	0.6	1.2	1.7	0.85
120	5	60:1	1.0	2.0	1.0	0.5
180	7.5	90:1	1.5	3.0	0.66	0.33
240	10	120:1	2.0	4.0	0.50	0.25
360	15	180:1	3.0	6.0	0.33	0.17
480	20	240:1	4.0	8.0	0.25	0.13
600	25	300:1	5.0	10.0	0.20	0.10
720	30	360:1	6.0	12.0	0.16	0.08
999	41.6	500:1	8.3	16.6	0.12	0.06

*TAPE CASSETTE: T-120

** STANDARD REAL-TIME VIDEO

RESOLUTION: 375 TV LINES MONOCHROME
240 TV LINES COLOR

Table 8-4 Time-Lapse Recording Times vs. Playback Speeds

pictures are being recorded intermittently. This permits recording many hours of scene activity on the tape cassette, maximizing the use of space, minimizing the expense, and eliminating the inconvenience of having to change the cassette every few hours. To review the tape, it can be scanned very quickly at normal playback speeds. When more careful examination of a particular series of frames is required, playback speed is slowed or stopped and a more careful and discriminating observation made.

To obtain the best synchronization from multiple cameras sequentially recorded by a VCR, it is recommended that these video signals be synchronized using a 2:1 sync generator. This technique provides picture stability, enhances picture quality, and prevents picture roll, jitter, tearing, or other disturbances. If random-interlace cameras are used, they should be externally synchronized.

In theory, present consumer-type VCRs could be used for security applications when real-time recording is required. However, most security applications require 24-hours-a-day operation without stopping, and continuous VCR operation has proved to be a strain on all consumer-grade VCRs. Consumer VCRs are not designed for continuous use, do not operate in a time-lapse mode, and are not a reliable choice for security applications. Time-lapse VCRs developed for the security industry are specially designed to withstand the additional burden of long-term recording over continuous periods; as a result, they have higher reliability and are more expensive than standard VCRs.

8.4 VCR OPTIONS

Most VCRs offer options such as

- built-in camera switcher,
- time/date generator,
- sequence or interval recording of multiple cameras on one VCR,
- interface with other devices, such as a cash register or ATM,
- remote control via RS-232 or other signals,
- 12-volt DC power operation for portable use, and
- frame/field selection.

8.4.1 Camera Switching/Selecting

A very useful option available on some VCR time-lapse recorders allows recording of multiple cameras and selected playback of numerically coded cameras. Figure 8-7 shows the technique for a 16-camera-input system using 8 cameras. This technique allows the multiplexing of up to 16 camera inputs onto one videotape, thereby reducing equipment cost by eliminating the need for one videotape recorder per camera input.

The VCR has the ability to separate recordings made from each camera by displaying only the frames taken by one particular camera. With this option, one particular camera can be viewed individually when reviewing the tape. This is particularly useful when many cameras are recorded on one VCR. Rather than sit through scenes from all the cameras when only one is of interest, the operator can select a particular camera to be viewed on the monitor. To assist in locating a specific frame, the operator can advance or back up (shuttle) one picture (frame or field) at a time.

In operation, during real-time or time-lapse video recording, the encoder inserts a binary code on the synchronizing portion of the video signal for every field, with each camera uniquely identified by a two-digit ID number. With this specific dedicated coding scheme, a pulse is generated once per frame and the video switcher then dwells on each position equal to a time set by the number of frames on a thumbwheel setting on the recorder control console. To recover the video image, a two-digit thumbwheel on the front panel selects the desired video camera from the frame store device. The high-speed switching and encoding technique permits as many as 16 cameras to be recorded on a single VCR. When the tape is played back in real time, any of the 16 camera scenes can be chosen for display. In Figure 8-7, camera scene 5 has been chosen for presentation. In this example, the pictures are updated every 0.266 seconds.

8.4.2 RS-232 Communications

Some VCRs interface with computer systems via an RS-232 port built into the machine. This electronic communications port enables the computer to talk to the VCR and control it. The VCR communicates back to the computer via the same RS-232 communications link.

A time-lapse recorder able to communicate via an RS-232 port by two-way data transmission can be used in large security system applications, since the recorder is controlled by a central computer and communicates its status back to the central control. This feature is important in large integrated security systems where there are a large number of critical and demanding security requirements, and its use eliminates guard responses.

The RS-232 port permits the recorder to be interfaced with the computer communication port via transmission paths such as telephone dedicated two-wire or wireless systems. The computer becomes a command post for remote control of recorder functions, such as real-time or time-lapse mode, time-lapse speeds, stop, play, record, rewind, fast-forward, scan, reverse-scan, pause, advance, and adjust. These remote functions are put at the fingertips of the security operator—whether in the console room, at the computer across the street, or at a distant location.

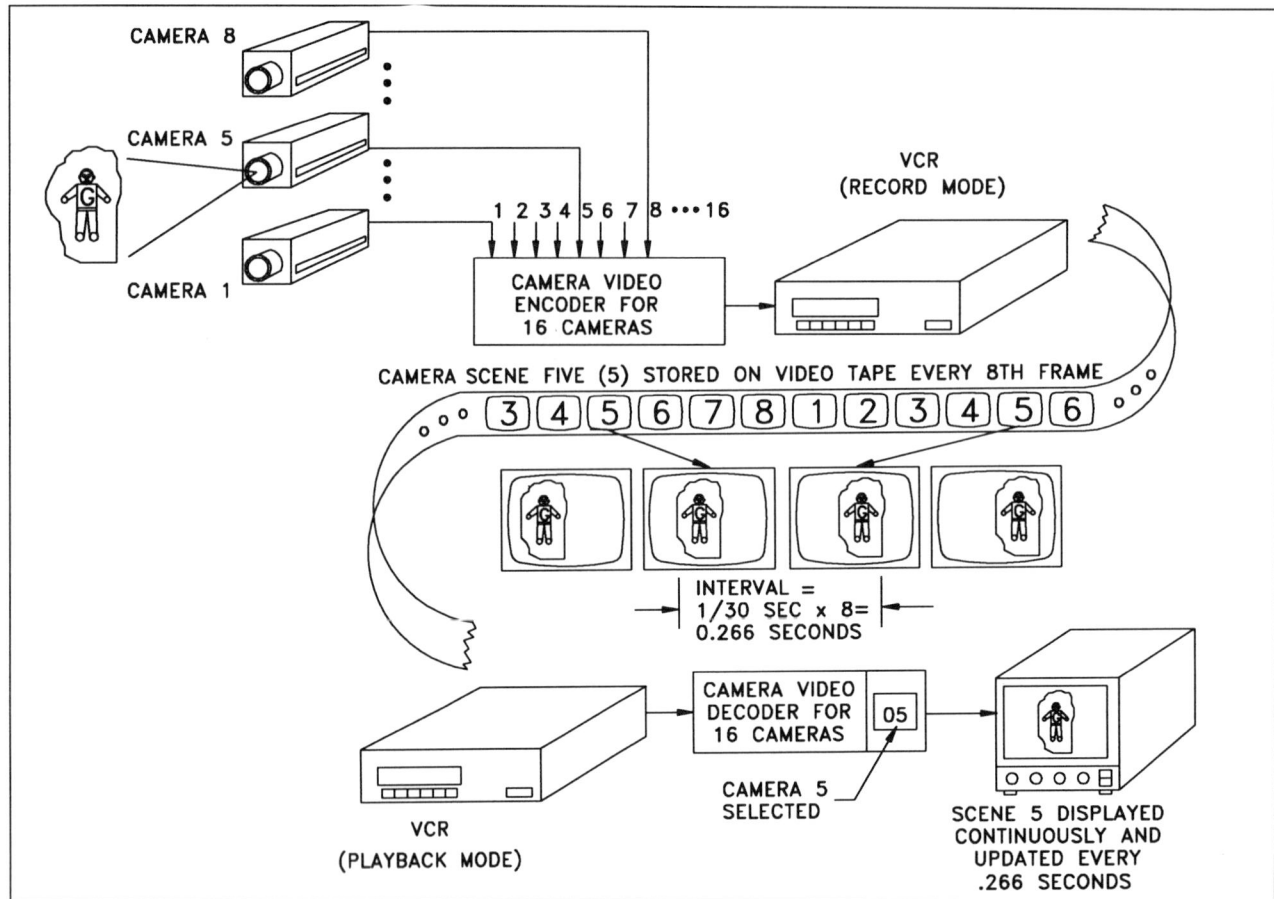

FIGURE 8-7   Multiplexing multiple cameras onto one time-lapse VCR

8.4.3 Scrambling

The widespread use of video recording for security purposes and the inherent highly sensitive security information recorded thereon generates a need for scrambling or encrypting of a videocassette recording. Equipment is available to scramble or encrypt the videotape signal to prevent unauthorized viewing of video recordings. This video-scrambling technology safeguards the video signal as it is recorded or transmitted, thereby protecting the information on the cassette. The encoder produces a secure signal totally unusable in its scrambled form. In one system, the encryption code changes constantly, thereby rendering even frame-by-frame decoding fruitless. The scrambling is done in the video signal, making it virtually impossible to reconstruct an intelligible picture. A password feature can be included, so that only personnel entitled to view the tape can gain access to it. In some systems, special codes can be added to a recording, thereby restricting descrambling to a scheduled time interval. The complete system consists of an encoder and decoder connected to the VCR at each location (Figure 8-8).

8.4.4 On-Screen Annotating and Editing

Some recorders have a built-in alphanumeric character generator to annotate information on the tape, such as time, date, day of the week, recording speed, alarm input, camera identifier, and time on/off status. For example, in a retail application the recorder can annotate the video frame with the cash register dollar amount to check a cashier's performance. In a bank ATM application, a video image annotated with the transaction number can help identify the person entering the transaction.

Another option is on-screen editing. The RS-232 interface permits the operator to add text to the videotape whenever additional identification is required, such as making notes of the action taken during an alarm condition or documenting information on a scene. Up to 8 lines with 20 characters per line for each video frame are typical capabilities. Usually the first two lines are dedicated (preprogrammed) and record the time and date or other preprogrammed data. The remaining lines are available for text entered by a security operator. In typical applications, this annotation provides the following:

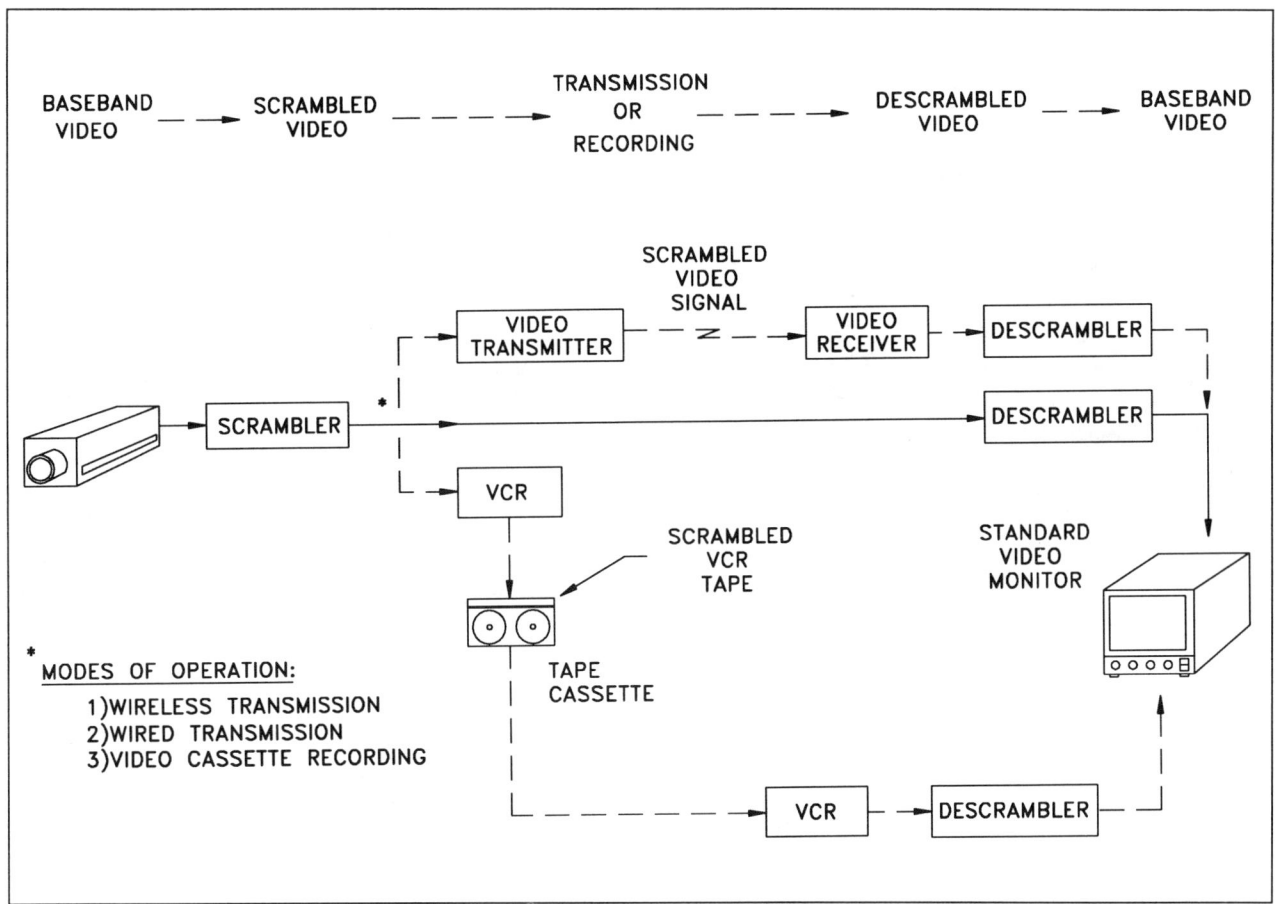

FIGURE 8-8   Video recording scrambling and encryption

- superimposed listing of cash register data covering the transaction seen on the videotape;
- a record of the identification data or personal identification number (PIN) of an individual using an ID card in an access-control environment;
- verification of the authenticity of ID cards used at remote ATMs, electronic gas pumps, grocery stores, cash-dispensing machine, and so on;
- data on action taken by security personnel as a result of alarm conditions from a particular CCTV camera.

8.5 VIDEO IMAGE/FRAME-STORAGE DISK

The real-time or time-lapse VCR provides a means for serially recording consecutive frames and scenes over a period of time ranging from seconds to many hours, and recording thousands of individual frames of video pictures on magnetic tape. A 2-hour VHS cassette records (stores) 216,000 frames of video (2 hr × 30 frames/sec = 216,000 frames). An inherent characteristic of all videocassette recordings is that the time required for locating a particular frame or sequence of frames on the tape can be lengthy;

the videocassette is certainly not considered a means for fast retrieval. If the videotape is not coded in some way to identify individual camera scenes as they are recorded (sequentially or by time and date), locating a particular scene taken with a particular camera at a particular time is usually impractical. As an improvement, if camera information and time and date are coded on the tape, equipment is available so that the operator can enter the camera number, time, and date, and the frames on the tape will be retrieved by the VCR. However the VCR equipment can still take considerable time to locate a specific frame since many minutes can be needed to move (shuttle) a particular frame to the playback head.

If a particular frame or series of frames or short real-time sequence of pictures needs to be stored or retrieved quickly, that is, in fractions of a second or seconds, a different recording technique is required. For example, an alarm signal occurs and the CCTV picture from the camera monitoring the location needs to be recorded instantly and then retrieved in a few seconds and perhaps the video image printed out on a hard-copy printer or displayed at remote monitoring locations so that action may be taken—all in a few seconds. A magnetic or optical hard disk can help perform this task (Figure 8-9).

FIGURE 8-9 Magnetic and optical disk recorders

In the disk-recording method, the magnetic or optical video picture is stored on a disk rotating at high speed. The picture is stored and identified by coding the signal to the specific camera and the time and date at which it was put on disk. At a later time the stored picture can be retrieved in random access at high speed. In one system, all the pictures are recorded on parallel, circular, or concentric tracks on the magnetic disk and accessed in parallel rather than serially, as required in VCR tape recording. The magnetic disks are all erasable and re-recordable. Most optical disks are write-once read-many (WORM) disks, but some are erasable. The video information is stored on the magnetic disks in either analog or digital form. At present all optical disks are recorded by converting the analog video signal into a digital signal and then recording the individual bits as digital information on the disk.

Hard-disk magnetic media for storing video images make use of microcomputer hard disks and store from a few hundred to several thousand monochrome video frames on one 20-megabyte disk. While there is some similarity between disk storage and the time-lapse video recorder in that single frames (or fields) of video are stored, the hard disk can provide rapid random access to the stored pictures.

Two generic systems have evolved: (1) microcomputer-based, digitized hard-disk video file, and (2) analog hard-disk video file.

8.5.1 Magnetic-Disk Image Storage

An example of a microcomputer-based digital hard-disk video file is a personal computer (PC) AT with a video analog-to-digital conversion board and software to code and store the video frames on a magnetic hard drive. A PC with a 20-megabyte hard disk drive stores about 200 pictures, with any picture retrievable in a few seconds.

An example of an analog hard-disk video file (requiring no computer) stores 2460 pictures on a 20-megabyte hard disk and retrieves any picture in less than 0.05 seconds. Storage requirements using the analog technique are about one-fifth that of digital.

8.5.1.1 Analog

There is a trade-off in digital versus analog picture storage on disks. Analog storage and retrieval has been available longer and has seen widespread consumer application in magnetic-tape VCR products. This acceptance has provided high-volume, high-reliability, and yet low-cost range of products. Analog storage excels in its ability to store large numbers of pictures per given media size (216,000 pictures on a 2-hour cassette). The disadvantage of cassette tape storage is its long access time to locate a specific frame, since videocassette tape is a serial medium and retrieval time is related to the location of the picture on the tape.

A technique has been developed to store the analog video data on a digital magnetic disk in the form of an analog magnetic signal (Figure 8-10). The magnetic analog signal represents the composite video signal received from the camera. In one revolution of the disk ($^1/_{60}$ of a second), one video field is stored. A full frame (two fields) is stored in $^1/_{30}$ second. Since a 20-megabyte Winchester drive has 2460 tracks, 2460 monochrome video pictures can be stored on a standard 20-megabyte hard drive in analog form. The picture (one field) can be accessed in 0.05 seconds. This is presently the fastest system available to store and access a video field, and because the medium is magnetic, it is erasable and reusable. This system is monochrome and has 250-TV-line horizontal resolution. By recording two fields on two tracks, approximately 500-TV-line resolution is achieved. A color version of the analog storage system is available and capable of recording 1250 frames on a 20-megabyte hard disk. Storage and retrieval times are 0.20 seconds.

All or a selected number of stored video images can be backed up using a standard VCR. Standard thermal video printers (or others) print a hard copy of any video picture.

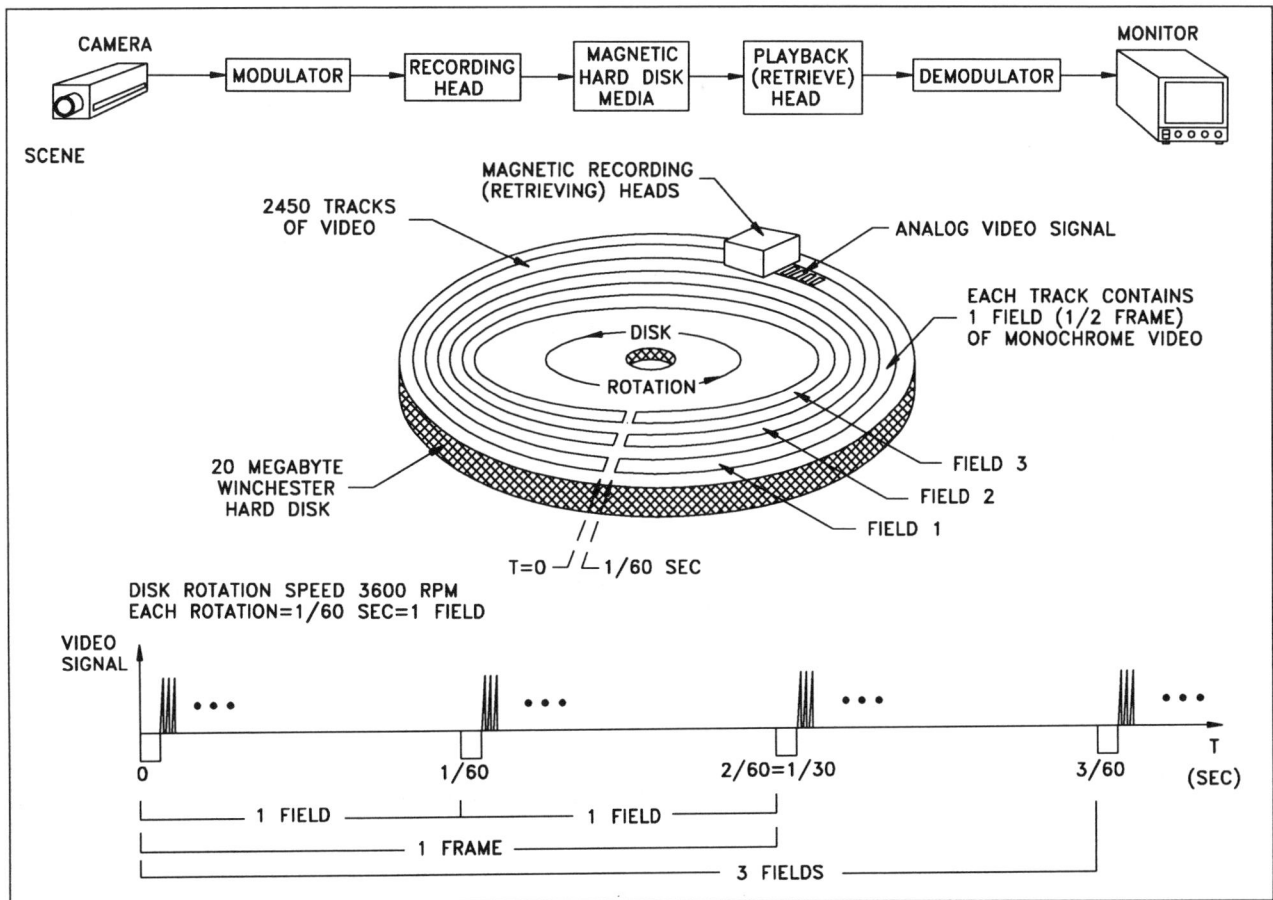

FIGURE 8-10   Analog video storage on a magnetic hard disk

8.5.1.2 Digital

A technique in widespread use for storing video images on magnetic hard disk is to digitize the video image (Figure 8-11). This method requires from five to ten times more storage space than analog storage. It requires five to ten times more time to store the pictures and correspondingly more time to retrieve them. Depending on the size of the magnetic hard disk, many thousands of high-resolution monochrome or color images can be stored on it. The following sections describe the technique and cite examples of some available hardware.

8.5.2 Optical-Disk Image Storage

When a large number (many thousands) of video frames (or fields) must be stored, an optical-disk medium is chosen. Optical storage media are durable, removable disks that can store hundreds of thousands of video pictures in a coded digital format. There are two generic systems available: (1) nonerasable WORM and (2) erasable. These two electrooptical storage systems are described in the following sections.

8.5.2.1 Write-Once Read-Many Disk

The optical-disk recording system provides a compact means to store many thousands of video images. Typical 5¼-inch WORM optical-disk systems store 10,000 to 20,000 color images in digital form (Figure 8-12).

The drive uses an 800-megabyte, double-sided, removable diskette, which is rugged and reliable. In some security applications, a WORM drive has a significant advantage over magnetic recording media because an image cannot be overwritten, eliminating the risk of accidental or intentional removal or deletion of video pictures.

Another attribute of the WORM system is that the picture file is removable and can therefore be secured under lock and key, or stored in a vault when the terminal is shut down or the system is turned off. Cost and reliability are also favorable. If one side of one removable cartridge stores 10,000 color images, the cost of removable cartridges is less than 20 cents per million bytes, or 0.5 cents per stored picture. Reliability is extremely high, with manufacturers quoting indefinite life for the cartridges and a minimum mean time between failure of greater than 10 years. The reason for this longevity is that nothing touches the disk itself except a light beam used to write onto and read from the disk.

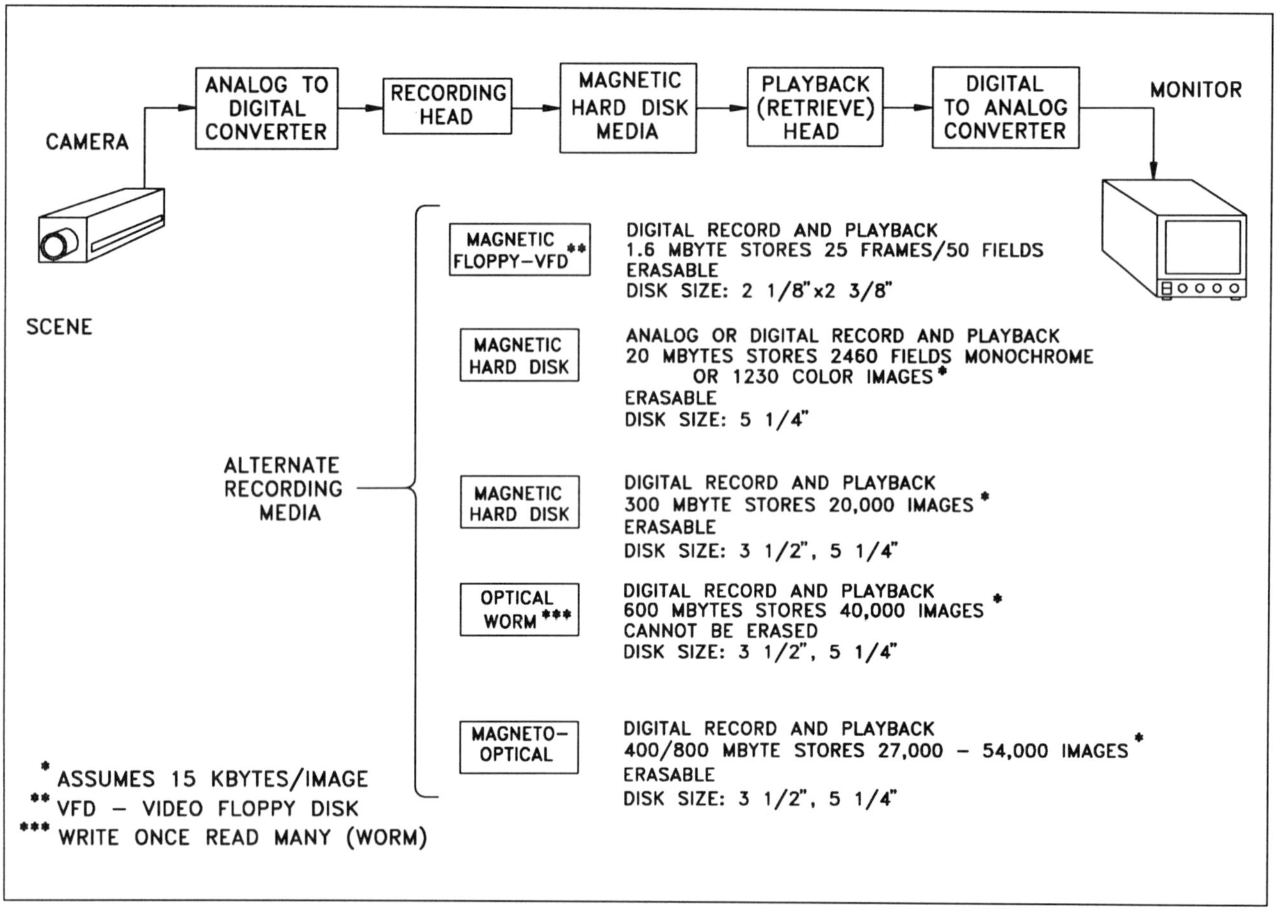

CAMERA → ANALOG TO DIGITAL CONVERTER → RECORDING HEAD → MAGNETIC HARD DISK MEDIA → PLAYBACK (RETRIEVE) HEAD → DIGITAL TO ANALOG CONVERTER → MONITOR

SCENE

ALTERNATE RECORDING MEDIA

MAGNETIC FLOPPY–VFD **
DIGITAL RECORD AND PLAYBACK
1.6 MBYTE STORES 25 FRAMES/50 FIELDS
ERASABLE
DISK SIZE: 2 1/8"x2 3/8"

MAGNETIC HARD DISK
ANALOG OR DIGITAL RECORD AND PLAYBACK
20 MBYTES STORES 2460 FIELDS MONOCHROME
OR 1230 COLOR IMAGES *
ERASABLE
DISK SIZE: 5 1/4"

MAGNETIC HARD DISK
DIGITAL RECORD AND PLAYBACK
300 MBYTE STORES 20,000 IMAGES *
ERASABLE
DISK SIZE: 3 1/2", 5 1/4"

OPTICAL WORM ***
DIGITAL RECORD AND PLAYBACK
600 MBYTES STORES 40,000 IMAGES *
CANNOT BE ERASED
DISK SIZE: 3 1/2", 5 1/4"

MAGNETO-OPTICAL
DIGITAL RECORD AND PLAYBACK
400/800 MBYTE STORES 27,000 – 54,000 IMAGES *
ERASABLE
DISK SIZE: 3 1/2", 5 1/4"

* ASSUMES 15 KBYTES/IMAGE
** VFD – VIDEO FLOPPY DISK
*** WRITE ONCE READ MANY (WORM)

FIGURE 8-11   Digital video storage on a magnetic hard disk

8.5.2.2 Erasable Optical Disk

While there have been steady improvements in magnetic-disk and WORM storage media, a new erasable optical-disk medium is now available. The images on these removable optical disks can be erased (as on present magnetic media) and overwritten with new images. They operate in the same way as magnetic disks, with an important exception: optical disks can store significantly larger amounts of video data per disk. One technique for erasable magneto-optical recording uses a laser to heat the magnetic material at a single point to its Curie temperature (the temperature at which the residual magnetization state of the magnetic film can be reversed), while the proper external magnetic field is applied (Figure 8-13).

This magnetic field polarity can later be "read out" by reflecting a low-power laser beam off the point. Each point represents a bit of stored information.

Each image stored on the magnetic or optical hard disk is uniquely identified and may be retrieved in random access in less than 1 second. This is a significant increase in storage capability: approximately 31 reels of data tape are equivalent to one single 5¼-inch-diameter optical disk—the size of an ordinary compact disc. This optical disk can store up to 800 megabytes of information—10,000 color video pictures (uncompressed).

While most optical disks used in security are WORM, erasable optical disks are in use. Erasable disks use the principle of magneto-optics to record the video information onto the disk in digital form. The video image data or other information are erasable, allowing the same disk to be reused many times, just like magnetic tape. The 5¼-inch

FIGURE 8-12   Write-once read-many optical-disk recorder

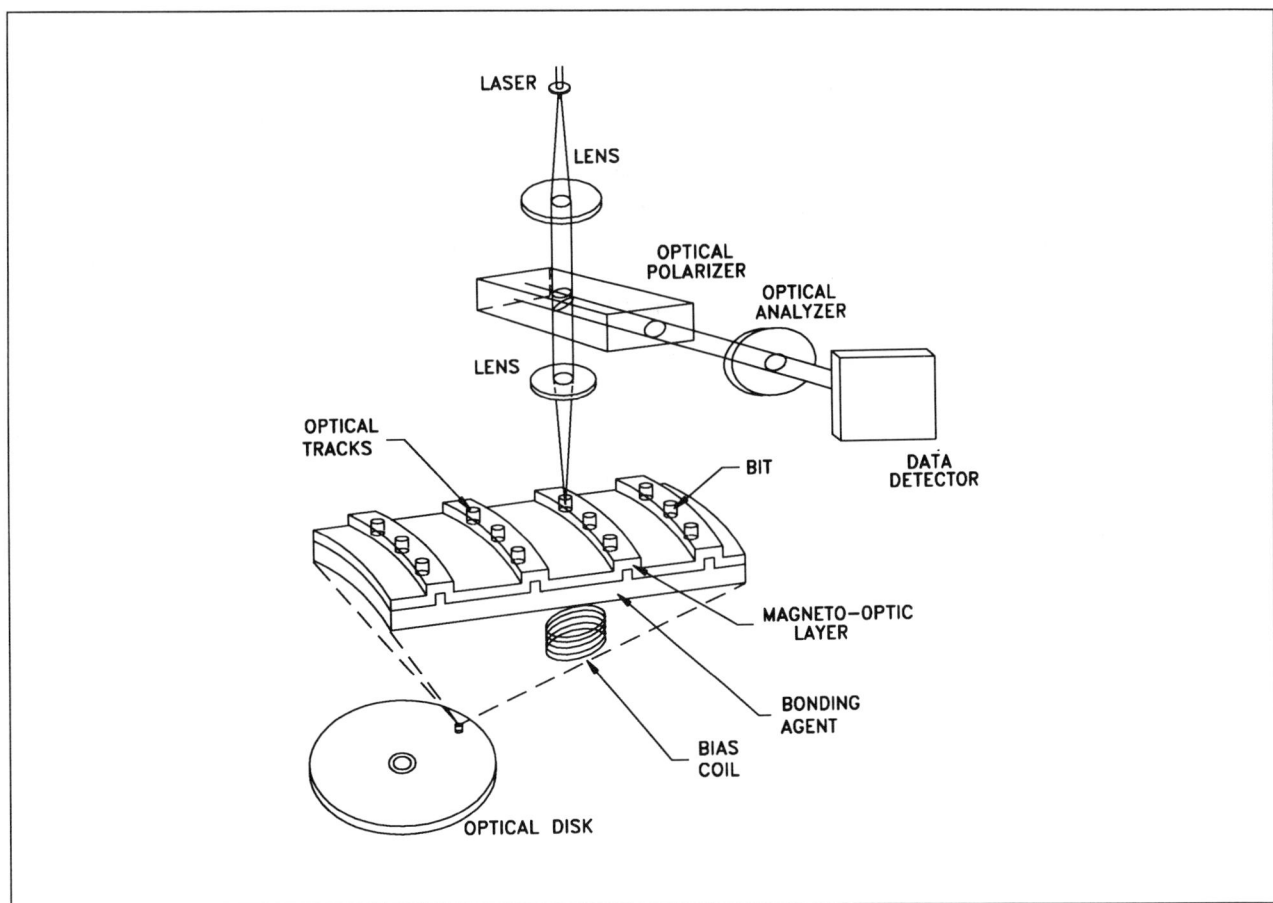

FIGURE 8-13 Erasable optical-disk recorder

optical disk with the magneto-optic disk holds 600 megabytes of information and stores up to 1000 times more information than the standard 5¼-inch floppy diskette used in computers. The most advanced Winchester hard disks stores 43 megabytes of information in a single square inch of media, while the magneto-optic disk stores 300 megabytes per square inch. Furthermore, magneto-optic disks are as portable as floppy disks, unlike most high-capacity 5¼-inch magnetic hard disks, which must be permanently housed in heavy drives. A removable hard-disk-drive cartridge is available.

Erasable magneto-optic disks offer more than high capacity and portability, however. They provide random access to video information in fractions of a second rather than in the minutes it takes with serial VCR magnetic tape. Reading, writing, and erasing the information on the optical disk are done using light energy and not magnetic heads that touch or skim across the recording material. Therefore, magneto-optical disks have a much longer life and a higher reliability than magnetic disks, as they are immune to wear and head crashes (usually catastrophic events in which sudden vibration or dust particles cause the mechanical head to bump into the recording material, thereby damaging it). In the case of the optical disk, the magnetic

layer storing the information is imbedded within a layer of plastic or glass, protecting it from dust, wear, and other problems that plague conventional magnetic recording techniques. Clearly, the optical disk will become the storage medium of choice in the future when large numbers of high-resolution video images need to be stored and retrieved for later use.

8.5.3 Still-Video Floppy Disk

A convenient technique for recording still video (SV) color images uses an electronic equivalent of a 35-mm camera and a small magnetic video floppy (VF) disk instead of film. (Figure 8-14).

The magnetic VF disk is similar to the 3½-inch personal computer disk, but smaller. This industry-standard disk stores up to 50 video field images in color. The source of the floppy disk images can be a video camera equipped with a floppy disk recorder or a 35-mm electronic video camera. Color prints can be produced from the floppy disk by the same method as from a real-time video image or recording.

The SV recorder lens, image sensor, camera electronics, recording electronics, and memory disk are all contained

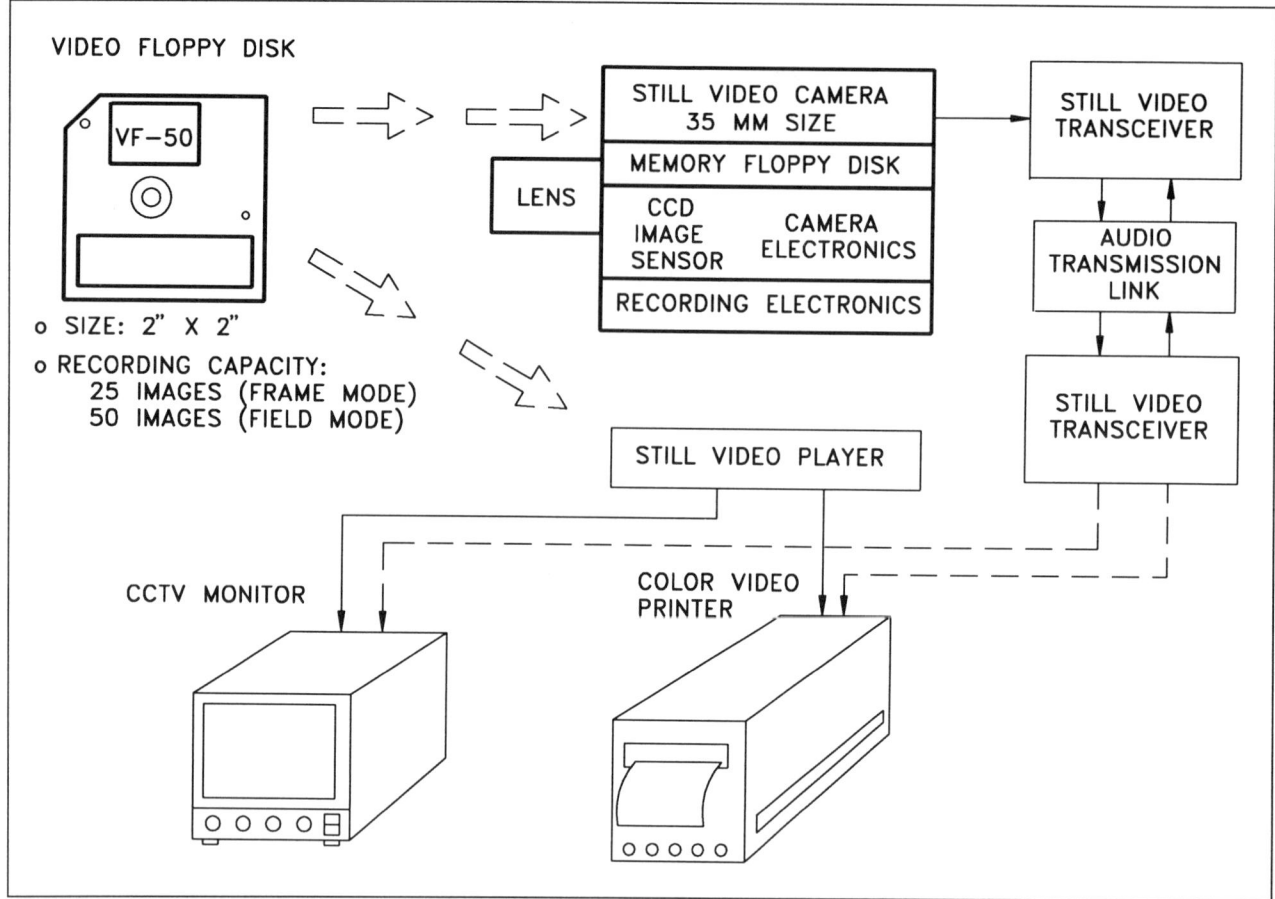

FIGURE 8-14 Still video floppy disk recorder

in a housing similar to a 35-mm camera. The camera is designed to be operated manually or by a remote control unit. The SV camera does not provide real-time video surveillance but rather operates as a still-frame camera, much like a film security camera but with lower resolution. The medium for recording the video image is the key to the system: a durable floppy disk in which the video images are electromagnetically stored via a recording head inside the camera. The industry-wide standard is a VF disk accepting both normal and high-band video signals. The small 2 × 2-inch disk can record 50 television images in field mode and 25 full frames in frame mode and be easily transported or mailed to remote locations. Images are captured by a high-resolution, ½-inch CCD sensor with 360,000 (or more) pixels, providing a high-quality color picture and horizontal and vertical resolution of 400 TV lines. These cameras are equipped with an automatic-focusing sensor to ensure a sharp image, and a variable electronic flash for correct exposure. The amount of light required for each image is calculated and controlled with the electronic-flash measuring device. This precise exposure is accomplished by adding a feedback loop to the conventional method of determining aperture size and shutter speed. The light

sensor samples the light reflected from the scene and the CCD reports back to the camera electronics how to set the lens aperture and flash intensity. The all-electronic camera has the capability of high-speed continuous shooting (up to 50 fields) at recording speeds up to five images per second. Each frame can be annotated with information such as year, month, hour, and minute.

To play back a picture, the VF disk is inserted into the SV player and any of the 50 fields (or 25 frames) can be selected randomly by the operator. As with other video images, the system output can be printed on any thermal-transfer or other hard-copy printer for a permanent record. The video still image can be transmitted over a standard telephone line or other audio-bandwidth link via a transceiver at each end. The VF disk, like other magnetic media, can be reused by erasing the images on it and recording new ones. The SV imaging system provides a convenient means for replacing sometimes-cumbersome film camera systems, which require film development and printing.

In another system, a lens, CCD camera, and a single-frame-store video board captures (digitizes) a single video frame. It is small (½ × 3 × 6-inch), low cost, and useful when only one frame needs to be recorded. The stored image can be displayed on a standard CCTV monitor, recorded on a VCR, or printed on a video printer. Image storage is initiated via a dry contact switch closure.

8.6 VIDEO HARD-COPY PRINTERS

Hard-copy printout of video monitor scenes, VCR recordings, or other recording media has become an important adjunct to many security systems. Monochrome and color printers permit good-to-excellent-quality reproduction of the scene on paper or film-based media. The hard copy is used by security personnel for apprehending an offender, responding to a security violation, or for permanent record.

Early models of thermal hard-copy printers produced crude facsimiles of the monitor picture with low resolution and poor gray-scale rendition. Today's advanced technology enables printers to produce excellent monochrome or color image prints with resolution approaching that of the camera, lens, and monitor. Monochrome prints are inexpensive: 5 cents a copy. Color prints cost 60 to 80 cents a copy, generally less than a photographic film equivalent.

Of the several monochrome and color printout techniques available for the security industry, thermal-transfer printers are the most popular, because of cost considerations. Color prints are made using the thermal plastic (wax) or dye-diffusion technique. Other printing methods include ink-jet (continuous or drop-on-demand) and laser. The thermal printer in widespread use with resolution from 250 to 500 lines is probably the best choice for reproducing monochrome images with reasonable continuous-tone printing. Table 8-5 summarizes the characteristics of video hard-copy printers.

8.6.1 Thermal

The thermal printer enjoys popular demand for printing monochrome and color video images because of its ruggedness, convenience, and reasonable price. Monochrome thermal printers cost from $1100 to $1600 and provide a fast means—8 seconds per print—for obtaining a hard-copy printout from any video signal. The typical video thermal printer (Figure 8-15) holds a roll of plastic wax–coated thermal paper sufficient to produce 120 3 × 5-inch video pictures.

The principle of thermal printing is shown in Figure 8-16.

The video signal from the camera is converted into a digital signal and stored in a random access memory (RAM) or frame-storage device. The video freeze-frame module captures the video image and "freezes" the picture as a snapshot of a moving scene. This temporary storage allows the printer to operate at a much slower speed than the actual real-time video image. After the picture (frame or field) has been captured by the freeze-frame electronics, the video signal from the camera is converted to an electrical drive signal for the thermal head located adjacent to the paper. Depending on the video drive signal level, the paper is locally heated, causing the wax on the paper to melt and turn black (or another color). Depending on the amount of heat applied, a larger or smaller dot is produced. As the video information is scanned across the slowly moving paper, the image is "burned in," thereby creating a facsimile of the video image. Scanning an entire monochrome video image one pixel at a time takes approximately 8 seconds. The printed video image is recorded on Mylar-treated paper, which resists fading (from sunlight) and tearing and serves as an excellent record. Since the video frame is stored in the printer until a new frame is captured, multiple copies can be made.

8.6.1.1 Monochrome

Several manufacturers provide models having different resolutions, speeds, and paper sizes. A printer at the low end of the resolution scale has 250 by 250 pixels with 16 shades of gray. This print provides adequate information for many security applications where fine detail is not required. When higher resolution is required, systems having 640 by 640 pixels with 64 levels of gray scale are available. These printers provide a horizontal resolution of 470 TV lines, at a printout speed of 9 seconds per picture. Picture quality from this type of system is as good as most television camera and lens systems are capable of producing (Figure 8-17).

Print paper rolls for these systems yield 120 to 180 pictures per roll; cost per picture ranges from 4 to 10 cents. These pictures represent high-quality monochrome hard copies and are almost equivalent to instant Polaroid photo prints because of the extremely dense pattern of the picture and the 64 gray levels.

Thermal printers are totally silent, since they are nonimpact type and require no ribbons or toners. Large-format printers are also available, which produce 8½ by 6-inch monochrome prints.

For highest resolution, a high-speed monochrome video thermal printer with excellent gray-scale rendition is available. The printer uses a direct thermal printing process and has a resolution of 1200 by 1000 pixels, equivalent to 300 dots per inch. High speed is achieved by using a fast frame-grabber board (digitizer). The printer is easy to use, and paper cost is approximately 25 cents a copy, or about one-half the cost of the film equivalent. A full picture with this high resolution is printed out in 26 seconds; duplicates can be made from the image stored in the frame-grabber memory. This expensive printer should be used only with the highest-resolution surveillance system, where all of the detail obtained from the camera must be recorded on hard copy without any loss in image quality.

VIDEO PRINTER TYPE	SHADES OF GRAY BITS( )	RESOLUTION (PIXELS)*	PRINT SIZE WxH (IN)	PRINTS PER ROLL	MAXIMUM PRINT TIME (SECONDS)	PRICE RANGE ($)	PRINT COST (CENTS)
THERMAL–COLOR (DYE–DIFFUSION)	256 (8) 256 (8)	640x480 1280x960	4x3 7.9x5.9	100 50	70 160	2,000	60
THERMAL–MONOCHROME	64 (6)**	250x250 640x640	5x3 6x8.5	120–180	8–17	800–1,400	4–10
THERMAL–MONOCHROME (HIGH RESOLUTION)	256 (8)	896x508 508x508	4x3	270	4	1,200–1,600	25
THERMAL–MONOCHROME (LARGE FORMAT)	16 (4) TO 64 (6)	300x300	8.5x7.0	–	30–50	2,000	25
INK JET	–	80–400	11x8.5	SHEET	20–30	500–1,000	2
LASER	2 (1)	300x300+	11x8.5	SHEET	2	1,000	2

* EQUIVALENT TV LINES = 0.75 PIXELS (EX: 640x0.75 = 470 TV LINES)

** 4 BITS DIGITAL AND 2 BITS DITHERING

+ EQUIVALENT TO 300 DOTS/INCH (STANDARD LASER PRINTER RESOLUTION)

Table 8-5   Characteristics of Video Hard-Copy Printers

FIGURE 8-15   Thermal video printer

8.6.1.2 Color

The increased use of color video cameras and systems in the security industry has motivated manufacturers to provide cost-effective solutions for printing color images.

In color video systems, the lens receives the color picture information and through the color camera converts the light image into three electrical signals corresponding to the red, green, and blue (R, G, B) color components in the scene. If these signals were presented on an RGB monitor, the color video image of the scene would appear on the monitor in a normal way. In a color printer, the three primary colors in the video signal, R, G, and B, must be reversed to obtain their complementary colors, cyan, magenta, and yellow.

The principal technology used is the thermal printer. The two types used for color hard copy are (1) thermal transfer printer (TTP) (thermal plastic wax) and (2) thermal sublimation printer (TSP) (dye diffusion). Both techniques can produce brilliant colors and excellent resolution.

In the color TTP, a plastic-wax, single-color-coated ribbon (the width of the paper roll) is inserted between the thermal print head and the paper (Figure 8-18). The ribbon is heated (locally) from behind, causing the wax-based ink coating to melt and the image to transfer to the paper.

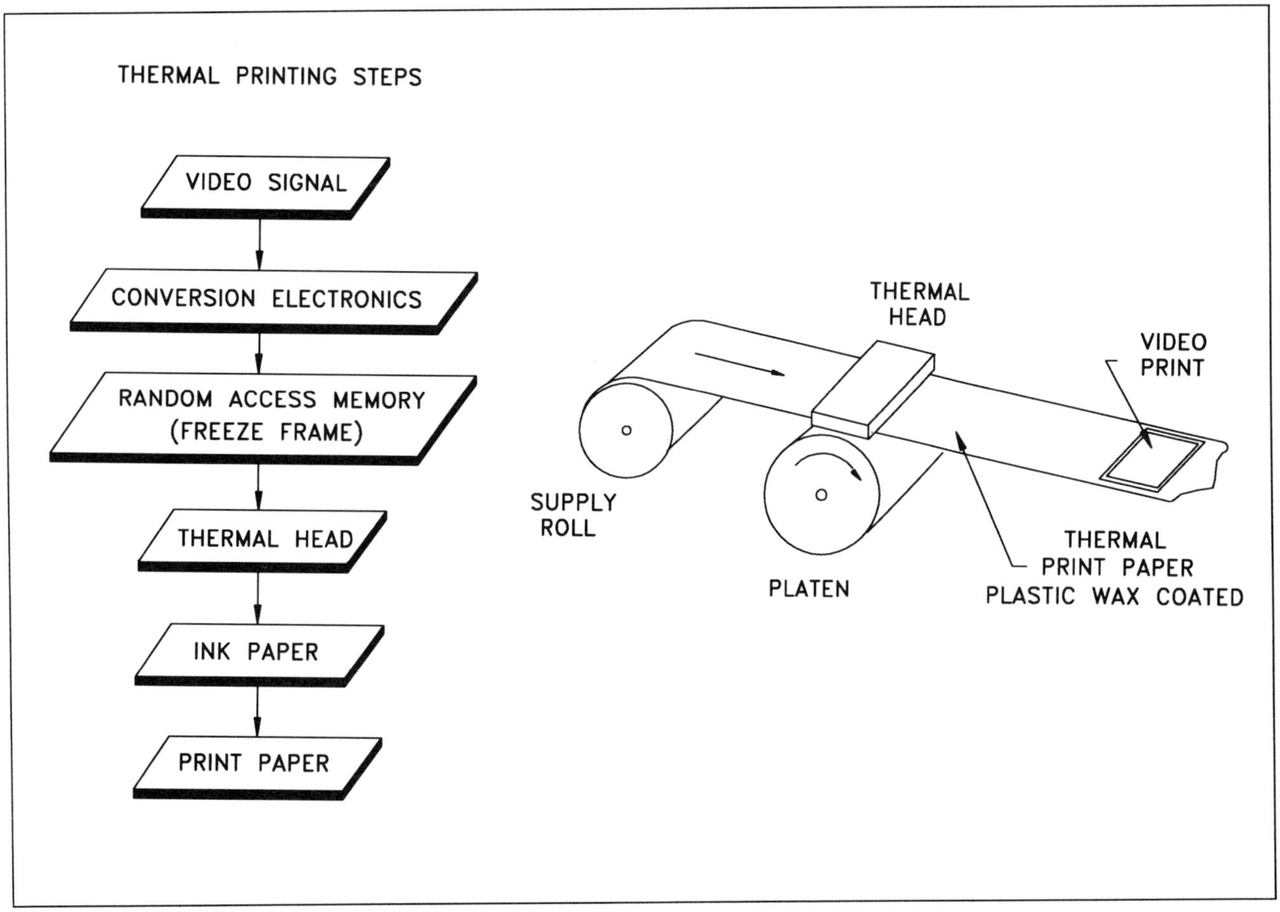

THERMAL PRINTING STEPS

```
VIDEO SIGNAL
      |
      v
CONVERSION ELECTRONICS
      |
      v
RANDOM ACCESS MEMORY
(FREEZE FRAME)
      |
      v
THERMAL HEAD
      |
      v
INK PAPER
      |
      v
PRINT PAPER
```

THERMAL HEAD

VIDEO PRINT

SUPPLY ROLL

PLATEN

THERMAL PRINT PAPER PLASTIC WAX COATED

FIGURE 8-16   Thermal printer technology

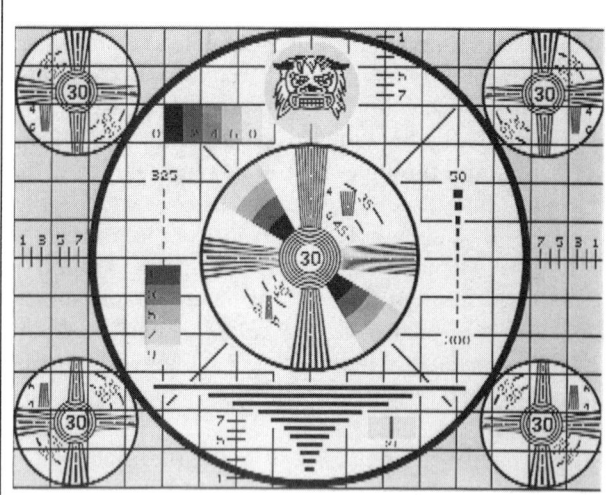

FIGURE 8-17 Monochrome thermal printer quality: (a) CCTV test pattern, (b) facial identification

Reproducing a satisfactory color image requires precisely engineered mechanical components so that the absolute registration between the three colors is printed. It also requires a high level of electronics technology to accurately combine the timing, signal level, and video fidelity to ensure a faithfully reproduced video image.

The TSP dye-diffusion printing media consists of the three-ink dye paper and hard-copy print paper. The dye-diffusion color printer operates through the use of a polyester-based substrate (donor element) that contains a dye and binder layer that when heated from the back side of the polyester, sublimates (becomes gaseous) and transfers to the paper, where the dye then diffuses into the paper itself (Figure 8-19).

The ink paper consists of a cartridge containing three-color sequential printing inks (yellow, magenta, and cyan).

One advantage of the dye-diffusion system over the wax-coated thermal plastic system is that it provides true continuous-tone copies, with the density of colors directly proportional to the heat supplied. Dye-diffusion thermal-transfer printers generally cost considerably more than their TTP counterparts.

Resolution for color video printers is typically 500 dots horizontal, and printout time approximately 80 seconds per print. Since each point (pixel) in a picture or resolution element in the color video image is composed of three separate colors, the actual detail resolution of the image is one-third the number of dots, or typically less than 200-TV-line resolution for the printed color image. While this is significantly less than the 500- or 600-TV-line resolution in the monochrome image, the addition of color to the print adds useful information. The print paper roll produces 3 by 4-inch pictures.

The full-color prints are produced in the thermal plastic color printer through the multiple passes of three ribbons having the colors cyan, magenta, and yellow.

The inking paper used is divided into three sections with different-colored ink; these three sections pass the thermal printer platen in sequence. As each color passes over the thermal head, an electrical signal proportional to the amount of the respective color in the video signal heats the head so that the ink of the required color is deposited on the paper. Depending on the amount of heat applied, a larger or smaller amount of ink from the paper will be transferred from the base film to the print paper. The first time the paper passes the head, yellow is deposited on it, then magenta, then cyan. By printing these three colors so that they are superimposed exactly on each other, the printer is able to produce a high-resolution print with excellent color rendition. By this principle, each dot on the final print copy is transferred from the base film ink layer to the print paper.

8.6.2 Ink Jet

There are two different types of hard-copy ink-jet printers. One is the continuous-jet printer, which uses a steady stream of ink droplets emanating from print nozzles under pressure. An electric charge is selectively applied to the droplets, causing some to be deflected toward the print paper and others away from the paper. The printout is the composite of all the individual dots in the image produced in this manner.

A simpler and more popular ink-jet printer is called the drop-on-demand printer. This printer forms droplets of ink in the nozzle and ejects them through appropriate timing of electronic signals, thereby producing the desired image on the paper. The majority of ink-jet printers produce a single dot size for each dot. Ink-jet printers have not yet found a significant market in the surveillance field and are not in wide use today because of their limited resolution and color reproduction.

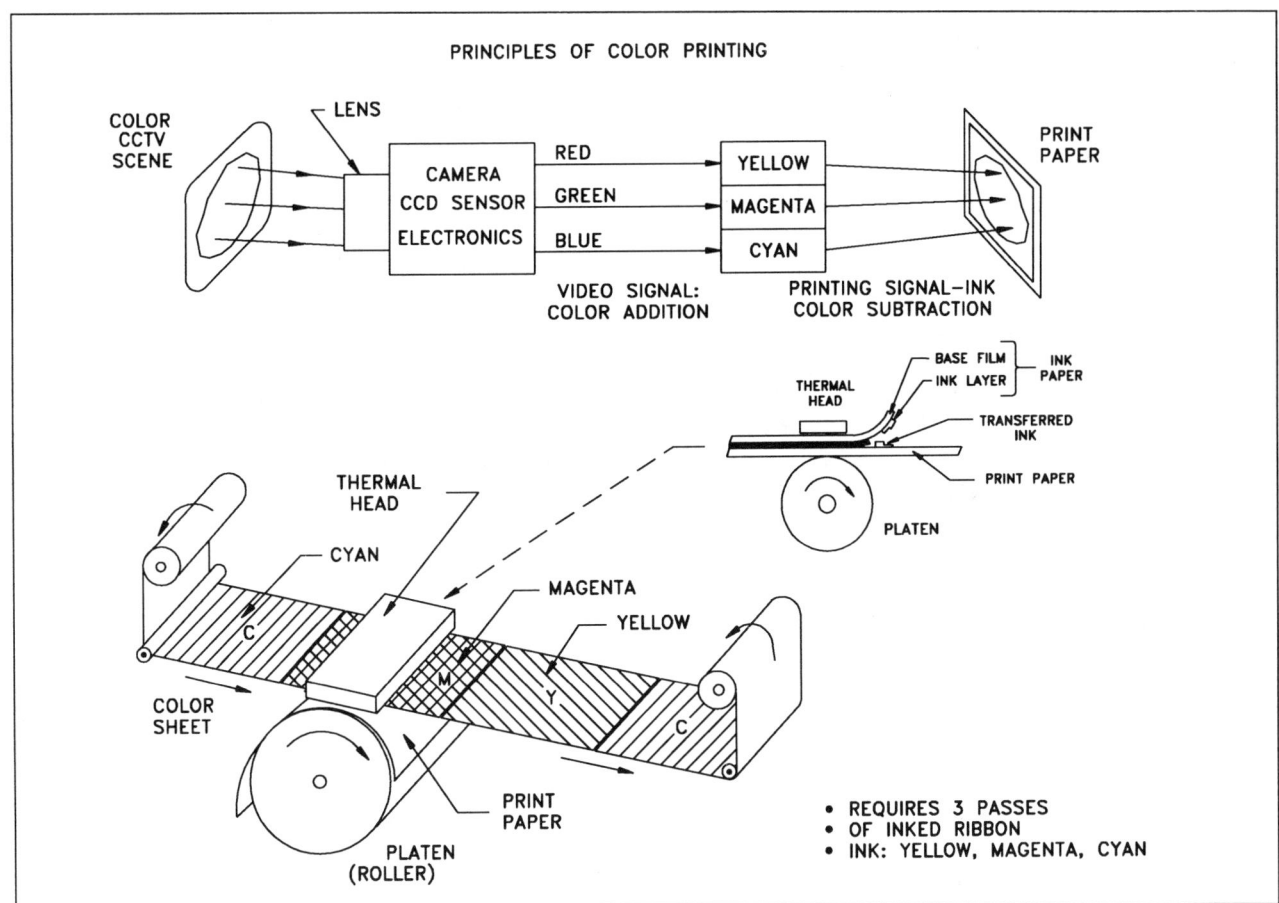

FIGURE 8-18 Plastic-wax color thermal transfer printer technology

8.6.3 Laser

Laser printers provide another alternative for producing hard-copy printouts. The common laser printer used in microcomputer desktop publishing does not produce high-quality monochrome images. It has reasonable resolution—300 dots per inch—but poor halftone (gray-scale) rendition.

8.6.4 High-Resolution Thermal Laser

Another class of laser printers using entirely different and more complex principles produces extremely high resolution, continuous-tone printed images. Typical systems have resolution of 500 to 600 pixels and 64 levels of gray scale, and require 60 to 80 seconds to print out. These printers carry a very high price tag and are normally used for printing from still images. Therefore they have not yet found their way into the surveillance field.

8.6.5 Film

Hard-copy video images can be printed on black-and-white or color photographic film, such as instant prints developed by Polaroid Corporation. The image is captured in a freeze-frame image storage device, displayed on a cathode ray tube, and recorded with a Polaroid film camera. While the resolution and rendition of the image is quite good, Polaroid film is more expensive and more difficult to work with than thermal paper. A further limitation is the small number of prints or slides available in a magazine (12 for prints and 20 for slides).

8.7 SUMMARY

There are several reasons why it is important to record CCTV monitor scenes on VCRs and other media.

1. The VCR (or other) taped image establishes an audit trail for the video scene. It can be viewed at a convenient

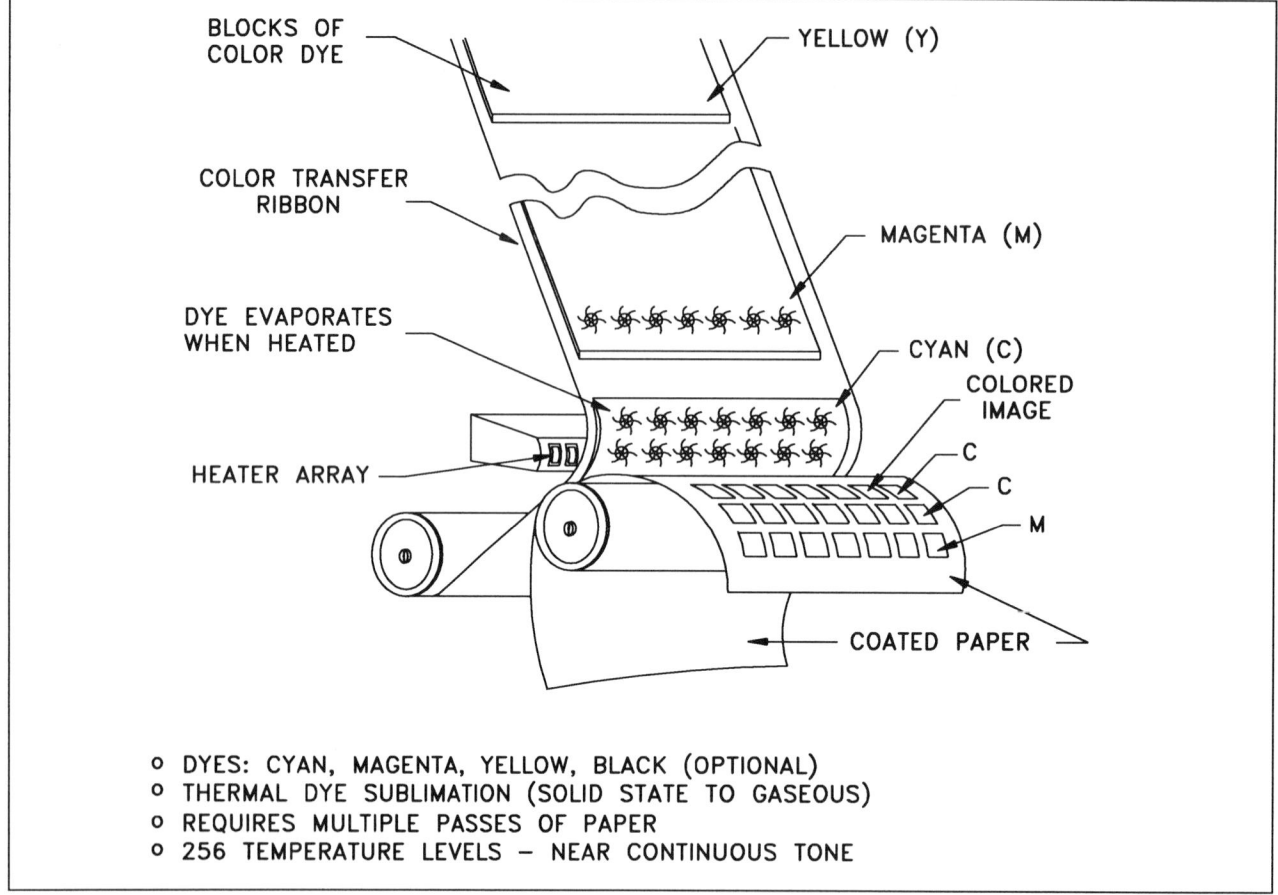

BLOCKS OF COLOR DYE

YELLOW (Y)

COLOR TRANSFER RIBBON

MAGENTA (M)

DYE EVAPORATES WHEN HEATED

CYAN (C)

COLORED IMAGE

HEATER ARRAY

C

C

M

COATED PAPER

o DYES: CYAN, MAGENTA, YELLOW, BLACK (OPTIONAL)
o THERMAL DYE SUBLIMATION (SOLID STATE TO GASEOUS)
o REQUIRES MULTIPLE PASSES OF PAPER
o 256 TEMPERATURE LEVELS – NEAR CONTINUOUS TONE

FIGURE 8-19 Dye-diffusion color printer thermal sublimation process

time by security, law enforcement, or corporate personnel to identify a person, determine the activity that occurred, or assess the responses of security personnel.

2. The video recording provides a permanent medium with which to establish credible evidence for possible dismissal of a person involved in criminal activity or suspected thereof, and for use in a criminal trial, civil litigation, or dismissal. Both film (still and motion-picture) and video recordings have been used to establish the guilt or innocence of a defendant.

3. The video recording provides a basis of comparison with an earlier set of tapes to establish if there was a change in condition at a particular location, such as moved or removed equipment or personnel patterns, including times of arrival and departure.

4. The video record offers the ability to instantly replay a scene. In contrast film in photographic systems must first be processed (with the exception of Polaroid film).

This single feature may be more important in pursuit situations than the higher resolution and quality obtained with film.

5. VCRs are an excellent tool for training and evaluation purposes. They serve as a source of feedback when evaluating employee performance. By reviewing the videotape, management can see which employees are working efficiently and which employees are not performing up to standards, without on-site supervision.

6. Magnetic floppy, magnetic hard disk, and optical hard disk recorders are used when a specific scene(s) must be retrieved quickly from a large database of stored images. Video printers are used when (1) a hard copy audit trail of a scene(s) is required for dismissal, courtroom, or insurance purposes, (2) a guard is to be dispatched with a hard copy of the individual to be apprehended or a scene to inspect, or (3) a hard copy is used to produce a photo ID or visitor's badge.

Chapter 9
Video Switchers and Consoles

CONTENTS

9.1 OVERVIEW

The function of the CCTV switcher in any multiple-camera security system is to connect a specific camera to a specific monitor (or recorder or other device) and display the video image in a logical sequence. The switched pictures on the monitor can be recorded on a VCR or printed on a video printer. In both small and large installations, the switcher component performs a vital function, which simplifies system use and maximizes the information presented to the security operator. In small security systems—having several cameras and one or two monitors—a switcher may not be necessary, since all camera scenes can be displayed on the monitors simultaneously. For a medium or large installation, where it is necessary to limit the number of monitors in the control console, a one-to-one camera-to-monitor correspondence is not practical. Physical space may be limited, and one security guard may not be able to view multiple monitors simultaneously. To view multiple cameras simultaneously on a single monitor, a combiner or splitter must be used (Chapter 13).

The following list presents the advantages of a single-monitor display of multiple cameras accomplished via sequencing techniques:

1. One monitor costs less than multiple monitors.
2. A single monitor occupies less space in the console than multiple monitors, even if the single monitor has a larger screen size than would be considered for multiple monitors.
3. Operator inattention or fatigue is less likely to occur with a single monitor.
4. There is proportionately less monitor maintenance required when the number of monitors is reduced.

While most asset-protection professionals recognize these advantages, there are some disadvantages:

1. When using one monitor, it is impossible to observe all camera locations simultaneously. This deficiency is especially important in situations involving continuous movement, or when it is important to view and observe activities at several locations simultaneously.
2. Since the switcher is sequencing from camera to camera, a long time can pass before a particular camera is seen again. In the case of four cameras, the operator will be viewing each camera only one-fourth of the time. There will be substantial periods of time during which the majority of the facility under CCTV surveillance will not be monitored by the console operator.

This may bring into question the cost-effectiveness of cameras whose dwell times (actual time viewing scene) are short and are seen for only a short period of time.

3. If there is a failure in the single monitor, no pictures will be displayed until it is replaced. Unless a spare monitor can be switched in, there will be some downtime before the new monitor is installed.

As a result, there is a balance between the simple system having a single monitor and one having multiple monitors. Each situation requires analysis to determine the optimum number of monitors and the type of switcher required.

The function of switching the video information from each of the cameras to the monitors can be divided into two basic categories: (1) single-output switching—switching one or more camera signals to a single output cable and into one or more monitors; (2) multiple-output switching—switching one or more camera signals to multiple output cables and into multiple monitors. All video switchers generally fall into one or the other of these two groups.

The two groups can be further categorized by considering the quality of the switching required and the location of the switcher in relation to the switching controls. When the camera-to-monitor distances are short (a few hundred feet), the switcher and the switching controls are one and the same and are located at the console. In installations having large distances between the cameras and monitor, the switcher and switcher controls are housed in two separate units, with the switcher located near the camera sites and the switching controls located near the console monitor.

The quality of switching, that is, how smoothly (clear, noiseless picture) the monitor picture from camera 1 can be changed to camera 2 and so on, is influenced by two related factors: (1) the type of signals to be switched—synchronous or asynchronous; and (2) the switching action itself—the manner or period in which the switchover occurs. Section 9.2.9 presents a more detailed discussion of the quality of switching. The many different types of video switchers to connect multiple cameras to a single or multiple monitors are described in the following sections.

Most sequential switchers, whether homing, bridging, looping, or alarm, have a three-position switch for each camera input. When one of these switches is in the up position, it is said to be in the Bypass mode. Any of the camera switches set in this position will cause the switcher to automatically skip the corresponding camera in the sequential switching cycle. The center position of these switches is called Automatic (Auto) mode. Any camera switch in this position will cause the switcher to automatically include the corresponding camera in the normal switching cycle. The down position of these camera switches can have several different functions.

In large CCTV security systems, microcomputer matrix switchers are used (Section 9.3). These large systems can connect hundreds of cameras to dozens of video monitors, recorders, or printers, either automatically via RS-232 communication links or through the operator. These computer-controlled switchers are software-programmable, so they can simultaneously switch multiple cameras to multiple output devices.

9.2 SWITCHER TYPES

Small-to-medium CCTV security systems use five basic switcher types: manual, homing, bridging, looping, and alarming. Using one or a combination of these five switcher types, cameras at multiple remote sites can be routed to multiple monitoring locations for direct observation, recording, or printing. The following sections describe the unique features of each switcher.

9.2.1 Manual

The simplest CCTV switcher is the manual switcher, with which the console operator manually chooses one camera from a number of cameras and displays it on a single video monitor. Front-panel push-button switches are activated manually by the operator to connect the individual camera to the monitor. Figure 9-1 shows the two types available: manual passive and manual active.

The basic difference between the two is that the manual passive switcher uses a simple contact switch, whereas the manual active switcher uses an electronic switch. Manual switchers are available to switch from 4 to 32 CCTV cameras.

9.2.2 Homing Sequential

The homing sequential switcher allows the continuous viewing of any normally sequenced television camera. The camera signal is connected to a single output monitor. This switcher has a three-position switch for each camera: Auto, Homing, and Bypass. Figure 9-2 shows a homing sequential switcher.

In the Automatic position, the switcher automatically selects and switches the video signal from one camera after another to the monitor, according to the sequence set by the security operator. The length of time each camera picture is presented on the monitor (dwell time) can be changed by the operator. The homing sequential switcher automatically sequences from one camera to the next camera, assuming the cameras have not been bypassed. When the specific camera control switch is pressed to the Home position, the camera input is continuously displayed on the single monitor and the switching sequence stops. All camera inputs are automatically electronically terminated by the homing switcher.

In summary, the three-position front-panel switches on the homing sequential switcher provide three separate

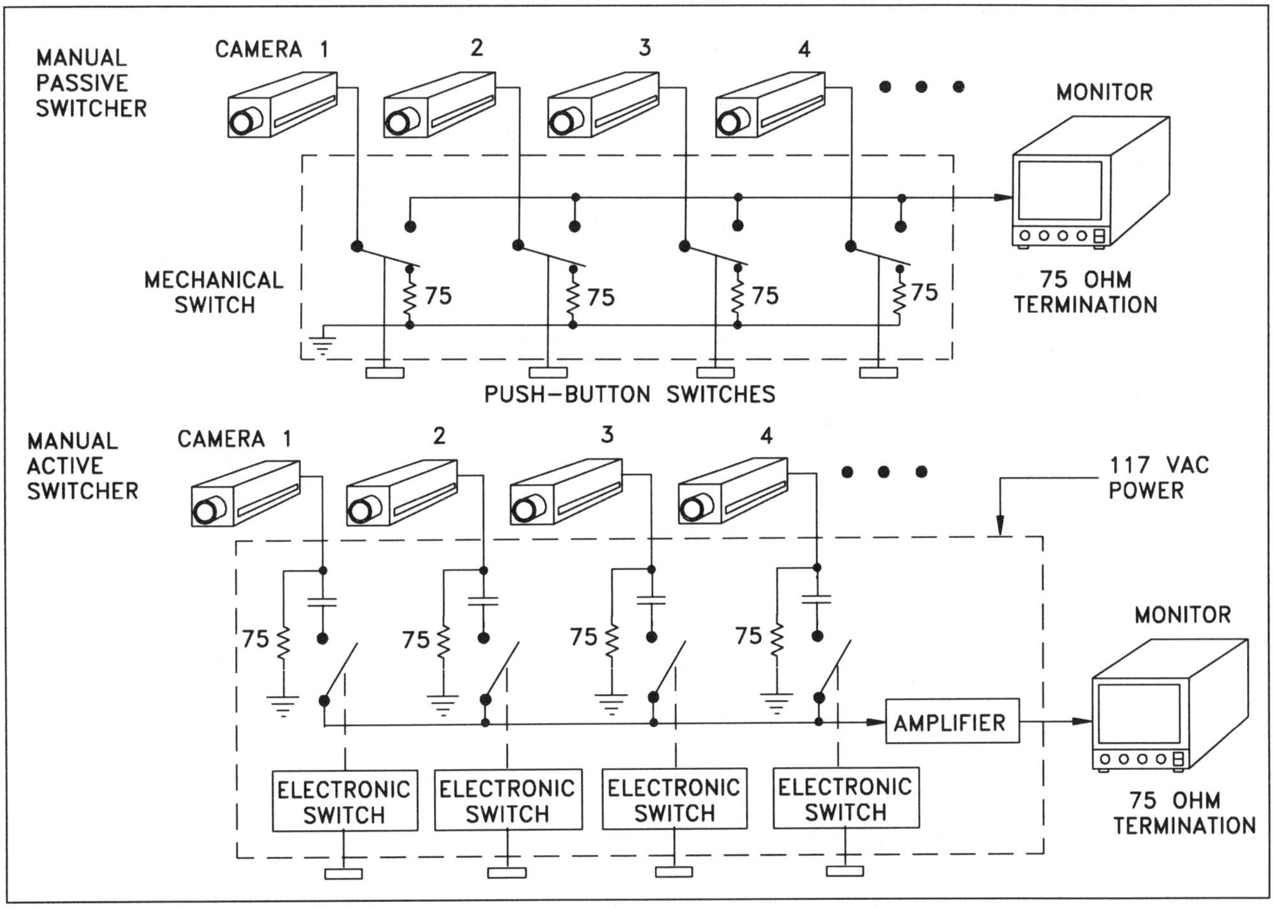

FIGURE 9-1 Manual passive and manual active switchers

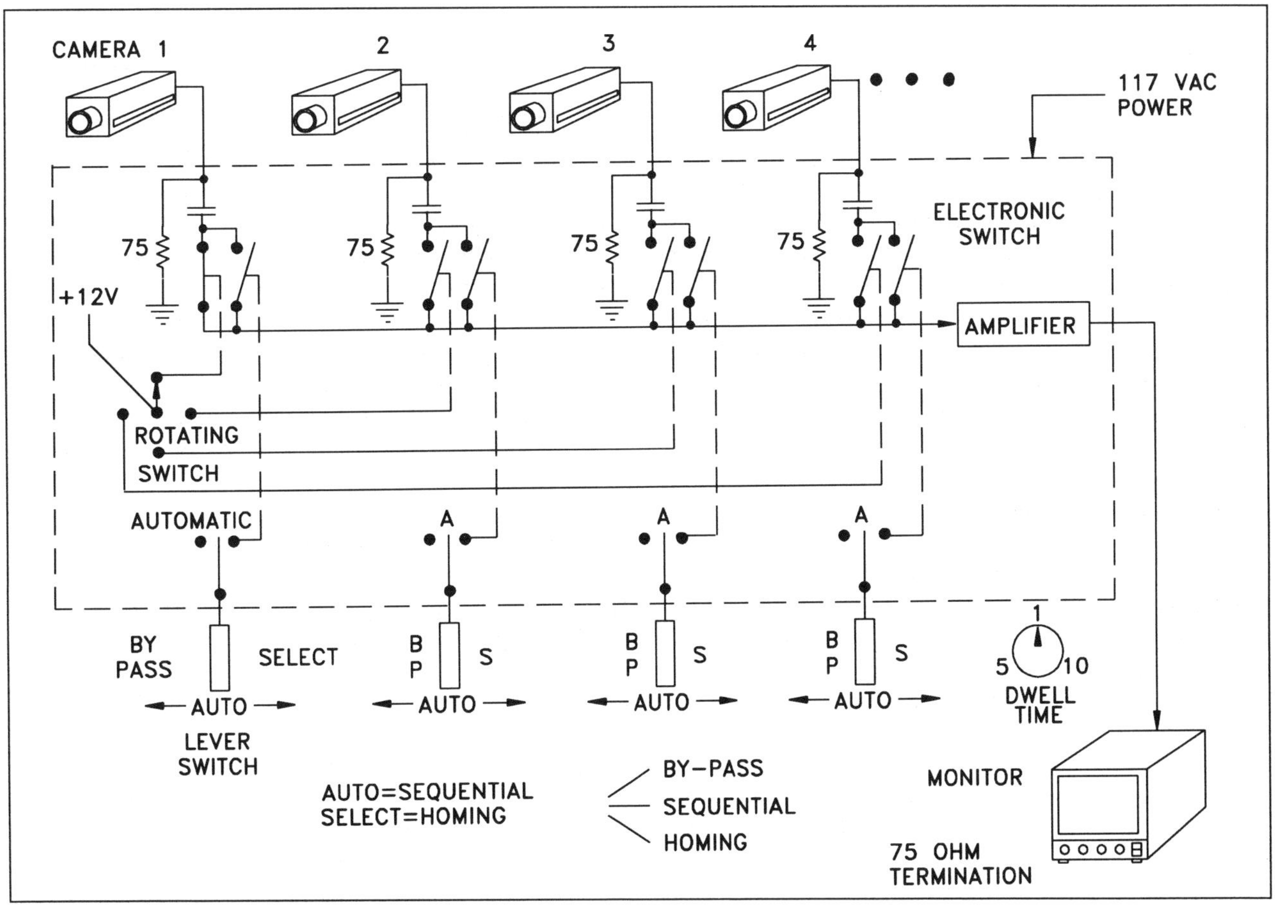

FIGURE 9-2 Homing sequential switcher

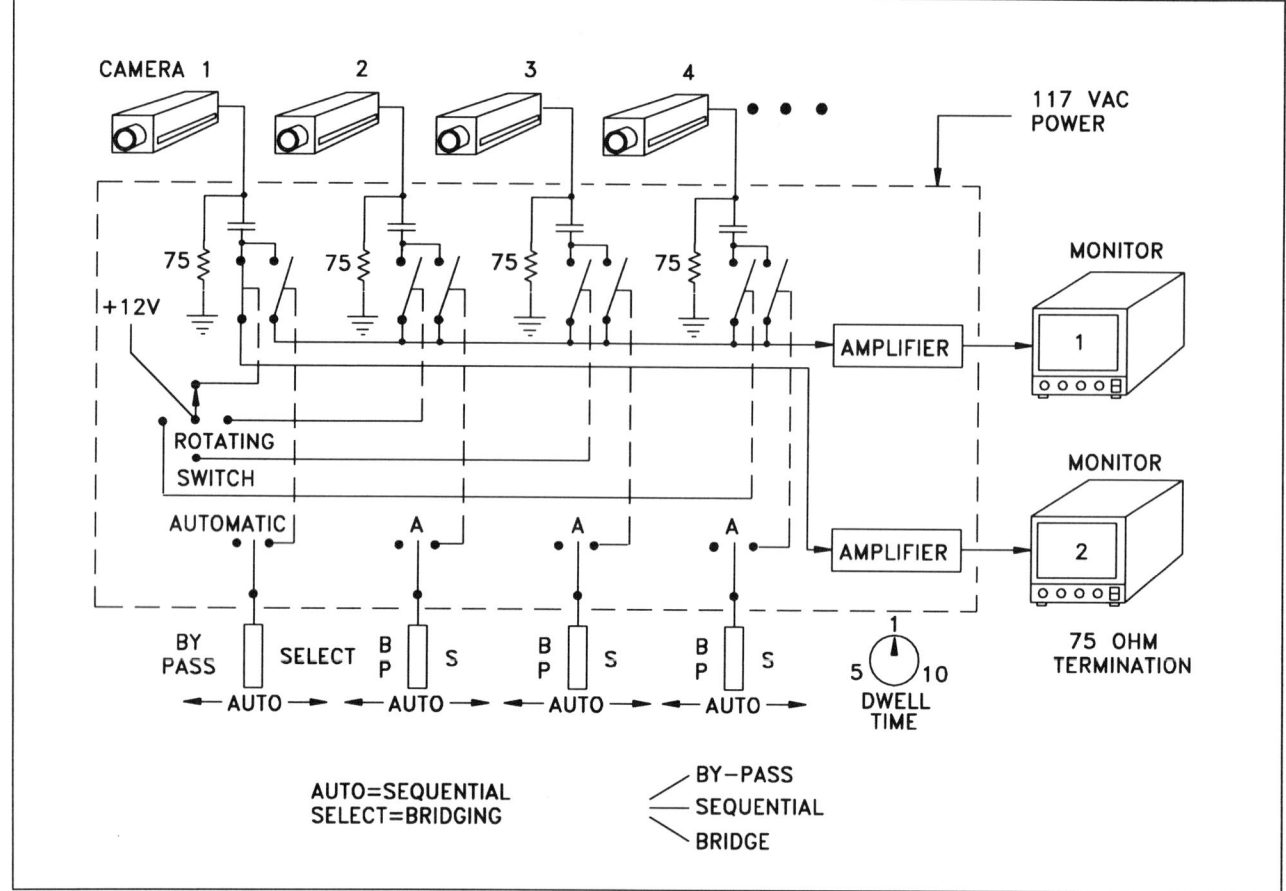

FIGURE 9-3 Bridging sequential switcher

operational functions: automatic sequencing, bypass, and homing (select). When a switch is set to Bypass, that particular camera is not displayed. When the switch is set to Homing, that camera picture is presented continuously on the monitor and in essence overrides the automatic sequencing function. This permits continuous observation of any particular camera at the operator's command. When the switch is in the Automatic position, all cameras are sequenced onto the monitor, one at a time.

9.2.3 Bridging Sequential

Bridging sequential switchers operate like the homing sequential switcher but have the additional feature that two monitors can display the CCTV cameras. Figure 9-3 shows the block diagram for a bridging sequential switcher.

Monitor 1 will always display the cameras selected for sequential viewing. Monitor 2 will display only the camera manually selected for detailed viewing. For instance, pressing a particular camera control switch to the down position puts the picture on the second or bridged monitor for detailed viewing, while the sequence of all cameras not

bypassed continues on the first monitor: that is, monitor 1 sees the switched sequence of cameras while monitor 2 sees a selected camera continuously. As with the homing sequential switcher, all camera inputs are properly terminated in 75 ohms.

The first monitor (the sequential monitor) functions as a homing sequential switcher. The bridging monitor displays whatever camera is manually selected. This allows the operator to maintain a system overview while viewing in detail the camera covering a scene of particular interest.

9.2.4 Looping-Homing Sequential

The looping-homing sequential switcher operates like the homing sequential switcher, with the additional feature that all camera inputs can be brought out to a second switcher or other device at another location (Figure 9-4).

This type of switcher has the ability to drive a second switcher, monitors, VCRs, and transmission devices for remote transmission, thereby providing video images at multiple locations for display or recording. Unlike other switchers, the looping-homing sequential switcher camera

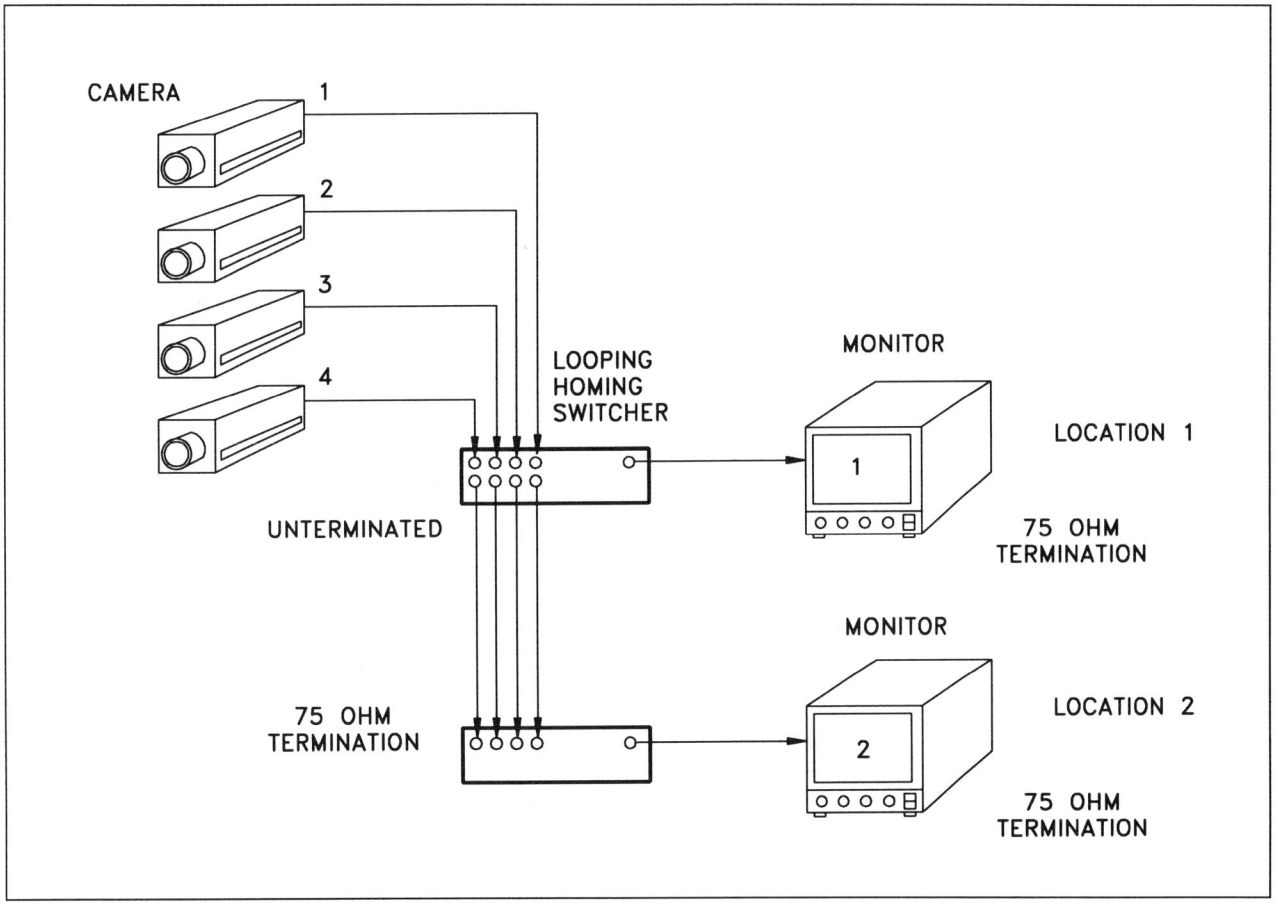

FIGURE 9-4 Looping-homing sequential switcher

inputs are not terminated, thereby allowing multiple devices to be connected to the switcher output. For proper operation, one of these devices, generally the last device in the line, is terminated in a 75-ohm impedance.

9.2.5 Looping-Bridging Sequential

The looping-bridging sequential switcher operates in the same manner as the bridging sequential switcher except that the looping feature is added. As with the looping-homing sequential switcher, the camera inputs are not terminated in the switcher.

Figure 9-5 shows the block diagram for looping-bridging sequential systems.

A looping switcher provides the ability to establish two independently controlled locations. Each station may select any camera for viewing without interfering with the operation of the other station.

9.2.6 Alarming

An alarming switcher automatically displays a picture on a monitor and/or starts a VCR each time it is activated by a camera motion detector or other alarm input (Figure 9-6).

These switchers are available in homing, bridging, looping-homing, and looping-bridging configurations.

When an alarm (or input) is received, a corresponding output signal is generated. This signal is transmitted to a VCR, which activates a switch closure in the VCR, keying the VCR to operate at a preselected speed.

9.2.7 Remote Sequential

The use of the homing, bridging, looping, and alarm versions of sequential switchers just described assumes that the distance between the camera locations and the monitor (control console) location is relatively short. In many installations, this is not the case and the cost becomes prohibitive to provide separate video coaxial cables from each camera to the distant monitor location. Remote sequential switchers overcome this problem. The remote sequential switcher consists of two parts, a control unit and a switching unit, and is available in all of the aforementioned versions to provide complete system design flexibility. Both units are connected by means of multiconductor cables, fiber optics, or, more commonly, a multiplexed frequency shift key (FSK) or RS-232 communications system (Figure 9-7).

The control unit is located near the monitor, and the switcher unit is located closest to the central location of all

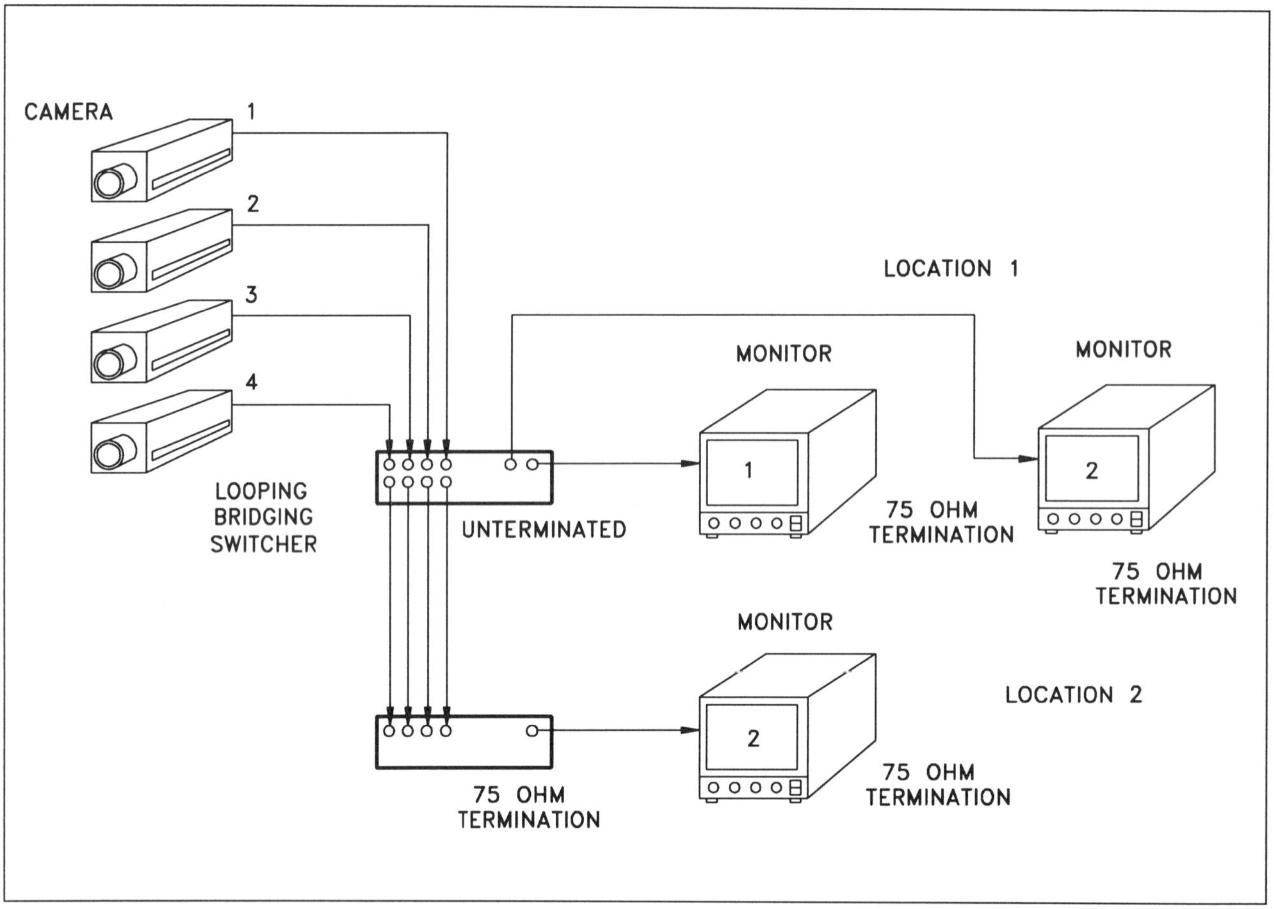

FIGURE 9-5 Looping-bridging sequential switcher

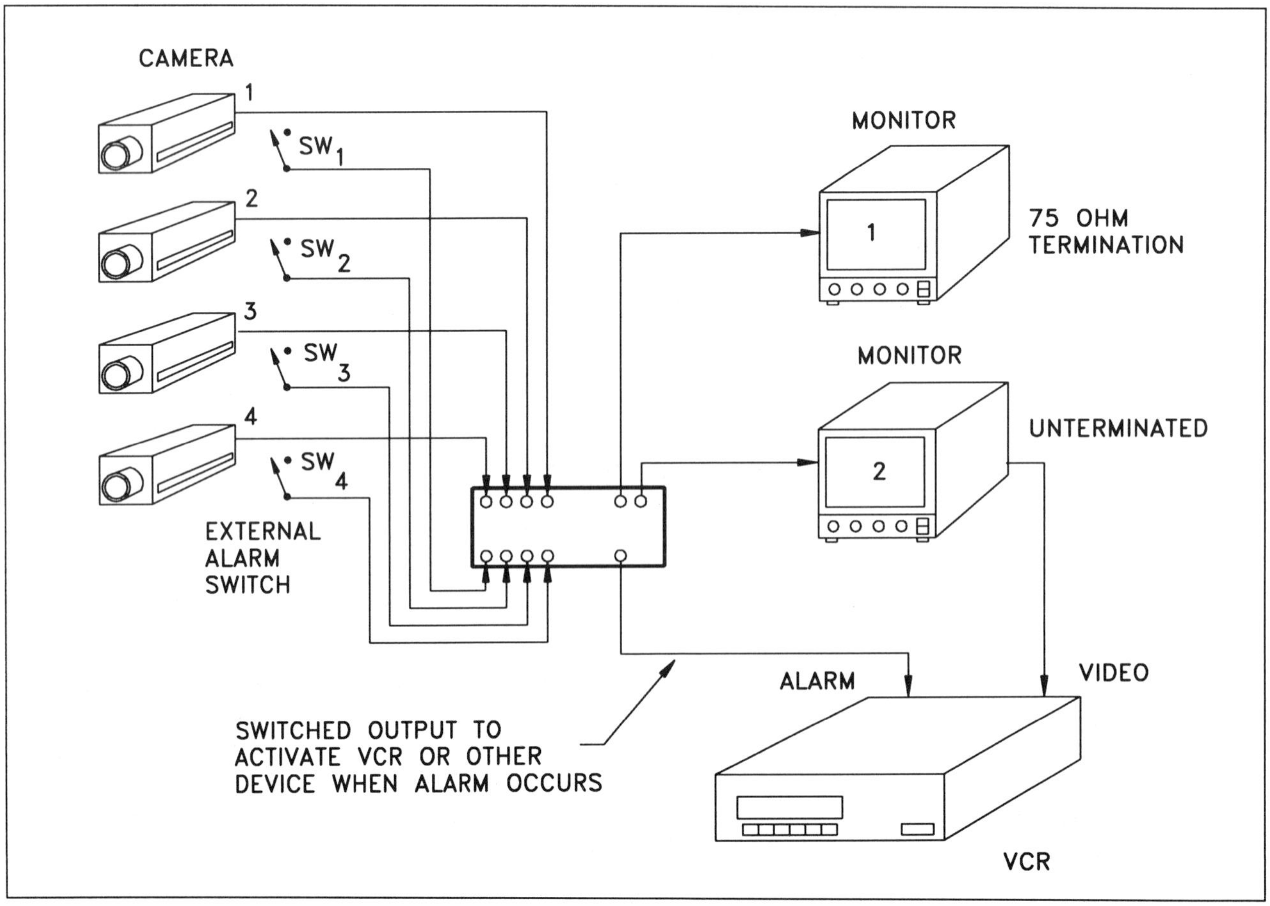

FIGURE 9-6 Alarming bridging sequential switcher

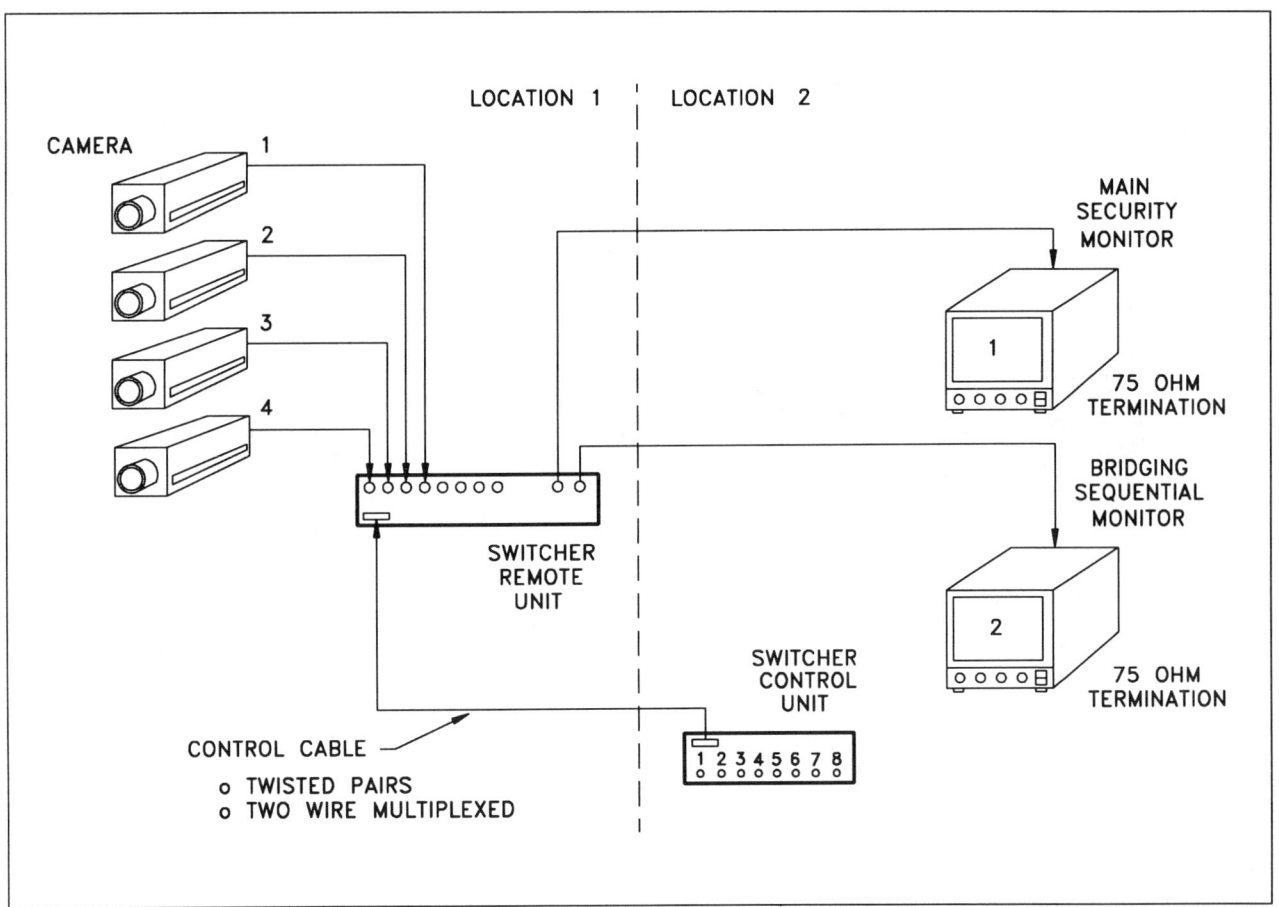

FIGURE 9-7 Remote homing sequential switcher

the cameras. The physical separation of the switching and control functions avoids the use of individual camera coaxial cables to the control console; each switcher requires only one or two coaxial cables for monitor input. The switching function is accomplished as follows. The control unit located at the monitor sends an audio, FSK-modulated signal via selected transmission medium to the remote unit located near the camera remote-switching unit. This audio control signal shifts between two frequencies according to the digital information transmitted, representing the desired switching function. The selected camera(s), after being switched in the remote switching unit, is then sent to the monitor via a single coaxial cable or other transmission medium. The remote bridging sequential switcher requires two output coaxial cables. The FSK-modulated signal technique is the most economical use of control between the camera and the monitor location.

9.2.8 Alarm-Programmed Sequential

The homing, bridging, and remote sequential switchers can be provided with an alarm feature. In the event of an external alarm switch closure caused by any type of sensor input, simple switch closure, IR source, video motion detec-

tor, or pressure transducer, the alarmed camera will automatically override the preselected video or will be auto- matically displayed on the second monitor, when a bridging type switcher is used. The automatic homing of the alarmed camera overrides any manually bridged display on the second monitor. The sequence of all cameras not bypassed continues on the first monitor. Simultaneously with this switching, an alarm contact within the switcher closes to operate a VCR, video printer or any other alarm-indicating equipment, if connected. Automatic alarm-programmed switchers are especially suitable for applications in which monitors are occasionally unmanned and use VCRs to record abnormal events. They are also particularly useful during off hours or over weekends when real-time or time-lapse recorders are used to monitor multiple cameras.

There are many reasons for recording video pictures from alarm inputs. Such alarms are caused by some event detected by sensors and trigger a TV response, such as displaying a camera scene on a monitor, activating a VCR, or changing a VCR mode from time-lapse to real-time. In a typical state-of-the-art system using alarm sensors of various kinds and CCTV motion detectors, the occurrence of the alarm event is processed via the sequential switcher or by a computer-controlled matrix switcher. For example, auto-

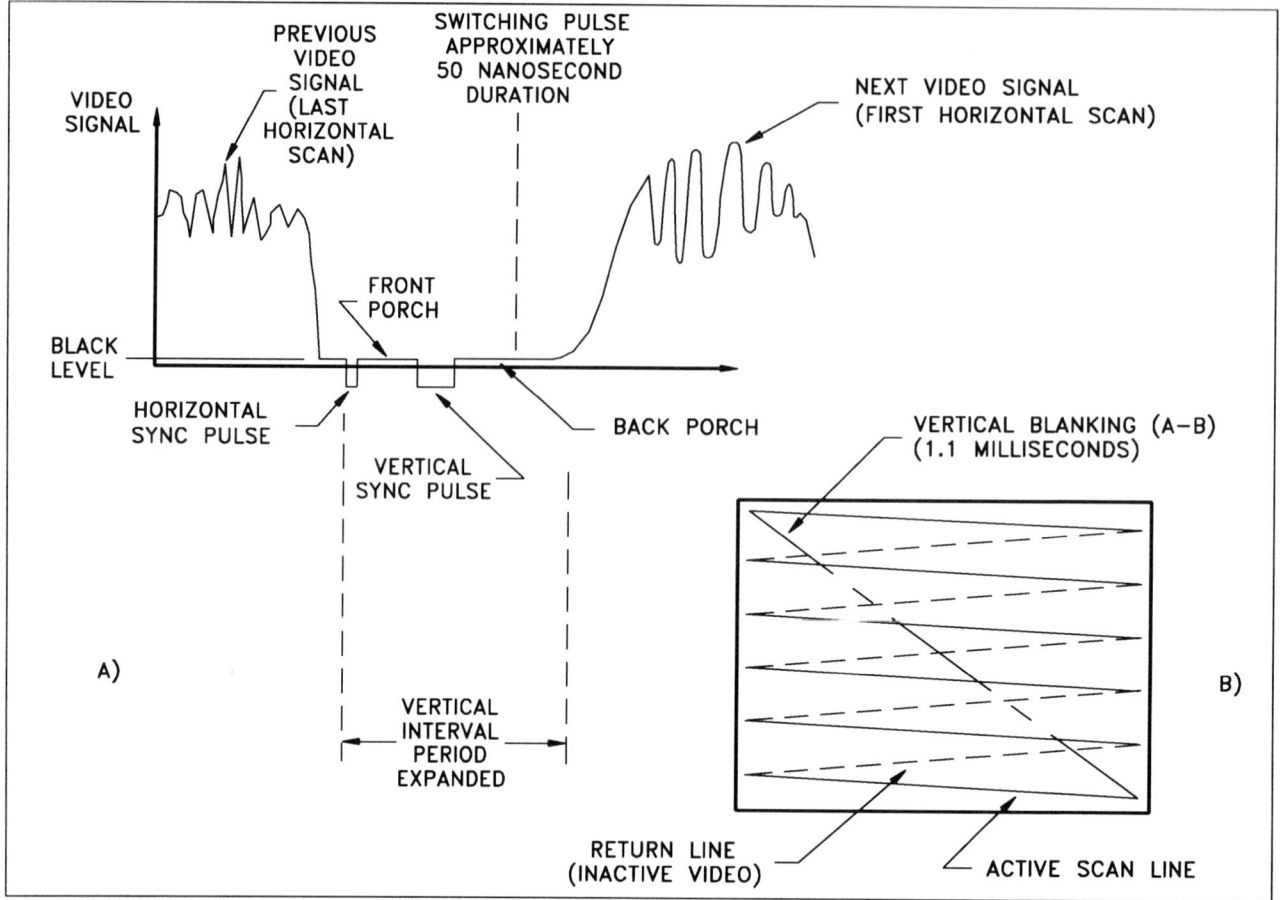

FIGURE 9-8 Vertical-interval switching

matic switching occurs to a specific video camera when a holdup alarm is activated by a bank teller. The alarm is received at the switcher, which identifies the camera providing coverage to the teller area affected, calls that camera up to the display or command monitor, and begins to record the event.

Another application is the use of a vehicle electromagnetic-loop detector or other alarm device, such as infrared or microwave, to signal the arrival of a person or vehicle at a controlled entry point or gate—and at the same time display the CCTV camera picture at that entry location. The scene showing the person is then recorded on a VCR. The output monitoring device could be a bell, light, or other signaling unit, which would notify a security guard to dispatch a guard to the scene or alert a guard at the scene. Even if there are multiple monitors affording the opportunity to observe all locations, the use of alarming switchers puts attention in areas where guard action is really required. The activation of the alarm signals a significant occurrence within the field of view covered by a particular camera.

9.2.9 Vertical Interval

There are two types of video signals that are switched: nonsynchronous and synchronous. Nonsynchronous signals involve the inherent discontinuity of timing pulses that cannot be corrected by special switching methods and show up as noise disturbances in the picture. Synchronous signals lend themselves to methods of switching where controlled transition maintains a degree of signal continuity and provides a clean, noise-free video picture during switching.

Timing, as related to the actual switching interval and the time and manner in which it takes place, is also an important factor in television camera switching. If there are long breaks in the video signal, streaks, a momentary black screen, or other picture irregularities can occur. When switching composite video signals containing picture information and synchronizing pulses, a break may occur during the synchronizing time and the synchronization signal may be completely lost. This results in picture rolling or tearing when the picture from the next camera appears and is an unacceptable state. The solution to ensure clean video

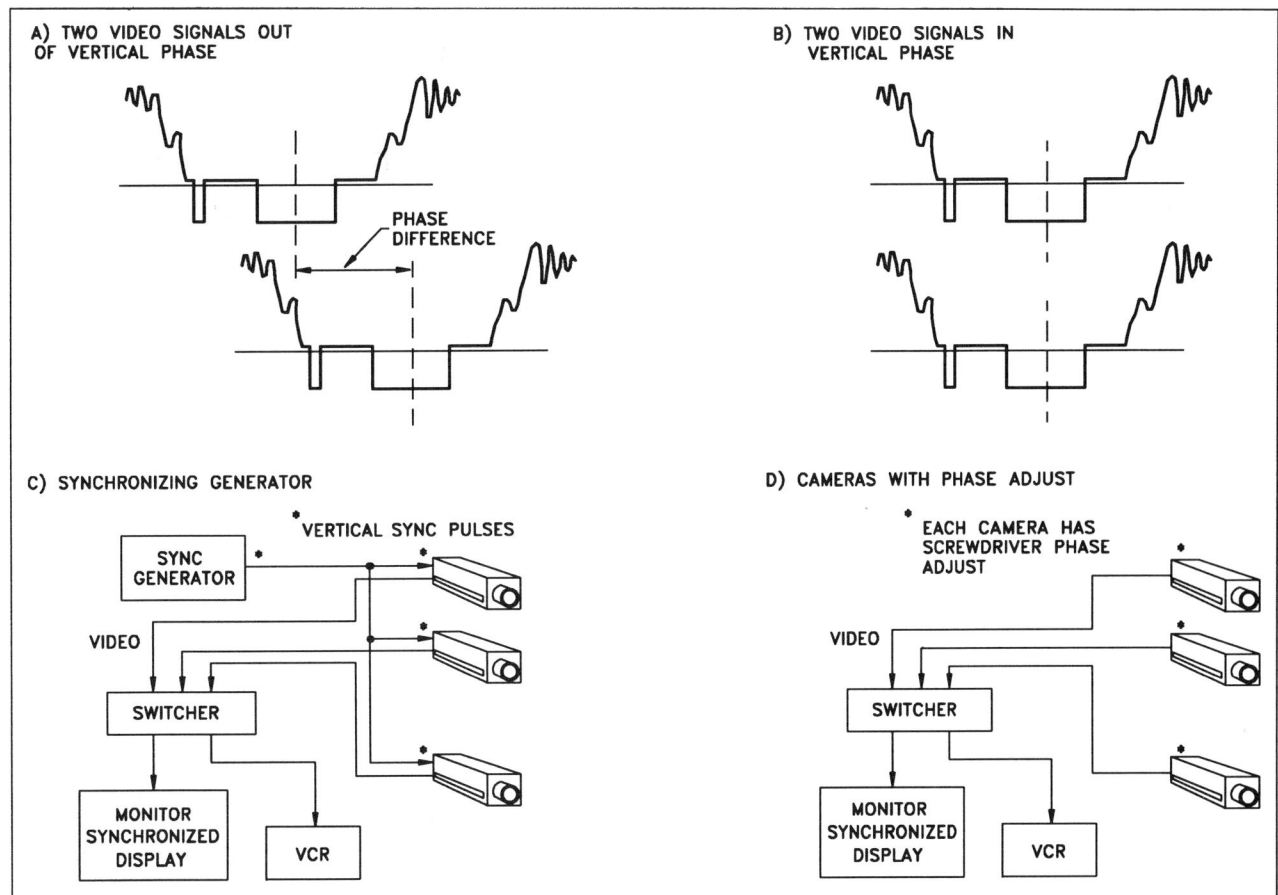

FIGURE 9-9   Sequential switching synchronization

camera switching is vertical-interval switching. With this method, the switching is allowed to occur only during the vertical interval in the television signal between picture frames (Figure 9-8), during which time no picture information is being transmitted.

Since no visible picture is displayed on the monitor during the vertical-interval switching time, switching during this period does not cause picture interruption or deterioration. This technique permits switching from one camera to the next with no noise or interruption of intelligence.

To understand vertical-interval switching, refer to Figure 2-16. The camera video signal is generated by periodically scanning the tube or solid-state camera sensor. For the solid-state sensor, the camera clocking signal scans from left to right and reads out the electrical information representing the light image on the sensor. When the clocking signal reaches the right side of the sensor, it returns to the left side and begins another scan. During the return time, the clocking signal is addressed down two rows of sensor pixels. After it completes 262 1/2 scans (one-half of the full frame), the clocking signal reaches the bottom of the sensor and returns to the top. The clocking signal then

scans the alternate pixel rows; after completing the second scan, the full sensor has been read out. The return trip represented by the dotted line (Figure 9-8) takes about 1.1 milliseconds and is referred to as vertical blanking, since during this time no video signal is generated. In summary, the picture information occurs during the left-to-right and top-to-bottom scanning, and both the horizontal and vertical synchronization occur during the right-to-left scanning and the vertical blanking in between scans.

In the typical CCTV surveillance application, cameras will not be synchronized. While they may have waveforms or signals like Figure 9-8, the time relationship between them is not synchronized nor in phase (Figure 9-9).

Since the synchronization pulses from each camera occur at different times, when the switcher switches from one camera signal to the next, a noticeably scrambled or distorted nonsynchronized image occurs as the monitor tries to adjust to the synchronization pulses of the new signal. While a temporarily distorted picture might be tolerable in some simple direct-viewing applications, in situations where there are many cameras or the information is recorded on a VCR, the result is unsatisfactory. Since VCR tape drives use the camera synchronizing pulses to synchronize the tape, it may take several seconds for the VCR to synchronize to the new camera signal, thereby recording useless video frames for several seconds each time the switcher is switched. The signals shown in Figure 9-9a are said to be out of phase, a situation that is correctable with at least two methods.

One technique for producing in-phase signals (Figure 9-9b) is to install a synchronizing generator, which provides a synchronizing signal to the cameras to assure that they are all in the same phase and operating at the same frequency (synchronized). As an alternative, some cameras can be adjusted so that the phase is the same for each camera. Phasing each camera to be the same does not produce a clean switchover, however. Even though the signals may be in phase, if the switching occurs during the video portion of the signal, there will be visible transient effects such as spikes and flashes and other deleterious effects on the monitor or recorder-reproduced picture. To eliminate this problem, the switcher is designed to switch during the vertical interval (Figure 9-8), hence the name vertical-interval switching.

In operation, the switcher circuitry detects the vertical interval in the signal and delays the actual switchover from one camera to the next to the time when vertical blanking is occurring. By using this method, no transient effects are visible on the monitor or in the recorded image, thus producing successful switching. While the vertical-interval switching technique may not be important in simple systems, it is extremely important in medium-to-large systems, and in any system using a VCR.

9.2.10 Choosing the Correct Switcher

The following paragraphs suggest the progression in switching complexity when starting with a simple security application and upgrading to a very complex application.

1. **Passive Switcher.** The manual passive switcher is the simplest switcher and has the ability (depending on model) to switch 4, 8, 16, or 32 cameras and display any one of them on a single monitor. The switcher is available in either passive or active type, which employ different kinds of electrical switches. In a simple application, any one of the input cameras can be presented on the single monitor, one at a time, through manual switching by the security guard. The only difference noticeable on the monitor between the active and passive switcher will be that the passive switcher monitor will display various noise transients while the active will provide a clean switchover between cameras.

2. **Sequential Switcher.** The sequential switcher is used when it is necessary to switch automatically from camera to camera so that the guard can observe all camera scenes periodically. This switcher could be used in the previous simple example, as well as in more complex applications, where more cameras are used. As in the manual active switcher, the electronic circuitry provides fast, clean switching with no transients on the screen and is available with camera dwell times of 1 to 50 or 60 seconds, depending on the adjustment made by the operator.

3. **Homing Sequential Switcher.** The homing sequential switcher has the additional feature of permitting the operator to stop and look at one particular camera picture continuously or sequentially display all the camera pictures with a dwell time set (adjusted) by the operator. This system permits the operator to continuously scan through all the cameras and simultaneously pick out one camera and view it continuously. In the sequential mode, the dwell time, or time for which any particular camera is viewed, is independently adjustable for each camera and provides the operator with flexibility to view different camera scenes for different periods of time. The homing sequential switcher provides the operator with three options and adjustments: (1) automatic switching, (2) timing, (3) and bypass control.

4. **Bridging Sequential Switcher.** The bridging sequential switcher has two separate outputs for two monitors. One output is for the programmed sequence of cameras; the second is for the continuous display of a single camera. Unlike the homing sequential switcher, the bridging sequential switcher provides this constant viewing of a selected input without giving up the overview of all the camera scenes provided by the sequential program. With the bridging sequential switcher, if the operator wants to observe a particular camera area continuously, he moves the switch to Select, thereby displaying that camera picture on the monitor continuously while simultaneously the other monitor continues to display the sequentially switched camera sequence, including the camera that is displayed on the second monitor continuously.

9.3 MATRIX SWITCHER—MICROCOMPUTER CONTROL CONSOLE

Microprocessors and microcomputers have revolutionized the security field. There are microcomputer-based control systems for preprogramming and automating—camera pan/tilt, preset camera pointing, and automatic switching functions—all in one console unit. Prior to microcomputer systems, these functions were provided by manually programmed dedicated systems, offering only limited flexibility. The new microcomputer-based systems have substantially expanded capabilities, ease of operation, and far greater flexibility in one single system.

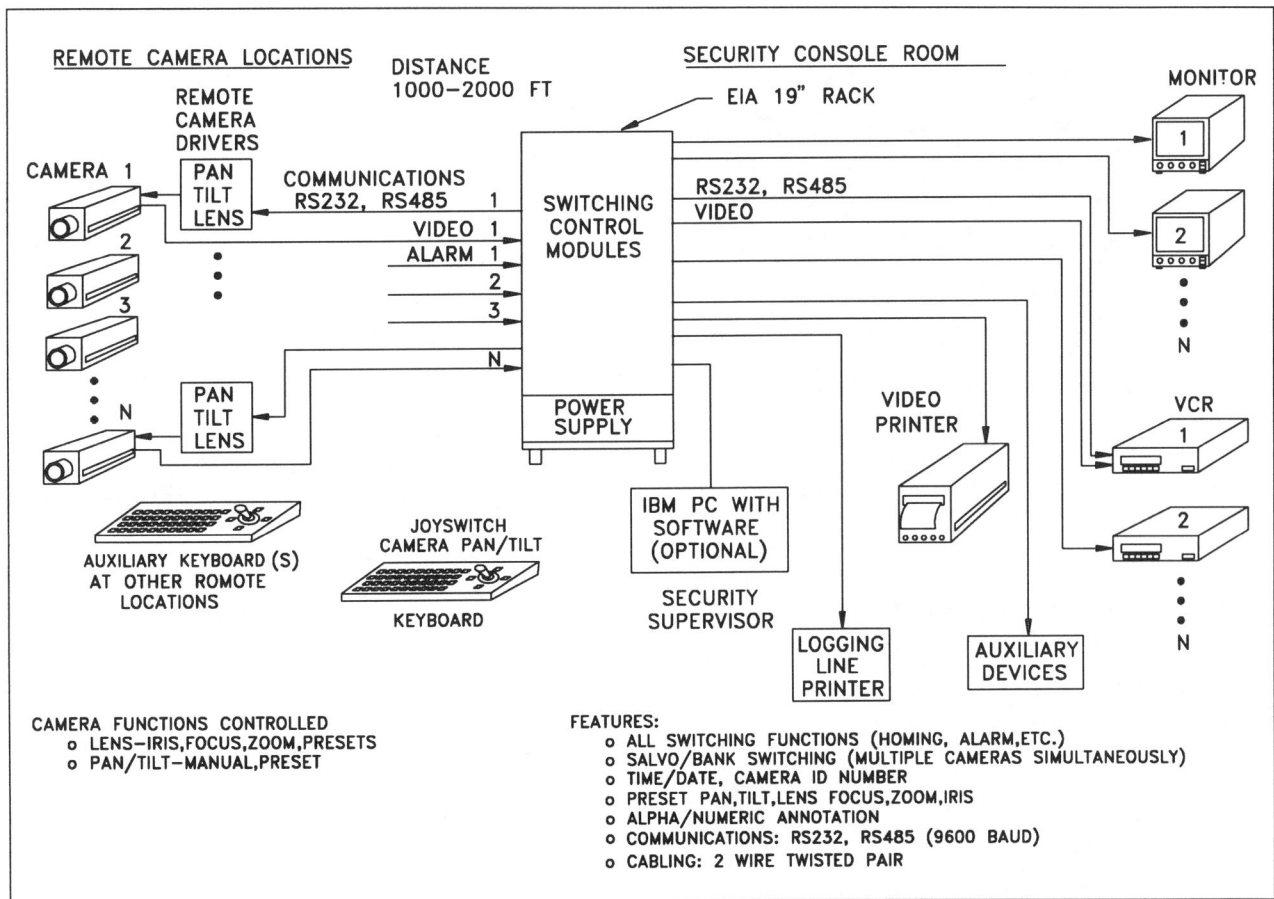

FIGURE 9-10   Configurable microprocessor-controlled video switching system

9.3.1 Specialized Switchers

Specialized switchers, in which several security functions have been integrated into a medium-size, dedicated microprocessor-controlled video security system, combine the functions of an alarming switcher, a digital motion detector, time/date generator, and video printer.

One system accepts up to eight camera inputs and allows the security operator to choose via manual or programmable automatic sequential switching. Each camera dwell time is adjusted by the user from 2 to 60 seconds. Each individual camera channel mode setting can be changed to permit any camera input to be processed as (1) monitor only, (2) monitor plus alarm, (3) monitor/alarm plus print, or (4) camera off. All of the eight camera video channels are processed simultaneously by the microprocessor central processing unit (CPU) to digitally detect movement in discrete target zones (user-programmable from 1 to 4) within each picture channel. The size, shape, and location of the target zones are user-adjustable, as are the sensitivity levels for each individual video channel. When motion is sensed in any of the target areas, there is an audible signal and a visible

alarm. If so adjusted, in 6 seconds it prints a hard copy of the picture on the monitor that caused the disturbance or the motion, annotated with date, time, and camera number. If there is motion or activity in more than one area, the device can store up to four images at one time in memory and subsequently print each one. While this system has capability similar to other microprocessor-controlled systems, it conveniently combines in one unit the switching capability, motion detection, and hard-copy printout.

9.3.2 Large Switchers

When a security system reaches a higher level of capability—24 to 32 or more cameras and associated monitors—it becomes more efficient to use a configurable microprocessor-controlled video switching and control system (Figure 9-10).

Several manufacturers produce systems that can switch and control up to 255 cameras and 32 monitors. These systems are built in modular form with removable printed-

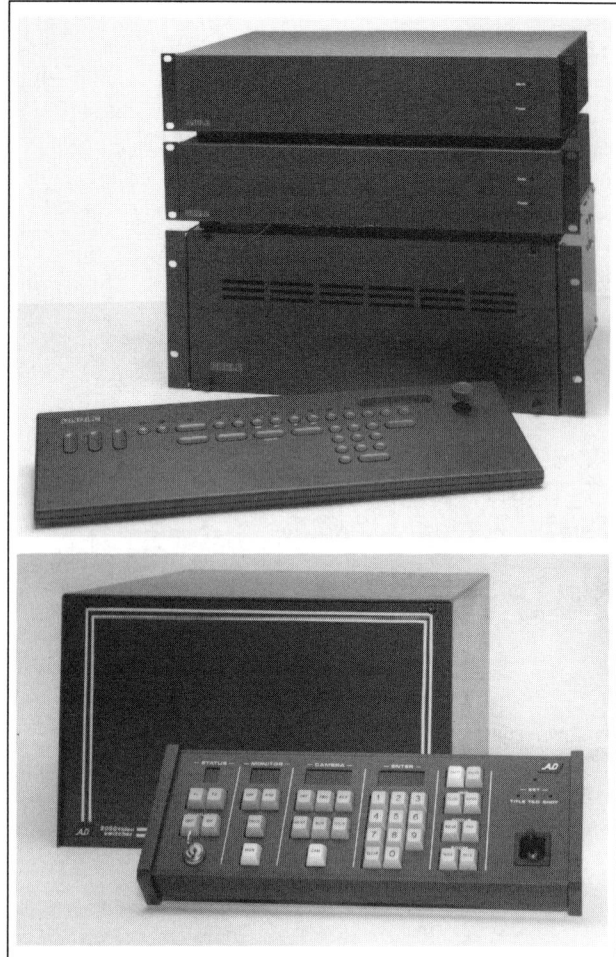

FIGURE 9-11   Microcomputer switching systems

circuit boards and rack-mounted modules, permitting the user to begin with the basic system and expand when necessary. The removable modules and plug-in units also provide on-line serviceability and reduce or eliminate system downtime. The control systems are divided into several subchassis or modules, including:

1. a keyboard and joystick desktop console;
2. a rack-mounted card cage chassis housing the multiple submodules for the switching and control functions;
3. a power supply; and
4. remote modules located near the cameras for driving the camera, lens, and pan/tilt hardware, as well as for communicating the information to the control unit.

The control unit contains the system software and microprocessor hardware. In some systems, customized switching programs can be included in the hardware using electrically programmable memories (EPROM). These solid-state memory devices allow storage of switching instructions to be used at a later time when automatic sequencing is desired. Most systems feature a time/date display on the

monitors, full identification of the camera ID number, and 20 to 40 alphanumeric characters for camera location information or other pertinent data. A typical titler card provides each video input with time and date, a three-digit identification number, and a user-programmable 40-character alphanumeric message displayed in two 20-character lines.

Communication from the console to the remote control camera module is via RS-232 or RS-485 communication protocol at a 9600-baud data rate. Distances between the control console and remote console can be 1000 to 5000 feet, with the data signal cable a single twisted-pair, 22-AWG shielded wire. Most equipment is housed in 5-to-7-inch-high EIA 19-inch rack modules, thereby removing most of the electronics from the desktop area except for the keyboard. Some systems have the ability to connect several keyboards to the same control system, thereby permitting control of the system from several locations.

All basic microprocessor-controlled systems have the capability for manual, homing, looping, sequential, auto alarming, bridging, and remote switching functions. A unique feature called salvo switching allows the operator to switch a selected bank of cameras into a bank of monitors as a synchronized group—all of the monitors switched together, in step, with one dwell time for each step. The unique salvo switching feature allows the operator to view all scenes in one general area, such as a single floor in a building, before switching to the next floor. This feature can significantly increase the monitoring ability of the security guard, since it automatically switches a logical array of cameras.

These systems can provide the same control over alarm functions as over the video network functions. The alarms are constantly monitored by the microprocessor control console. If one or more of the alarms is activated, the system automatically switches in the camera nearest the alarm and displays its video scene on the appropriate monitor. The types of alarms accommodated include switches, infrared sensors, and television motion detectors. Other alarm signals can be monitored via an audible tone alert and a VCR activated to real-time from time-lapse to record the alarm, all automatically. The operator has the ability to bypass or restore cameras and alarms at will. Individual camera dwell times and sequencing times can be set by the operator on all cameras.

For the purpose of recording particular actions or alarms occurring on the system, as well as the system status, microprocessor systems have a printer port for logging a printed record of the systems' status as changes occur. In large systems, it is possible to program camera, monitor, recorder, and other system functions and hardware via the use of a PC and a software package. By this means, the system can be customized to suit a specific security application. To provide additional safeguards, system passwords can be programmed and lockout tables used, thereby limiting access of unauthorized personnel. In addition to the

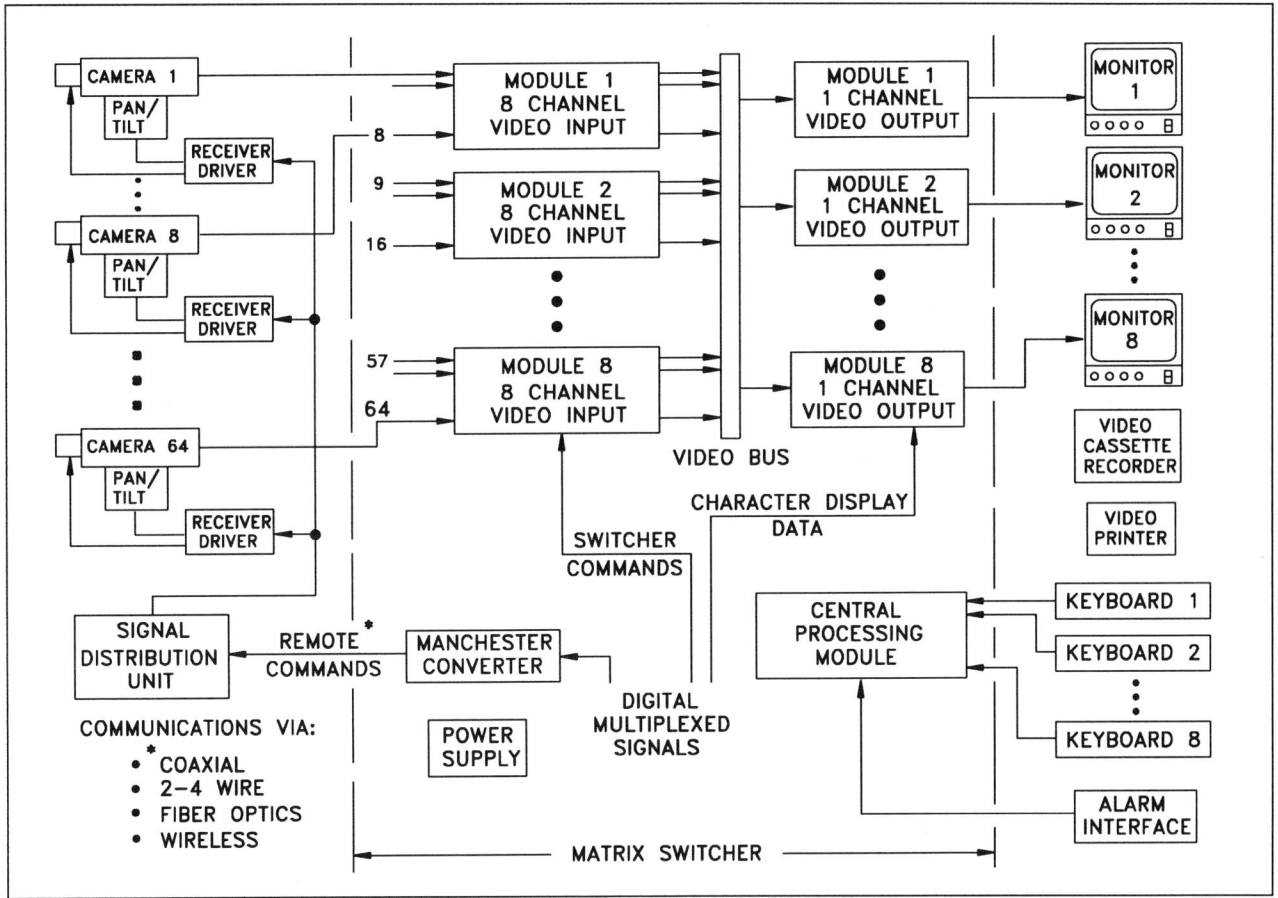

FIGURE 9-12 Matrix switcher block diagram

operational switching sequences normally entered from the PC keyboard, complex switching sequences can be programmed off-line using the PC and then downloaded to the microprocessor control system. Examples of such complex switching might include many pan/tilt presets for camera pointing position, and lens iris, zoom focal length, and focus settings.

These functions are accomplished via receiving modules located at the camera sites, which receive the control functions via the control unit and RS-232 communications cable. The computer-controlled console can memorize preset pan/tilt and camera/lens settings such as azimuth, elevation, shutter settings, and zoom lens parameters. By this means a camera is redirected and focused on a particular part of its total FOV for the purpose of perhaps identifying someone passing through an area or access location at a specific time. This function is accomplished manually by the operator by selecting a specific camera and preset number or automatically preprogramming so that if an alarm occurs at a location in the scene, the camera automatically goes to the preset condition. Simultaneously, a VCR is activated into real-time mode and records the activity at the designated preset camera position. A unique feature in one system provides a dither (small movement),

which protects intensified tube cameras from producing a burned-in image (not required for solid-state sensors). It does this by periodically panning the camera a few degrees left and right, thereby changing the image seen by the camera and thus avoiding any degradation of the camera tube.

In some applications in which an audio signal must be switched, a separate microprocessor system is used to switch the audio signal simultaneously with the video from a camera. This provides full audio and video communications between the guard console operator and another person at the remote camera location.

Figure 9-11 shows examples of state-of-the-art microcomputer switching systems used in large security applications having hundreds of cameras and dozens of monitors, VCRs, video disk recorders, and printers.

These microcomputer switchers require special knowledge and experience to specify and configure, but when properly chosen, they provide a powerful, cost-effective solution.

Figure 9-12 shows a block diagram of a typical matrix switcher used in a large security installation.

All cameras, lenses, pan/tilt platforms, monitors, recorders, and printers are controlled, monitored, and switched via the central matrix switcher. The matrix switcher communicates control functions to the various hardware via RS-232 or RS-485 protocol or time-multiplexed signals. Video signals from the cameras are transmitted from the remote locations via individual coaxial, two-wire, fiber-optic, or wireless channels. The matrix switcher has a separate video input connector for each camera. It likewise has a separate output connector for each monitor, recorder, or printer device. When required, two-way intercoms can be added to provide simultaneous audio and video switching.

To bring the matrix switcher and camera and monitoring equipment on-line, the matrix switcher must be "configured" or programmed according to the hardware connected to it and the required functioning of the system. This can take hours or days to accomplish and requires a detailed plan with methodical procedures.

9.4 SUMMARY

The heart of a good security system is a highly functional video switching control unit. The cameras, monitors, and recorders have the ability to view, monitor, and record visual intelligence. During the design phase of any security system, management and security personnel must decide what information needs to be displayed, recorded, and printed—and then choose the switching system to accomplish it.

Chapter 10
Camera Pan/Tilt Mechanisms

CONTENTS

10.1 OVERVIEW

One of the many camera and lens accessories available to complete a CCTV security system is the pan/tilt mechanism. The pan/tilt mechanism rotates and tilts the camera to point it in a specified direction. These electromechanical platforms are available for lightweight, indoor, small-camera applications as well as for large, outdoor, heavy camera/lens installations. Pan/tilt mechanisms are designed to operate in manual or automatic mode, using a remote control joystick mounted on a control console. Figure 10-1 shows examples of indoor and outdoor pan/tilt units.

While the CCTV lens and camera can see and point the way the human eye does, but at a remote distance, security circumstances arise when the camera/lens combination must be redirected in order to see a specific activity or scene. The console-mounted pan/tilt controller remotely directs the pan/tilt unit and the camera pointing direction, rotating it in azimuth (panning) and elevation (tilting), like a person moving his head and looking in different directions. Most pan/tilt units provide from 270- to 360-degree azimuth panning and from 0- (horizontal) to 90-degree (directly downward) tilting. To control the pan/tilt unit, the security operator uses a lever switch, joystick, or thumb trackball at the control console. Figure 10-2 shows some typical pan/tilt controllers.

The joystick is moved forward or backward, left or right corresponding to the vertical (tilt) and horizontal (pan) movements of the camera/lens unit. Likewise, the thumb

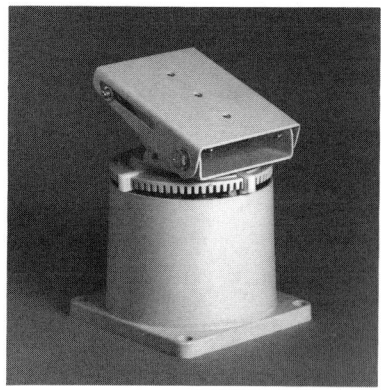

INDOOR

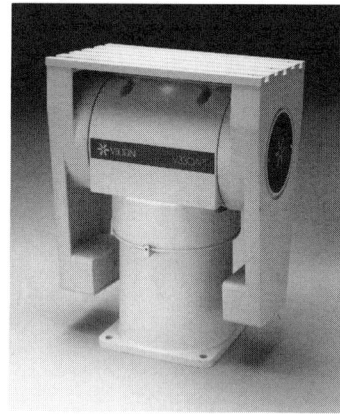

OUTDOOR

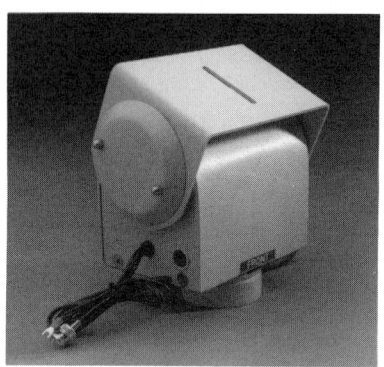

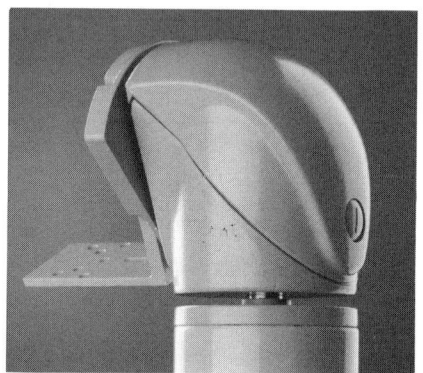

FIGURE 10-1 Indoor and outdoor pan/tilt mechanisms

A) FOUR MOMENTARY SWITCHES

B) FLAT DISK

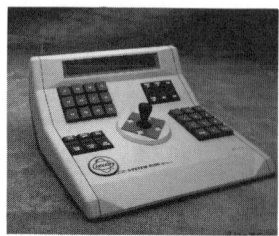

C) JOYSTICK

PAN/TILT SPEED DIRECTLY
PROPORTIONAL TO AMOUNT
JOY STICK IS PUSHED OFF
CENTER POSITION

FIGURE 10-2 Pan/tilt controllers

trackball is moved using the thumb to initiate pan and tilt. For the disk and four-button design, left and right initiate panning, and top and bottom initiate tilt. The CCTV lens and camera (and housing if used) are secured to a small platform on the top or side of the pan/tilt mechanism. The platform and camera move in accordance with the control signal sent by the control unit. By watching the television monitor, the security officer can orient the camera to view any area within the pan/tilt rotation range.

Pan/tilt platforms are available in a wide range of sizes and functions, from lightweight pan-only units (scanners) for indoor applications to heavy-duty, environmentally protected pan/tilts for outdoor applications. Outdoor ruggedized units are environmentally enclosed and are intended to withstand severe conditions of temperature, precipitation, dust, and humidity.

There are three primary applications for pan/tilt units: (1) monitoring a fixed or point site, (2) monitoring an area site, and (3) monitoring a volume site. A point site might correspond to a bank teller, where the area of interest is the teller, customer, and the business transaction. An area site might be a building lobby, where the activity in the lobby is of interest. A volume site might be a warehouse, where many vertical bays (levels) must be under surveillance at different times. For the fixed or point site, a single fixed camera and lens are suitable to view the entire area. In the area and volume cases, a single fixed camera cannot see the entire area of interest, so a panning and tilting camera and a zoom lens are required.

Applications that use pan/tilt mechanisms can be roughly classified into indoor and outdoor types, mounted in either an overt configuration—without a concealing dome or housing—or a covert application, where the pan/tilt mechanism and lens/camera unit are hidden in a dome, hidden behind some visually opaque window, or camouflaged by some other means. A common type used in indoor applications consists of a panning platform that rotates the camera via direct operator control or automatically pans from side to side to cover the specific area. A second type is the pan/tilt unit in which panning and tilting are done manually (under operator control) or automatically. Both types can be augmented with a remotely controlled zoom lens to change the FOV and therefore the instantaneous camera coverage within the full range of the scanner or pan/tilt unit.

In outdoor applications, the pan/tilt mechanism and the housing containing the camera/lens must be designed to withstand wind loading, precipitation, dust, dirt, and all types of vandalism. In an outdoor application, the lens and camera can be enclosed in an environmentally shielded housing and mounted on an environmental outdoor pan/tilt mechanism; or the lens, camera, and indoor pan/tilt mechanism can be installed in an environmentally enclosed housing or dome. The advantages and disadvantages of these two techniques are discussed later.

Complete CCTV lens/camera, pan/tilt/housing systems have been introduced, in which all these parts are integrated into a custom housing or dome and controlled from a remote console location. While this combination provides an efficient, small, integrated system, it is often more expensive than using the simpler modular design. Typical systems are discussed in the following sections.

In a parking lot or building exterior, pan/tilt lens/camera systems are mounted on a pedestal or on a wall or ceiling, with mounting brackets.

For covert applications, pan/tilt zoom systems are available with only a small lens port; all pan/tilt and zoom camera/lens hardware is hidden behind a wall or ceiling.

The transmission of video and the control of pan/tilt mechanism movement and lens function is accomplished through separate video and control cables, or through a multiplexed communication channel in which the video signal and the controls are combined on a single coaxial cable, thereby eliminating additional wiring (Figure 10-3).

The choice of method is determined by the specific application. In small CCTV systems, separate video and control cables are usually less expensive, since additional modulation, demodulation, and transmitter/receiver units are required when the control signals are multiplexed with the video signals. In larger installations, multiplexing has an advantage, since long cable runs are more expensive than extra electronic equipment. Consult the manufacturer and identify the factors specific to the application and the features of the equipment to determine what technique to use.

10.2 SPATIAL COVERAGE REQUIRED

The intelligence the security guard obtains from the monitor picture is only as good as the information the camera/lens combination sees. It is therefore important in any security application to determine the viewing requirements of each camera/lens to select the optimum combination for transmitting maximum intelligence to the operator. One of the questions to ask in any CCTV security system is whether a camera will be mounted on a fixed mount—that is, pointed in only one direction—or on a pan, tilt, or pan/tilt platform, so the camera can be pointed in different directions at different times (Figure 10-4). A good analogy is the seeing ability of a human: When we look in a single direction, we see all of the activity in that direction all of the time. When we move our eyes or head in another direction within a large FOV, we cannot see the entire scene at one time, and therefore cannot see all of the activity in the entire FOV that our eyes and our head can see in totality. We and the television camera in essence time-share the viewing of the scene, spending a fraction of time looking at any particular portion of the total scene area. In order to

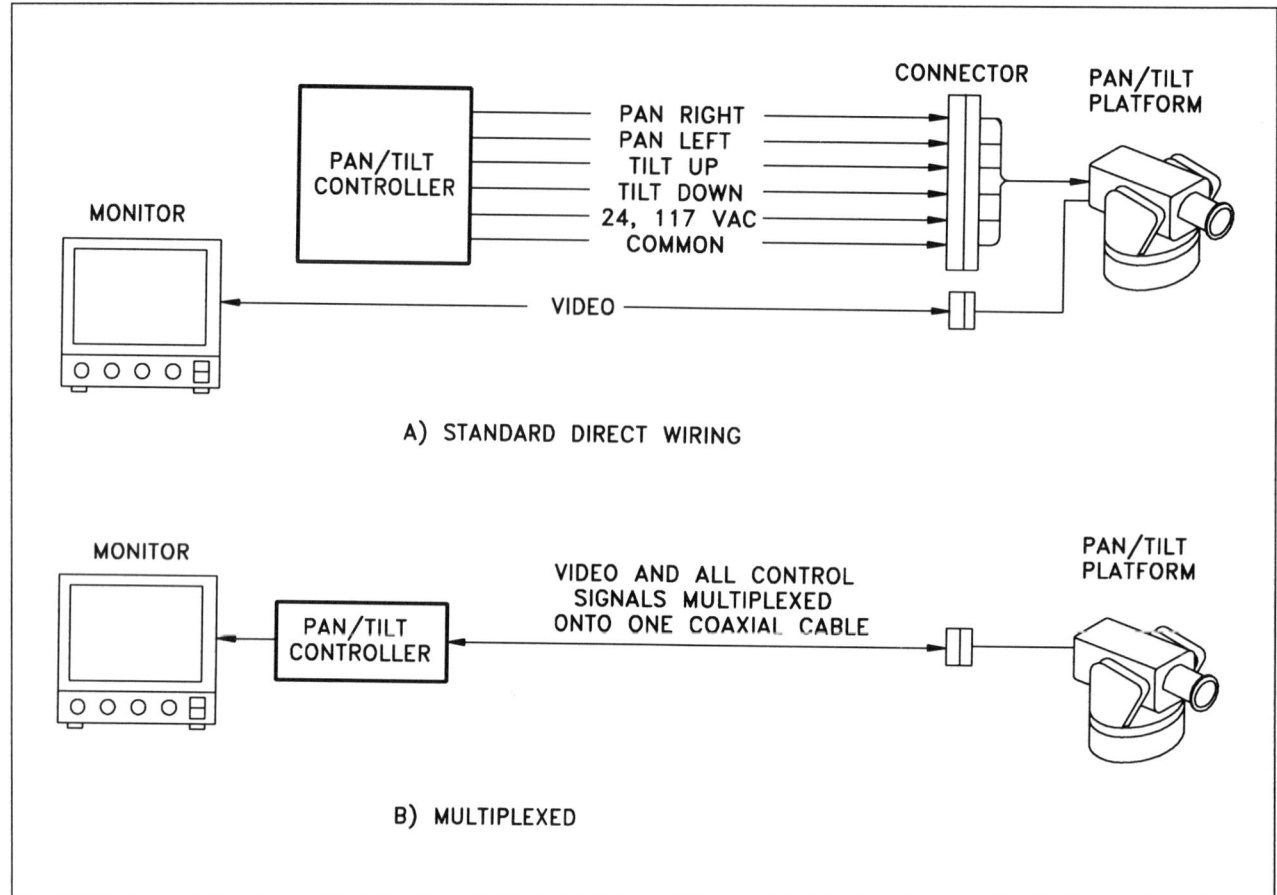

FIGURE 10-3 Direct and multiplexed pan/tilt video and control cable

analyze this viewing requirement, CCTV applications are divided into three types: point (fixed), area, and volume.

10.2.1 Point Scenes

A typical security surveillance point scene might be a supermarket checkout line, where the information of interest is the merchandise being rung up, the amount displayed on the cash register, the cashier, and the customer; or the scene might be a bank teller station, where the crucial intelligence is the identity of the teller. In such scenes, all of the activity occurs in a relatively small portion of the FOV. Although a single camera with an FFL lens may be satisfactory, a better solution might be a fixed two-lens, one-camera system, with one lens viewing the amount rung up on the cash register and the other viewing the cashier, customer, and merchandise. Another solution is a zoom lens with some panning and tilting motion. This would be necessary for instance in a jewelry store (or department store), where the counter area is viewed with a wide-angle lens and a small piece of jewelry is viewed with a narrow-angle lens. The pan/tilt is necessary with a zoom lens because the high-magnification (telephoto) position results in a very small FOV, so the camera must be pointed at the exact center of the target. Figure 10-5 illustrates two conditions of such a cash-ier scene, in which the zoom lens views the entry area in the wide-field mode and can identify monetary denominations in the telephoto position, with the pan/tilt mechanism pointing the camera/lens.

In the wide-field scene, the article of interest is not normally centered in the FOV. In this situation, the operator must redirect the camera/lens combination via the pan/tilt control to center the object of interest and make it observable in the high-magnification, telephoto position.

10.2.2 Area Scenes

A second broad classification of surveillance scenes includes a not-yet-known area of activity within a relatively large location. A typical application might be to observe activity at one or more doors in the front lobby of a building. In this situation (Figure 10-6), a camera mounted on a pan/tilt mechanism with zoom lens and located in one corner or on one wall of the lobby can accomplish all necessary surveillance.

If only a casual observation of the room is required, a fixed camera and lens may be sufficient; however any detailed activity will probably be missed. To identify personnel in the lobby or identify a person or some specific activity at one of the distant locations in the lobby, it is necessary to

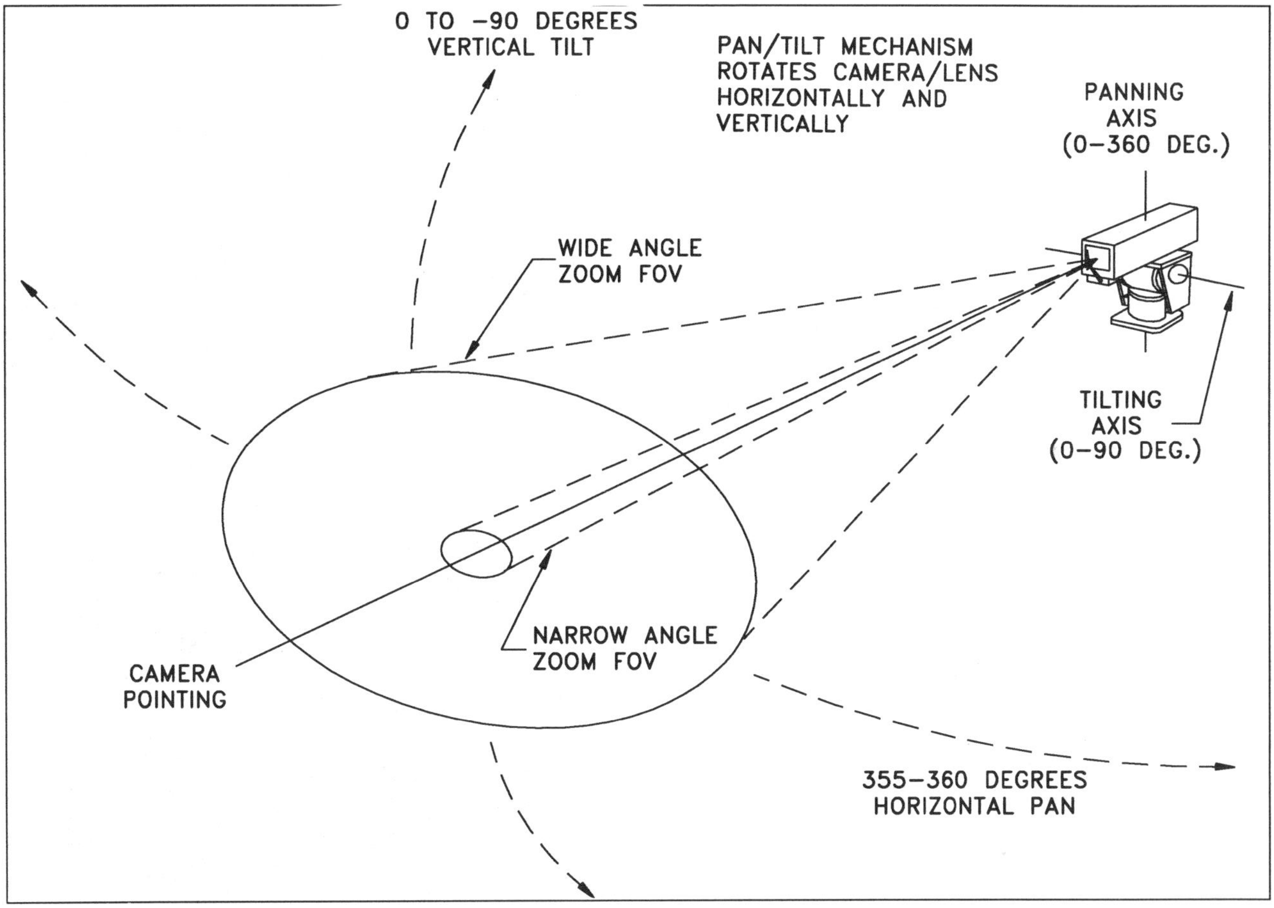

FIGURE 10-4 Spatial coverage using pan/tilt systems

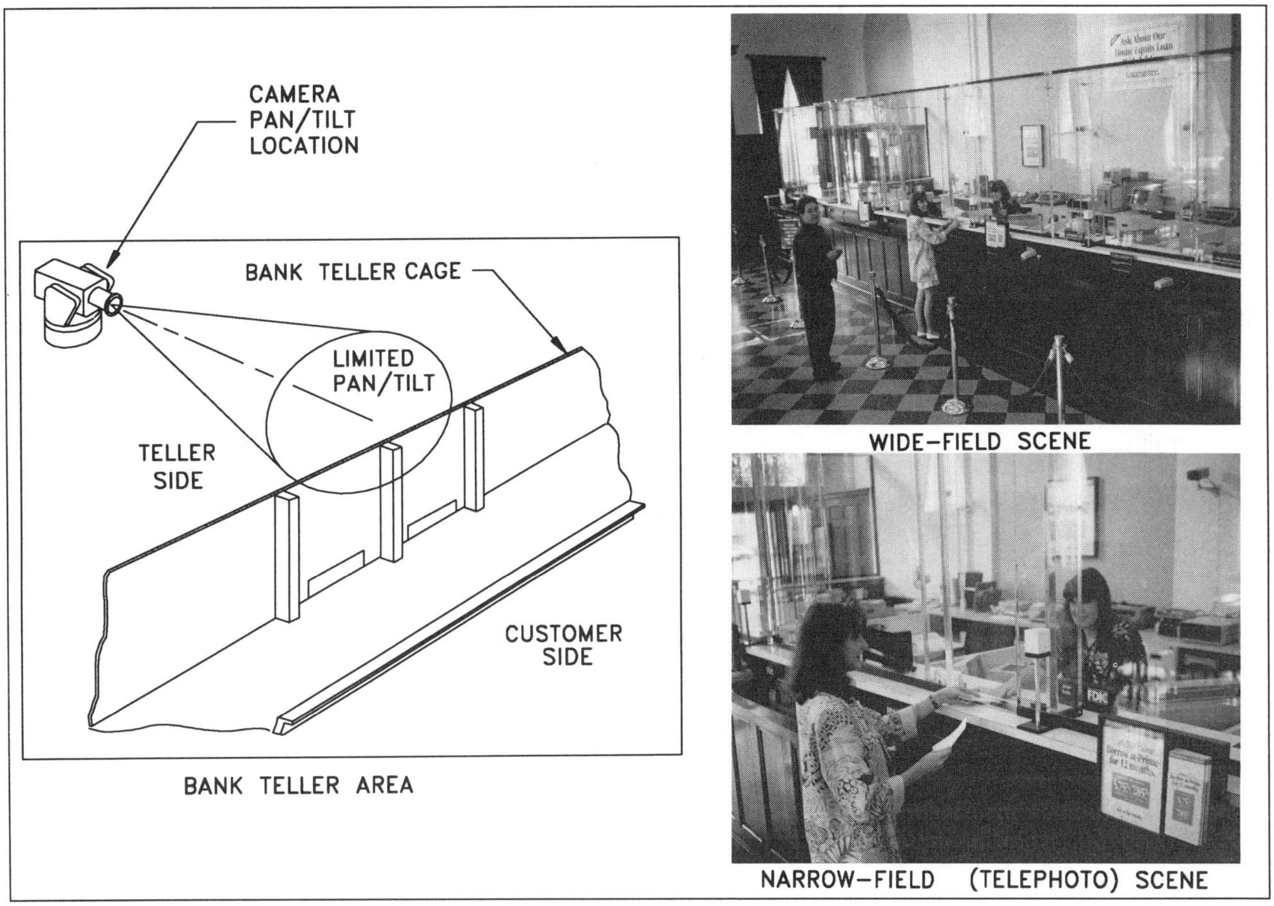

FIGURE 10-5 Point scene using pan/tilt

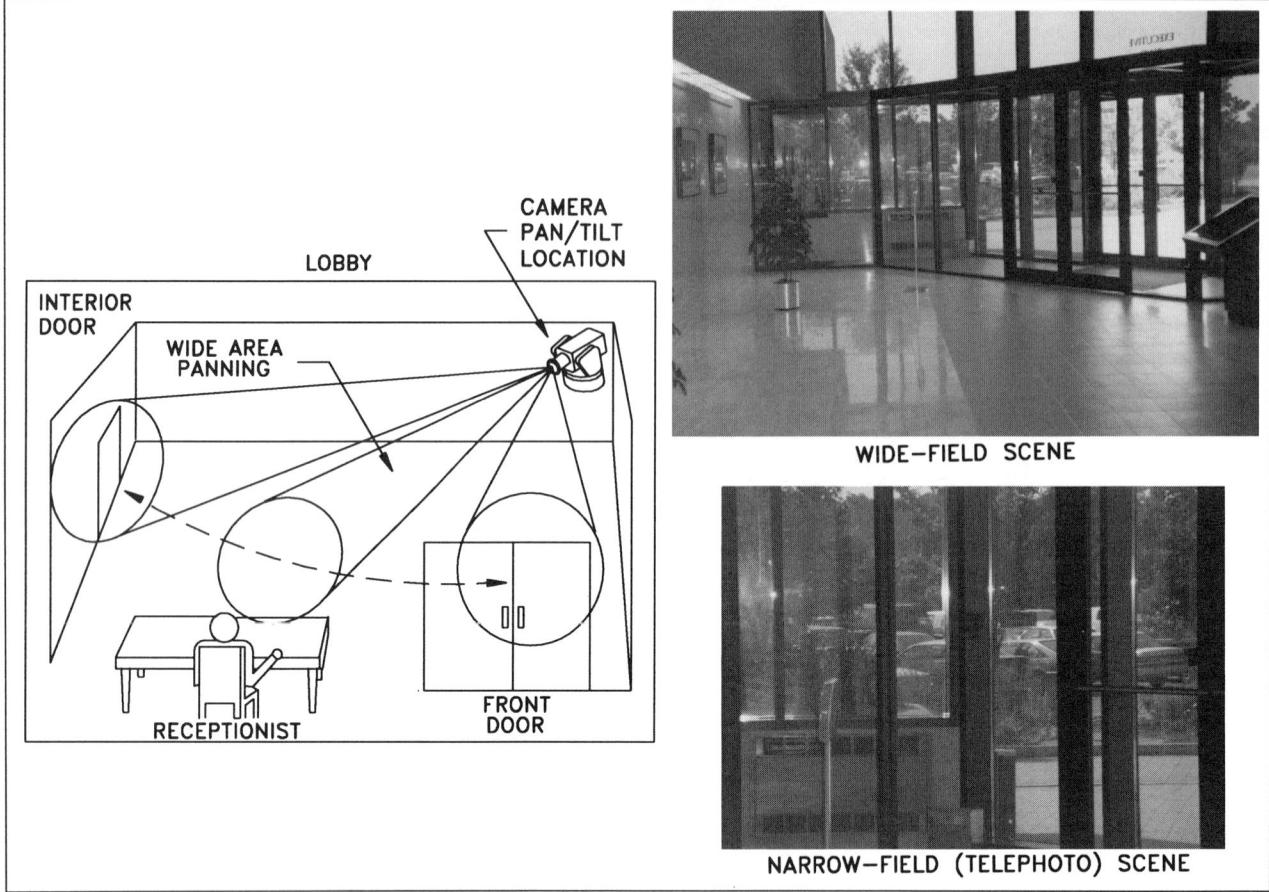

FIGURE 10-6 Area scene surveillance using pan/tilt

have a pan/tilt and zoom lens combination operated by remote control. This is called an area application, since the camera pointing must be directed to different locations in the room, as selected by the security personnel. When a particular activity must be viewed, the operator points the camera to the new location, directs the zoom lens to zoom to the telephoto position, thereby magnifying the area of activity on the monitor and leading to identification and further action. While this scenario is similar to the point scene, this scene is significantly larger than a point scene, often requiring a faster pan/tilt unit in order to acquire the target quickly.

10.2.3 Volume Scenes

The third category involves very large areas of surveillance, typically a large warehouse or outdoor parking area in which essentially a volume of space must be surveyed. In a warehouse, there may be many aisles to be viewed, high stacks of material, ground-level activity, or widely dispersed, elevated areas of interest. For this application, the best pan/tilt and zoom lens hardware is required, to provide large ranges of magnification for wide-angle (low-magnifi-

cation) and narrow-angle (high-magnification) conditions, and to rapidly point the camera in all directions, depending on the activity. Figure 10-7 illustrates the expanded requirements of this volume scene.

Some warehouse and supermarket installations have the camera/lens and pan/tilt mechanism on a roving track-mounted mechanism, so that the combination can move physically throughout the facility to gain a more advantageous perspective on the activity. This technique is usually confined to indoor applications, since the track mechanism is difficult to protect in an outdoor environment. Since the roving camera cannot be in position to view all areas, it does not provide full-time coverage. In an outdoor parking lot environment where the area to be covered is very large, multiple camera systems on pan/tilt platforms with zoom lenses are used.

10.3 PANNING AND TILTING

There are basically two camera-mount configurations to point the camera in different directions: (1) panning (or scanning) and (2) panning and tilting. The simplest camera movement consists of horizontal scanning, in which the

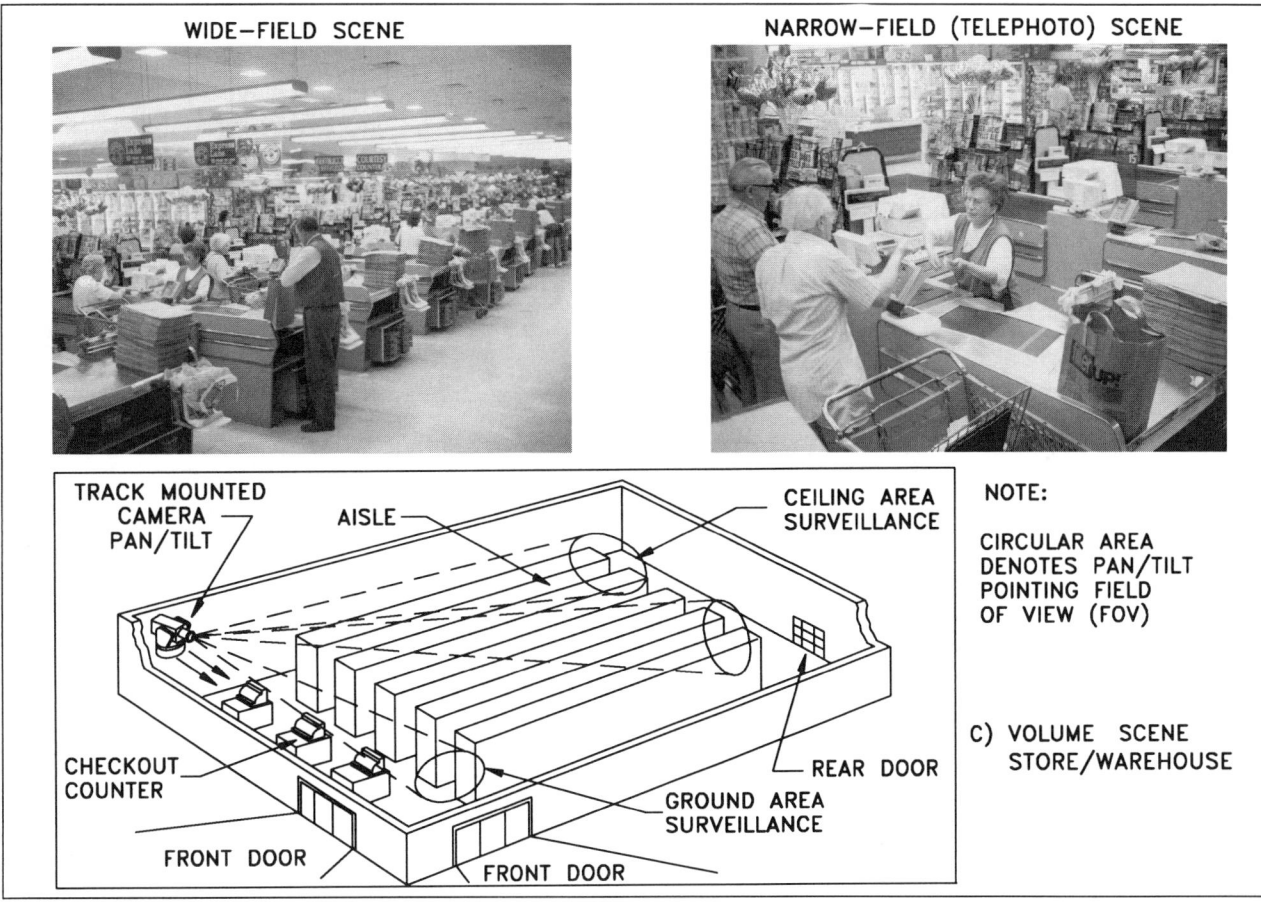

FIGURE 10-7 Volume scene using pan/tilt

camera is mounted on a rotating platform that moves from left to right (and vice versa) to scan a scene (Figure 10-8).

Most systems scan through a range of 355 degrees, with some capable of full, continuous 360-degree scanning. Panning systems are used in front-lobby applications in which all activity of interest occurs in a single horizontal plane: the camera must move only from left to right and stop at some particular location in the plane. The systems available include automatic panning, in which the camera/lens is rotated automatically through a specified angle at a constant rate, thereby allowing a guard at a monitor or VCR to see and record activity occurring in the area. Even in this simple situation, however, when the camera is pointed in a particular direction (say, the right-corner door), activity occurring in the center or left part of the scene is not under surveillance. The application must have suitable scanning speed available and camera/lens angular field of coverage, so the combination can be pointed in the direction of activity quickly, and provide the protection required. Panning systems can be set to scan a small sector of the entire scanning range and scan at different speeds from 8 to 80 degrees per second, thereby giving the operator the ability to optimize viewing of the area.

10.4 PAN/TILT MECHANISM

When expected activity is anywhere in a horizontal or vertical location, it is necessary to use the pan/tilt camera platform. The pan/tilt system points the camera and lens in two different axes of rotation, one around the horizontal axis to tilt and the other around the vertical axis to pan (Figure 10-9).

The panning action was described in Section 10.3, with the same adjustable parameters: speed of scanning and angular coverage. As with panning systems, there are equipments available to provide a full 360-degree panning, so that if activity occurs at one end of the FOV, the lens/camera can continue rotating to provide full coverage.

To accomplish vertical movement or pointing, a second motor and drivetrain are required to raise or lower the camera/lens angle of view. The angular FOV provided by such a system usually ranges from 0 degrees to −90 degrees, thereby covering activity in most scenes of interest. A joystick, thumb trackball, or four-button setup (Figure 10-2) is used to move the camera-pointing angle horizontally and/or vertically. Moving the joystick in between horizontal and vertical activates both drives. Various manufacturers

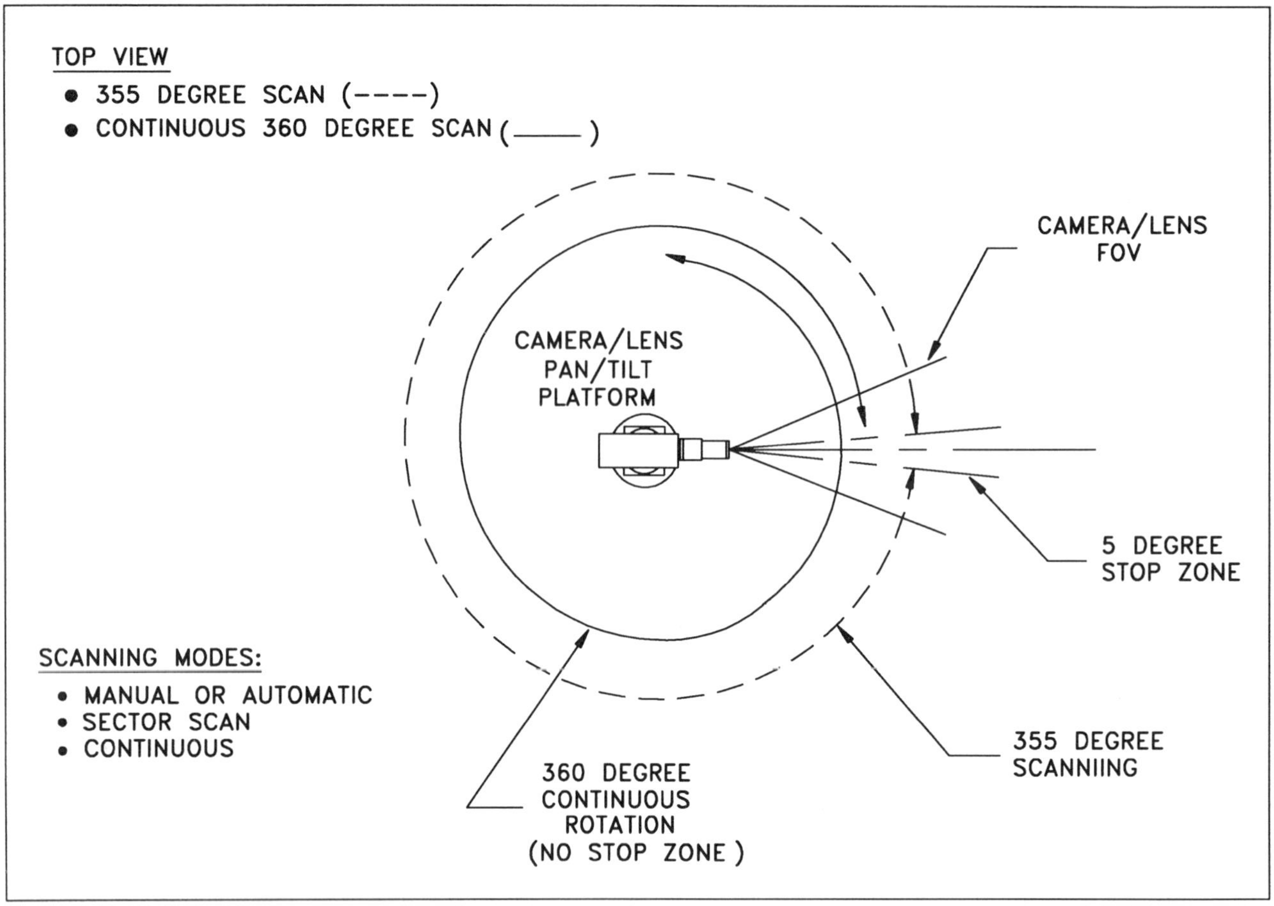

TOP VIEW

- 355 DEGREE SCAN (– – – –)
- CONTINUOUS 360 DEGREE SCAN (——)

CAMERA/LENS FOV

CAMERA/LENS PAN/TILT PLATFORM

5 DEGREE STOP ZONE

SCANNING MODES:

- MANUAL OR AUTOMATIC
- SECTOR SCAN
- CONTINUOUS

360 DEGREE CONTINUOUS ROTATION (NO STOP ZONE)

355 DEGREE SCANNIING

FIGURE 10-8 Horizontal scanning function

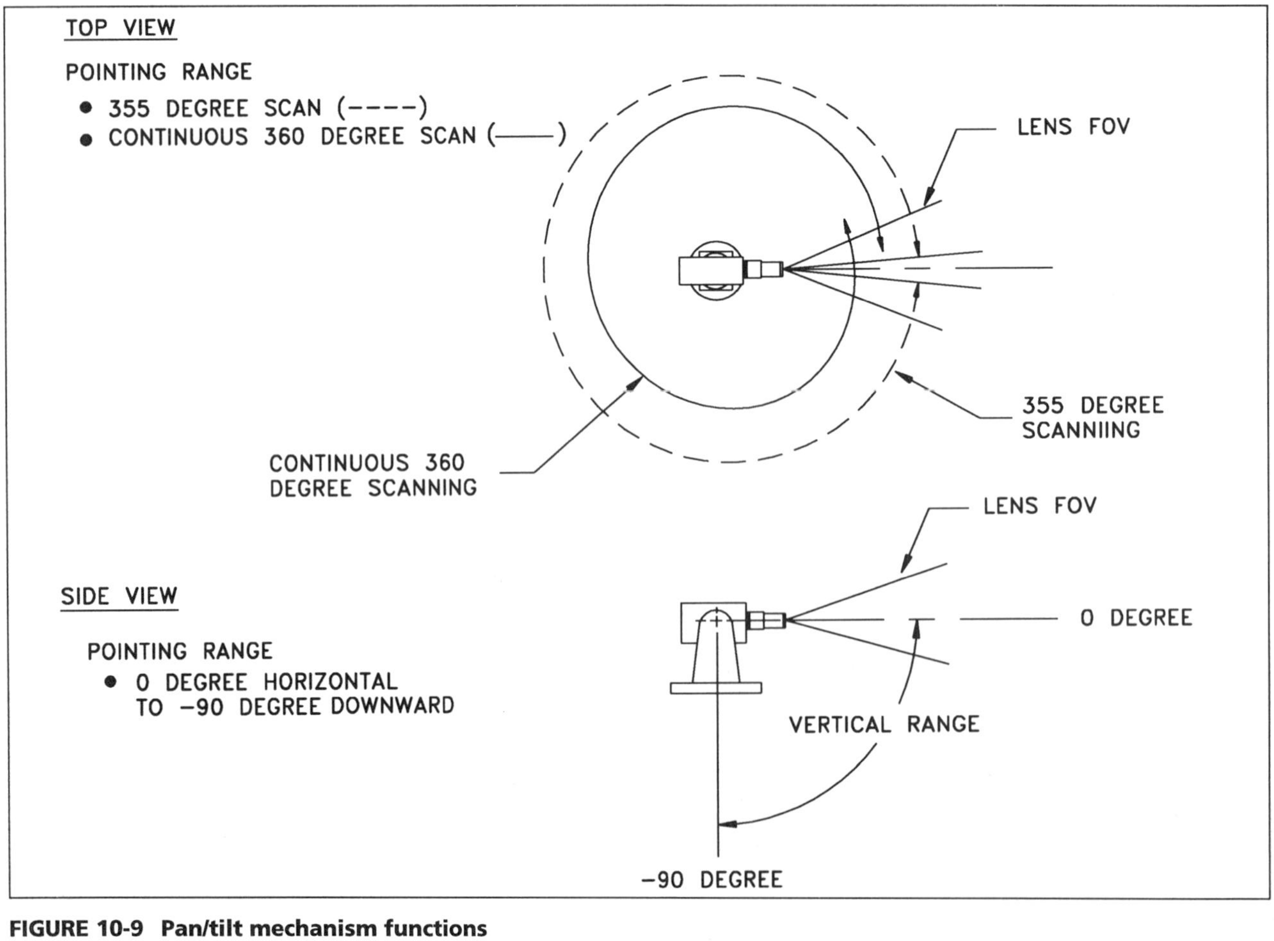

TOP VIEW

POINTING RANGE

- 355 DEGREE SCAN (– – – –)
- CONTINUOUS 360 DEGREE SCAN (——)

LENS FOV

355 DEGREE SCANNIING

CONTINUOUS 360 DEGREE SCANNING

SIDE VIEW

POINTING RANGE

- 0 DEGREE HORIZONTAL TO –90 DEGREE DOWNWARD

LENS FOV

O DEGREE

VERTICAL RANGE

–90 DEGREE

FIGURE 10-9 Pan/tilt mechanism functions

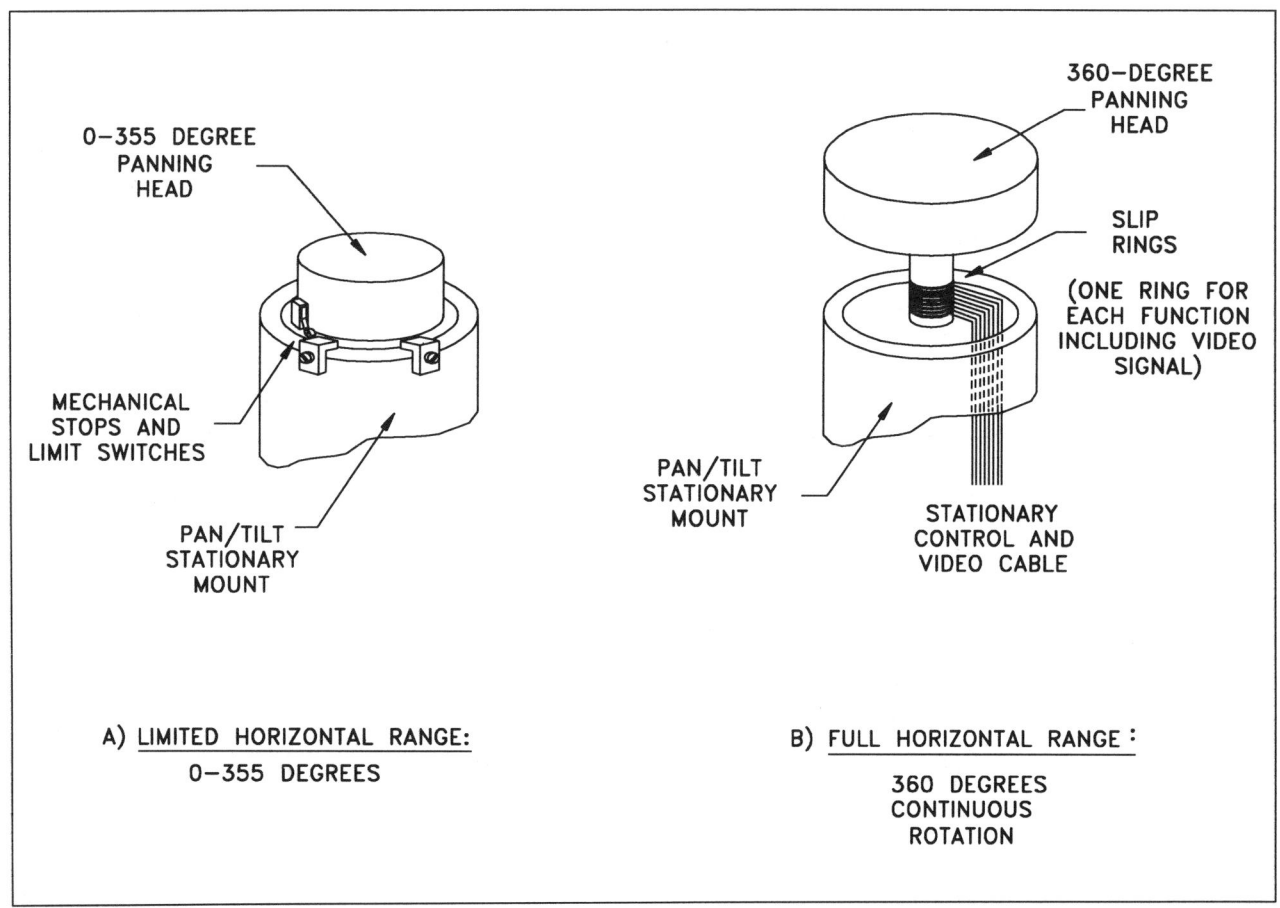

FIGURE 10-10   Azimuth limit stops vs. slip rings

have incorporated a proportional joystick control, in which pan/tilt mechanism speed of movement is proportional to the amount of joystick deflection from its center off position. This is particularly useful when the system has variable-speed capability, so the security operator can point the camera to the direction of activity as quickly as possible.

When CCTV is integrated with microcomputer-controlled hardware, the security operator can program various preset lens/camera pointing angles, so that later he has to push only a single switch at the console to automatically direct the pan/tilt platform. This is useful when an alarm sensor can activate the camera pointing to the preset direction, pointing the camera at the target in the shortest possible time. Other programmable presets include zoom lens focal length and focus, thereby optimizing the scene at which the camera is pointed.

Pan/tilt mechanisms have specific ranges of azimuth and elevation over which they can operate. In the tilt (elevation) axis, electromechanical switches sense when the tilt platform and shaft have rotated to the upper or lower extreme (maximum range), and they open to de-energize the tilt drive motor—otherwise the mechanism could be damaged. In the horizontal (azimuth) direction, similar limit switches (or other sensing mechanisms) are incorporated, so that

the horizontal travel can be limited to a specific angular sector or go to its extreme horizontal range limits. In most pan/tilt mechanisms, this maximum range is usually 355 degrees. These mechanisms cannot go beyond 355 degrees and must therefore come back around if they are to follow the movement of a person or object that is moving out of the angular range of the horizontal panning direction (Figure 10-10).

To overcome this limitation and allow continuous camera and lens rotation (360 degrees and beyond), slip rings are used. By making all electrical connections from the stationary platform to the moving pan/tilt mechanism and camera lens assembly through slip rings, the moving platform is able to rotate continuously in any angular direction without having to stop and back up in order to look at a particular location. Limit switches are incorporated if the range of rotation must be reduced, but no mechanical stops are installed if the slip ring mechanism is to provide full angular travel (continuous 360-degree movement).

Clutches are used in pan/tilt drives in between the motor and gear-train drives so that if the output shaft encounters resistance, the motor will not stall and burn out—the motor will keep running until the overload is removed. Some pan/tilt mechanisms perform a similar function: when the motor stalls, they sense the overload via the increased electrical current to the motor and then reduce the current or turn it off, so that no harm is done to the motor or drivetrain.

Once the motors driving the pan and tilt axes have turned off, brakes or antidrive backdrive gears must be used to prevent the platforms from "coasting," that is, continuing to move under the load of the camera or other loading factors. Brakes can be energized when the motor turns off so that the platforms come rapidly to a complete stop. Without brakes, the platform will coast until the motor has come to a complete stop. If the angle of overshoot is small, this may be an acceptable solution. However, once the platform has stopped, a sufficiently large gear ratio and the correct gear type must be used so that the load cannot continue to drive the motor, allowing the platform to move and rotate.

Most systems use microswitches or other electromagnetic devices to determine when the motor has driven the platform to a particular angular location or to the limit of travel. An installer can move or adjust these switches to change the range of travel. They usually operate satisfactorily for long periods of time, but they do need periodic adjustment and can cause some operational problems. Newer designs utilize electronic positioning, in which a reference positioning potentiometer is installed to provide a basis for determining the angular position of the horizontal or vertical platform at any moment. The primary use for the potentiometer feedback is to direct the electronic control system to program preset camera/lens parameters. These parameters include horizontal pan angle, vertical tilt angle, zoom lens FL, focus, and iris settings. With modern electronic control systems, the security operator can pan, tilt, and zoom to a particular location, set parameters for the system to memorize as a preset combination, and assign a unique code for that condition. To accomplish this, the system's electronics read and memorize the potentiometer's shaft position (by measuring a voltage) on the pan/tilt and camera motors. During operation, if an action occurs at the preset site for that code, the operator can manually call for that combination; or if an alarm occurs in that area, the system can automatically call up the particular code and view the site. When the code is initiated, the pan/tilt points the camera in the direction of the activity, the zoom lens adjusts to the preset FL and focuses, and the operator sees a scene in the quickest possible way. This operating procedure in these sophisticated integrated systems reduces human stress, effort, and error. This concept has been expanded to multiple cameras and multiple sites, allowing the operator to program different presets and codes so that the system will "tour" the sites. This feature optimizes both human and mechanical efficiency: it eliminates the need for the console operator to hunt for particular targets or activities.

10.5 MECHANICAL CONFIGURATIONS

There are several mechanical configurations for pan/tilt platforms, with options for camera/lens location and the rotation axes of the pan/tilt mechanism. Figure 10-11 illustrates the two basic configurations; a brief analysis of their advantages and disadvantages follows.

Historically, the most popular camera/lens location has been above or over the center of the pan/tilt mechanism; the majority of manufacturers produced this type. A second configuration, the side-mounted camera/lens, places the camera and lens to one side of the vertical panning axis, at the same height as the tilt motor drive and hardware. This side-mounted configuration is designed to reduce the wind and mechanical loading on the vertical drive motor and gear mechanism, thereby improving its performance, life, and reliability. This configuration is more popular in Europe.

10.5.1 Above-Mounted

In the over-the-center-mounted configuration, the system is balanced horizontally when the camera/lens is horizontal, but when the camera/lens is rotated up or down, the weight of the camera/lens combination becomes unbalanced. To improve balance, additional weights are added to the tilting arm, so that undue stress is not imposed on the motor, bearings, and gear train. This additional weight must be driven by the vertical tilt motor in addition to the camera/lens/housing load. As the camera tilts up or down from its normal horizontal position, the weights and/or springs counterbalance the load.

10.5.2 Side-Mounted

The side-mounted configuration requires no counterweights or springs if the camera is symmetrically mounted on the tilt arm located to the side of and in line with the horizontal (tilt) rotating axis. While this configuration is not symmetrical in the vertical plane—the camera is off to one side of the vertical (panning) axis—it nevertheless results in a lower imbalance on the horizontal tilting axis, thereby providing less wind loading and less unbalanced forces on the mechanism.

Because counterbalancing weights and internal springs are not needed in the side-mounted configuration, the horizontal axis remains approximately balanced, thereby reducing the power requirement of the motor and gear train assembly. The counterweights and springs are not necessary because, as shown in the side-mounted design, the camera, lens, and housing are closer to the center of the axis of rotation. Also, if the side-mounted system must

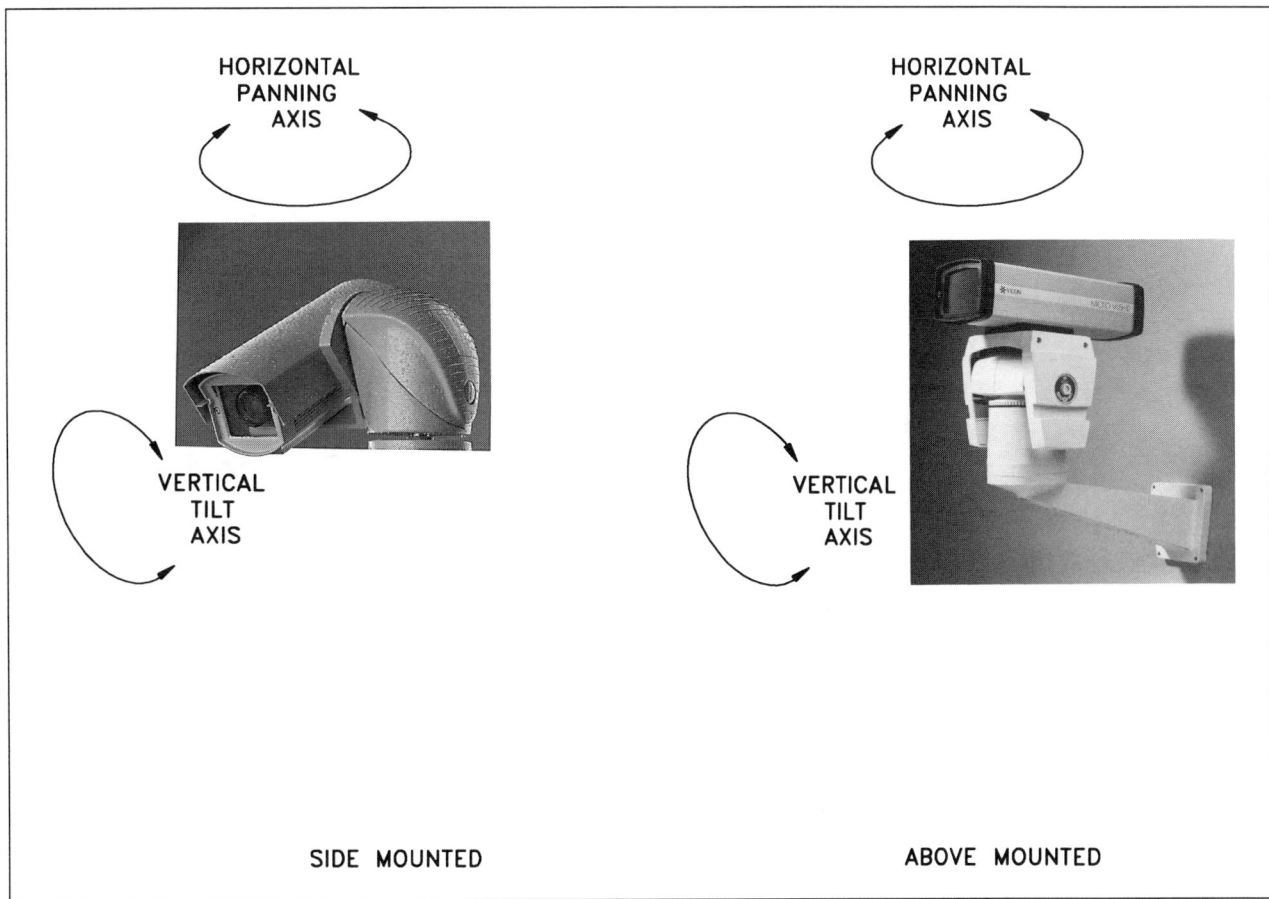

FIGURE 10-11 Above- and side-mounted pan/tilt configurations

be inverted (say, when mounted on a ceiling), no changes are required besides repositioning the arm. To invert a top-mounted system, the springs and weights usually have to be removed.

10.6 INDOOR

Indoor panning or pan/tilt platforms are lightweight and require no environmental protection. They can be mounted overtly and directly with wall or ceiling brackets (Figure 10-12a).

In a friendly environment, the camera and lens are mounted directly to the pan/tilt platform with no exterior housing, and the cables are suitably dressed to the mount. In a hostile environment, the camera and lens are enclosed in an indoor environmental housing to protect them from vandalism or other abuse.

The availability of small solid-state CCTV cameras and their accompanying small lens packages has made it possible to put lenses and cameras into small dome housings. Figure 10-12b shows an attractive pan/tilt assembly enclosed in a small hemispherical dome mounted in the ceiling, which results in an architecturally aesthetic installation.

In these systems, the pan/tilt and camera/lens are all integrated and mounted inside the dome housing; then most of the mechanism is mounted above the ceiling, providing an attractive, low-profile housing. Operationally, locating the pan/tilt and camera/lens inside a smoked plastic dome provides additional security and deterrence, because personnel at floor level cannot determine where the camera is pointing at any particular time, which may deter offenders from carrying out harmful overt or covert activities. Figure 10-12c shows a larger integrated pan/tilt camera system suspended in a dome below ceiling level.

10.7 OUTDOOR

The major difference between the indoor and outdoor pan/tilt hardware results from the outdoor environment. These adverse outdoor conditions manifest themselves in wind loading, precipitation, temperature extremes, humidity, dust, dirt, corrosive atmosphere, and higher levels of vandalism. To successfully cope with these additional burdens, the pan/tilt mechanism and housing must be more rugged and have higher environmental resistance.

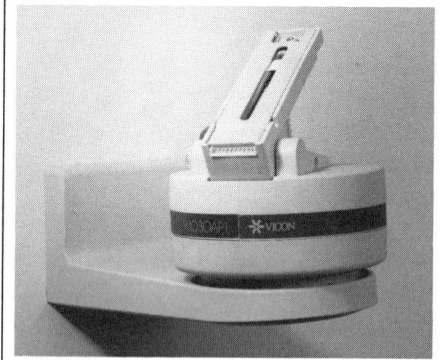

BRACKET
MOUNTED

A) WALL

CEILING
MOUNTED
7" DIA.
HEMISPHERE

B) SMALL DOME

BELOW THE
CEILING
20" DIA.
HEMISPHERE

C) DOME

FIGURE 10-12 Indoor pan/tilt platforms

Compared with top-mounted configurations, the inherent symmetry in the side-mounted configuration reduces the turning force required to overcome the retarding forces when the camera must be tilted (Figure 10-13).

In the overhead design, the center-of-gravity distance of the camera/lens/housing load is 5 to 10 inches away from the axis of rotation; in the side-mounted design, the same components are on the horizontal (tilt) rotation axis. While both systems operate satisfactorily in outdoor environments, the motor and gear train size required in the side-mounted design is smaller. Although the top-mounted design is the most popular in the United States, the European side-mounted type is gaining acceptance.

10.7.1 Wind Loading

Wind loading must be considered for all outdoor pan/tilt camera systems. Wind loading is the force exerted on the camera housing assembly and pan/tilt mechanism that can cause it to move or rotate in an unwanted direction, cause the entire system to vibrate, or prevent proper motion (or any motion) of the system. One of the deleterious effects to the CCTV picture caused by wind loading is vibration, which produces a blurred or unfocused image on the monitor. In a more serious case, sufficiently high wind loading on an inadequately designed system will cause mechanical failure of the gear drives or motors. To minimize wind loading, the system should be as small as possible and all components located as close as possible to the center of rotation of the pan and tilt axes. In Figure 10-13 the wind comes from one direction and produces an unbalanced force on the pan/tilt mechanism and camera housing, causing it to rotate or move in an unwanted direction. For example, if the camera/lens/housing load is energized to rotate upward (tilt up) and the unbalanced wind force would rotate the camera downward, the motor and gear train must have sufficient torque to overcome the counteracting wind loading force (torque) and the camera/lens/housing torque. A well-designed system will have sufficient motor/gear torque to overcome the wind and the load, including a safety factor.

The long-term effects of wind, snow, and ice loading on the pan/tilt mechanism are wear and tear and additional maintenance. Even in a well-designed system, the frequent motions and starting and stopping of the motor, gears, and camera assembly require periodic maintenance. This takes the form of lubricating the gear train and/or replacing a motor, clutch, or gear assembly. So careful consideration must be given to the pan/tilt design, choosing the correct

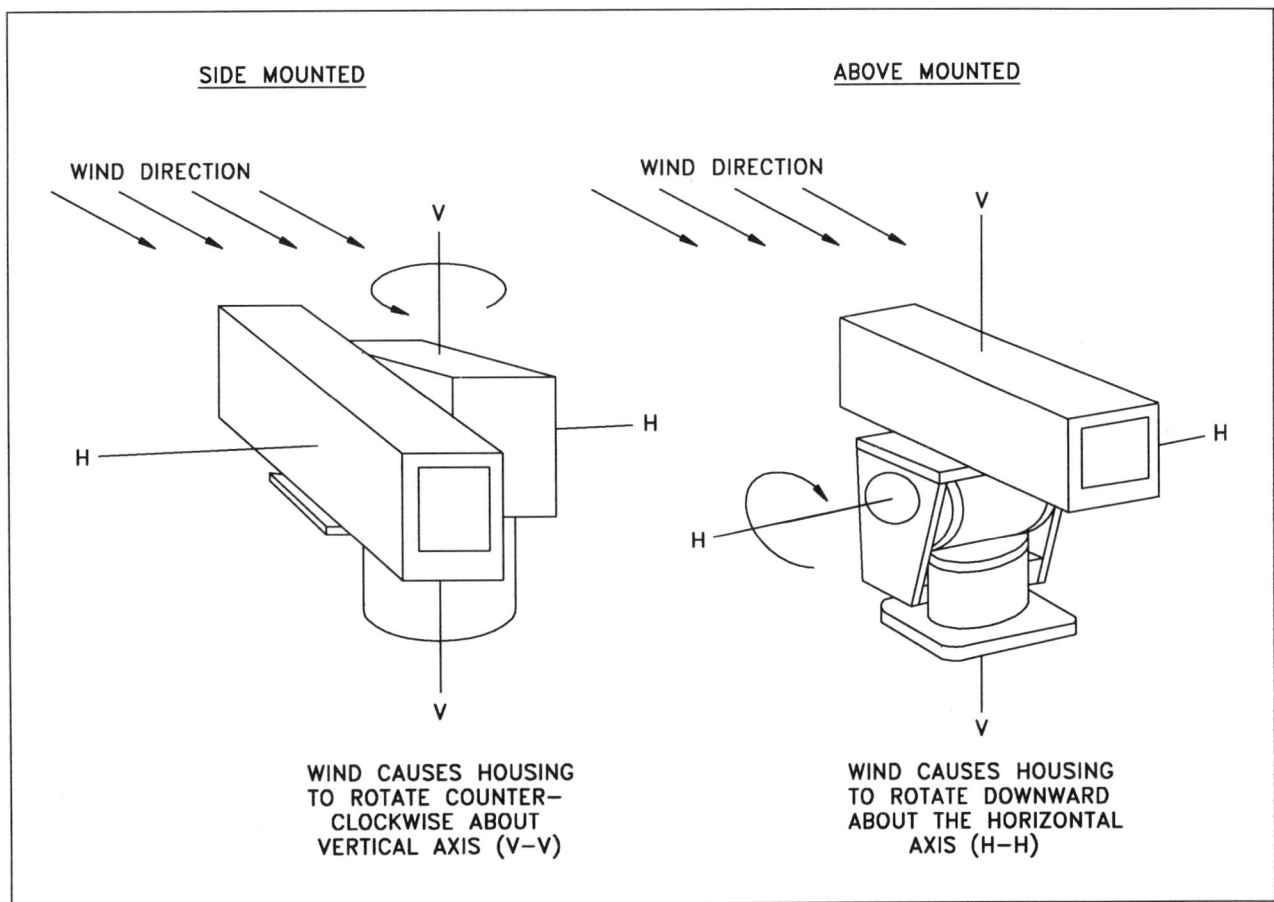

FIGURE 10-13 Wind loading on pan/tilt and camera housing

application, and recognizing the effects of wind, snow, and ice loading. In most outdoor applications, adding an extra 5 or 10 pounds to the load rating factor of the pan/tilt system is a worthwhile investment.

10.7.2 Precipitation

For long life and reliable service, outdoor pan/tilt units should be completely watertight and dustproof, so that rain, snow, dust, dirt, or corrosive materials do not find their way into the bearings, gear train, or motor drive assembly. It is difficult to determine the quality of the environmental pan/tilt design from the manufacturer's specification. It is usually necessary to rely on the manufacturer's reputation, as well as information obtained from installers and users, as to the quality and durability of the equipment. There are some guidelines, however:

- Sealed ball bearings are preferred over sleeve bearings.
- All bearing shaft ends should be protected with environmental seals to prevent entry of liquids, dirt, or corrosive materials.
- The pan/tilt mechanism should be constructed of non-corrosive materials, such as die-cast aluminum, magnesium, or outdoor plastic.

- Aluminum and magnesium housing materials should be painted with a durable outdoor enamel or epoxy paint. Any unpainted aluminum surfaces should be anodized.
- Suitable outdoor plastic types include polycarbonate, ABS, and PVC. While many types of plastic can withstand outdoor weathering conditions as well as or better than metal, unsuitable plastic parts may crack or deteriorate due to (1) breakdown of the plastic by the sun's rays (particularly ultraviolet radiation), corrosive liquids, or gases; (2) cracking due to large variations in temperature excursions; or (3) cracking caused by ice forming between several parts and then expanding.

While the materials used by the manufacturer may be an indicator of performance, the actual reliability and maintenance requirements of the equipment can only be determined through manufacturers' test results, if available, or dealer and end-user comments. If plastic material is used, flame-retardant types are absolutely essential.

As mentioned previously, snow and ice cause an additional loading factor on the pan/tilt mechanism, which must be considered in the design. Ice forming between the stationary mount and the panning or tilting mechanism can prevent the motor from "breaking loose" and providing the desired motion. In inclement, freezing weather, it is often advisable to operate pan/tilt mechanisms regularly, to produce heat from the motors and keep ice from forming at these critical locations.

Additional torques that the pan/tilt mechanism must overcome in subfreezing climatic conditions include those from ice and snow loading. This loading takes two forms: (1) the additional weight of snow or ice that the camera housing must carry, which the motor gear train must try to lift; and (2) freezing ice or snow at the location of the turning shaft (bearing), which must be dislodged. The motor and gear train must overcome this additional force to rotate and point the camera.

10.7.3 Temperature

Most pan/tilt mechanisms have been designed to operate over temperature ranges found in most parts of the world. However, installations in very severe cold or hot weather require special pan/tilt mechanisms for reliable operation.

In desert and tropical locations, where the temperature may reach 130° Fahrenheit (72° Centigrade), special motors with high-temperature insulation on their windings are required. Operating conventional motors under higher temperatures shortens the life of and burns out the motor. Likewise, other electrical components, such as motor capacitors and circuit boards, must be rated for higher temperatures. An additional consideration for these hot environments is expansion of mechanical parts, which could cause binding or jamming and render the system inoperable.

In extremely cold climates—40°F (−10°C) or lower—the grease and lubricants used in the gear train must be chosen so that they do not become too viscous (hard), thereby preventing rotation of the motor and gear train. Most other electrical components in the system operate without any problem under very cold conditions. It is important that the pan/tilt mechanism be properly sealed at the bearing and cable entry/exit locations so that moisture does not enter the mechanism and freeze.

10.7.4 Dust, Dirt, and Corrosion

While pan/tilt mechanisms usually operate 10 feet or more above the ground, some locations have sufficient dust, dirt, sand, or corrosive atmosphere to require special attention. Although a well-designed outdoor pan/tilt mechanism will have seals to prevent these contaminants from entering the bearings and internal parts of the mechanism, it is important that they have been designed and are operating properly. Of particular importance is the seal at the electrical cabling entry and exit locations.

When an installation is in a sand-blown environment, additional protection of the housing is required, in the way of more durable finishes to prevent damage from blowing

sand or other particulate matter. Fine sand also causes bearing surfaces to wear out sooner.

When an installation is in an area containing a corrosive chemical atmosphere, the pan/tilt housing manufacturer must be provided with the specific information on the known chemicals active in the area. Some active chemicals include sulfuric, nitric, and hydrochloric acids, salt spray, and other substances. While many plastics have excellent corrosive resistance to these chemicals, aluminum and steel, which are in widespread use for pan/tilt housings, do not have good resistance against most of these hazardous chemicals and salt spray.

10.7.5 Outdoor Domes

One technique used to eliminate the entire outdoor pan/tilt mechanism environmental problem is to house the camera, lens, and pan/tilt mechanism in a plastic dome. Figure 10-14 illustrates one such configuration, which is 18 inches in diameter and contains a pan/tilt mechanism inverted and suspended inside the dome, with a zoom lens and CCTV camera.

The enclosure of the camera/lens and pan/tilt mechanism in the dome totally eliminates the precipitation, wind loading, dust, and dirt problem. Yet it has one disadvantage: viewing through a curved dome is inferior to viewing through a plane (flat) window on a standard outdoor housing. A second viewing disadvantage is that the dome can be cleaned only manually, whereas an outdoor housing with a flat front window can be cleaned periodically with a windshield wiper. Water droplets on the dome also tend to reduce visibility. In any application, the dome's complete protection of the mechanism must be weighed against its inferior viewing ability compared with the standard outdoor housing pan/tilt mounting. The small dome pan/tilt design shown in Figure 10-12b is adaptable for use in outdoor applications under building eaves or passageways.

10.8 PAN/TILT—COMPONENTS VERSUS INTEGRATED SYSTEM

The majority of pan/tilt camera/lens installations consist of components obtained from several different manufacturers, all assembled by the dealer or installer and made to operate as a complete system. Considering the wide variety of applications, this is a practical solution for many installations. However, sometimes a specialized application requires an integrated camera/lens, pan/tilt, housing, and mounting system from a single manufacturer. Special parameters might include higher scanning speed, wider angular coverage, or an unusual combination of lens/camera characteristics. Figure 10-15 shows an example of an integrated camera head system.

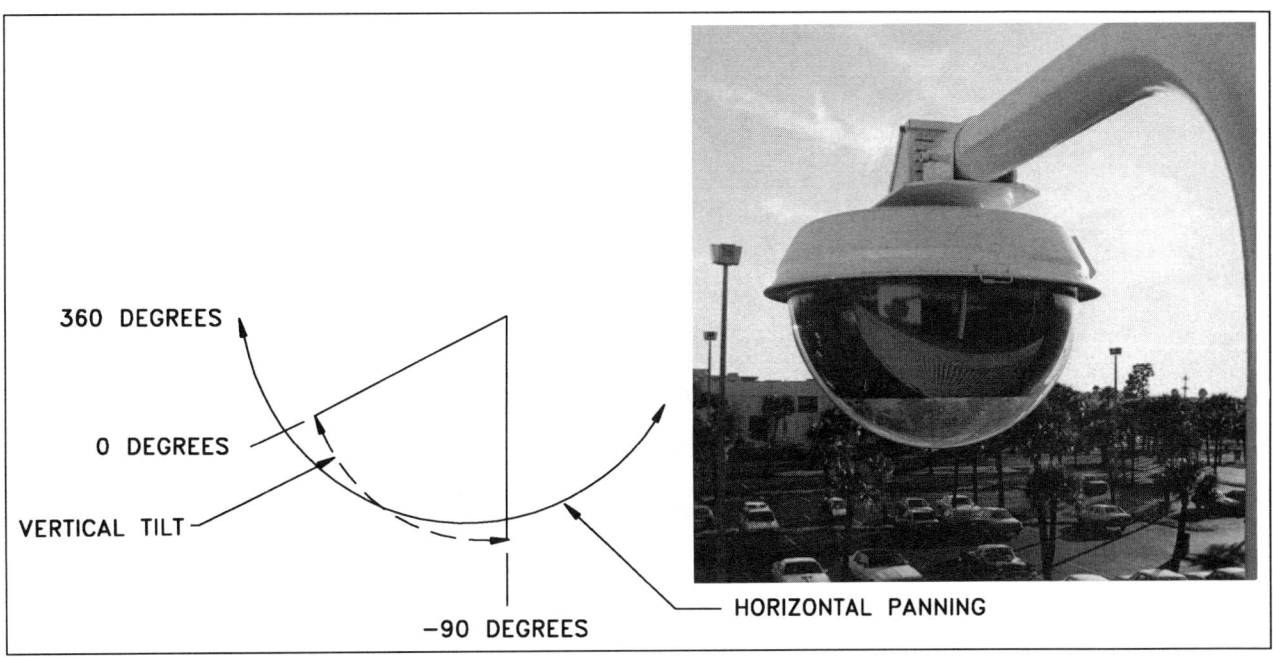

360 DEGREES

0 DEGREES

VERTICAL TILT

−90 DEGREES

HORIZONTAL PANNING

FIGURE 10-14 Pan/tilt mechanism in plastic dome

The system has a maximum scanning rate of 80 degrees per second and is capable of panning 360 degrees continuously (using electrical slip rings), all housed in a 12-inch-diameter ceiling-mounted dome. With 360-degree continuous horizontal scanning, the lens/camera does not have to come back 360 degrees in order to follow a moving target. To achieve these results, the moving parts are small, with low masses and moments of inertia. Only through advanced engineering and packaging are these fast scan rates obtained.

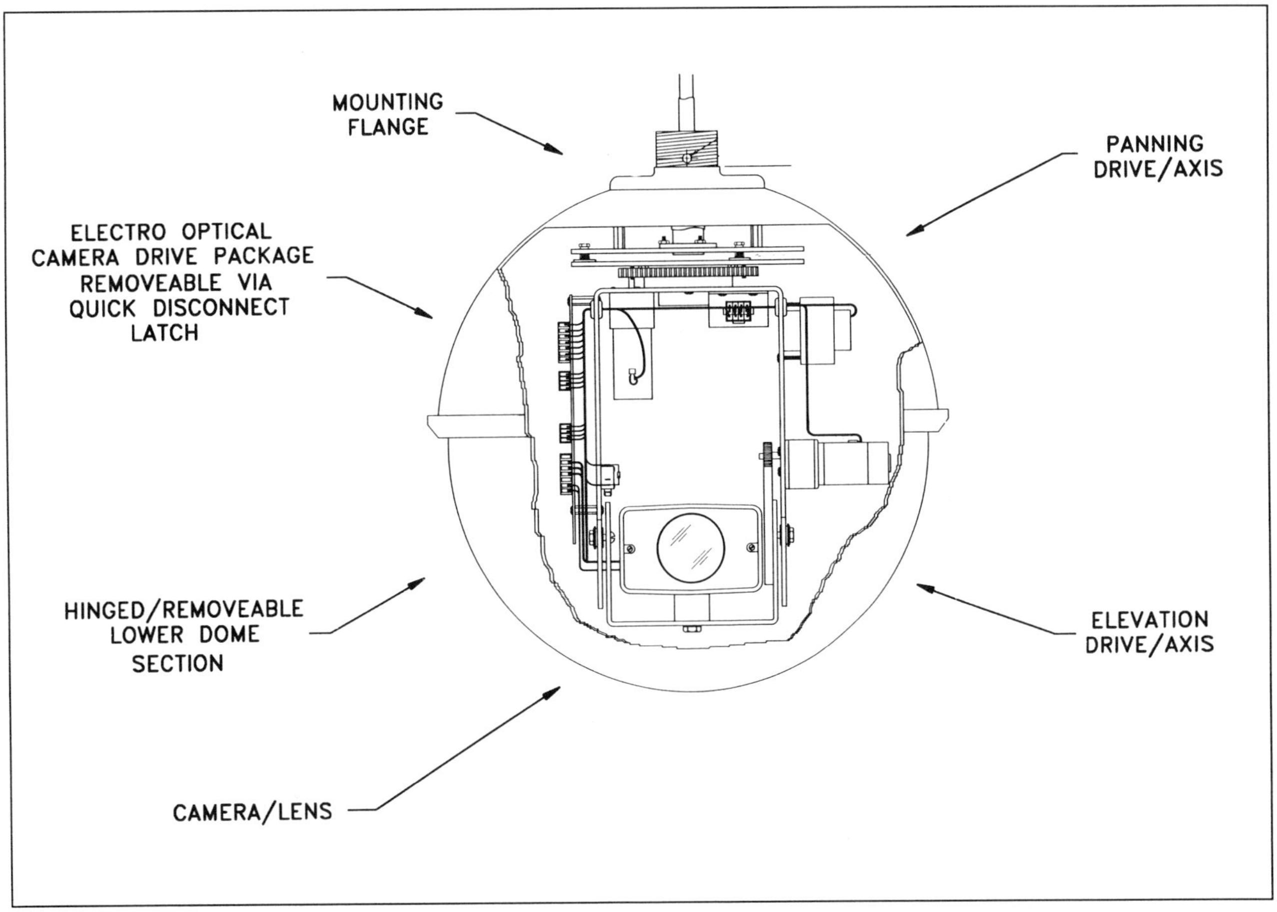

MOUNTING
FLANGE

PANNING
DRIVE/AXIS

ELECTRO OPTICAL
CAMERA DRIVE PACKAGE
REMOVEABLE VIA
QUICK DISCONNECT
LATCH

HINGED/REMOVEABLE
LOWER DOME
SECTION

ELEVATION
DRIVE/AXIS

CAMERA/LENS

FIGURE 10-15 Integrated pan/tilt camera head configuration

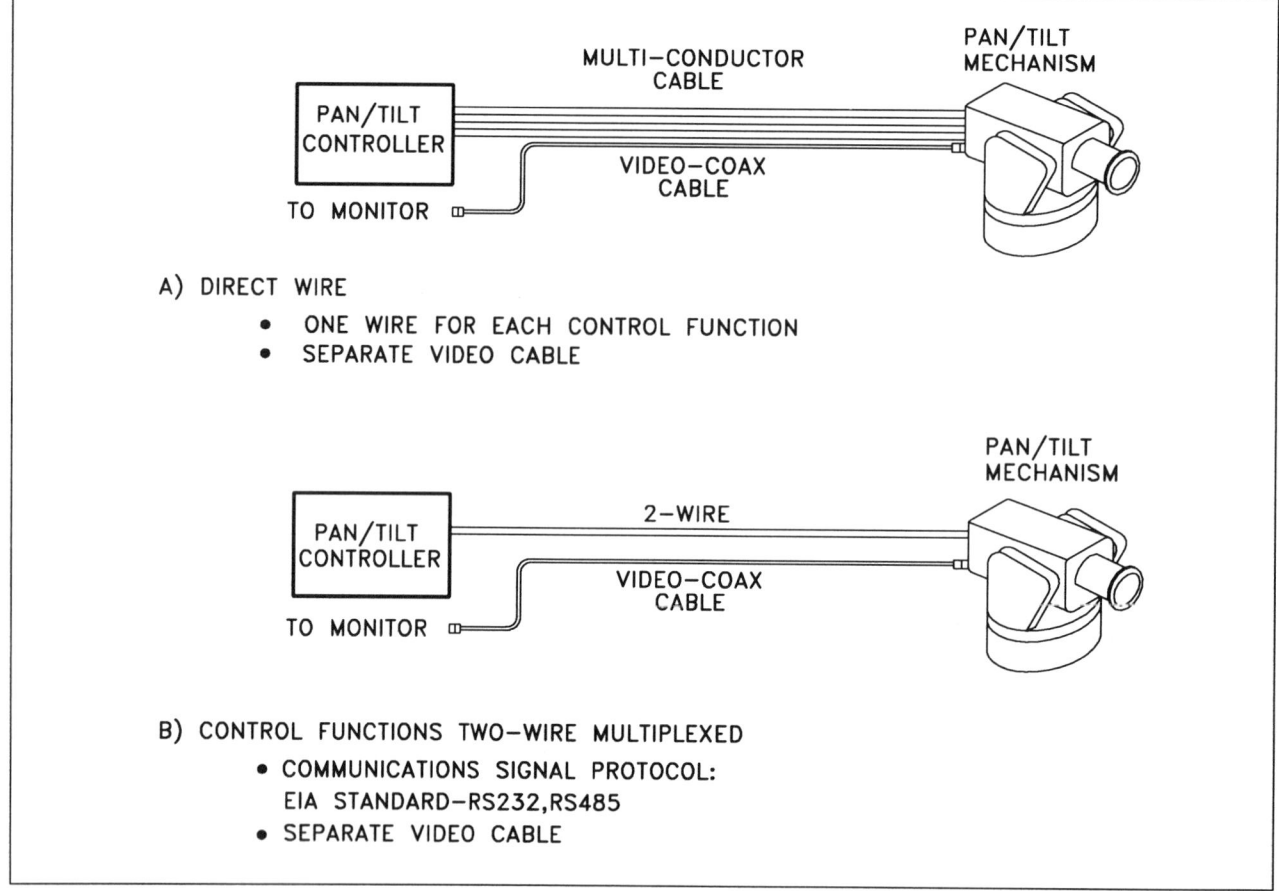

A) DIRECT WIRE
* ONE WIRE FOR EACH CONTROL FUNCTION
* SEPARATE VIDEO CABLE

B) CONTROL FUNCTIONS TWO-WIRE MULTIPLEXED
* COMMUNICATIONS SIGNAL PROTOCOL:
  EIA STANDARD-RS232,RS485
* SEPARATE VIDEO CABLE

FIGURE 10-16 Pan/tilt control using direct-wire and multiplexing

The obvious advantage of this system is that if an incident occurs anywhere within the dynamic FOV of the pan/tilt system, the camera/lens can be pointed in any direction in the shortest possible time, while the lens zooms and focuses on the target. Computer systems with camera-pointing preset capabilities can take advantage of these fast pan/tilt designs (see Section 9.3). While some pan/tilt mechanisms alone have high scan rates—up to 80 degrees per second—when combined with the additional inertia of the camera and lens (obtained from different manufacturers), these fast scan rates cannot be maintained.

10.9 PAN/TILT/ZOOM LENS CONTROL SIGNAL COMMUNICATION

There are basically four techniques for a security console controller to communicate with and control the remote pan/tilt zoom unit.

1. direct-wire—multiconductor
2. two-wire—multiplexed controls plus coax
3. single-cable—multiplexed video and controls on coax
4. wireless controls and video

10.9.1 Direct-Wire—Multiconductor

The simplest control of the pan/tilt/zoom lens mechanism is via direct-wire, using one wire for each control function and a separate video coax cable (Figure 10-16a).

This straightforward system is in widespread use for many small or short-run (under a thousand feet) installations. This technique requires no additional driver electronics for transmitting the control signals and no additional receiver electronics at the camera end. The controller consists of switches that control all functions, set manually by the operator or memorized by the system for automatic operation. Wire size must be large enough to minimize voltage drop to the motors and electronics.

10.9.2 Two- and Four-Wire—Multiplexed

For longer distances or when there are many different camera sites, a significant reduction in number of conductors and wire runs is accomplished by multiplexing (time-sharing) the control signals at the control console onto two

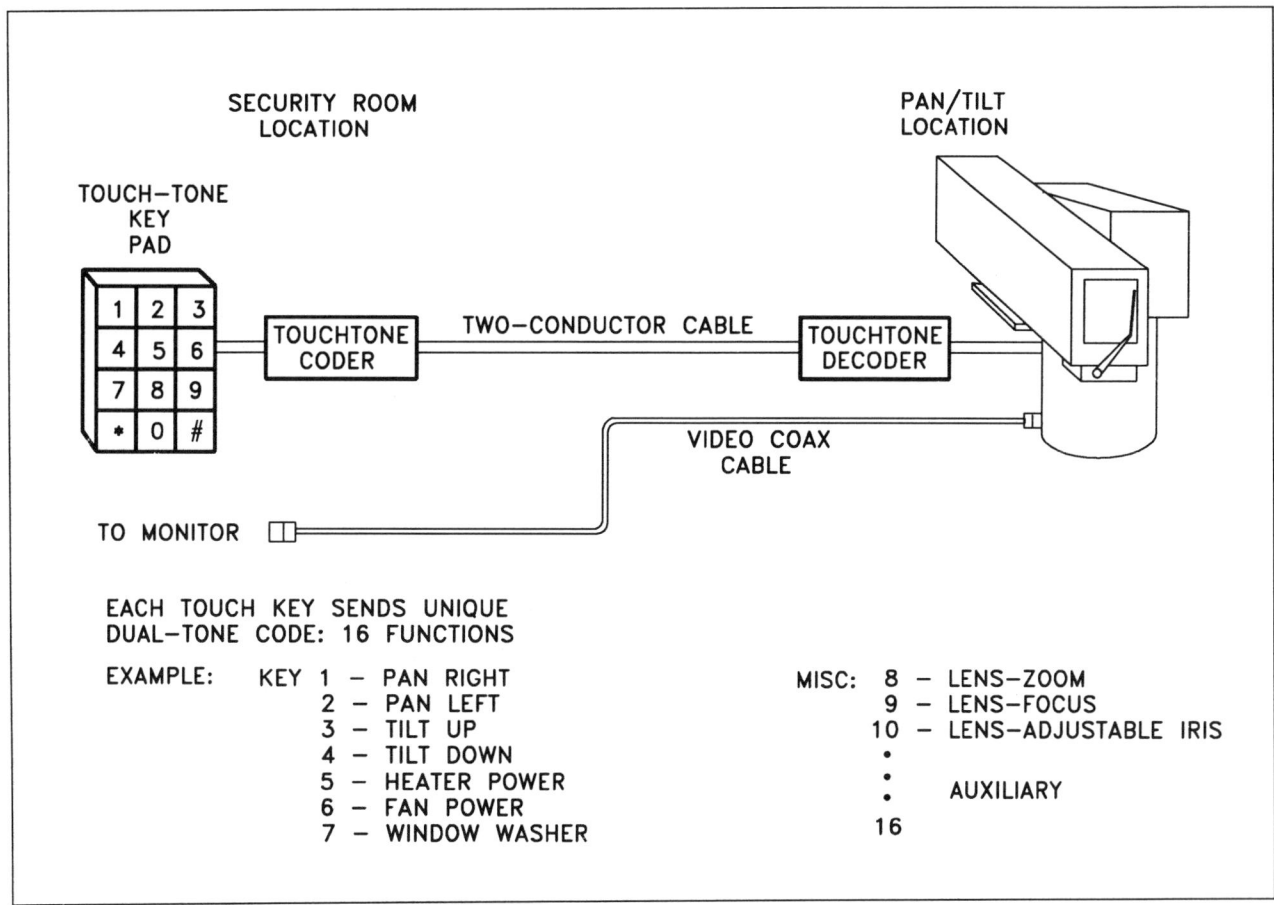

FIGURE 10-17 Pan/tilt control using telephone Touch-Tone

wires, sending them to the camera site, and then demultiplexing them or separating them again to provide the signals necessary to drive the pan/tilt/zoom unit (Figure 10-16b). Since the two wires need carry only communications information and not current to drive the motors, any long-distance two-wire communication system suffices. Two popular transmission codes (protocols) are the EIA RS-422 and RS-485. The video signal is transmitted on a separate coaxial cable.

A technique available to control a pan/tilt/zoom camera head system via two wires uses the telephone Touch-Tone communications system. Using this reliable design takes advantage of the world's most advanced large-scale integration technology, developed by AT&T. With this technique, a single twisted cable can route all command tones from the controlling site to a camera site and control the pan, tilt, zoom, iris, focus, and many optional auxiliary functions via a single simple Touch-Tone keyboard (Figure 10-17).

Touch-Tone remote control allows substantial savings in cable runs and the labor required to install them, and it eliminates the fear factor associated with conventional pan/tilt systems when an unknown communications scheme is utilized. Some of the optional auxiliary functions available are camera power (via an AC convenience outlet on the unit); enabling of autopan; remote switching of AC or DC accessories, such as a heater, blower, wiper, or tamper switch; and so on. The system can accommodate additional keypads at a later time, allowing easy expansion to a multiuser system. By using the FCC-approved telephone interface unit, these products can be interfaced to any telephone network.

10.9.3 Single-Cable—Multiplexed

Several companies manufacture systems that multiplex or time-share the control signals on the same video signal cable, thereby allowing video to be transmitted from the camera to the monitor console site and camera control signals to be transmitted from the security console to the camera site, all on one coaxial cable (Figure 10-18).

For direct wiring, this is the most efficient, minimum-cost technique, since only a single coaxial cable is required. The technique is cost-effective when the distance between camera and monitor is long, and when there are multiple sites. The system requires a special multiplexer that combines the

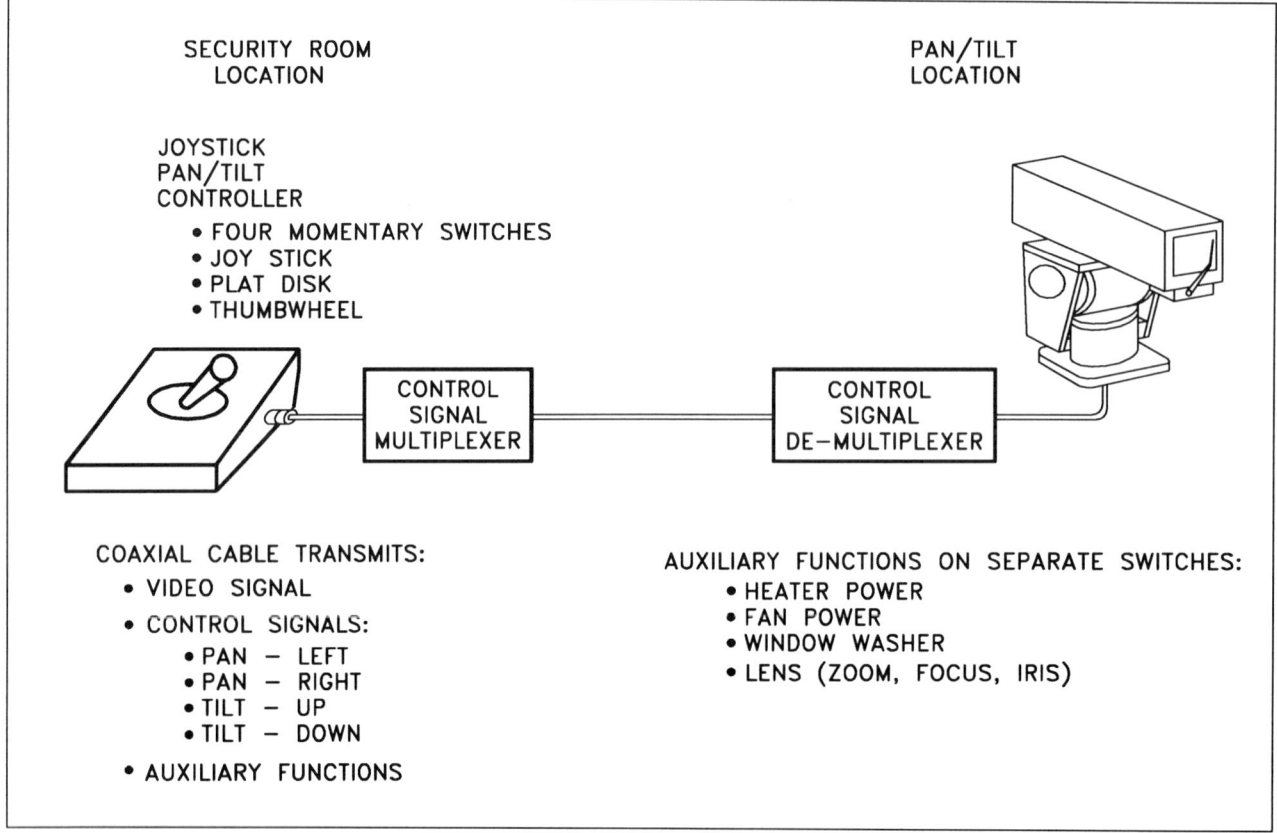

SECURITY ROOM
LOCATION

PAN/TILT
LOCATION

JOYSTICK
PAN/TILT
CONTROLLER
- FOUR MOMENTARY SWITCHES
- JOY STICK
- PLAT DISK
- THUMBWHEEL

CONTROL
SIGNAL
MULTIPLEXER

CONTROL
SIGNAL
DE−MULTIPLEXER

COAXIAL CABLE TRANSMITS:
- VIDEO SIGNAL
- CONTROL SIGNALS:
  - PAN − LEFT
  - PAN − RIGHT
  - TILT − UP
  - TILT − DOWN
- AUXILIARY FUNCTIONS

AUXILIARY FUNCTIONS ON SEPARATE SWITCHES:
- HEATER POWER
- FAN POWER
- WINDOW WASHER
- LENS (ZOOM, FOCUS, IRIS)

FIGURE 10-18   Multiplexed pan/tilt control using single cable

video and control signals at the camera and the monitor ends. An advantage of multiplexing the control signals onto the video signal is that additional transmission or control signals can be added to the system without adding new cable. These additional functions can include lens controls, alarm functions, or tamper switches.

In all three techniques just described, fiber-optic transmission is an alternative to copper wire. Several manufacturers have equipment that transmits the control signals, alarms, and other signals—even in some cases the video signal—on a single fiber-optic channel. As mentioned in Chapter 6, the fiber-optic advantages include noise immunity, long transmission distance, absence of ground loops, higher security, and reliable operation from different building sites in harsh environments.

10.9.4 Wireless

Control signals can be transmitted from the console to the camera location via wireless remote control communication. The control signals are multiplexed onto a single channel and transmitted on RF, microwave, or light-wave (visible or infrared) communication links. In extreme security environments (such as military or nuclear sites),

wireless transmission of video, command, and control signals is used as a backup to a hard-wired (copper or fiber-optic) system.

10.10 PAN/TILT MOUNTING

Many manufacturers produce attractive mountings for indoor and outdoor camera/lens and pan/tilt systems (Figure 10 19).

For indoor applications, the pan/tilt unit is securely attached to a wall or ceiling mounting bracket. The electrical cables connected from the camera, housing, and the pan/tilt mechanism are secured to the bracket and directed into the wall or ceiling. The system should be located with enough clearance between the rear of the camera housing (including extended cables) and the wall so that there is no interference when the unit pans over its full range. Also, the unit should be far enough below the ceiling so that when the camera is pointed down to its maximum depression angle, the rear of the camera and cable do not interfere and collide with the ceiling.

Figure 10-20a illustrates an outdoor pan/tilt unit in standard configuration mounted on a building roof edge

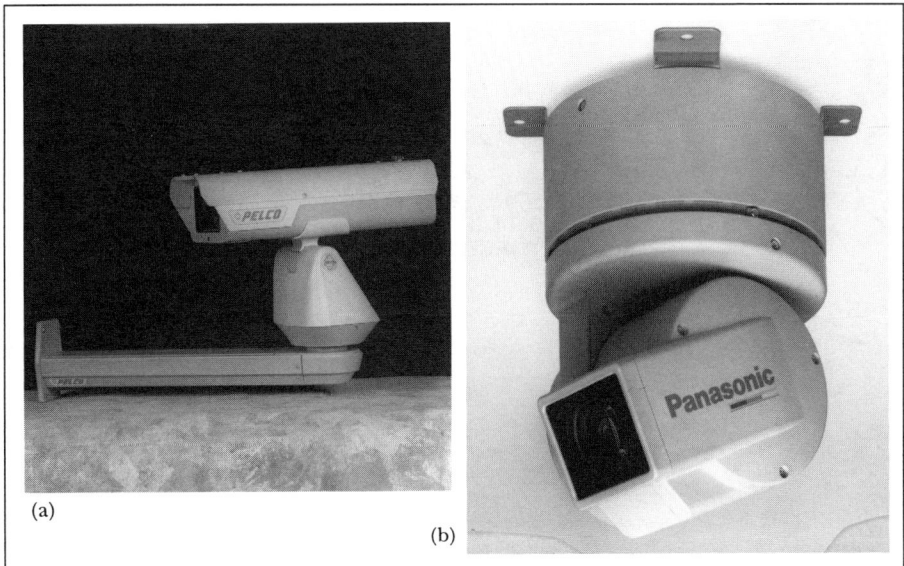

FIGURE 10-19 Pan/tilt mounting assembly: (a) wall, (b) ceiling

and capable of scanning 270 degrees horizontally to view a parking lot.

With such a large angular FOV to cover (an entire parking lot), this solution should be used where only sporadic activity is monitored, since panning with a standard unit from one end of a building to the other would typically take 10–15 seconds, which may be unacceptable. If a fast-scan unit is used, the panning time can be reduced to a few seconds, which might be acceptable.

Figure 10-20b shows a camera pan/tilt assembly mounted in a clear plastic dome housing on a pedestal mount, which provides wide-angle surveillance at an entryway and parking lot. Mounting the camera away from the building on a pole provides good viewing of the entire entry

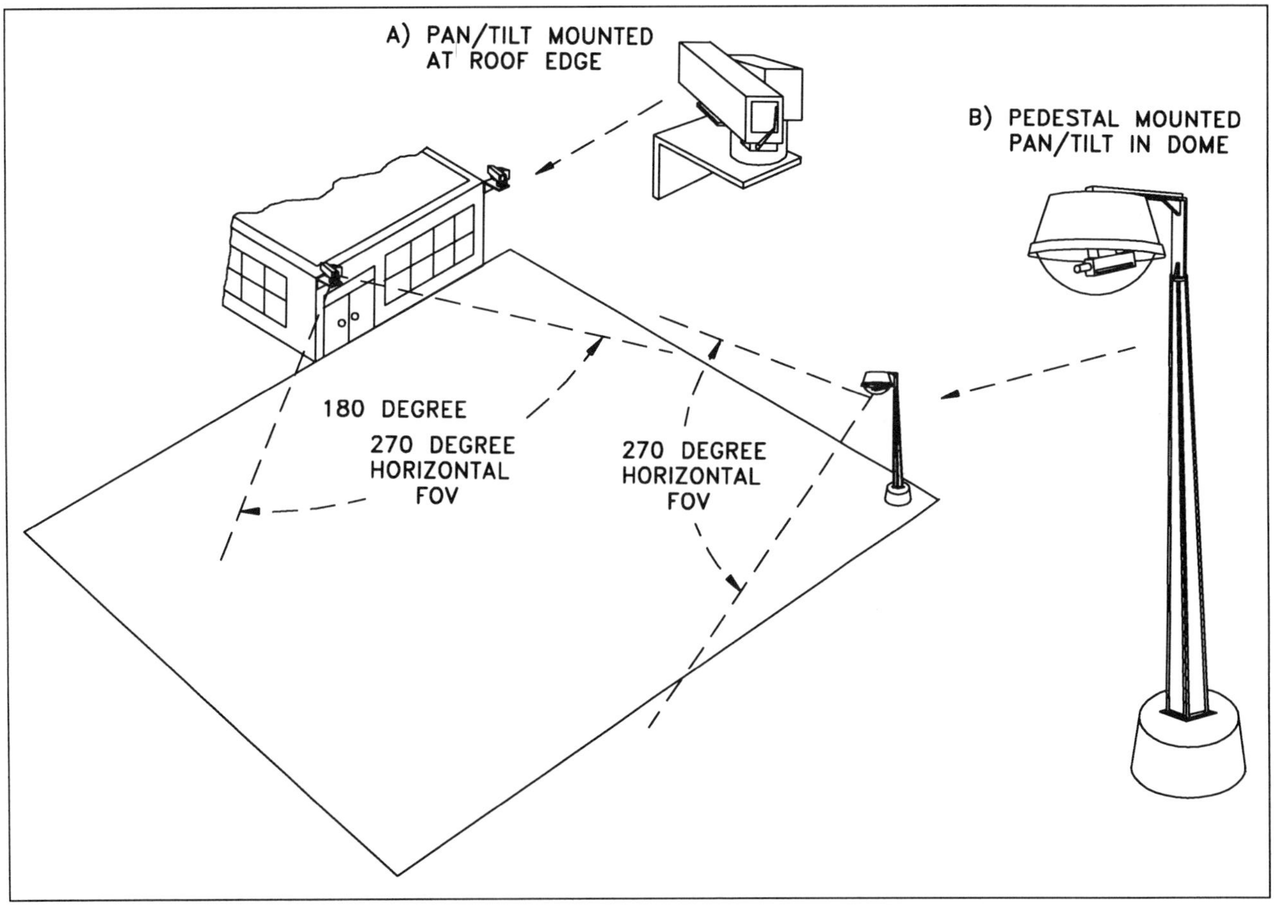

FIGURE 10-20 Outdoor pan/tilt installation

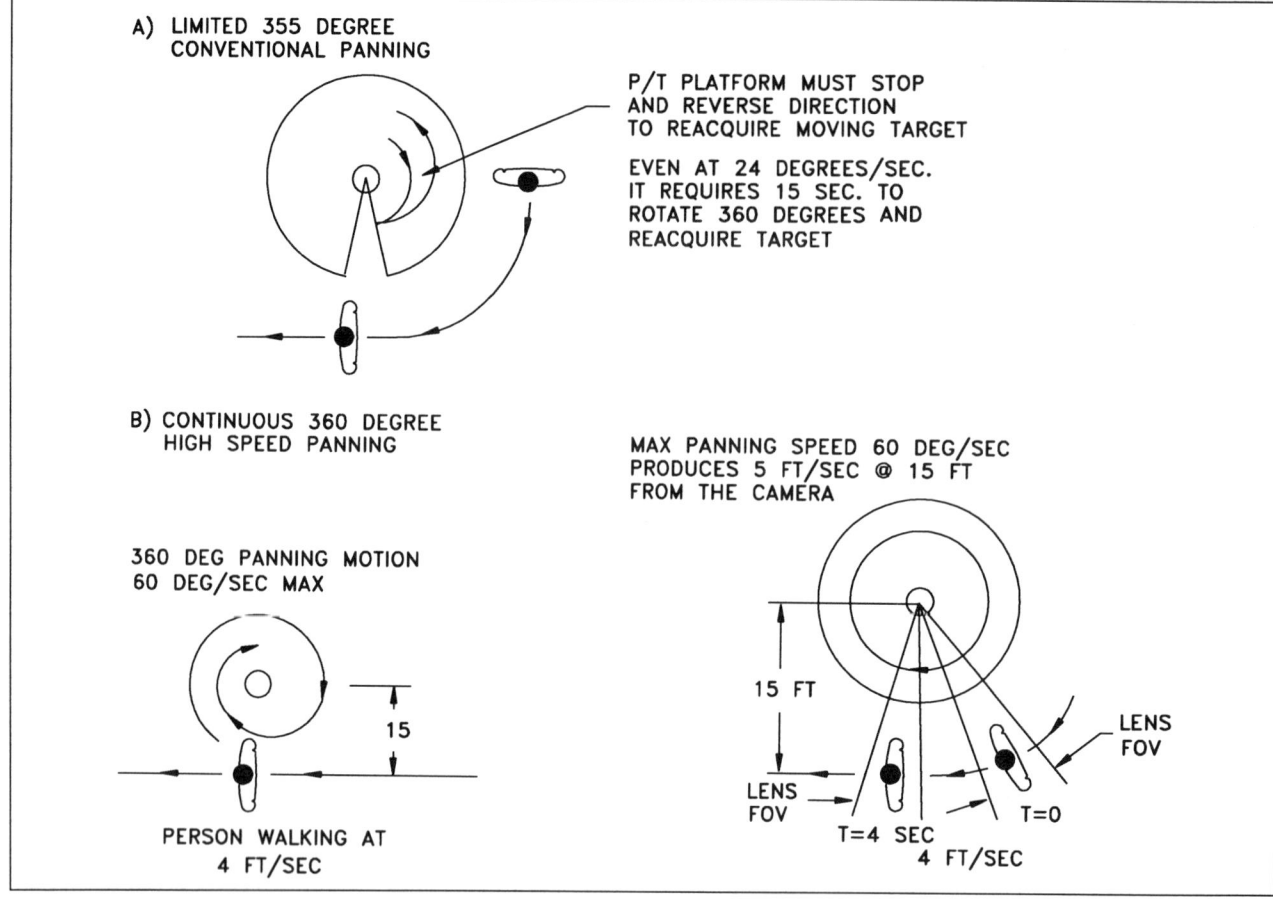

A) LIMITED 355 DEGREE CONVENTIONAL PANNING

P/T PLATFORM MUST STOP AND REVERSE DIRECTION TO REACQUIRE MOVING TARGET

EVEN AT 24 DEGREES/SEC. IT REQUIRES 15 SEC. TO ROTATE 360 DEGREES AND REACQUIRE TARGET

B) CONTINUOUS 360 DEGREE HIGH SPEED PANNING

MAX PANNING SPEED 60 DEG/SEC PRODUCES 5 FT/SEC @ 15 FT FROM THE CAMERA

360 DEG PANNING MOTION 60 DEG/SEC MAX

15

PERSON WALKING AT 4 FT/SEC

15 FT

LENS FOV

LENS FOV

T=4 SEC   T=0

4 FT/SEC

FIGURE 10-21   Relationship between pan/tilt speed and moving target

area with a single camera. The presence of the camera system serves as a deterrent to crime while it captures the necessary visual information for possible apprehension and prosecution. The same scanning limitations as described in the previous system apply.

10.11 THE VALUE OF SLIP RINGS

Most pan/tilt mechanisms use a mechanical stop at each end of the horizontal and vertical panning and tilting motion to prevent the wires connected to the moving camera/lens assembly from getting twisted (the wire ends are terminated in the stationary wall mount). This means that the camera cannot scan more than 355 degrees horizontally before it must stop and then pan in the opposite direction. Even at a 24-degree-per-second pan speed, nearly 15 seconds is required to reacquire a subject or target that is moving past the end of the panning range (see Figure 10-21a). During most of this 15 seconds, the target is out of sight of the camera and probably lost.

In the continuous panning system using slip rings (Figure 10-21b), the camera/lens combination rotates continuously—beyond 360 degrees—without any concern for

twisted wires, since the electrical signals and power pass through stationary slip rings and contacts on each slip ring. No matter where the target moves in the lens FOV, the pan motion can continue: the subject never leaves the FOV. There are no restricting mechanical stops to limit the pan/tilt unit's rotation. Since there are no cables and harnesses to flex back and forth as the camera/lens platform pans or tilts, there is also a significant improvement in reliability.

An example in which panning speed and continuous motion (more than 360 degrees) is important follows: A person walks past a pan/tilt unit 15 feet away the dome (Figure 10-21b). If the person is walking at a normal rate of about 4 feet per second and the dome is panning at a rate of 1 foot per second (12 degrees per second), the monitor scene at 15 feet is moving at a rate of 3 feet per second. The subject is quickly lost because the pan/tilt cannot pan fast enough to follow the subject. With a high-speed 60-degrees-per-second panning system, a target at 15 feet from the camera produces a picture going by at a rate of 5 feet per second (1 foot per second faster than the target) and the subject is not lost. In this example, the panning speed would be reduced to 4 feet per second to keep the target in the center of the picture.

FIGURE 10-22 Low-profile pan/tilt in dome housing

Another system providing an integrated package for pan/tilt/zoom is designed into a low-profile section of a hemisphere, protruding only 7 inches below and $7\frac{1}{2}$ inches above the ceiling (Figure 10-22).

The system has a maximum panning speed of 18 degrees per second and is capable of a full 360-degree rotation through the use of slip rings. The maximum tilt speed is 12 degrees per second. A single joystick is used to control the pan and tilt motion and contains a latching on/off push-button control for auto-pan. It has a 12.5-to-75-mm f/1.2 zoom lens providing a 6:1 zoom ratio. Other lens options include 8.5, 16, 35, and 70-mm FLs with manual or automatic iris. The system contains a unique camera-mirror design requiring less space for tilt, permitting the low profile below the ceiling. An inner black dome hides the camera from view to create a dome effect. The clear outer dome does not add any appreciable light loss to the optical path that the light must take (optical train), thereby providing pictures of almost equal quality to systems having no dome. The monitor screen can be annotated with time/date, the name of a particular location (or facility) or alarm point, camera identification, and so on.

10.12 HIGH-SPEED PAN/TILT SYSTEMS

The availability of small cameras and lenses has caused a major change in the design of new pan/tilt mechanisms. The new units are much smaller and provide a higher scan rate because of the lower camera, lens, and enclosure weight and inertia.

Figure 10-23a illustrates a very small pan/tilt unit capable of moving at an angular rate of 60 degrees per second. This system is suitable for mounting on a wall, in a ceiling, or within a small hemispherical dome. It receives control signals via a two-wire, Touch-Tone, or rotary-dial telephone system, which simplifies communications to and from the camera unit. Since it operates on the Touch-Tone system, it can function from any two hard-wired or wireless remote sites having a telephone-type communication system.

Another system available is a high-velocity, rate-proportional-control, digital-tracking system designed for high-speed positioning of a camera and lens load (Figure 10-23b). The system uses slip rings to permit 360-degree continuous rotation, scan speeds from 0 to 80 degrees per second, and tilt speeds from 0 to 25 degrees per second. At these high speeds, dynamic braking is used to provide a plus-or-minus 0.5-degree accuracy of stopping at any speed. Zoom lenses used with this design include the option of a 6:1 or a 10:1 zoom ratio. The system uses a solid-state CCD camera for long life and is often mounted in a hemispherical dome to protect and disguise it.

The system is operated via keypad and joystick control, which is a rate-proportional device: the farther the joystick is moved from the center off position, the faster the speed of the function to be performed. Move the joystick forward or backward and the pan/tilt unit tilts. Move it to the left or right and the unit scans (pans). Move the joystick at an angle between the two and the pan/tilt unit pans and tilts at the same time. Turning the joystick knob clockwise or counterclockwise initiates the lens zoom function. Preset

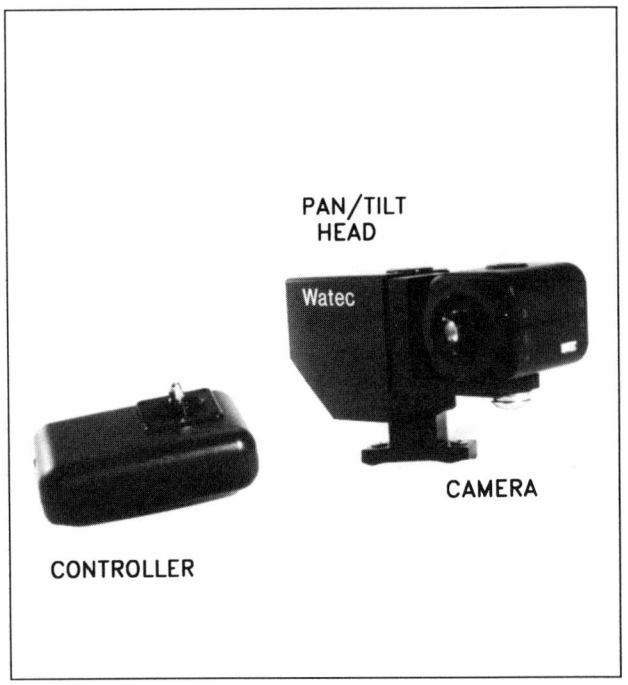

FIGURE 10-23 High-speed pan/tilt systems

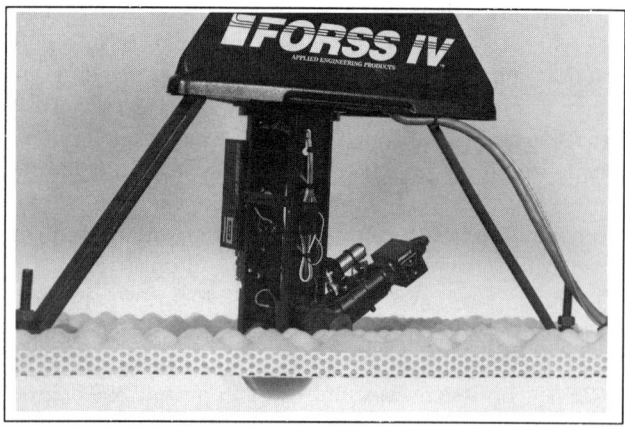

FIGURE 10-24 Low-profile pan/tilt mirror system

camera and lens parameters are entered into the unit via the keypad. By pressing a two-digit number on the keypad, any one of 99 presets can be executed. The information contained in each camera preset location includes the pan angle, tilt angle, zoom FL, iris opening, focus setting, and camera on/off. This system controls up to 32 cameras, drives two or four output monitors, and is capable of processing alarm inputs to direct the camera at the alarm location.

A unique pan/tilt device capable of high-speed pan and tilt is shown in Figure 10-23c. This system can pan at a rate of 18 degrees per second and tilt at a rate of 12.5 degrees per second. It is small and lightweight, which enables it to scan quickly. Control is via joystick, providing proportional-rate control of pan, tilt, and zoom.

Another system uses a very fast pan/tilt mechanism pivoted about the front lens, so that only a 5-inch dome is required for concealment (Figure 10-23d). The lightweight, integrated assembly weighs less than 10 pounds so that it can be installed above the ceiling. The small dome minimizes dome intrusion into the room without sacrificing function. The unit features a very fast panning speed of 180 degrees per second and tilting speed of 80 degrees per second. Braking accuracy is plus-or-minus 0.5 degrees. The system is interconnected to the control console and monitor by a single coaxial cable. The resolution is sufficient to read serial numbers of currency at a distance of 12 feet through the 5-inch dome.

Another unique pan/tilt system uses a vertically mounted camera and pinhole lens, which views the scene through a 45-degree mirror (Figure 10-24). The camera, lens, and mirror assembly are rotated on a vertical axis, as shown. This design provides an aesthetic package for wall mounting on interior or exterior locations with the ability for rapid panning and tilting. Since the panning is accomplished by rotation about the camera/lens/mirror axis, only a small volume is required to house the entire system. Because the vertical scanning—typically 30 degrees—is less

than that obtainable with a standard pan/tilt system, it occupies less volume.

10.13 COVERT MIRROR PAN/TILT SYSTEMS

A panning-mirror scanning system having no similarity to those previously discussed is shown in Figure 10-25. It consists of a motorized camera/lens/mirror combination to provide horizontal panning. The only part of the system in view of the observer is the small mirror, which scans horizontally to provide viewing in any azimuth direction. The camera, lens, and panning system are mounted above the ceiling level. The use of slip rings provides full 360-degree continuous rotation, with its attendant advantages.

A more comprehensive pan/tilt/zoom system using the covert mirror principle is shown in Figure 10-26. In this system, the small mirror is the only element seen by the observer. The camera, zoom lens, and pan/tilt mechanism are located above the ceiling, completely hidden from view. Panning is accomplished by rotating the mirror and camera/lens mechanism about the vertical axis. Tilting is accomplished by moving the camera, which translates linear motion into tilting motion. Zooming is accomplished via a zoom-pinhole-lens combination between the viewing mirror and camera. This system does not have the full tilting range capability of the normal pan/tilt system, nor the image quality and optical speed of standard lenses, but it nevertheless provides sufficient tilting for many applications and is a truly covert pan/tilt/zoom CCTV camera system.

10.14 GIMBAL MOUNTING

A camera/lens mounting configuration used to compensate for motion in three mutually perpendicular axes—pitch, roll, and yaw—is used on moving vehicles, aircraft, and ships for pointing and image stabilization, as shown in Figure 10-27.

The stabilized gimbal-mounted camera is used extensively in helicopter CCTV surveillance systems by law-enforcement agencies. Unlike a stabilized lens (see Chapter 4), which compensates for unwanted vibration and rotation, the stabilized gimballed platform can compensate for unwanted vibration and rotation over its full range of elevation and azimuth. The outermost gimbal compensates for front-to-back pitching (tilt) and the inner gimbal for side-to-side rolling. When these two motions are compensated for, the center rotating platform is stable (maintained horizontal). Azimuth pointing (panning) is achieved by rotating the camera/lens platform. Since the vehicle is usually pointed in the general direction of the target, the two gimbals and platform angular travel are limited to 90 degrees or less. The two-gimbal camera/lens combination is balanced and centered on all three axes to produce a

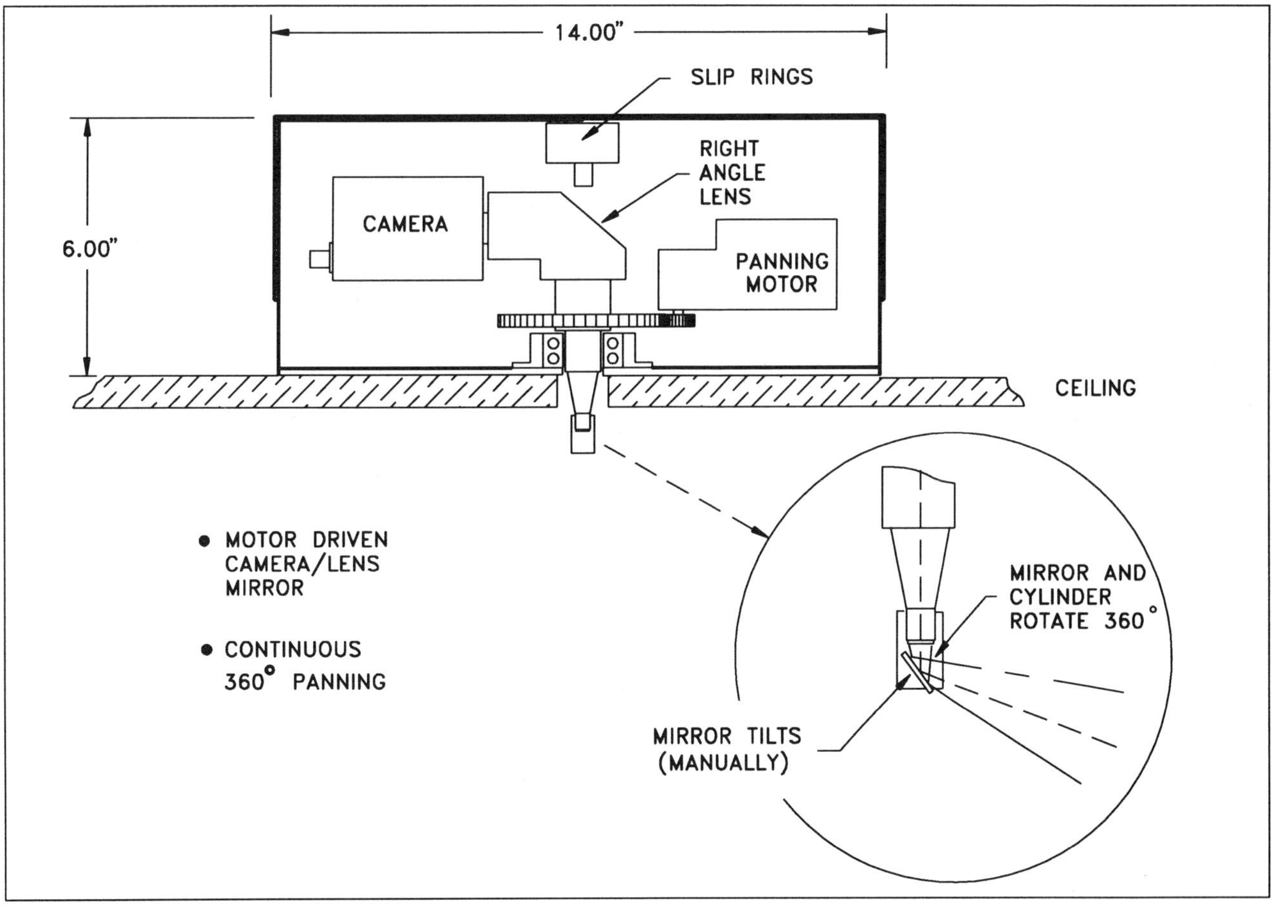

FIGURE 10-25   Covert mirror panning system

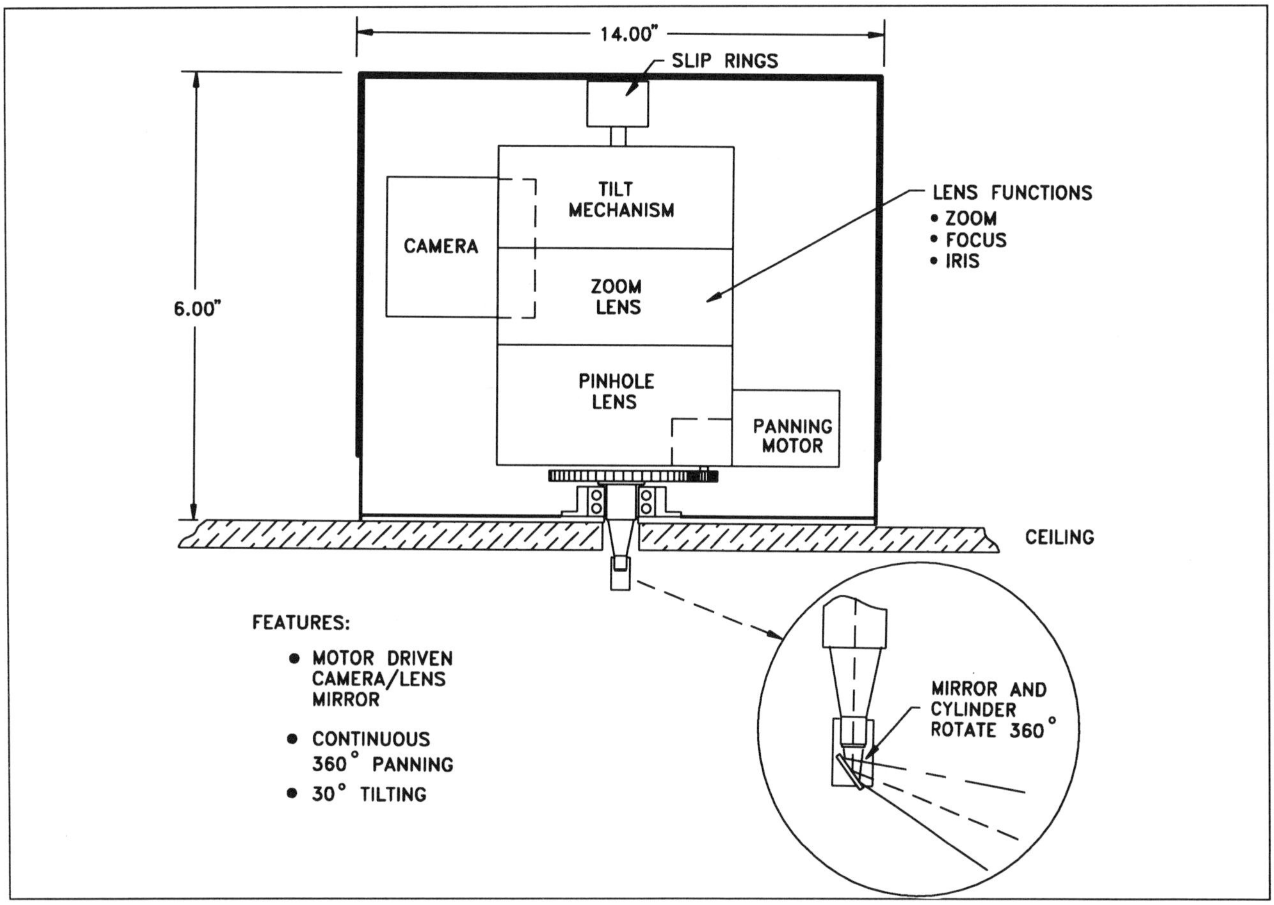

FIGURE 10-26   Covert mirror pan/tilt/zoom system

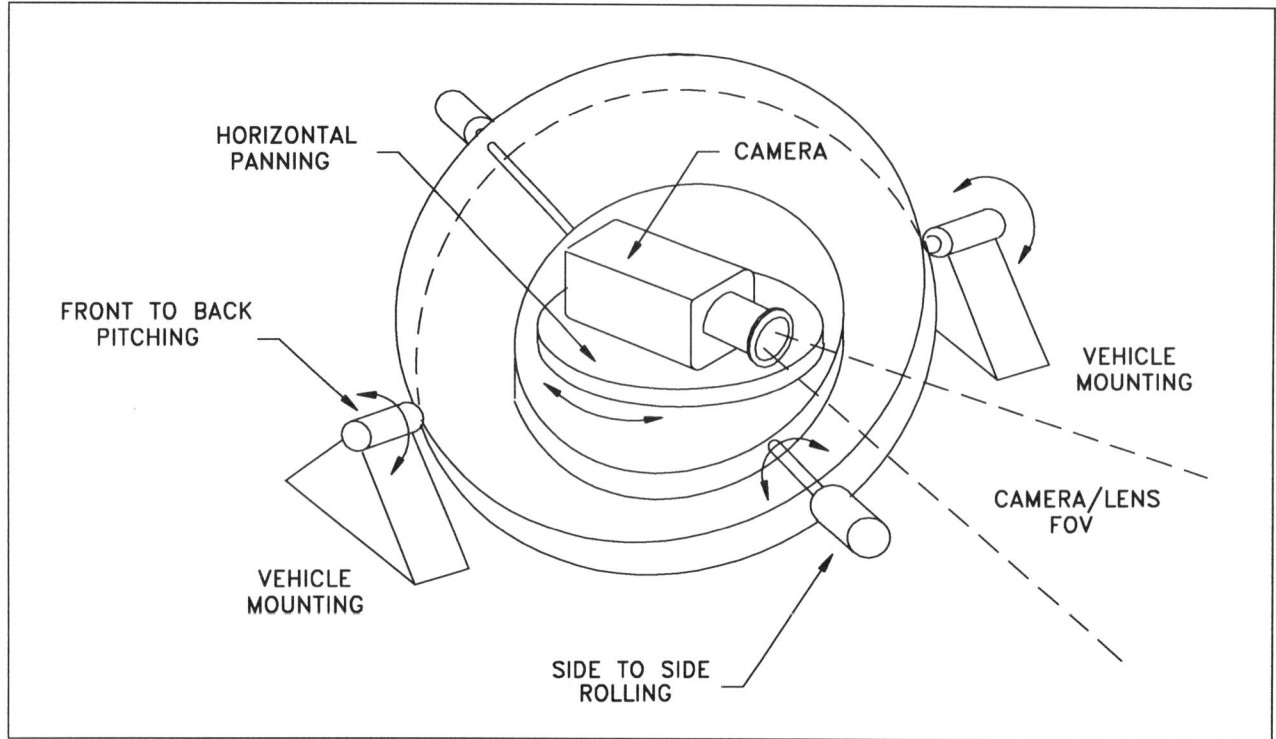

FIGURE 10-27 Gimballed platform

completely symmetrical loading on the vertical (pitch), horizontal (yaw), and roll axes which aids in stabilization.

The gimballed configuration is not in widespread use because of its high cost, but it must be used when the CCTV camera package is to be mounted on an unstable platform.

10.15 SERVICING

Modern integrated CCTV security systems are assembled from many different components to provide the final result: (1) the picture on the monitor, (2) the image recorded on the VCR, or (3) the hard-copy printout on the video printer. The part of the system with the lowest reliability and most responsible for downtime and problems is the pan/tilt mechanism. Components causing problems include (1) kinked cabling at the pan/tilt mechanism, (2) burned-out motors, (3) failed gear trains, and (4) broken mechanical stops. This section addresses some of these problems, identifies methods to prevent many of them, and outlines procedures for good system design.

10.15.1 Cabling

The pan/tilt mechanism component requiring most servicing is the cabling at the interface between the moving head and the stationary mount. This constantly flexing and moving cable will eventually break and fail. Most pan/tilt

systems contain multiple conductors for control, power, and the video signal. These conductors flex every time the camera is pointed to a new direction or if wind causes the cable to vibrate. To maximize the lifetime of these cables, they must be specified and installed correctly. The cables must be dressed (tied or clamped) to the housing and support structure at proper locations, so that over the full motion of the pan/tilt head there is no undue stretching or flexing, or any interference between the cable and any of the moving or stationary parts. Since installations are often in high or difficult-to-reach locations, failure often occurs because the cables have not been installed or dressed properly.

Different grades of wire are suitable for outdoor pan/tilt system use, each with different qualities and lifetimes. The outdoor cable types for pan/tilt mechanisms include coiled cables, in which the coaxial and control cables are enclosed in a single coiled cable that looks like a standard telephone headset cord, protecting them from the environment and overbending.

To minimize downtime in the event of a cable failure, the flexible cable between the pan/tilt mechanism and the stationary bracket should be installed so that a failed cable can be replaced quickly.

Several manufacturers use slip-ring (Section 10.8) assemblies, where all cables feed through the center shaft of the pan/tilt unit so that no flexing cables are on the outside. All cable flexing is done internally, which by design makes for a more reliable system. The cable is always protected from damage during installation or the environment and will not deteriorate and become brittle, a cause of premature failure. Slip rings transfer all power, control, and video signals at the pan/tilt camera site to the transmission cable, which provides a significant improvement over the other cabling methods. These slip rings are concentric rings mounted around the shaft with electrical contacts that slide on the rings, providing transmission of the electrical voltage and current from the stationary mount to the rotating shaft and vice versa. Slip-ring technology has been used in other industries for many years and is now being used in the security industry because of the increased need for reliability. This technique is more expensive than using flexible cables during initial installation. However, when installation plus maintenance costs over the life of the equipment and the possible cost of loss of security are taken into account, slip rings are a good investment. Properly designed slip-ring assemblies provide many years of reliable operation in addition to the extra security of continuous 360-degree rotation.

10.15.2 Mechanical

Most pan/tilt systems use motors and gear trains, which operate satisfactorily over long periods of time without attention under "normal" conditions. Unfortunately, many installations are not normal, such as when careless maintenance or inadvertent abuse causes premature failure. While most manufacturers test their designs and equipment thoroughly at the factory and the installers install them properly, often under adverse conditions, sometimes hazardous circumstances on tall buildings or poles inflict abnormal abuse on the systems. These can include inadvertent mechanical overload on the pan/tilt axes during installation or at some other time, which may cause permanent damage to the gear train or motor. Factors such as breakaway, when a unit has been frozen by ice, can sometimes cause overheating of the motor and eventual damage and failure. While the system is being used under this condition, the operator may not be aware that an overload is occurring and will unknowingly damage it.

Sometimes a mechanical stop will go out of adjustment and the motor will drive the unit beyond its end travel and continue driving, possibly overheating the motor or causing damage to the gear train. The operator may not be aware of this condition and attempt to continue operating with a damaged system. Some manufacturers have designed out these possible failure modes, providing a more reliable system. The use of potentiometers, optical, or magnetic encoders to determine the pan/tilt pointing direction (azimuth and elevation), as well as using this signal to shut off

the motor, provide improved reliability over mechanical means. As other components in the CCTV security system achieve longer mean time between failure, pressure will be placed on pan/tilt mechanism manufacturers to produce more reliable, longer life systems in which the design and component reliability will ensure life compatibility with the other components of the CCTV system.

10.15.3 Environmental

Precipitation in the form of water, snow, sleet, or ice causes deleterious long-term effects on pan/tilt mechanism parts. All components, such as mechanical hardware or brackets, are not always corrosion-resistant; as they age, they corrode and become loose or fail. A typical failure occurs in a loosened shaft mechanical stop which fails to stop the motor (via the microswitches installed), thereby causing damage to the motor and gears. Depending on the design, water entering between a housing and the shaft, as well as accumulation of dirt with the water or grease, can jam the drive, thereby causing failure or intermittent, rough operation of the unit. While close scrutiny of the mechanical design suggests some information on its future performance, feedback from dealers, installers, and end users provides better information with which to make an intelligent choice.

It is difficult to tell quantitatively how much dirt or dust or corrosive material in the pan/tilt environment will eventually cause deterioration in its performance. The best approach is to perform periodic maintenance and inspection to determine if there is a problem. Manufacturer's lubrication or inspection recommendations should be followed, and any questions should be addressed to the manufacturer for additional preventive maintenance.

10.16 SUMMARY

A crucial requirement of any CCTV security surveillance system is its ability to see as large an FOV as possible with as much clarity (resolution) as possible. Since these two parameters work against each other (wide FOV means low resolution and vice versa), a pan/tilt camera/lens platform with a zoom lens solves some of these problems. When a pan/tilt/zoom system is used, dead zones are created, and there is less than 100% viewing of the entire system FOV. New fast-scan systems reduce the dead zones, but at significant increase in cost. To ensure reliable operation of the pan/tilt units, preventive maintenance should be performed at regular intervals.

Chapter 11
Camera Housings and Accessories

CONTENTS

11.1 OVERVIEW

The two primary functions of any environmental housing are to protect the camera and lens from vandalism and protect them from the environment. Some housings need to be attractive and unobtrusive. To meet these requirements, indoor and outdoor housings are fabricated from steel, aluminum, or high-impact plastic. The CCTV housing enclosure should be large enough to contain the lens and camera, with access via a hinged or removable cover. Figure 11-1 shows two examples of standard rectangular indoor and outdoor CCTV housings in common use today.

There are many varieties of CCTV camera housings available for indoor and outdoor security applications. The shapes and forms they take include (1) rectangular housings (the most common and functional, mounted on brackets); (2) wall or ceiling housings; (3) round domes mounted on a ceiling, wall, or pylon; (4) triangular housings, mounted in a corner, along the ceiling-wall interface, or in an elevator; and (5) housings mounted on the ceiling at an angle.

While environmental housings were first used for functional purposes, there is an increased demand for aesthetically designed housings to match the decor of a building interior or exterior, as specified by the architect. Housings used on facility properties are designed to match landscaping and grounds and/or specific lighting conditions. Many security installations require discreet CCTV surveillance equipment, not eye-catching or obtrusive, but blending in with the surrounding environment.

Indoor housings are used to protect the lens, camera, and wiring from vandalism. The housing, camera, and lens are often within reach of personnel who could otherwise damage or remove the equipment. Of particular concern are high-risk locations such as jail cells and public-access locations, which require more rugged housing materials: stainless steel or high-impact polycarbonate plastic materials. The only adverse indoor environment the camera/lens needs protection from is dust, dirt, or some corrosive atmosphere.

Outdoor environmental housings, which can be subject to wind loading or ice buildup, should be no larger or heavier than required to house the camera, lens, and associated

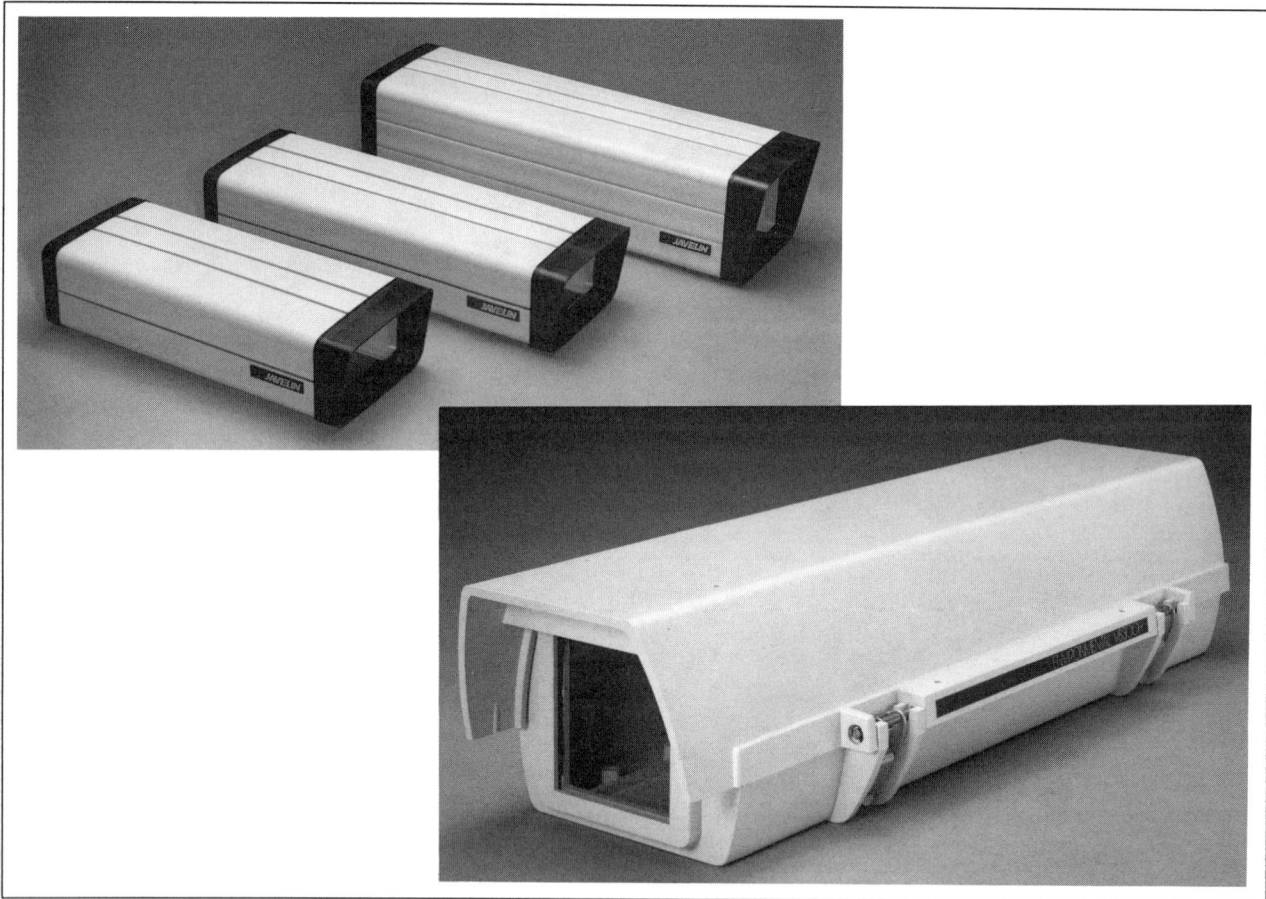

FIGURE 11-1 Standard indoor and outdoor camera housings

wiring and accessories. They should be constructed to withstand the harsh outdoor environment and added abuse from vandalism or attack.

Housings are used to protect vital electronic CCTV surveillance equipment; consequently, the material used for their construction must be chosen carefully. Underwriters Laboratories (UL) has developed guidelines for minimum fire-safety requirements and suggested tests and ratings for fireproof or fire-retardant designs. Likewise, the Electronic Industries Association (EIA) is developing guidelines for improved interchangeability among manufacturers' products. These guidelines relate to mechanical design parameters, such as mounting-hole locations and electrical-cable entry and fitting requirements.

The following sections describe rectangular, dome, and other special indoor and outdoor housings, as well as accessories such as heaters, fans, thermostats, and windshield wipers and washers.

11.2 INDOOR HOUSINGS

Indoor housings must protect the camera and lens from pollutants such as dust and other particulate matter, a corrosive atmosphere, and tampering or vandalism. Indoor

housings are constructed of painted or anodized aluminum, painted steel, brass, and plastic. The material for plastic housings should be flameproof or flame-retardant, as designated by local codes and UL recommendations. The housings must have sufficient strength to protect the lens and camera, and be sturdily mounted onto a fixed wall or ceiling mount or recessed in a wall or ceiling. The lens should view through a clear window made of safety glass or plastic (preferably polycarbonate). The electrical input/output access locations should be designed and positioned for easy maintenance. For easy access and servicing of internal parts, the top half of the housing should be hinged or removable. In some designs, the entire camera/lens assembly is removable for servicing. Figure 11-2 shows a typical indoor housing with its interior exposed.

The common rectangular housing is available in many sizes and is the least expensive. For vandalism protection, many housings are available with locks or tamper-proof hardware, which makes the cover removable only with a special tool. This prevents tampering with the lens and camera. In very high risk areas, welded stainless-steel housings with thick polycarbonate windows ($\frac{3}{8}$- or $\frac{1}{2}$-inch) and high-security locks are used. Some housings are designed to provide concealment and improved aesthetics by recessing them into the wall or ceiling. The four housing types

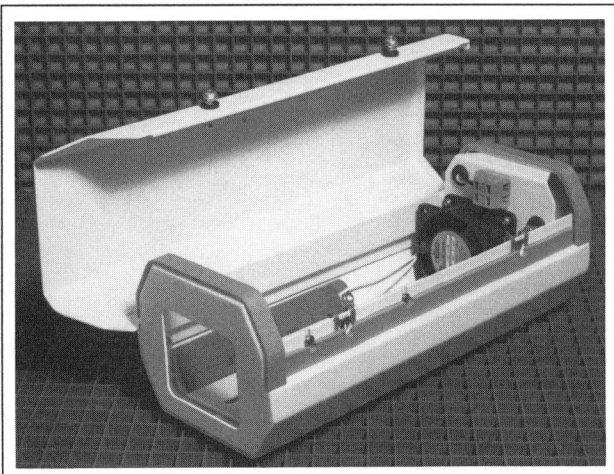

FIGURE 11-2 Indoor housing showing interior

that account for most security installations are (1) rectangular, (2) dome, (3) triangular, and (4) recessed.

11.2.1 Rectangular

The most popular type of housing is the standard rectangular design, since it can be fabricated at low cost, is sturdy, and is available from many manufacturers in many sizes and attractive styles.

Under normal circumstances, indoor housings do not require any special corrosion-resistant finishes. The housings are made from painted or anodized aluminum, painted steel, or high-impact plastic, such as polyvinyl chloride (PVC), acritile buterated styrene (ABS), or polycarbonate (such as Lexan).

Accessibility to the camera/lens assembly for installation and servicing is important. CCTV surveillance cameras are always mounted near ceiling height, on a pedestal, or at some elevated location requiring service personnel to be on ladders or power lifts. The housing design must permit easy access and serviceability under these conditions. Manufacturers provide one of several means to gain access to the housing: (1) removable top cover, (2) hinged top cover, (3) top cover or camera/lens on slide, (4) removable front and/or rear cover, (5) hinged bottom cover (dome), or (6) top cover on slide (Figure 11-3).

11.2.2 Dome

A second category of indoor housing consists of a round or hemispherical clear or tinted dome, in which a camera, lens, and an optional pan/tilt mechanism are housed. The ceiling-mounted hemispherical dome and the below-the-

ceiling and wall-mounted domes on brackets look totally different from the rectangular housing and often blend in better with architectural decor. Since they look like a lighting fixture, they are less obtrusive than rectangular housings. Since the hemispherical dome is circularly symmetrical, it can be in a fixed position and the CCTV camera pointed in any direction to view the scene. A pan/tilt unit used in a dome can rotate and tilt the camera and lens while still remaining inside the confines of the dome. This contrasts to cameras inside rectangular housings: if the camera moves, the entire housing assembly has to move as a unit.

If the dome is tinted so that the person down at floor level viewing the dome cannot see the camera and lens, it is possible to point the camera in any direction without the observer seeing it move. This capability can act as an additional security deterrent, because the observer does not know when he or she is under surveillance.

There are three different types of plastic dome materials through which the lens can look: (1) clear, (2) coated (with semitransparent aluminum or chrome), and (3) tinted or smoked. When the dome housing is used for protection only and its pointing direction need not be concealed, the clear plastic dome is the best choice, since it produces only a small (10 or 15%) light loss. If the camera's pointing direction is to be concealed for additional security, a coated or tinted dome is required. The aluminized dome is the earliest version of the coated dome; it attenuates the light passing through it by approximately two f-stops (equivalent to approximately a 75% light reduction). While this type of dome is still in use, the preferred dome material is a smoked plastic or tinted plastic, which attenuates light approximately one f-stop, or 50%.

In contrast to rectangular housings using flat plastic or glass windows with excellent optical quality and transmission, all dome systems add some degree of optical distortion to the video picture. In high-quality domes, the image distortion is almost negligible, but in many systems the distortion or loss in resolution is noticeable and even objectionable. In any dome-housing application, the camera/lens must view through the surface of the dome perpendicularly, as shown in Figure 11-4a.

Under this condition, there is at least a symmetry of distortion, that is, the primary effect is that of a weak lens producing a small change in the focal length of the total lensing system, which may not be objectionable or noticeable. If the camera/lens pointing axis is not perpendicular to the dome surface (Figure 11-4b) and looks at an oblique angle through the dome housing material, noticeable distortion will occur; for example, images may appear elongated vertically or horizontally. If the dome and camera are in a fixed position with respect to one another, the distortion is generally less noticeable than if the lens is scanning or tilting while the dome remains still. Figure 11-4c shows two widely used dome housing configurations.

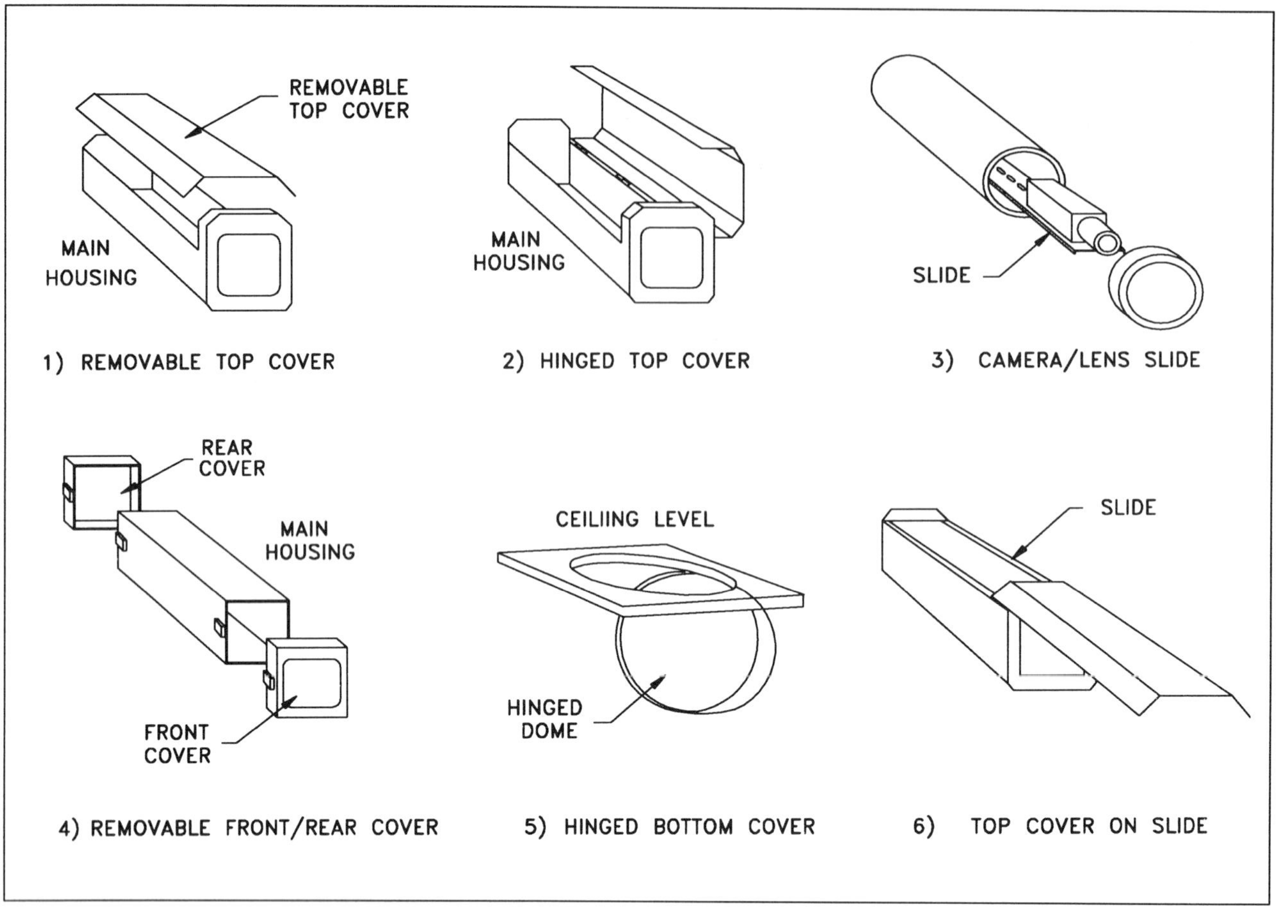
REMOVABLE
TOP COVER
MAIN
HOUSING
MAIN
HOUSING
SLIDE
1) REMOVABLE TOP COVER
2) HINGED TOP COVER
3) CAMERA/LENS SLIDE
REAR
COVER
MAIN
HOUSING
FRONT
COVER
CEILIING LEVEL
HINGED
DOME
SLIDE
4) REMOVABLE FRONT/REAR COVER
5) HINGED BOTTOM COVER
6) TOP COVER ON SLIDE

FIGURE 11-3 Camera housing access methods

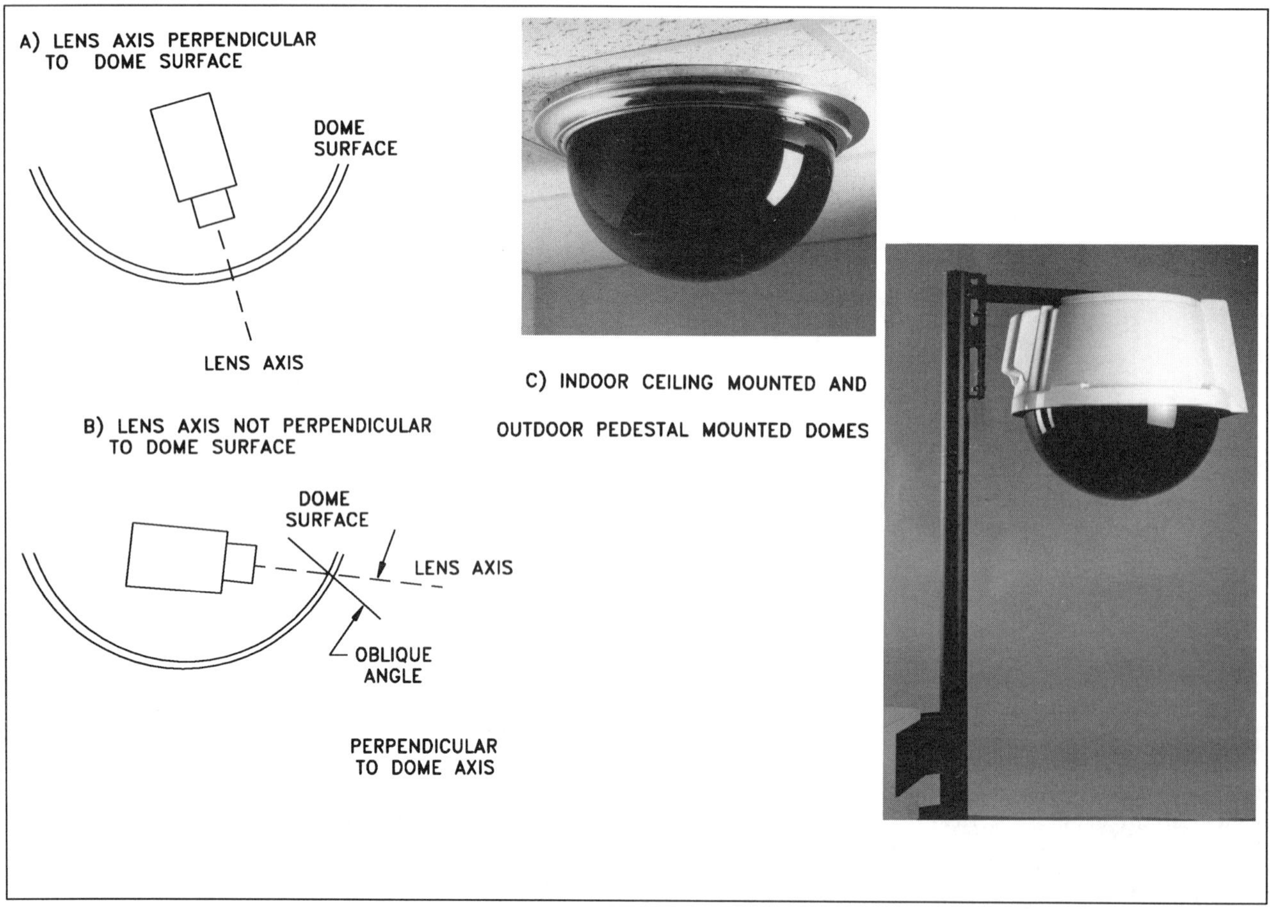
A) LENS AXIS PERPENDICULAR
TO DOME SURFACE
DOME
SURFACE
LENS AXIS
B) LENS AXIS NOT PERPENDICULAR
TO DOME SURFACE
DOME
SURFACE
LENS AXIS
OBLIQUE
ANGLE
PERPENDICULAR
TO DOME AXIS
C) INDOOR CEILING MOUNTED AND
OUTDOOR PEDESTAL MOUNTED DOMES

FIGURE 11-4 Dome camera housings

A) STAINLESS STEEL
   WIDE FOV

B) HIGH IMPACT PLASTIC
   WIDE FOV

C) STEEL
   ADJUSTABLE MIRROR
   MEDIUM FOV

FIGURE 11-5   Triangular ceiling/corner mount housings

11.2.3 Triangular

The triangular housing is mounted in the corner of a room at the intersection of two walls and the ceiling. Typical locations may be in a small room or lobby, an elevator, a stairwell, or a jail cell. Figure 11-5a shows a design used with a wide-angle (90-degree FOV) lens installed extensively in elevators.

The camera is installed in the triangular housing at an oblique angle, pointing down from the ceiling horizontal toward the area of interest; the camera viewing window is perpendicular to the lens. One version has a stainless-steel housing and polycarbonate window, which would be most suitable for an elevator cab (see Section 11.5.2). The hinged front cover is secured with a lock and swings out for servicing or removal of the camera assembly. Another version is a high-impact plastic housing used extensively in small rooms (Figure 11-5b).

A unique corner camera housing is shown in Figure 11-5c. This design fits into the corner of a room at ceiling height and contains a camera and lens pointing up toward the ceiling; a first-surface mirror reflects light to the lens, allowing the camera to view the scene. The mirror is adjustable in azimuth and elevation so that the camera can point to different areas in a room. Since the mirror reverses the scene image, the camera scan must be reversed to make it normal (see Section 13.4). The system works well with lenses having 75-mm narrow-angle to 6-mm wide-angle FLs. On a $\frac{1}{2}$-inch CCD sensor, this corresponds to a horizontal angular coverage of 5 to 56 degrees, respectively. This attractive housing is suitable for front lobby installations and other areas where an unobtrusive corner mount design is required.

11.2.4 Recessed in Ceiling or Wall

Recessed, partially concealed housings are often mounted in ceilings and walls. Figure 11-6 shows examples of these housings, including rectangular, triangular, and dome-shaped.

The round, semicircular, and tapered housings shown in Figure 11-6 offer design flexibility, since the camera and lens can be pointed in any horizontal direction while the square or rectangular ceiling tile remains in place. These housings are used where a low-profile, but not covert, type of CCTV surveillance camera is required. In ceiling installations, most of the housing, camera, lens, and accessories are mounted and suspended above the ceiling level. The only portion showing below ceiling level, and observable by the person below, is a small part of the housing and the window through which the camera lens views. The cameras and lenses are accessible from below ceiling level by unlocking a cover that swings down or by gaining access to the portion of the housing above the ceiling from an adjacent ceiling tile. It is important for all ceiling-mount housings to be permanently attached with a safety chain or cable to a structural member of the building above the ceiling, so that if the hanging ceiling support fails, the housing and contents do not fall to the floor, possibly injuring personnel.

With the increased use of CCTV security in public locations, be they industrial, private, or government buildings, more attention is being given to the decorative and aesthetic features of the housing. Figure 11-7 shows examples of decorative protective housings for installations requiring an aesthetic design.

These housings are often available in finishes of brass, gold, and chrome, with satin, textured, or highly polished

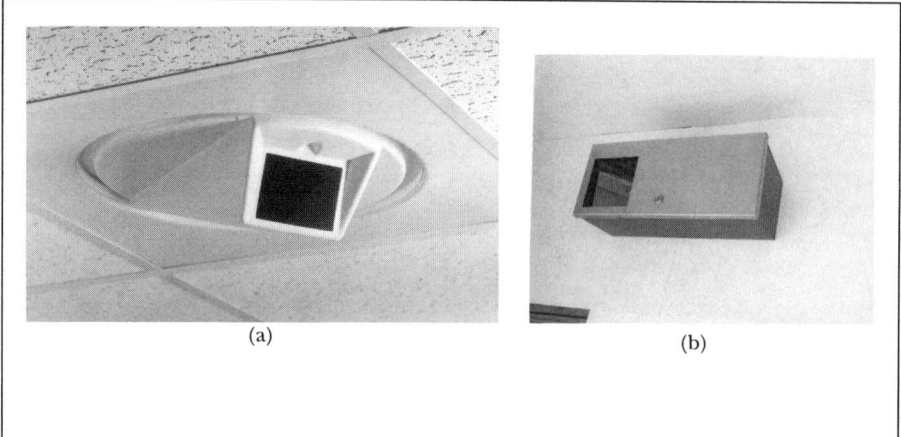

FIGURE 11-6 Recessed ceiling (a) and wall (b) housings

custom paint colors and textures, as well as clear or colored plastics. Some manufacturers offer special shapes and custom configurations for matching specific architectural designs.

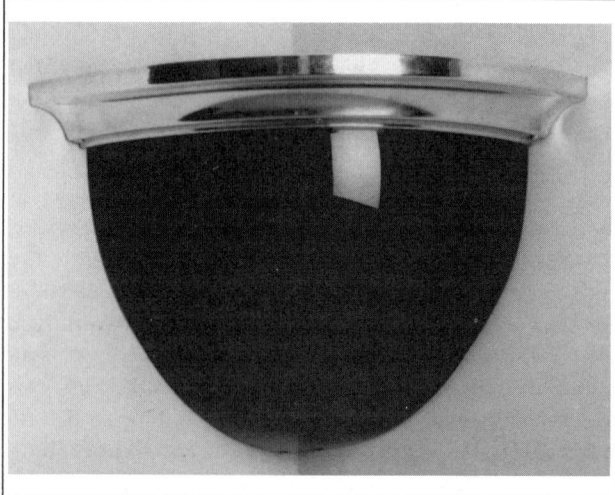

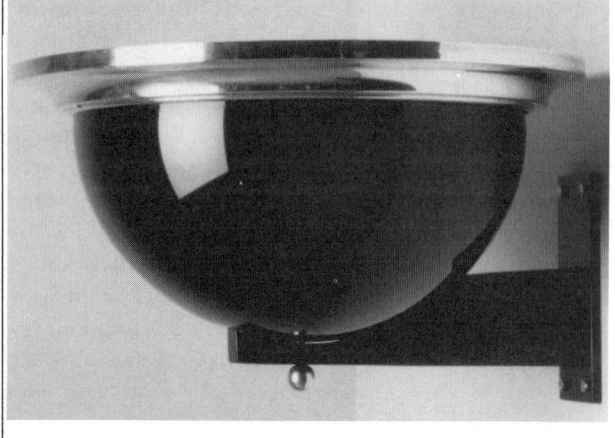

FIGURE 11-7 Aesthetic and decorative housings displaying corner and wall mounting

11.3 OUTDOOR HOUSINGS

Outdoor housings must protect the camera from vandalism as well as adverse environmental conditions. The vandalism encountered can range from rocks or sticks thrown at the housing to gunfire. Since these security housings are mounted on pedestals or building walls, they are prime targets.

In outdoor installations, environmental factors such as precipitation—rain, hail, snow, sleet, ice, and condensing humidity—require that the camera be mounted in a protective enclosure. The types of particulate matter encountered outdoors include dirt and dust, sand, fly ash, soot, and any other material local to a particular site.

In outdoor locations, the presence of a corrosive atmosphere, whether natural or artificial, can cause rapid deterioration, failure, and premature replacement of the camera and lens if not properly protected. These substances include industrial chemicals, acids, and salt spray. Outdoor housings should have external finishes that withstand the atmosphere in which they are to operate. In hot climates, a bright aluminum or white finish is desirable, to reflect most of the heat.

11.3.1 Outdoor Housing Types

Outdoor housings share many of the same requirements as indoor housings. Accessibility to the camera and lens during installation and maintenance are more important, since outdoor CCTV equipment is often mounted high above the ground and serviced under adverse conditions.

11.3.1.1 Rectangular

In outdoor applications, the rectangular plastic or metal housing is the most popular choice. These housings are easily mounted from a bracket on a building, wall, or pole, or hung from a building overhang to provide a solid mounting.

11.3.1.2 Dome

Domes are mounted on an individual pole or pylon, under the eaves of a building, or on a bracket mounted off the wall of a building. The portion of the dome that is being used for viewing has its surface pointing downward and tends to be self-cleaning; however, water droplets on the surface will reduce visibility (see Section 10.14.3).

11.3.1.3 Triangular

The triangular housing is mounted outdoors in areas where two walls meet, resulting in a very unobtrusive installation. When mounted in hot or cold environments, the housings in Figure 11-5 are provided with a heater or fan and thermostat.

11.3.2 Outdoor Design Materials

Outdoor housings are manufactured from aluminum, painted steel, stainless steel, and outdoor-type plastic. It is important that plastic outdoor housings be fabricated from ultraviolet-inhibiting materials, to prevent the housing from deteriorating due to sunlight. Plastics not treated will crack, and colors will fade. High-quality baked-enamel, painted-steel, and stainless-steel housings will last many years. Where long-lasting, high-security, vandal-proof housings are required, stainless steel is the choice because it does not rust or corrode and is extremely tough.

Aluminum is a good choice for an outdoor housing when it is properly finished in a baked polyurethane enamel paint, anodized, or anodized and painted, which is the most durable finish. Aluminum and steel housings should not be used when a corrosive atmosphere is expected. Stainless steel and special plastics are the best choice for a salt-spray environment. Consult the housing or materials manufacturer for the proper choice.

11.4 INDOOR/OUTDOOR ACCESSORIES

Like the indoor housing, the outdoor housing protects the camera and lens from vandalism, and it also keeps out the outdoor environment. Many housings are provided with key locks to prevent opening the housing and tamper switches to activate an electrical alarm if the housing is opened. The environmental factors from which the camera and lens must be protected include precipitation, blowing sand and dirt, corrosion, and temperature extremes.

In warmer climates where the temperature does not drop below freezing, only a fan and thermostat are required to maintain suitable operating temperatures for the camera and lens. The thermostat is designed to automatically turn on the fan when the temperature in the interior of the housing rises above some value—usually between 90 and 100°F (32 to 38°C)—and turn it off when it falls a few degrees below the set temperature. In cold climates, a heater and thermostat are used to keep the lens and camera above about 45 to 55°F (7 to 13°C). The heater prevents condensation on the window and lens and keeps the automatic-iris mechanism and camera operative. In freezing weather, it prevents moisture from freezing within the environmentally enclosed housing. The thermostat applies power to the heater when the temperature goes down below the dew point. When the camera-lens housing is located in an interior close-to-the-ceiling environment or in an outdoor warm environment, a thermostatically controlled fan is used to cool the camera/lens combination. The fan should contain a removable filter that can be cleaned or replaced periodically.

Another accessory is the window washer and wiper. If the housing is rectangular and pointing down at 15 or 20 degrees or more, it is generally unnecessary to provide the housing with a window wiper and washer, as rain will run off the window, along with dirt, and allow proper viewing. If, however, the housing is located in a dusty environment or is in a more horizontal direction, it is advisable to include a window washer/wiper assembly. This assembly is mounted below and in front of the window and operates like an automobile washer/wiper system. The wiper motor and liquid washing pump is energized remotely from the control console.

Most environmental housings, indoor or outdoor, are supplied with plastic or safety (tempered) glass windows for the lens to view through. For indoor applications, these windows may be acrylic, polycarbonate, or glass, depending on the design. For outdoor applications, they are usually polycarbonate or safety glass. The choice of acrylic versus polycarbonate depends on whether the application is to be maximally tamper-proof or only moderately so, and whether the housing is used indoors or outdoors. Acrylic cracks or breaks when it is exposed to temperature extremes or if it is hit. Polycarbonate, in contrast, will only be dented under impact (it will not crack) and will remain optically clear under normal cleaning action and withstand outdoor weathering. For maximum resistance to scratching and abrasion from cleaning, mar-resistant polycarbonate (and acrylic) material is available and is highly recommended. Window thicknesses range from $\frac{1}{8}$-inch for light duty to $\frac{1}{4}$-inch for normal service to $\frac{3}{8}$- to $\frac{1}{2}$-inch for high security.

11.5 SPECIALTY HOUSINGS

There are security applications in which cameras must be located in very hostile environments. To protect the camera and lens from damage and downtime, manufacturers offer

FIGURE 11-8   Correctional-facility housings

housings that can withstand high mechanical impact from hand-thrown or fired projectiles, extreme high temperature, dust, sand, liquid, and explosive gas. The following housings have unique characteristics for solving these extreme security or special environmental applications.

11.5.1 High-Security

There are numerous armored camera/lens enclosures for installation in correctional institutions. Figure 11-8 illustrates several high-security housings designed specifically for mounting in jails and detention and holding cells, which provide maximum protection from vandalism. The housings have no exposed hardware and all use heavy-duty locks with keys that can be removed only when the lock is in the locked position.

The housings are fabricated from 10-gauge (0.134-inch-thick) welded steel. The window material is $\frac{1}{2}$-inch polycarbonate, with an abrasion-resistant finish. These housings withstand blows and impacts from hammers, rocks, and some firearm projectiles, without penetrating or destroying the integrity of the housing.

11.5.2 Elevator

Figure 11-9 illustrates two examples of hardened camera/lens housings designed specifically for elevator applications. The photograph of the elevator interior illustrates that the full interior of an elevator can be monitored from one wide-angle camera/lens system.

The first housing is a moderate-size (12 inches high), high-security housing of welded stainless steel with a polycarbonate window. The tamper-proof camera assembly is complete with a wide-angle, 90-degree FOV lens and stainless-steel housing. In this configuration, the camera can be tilted +10 degrees for minor adjustments of the vertical FOV. The high-security housing has a hinged, lockable cover for easy, controlled access to all internal parts, and a tough mar-resistant polycarbonate (Lexan) viewing window. All mounting and electrical access holes are inaccessible to the public, located on the rear and top surfaces. Power is supplied to the camera, and video comes out from the camera through one of several access holes in the rear and top. The installation meets codes that require unbroken fire walls. The housing accommodates most CCD solid-state cameras and wide-angle manual- or automatic-iris lenses. A

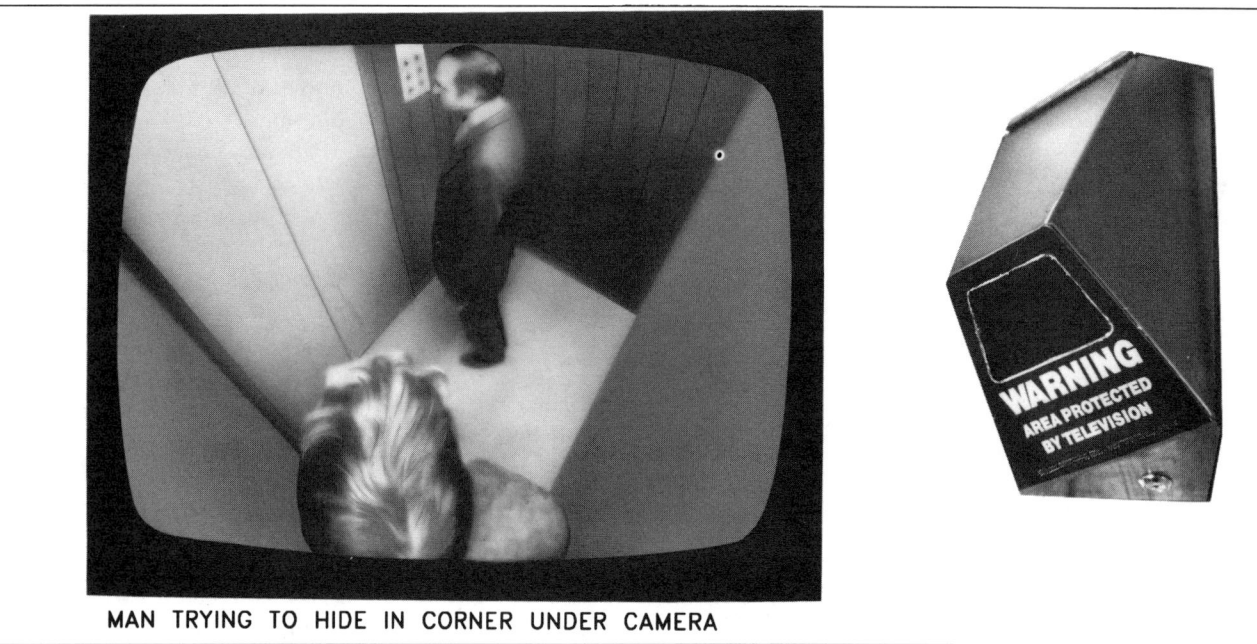

MAN TRYING TO HIDE IN CORNER UNDER CAMERA
WARNING
AREA PROTECTED
BY TELEVISION

FIGURE 11-9 Elevator-cab housings

larger housing uses an infrared lighting source to obtain a useful picture under completely dark, unlighted conditions. A smaller housing 6 inches high can protect all small $\frac{1}{3}$- and $\frac{1}{2}$-inch-format CCD cameras and associated wide-angle lenses. Figure 11-10 illustrates camera viewing parameters for elevator-cab surveillance cameras.

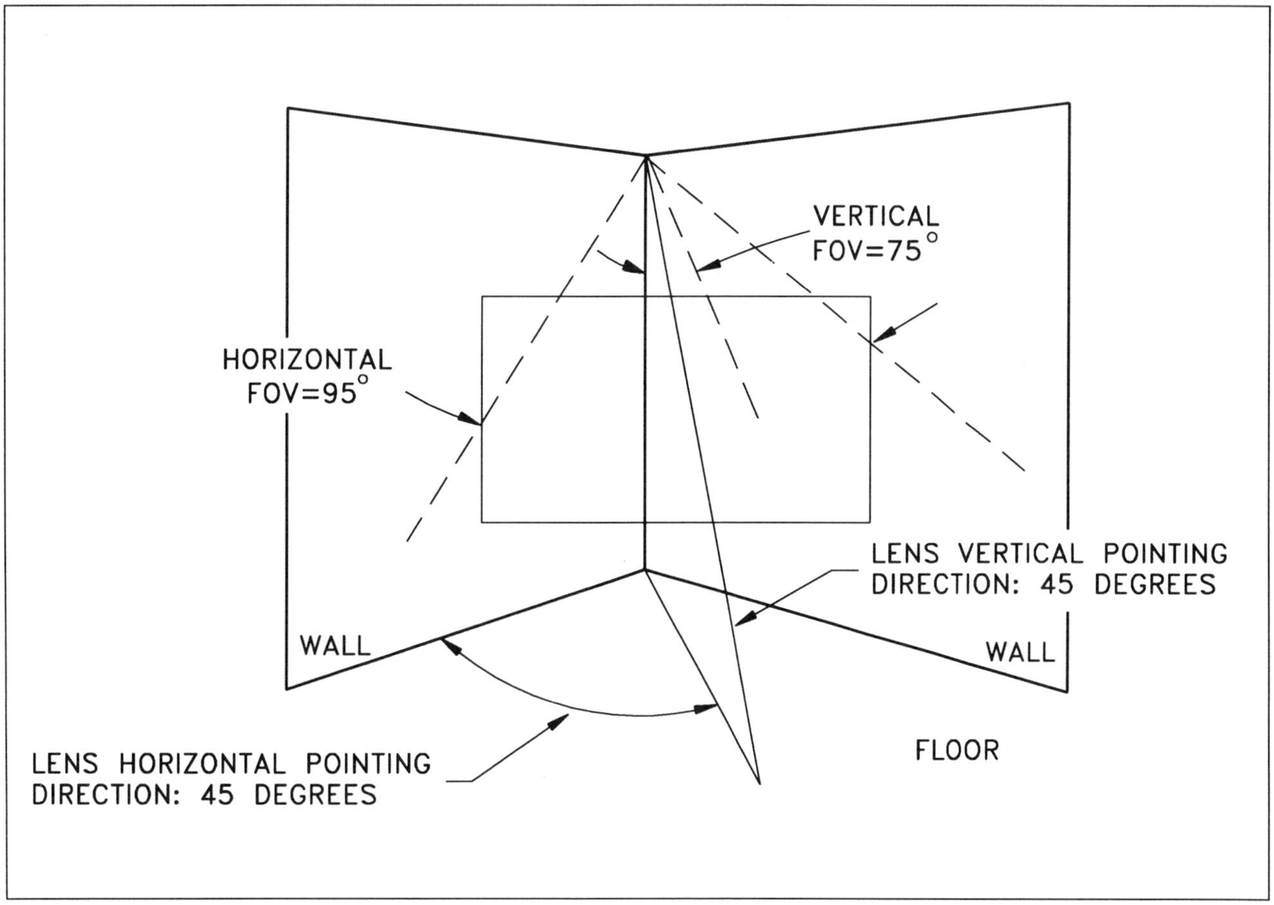
VERTICAL
FOV=75°
HORIZONTAL
FOV=95°
LENS VERTICAL POINTING
DIRECTION: 45 DEGREES
WALL
WALL
FLOOR
LENS HORIZONTAL POINTING
DIRECTION: 45 DEGREES

FIGURE 11-10 Elevator-cab-housing parameters

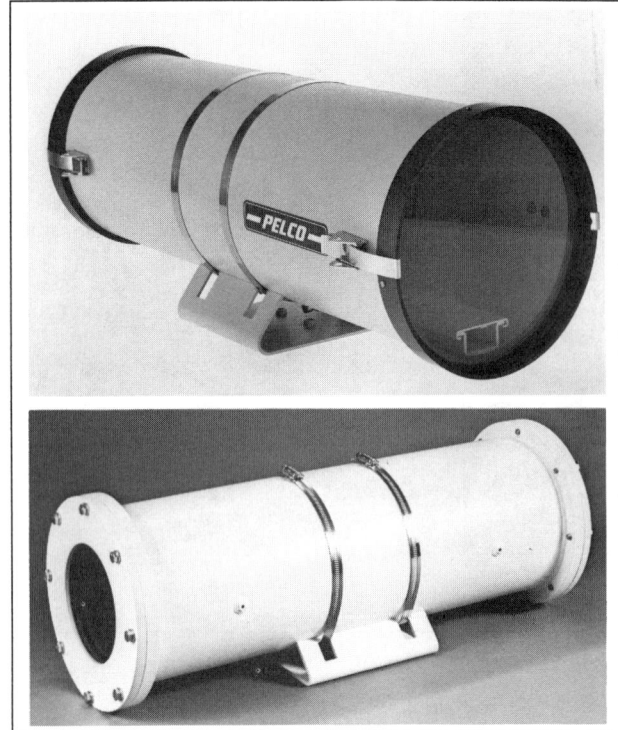

FIGURE 11-11   Dust-proof (a) and explosion-proof (b) housings

11.5.3 Dust-Proof and Explosion-Proof

Figure 11-11 illustrates examples of dust-proof and explosion-proof environmental housings.

The dust-proof housing is similar to many other camera housings except that it is totally sealed to the outside atmosphere and therefore can be used in sandy and dusty environments. When fabricated from stainless steel, these housings withstand the effects of corrosive environments. The window material is tempered glass to provide safety and maximum resistance to abrasion and corrosion. To provide some cooling of the camera and lens, a fan is used to circulate the air inside the housing. A sun shield above the camera housing protects it from direct solar radiation. The housing is provided with air fittings so that an external filtered compressed-air supply can be used (if available) to maintain moderate operating temperatures. These housings are not considered explosion-proof.

Explosion-proof housings are designed to meet the rigorous safety requirements of explosion-proof and dust-ignition-proof electrical equipment, for installation and use in hazardous locations. Many housings meet the requirements of the National Electric Code Class 1, Division 1, and Class 2, Division 1, and are certified per the requirements of UL 1203 specifications and procedures. These housings are generally of heavy-wall, all-aluminum construction and are available in 6-, 8-, and 10-inch diameters to accommodate most camera/lens combinations. They are fitted with explosion-proof, sealable fittings for electrical power/control input and video signal output. Optional sun shrouds are available for operation in hot environments.

11.5.4 Pressurized and Water-Cooled

Pressurized housings are used in hazardous atmospheres. They meet these requirements by purging with an inert gas and pressurizing in accordance with National Fire Protection Association specification Number 946 (Figure 11-12a).

The optimum pointing direction for the lens and camera is 45 degrees with respect to both adjacent walls and 45 degrees down from the ceiling horizontal plane. With a wide-angle, 90-degree horizontal FOV, the entire elevator cab is viewed with no hidden areas, providing 100% television coverage of the cab area. Since the housing is exposed to the public, it must be securely locked and manufactured of tamper-proof materials, such as steel or stainless steel, with a polycarbonate (Lexan) window.

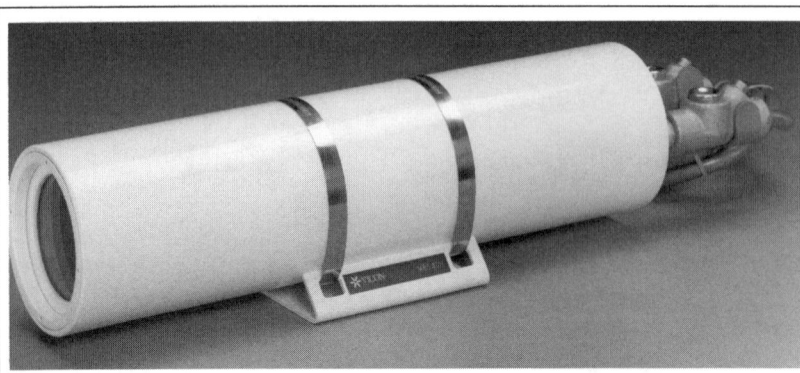

FIGURE 11-12   Pressurized (a) and water-cooled (b) environmental housings

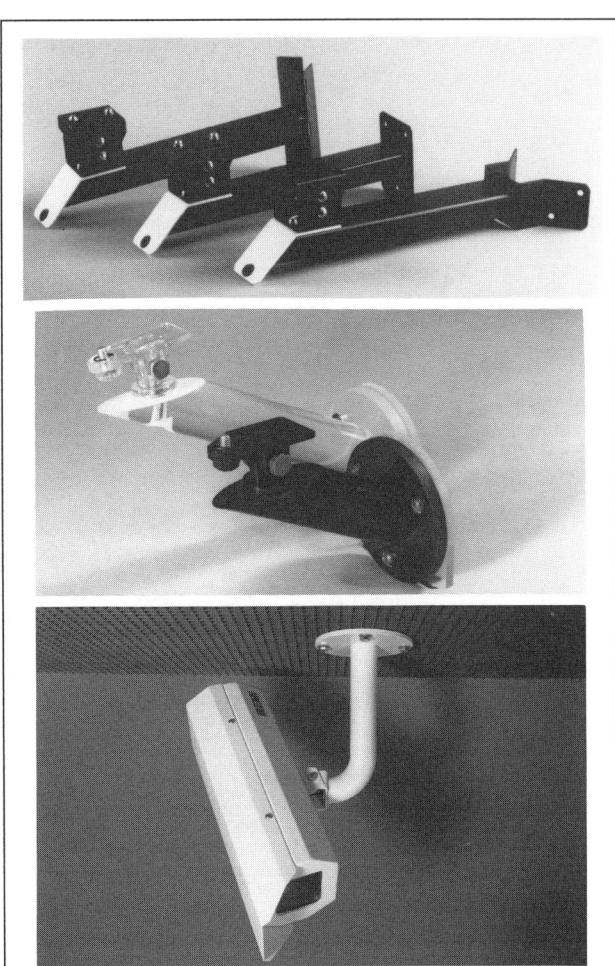

FIGURE 11-13 Camera housing brackets and mounts

The housings are fabricated from thick-walled aluminum with corrosion-resistant finishes. The window is $\frac{1}{2}$-inch-thick tempered and polished plate glass. These housings can be back-filled (purged) with low-pressure nitrogen gas to a pressure of 15 pounds per square inch (psi). Nitrogen is completely inert and prevents an explosion from occurring if there is any spark or electrical malfunction in the housing. The housings have hermetically sealed O-ring seals located between the access cover and the housing. All electrical terminations are made and brought out through hermetic seals. To purge the housing, the access cover is mounted and secured, and the housing is filled with dry nitrogen to a pressure of 15 psi by means of a filling valve and pressure-relief valve. The purge is then closed and the nitrogen filling tube removed. These housings are significantly more expensive than standard housings, since they must be designed to be hermetically sealed to provide a positive pressure of 15 psi differential pressure, and to withstand an explosion.

Water-cooled housings are designed for use in extremely hot indoor or outdoor locations. They require a constant supply of cooling water for proper operation (Figure 11-12b). A 1-inch-thick water jacket built into the housing

effectively shields the camera/lens from the outside environment. Depending on the application, the housings are made from aluminum or stainless steel. An internal fan provides constant air circulation within the housing, aids in efficient heat transfer to the water jacket, and prevents heat buildup. The housing is supplied with a $\frac{1}{4}$-inch-thick Pyrex heat-resistant window for operating at temperatures up to 550°F (288°C). Consult the manufacturer to obtain recommendations for the specific operating environment.

11.6 CAMERA HOUSING MOUNTS AND BRACKETS

A large variety of brackets and mounts are available to secure camera, housing, and pan/tilt platforms safely to walls, ceilings, poles, pedestals, and other structures. Figure 11-13 shows some examples of these brackets and mounts.

There are many manufacturers of housings and brackets, and not all are compatible. Whenever possible, the housing and bracket should be purchased from the same manufacturer to avoid extra costs for reworking parts that don't interface properly.

11.7 HOUSING GUIDELINES AND SPECIFICATIONS

The EIA has written a guideline of recommended design parameters for housing manufacturers for hole configurations on mounting brackets and housing mountings. At present, not all manufacturers use the same mounting-hole configuration. The EIA has recommended guidelines for the electrical input/output wiring and connector configurations, so that there is interchangeability between manufacturers and so that safe procedures are followed by manufacturers and installers. Local building codes and UL codes specify the minimum requirement for electrical enclosure materials (plastic housing materials); they should be consulted to be sure materials are suitable. The purchaser must be aware of the requirements for each application and look carefully at the manufacturer's specifications to determine the most suitable housing.

11.8 SUMMARY

The security camera housing, while not contributing directly to the CCTV picture, nevertheless plays an important role in protecting the camera and lens and ensuring that it will be in a controlled environment to maximize long life and picture quality under varying conditions. The housing protects two of the most important parts of the CCTV system. The housing does influence the ability to obtain optimum performance from the camera/lens or to degrade its performance, if not properly chosen. With the large number of housing manufacturers to choose from, there is a housing configuration for almost any application.

Chapter 12
Video Motion Detectors

CONTENTS

12.1 OVERVIEW

The useful information on a CCTV security monitor often comes from motion within the scene—a moving person, vehicle, or object or some activity involving motion. Irrespective of the number of security monitors, it is useful to have a device to alert the guard to motion or activity in a scene, such as an alarm. Medium-to-large CCTV installations generate many camera scenes, which must ultimately be displayed on monitors, but it is difficult for a security guard to watch multiple monitors over long periods of time.

A technique is necessary to reduce the number of monitors the guard must view but at the same time increase his or her ability to react to real threats.

The video motion detector (VMD) plays an important role in meeting these two requirements. The VMD electronically analyzes and monitors CCTV camera images to detect changes (motion) that are judged large enough to warrant an alarm. They potentially provide an electronic alternative to sitting and staring at the monitors, and they notify the guard immediately of situations requiring attention. VMD systems operate to detect changes in a specified area within the camera FOV. They do this by comparing the light levels of camera pixels from one television frame to the next, looking for changes considered significant. In the simpler, lower-cost analog VMD systems, large areas in the incoming frame are compared with those of a previous reference frame. This type of system works reasonably well indoors, where there are few changes in the scene, where lighting is constant, and where a simple area zone provides adequate coverage. Analog systems are, however, susceptible to false alarms caused by lighting changes or camera vibration and are therefore not recommended for most outdoor applications. In outdoor applications, the more powerful digital VMD is used. This computer-based VMD can analyze thousands of picture zones and operate with low false-alarm rates, even under severe light-level changes.

The VMD is a sensitive and valuable CCTV security tool, since it provides security personnel the visual information taken at the intrusion location when there is motion in the lens FOV. The intrusion scenario can be recorded on a VCR and printed on a hard-copy thermal video printer.

A primary function of the VMD is to allow the security force to make optimum decisions about an intrusion in a mini-

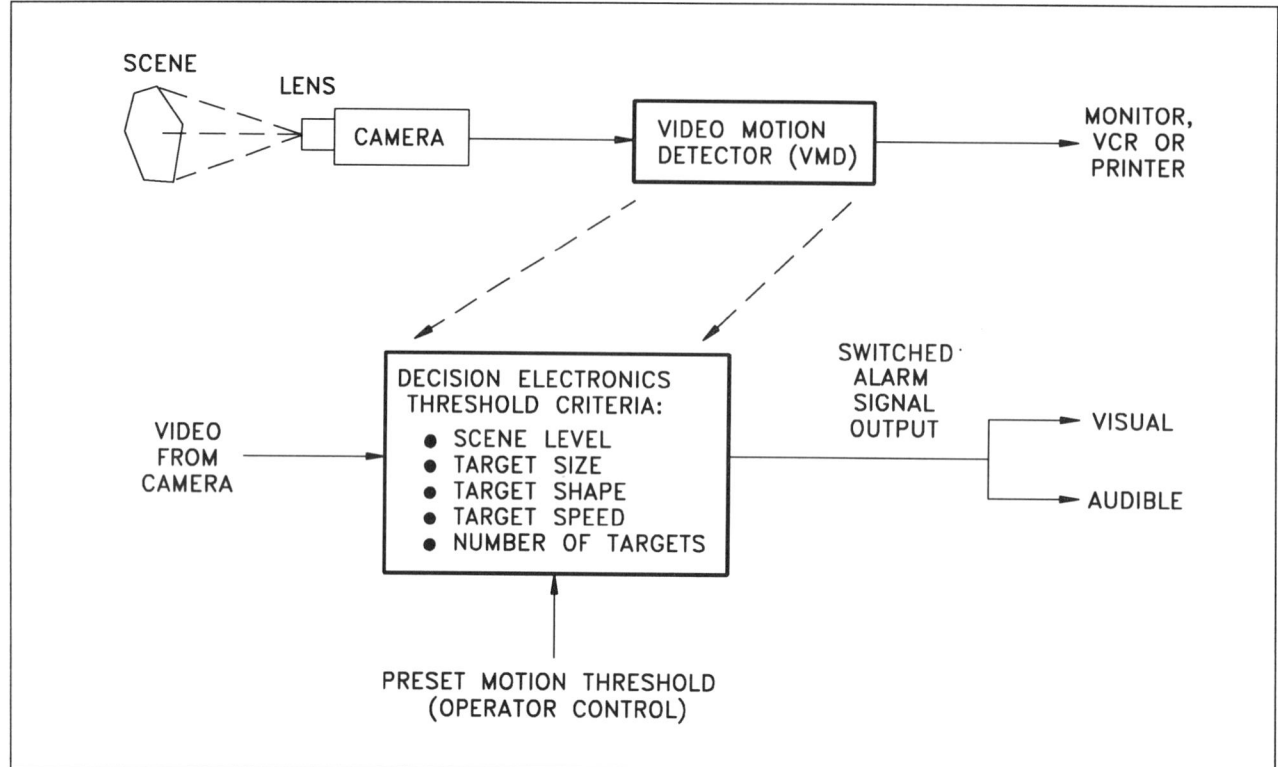

FIGURE 12-1 CCTV motion detection system

mum amount of time. Intruders are often very sophisticated in their technique, making the guard's response more complex. Since the intrusion scenario works to the advantage of the intruder, because he can spend time planning it as well as anticipating the response agency's action under pressure, the offender benefits from a hasty decision.

The following four ingredients should be part of any security system philosophy: (1) surveillance, (2) detection, (3) assessment, and (4) response. The VMD system hardware must provide the surveillance, detection, and assessment, and provide accurate, detailed, and concise information to the guard force, allowing the force to respond optimally. As a free by-product, the VMD also makes available a training tool to practice and perfect the philosophy. To achieve detection probabilities of greater than 50% in any moderate-to-large security system, the integrated CCTV system must operate with an automated VMD detection system.

12.2 FUNCTIONAL OPERATION

A CCTV camera output provided with appropriate VMD processing electronics can make the camera operate as an alarm sensor. The VMD processing electronics memorize the instantaneous television picture, and then if some part of the picture changes by a prescribed amount, the system generates an alarm signal to alert a guard or activate a VCR.

Two VMD types of processing electronics have been developed, analog and digital; the digital form provides more capability and reliability but costs more. The analog VMD detects the change in light level in one or a small number of scene locations (zones). The digital VMD electronically analyzes hundreds or thousands of zones in the CCTV signal and provides security information and capabilities such as: location in the picture where a motion or intrusion has occurred, various audible and visible alarm signals, a graphic monitor map showing the motion path, and a record of the intrusion using a VCR or video printer. Figure 12-1 shows a block diagram for the VMD system and illustrates in principle the information it provides.

In normal operation when there is no motion or change in a scene, the VMD takes the video signal from the camera, stores the video frame (containing no motion), continually updates and memorizes the subsequent frames, and compares them to the previous frame to see if there is a difference in the new frame. If there is no motion, there is no alarm. If there is a difference of measurable and defined value, then an alarm is declared and an output produced.

Surveillance of any scene is achieved by the use of conventional CCTV cameras and lenses positioned throughout the area of interest, at locations that permit recognizing an intruder or movement within the lens FOV. Cameras should be positioned so they can view all activity and targets of interest, and should be designed so they cannot be tampered with unnoticed. Environment is a major factor for choosing the VMD in outdoor applications. The digital

VMD can tolerate some camera vibration, but the camera should be mounted as securely as possible. The VMD can also tolerate large-area scene motion and light-level changes as might occur when a cloud passes in front of the sun, without causing a false alarm. Some digital VMD systems can subtract out or ignore inherent scene motions such as waving flags, leaves, or trees, so that they will not be a source of false alarms. Some have the ability to selectively sensitize and desensitize certain portions of the scene in order to prevent false alarms. Desensitizing a part of the scene where inherent motion and no real activity is expected (such as leaves rustling on trees) reduces the chance of false alarms.

12.2.1 Surveillance

CCTV surveillance is accomplished via the use of cameras and lenses located and positioned for maximum viewing (intelligence gathering) of an area. CCTV can act synergistically with other alarms as remote eyes, to present a visual image of an area, as well as the source for an alarm input. Since the camera, transmission means, and monitoring equipment may already be included in the security system, the addition of the alarm function using the CCTV system requires only the VMD unit to analyze the television scene. The addition of the VMD to the CCTV security system is an additional cost, which can be small or large depending on the application: indoor or outdoor, analog or digital. Use of the VMD provides a significant increase in security. The correlation of the information in the CCTV scene as viewed by the operator and recorded on a VCR, along with the alarm output from the VMD, provides a strong and effective means for assessing the cause of the video motion.

Monitoring a large target such as a parking lot presents multiple possibilities. Three techniques have been used to monitor these areas with a VMD: (1) a wide-angle lens, (2) multiple cameras, and (3) dual-lens, split-screen. When a wide-angle lens is used, the alarm source (intruder) appears small on the monitor screen and a guard does not detect the intruder, especially if the intruder takes cover quickly. The VMD can detect the intruder and register an alarm. With multiple cameras, the parking lot FOV is divided among the cameras, each viewing a section. Each camera must use a separate VMD. Likewise, the split-image optics divides the FOV and displays and combines views on the monitors, with each camera using a VMD.

If the system includes pan/tilt equipment, the guard must pan, tilt, and zoom the camera/lens to locate the alarm source. This is not a simple task, and in the time required for the guard to perform it, the intruder may be gone. If this occurs, a responding force becomes necessary. In more sophisticated systems, the location of the motion in the picture is used to point the pan/tilt platform in the direction of the motion, to speed reaction time.

12.2.2 Detection Probability

A guard monitoring a medium-to-large CCTV security system must view many monitors that display either: (1) sequenced scenes, (2) several monitors—one for each camera, or (3) monitors with split screens. To assure a high probability of detection, the lens magnification must be such that an intruder is displayed on the monitor magnified enough so that the guard can easily see him. Using multiple cameras provides the necessary coverage to detect the intruder. The dual lens, with one lens a wide-angle (or zoom lens) and the other a telephoto (Chapter 4) is a good solution.

For a guard's response to an intrusion to be effective, the guard must first know that he is responding to a real intrusion, and second, its location and nature. The VMD's function is to display only intrusion alarms on the video monitor without any human intervention. The guard then assesses the alarm by viewing the monitor.

The system must give instantaneous information as to the exact location and nature of the disturbance and must (1) respond to small changes (motion) in the camera/lens FOV, (2) activate an alarm output on the monitor to alert the guard that an intrusion has occurred, and (3) display the alarmed scene on the monitor, accompanied by an audible alarm and activation of a VCR and/or video printer. The displayed scene should show the location within the scene that has been disturbed and give immediate information to security personnel as to the precise location, movement, and nature of the alarm. If an intruder is hiding, a flashing pattern on the monitor should show the path of the intruder from entry of the scene to the point where he is hiding.

The guard has a difficult job to perform. He is responsible for monitoring scenes that after a short time are no more interesting than a television test pattern, and he will have a tendency to become mesmerized. One question arises: How large should the intruder's image on the monitor be to attract the guard's attention? A second question: Are there sufficient cameras so that an intruder is not missed? If more cameras are added, more scenes must be sequenced and watched, with each scene viewed for a shorter time, making it more difficult for the security guard. This is true in both the single- or multiple-monitor approach. A second guard may be required to view the increased number of camera scenes and the higher rate of switching. The third method is to provide split screens showing multiple scenes on a single monitor. The guard must pay attention to the increased number of scenes, but the fewer screens reduces the risk of missing activity in a scene.

Intrusion-detection probability is controlled by the placement of cameras and is a system design parameter. With proper camera placement and reliable equipment, target-detection probabilities are 95 to 99%. Alarm assessment takes place in the time it takes for the operator to view the scene and identify the cause. When a VMD is used, the security operator does not have to locate the movement on the screen, since the cause of the alarm is indicated by the brightened flashing map on the monitor. If it is an intruder, the guard responds accordingly, knowing exactly where the intruder is and who he will be confronting. If it is not an intruder, he can press the reset button and go on to the next alarm.

VMD alarms are valuable not only because they can cue a CCTV response but also because they are an independent source of vital input. There may be particular situations where a specific activity within an area covered by the camera would be difficult to detect with other conventional forms of alarms. It is often important to know not only that an intrusion occurred in a certain space or area but also the path the intruder took. VMDs with enhanced mapping display capability can provide this information.

12.2.3 Motion Assessment

Assessment is the ability of the console operator to identify and evaluate the cause of the alarm. This judgment call is one of the most important decisions, for two reasons: (1) if a real intrusion occurs, the guard's assessment must be rapid and accurate and depend on a visual judgment; (2) if the alarm is not a valid intrusion, the guard must be able to make that decision rapidly and accurately, which again requires visual observation of the cause of the alarm, and then cancel it.

In some VMD systems, a random-access memory (RAM) module stores the alarmed locations in a separate RAM alarm map memory (AMM). Upon alarm, the contents of the AMM are displayed on the alarmed video monitor scene as a flashing, highlighted array of alarm points. This feature is a key to quick, accurate assessment of all alarms. The AMM enables the operator to determine instantly the exact location where the disturbance or intrusion has occurred, and provides a quick, precise evaluation of the alarm, so he can provide the appropriate response. To clear the alarm condition after a response has been made, the operator presses a reset switch and the monitor returns to the normal blank condition. This accurate, rapid assessment optimizes the use of the response force. If a second or additional alarms occur prior to resetting, the alarm scenes are displayed with their alarm maps in sequence on the master monitor, at a selectable rate.

When a large number of cameras are alarmed simultaneously, an assessment problem can occur. By the time the guard views the last camera, the intruder most likely has left the scene and only the map remains. The digital VMD effectively controls the situation by providing a video output to record all alarmed camera signals. This is done automatically while the guard watches the monitor. The video frames (scenes) are sent to the VCR at a rate of 30 per second. The pictures are recorded—one from each camera—in sequence and continue until the operator resets

the equipment. When a guard realizes a multiple-intrusion attempt is in progress, he can play back the VCR video into the monitor and replay the intrusion with the alarm map to determine the cause of the alarm in the scene. Using this technique, the alarm assessment capability is extremely high. The guard need not leave the console during an alarm condition unless he must respond to, or direct a response to, a real intrusion. He can observe the progress of the intruder into the area by observing the monitor as the intrusion map is generated.

12.2.4 Guard Response

Security protection results from the combination of precision detection and optimum response. Optimum response implies that much thought and planning be expended to bring about an effective response. In an optimal system, security personnel are ready to respond to an alarm input displayed on the CCTV monitor activated by a VMD or other alarm. When an intrusion occurs, the system quickly makes an analysis of the situation, displays the camera view in the alarm condition, and any other CRT display, instructing the guard of his function and the optimum response.

Without visual information, security personnel do not know what caused the alarm, resulting in a total lack of security. Security systems generally deteriorate in this area, but this is the point where the VMD provides the crucial input. Here the human factor is summoned to determine whether the alarm is real or false. When an alarm is interpreted as real, a common labor-intensive practice is to dispatch a guard to visually observe the zone that caused the alarm. Upon his arrival at the scene, the probability is high that he will not be able to make an assessment (because the intruder is no longer in the area). The only certainty is there was no positive security. An additional detriment to all "nonassessable" alarm situations is that the responding guard becomes conditioned to assuming that nonaddressed alarms are not intrusions—again, no dependable security.

12.2.5 Training

In the intrusion scenario, when an alarm occurs the console operator is called upon for the first time to evaluate the alarm on a previously blank CCTV monitor. The monitor displays the intruder and the exact location within the scene by some flashing indicator superimposed on his exact location.

CCTV VMD and VCR systems are used to test a security plan and response, train management and security personnel, and evaluate guard and system performance. A system using the VMD permits security personnel to train before an actual event; and when an intrusion does occur, the system can immediately recall the decisions to form an instant plan of action, which directs the efforts of the response force in an optimum way. This important training improves the plan, the guard response time and method, and overall security.

The ideal VMD system would give a 100% probability of detection of intrusions, zero false-alarm rate, zero nuisance alarms, and zero equipment failure. This system has not yet been developed, but several companies have achieved products coming close to these goals.

12.3 INDOOR VERSUS OUTDOOR

Before any VMD can be applied to a particular application, its location—indoor or outdoor—must be considered. In an indoor application, the light-level changes are usually predictable or at least not very significant. Successful VMD operation depends on recognizing light-level changes in specific parts of the scene (caused by an intrusion or disturbance) in contrast to overall scene light-level changes caused by changing lighting conditions. These two phenomena must be differentiated to avoid undue false alarms. In indoor lighting applications where the light level is controlled by the user, a simpler analog VMD system is used.

The protection of outdoor areas presents the most difficult problem in facility security. All sensing devices are plagued by false alarms due to the unpredictable nature of natural phenomena and intentional artificial alarms. Seismic sensors produce false alarms due to vibrations caused by wind, vehicles, and other objects. Microwave sensors produce false alarms due to moving animals, blowing papers, or leaves. An effective outdoor security system is best augmented by CCTV cameras viewing the actual scenes to filter out and recognize false alarms. Although an alarm denotes that a certain area has been disturbed, without a visual image, little information is provided as to the nature of the alarm or the precise location at which it occurred. Without a video image, security personnel must be sent out to investigate and determine the nature of an alarm. Since outdoor areas to be monitored are often large, in many cases by the time a security guard responds to the alarm, the disturbance has disappeared.

Since the VMD makes its decision based on the scene the camera is viewing, it is important that lighting of the camera site be adequate. The VMD equipment must be able to compensate for variations in average scene lighting occurring during daylight hours as well as when auxiliary artificial lighting is provided during nighttime operation. VMD systems operate with scenes illuminated by visible or infrared lighting.

In outdoor applications, the environment is not as controllable: significant light-level changes are caused by sunlight, cloud variations, lightning, and many different types of objects passing through the camera/lens FOV. Most digital VMD systems operate well in most outdoor condi-

tions, but under adverse environmental conditions of heavy snow or rain, they lose some of their capability, and alternative systems using other sensors should be relied upon. The VMD used in an outdoor environment has a significantly higher potential for false alarms due to these unpredictable lighting changes and moving clutter, such as leaves and branches. The digital VMD must have outdoor algorithms that correctly account for rapid changes in overall scene brightness and illumination, as well as area changes in illumination caused by rapidly moving phenomena. Likewise, consideration must be given to updating the sampling of each zone set up for intrusion. If there is movement in the scene, it must be detected while the movement is still in the scene. Therefore, if updates of the scene occur at too slow a rate, an object at a distance may elude detection.

To determine whether a target is of interest or a false alarm, the equipment must be able to distinguish its size, speed, shape. In outdoor applications, a digital VMD is the only solution. Figures 12-2 and 12-3 show block diagrams of analog and digital VMDs.

Since there is a significant cost difference (between 10 and 30 times) between the analog and digital VMD, careful consideration must be made in choosing the system.

12.4 ANALOG VERSUS DIGITAL

Analog VMDs have been available for many years and provide a low-cost video device to detect simple motion in a CCTV scene. They only operate reliably in indoor, well-controlled environmental and lighting conditions and should not be used for outdoor applications. The analog VMD is connected into the video system as shown in Figure 12-4.

The simplest VMD uses analog subtraction. The reference frame and the frame in which motion has occurred are subtracted and an alarm declared, depending on the amount of signal difference between frames. This analog system, while acceptable for most indoor applications, is prone to false alarms and is not suitable for outdoor applications. The digital VMD should be used in outdoor applications.

The more sophisticated and expensive digital VMD systems use elemental detection zones, in which the scene is divided into a large number of zones (hundreds to thousands) and converted into a digital signal. The processor analyzes these individual zones and makes a decision whether or not an alarm is present. With these microprocessor-based systems, many parameters are analyzed, thereby forming a more reliable basis for an alarm signal decision. Light-level changes in these digital VMD systems are compared with the previously stored values ratiometrically—that is, on a percentage basis. Ratiometric thresholding causes the system to cancel out any gross change in the

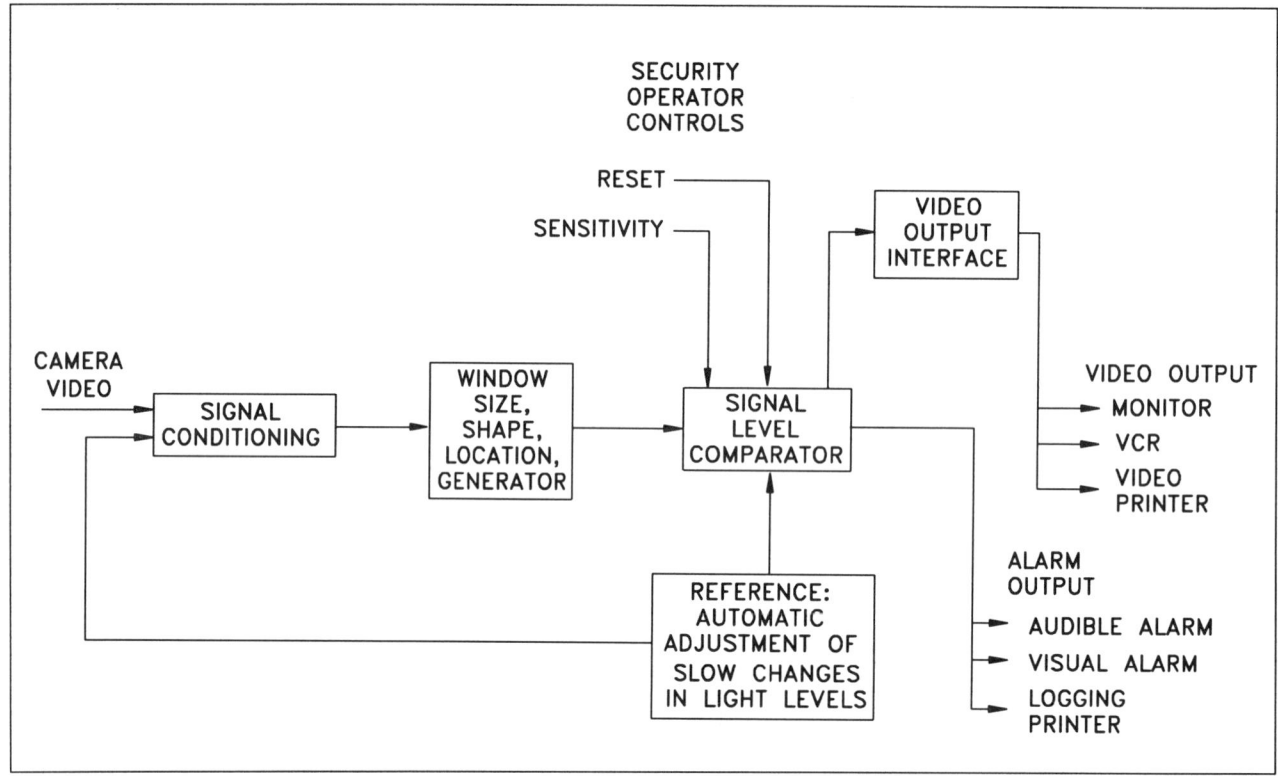

FIGURE 12-2 Analog VMD block diagram

scene lighting, so that an alarm decision is made strictly on an incremental basis, for a small portion of the total picture area.

The digital electronics in the digital VMD subdivide the television scene into many small elemental zones—as many as 10,000—and make a zone-by-zone comparison (subtrac-

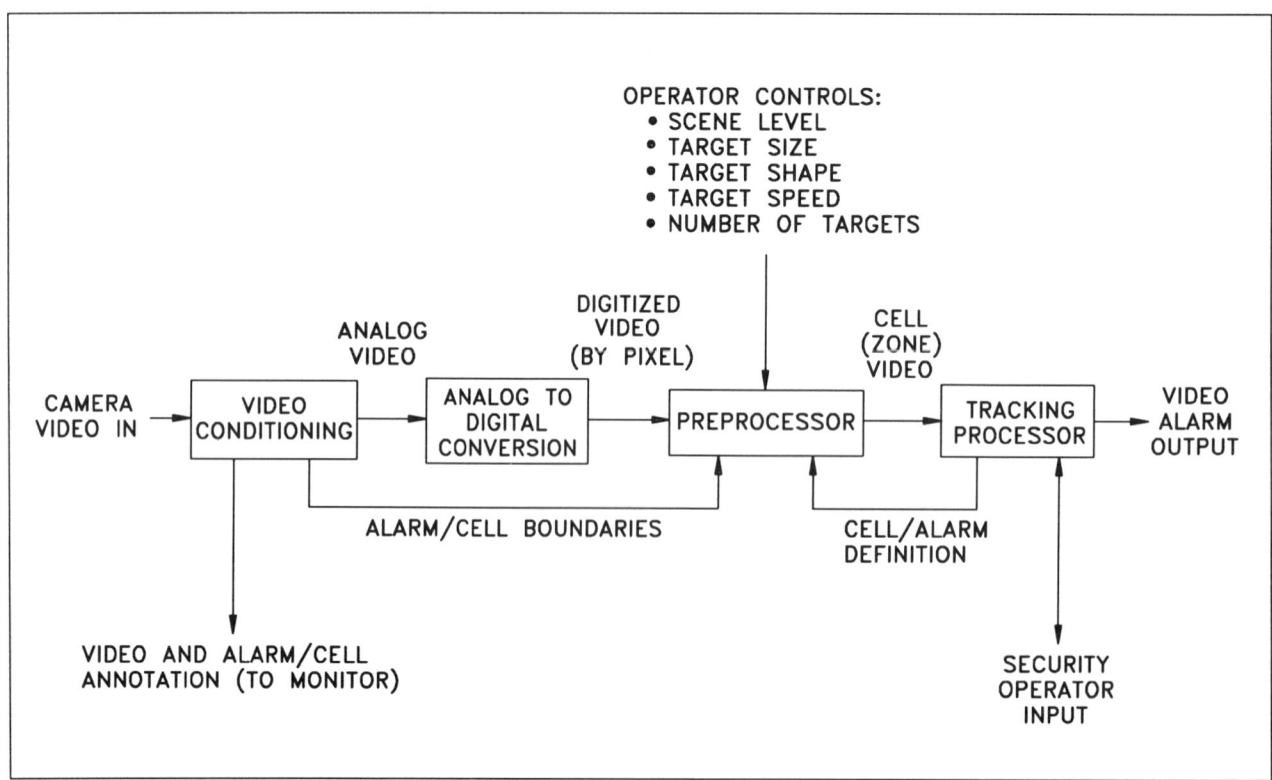

FIGURE 12-3 Digital VMD block diagram

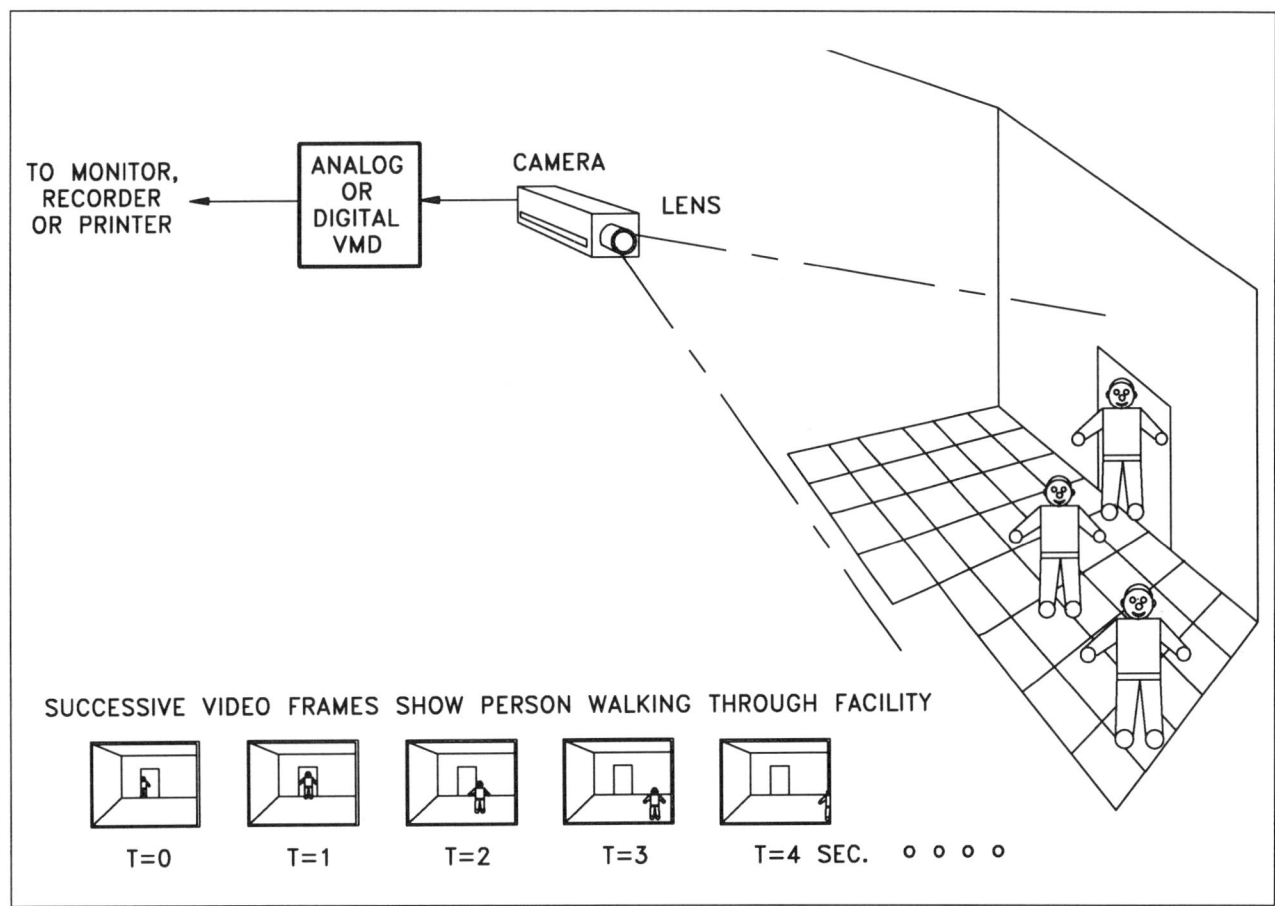

FIGURE 12-4 VMD in the CCTV security system

tion) of the nonmoving or steady scene with the motion scene, and goes into an alarm mode when a threshold is detected in any one or a multiple of these zones. By converting the signal from analog to digital and dividing it into many zones, a much more sensitive device results. This technique allows discrimination between real targets and false alarms and other scene lighting variations and provides a more reliable system for outdoor use.

The user-selected zones are positioned over specific areas where motion is expected. These zones may cover assets to be protected, entry or exit points, parking lot slots, perimeter areas, and perimeter fence lines. Each zone may be set with a different sensitivity appropriate to the percentage change required to trigger the alarm in that zone. The larger the percentage required to cause an alarm, the less sensitive the system is to contrast changes and the less likely it is to produce false alarms. The digital VMD is more sensitive than the large-area-detection analog VMD.

12.5 ANALOG HARDWARE

All analog VMDs have an adjustable detection-of-motion zone (DMZ), which is a selected portion of the monitor screen. Any movement (change of light level) in the scene within the DMZ automatically triggers any of four alarms: (1) an internal audible alarm, (2) a front-panel signal light, (3) an AC outlet that can activate an AC-operated signaling device, or (4) an isolated terminal relay contact to activate a VCR, printer, bell, or other security device.

The analog VMD operates by analyzing the analog video signal from the camera and determining whether the scene has changed. The system "memorizes" the value of a standard reference scene depicted within the DMZ and compares it with a value in the current real-time scene. If the two values are the same within the active DMZ, electronic circuitry declares that there has been no motion and no alarm is declared. On the other hand, if there has been a scene change caused by someone intruding into the scene or an object moving or some other light-level disturbance, providing the change is larger than a prescribed amount, typically 10 to 25%, then electronic circuitry decides that a change has occurred, there has been motion in the alarmed area, and an alarm signal is produced. This alarm signal is used to produce an audible or visual alarm, turn on a VCR, or activate a video printer. The VMD operates independently from the video monitor or any other recording equipment and in no way interferes with it. Table 12-1 summarizes the parameters of several commercially available analog VMD systems.

On most analog VMD equipment, the size, shape, and location of the active area in the entire scene is adjusted with front-panel controls. The DMZ size and configuration chosen depends on the requirements of the surveillance application. Figure 12-5 illustrates some examples of DMZ shapes available, including split-screen, square or rectangle, L-, C-, and U-shaped.

The areas of sensitivity are chosen to surround a location in the scene where motion is expected. The DMZ enables the operator to select (sensitize) specific portions of the camera scene area, while the entire scene is always displayed. An alarm occurs only if there is motion in the DMZ itself. Depending on the equipment, DMZ is represented on the video monitor screen by a brightness-enhanced window (or a brightness-enhanced frame), adjustable via the front-panel controls. After initial setup, the brightened window (or frame) may be switched off so that the scene looks normal to the operator. The active DMZ on the screen can be set up to cover an area anywhere from 5 to 90% of the viewed picture width and height. The VMD system sensitivity is usually set to respond to a 25% change in video signal level, in 1% of the picture area occurring within a time period of several frames.

12.6 DIGITAL SYSTEMS

While analog VMDs have been in use for security applications for many years, they have been successful only in indoor applications where lighting has been well controlled. In outdoor applications, a far more complex digital electronic system is needed to provide reliable security. The VMD must take into account the many variations of lighting, type of target movement, and electrical background disturbances caused by external sources and noise in the system. In the past, these sophisticated expensive systems have been used in large government facilities and nuclear power plants. With lower cost and more powerful computers, digital VMDs are now more widespread in commercial installations.

The CCTV cameras are connected to the digital VMD processing unit, which converts the analog CCTV signal into a digital code. For each camera, a specific detection pattern or area is selected, or already built into the processor. The detection pattern is part or all of the camera image scene within which specific sample points are designated. Depending on the manufacturer, the sample points vary in number and location. At a designated rate, the sample or reference image from a specific camera is converted from the analog to digital format, and the digital values are stored in temporary memory in the VMD unit. This reference or base image is updated at variable rates to compensate for small changes in the scene that do not constitute alarm events.

At programmable rates, camera images at a later time are converted to a digital format and electronically compared with the stored reference image. If there has been movement in the scene or any variation in a significant number of sample points over some range, an alarm may be triggered. If some harmless objects, such as a small animal or bird, passes through the scene, no alarm will occur. If, however, there is movement within the scene—such as a person entering or a window opening or closing—the VMD will be triggered. The number of sample points and the amount of change within the areas to produce an alarm output depend on the particular manufacturer, model, and control settings.

Depending on the design, a VMD can process 1, 10, 16, 32, or 64 cameras and sample them serially: that is, camera 1, then camera 2, and so on, and then back to camera 1. Some systems sample and process multiple cameras simultaneously, analyze and respond to multiple alarms. When a VMD detects an alarm event, its output can be used for multiple functions. It can display the alarmed camera on a monitor, alert a guard with a visible or audible signal, record the alarm on a VCR, send the alarm signal to a remote site, or activate a time-lapse VCR with an alarm input to change its recording mode from time-lapse to real-time. Table 12-2 summarizes the parameters of several commercially available digital VMD systems.

12.6.1 Mode of Operation

Figure 12-6 shows a block diagram of a generic digital VMD. The digital VMD digitizes the frame from each camera into a large number of zones corresponding to exact locations on the monitor screen. The number of digitized zones varies from hundreds to many thousands. The system assigns an absolute gray-scale value (light level) to each zone and stores the digitized gray-scale value and location in RAM. This procedure is carried out for each television camera channel. The number of analog gray-scale levels for a good analog television picture is 10. The digital VMD can digitize the picture into 16 to 256 gray-scale levels, thereby storing (memorizing) the image scene very accurately. After this reference scene has been memorized in RAM, the digital VMD digitizes subsequent camera frames and compares them zone by zone to the stored values. If the stored levels at any location differ by one or two levels in gray scale—between the stored frame and the live frame—an alarm condition exists.

One digital VMD digitizes the scene by creating up to 16,000 individual zone locations per scene in up to 16 camera scenes. With this high resolving power, the system can detect an intruder occupying as little as 0.01% of the area. When using 16 cameras, the system can detect an object occupying 0.006% of the total monitored area. The digital VMD system operates normally with a blank monitor. When a camera receives or detects motion, an audible alert is sounded and the disturbed scene appears on the monitor. VCRs are activated for recording the intrusion

MANUFACTURER MODEL	FRONT PANEL CONTROLS	NUMBER OF CAMERAS MONITORED	ALARM ALERT		ADJUSTABLE DETECTOR AREA				WINDOW OUTLINE	WINDOW SENSISTIVITY
			AUDIO	VISUAL	SIZE	SHAPE	POSITION	WINDOW		
AMERICAN DYNAMICS AD1461	AUTO RESET MANUAL RESET WINDOW	1	YES 2 TONE	YES FLASHING LED	YES	YES	YES	RECTANGULAR	YES	ADJUSTABLE
JAVELIN J314MD	MODE, ALARM, RESET ALARM WINDOW FRAME BRIGHTNESS SENSITIVITY, HORIZ POSITION VERT POSITION	1	YES 2 TONE	YES FLASHING LED	YES	YES	YES	RECTANGULAR L, ⌐, ⊐, ⊔, ⊓,	YES	ADJUSTABLE
BURLE TC8210MD	MODE, ALARM, RESET ALARM SENSITIVITY, HORIZ POSITION VERT POSITION	1	YES	YES FLASHING LED	YES	YES	YES	% OF SCREEN 4% MINIMUM TO FULL SCREEN	—	ADJUSTABLE

Table 12-1   Representative Analog VMD System Parameters

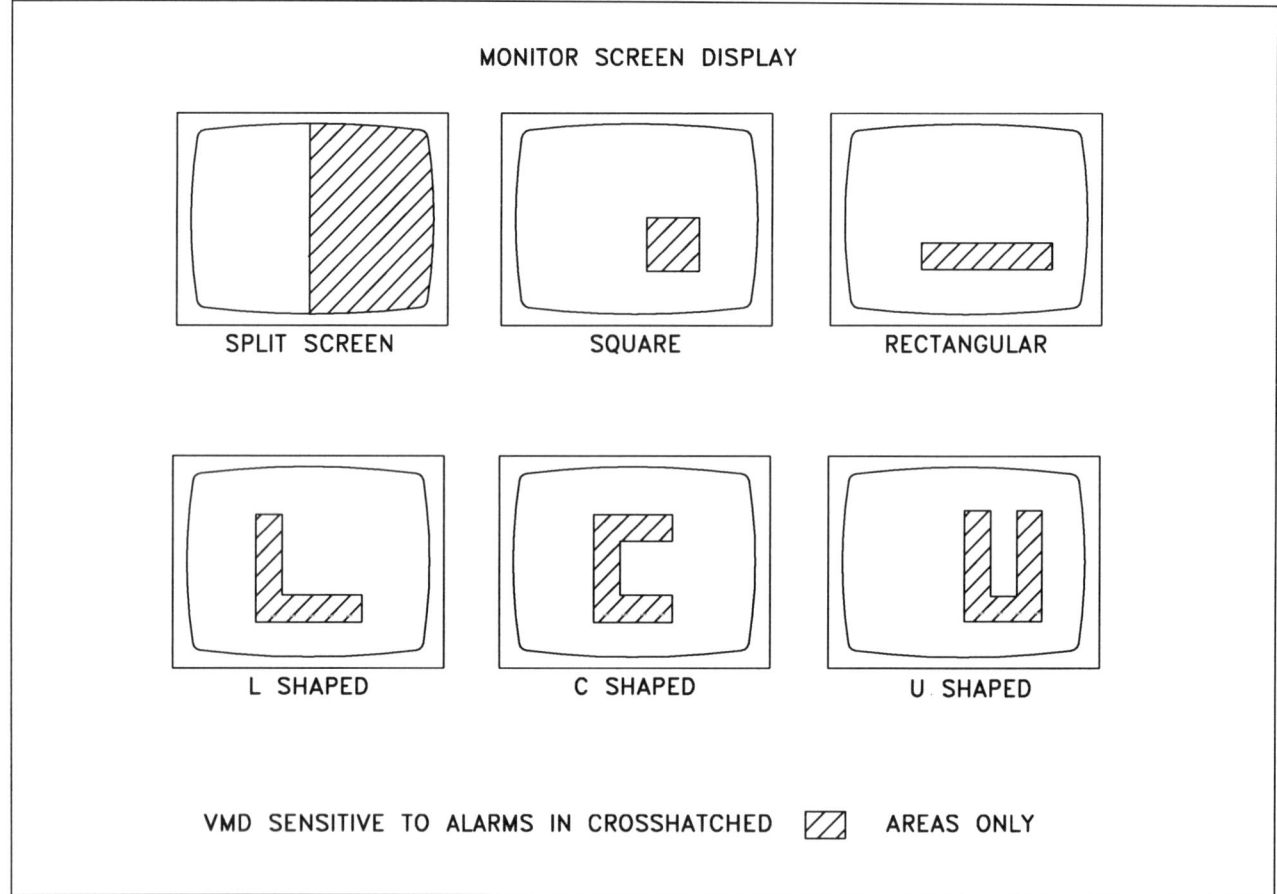

FIGURE 12-5   Analog VMD sensitized shapes

scene or for reviewing the alarmed scene at a later time. When the digital VMD displays the picture on the monitor, the guard sees the intruder in the scene even though he occupies only a small portion. The guard will also know where the intruder is, even if he is hidden from camera view, since the system displays the intruder's path on the monitor. This display is accomplished by displaying bright flashes on the monitor at all locations the intruder has passed through. The guard now knows not only which scene was intruded upon but also the exact location of the intruder in that scene at that instant. He can therefore concentrate immediately on what decision to make and what action to take.

An auxiliary display useful with VMD systems is an illuminated graphic display consisting of an overlay that is a plan view diagram of the entire monitored site. The map overlay shows the location of each camera and alarm sensor and flashes on the display when an intrusion occurs. To ensure that no intrusion is missed, particularly if there are simultaneous intrusions or motions in the scenes, VCR recording is used. The VCR records the video scene, the intruder, his track through the scene, as well as a graphic alarm map if available. In the event of multiple video alarms in a single VCR system, the recorder is set to record one alarm scene

for a predetermined time interval and then switch to the next alarm scene. If a nonvideo sensor detects an alarm, the system activates the appropriate camera(s) and the VCR. The displayed information enables the console operator to assess the situation rapidly and accurately and report any diversionary tactics. Present digital VMD equipment is able to detect 20 times the number of intrusions as those detected by a guard looking at the video monitor without the benefit of the digital VMD. This digital VMD system is not easily mesmerized!

12.6.2 The Storing Process

The digital VMD analyzer detects the alarm condition by storing the scene in solid-state RAM. In one system, the storage process takes approximately 33 milliseconds and consists of sampling the picture scene (up to 16,384 discrete locations) that are spaced throughout the scene. At each location the brightness is measured (one of 16 to 256 different gray-scale levels). The address (pixel location in the scene and camera) is stored with the brightness number. This occurs for all zones in the scene. After the brightness and location information are stored, a comparison

MANUFACTURER / MODEL	SAMPLE PIXELS (CELLS)	CELL GRAY SCALE LEVELS	SENSITIVITY (PER CHANNEL)	REFRESH RATE	DWELL TIME	PROGRAMMING	RECORDER ACTIVATION	OUTPUT ALARM/ VIDEO	INDOOR/ OUTDOOR
AMERICAN DYNAMICS AD4000			ADJUSTABLE 8 LEVELS			LIGHT PEN	YES	3 VIDEO , 3 ALARM RELAYS , 8 TTL	YES/YES
AITECH 816	30,720	64 (6 BIT)	8 LEVELS	4/15–60/15 SEC	NORMAL:1–60 SEC ALARM:1–10 SEC	ON SCREEN MENU DRIVEN PROMPTS FUNCTION KEYS	YES		
BURLE INDUSTRIES TC214 TC218	1024 DETECTION POINTS		ADJUSTABLE 28 LEVELS PER TIME/ZONE			ON SCREEN MENU DRIVEN			YES/YES
COMPUTING DEVICES DAVID	64,000	256				MENU DRIVEN			YES/YES
DIGI–SPEC DS–32	262,144		INDIVIDUAL CAMERA		1–99 SEC		YES		YES/YES
FOR–A SVS–860	3328 CELLS (64x52)		ADJUSTABLE	0.1–5 SEC/FIELD		LIGHT PEN, INACTIVE AREA CELLS ERASED OR ADDED		6 VIDEO	
I–DEN IS–80	3328 (64x52)		MOTION,SPEED DIRECTION, DISTANCE TRAVELED					3 VIDEO LED PC AT BUZZER	
PRESEARCH		256							
SAS–TEC VSM 210	1024 (32x32)		TARGET SIZE MOTION PRO– GRESSION AND DIRECTION			AREA EXCLUSION 8 BUTTON KEY– BOARD. SOFTWARE CONTROLLED . MENU DRIVEN		LED AUDIO 64 VIDEO CONTROLLABLE	YES/YES
SONY YS–S100	2x288		PROGRAMABLE	0,3,5,15,30 SEC 1,2,4,MIN	5,10,20,30 SEC 1,3,5, MIN				
VISION RESEARCH DIGIPLEX 4000			PROGRAMABLE EACH CAMERA	ADJUSTABLE		LIGHT PEN MASKING– INACTIVE CELLS	YES	LED, AUDIBLE SOUND	YES/YES
VISION SYSTEMS ADPRO 1650 VMD–1 VMD–10									YES/YES

Table 12-2   Digital VMD System Parameters

MANUFACTURER / MODEL	FRONT PANEL CONTROLS	NUMBER OF ZONES	TARGET/CONTROL PARAMETERS	ON-SCREEN ALARM GRAPHICS (TYPE)	NUMBER OF CAMERAS MONITORED	ALARM INPUTS	COMMUNICATION PORT(S)
AMERICAN DYNAMICS AD4000	ARM/DISARM TARGET AREA (ZONES), SENSITIVITY, 18 PARAMETERS TO OPTIMIZE	2800	ZONES OF MOTION DISPLAYED TO SHOW TRACK OF INTRUDER	CAN DEACTIVATE ZONES	8	8	
AITECH 816	ON SCREEN MENU: WINDOW, SCHEDULES	8 WINDOWS: SENSITIVITY, LOCATION, SHAPE, SIZE	ALARM INDICATORS: MAIN-LED, AUDIBLE CAMERA ID NUMBER, MOTION ALERT, VIDEO CHARACTER GEN, ALPHA-NUMERIC	8 WINDOWS/CAMERA SHAPE, SIZE AND LOCATION PROGRAMABLE	8 EXPANDABLE TO 16	8	25 PIN D FULL DUPLEX SERIAL PORT 1200 BPS
BURLE INDUSTRIES TC214 TC218	TRACKBALL	32	ALARM STATUS AREA OF INTRUSION TIME/DATE, CAMERA ID	USER DEFINED	4,8,		RS232
COMPUTING DEVICES SENSTAR DAVID 200			ZONE DEFINITION AREA OF INTRUSION		12		
DIGI-SPEC DS-32	SENSITIVITY	720	MOTION DISPLAYED IN REAL TIME	ACTIVE AND INACTIVE AREAS PROGRAMABLE	2-32(MODULAR)		RS232
FOR-A SVS-660	SENSITIVITY	3200	AREA OF INTRUSION	VERT. AND HORIZ USER DEFINED	6		
I-DEN IS-80	SENSITIVITY, 8 CHANNEL	3328		USER DEFINED	8-96		
PRESEARCH	KEYBOARD		YES	USER DEFINED	64		
SAS-TEC VSM 210	8 BUTTON KEYBOARD	1024	TIME/DATE CAMERA ID NUMBER 2 LINES ALPHA/ NUMERIC	USER DEFINED	1024		
SONY YS-S100	SENSITIVITY, DWELL, REFRESH	288	TIME/DATE CAMERA ID NUMBER		8	8	RS232,RS485
VISION RESEARCH DIGIPLEX 4000		4000	TIME/DATE		8		
VISION SYSTEMS ADPRO 1650		540			2-16 STANDARD 128 WITH PRO 128	UP TO 16 16	
VMD-10						10	

Table 12-2 Digital VMD System Parameters (continued)

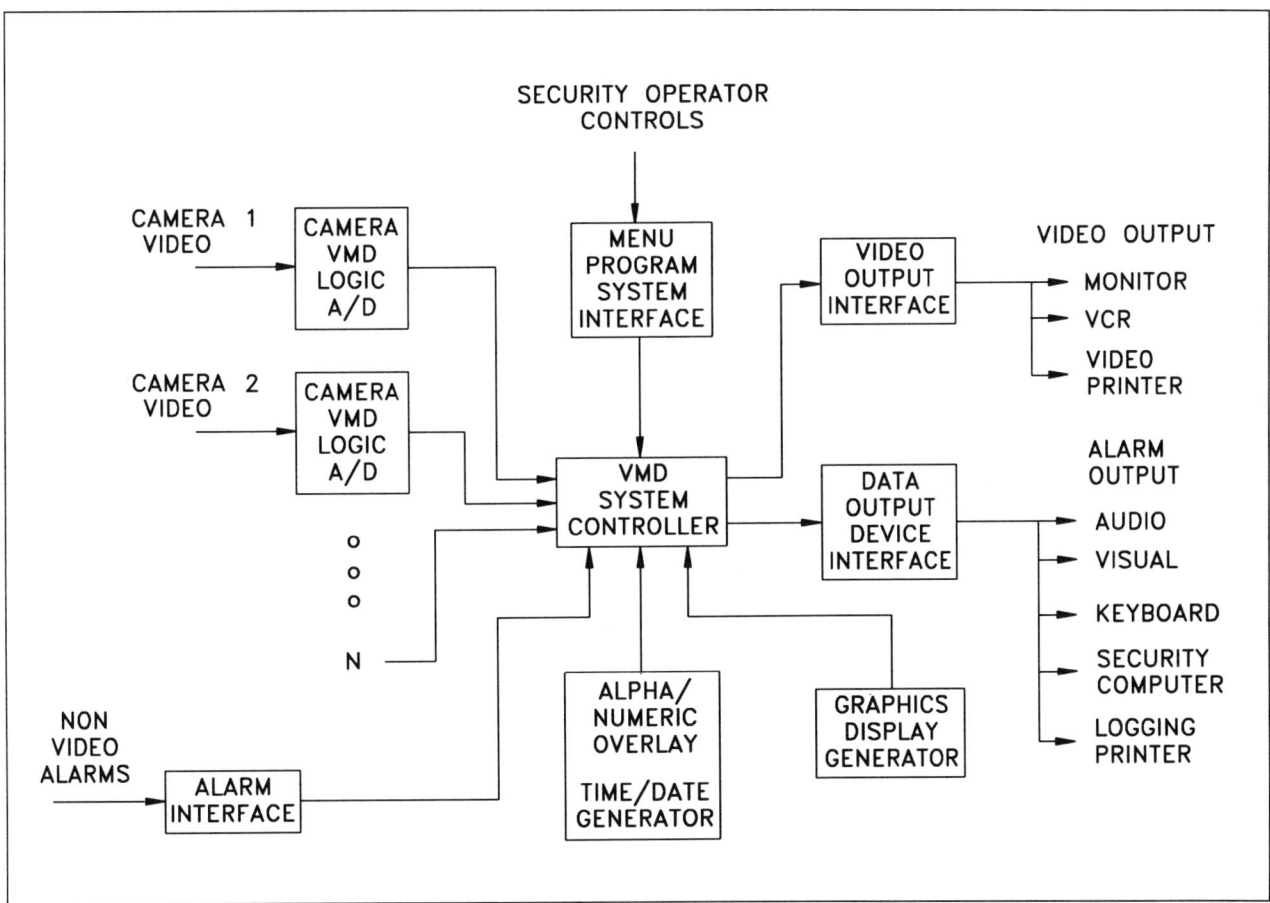

FIGURE 12-6 Digital VMD block diagram

process is initiated that compares the present live picture from the camera (which the camera generates 30 times a second) to the stored picture. Whenever there is a brightness discrepancy in any zone, the address of that particular zone location is also stored with its brightness value. Zone locations where these differences are caused by electrical noise or ambient scene motion such as blowing leaves, trees, or flags are processed out and are not considered alarms. All scene areas where detection is not desired are removed or masked out.

When a sufficient number of zones changes, an alarm is processed. The comparison process occurs across the entire scene 30 times a second. The alarm condition is established by counting the number of locations in disagreement; if a preset threshhold count is reached (any number, but generally 1 in 8 counts), the system then alarms. The count is cleared each time a new storage process takes place. The memory is refreshed on a preset basis and ranges from 1/15th of a second to many seconds. Memory refresh prevents normal changes, such as scene lighting, moving clouds, or electronic drifts in the camera from being interpreted as alarm conditions. The camera viewing the intrusion scene is automatically switched to the monitor (any standard CCTV monitor) and the scene displayed. The

monitor is usually blank prior to an alarm, since there is no reason to display the scene if no activity is occurring.

Some digital VMDs monitor up to 32 separate CCTV cameras by sampling (time-sharing) each camera sequentially. Each camera can have a separately adjustable sensitized alarming area, thereby optimizing each camera to the scene it views. Likewise, the number of sensitive zones in each camera is chosen independently to match the scene requirement. If one camera views a large-area scene looking for small intrusions, the operator can make the alarming zone small for this first channel. If another camera views a small area scene looking for large intrusions, the operator can make the alarming zone large for this channel, and so on.

12.6.3 Setup

Equipment setup procedures differ from manufacturer to manufacturer, but there are some common parameters and controls that must be determined and set when initially installing the VMD system.

1. Channel Mode Control. A three-position switch selects the mode for each CCTV camera channel. In the down position—INHIBIT—the channel is disabled and no alarms are registered. In the middle position—NORMAL—the cameras are ready for motion detection and alarming. In the up position—SET—the console operator can manually select any camera on the alarm monitor. When released from the SET position, the switch returns to the NORMAL mode.

2. Alarm Area Control. The alarm area control lets the operator manually adjust the position and size of the alarmed area zone. These adjustments can desensitize areas of the camera's FOV where normal movement would cause an unnecessary alarm. For example, in an outdoor scene where a flag is constantly waving, the desensitized area would appear on the monitor but movement within that area would not cause an alarm.

3. Refresh Control. The refresh rate refers to the time interval during which the reference frame memorized in RAM is stored, before it is again updated. Systems use refresh rates varying from 1/30 second up to several seconds. The operator selects the refresh rate, which is normally a function of the number of cameras and the kinds of alarms expected in the scenes.

4. Ranging Control. Most systems allow adjustment of the electronic analog dynamic range of the analog-to-digital (A/D) converter. The function of the A/D converter is to change the cameras' analog electronic video signal to digital values. To provide the best scene resolution for each camera, the operator adjusts the range of white to black level in the digitized video signal.

5. Masking Control. The masking control allows the operator to enter scene areas on the monitor screen for which no alarming will occur. It is entered by inserting rectangular, square, or other masked areas. In some systems the operator enters the masking with a light pen. The light pen permits irregular shapes to be desensitized merely by drawing around the object in the CCTV monitor scene.

12.7   DIGITAL HARDWARE

There is no standardization for the design and specifications of digital VMD systems. For this reason, the features of representative VMD systems and specific attributes are described in the following sections. The choice of these two systems is not an endorsement of the products, but rather an indication of what is representative in the field.

12.7.1   Single-Channel

The Adpro Pro 1650 operates with 2 to 16 monochrome or color CCTV cameras in an outdoor or indoor environment. Each CCTV camera is electronically monitored and alerts the guard when an alarm occurs. There are 540 detection zones per camera, ratiometric thresholding is used, and 16 nonvideo alarm detectors or groups of detectors can input alarms to the system. Any one of these detectors may turn on its associated camera and control up to 16 alarm outputs, a VCR channel, and an event-logging printer. For a very large facility, the Pro 128 can network with up to 8 Pro 1600s. This would comprise 128 cameras concurrently and 8 times (1024) the number of sensor inputs and alarms.

12.7.2   Multiple-Channel

Two systems by American Dynamics, the AD 1465 and AD 4000, are described here. The AD 1465 monitors up to 16 cameras and samples 10,752 picture zones per channel. These picture zones or cells are arranged in 896 addressable cells, contiguously covering over 85% of the picture area. The user can group the cells into 1 to 7 independent detection zones plus a no-detection zone. The cells need not be contiguous, that is, any number of the 896 cells in any location may be assigned to zone 1, any number of the remaining cells to zone 2, and so on. Zones may be of any shape and be divided into separate areas to accommodate unique detection requirements. Zones can be individually turned on or off to accommodate entrance, hallway, parking area, or other locations. Two examples of zones being turned on or off individually are the following: (1) A zone encompassing a gate or doorway can be turned off during shift changes while other zones in the same scene can remain active to alarm and alert an operator of unauthorized intrusions. (2) A zone encompassing a file cabinet can be left off during normal working hours and turned on overnight. Independent 16-step zone sensitivity, signal integration (retention), plus multilevel digital filtering maximize motion alarm detection and minimize false alarms. Periodic automatic rebalancing minimizes the effect of slow light changes, such as those occurring between daylight and nighttime conditions.

In operation, a cell is activated by the changes in the video content of successive picture fields. A higher retention setting delays automatic rebalancing to optimize detection of slow changes or slow-moving objects. Both the video change (sensitivity) and the rebalancing time (retention) assigned to a zone can be adjusted to optimize detection and minimize false alarms for that zone. Any activated cell in a zone alerts (activates) that zone and channel.

The system has an integral video switcher with dual video outputs and RS-232 port to allow the VMD to function as a standalone system. An audio output is available to warn the operator of an alert, and a relay closure can start a VCR for recording alerted channels. The RS-232 ports provide both a control input and an alarm output. They permit remote system control via a separate control keyboard, a data terminal, or a computer.

Either of two on-screen alert presentation modes may be selected: normal or trace. In the normal mode, a bright dot

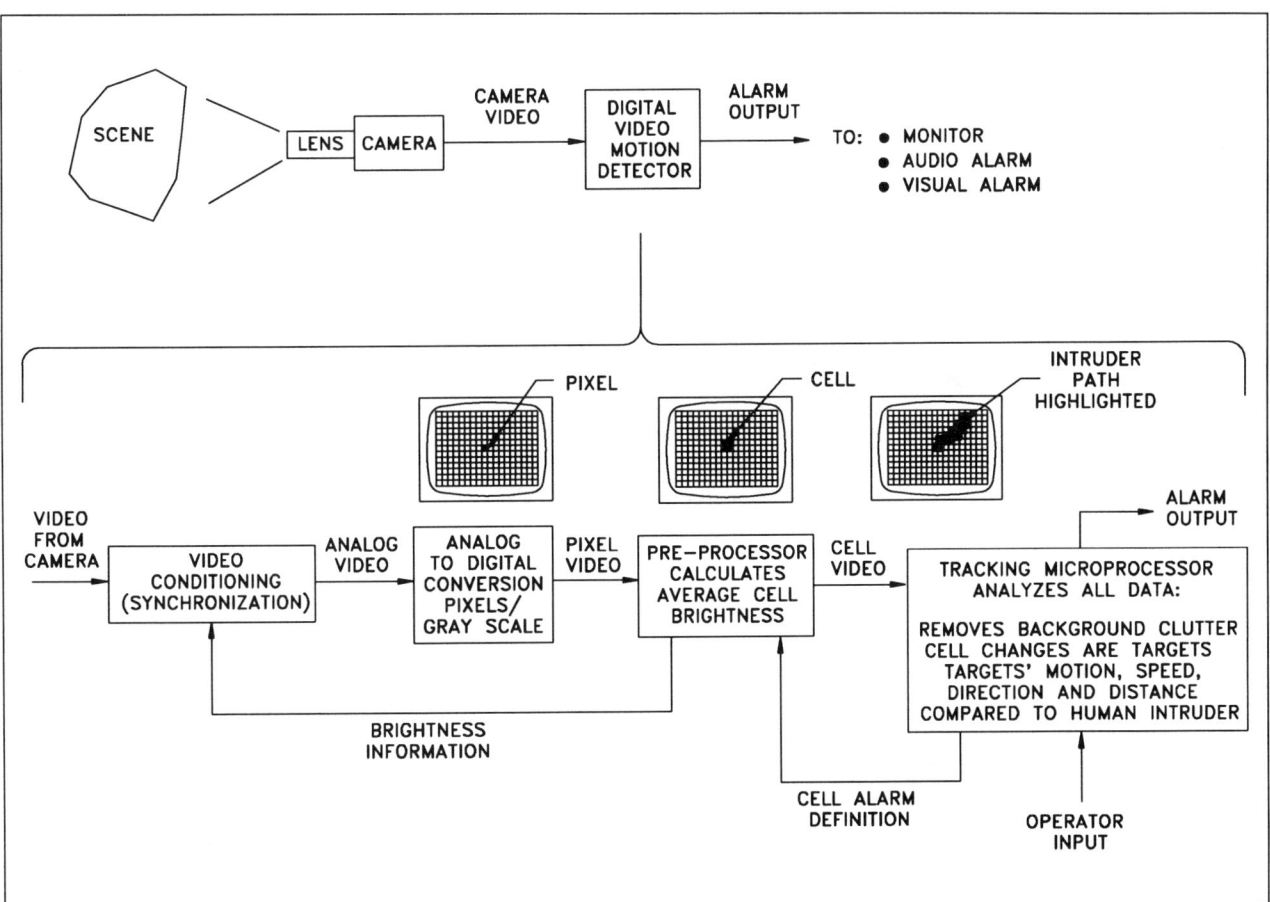

FIGURE 12-7 Block diagram for digital VMD in security system

is displayed in the picture on the alarm monitor at the center of each activated cell. With manual reset, this dot remains lighted until the channel is reset. With automatic reset, each dot disappears 16 seconds after the cell was first activated. Thus an intruder moving into a zone will cause a series of dots to appear as he first activates cells and leaves a trail of dots through the zone or to the point in the zone where he stopped or hid.

In the trace mode, a bright dot is displayed in the picture on the alarm monitor at the center of each activated cell as in the normal mode. In addition, each illuminated dot emits a quick burst of flashes 8 seconds after it is activated. With manual reset, this results in a continuous moving trail of flashes at 8-second intervals along the path of intrusion. With auto-reset, a single burst of flashes occurs before each set is automatically reset. These flashes can assist the operator in determining the size, direction, and location of an intrusion. Figure 12-7 shows the system block diagram for the VMD in a security system.

A more comprehensive VMD digital system is represented by the AD 4000. A high-speed microprocessor analyzes detected motion for size, position, and rate of movement to discriminate against undesired targets and verify a valid intrusion before the system signals an alarm.

Verified intrusions initiate audio and visual alarm signals. Video from alarmed cameras is connected to outputs for an alarm monitor, a VCR, and a freeze-frame recorder to monitor and record the track and position of intrusions. Independent output relays provide control of external devices. A built-in sequential switcher provides normal system viewing of all cameras by separate video output.

For ease of operation, user-defined detection of active areas is initiated using a light pen. Via the light pen, target-detection areas for each camera are easily established, with up to 2700 detection zones possible. Zones can be individually deactivated while observing the picture to eliminate detection of areas where insignificant or acceptable motion could cause some false alarms. The system has the ability to perform target discrimination. Each camera module is programmable to optimize target discrimination based on a combination of anticipated characteristics, such as size, rate of movement, and indoor/outdoor scenes. In order to see the intrusion track and position display, zones where motion has been detected are highlighted on the video displays. There are two user-selectable operating modes: instant and processed. In the instant mode, an alert is issued when motion is first detected; the video from the camera is sent to the video outputs for inspection and recording. If the system analysis verifies a valid intrusion, it sends an alarm signal. In the processed mode, the system signals an alarm and puts the video from the camera on the video outputs only after system analysis verifies a valid intrusion.

To prevent lost pictures at the time of alarm, the AD 4000 uses a freeze-frame recorder and captures the pictures frozen at the instant of alert and alarm. Thus the action often missed by an operator at the time of an alarm can be reviewed and documented by a VCR recording of the frozen pictures.

Processed and instant operating modes are user-selectable. In both modes, when motion is detected in active detection zones in a camera picture, system analysis is initiated immediately to verify a valid intrusion. In large installations, multiple AD 4000 systems can be combined to increase camera capability from 8 to 32.

12.7.3 Image Capture

The DAVID VMD operates with up to 12 video cameras. The A/D converter digitizes each frame into 6400 pixels. Each pixel has 256 gray-scale levels to detect small changes in brightness or contrast. The digital processor computes the average brightness in real time, in each detection cell. The system microprocessor analyzes the cell data and removes background clutter and identifies any changes in the cells as targets to be tracked. The target's motion, speed, direction, and distance traveled are analyzed to see if they match the characteristics of a human intruder. When a human intruder is identified, on-screen graphics highlight his position and an alarm is signaled.

DAVID is menu-driven, thereby providing an easier setup for the operator. Special setup graphics define the camera zones to be monitored. Target discrimination is based on target size, contrast, speed, and direction. Target tracking is used to verify detection before declaring an alarm, resulting in a low false-alarm rate. The operator sets up sharply

defined detection zones configurable for each camera, which may be tailored to reflect the optical differences between near and distant areas and act as distance compensation.

12.8 SUMMARY

The use of a VMD significantly increases the security level and reduces the human error in any security system. The choice of the optimum VMD for a specific application requires that the security designer understand the equipment capabilities and limitations and match them to the problem. Of highest importance is whether the VMD can properly react to the changing lighting conditions in the television scene and generate meaningful alarm information and reject false alarms. The present state of the art indicates that analog VMDs can operate acceptably only in well-controlled indoor environments, while digital VMDs can operate in all indoor environments and do well in most outdoor environments. Because of the variety of approaches and differences in digital VMD equipment, characteristics of systems manufactured by leaders in the field must be considered on their own merits. Analyzing the systems described exposes the designer to some of the features available and permits asking the manufacturer sensible questions to determine suitability for the problem to be solved. Some helpful comments and hints follow:

- Analog or digital VMDs are suitable for indoor applications.
- Digital VMDs should be used for all outdoor applications.
- The VMD should be able to switch video to a VCR and produce a hard-copy video printout. Once the VMD system is set up, most of the decision making should be automatic.
- Following initial setup, alarm declaration should be automatic, using a menu-driven program.

Chapter 13

Electronic Image Splitting, Compression, Reversal, and Video Annotation

13.1 OVERVIEW

Many video accessories can significantly enhance the effectiveness of a CCTV security system, including (1) electronic image-splitting and -compression devices, (2) data annotation equipment, and (3) video scan-reversal units.

An electronic image-splitting device is a control unit interposed between the camera (or switcher) and the monitor, VCR, or video printer. It combines parts of the scenes of two or more cameras and displays them on one monitor.

The use of the inexpensive image splitter increases security into two ways: (1) the screen splitter reduces the number of recorders required, since it combines several scenes onto one video signal, and (2) it permits the guard to view a monitor with two or three scenes on it rather than one.

Image-compression devices are used to compress and display two or more full video scenes on one monitor. Devices are available for displaying as many as 32 full video scenes on a single monitor, thereby reducing the number of monitors required in the security console. This feature is particularly important in larger installations, where a guard cannot effectively view more than a half dozen monitors at a time.

Video annotation equipment represents a class of accessories that display alphanumeric characters on a video monitor to supply information such as the time, date, camera number, camera scene information (hallway, lobby, etc.), alarm source, and other instructional information for the guard.

The scan-reversal unit reverses the video picture horizontal orientation, as required in some applications using mirror optics. An example of this is a mirror used with a fixed-focal-length (FFL) lens to redirect the camera viewing angle.

13.2 Electronic Image Splitting/Compression

Chapter 4 described several image-splitting optics that view two or three independent scenes from two or three different lenses and superimpose the scenes onto the same camera and monitor. In the case of two lenses, only one-half

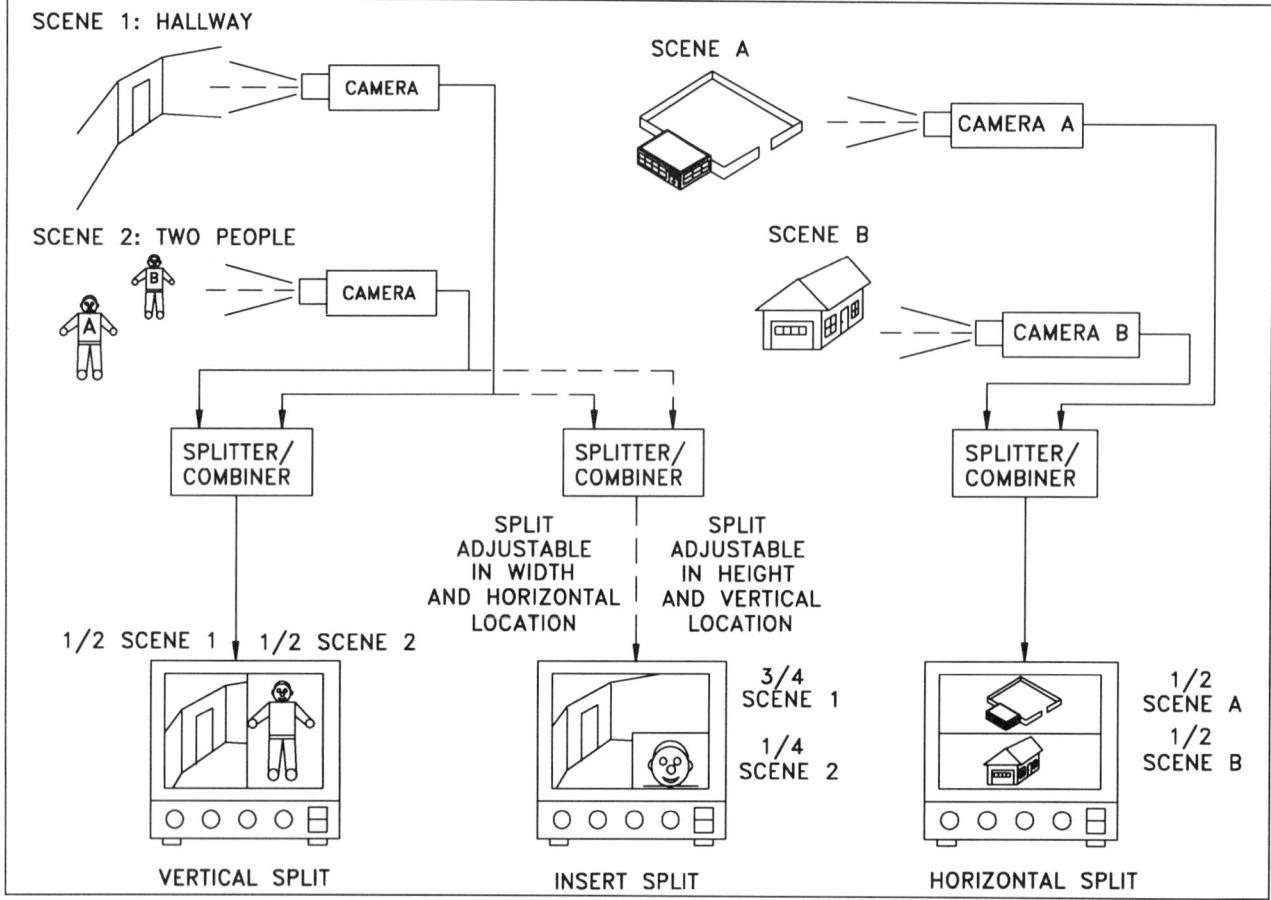

FIGURE 13-1 Monitor picture from two-camera image splitter

of each of the original scenes is used (for three lenses, only one-third of each is used). The combined image can be recorded on a single standard VCR or printed on a single video printer.

Several manufacturers produce image-compression equipment that combines the full video signals from 4, 9, 16, or 32 individual cameras and displays them side by side or one above the other on a single monitor. This compression equipment displays the entire camera scene on the monitor but at a cost in resolution (loss of half the horizontal resolution and half the vertical).

13.2.1 Video Image Splitting

Many manufacturers produce electronic image-splitting (actually combining) devices, which take two video monochrome or color camera signals and display a part of each picture on one monitor. The vertical split case shows two camera scenes: one-half picture from camera 1 and one-half picture from camera 2, each displayed side by side.

In the horizontal split case, the same two camera scenes are shown on a single monitor, with the image splitter displaying one-half of scene A on top and one-half of scene B on the bottom. Manufacturers produce devices to insert a rectangular section of one picture into another. This insert split shows a part of the camera 2 scene inserted into the picture of camera 1. The inserted picture is adjustable in size and can be bounded by an adjustable gray, black, or white border. The inserted picture can be positioned anywhere on the monitor screen. It can be adjusted to make a full horizontal or vertical split, make an insert in any corner, or make a "floating" insert (Figure 13-1).

Almost all electronic-splitting systems require that the two video signals be synchronized. Figure 13-2 shows the connections for synchronizing the two cameras in the system.

One camera can be unsynchronized but should have a vertical and/or horizontal output drive signal to synchronize the second camera. This requires that the second camera be able to be synchronized by a horizontal or vertical sync signal, or a composite video signal. Alternatively, the two video cameras can be synchronized by some external synchronizing source. The best synchronization between the two cameras is obtained when the driving camera has a 2:1 interlace. Normally, any random or 2:1-interlaced camera is used as a master camera and any

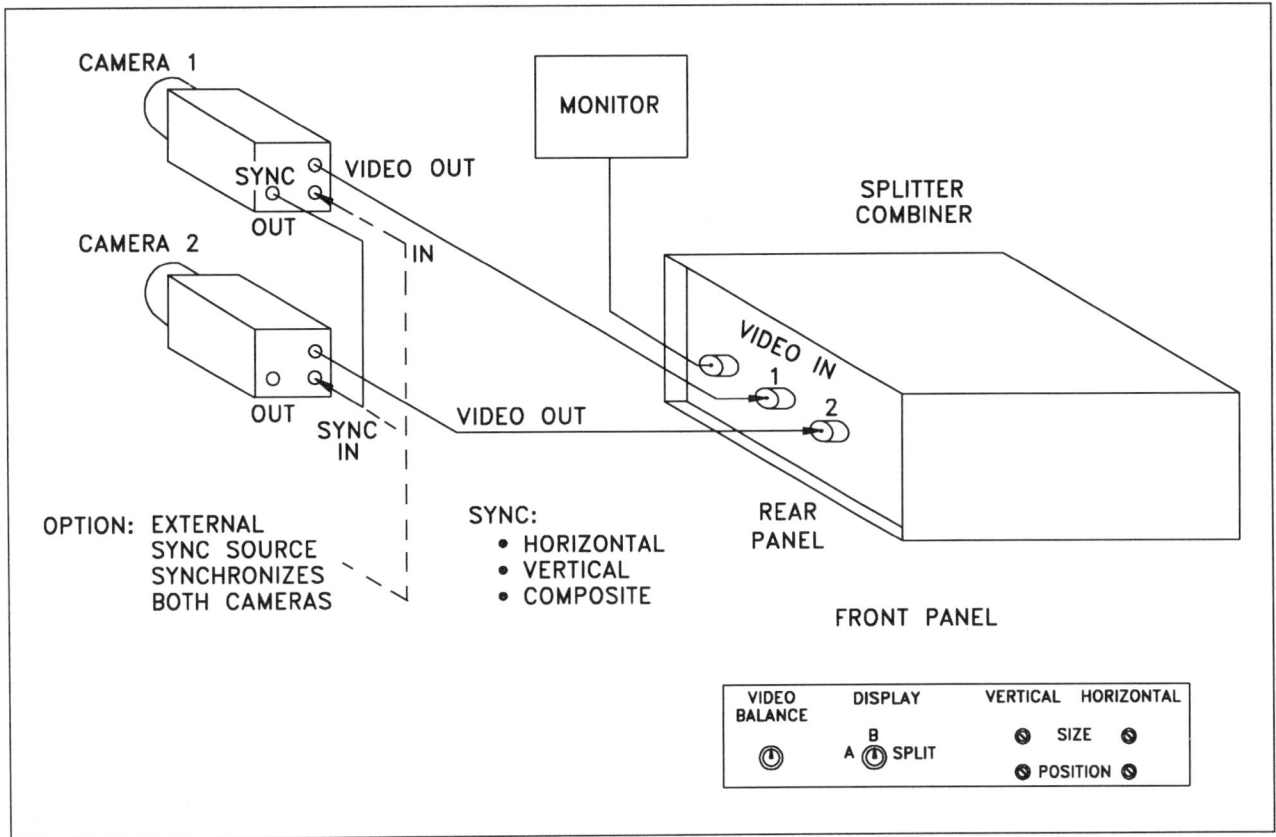

FIGURE 13-2 Interconnecting diagram for electronic-splitting system

camera accepting horizontal, vertical, or composite video signal drive is used as a slave camera.

The adjustable front-panel controls on a video splitter include vertical position, vertical size, horizontal position, horizontal size, and video level balance. There is also a three-position switch so that the full camera 1 scene, the full camera 2 scene, or the split image can be displayed. Different manufacturer models provide controls to produce no outline along the split, black outline, gray outline, or white outline. More than one splitter-inserter may be cascaded in series to display video from more than two cameras on a single monitor. Three cameras require two splitters, four cameras require three splitters, and so on. The splitter-inserter selects only a portion of each picture: *it does not compress* the picture into a smaller area. When camera scenes must be transmitted to remote guard locations, the splitter-inserter technique reduces the number of expensive coaxial-cable, fiber-optic, microwave, RF, or infrared video links. Eliminating one or more of these links can reduce system cost considerably. A cost-saving retail-store splitter application is shown in Figure 13-3.

The monitor displays a quarter-split, three-quarter split picture, where the three-quarter image provides information from a wide-angle lens viewing an entire cash register area and the one-quarter split shows a close-up of the merchandise on the conveyor. The video output from the screen splitter is standard composite video and is accepted by any standard monitor, real-time or time-lapse VCR, or video printer and requires no special processing for transmission, display, recording, or printing.

13.2.2 Video Image Compression

A second generic type of image combining is accomplished by *compressing* the video images from four or more cameras and combining and displaying them on a single monitor. Equipment from manufacturers is available to compress 4, 9, 16, and 32 full video images onto a single video monitor. In the case of a quad combiner (using four cameras), the complete original picture is compressed into one-fourth its original size (half-width, half-height). The resolution is likewise reduced to one-fourth of its original ($\frac{1}{2}$ H × $\frac{1}{2}$ V = $\frac{1}{4}$). Most video-compression systems are capable of combining and displaying either monochrome or color cameras.

The image-splitting lenses described in Chapter 4 display only a portion or fraction of each individual original scene. Likewise, the electronic splitting devices described in Section 13.2.1 display only a portion or fraction of each scene. In the video compressor, through the use of computer technology, the entire video picture is compressed into a

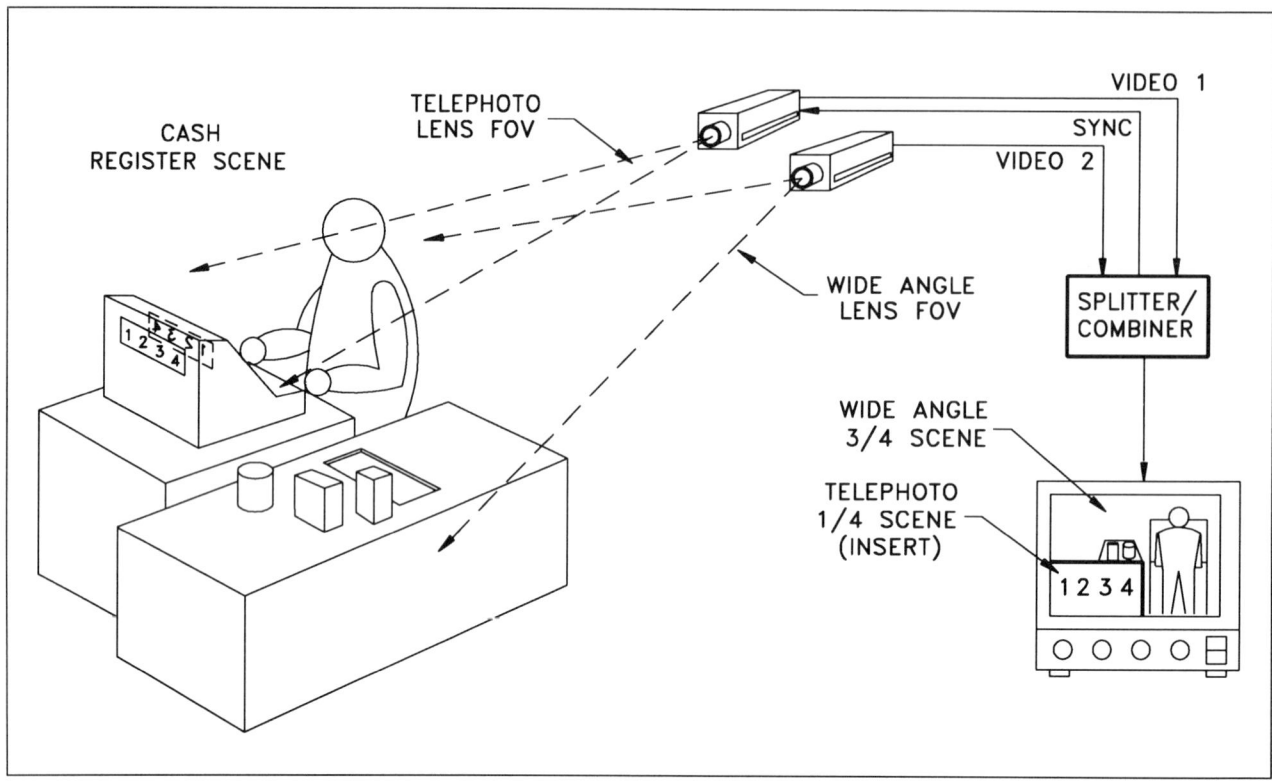

FIGURE 13-3   One-quarter/three-quarter image split

smaller size so that multiple *complete* video images are displayed on the same monitor screen.

Obviously, the more images displayed on a single monitor, the less resolution available for each individual scene. For instance, if four video scenes are displayed on one monitor, the horizontal and vertical resolution are each decreased by a factor of 2 and the entire picture resolution is reduced by a factor of 4. The decrease in resolution is proportional to the number of scenes on the screen; therefore, the resolution obtainable from the 9-, 16-, and 32-way splits is correspondingly reduced by a factor of 9, 16, and 32, respectively. This decrease in resolution can be compared with using a lens with less magnification and should be considered as a system-parameter trade-off rather than a limitation. The advantage of video image compression is that more than one camera scene can be displayed on a single monitor. Often there is sufficient resolution in the video scenes so that the image can be compressed and viewed on a single monitor and still provide useful communication and intelligence to the security guard. When the system is connected into a video motion detector (VMD) or some other camera-alarming device, or if the guard can see motion in one of the pictures, the guard can switch the alarmed camera scene to full screen presentation. This means that the guard has the ability to view many camera scenes on one monitor and then display only that scene requiring his attention.

The following paragraphs describe special features and characteristics available in image-compression hardware. By far the most popular version is the quad (four-way) combiner/compressor, displaying four individual camera scenes on one monitor. In contrast to the electronic image-splitter/inserter equipment described in Section 13.2.1, all combined video scenes displayed on the image-compression equipment are full video scenes as they appear at each camera output.

13.2.2.1 Quad Combiner

Combining four full-camera pictures on one monitor is accomplished using a quad compressor (Figure 13-4). The quad system takes four standard synchronized or nonsynchronized CCTV camera signals, compresses the pictures, and displays them on a single monitor. Earlier models required that all cameras be synchronized, but most systems now accept inputs from random or 2:1-interlaced cameras, VCRs, switchers, and other devices, and no external synchronization is required. Specific design details differ depending on manufacturer, but the results are essentially the same. The camera signals are digitized and stored temporarily in random-access memory (RAM), usually a 256×256 or a 512×512 array, with 64, 128, or 256 gray-scale levels and compressed and combined into one frame. Depending on operator control, either the full-screen (select) pictures

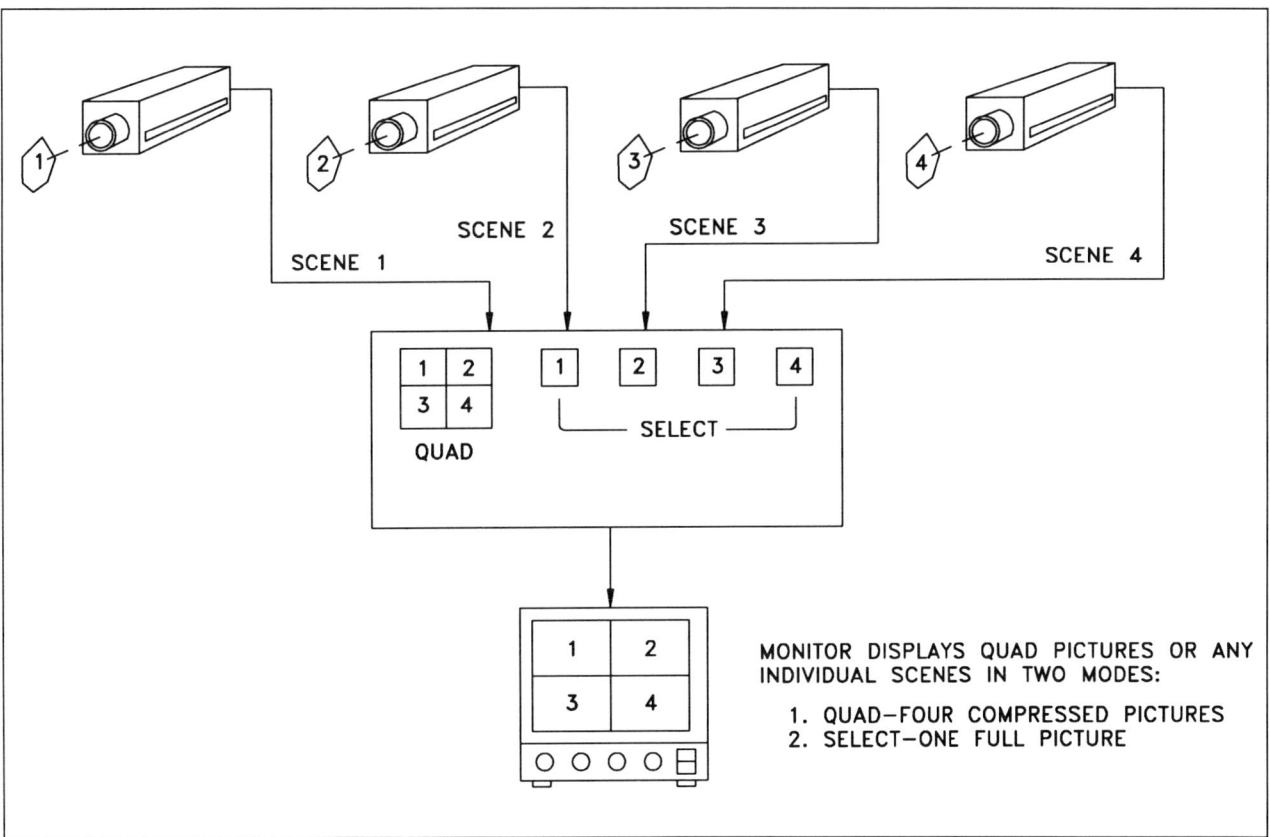

FIGURE 13-4 Quad combiner system

or the digitized compressed pictures (quad) are displayed on the monitor. The quad units are provided with individual channel-gain and level-control circuits to automatically balance the brightness and contrast of the four individual cameras. If an alarm occurs on any of the individual cameras, that camera can be displayed full-screen for more scrutiny. Some quad systems can be controlled remotely via an RS-232 communications link. Figure 13-5 shows the typical quad pictures.

Figure 13-5 shows diagrammatically the five different scenes that the quad system can display. In the Select mode, the full-screen images from camera 1 output, camera 2, camera 3, and camera 4 can be selected. In the Quad mode, each individual picture on the monitor is a full-camera scene reduced in size (compressed).

By simultaneously displaying four camera views in one-quarter-screen pictures, the quad compression system keeps the operator current on events occurring in the four camera scenes. When activity occurs in one picture, the guard enlarges it from the quadrant size into a full screen.

Some image-compression equipment provides a means for "freezing" the picture, thereby holding it stationary on the monitor for detailed examination. This permits the security operator to view a single scene in more detail over a period of time, until it is released by the operator. In this mode a VCR recording can be made of the full-screen or quad pictures and frozen snapshots can be made on the VCR or video printer for documentation or analysis.

13.2.2.2 Nine- and 16-Camera Combiner

While the majority of image compression is accomplished using a quad combiner, some applications can take advantage of displaying a larger number of cameras on a single monitor. Figure 13-6 diagrammatically shows a 9- and a 16-scene combiner.

As in the case of the quad combiner, a single-camera full-screen scene can be displayed when activated by an alarm or by the guard. A front-panel control lets the user select any four inputs in a two-by-two display (quad), any 9 inputs in a three-by-three display, and any 16 inputs in a four-by-four display. The displays or inputs within any display can be changed with a push-button switch on the front panel, or by remote serial control signals via a standard RS-232 link. As with quad units, these systems accept synchronized or nonsynchronized video signals.

13.2.2.3 Picture-in-a-Picture

A very useful variation of the image-compression technique is the compressed picture inserter or picture-in-a-picture (Figure 13-7).

With this unit, a camera picture is compressed and inserted as part of a full-frame picture. Via a front-panel control, the compressed picture can be interchanged with the full picture. Some units have multiple camera inputs to choose from. Most units display monochrome or color

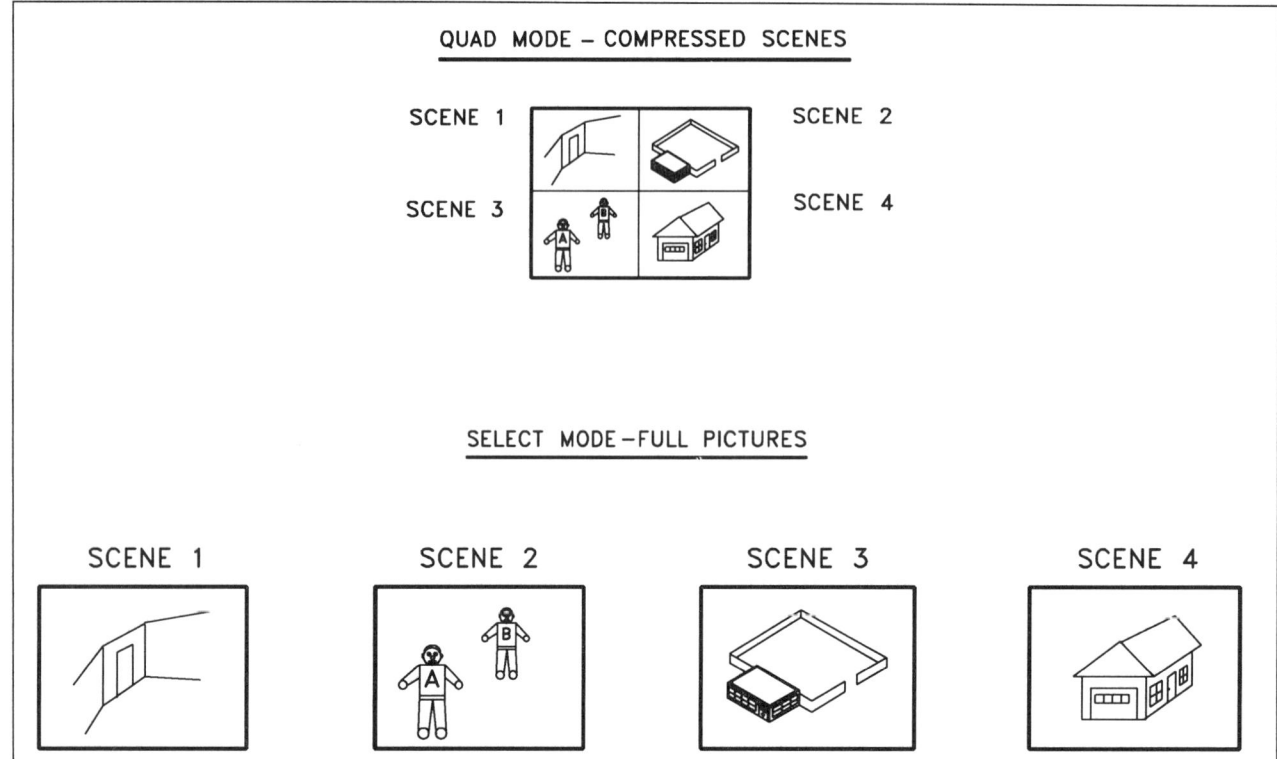

FIGURE 13-5 Scenes obtainable with quad combiner

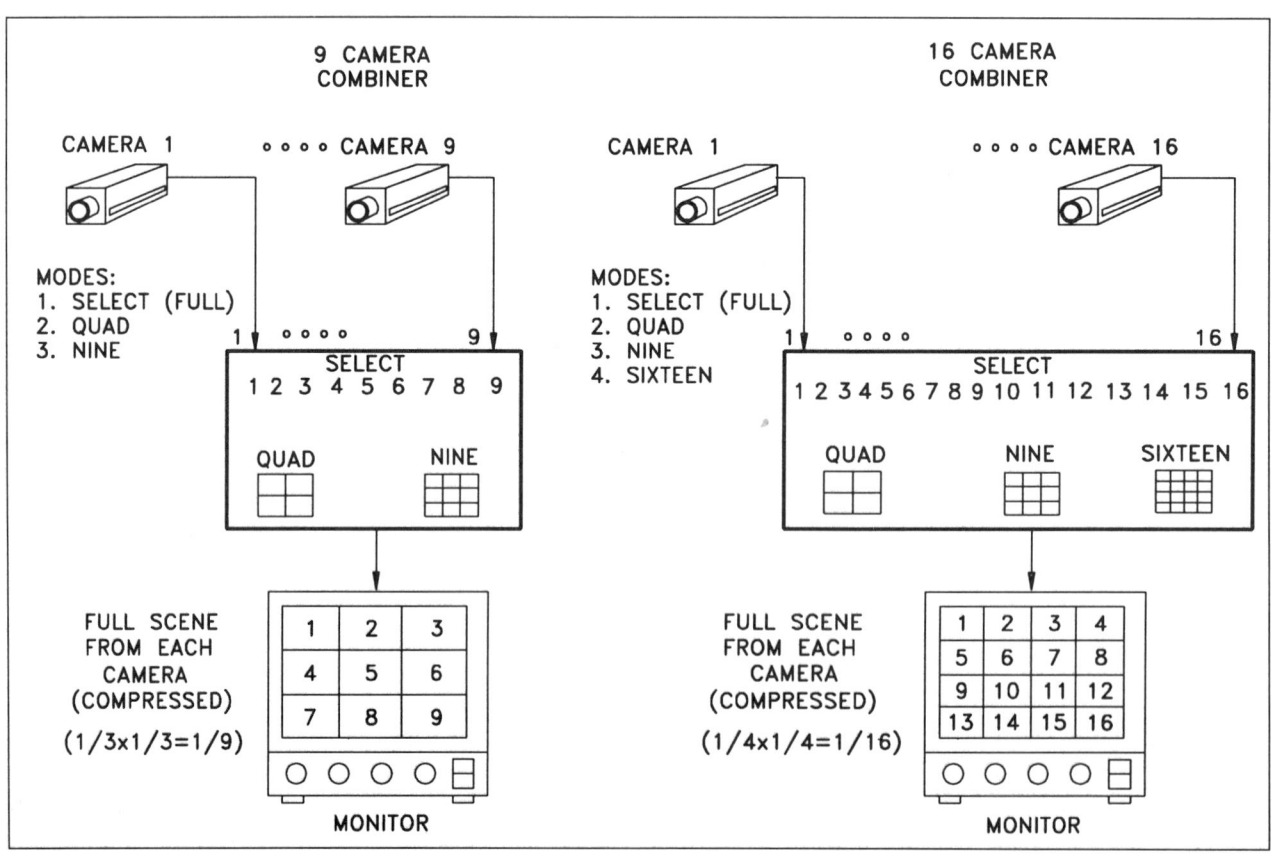

FIGURE 13-6 Nine- and 16-scene combiner

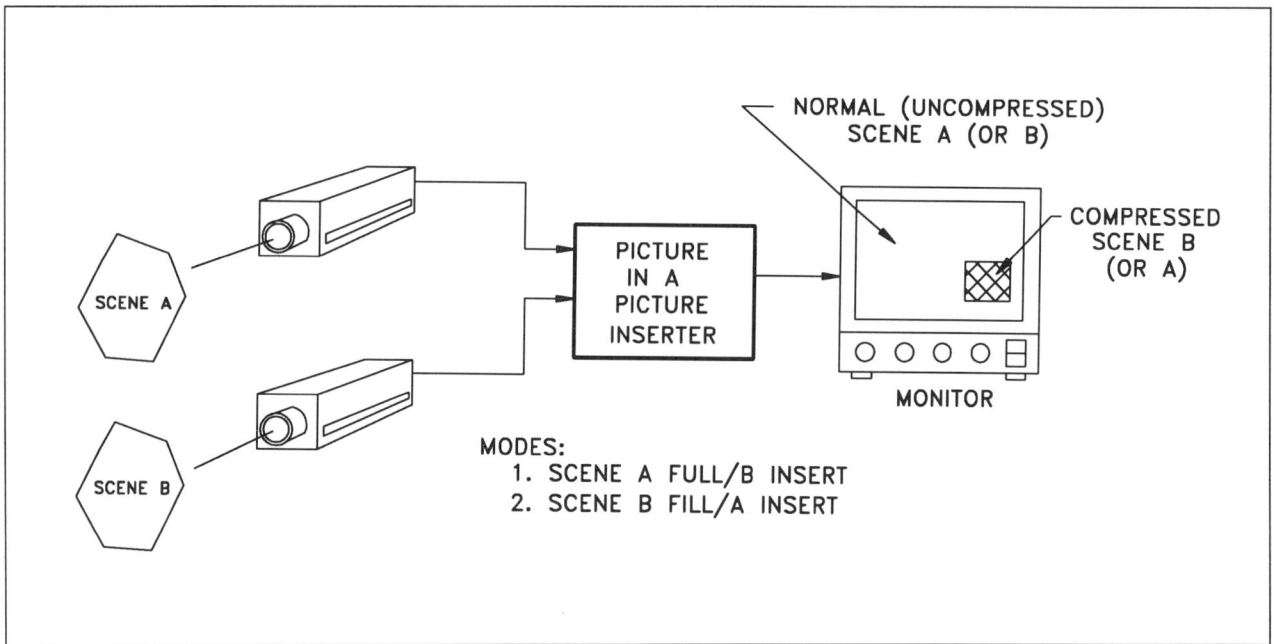

FIGURE 13-7 Picture-in-a-picture inserter

pictures on the main image and monochrome on the compressed inserted picture, but some display color on both. Freeze-frame of the compressed picture is available on most units.

13.3 DATA INSERTER—VIDEO TYPEWRITER

It is often necessary to superimpose alphanumeric data on a video display for security applications. The information contained in these displays may include camera number, time and date, computer terminal transaction number, cash register amount, alarm instructions, and any other message providing the guard with additional useful information and/or evidence presentable in a courtroom for prosecution. The equipments available to provide these functions vary from very simple time/date generators to comprehensive typewriter keyboards or video typewriters that can store programmed messages and are displayed on the screen upon command or automatically. This section describes various alphanumeric generators used in many security systems.

13.3.1 Time/Date Generator

The most commonly used video annotation equipment is the time/date generator. Time and date annotation are required when a video recording from a VCR is used in a courtroom procedure. The time/date unit is connected in series with the camera and monitor, functionally adds the time and date to the video signal passing through it, and displays the picture and the time/date on the monitor (or

records it on the VCR). Since both the scene and the annotation are on the same composite signal or VCR frame, they are permanently correlated in time, which is important in a security application where prosecution relies on the correlation of two events or connection between an event and an individual. The days of the month and year are preprogrammed in memory, and the system even remembers to put in February 29 during a leap year. The time and date annotation can be positioned anywhere within the effective picture area. Most units provide a backup battery so that if main power is lost during a power failure the time/date clock will retain correct time and date for many hours. Typical battery backup time is 30 to 100 hours. Once the unit has been set up and the time and date adjusted, it requires no additional attention.

Most units provide year, month, day of month, hours, minutes, and seconds in the annotation. The seconds annotation may be suppressed in the digital readout. Most units provide a variable alphanumeric readout intensity relative to picture intensity via a front-panel control. For maximum picture usability and optimum readout of the numeric time/date information on the display, the ability to put a border of black, gray, or white around the time and date, block it, or provide no edges around the time/date digital display is provided.

To facilitate easy setup, the equipment has user-friendly on-screen prompts to set the time and date, position the digital information on the screen, and set the character brightness. For commercial/military applications, most units have a 12-hour AM/PM designator mode or a 24-hour mode. One manufacturer provides a user-selectable built-in daylight savings time corrector. Various manufacturers make equipment operating from a 12-volt DC supply for use

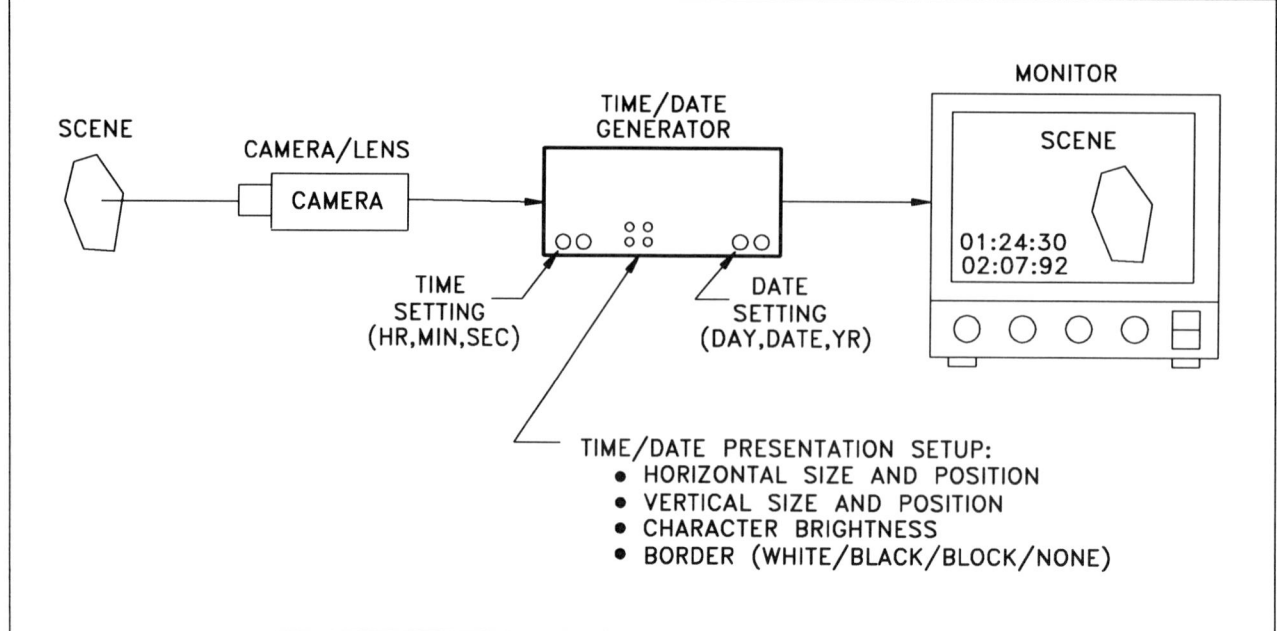

FIGURE 13-8   Time/date generator

with portable cameras and VCRs. Figure 13-8 illustrates a typical time/date generator, its screen presentation, and setup controls.

13.3.2 Camera ID Inserter

When multiple cameras are used, it is useful to annotate a video monitor display with the camera number, corresponding to a particular scene or area in the facility. This function is performed by extra camera circuitry provided by the camera manufacturer or by a separate camera ID inserter unit. As with the time/date generator, the camera ID unit, when provided separately, has the video looped through it so the camera ID number is superimposed onto the video picture. When the camera manufacturer supplies the ID unit, it is an internal auxiliary board module having a slide ("dip") switch used to set the camera ID number.

13.3.3 Message Generator

When more-comprehensive annotation of a video display is required—information such as time/date, camera number, title of data display, message display, and so on—more versatile equipment is needed. This "video typewriter" unit is connected between the video camera and monitor and is programmed via front-panel controls, from a keyboard, or through an RS-232 interface. Most units can generate several lines of alphanumeric characters, providing full text and numerals on the monitor. Figure 13-9a shows an example of a video annotator.

Figure 13-9b shows a simple three-line display. The system can generate an on-screen menu. In another mode the system can generate a systems diagnostics and illustrate all the alphanumerics it can display. In the menu mode, the user can set the clock, ID number, line position, line text, and so on, all via the menu and keyboard. Once the data have been entered, the keyboard may be removed; the equipment retains the information for call-up via front-panel switches on the control unit.

The system in Figure 13-10 can display 16 alphanumeric, characters with a maximum of 15 blanks between characters on the monitor screen. The characters displayed are selected from a built-in programmable read-only memory (PROM), which includes a selection of 28 different characters. The unit also accepts external data, which can be superimposed onto the CCTV picture via an RS-232 command signal. Built-in digital switches are used to change the character size in four steps in both the horizontal and vertical directions. Character position on the display can be varied horizontally and vertically. As with the time/date generators, the display characters can be presented in black or white for optimum readability. A black, white, or off-white mask can be selected and superimposed to surround the characters. The intensity of the character display is controlled by the operator to maximize readability.

A very comprehensive system displays clear legible characters in four sizes from a 32-page memory. It can also display 256 continuous lines of text and is capable of extensive text editing. The system operates with monochrome or color CCTV; in color mode, it offers 512 color selections, with a working palette of 14 colors at any time. It can provide nine different speeds for text roll and crawl, up or down the screen. It has adjustable edge and shadow con-

FIGURE 13-9 Video screen text annotation

trols for the characters, to optimize contrast with the video scene.

13.3.4 Transaction Inserter

Character generators can be dedicated to specific hardware by adding specific messages to inputs or commands. Several applications for which specific equipment has been de-

signed include transactions for (1) automatic teller machines, (2) point-of-sale (POS) terminals, 3) cash registers, (4) gas station terminals, (5) credit card verifiers, and (6) electronic scales. These dedicated text inserters combine digital text with video images for display on a monitor, recording by a VCR or printing on a video printer.

Most of these specialized systems operate with peripheral equipment via a standard RS-232 interface or other communication protocol. In the case of cash register, POS, and ATM equipment, the protocol sent by these equipments is interpreted by the message generator, converted into alphanumeric characters, and displayed on the screen. To operate properly with each peripheral device, the unit must match the unique protocol of the equipment. Some annotation equipment is versatile, so that it can be set in the field to match the protocol of the various equipments. Other annotation equipment is designed to be dedicated to specific devices. In one system, up to 25 lines of 80 characters per line can be displayed on each screen. The characters include upper- and lowercase letters, numerals, and special symbols. These message generators and annotators can be programmed to display only certain events, such as voids, no sales, cash refunds, and so on.

All the information displayed on the monitor—the video picture and the message and transaction—can be recorded on a video time-lapse recorder to show an audit trail of business transactions, including data on what cash was exchanged, and the time and date. By combining alphanumeric text with a video screen, a concise, easily reviewable record is obtained, offering permanent, positive, and complete identification.

13.4 VIDEO CAPTURE

The video-capture technique uses a video digitizing board capable of storing multiple video frames, usually from one to four (Figure 13-11).

When first switched on, it stores the first four video frames it receives. When the fifth frame arrives, it stores it while simultaneously discarding the first frame, keeping frames two, three, four, and five, and so on. By this technique, when an alarm at the camera location occurs, four video frames *prior* to the alarm are stored in the device. These four scenes and all subsequent ones are available for display and recording.

13.5 SCAN-REVERSAL UNIT

Most monochrome CCTV tube cameras can be modified easily to electronically reverse the scan on the tube so that if a mirror is located in front of the camera lens—which reverses the picture from right to left—the tube deflection

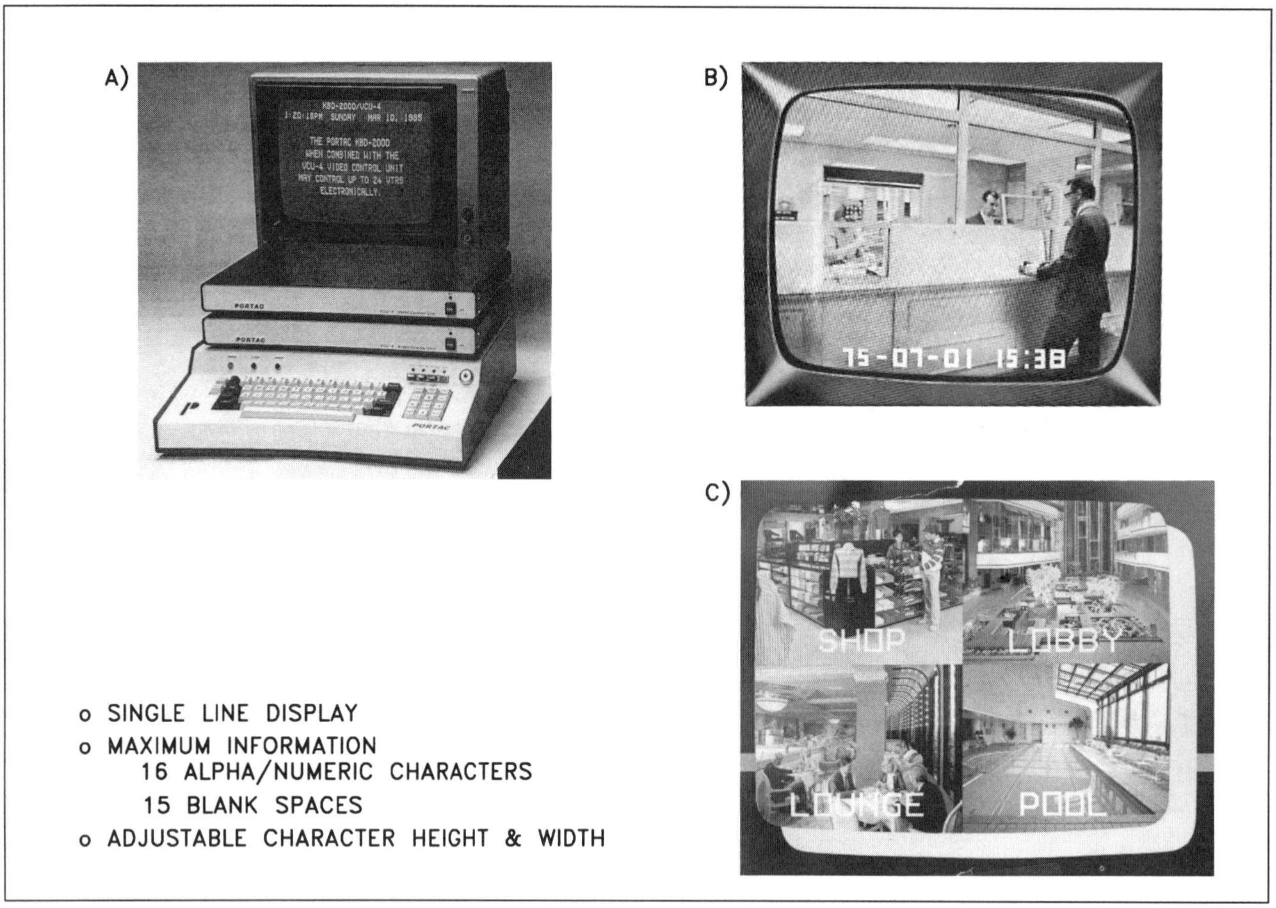

- SINGLE LINE DISPLAY
- MAXIMUM INFORMATION
  - 16 ALPHA/NUMERIC CHARACTERS
  - 15 BLANK SPACES
- ADJUSTABLE CHARACTER HEIGHT & WIDTH

FIGURE 13-10 Alphanumeric video annotation

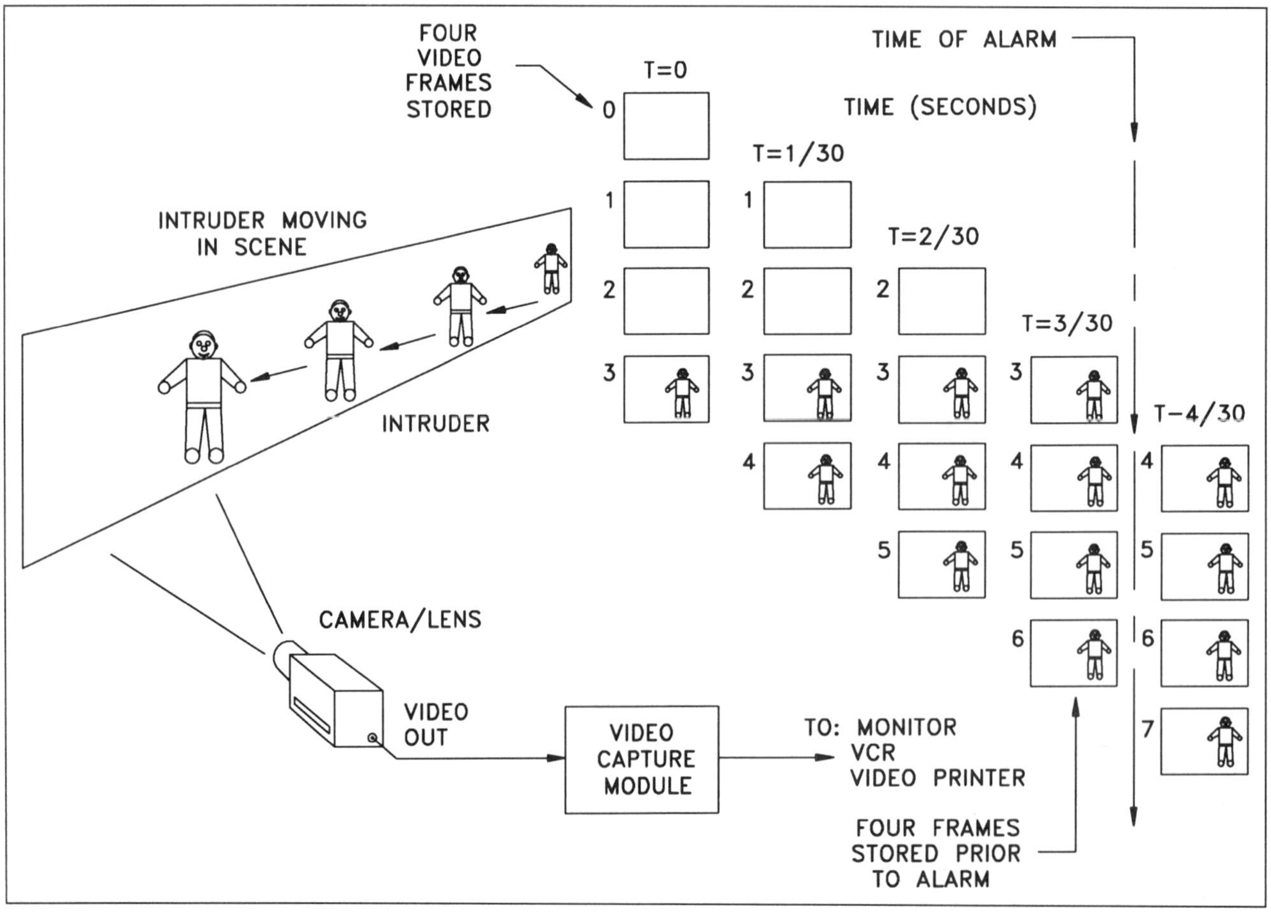

FIGURE 13-11 Four-frame video-capture module and diagram

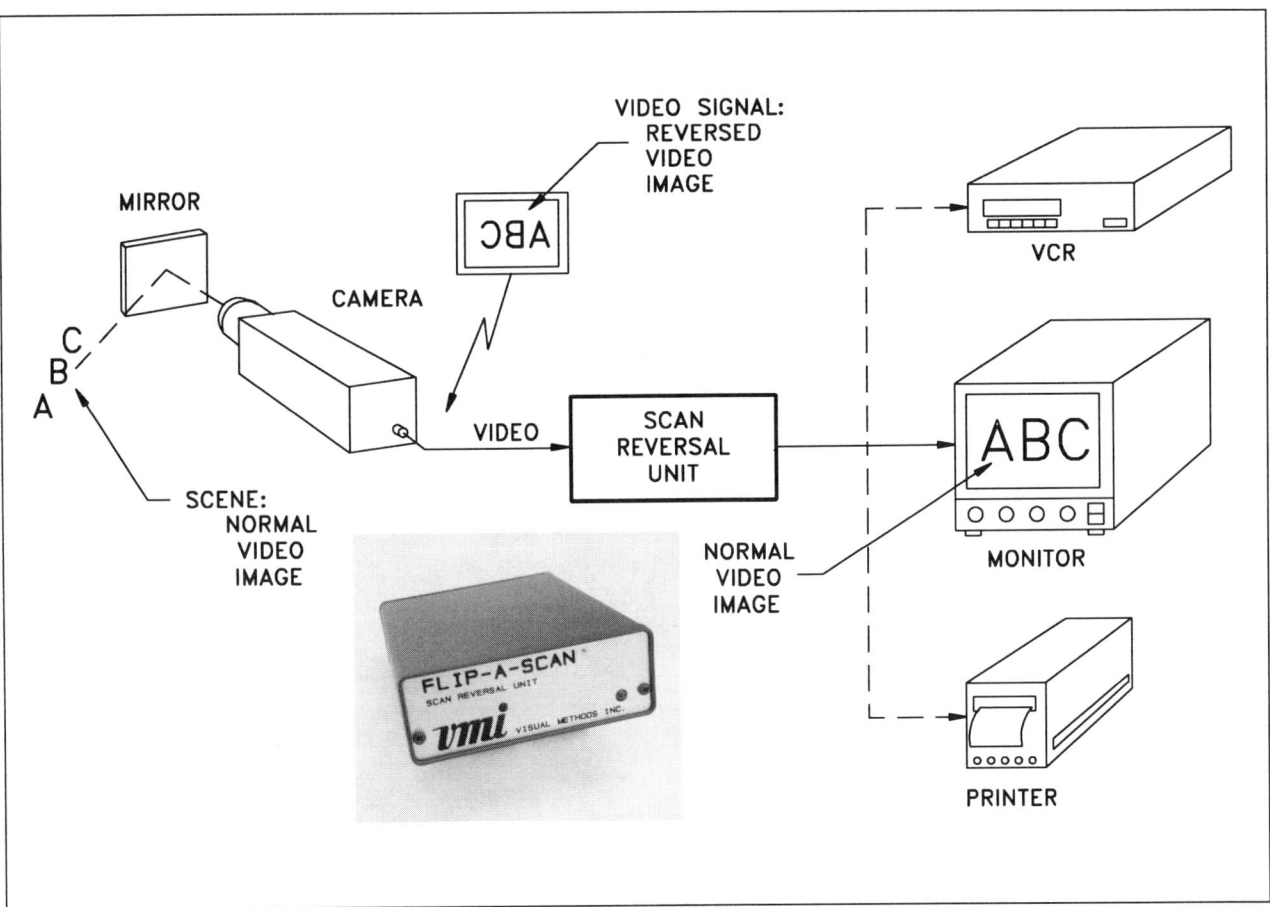

FIGURE 13-12   Scan-reversal requirements

coils can be reversed to produce the correct left-to-right orientation. While this is true for the monochrome tube camera, it is not true for the color tube camera or CCD solid-state-sensor cameras. A device called a scan reversal unit (SRU) has been developed for monochrome cameras to reinstate the correct left-to-right monitor display (Figure 13-12).

This procedure is necessary in all CCD camera installations where the camera sees a reversed image of the subject because it is viewing via a single mirror in front of the lens. The monochrome SRU stores one line of digitized video signal at a time in a static RAM and then reads it out in a reversed sequence during the storage of the next line. By using dual memory chips, each working alternately, the SRU displays the whole picture without any noticeable delay. The units available have approximately 600-line horizontal resolution, which is sufficient for most available monochrome CCD cameras. The SRU concept can also be extended to a color system.

13.6 ALARMS OVER COAXIAL CABLE

In most CCTV security applications, alarm information generated by nontelevision means—that is, not by VMDs but by contact closures or some other means—are transmit-

ted to the console monitor area via separate cable or transmission means. This procedure entails additional cost to transmit these alarms over the full camera-to-monitor distance. Equipment is available to multiplex the alarm onto a coaxial cable that is also transmitting the video signal, and then demodulate it at the monitor end to extract the alarm information, and process it as a normal alarm. This "piggyback" transmission is accomplished by multiplexing the coded alarm signal onto the video signal during the first 31 lines in the video scan (see Section 5.2). By encoding this alarm information during a nonvideo-signal time segment, the alarm information in essence "rides free" on the video signal.

In one configuration, up to 15 alarm points are transmitted on a single coaxial cable by individually coding each alarm point. The alarm inputs are digitally encoded (using a dip switch) and transmitted on the selected line pair of the vertical interval of the video signal, along with a 5-bit dip switch selectable error code to ensure signal integrity. The receiver board at the console decodes the signal information. The transmitter and receiver line and error-code dip-switch selections must be set identically. When a signal code match at the receiver judges the data as good, it drives an alarm output signal. The system has the capability to encode and decode (selectively) up to 15 alarm points and

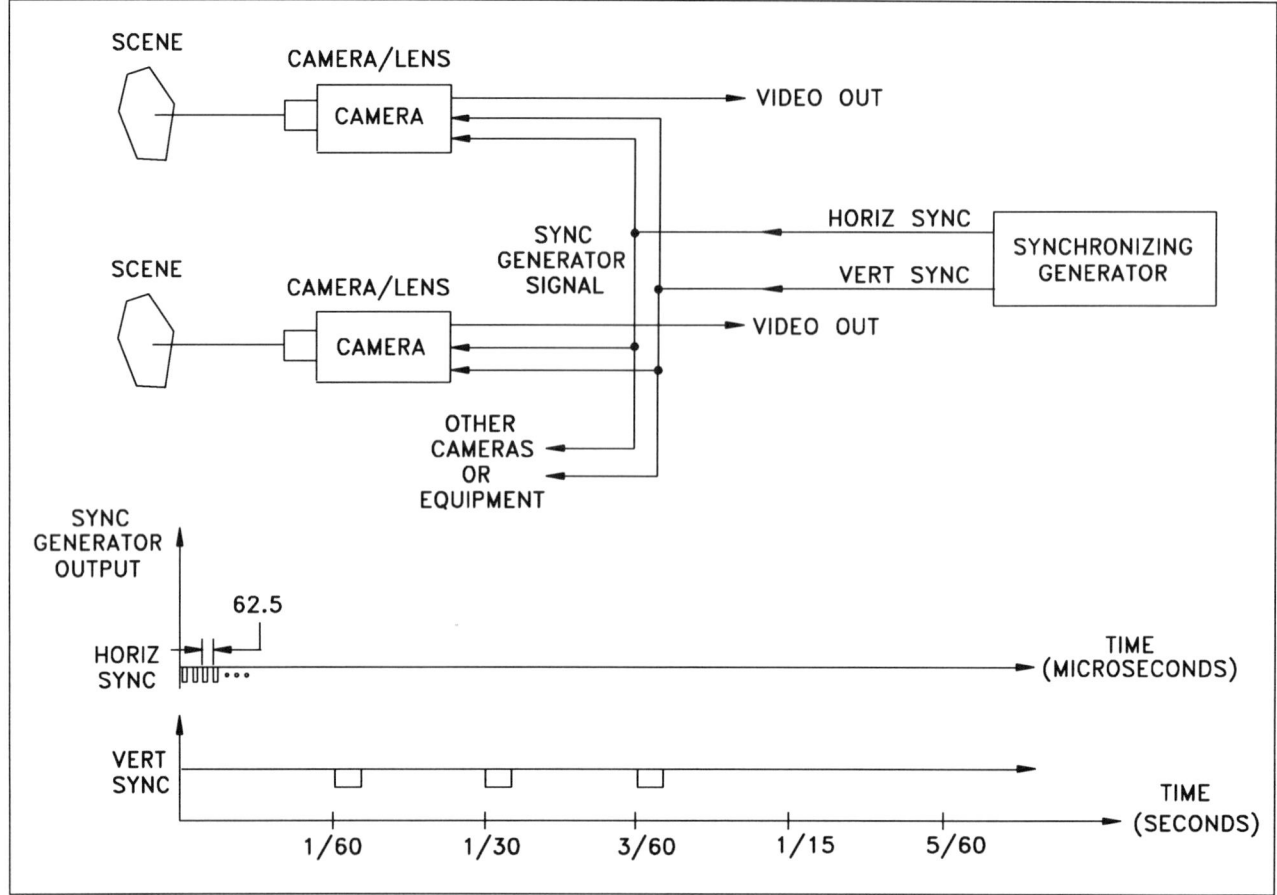

FIGURE 13-13   Video signal synchronizing generator system

transmit them down one coaxial cable. The system also works with other devices (using video lines in the vertical interval time) to transmit pan/tilt and lens control information. The encoded vertical interval data line can be set (chosen) from line 1 to line 31 in the video picture or can be put into the active video portion of the picture for visual detection of alarm conditions. The digitally encoded signal is VCR recordable. When the recorded signal is played back into a receiver unit, the alarm outputs will reflect the recorded alarm data. The addition of alarms to the video signal in no way alters its ability to be transmitted over coaxial-cable, two-wire, fiber-optic, microwave, RF, or IR laser transmission means. Combining the alarm point encoding with the video signal results in increased efficiency and lower cost of data and video transmission.

13.7 SYNCHRONIZING GENERATOR

In security systems having multiple cameras and monitors in multiple buildings and locations, including other peripheral equipment such as switchers, VCRs, printers, and time/date generators, it is often necessary to synchronize all cameras and devices from one timing source to ensure optimum picture quality. To accomplish this efficiently, a device called a video synchronizing generator is used. This device produces an accurately clocked set of synchronizing pulses and waveforms including vertical and horizontal sync, with suitable signal amplitude and drive power to synchronize many cameras and devices over long distances. Figure 13-13 shows how the synchronizing generator is connected into the security system.

13.8 SUMMARY

The aforementioned devices enhance the intelligence-gathering ability of video systems and help the guard be more effective. Video image-splitting,-combining,-compressing and picture-in-a-picture devices are powerful tools for increasing the amount of intelligence a CCTV system can provide. They should be used when more camera scenes must be monitored than can be displayed or viewed by available guard personnel.

The use of video annotating devices has become important in video monitoring and crime prosecution. Video annotation of time and date are absolutely essential in courtroom prosecution.

The video capture technique for viewing alarm scenes prior to the alarm illustrates how important CCTV can be for intrusion detection and false-alarm verification. Transmission of video, alarm, and control signals on the same cable (link) can save significant costs in retrofit and new installations. The video sync generator insures clean, noise-free synchronization of multiple cameras, monitors, recorders, and printers. There are many special accessories available from small and large manufacturers to enhance CCTV effectiveness.

Chapter 14
Covert CCTV

CONTENTS

14.1 OVERVIEW

Overt CCTV security equipment is installed in full view of the public; it is used to observe action while also letting subjects know that they are under surveillance. This technique often has the effect of deterring crime. Covert CCTV ideally operates so that the offender is not aware of the surveillance; it produces a permanent VCR record for later use in confronting, dismissing, or prosecuting the offender. Although overt CCTV security installations are very useful in apprehending offenders, in special situations, investigators, police officials, government agencies, retail operations, and industry security personnel require covert or hidden systems.

Covert CCTV cameras and lenses have become commonplace. Since the camera and lens are hidden, unsuspecting violators are often viewed, recorded, and even apprehended while committing the act. Although this hidden camera uses small optics, it can produce a high-quality CCTV picture.

Covert and overt CCTV are often used together to foil a professional criminal. The criminal, seeing the overt system, defeats or disables the overt cameras, but the covert cameras can still record the activity.

An unrelated reason for using covert CCTV is to avoid changing the architectural aesthetics of a building or surrounding area.

Covert CCTV cameras are concealed in common objects or located behind a small hole in an opaque barrier (such as a wall or ceiling). Cameras are camouflaged in common

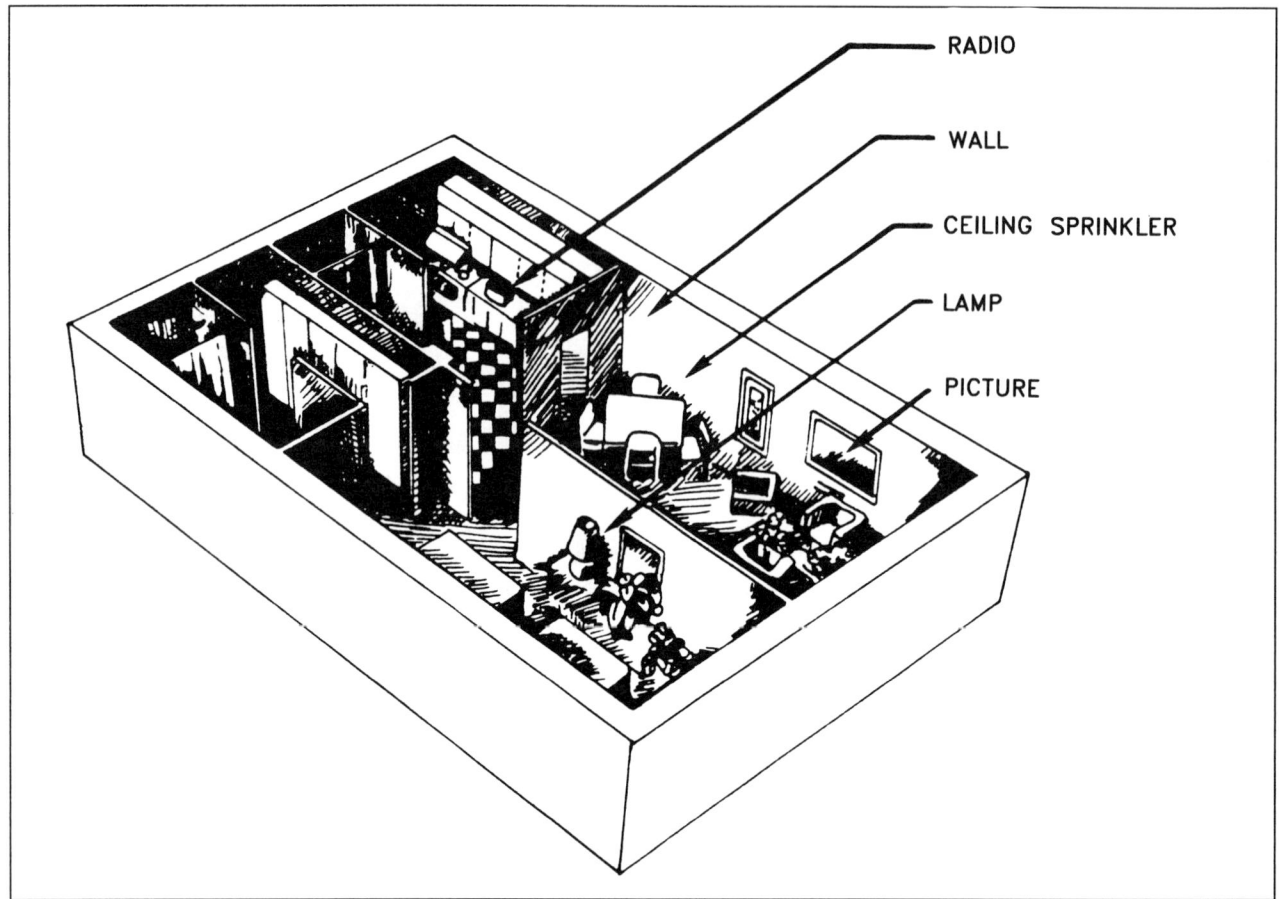

RADIO

WALL

CEILING SPRINKLER

LAMP

PICTURE

FIGURE 14-1   Covert CCTV lens/camera environment

objects such as lamps and lamp fixtures, table and wall clocks, radios, or books. A very effective covert system uses a camera and lens camouflaged in a ceiling-mounted sprinkler head.

Covert CCTV principles, techniques, and unique pinhole lenses and cameras are described. Lenses with a small front lens diameter, so that the lens and camera can view the scene through a 1/16-inch-diameter hole, are analyzed. Most of these lenses have a medium-to-wide FOV, from 12 to 78 degrees, to cover a large scene area, but still permit identification of persons and the monitoring of activities and actions. Special pinhole lens variations, including right-angle, automatic-iris, sprinkler-head, and fiber-optic are described, as well as small pinhole cameras combining a mini-lens and sensor into a small camera head and other complete miniature cameras.

In a low-light-level (LLL) application, a charge-coupled device (CCD) camera with a very sensitive sensor and IR light source or an image intensifier is used. Since many covert installations are temporary, wireless transmission systems are used to send the CCTV camera signal to the monitor, VCR, or video printer.

14.2 COVERT CCTV TECHNIQUES

CCTV lens and camera concealment is accomplished by having the lens view through a small hole, a series of small holes, or from behind a semitransparent window. Figure 14-1 shows a typical room in which covert television surveillance is installed.

A number of suitable lens and camera locations include the ceiling, a wall, a lamp fixture, a clock, or other articles normally found in the room. CCTV cameras are installed in one or more locations in the room, depending on the activity expected. Covert CCTV systems pose some unique optical problems compared with overt systems using standard lenses. Since the diameter of the front lens viewing the scene must, by necessity, be small in order to be hidden, the lens is designed to be optically fast, collecting and transmit-

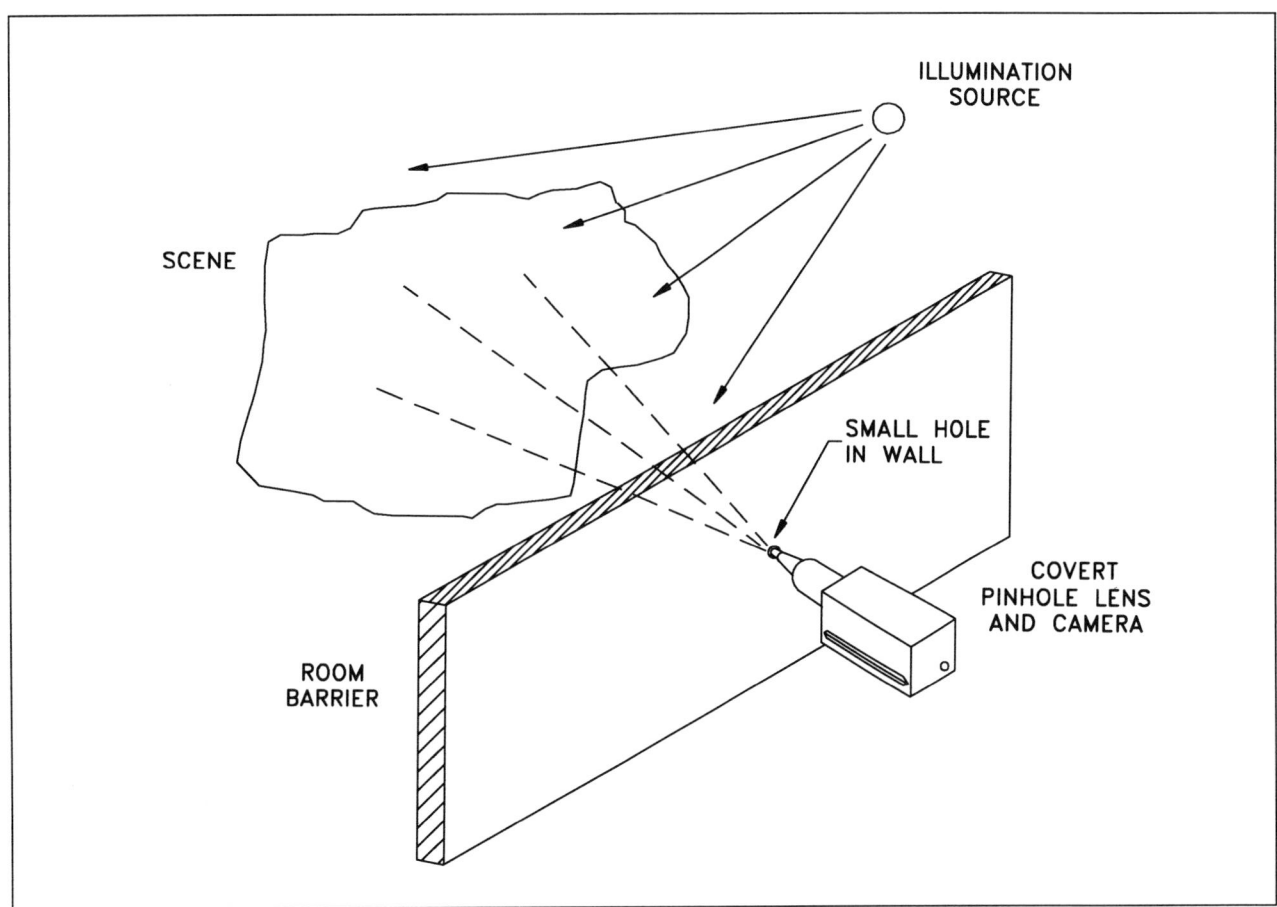

FIGURE 14-2   Covert CCTV surveillance

ting as much light as possible from the reflected scene to the television sensor. As a consequence, small-diameter lenses, called pinhole lenses, are used. (The term *pinhole* is a misnomer, as these lenses have a front diameter anywhere from ⅛ to ½ inch.)

There are several misconceptions regarding the factors determining a good pinhole camera or lens system for covert applications. Since the lens, camera, VCR, and system installation are a significant capital investment, an understanding of what constitutes a good system is important. Figure 14-2 shows the covert security problem.

The lens/camera must receive reflected light from an illuminated scene, the lens must collect and transmit the light to the camera sensor, and then the camera must transmit the video signal to a video monitor and/or VCR and video printer. Most covert pinhole lenses are designed for ⅓-, ½-, and ⅔-inch camera sensor formats. In indoor applications, the light sources are typically fluorescent, metal-arc, or tungsten types. Outdoor light sources include sunlight in the daytime and mercury, metal-arc, tungsten, sodium, or xenon lighting at night.

Figure 14-3 shows two basic configurations for pinhole lenses and cameras located behind a barrier. The hole in the barrier is usually chosen to be the same diameter (d) or smaller than the pinhole lens front lens element. When space permits, the straight-type installation is used. In confined or restricted locations with limited depth behind the barrier, the right-angle pinhole lens/camera is used. In both cases, to obtain the full lens FOV, it is imperative that the pinhole lens front lens element be located as close to the front of the barrier as possible to avoid "tunneling" (vignetting). When the pinhole lens front element is set back from the barrier surface, the lens is, in effect, viewing through a tunnel, and the image has a narrower FOV than the lens can produce. This appears on the monitor as a porthole-like (vignetted) picture.

An important installation problem often initially overlooked is the lens pointing angle required to see the desired FOV (Figure 14-4). Many applications require that the lens/camera point down at a shallow depression angle (30 degrees) from the ceiling (Figure 14-4a). This is accomplished by using the small-barrel, slow-taper lens. This feature allows pointing the small-barrel lens over a larger part of a room than the wide-barrel lens. Not all lenses can be mounted at a small angle to the ceiling because of the lens barrel shape (Figure 14-4b). Lenses having a large barrel

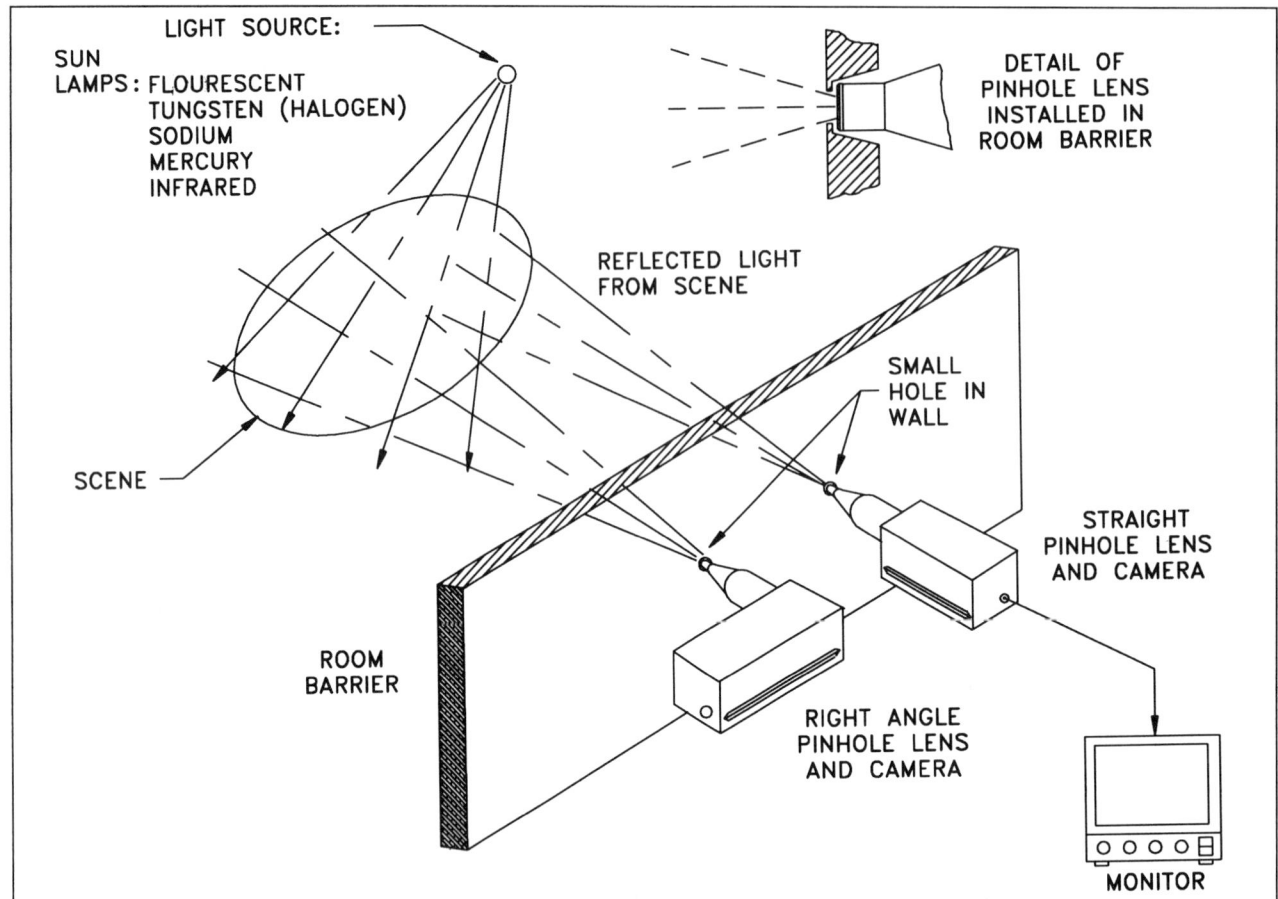

FIGURE 14-3   Straight and right-angle pinhole lens installation

diameter and fast taper at the front cannot be mounted at the shallow angles required. The small-barrel, slow-taper design permits easier installation than the fast-taper, since less material must be removed from the barrier and the lens has a faster optical speed, since it is larger and collects more light.

Figure 14-4 illustrates this installation problem. It shows a small hole on the scene side of the barrier and some material cut out of the barrier behind it to permit the front lens element to be located close to the front of the barrier surface. A pinhole lens having a small front diameter is simple to install. The smaller tapered barrel can be mounted at a smaller angle to the barrier than the wide-barrel lens. This feature allows pointing the small-barrel lens over a larger part of a room than the wide-barrel lens.

14.3   COVERT LENS/CAMERA TYPES

This section describes the widely installed pinhole CCTV lenses/cameras used for covert security applications. It includes standard pinholes, a compact pinhole lens kit, and mini-lenses. It also includes fiber optics, sprinkler-head configurations, and uniquely configured covert camera/lens combinations.

14.3.1   Pinhole Lenses

Figure 14-3 shows how pinhole lenses and cameras are mounted behind a wall, with the lens viewing through a small hole in the wall. Most are designed for $\frac{1}{3}$-, $\frac{1}{2}$-, or $\frac{2}{3}$-inch format cameras and have a manual- or automatic-iris control to adjust the light level reaching the camera. Figure 14-5 shows several samples of the generic pinhole lens types available. The right-angle version permits locating the camera and lens inside a narrow wall or above a ceiling.

The optical speed or f-number (f/#) of the pinhole lens is important for the successful implementation of a covert camera system. The lower the f-number, the more light reaching the CCTV camera and the better the television picture. The best theoretical f-number is equal to the focal length (FL) divided by the entrance lens diameter (d):

$$f/\# = \frac{FL}{d}$$

(14-1)

In practice the f-number obtained is worse than this number because of various losses caused by imperfect lens transmission, in turn caused by reflection, absorption, and other lens imaging properties. For a pinhole lens, the light

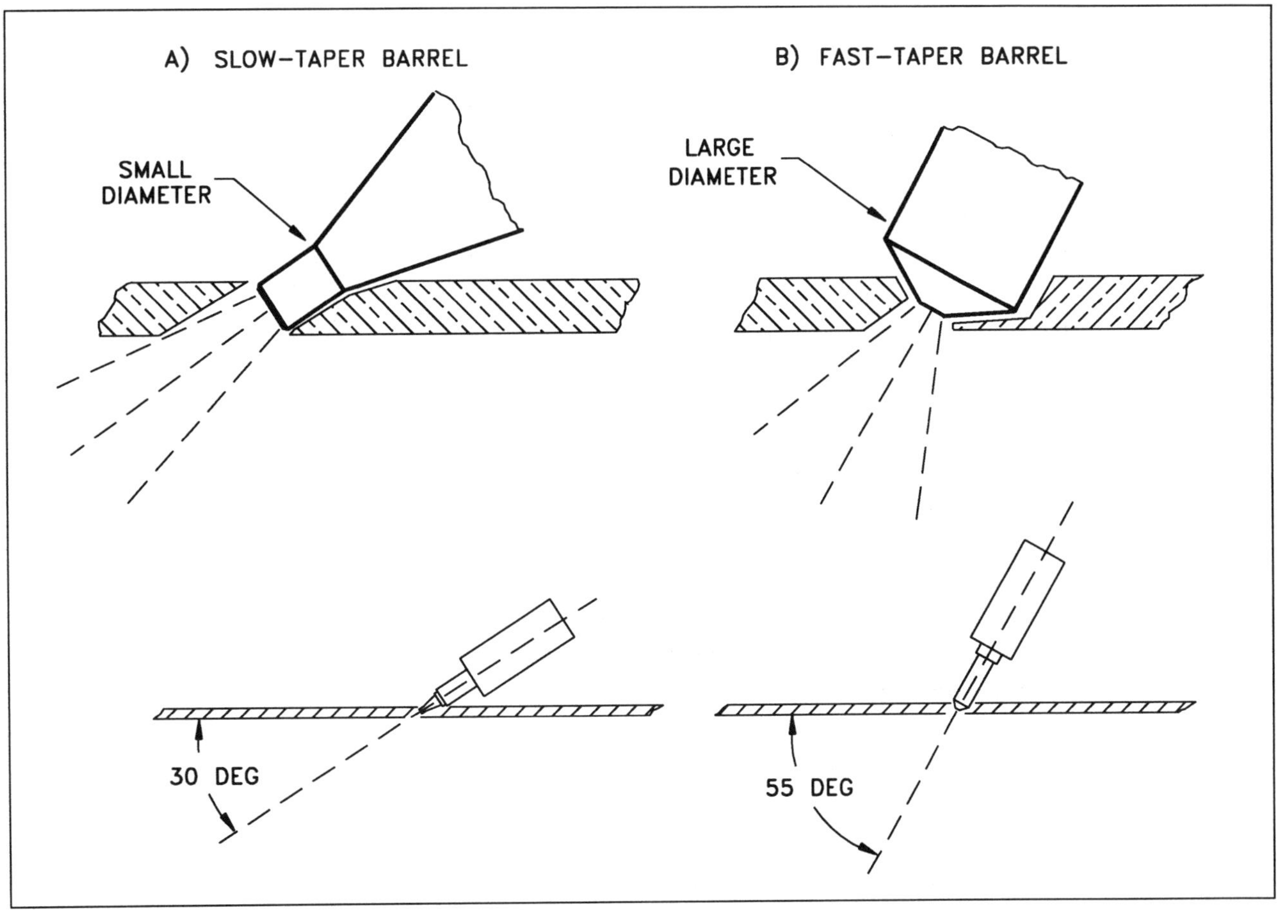

FIGURE 14-4 Short- vs. long-tapered pinhole lenses

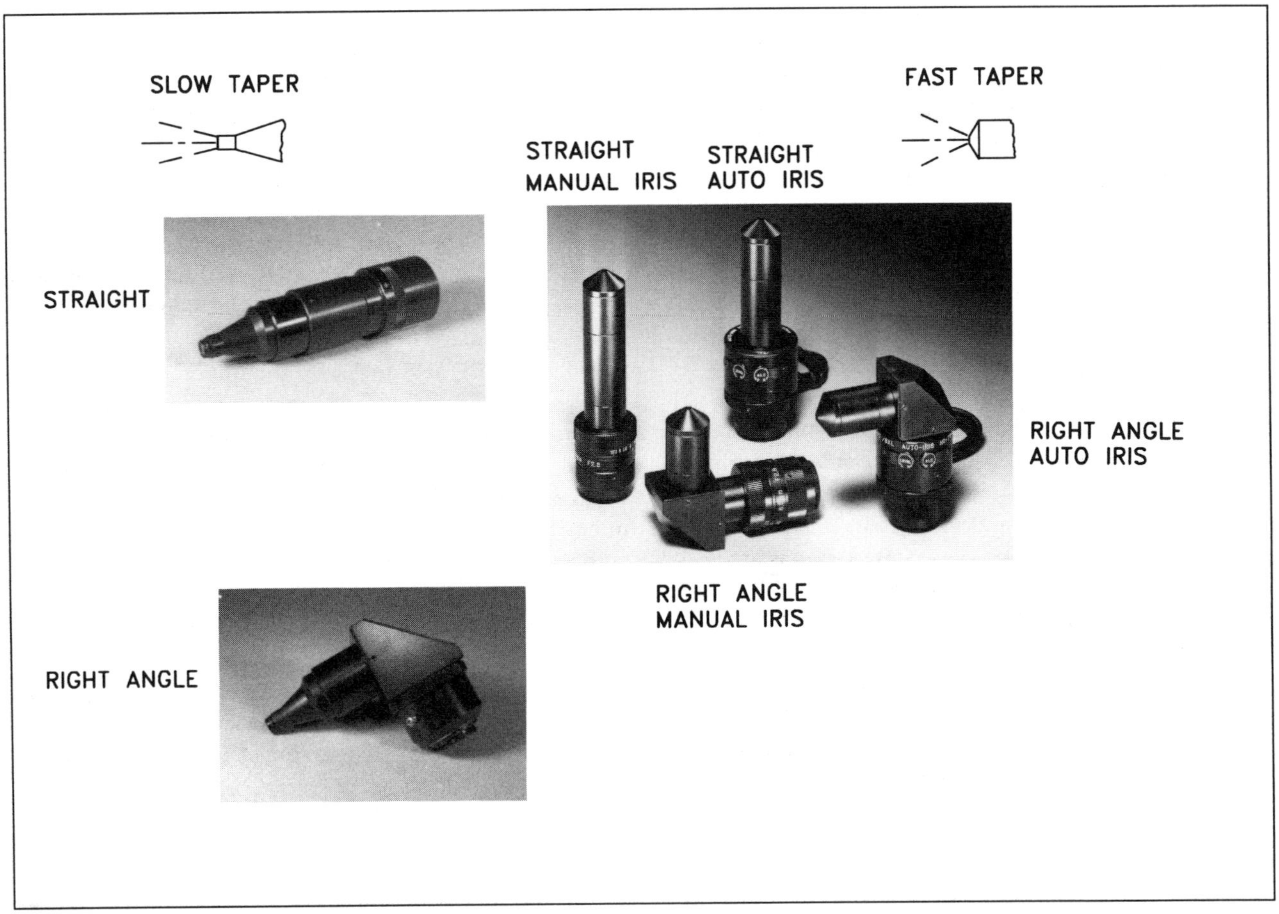

FIGURE 14-5 Standard straight and right-angle pinhole lenses

| FOCAL LENGTH (MM) | f/# | ANGULAR FIELD OF VIEW (FOV) IN DEGREES CAMERA FORMAT (INCH) | | | | | | TYPE |
| | | 1/3 | | 1/2 | | 2/3 | | |
		HORIZ	VERT	HORIZ	VERT	HORIZ	VERT	
4.0	2.5	53.4	43.1	76.3	61.6	--	--	STRAIGHT
5.5	3.0	38.7	31.0	60.4	47.2	77.3	61.9	STRAIGHT
5.5	3.0	38.7	31.0	60.4	47.2	77.3	61.9	RIGHT ANGLE
6.2	2.0	39.2	30.1	56.0	43.0	--	--	STRAIGHT
8.0	2.0	29.4	22.0	43.6	33.4	58.7	44.0	STRAIGHT
8.0	2.2	29.4	22.0	43.6	33.4	58.7	44.0	RIGHT ANGLE
9.0	3.5	26.3	20.2	38.8	30.1	52.6	40.3	STRAIGHT
9.0	3.5	26.3	20.2	38.8	30.1	52.6	40.3	RIGHT ANGLE
11.0	2.3	21.6	16.1	32.4	24.6	43.1	32.2	STRAIGHT
11.0	2.5	21.6	16.1	32.4	24.6	43.1	32.2	RIGHT ANGLE

Table 14-1   Covert Pinhole Lens Parameters

getting through the lens to the camera sensor is limited primarily by the diameter of the front lens or the mechanical opening through which it views. For this reason, the larger the entrance lens diameter, the more light getting through to the camera sensor, resulting in better picture quality, all other conditions remaining the same. The amount of light collected and transmitted through a lens system varies inversely as the square of the lens f-number: if the lens diameter is increased (or decreased) a small amount, the light passing through the lens increases (or decreases) by a large amount. Specifically, if the lens diameter is doubled, the light throughput quadruples. As an example, an f/2.0 lens transmits four times as much light as an f/4.0 lens. The f-number relationship is analogous to water flowing through a pipe: if the pipe diameter is doubled, four times as much water flows through it. Likewise, if the f-number is halved, four times as much light will be transmitted through the lens.

Many types of covert lenses are commercially available for CCTV security applications. Table 14-1 summarizes the characteristics of most manual- and automatic-iris pinhole lenses.

Most of these lenses are designed for $1/3$-, $1/2$-, and some for $2/3$-inch sensor formats, since these cameras are small, are in widespread use, and provide excellent image quality and resolution: 380 TV lines for $1/3$-inch and 450 to 570 TV lines for $1/2$- and $2/3$-inch format cameras. Some pinhole lenses have a very small entrance aperture, that is, 0.10 inch (2.5 mm), and are therefore optically slow (f/3.5–f/4.0) by design. A lens with a FL of 9 mm and a 2.5-mm aperture has, at best, a theoretical f-number of:

$$f/\# = \frac{9 \text{ mm}}{2.5 \text{ mm}} = 3.6$$

Other lens losses within this type of lens give an overall optical speed of approximately f/4.0.

A covert lens with an 11-mm FL and a 6-mm aperture has a theoretical f-number of:

$$f/\# = \frac{11 \text{ mm}}{6 \text{ mm}} = 1.83$$

Other lens losses result in an overall optical speed of approximately f/2.0.

The 9-mm lens with the smaller aperture works well if there is sufficient light. An advantage of the 6-mm-aperture lens is that it can be used in applications where a larger hole, that is, 6-mm diameter adequately conceals the lens and there is insufficient light available for the 9-mm FL lens with the 2.5-mm hole. The most important characteristics of a pinhole lens are (1) how fast is the optical speed—that is, how low is the f-number (the lower the better); and (2) ease of installation and use. When covert operation is required in locations having widely varying light-level conditions and when a silicon or Newvicon tube, CCD solid-state or other intensified LLL camera is used, a pinhole lens with an automatic iris controlling the light reaching the camera sensor is necessary. Some new shuttered CCD cameras may tolerate the use of manual-iris lenses. Check with the manufacturer for the light ranges over which the camera will operate. Figure 14-5 shows straight and right-angle pinhole lenses with manual and automatic irises capable of controlling the light level reaching the camera sensor over a 35,000-to-1 light-level range.

14.3.2 Convertible Pinhole Lens Kit

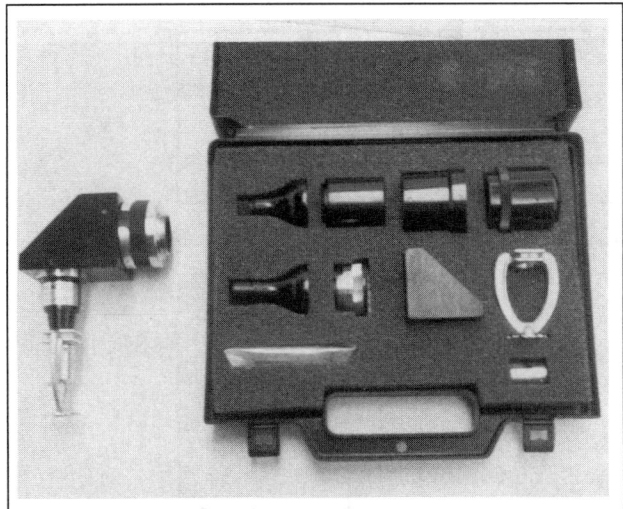

FIGURE 14-6 Pinhole Lens Kit and right-angle sprinkler lens assembled from kit

A generic characteristic of almost all pinhole-type lenses is that they invert the video picture and therefore the camera must be inverted to get a normal right-side-up picture. Some right-angle pinhole lenses reverse the image right to left and therefore require an electronic scan-reversal unit (Section 13.5) to regain the correct left-to-right orientation. Some pinhole lenses have a focusing ring or the front element of the lens can be adjusted to focus a sharp image on the camera sensor.

Pinhole lenses have been manufactured for many years in a variety of focal lengths (3.8, 4, 5.5, 6, 8, 9, 11 mm), in straight, right-angle, and manual- and automatic-iris configurations. The FL of most of these lenses can be doubled to obtain one-half the FOV by using a 2X extender. Pinhole lenses with 16-mm and 22-mm FLs are achieved by locating a 2X magnifier in between the 8-mm and 11-mm lenses and the camera. This automatically doubles the f-number of each lens (only one-fourth of the light transmitted). In many applications, the required FLs and configuration are not known in advance, and the user (or dealer) must have a large assortment of pinhole lenses, or take the risk that he or she will not have the right lens to do the job. This dilemma has been solved with the availability of a pinhole lens kit (Figure 14-6).

With this kit of pinhole lens parts, eight different FL lenses can be assembled in either a straight or right-angle configuration in minutes. An additional four combinations can be assembled for a disguised sprinkler-head covert application (Section 14.3.5). All lenses have a manual iris (an automatic iris is optional). Table 14-2 lists all the lens combinations for this versatile pinhole lens kit.

Tables 14-3, 14-4, and 14-5 tabulate the scene areas (width and height) as viewed with the popular pinhole lenses, on $\frac{1}{3}$-, $\frac{1}{2}$-, and $\frac{2}{3}$-inch sensor format cameras.

FOCAL LENGTH (MM)	f/#	CONFIGURATION	IMAGE ORIENTATION	COMMENTS
11	2.3	STRAIGHT	NORMAL	PINHOLE LENS
8	2.0	STRAIGHT	NORMAL	PINHOLE LENS
11	2.5	RIGHT—ANGLE	REVERSED	PINHOLE LENS
8	2.2	RIGHT—ANGLE	REVERSED	PINHOLE LENS
22	4.6	STRAIGHT	NORMAL	PINHOLE LENS
16	4.0	STRAIGHT	NORMAL	PINHOLE LENS
22	5.0	RIGHT—ANGLE	REVERSED	PINHOLE LENS
16	4.4	RIGHT—ANGLE	REVERSED	PINHOLE LENS
11	2.3	STRAIGHT	NORMAL	SPRINKLER HEAD
22	4.6	STRAIGHT	NORMAL	SPRINKLER HEAD
11	2.5	RIGHT—ANGLE	REVERSED	SPRINKLER HEAD
22	5.0	RIGHT—ANGLE	REVERSED	SPRINKLER HEAD

Table 14-2 Pinhole Lens Kit Combinations and Parameters

1/3 INCH SENSOR FORMAT LENS GUIDE						
PINHOLE LENS FOCAL LENGTH (MM)	CAMERA TO SCENE DISTANCE (D) IN FEET WIDTH AND HEIGHT OF AREA (WxH) IN FEET					
	5	10	15	20	25	30
	W x H	W x H	W x H	W x H	W x H	W x H
5.5	4.0x3.0	8.0x6.0	12.0x9.0	16.0x12.0	20.0x15.0	24.0x18.0
8.0	2.8x2.1	5.5x4.1	8.4x6.3	11.0x8.2	14.0x10.5	16.5x12.3
9.0	2.5x1.9	4.9x3.7	7.5x5.7	9.8x7.4	12.5x9.5	14.7x11.1
11.0	2.0x1.5	4.0x3.0	6.0x4.5	8.0x6.0	10.0x7.5	12.0x9.0
16.0	1.4x1.1	2.8x2.1	4.2x3.3	5.6x4.2	7.0x5.5	8.4x6.3
22.0	1.0x.8	2.0x1.5	3.0x2.4	4.0x3.0	5.0x4.0	6.0x4.5

Table 14-3 Pinhole Lens Guide for ⅓-Inch Format Camera

Several points should be considered when using standard, fully assembled pinhole lenses or pinhole lenses made from the pinhole lens kit:

1. Straight lenses invert the picture; therefore, the camera should be mounted in an inverted orientation.
2. Some right-angle lenses will show a right-to-left picture orientation instead of left-to-right, as with normal lenses. A scan-reversal unit will correct the problem. Check with the manufacturer.
3. The straight pinhole lens with the sprinkler-mirror attachment displays a right-to-left picture. Use an electronic scan-reversal unit to correct the problem. The right-angle sprinkler-mirror version displays a correct left-to-right picture.

As an example of choosing a pinhole lens and camera to view a scene 6 feet high by 8 feet wide at a distance of 15 feet using a ½-inch format camera, use Table 14-4 and choose an 11-mm FL lens. As another example, the scene area displayed on the monitor with an 8-mm lens on a ⅔-inch format camera in a ceiling at a distance of 20 feet is an area 22 feet wide by 16.4 feet high (Table 14-5).

Note that the FOV when using any of the medium-to-long-FL lenses is independent of the hole size through which the lens views, providing the hole produces no tunneling. Viewing through a wall with a wide-angle 4-to-8-mm FL pinhole lens may require a cone-shaped hole or an array of small holes to prevent tunneling (vignetting) of the scene image.

1/2 INCH SENSOR FORMAT LENS GUIDE						
PINHOLE LENS FOCAL LENGTH (MM)	CAMERA TO SCENE DISTANCE (D) IN FEET WIDTH AND HEIGHT OF AREA (WxH) IN FEET					
	5	10	15	20	25	30
	W x H	W x H	W x H	W x H	W x H	W x H
4.0	8.0x6.0	16.0x12.0	24.0x18.0	32.0x24.0	40.0x30.0	48.0x36.0
5.5	5.8x4.4	11.8x8.8	17.4x13.2	22.2x17.6	29.0x22.0	34.8x26.4
6.2	5.2x3.9	10.4x7.8	15.6x11.7	20.8x15.6	26.0x19.5	31.2x23.4
8.0	4.0x3.0	8.0x6.0	12.0x9.0	16.0x12.0	20.0x15.0	24.0x18.0
9.0	3.6x2.7	7.2x5.4	10.8x8.1	14.4x10.8	18.0x13.5	21.6x16.2
11.0	2.9x2.2	5.8x4.4	8.7x6.6	11.6x8.8	14.5x11.0	17.4x13.2
16.0	2.0x1.5	4.0x3.0	6.0x4.5	8.0x6.0	10.0x7.5	12.0x9.0
22.0	1.5x0.8	2.9x2.2	4.4x3.3	5.8x4.4	7.3x5.5	8.7x6.6

Table 14-4 Pinhole Lens Guide for ½-Inch Format Camera

2/3 INCH SENSOR FORMAT LENS GUIDE						
PINHOLE LENS FOCAL LENGTH (MM)	CAMERA TO SCENE DISTANCE (D) IN FEET WIDTH AND HEIGHT OF AREA (WxH) IN FEET					
	5	10	15	20	25	30
	W x H	W x H	W x H	W x H	W x H	W x H
5.5	8.0x6.0	16.0x12.0	24.0x18.0	32.0x24.0	40.0x30.0	48.0x36.0
8.0	5.5x4.1	11.0x8.2	16.5x12.3	22.0x16.4	27.5x20.5	33.0x24.6
9.0	4.9x3.7	9.8x7.4	14.7x11.1	19.6x14.8	24.5x18.5	29.4x22.2
11.0	4.0x3.0	8.0x6.0	12.0x9.0	16.0x12.0	20.0x15.0	24.0x18.0
16.0	2.8x2.1	5.6x4.2	8.4x6.3	11.2x8.4	14.0x10.5	16.8x12.6
22.0	2.0x1.5	4.0x3.0	6.0x4.5	8.0x6.0	10.0x7.5	12.0x9.0

Table 14-5   Pinhole Lens Guide for ⅔-Inch Format Camera

Mini-lenses and a mini-lens camera kit consisting of five interchangeable mini-lenses and a very small CCD camera are described in the next section.

14.3.3 Mini-Lenses

Mini-lenses are small FFL objective lenses used for covert surveillance when space is at a premium (Figure 14-7). The lenses shown have focal lengths of 3.8, 5.5, 8, and 11 mm. They have front-barrel diameters between ⅜ and ½ inch, making them easy to mount behind a barrier or in close quarters. Because these small lenses have no iris, they should be used in applications where the scene light level does not vary widely, or with electronically shuttered cameras. Mini-lenses, like other FFL lenses and unlike standard pinhole lenses, do not invert the image on the camera.

Since the small and short (less than ⅝ inch long) mini-lenses have only three to six optical lens elements, fast optical speeds of f/1.4 to f/1.8 are realized. Pinhole lenses, on the other hand, are 3 inches to 5 inches long, and have as many as 10 to 20 optical elements and optical speeds of f/2.0 to f/4.0. This makes the mini-lens approximately five times faster (able to collect five times more light) than the pinhole lens.

14.3.3.1 Off-Axis Optics

A useful variation of the mini-lens is one that is mounted with its optical axis laterally offset from the camera-sensor axis (Figure 14-8). This offset configuration allows the camera to view a scene at an angle away from from the camera pointing-axis. The physical amount the optics must be moved to produce a large offset angle is only a few millimeters, which is easily accomplished with this special mini-lens and its modified mount. The offset angle is cho-

sen so that, with the camera parallel to a mounting surface, the entire lens FOV views the scene of interest without viewing the mounting surface. This angle is 20 degrees for the 8-mm lens and 14 degrees for the 11-mm when using a ½-inch format camera; it is 25 degrees and 19 degrees, respectively, for the same lenses when used on a ⅔-inch camera. This technique has a direct benefit when a camera/lens is mounted flat against a wall or a ceiling or other mounting surface (Figure 14-8).

14.3.3.2 Optical Attenuation Techniques

Since mini-lenses do not have an iris, they should be used when the lighting conditions are fairly constant and do not exceed the dynamic range of the camera. If the scene is very brightly illuminated with an intense artificial light or the sun, several techniques can be used to attenuate the light to the lens/camera (Figure 14-9).

The first technique is to mount the mini-lens behind a light-attenuating filter (Figure 14-9a). This may take the form of a gray, neutral-density filter, a partially aluminized film, or a tinted/smoked glass or plastic material. Neutral-density filters are available from photographic supply stores. This technique uniformly attenuates the light across the full aperture of the lens. A second technique shown in Figures 14-9b through 14-9e is to mount the mini-lens behind a small hole, a pattern of small holes, a slit, or other hole(s). This is accomplished by either mounting a small cap with the hole(s) (Figure 14-9b) on the lens, or mounting the lens behind a hole(s) in the barrier (Figure 14-9c to 14-9e). The light level reaching the camera sensor can be set initially by locating the lens behind a hole smaller than the mini-lens diameter. This technique attenuates the light reaching the lens but does not do it uniformly. For medium-FL lenses (11 mm and above), almost any shape hole results in a satisfactory image on the sensor. When the

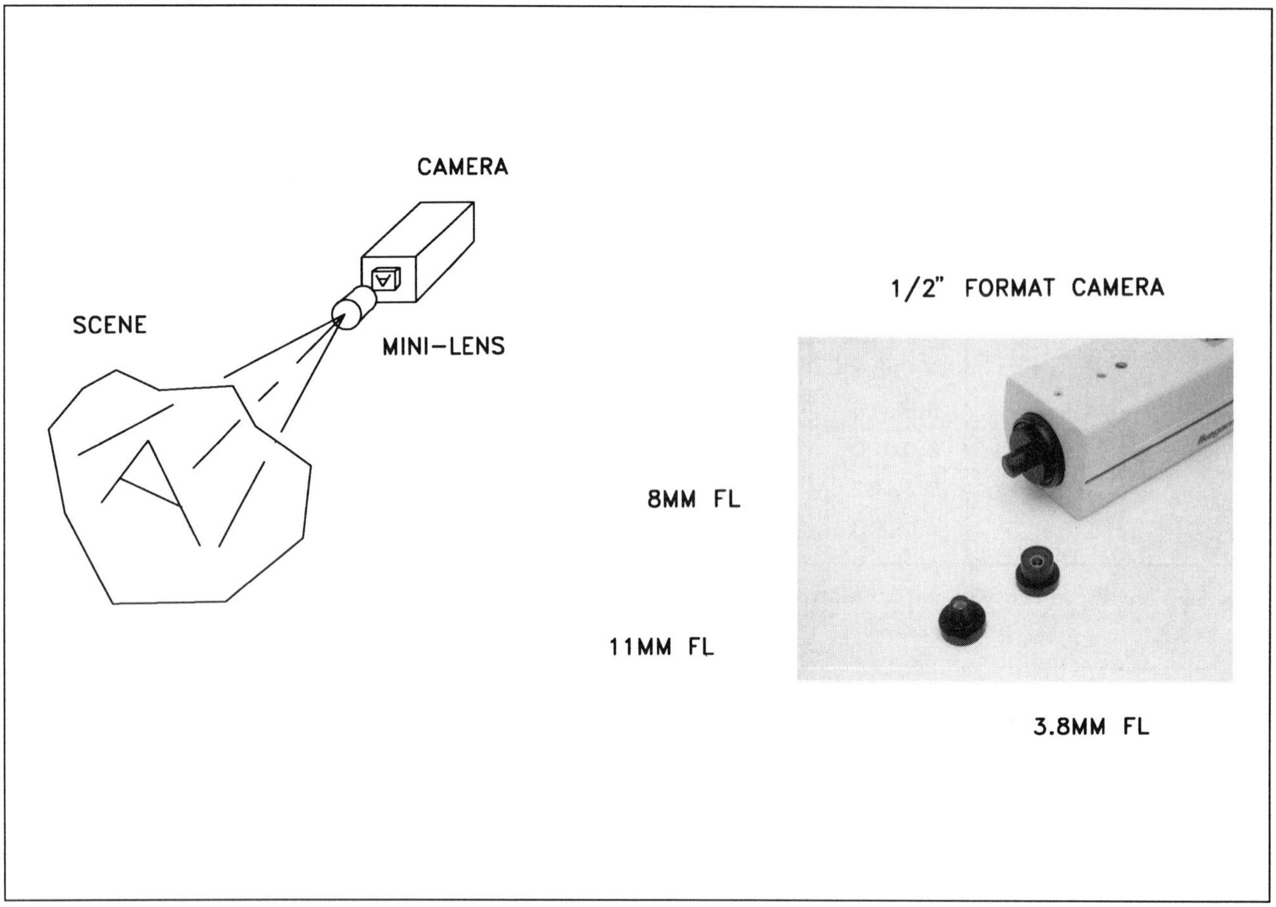

FIGURE 14-7   Mini CCTV lens and optical diagram

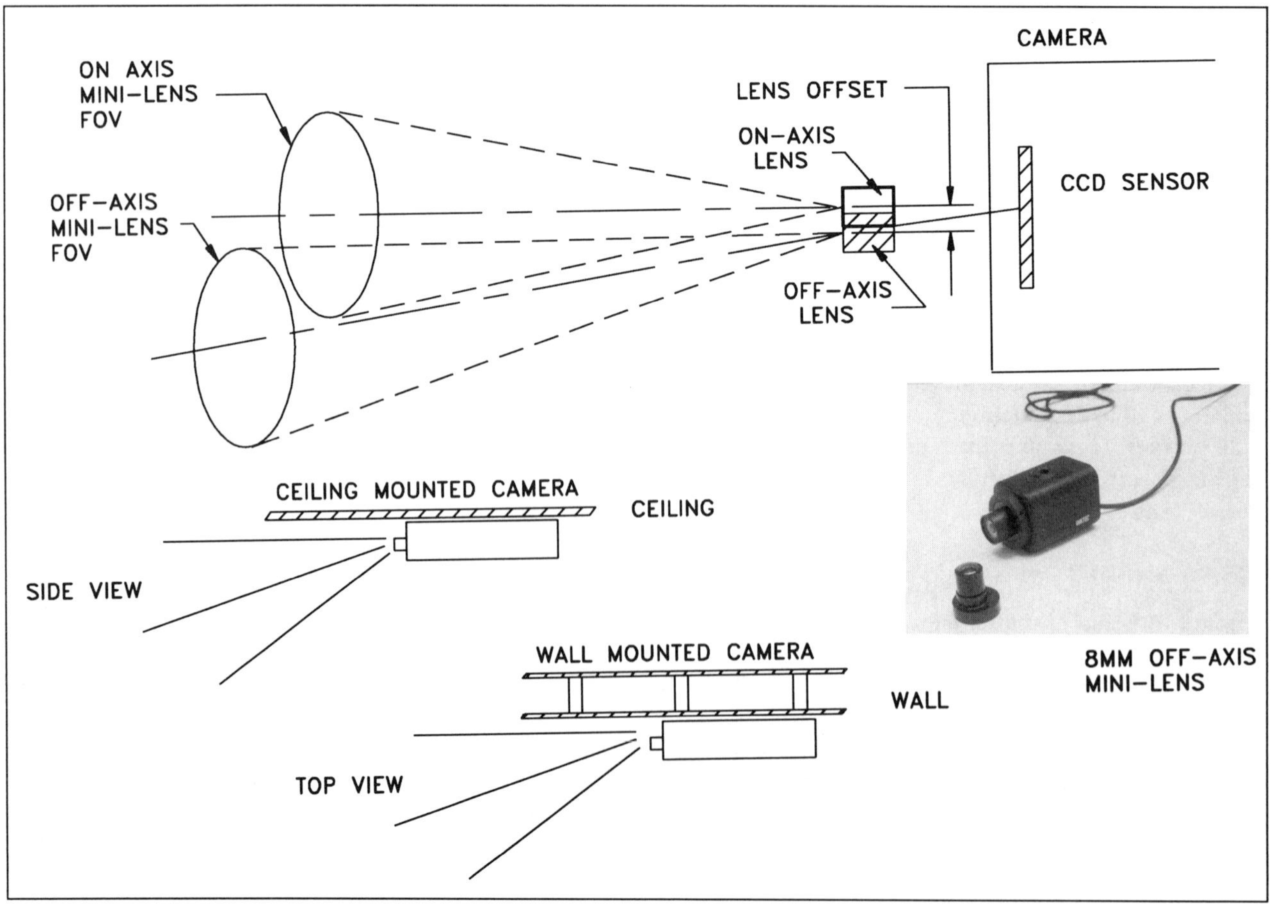

FIGURE 14-8   Off-axis optics configuration

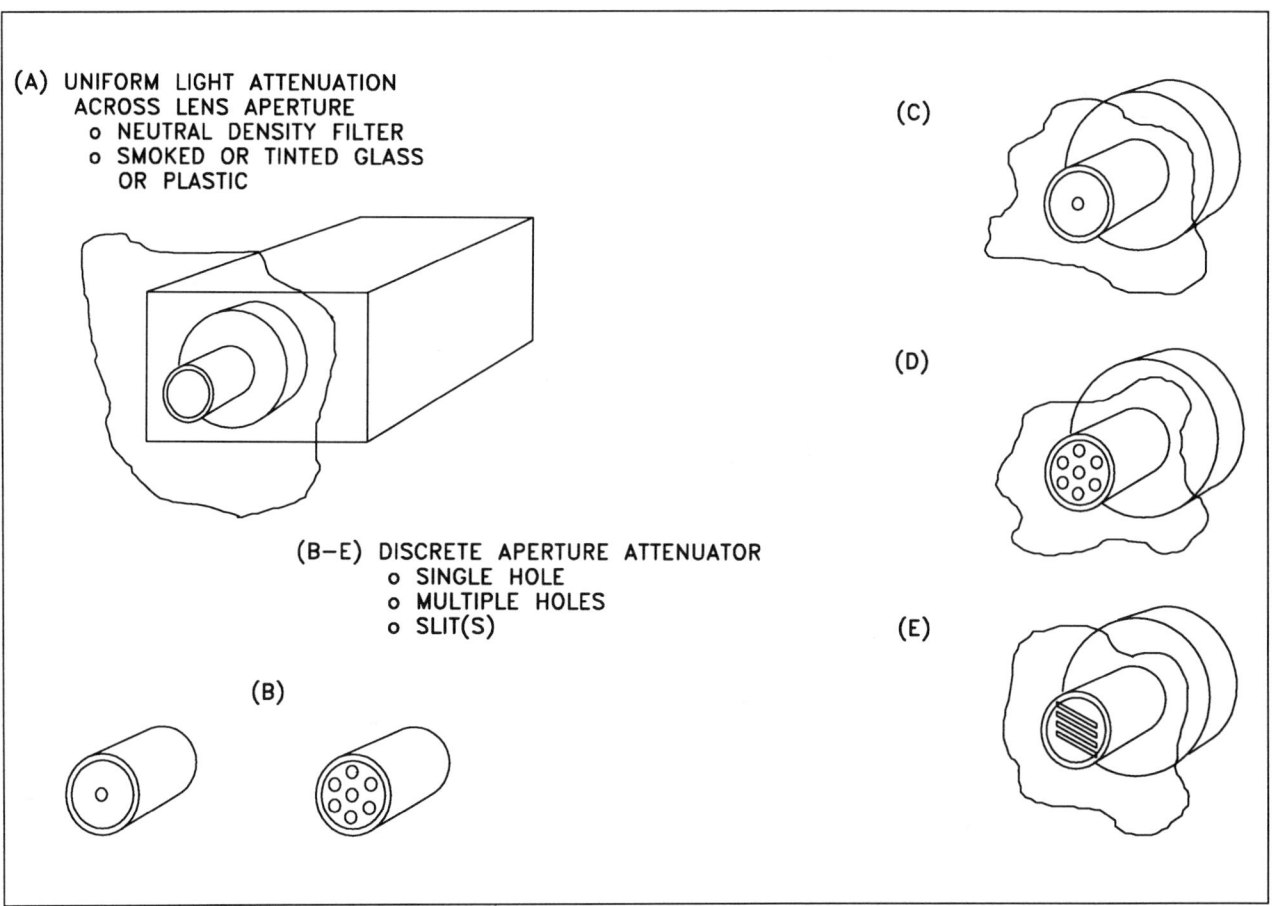

(A) UNIFORM LIGHT ATTENUATION
ACROSS LENS APERTURE
o NEUTRAL DENSITY FILTER
o SMOKED OR TINTED GLASS
OR PLASTIC

(C)

(D)

(B–E) DISCRETE APERTURE ATTENUATOR
o SINGLE HOLE
o MULTIPLE HOLES
o SLIT(S)

(E)

(B)

FIGURE 14-9   Lens optical attenuation techniques

11- or 22-mm mini-lens or pinhole lens is mounted behind a viewing barrier, a central hole as small as $^{1}/_{16}$th of an inch is suitable for producing a full image of the scene, providing sufficient light is available for the camera. When the 4-, 5.5-, or 8-mm mini-lens or pinhole lens views through a small hole, an undesirable porthole effect occurs, which is eliminated by having the lens view through a central hole and a series of concentric holes located around the central hole. The hole pattern must extend to the outer limits of the lens so that the full FOV of the lens is maintained. These concentric holes enable the lens to have peripheral vision or wide-angle viewing, and they eliminate vignetting. Figures 14-9b and 14-9d show two examples of this extended hole pattern. Either technique can provide attenuations required for sunlit or brightly illuminated scenes.

14.3.3.3 Mini-Camera/Mini-Lens Combination

A high-sensitivity pinhole camera results when a very fast mini-lens—f/1.4 to f/2.0—is coupled directly with the CCTV camera sensor.

Figure 14-10 illustrates a mini-lens camera kit with three standard on-axis mini-lenses having focal lengths of 3.8, 8, and 11 mm and two off-axis mounts for the 8- and 11-mm

FL lenses, and a very small, sensitive, high-resolution monochrome CCD camera.

The complete camera is only 1.38 × 1.38 × 2.2 inches long. The 11-mm FL lens extends 0.3 inch in front of the camera. The camera operates directly from 12 volts DC, requires only 2.5 watts of power, and produces a standard composite video output.

The small lens size and direct coupling to the camera sensor do not leave room for a manual or automatic iris. The camera has excellent electronic light-level compensation, but optimum performance is achieved if the lighting is fairly constant. The camera does have an automatic-iris input connector. Under bright light conditions, attenuation techniques shown in Figure 14-9 are used.

14.3.4 Comparison of Pinhole Lens and Mini-Lens

To compare different pinhole and mini-lenses with respect to their ability to transmit light to the camera sensor, a light power factor (LPF) is defined, with a slow pinhole lens (f/4.0) as a base reference. Table 14-6 summarizes the optical speed (f-number) and LPF for standard pinhole and mini-lenses.

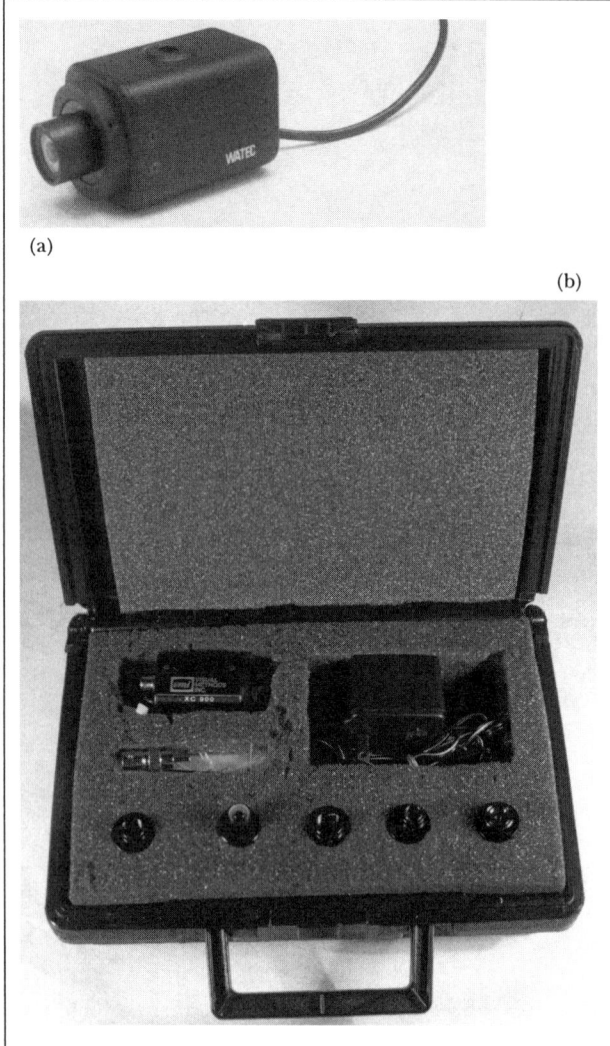

(a)

(b)

FIGURE 14-10   Mini-camera/mini-lens combination (a) and kit (b)

standing at floor level to detect or identify the lens and camera. Figure 14-11a shows the sprinkler pinhole lens attached to a standard CCTV camera mounted in a ceiling.

The covert surveillance sprinkler installed in the ceiling in no way affects the operation of the active fire-suppression sprinkler system; however, it should not be installed in locations that have no sprinkler system, so as not to give a false impression to fire and safety personnel.

The only portion of the lens system visible from below is the standard sprinkler head and the small ($3/8 \times 5/8$-inch) mirror assembly. In operation, light from the scene reflects off the small mirror, which directs it to the front of the pinhole lens. The 11- or 22-mm pinhole lens in turn transmits and focuses the scene onto the camera sensor. In the straight version, the image comes out reversed. In surveillance applications, this is often only an annoyance and not really a problem; an electronic scan-reversal unit will correct this condition. The right-angle version (Figure 14-11b) corrects this condition and produces a normal left-to-right image scan. The small mirror can be adjusted in elevation to point at different scene heights. To point in a particular azimuth direction, the entire camera-sprinkler lens assembly is rotated, with the mirror pointing in the direction of the scene of interest. When installed, most of the pinhole lens and the entire camera is concealed above the ceiling, with only a modified sprinkler head, a small mirror, and small lens in view below the ceiling. For many applications, this stationary pinhole lens pointing in one specific direction is adequate. For looking in different directions, the camera, sprinkler head and moving mirror assembly are made to pan (scan) via a motor drive. A scanning version of the sprinkler concept has a remote-control, 360-degree panning capability (Figure 14-12).

A motor drive scanning system provides remote panning capability. The sprinkler and small mirror assembly protrude below the ceiling and the camera and lens are above the ceiling, thereby making it covert to any observer below.

The f-number is usually critical in most applications with low light levels and where auxiliary lighting cannot be added. In Table 14-6, note the significant difference in light passing through the mini-lenses as compared with the pinhole lenses. A camera/lens using an f/1.8 mini-lens transmits almost five times as much light to the camera sensor as an f/4 pinhole lens. The f/1.4 mini-lens transmits more than eight times as much light as the f/4 pinhole lens.

14.3.5   Sprinkler-Head Pinhole Lenses

A very effective covert system uses a camera and lens camouflaged in a ceiling-mounted sprinkler head. Of the large variety of covert lenses available for the security television industry (pinhole, mini, fiber-optic), this unique, extremely useful product hides the pinhole lens in a ceiling sprinkler fixture, making it very difficult for an observer

14.3.6   Mirror on Pinhole Lens

Large plastic domes are often used to conceal a panning, tilting, and zooming television surveillance system from the observer (Section 11.2.2). The purpose for concealing the camera and lens in the dome is so the observer cannot see the direction in which the camera lens is pointing. Using this subterfuge, one camera system can scan and view a large area without the observer knowing at any instant whether he or she is under observation. Most domes are from 7 to 24 inches in diameter and drop below the ceiling by 6 to 12 inches, which does not add to the decor of the environment. The requirement that the CCTV lens view through the dome results in a typical light loss of 50%, and the image takes on some distortion because of the imperfect optical quality of the dome. The amount of light loss and distortion depends on the particular dome used and

FOCAL LENGTH (MM)	f/#	LENS TYPE	CONFIGURATION	LIGHT POWER FACTOR (LPF)*	ANGULAR FOV ON 1/2 INCH FORMAT (DEGREES)		COMMENTS
					HORIZ	VERT	
3.8	1.4	MINI	STRAIGHT	8.16	78.1	63.6	ULTRA WIDE–ANGLE
8.0	1.6	MINI	STRAIGHT	6.25	43.6	33.4	LONG TAPER
11.0	1.8	MINI	STRAIGHT	4.94	32.4	24.6	LONG TAPER
3.8	1.4	PINHOLE	STRAIGHT	8.16	78.1	63.6	ULTRA WIDE–ANGLE
3.8	1.4	PINHOLE	RIGHT–ANGLE	8.16	78.1	63.6	ULTRA WIDE–ANGLE
5.5	3.0	PINHOLE	STRAIGHT	1.78	64.7	51.1	WIDE–ANGLE
5.5	3.0	PINHOLE	RIGHT–ANGLE	1.78	64.7	51.1	WIDE–ANGLE
6.2	2.0	PINHOLE	STRAIGHT	4.00	56.0	43.0	SHORT, WIDE–ANGLE
8.0	2.0	PINHOLE	STRAIGHT	4.00	43.6	33.4	LONG TAPER
8.0	2.2	PINHOLE	RIGHT–ANGLE	3.31	43.6	33.4	LONG TAPER
9.0	3.5	PINHOLE	STRAIGHT	1.31	38.8	30.1	SHORT TAPER
9.0	3.5	PINHOLE	RIGHT–ANGLE	1.31	38.8	30.1	SHORT TAPER
11.0	2.3	PINHOLE	STRAIGHT	3.02	32.4	24.6	
11.0	2.5	PINHOLE	RIGHT–ANGLE	2.56	32.4	24.6	
16.0	4.0	PINHOLE	STRAIGHT	1.00	22.0	16.5	
16.0	4.4	PINHOLE	RIGHT–ANGLE	.83	22.0	16.5	
22.0	4.6	PINHOLE	STRAIGHT	.76	16.2	12.3	ALSO SPRINKLER
22.0	5.0	PINHOLE	RIGHT–ANGLE	.64	16.2	12.3	ALSO SPRINKLER

* INCREASE IN LIGHT LEVEL REACHING SENSOR BASED ON USING VALUE OF 1 FOR AN f/4 PINHOLE LENS

Table 14-6 Pinhole Lens and Mini-Lens Light Transmission Comparison

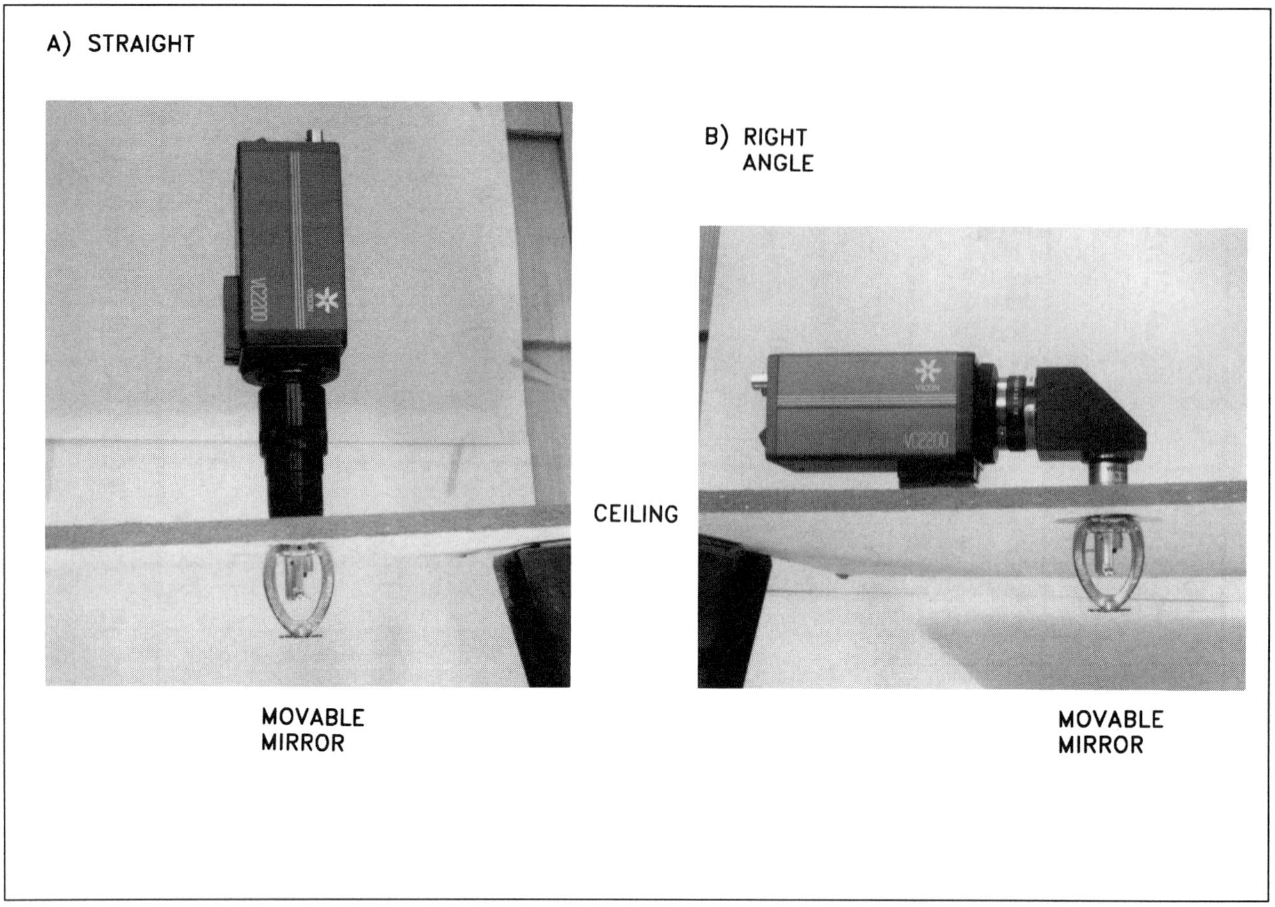

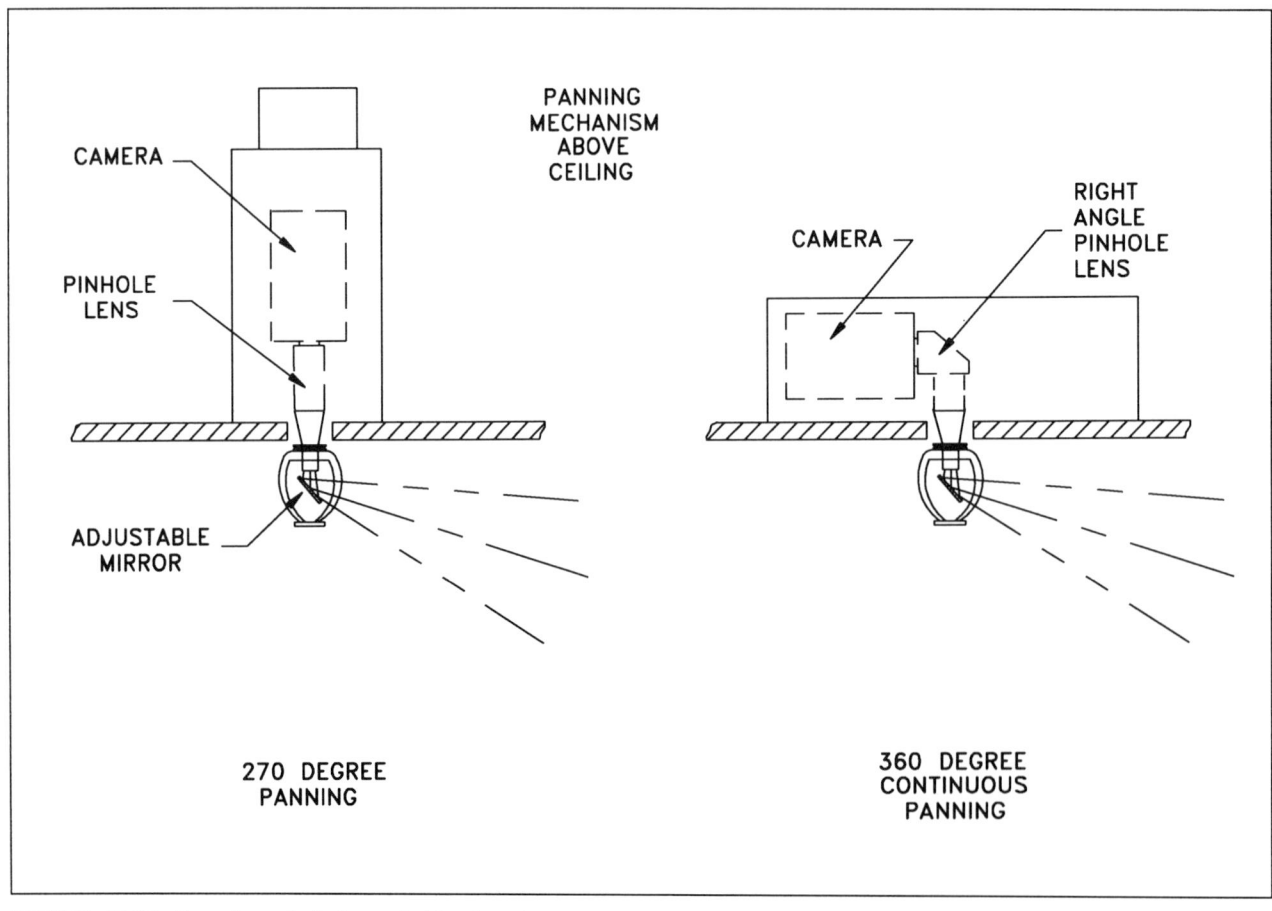

A) STRAIGHT

B) RIGHT ANGLE

CEILING

MOVABLE MIRROR

MOVABLE MIRROR

FIGURE 14-11   Sprinkler-head pinhole lens

CAMERA

PINHOLE LENS

ADJUSTABLE MIRROR

PANNING MECHANISM ABOVE CEILING

270 DEGREE PANNING

CAMERA

RIGHT ANGLE PINHOLE LENS

360 DEGREE CONTINUOUS PANNING

FIGURE 14-12   Panning pinhole sprinkler head

the part of the dome being looked through. A more aesthetic and covert camera/lens assembly is made up of a camera, pinhole lens and small mirror.

If the sprinkler-head assembly is removed from the right-angle lens shown in Figures 14-11 and 14-12, all that protrudes below the ceiling is a small mirror approximately $3/8 \times 5/8$ inches. This technique provides a very low profile and is difficult for an observer to detect at ground level. The pinhole/mirror system provides an alternative to some dome applications. The system can be fixed or have a 360-degree panning range, or a limited pan, tilt, and zoom capability depending on the design.

Two advantages of the moving mirror system over the dome are no large protruding dome suspended below the ceiling and easier installation, since only a small hole about $3/4$ inch in diameter is required to insert the lens and mirror through the ceiling. The small mirror scanning system has one limitation in that it cannot view the scene directly below its location. The dome system has two advantages over the scanning mirror: (1) the dome serves as a deterrent, since the observer sees the dome and believes a camera is active in it but does not know at any instant whether the camera is looking at him, and (2) the added capability of full-range zoom optics.

14.3.7 Fiber-Optic Lenses

For covert CCTV applications, all of the previous lenses find wide application when the barrier between the scene side and the camera/lens side is only a few inches (Figure 14-3), that is, the pinhole or mini-lens and camera can be mounted directly behind the barrier.

Difficult television security applications are sometimes solved by using coherent fiber-optic-bundle lenses. What if the camera must be located 6, 8, or 12 inches behind a thick concrete wall? What if a lens is on the outside of an ATM and the camera is 3 feet away, inside the building?

Fiber optics are used in surveillance applications when it is necessary to view a scene on the other side of a thick barrier or inside a confined area. The lens is installed behind a thick barrier (wall) with the objective lens on the scene side, the fiber-optic bundle within the wall, and the camera located on the protected side of the barrier. The lens viewing the scene can be a few inches or a few feet away from the camera.

There are three optical techniques to "lengthen" the camera's objective lens; two involve using a small-diameter "extender" that can be inserted into a hole or other enclosure in front of the camera sensor. The third uses a flexible fiber-optic bundle to transfer the image. The first of these methods is the rigid coherent fiber-optic conduit; the second is the borescope lens. These two special lenses can extend the objective lens several inches to several feet in front of the camera sensor.

The rigid fiber version is a fused array of fibers and cannot be bent. The flexible fiber version has hairlike fibers loosely contained in a protective sheath and can be flexed and bent easily. These fiber-optic lenses should not be confused with the single or multiple strands of fiber commonly used to transmit the time-modulated television signal from a camera to a remote monitor site over a long distance (hundreds of feet or miles; see Chapter 6). The coherent fiber-optic lens typically has 200,000 to 300,000 individual fibers forming an image-transferring array.

By combining lenses with coherent fiber-optic bundles, long, small-diameter optical systems are produced, which require drilling only a $1/4$ to $1\,1/4$-inch hole to position the front lens near the front side of the barrier. A small aperture hole (dependent on light-level available) is drilled completely through at the barrier surface and the camera and lens are connected on the opposite, protected side (Figure 14-13).

This lens/camera system solves many banking ATM and correctional-facility security problems.

Rigid fiber-optic lenses (image conduits) are $1/4$ to $1/2$-inch in diameter and from 6 to 12 inches long. Flexible fiber-optic lenses are $1\,1/4$ inches in diameter and 39 inches long. The fiber-optic pinhole lens is available with manual or automatic iris for $1/3$-, $1/2$-, and $2/3$-inch television formats.

A minor disadvantage of all fiber-optic systems is that the picture obtained is not as "clean" as that obtained with an "all-lens" pinhole lens. There are some cosmetic imperfections that look like dust spots, as well as a slight geometrical pattern caused by fiber stacking. These imperfections occur because several hundred thousand individual hairlike fibers make up the fiber-optic bundle, some of which are not perfectly transmitting. For most surveillance applications, the imperfections do not result in any significant loss of intelligence in the picture.

The rigid fiber-optic pinhole lens and the flexible fiber-optic lens are available with adjustable-focus and manual or automatic iris for $1/3$-, $1/2$-, or $2/3$-inch formats.

Figure 14-14 shows complete rigid and flexible fiber-optic lenses.

14.3.7.1 Configuration

A fiber-optic lens consists essentially of three parts: (1) an objective lens that focuses the scene onto the front end of the fiber-optic bundle, (2) a rigid conduit or flexible fiber coherent optic bundle that transfers the image a substantial distance (several inches to several feet), (3) and a relay lens at the output end of the fiber bundle that re-images that output and focuses its image onto the camera sensor (Figure 14-15).

The function of the objective lens is to image the scene onto the front surface of the fiber-optic bundle. This objective lens can be like any of the FFL, zoom, pinhole, manual-, or automatic-iris lenses described in Chapter 4. They may have a short FL (wide angle), medium FL, or long FL (telephoto). The only requirement of these objective lenses

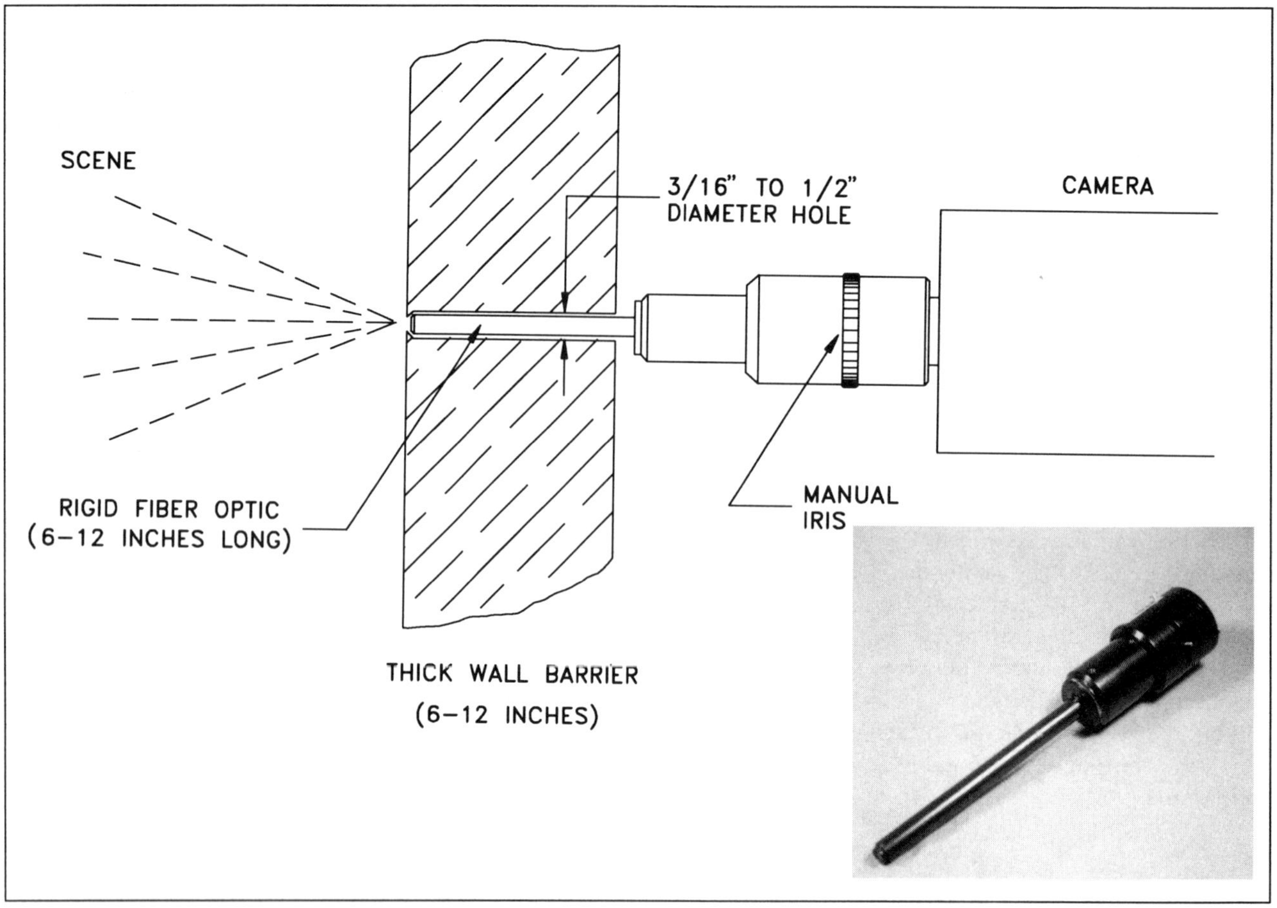

FIGURE 14-13   Fiber-optic pinhole lens installation in thick wall

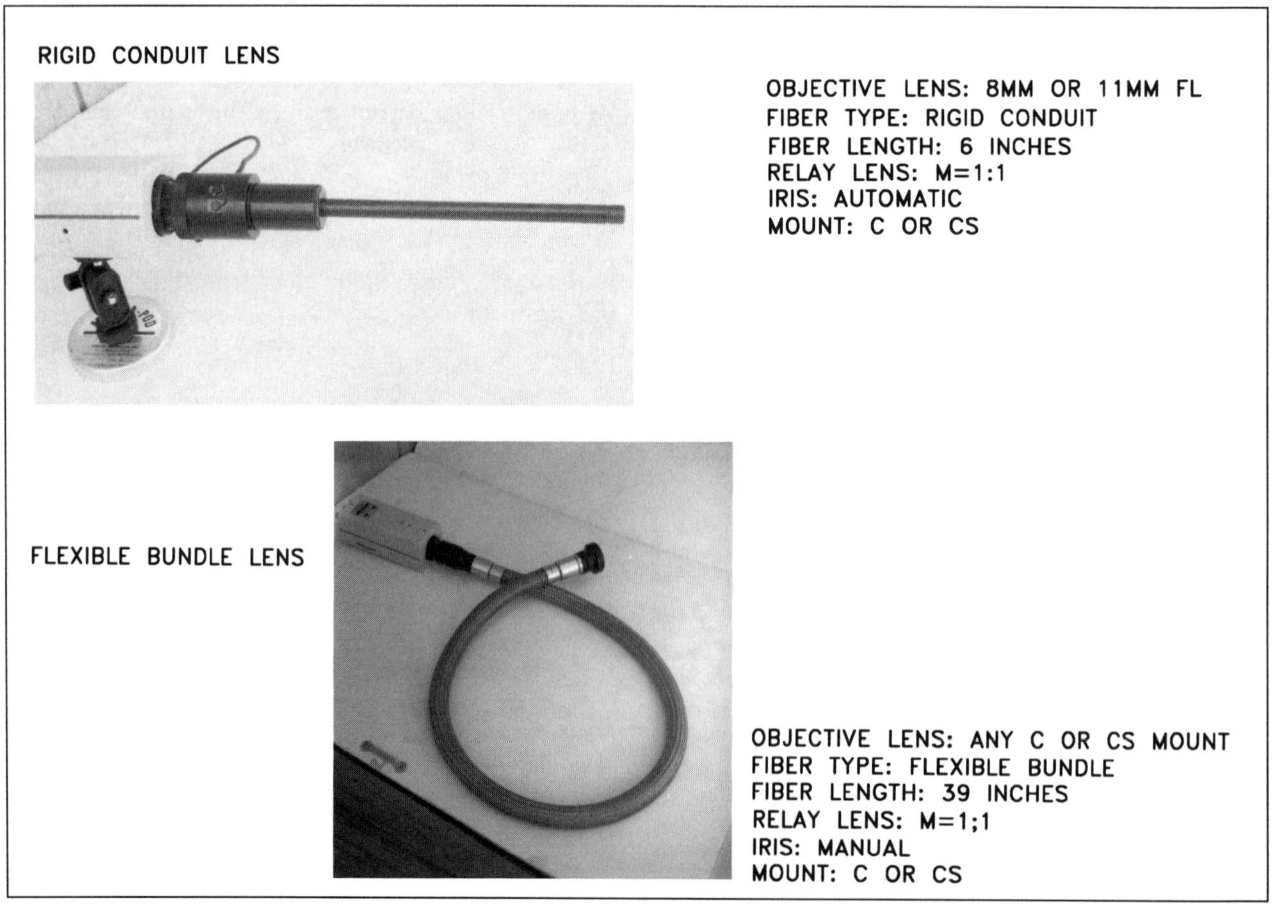

SCENE

3/16" TO 1/2"
DIAMETER HOLE

CAMERA

MANUAL
IRIS

RIGID FIBER OPTIC
(6–12 INCHES LONG)

THICK WALL BARRIER
(6–12 INCHES)

FIGURE 14-14   Rigid and flexible fiber-optic lenses

RIGID CONDUIT LENS

OBJECTIVE LENS: 8MM OR 11MM FL
FIBER TYPE: RIGID CONDUIT
FIBER LENGTH: 6 INCHES
RELAY LENS: M=1:1
IRIS: AUTOMATIC
MOUNT: C OR CS

FLEXIBLE BUNDLE LENS

OBJECTIVE LENS: ANY C OR CS MOUNT
FIBER TYPE: FLEXIBLE BUNDLE
FIBER LENGTH: 39 INCHES
RELAY LENS: M=1;1
IRIS: MANUAL
MOUNT: C OR CS

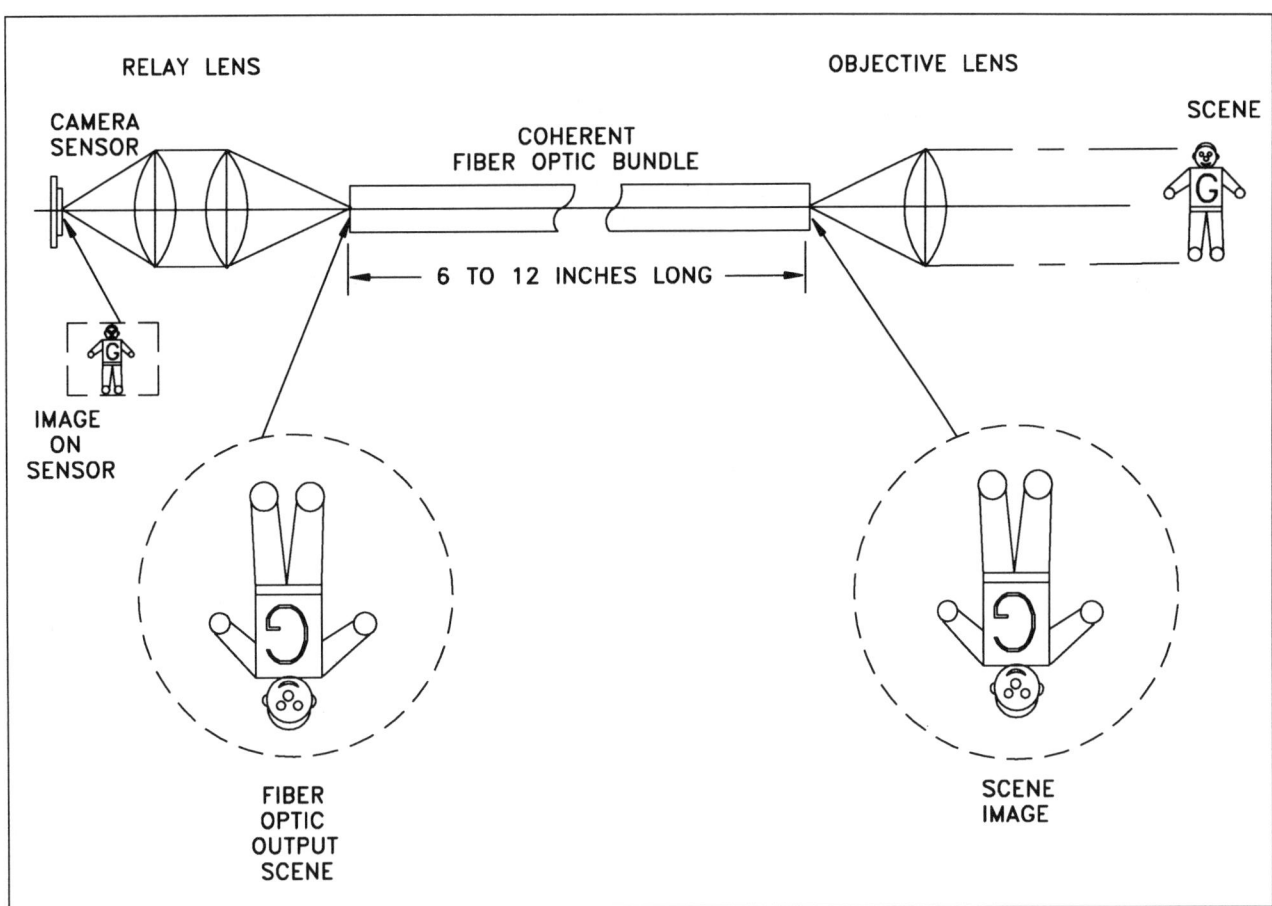

FIGURE 14-15 Fiber-optic lens configuration

is that they produce an image large enough to fill the full aperture (cross-sectional area) of the fiber-optic bundle. Fiber-optic bundles are available with cross-sectional areas suitable for $\frac{1}{3}$-, $\frac{1}{2}$-, and $\frac{2}{3}$-inch lens and camera formats. This means that they have fiber bundle diameters of 5.5, 8, and 11 mm, respectively (see Chapter 5).

The coherent fiber-optic bundle consists of several hundred thousand closely packed glass fibers to coherently transfer an image from one end of the fiber to the other, several inches to several feet (Figure 14-16).

After traversing the fiber bundle, the image is transferred by means of a relay lens to the camera sensor. As shown in Figure 14-16, fiber 1 transmits point 1 of the image from the objective lens down the fiber to a corresponding point 1 on the exit end of the fiber bundle. Likewise, all of the remaining points of the entrance image are transferred in an exact one-to-one correspondence to the exit end of the fiber bundle, thereby producing a coherent image. *Coherent* means that each point in the image on the front end of the fiber bundle corresponds to a specific point at the rear end of the fiber bundle. In order to produce an image having a resolution similar to CCTV sensor quality (400 to 500 TV

lines), the individual fibers used are approximately 12 microns (0.00047 inch) in diameter.

14.3.7.2 Rigid Fiber Pinhole Lens

In the case of the rigid fiber-optic bundle, the individual fibers are fused together to form a rigid glass rod or conduit. The diameter of the rigid fiber-optic bundle is approximately 0.4 inch for a $\frac{2}{3}$-inch format sensor, 0.3 inch for a $\frac{1}{2}$-inch format, and 0.2 inch for a $\frac{1}{3}$-inch format. The rod is usually protected from the environment and mechanical damage by a rigid metal tube. For the $\frac{2}{3}$-inch format, the outside diameter is about 0.5 inch. It should be noted that the image exiting the fiber-optic lens is inverted with respect to the image produced by a standard objective lens. This inversion is corrected by inverting the camera.

A rigid fiber-optic pinhole lens is used when a camera needs to view through thick barriers (such as concrete), where the pinhole lens cannot be installed near the viewing side of the wall (Figure 14-13). The fiber-optic lens with an 8-mm FL has a 58.7-degree horizontal FOV on a $\frac{2}{3}$-inch format camera, and the 11-mm has a 43.1-degree FOV. On a $\frac{1}{2}$-inch camera, the 8-mm FL lens has a 43.6-degree FOV

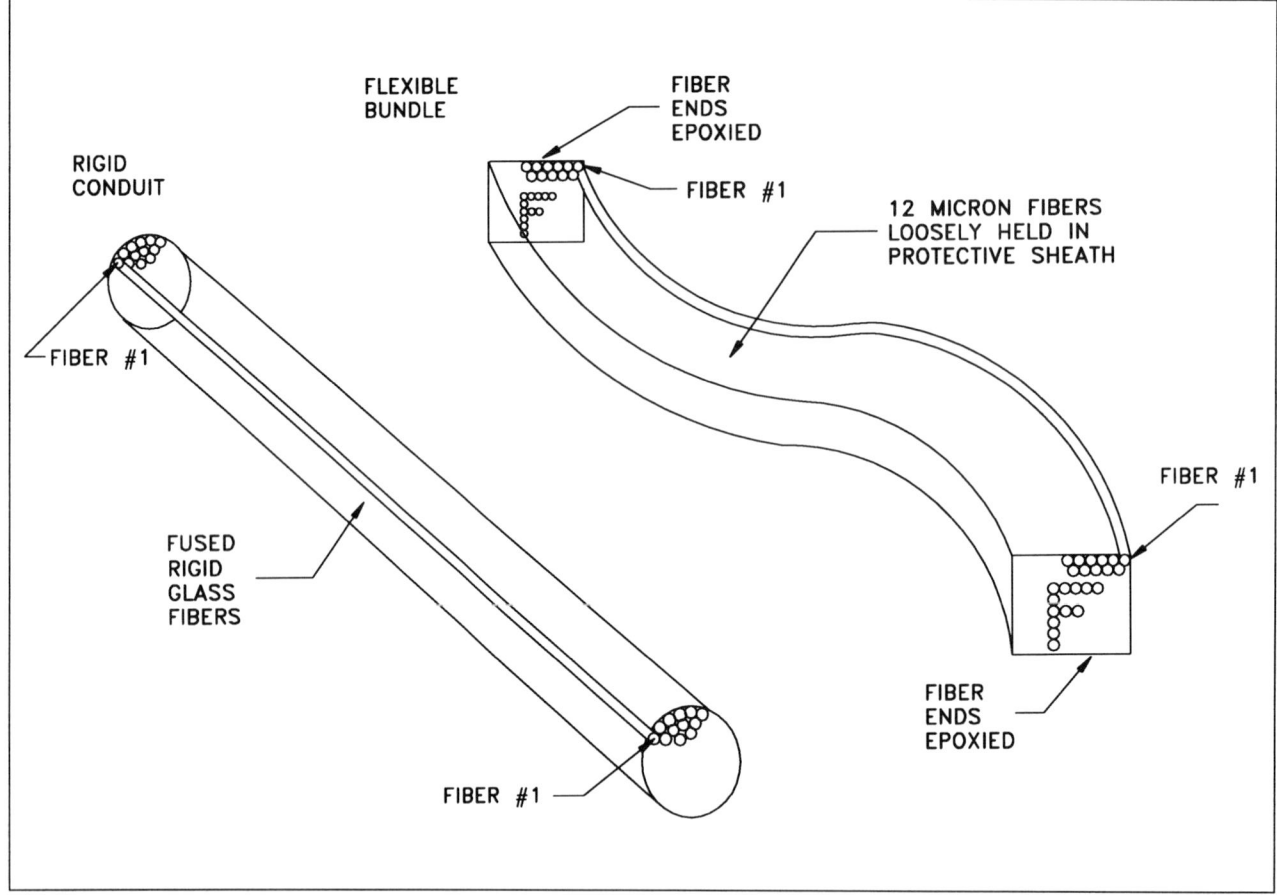

FIGURE 14-16   Fiber-bundle construction

and the 11-mm a 32.4-degree FOV. They both have a diameter of less than 0.5 inch, and are available in lengths of 6, 8, and 12 inches. The fiber-optic lens speed is between f/4 and f/8, depending on the fiber length—slower than the standard, all-lens-type pinhole lens. Figure 14-14 shows the 8-mm rigid fiber-optic lens.

14.3.7.3 Flexible Fiber

When the utmost flexibility (angular movement) between the front objective lens (pinhole or otherwise) and the camera is required, an alternative to the remote-head CCD camera is a coherent flexible fiber-optic bundle (Figure 14-14). The one significant advantage this lens has over a remote CCD is that there are no electrical voltages within 39 inches of the front lens, which may be important in some applications. The 39-inch-long fiber-optic bundle is encased in a protective, flexible, braided stainless-steel or plastic sheathing to provide environmental protection (from adverse weather, corrosive environment, or mechanical abuse). It can be twisted through 360 degrees with no image degradation. It, too, has spots like the rigid fiber-optic. The flexible fiber-optic lens has a 180-degree "twist" built into it and therefore does not invert the picture.

In the case of the flexible fiber-optic bundle, the individual fibers are fused together only at the ends, but are free to move in the length between the ends. The fragile fibers—one-fourth the diameter of a human hair—are protected from damage by encasing them in a loosely fitting plastic covering and a flexible stainless-steel or high-impact plastic braided sheath. End ferrules seal and terminate the bundle for mounting to the objective and relay lenses. For a ⅔-inch format camera, the outside diameter of the flexible fiber-optic assembly is approximately 1.25 inches. This diameter and construction permit the flexible fiber-optic to bend with a minimum bending radius of approximately 4 inches.

The flexible fiber-optic bundle front has a C mount and accepts any pinhole, C-, or CS-mount lens. The relay lens at the rear of the fiber bundle re-images the scene transmitted by the fiber onto the camera sensor. The back end of the assembly terminates in a male C mount, suitable for any C- or CS-mount camera.

14.3.7.4 Image Quality

As shown in Figure 14-16, the fiber-optic bundle is assembled from several hundred thousand individual glass fiber-optic strands. Although high technology and careful assembly techniques are used throughout the fiber bundle manufacturing process to achieve maximum, uniform optical transmission, there are small variations in transmission from one fiber to another and some broken fibers. The

```
FIBER: FLEXIBLE: 39 INCHES LONG
OBJECTIVE LENS: 25MM FL, F/1.4
RELAY LENS: M=1:1
OVERALL F/#: 4..0
```

```
FIBER: RIGID: 6 INCHES LONG
OBJECTIVE LENS: 8MM FL, F/1.6
RELAY LENS: M=1:1
OVERALL F/#: 6.0
```

FIGURE 14-17 Resolution and image quality from fiber-optic lenses

result is that in almost all fiber-optic systems, the picture obtained is not as "clean" as that obtained with an "all-lens" pinhole lens. There are some cosmetic imperfections that look like dust spots (actually non- or partially transmitting fibers), as well as a geometric pattern caused by packing the fibers during manufacture. These imperfections occur because there are several hundred thousand individual hair-like fibers comprising the fiber-optic bundle, and some of them are not transmitting perfectly. For many applications these imperfections do not result in any loss of picture intelligence, making the lens system adequate for identification of people, actions, and other information. Fiber-optic lenses have a resolution of 450 to 500 TV lines, similar to a standard ½- or ⅔-inch camera system. Figure 14-17 shows two examples of images produced from rigid and flexible fiber-optic lenses.

The photographs were taken directly from a 9-inch monochrome monitor using a CCD solid-state camera with resolution of 570 horizontal TV lines. Figure 14-17a shows the typical resolution and image quality obtainable from a 1-meter, flexible fiber-optic lens: approximately 450 TV lines horizontal and 350 vertical. The spots are caused by

partially transmitting or nontransmitting fibers. Figure 14-17b shows the same image obtained with an 8-inch rigid fiber-optic lens. The vignetting at the corners of the image was caused by the relay lens, not the fiber bundle. Note the spots and honeycomb pattern in the rigid fiber-optic monitor picture. The honeycomb is caused by the fiber-stacking procedure and consequent heat fusing of the rigid bundle.

14.3.8 Borescope Lenses

The borescope lens viewing system is a long thin tube housing with multiple relay lenses used to view inside objects (such as safes) or through barriers. Borescope sizes range from 12 to 30 inches long, and ⅛ to ⅜ inch in diameter (Figure 14-18). Special mini-borescopes are available with 1-mm to 2-mm outside diameters, 2 to 6 inches long.

Borescopes are constructed from stainless-steel tubing and contain an "all-lens" optical system. The long lengths and all-lens design mandate that such lenses have very high f-numbers: they are optically slow. Typical designs have an

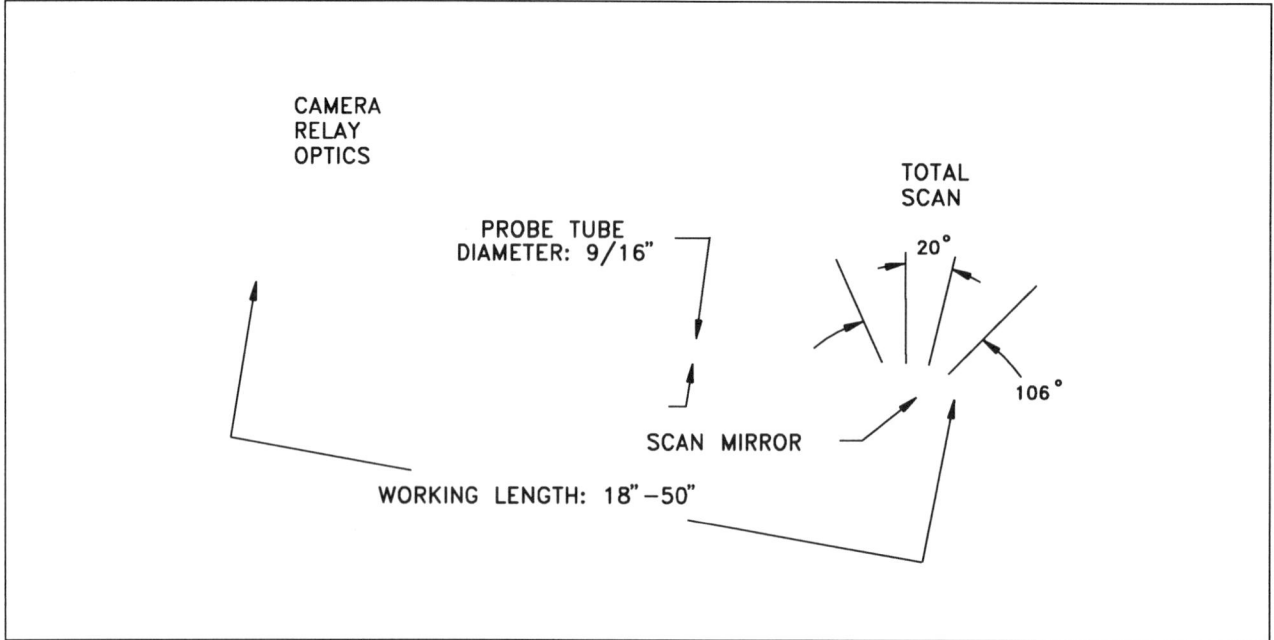

FIGURE 14-18 Borescope lens viewing system

f-number between f/15 and f/40. By comparison, an f/5 lens transmits 16 times more light than an f/20 lens. The borescope must be used with high levels of lighting or an LLL camera (Chapter 15).

14.4 SPECIAL COVERT CAMERAS

14.4.1 PC-Board Cameras

The miniaturization of ⅓- and ½-inch CCD camera electronics has generated a new family of small single and dual printed-circuit-board (PC-board) surveillance cameras. Three flat PC-board and compact sealed flat cameras are shown in Figure 14.19.

Figure 14-19a shows two monochrome ½-inch format monochrome cameras with a C-mount and automatic-iris option. Figure 14-19b shows a small (1.75 × 2.75-inch) ⅓-inch format PC-board CCD camera with a mini 8-mm FL lens. Other interchangeable lenses—3.8, 5.5, and 11 mm FL—are available.

Figure 14-19c shows a compact flat camera sealed in an epoxy case with pin terminals at the rear. The ⅓-inch format camera has 380-TV-line resolution and 0.2-fc sensitivity. All cameras are powered by 12 volts DC.

14.4.2 Remote-Head Cameras

The small size of mini-lenses and solid-state CCTV sensors permits the construction of an extremely small covert lens-sensor head by remoting the lens and sensor from the camera electronics via a small electrical cable. The cable link between the camera head and the camera electronics can vary between a few inches and 100 feet. Figure 14-20 shows a monochrome ⅔-inch format CCD remote-head camera with an 18-inch ribbon connecting the sensor-lens with the camera electronics. The lens is an 11-mm FL, f/1.8 lens. Figure 14-20b illustrates a small color CCD remote-head camera using a ½-inch format sensor having a lens-sensor-head size of 0.69 inch diameter × 2.25 inches long and weighing only 0.64 ounce. The lens shown is a 7.5-mm FL, f/1.6 lens.

The monochrome camera has a resolution of 450 TV lines and a light sensitivity of 0.1 fc (minimum scene illumination). The color camera has a resolution of 460 TV lines and a sensitivity of 1.0 fc (minimum scene illumination).

14.5 INFRARED COVERT LIGHTING

Clandestine surveillance augmented with invisible IR covert lighting can significantly increase the usefulness of covert CCTV installations. Since covert CCTV is intended to be hidden from its target, if the covert CCTV system can operate in near or total darkness, the person under surveillance will not be aware that he is under observation. By augmenting the CCTV system with an IR light, invisible to the human eye but not to the camera, the system can obtain a video picture as good as that under normal visible daylight conditions. Silicon-tube, solid-state CCD, and LLL cameras are sensitive to and can "see" with this IR lighting. The quality of the picture depends on the type of lamp or LED used, its power level and beam angle (see Chapter 3), and

A) MONOCHROME–PC BOARD

C MOUNT

5MM ATUOMATIC IRIS LENS

POWER: 12 VDC

B) MONOCHROME–PC BOARD

CCD SENSOR FORMAT: 1/3
SENSITIVITY: 2 LUX WITH F/1.8
SIZE: 1 13/16"X2 1/4"X1"
POWER: 12 VDC

C) MONOCHROME–EPOXY PC BOARD

CCD SENSOR FORMAT: 1/3"
HORIZ. RESOLUTION: 380 TV LINES
SENSITIVITY: 0.2 LUX WITH f/1.8
SIZE: 1.58"X1.58"X1"
POWER: 12 VDC

FIGURE 14-19 Flat PC-board cameras

A) MONOCHROME CAMERA
WITH RIBBON CABLE

SENSOR: 2/3" CCD
HEAD SIZE: 2"X2"X0.88"
HORIZ. RESOLUTION: 450 TV LINES
CABLE LENGTH: 18"

B) COLOR CAMERA WITH
MULTI–CONDUCTOR CABLE

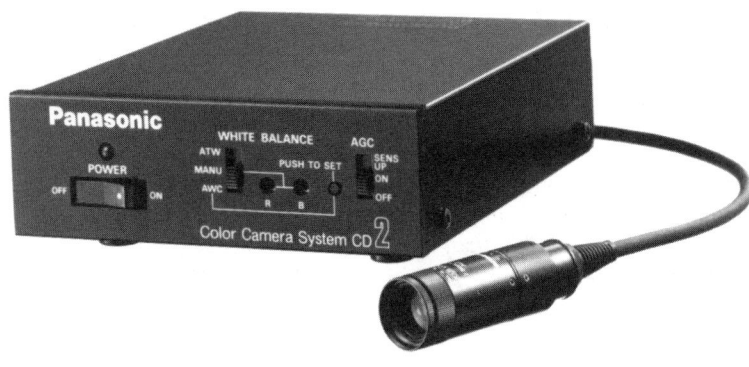

SENSOR: 1/2" CCD
HEAD SIZE: 0.69"DIA.X2 1/4"
HORIZ. RESOLUTION: 460 TV LINES
CABLE LENGTH: 30METERS

FIGURE 14-20 Remote-head CCTV cameras

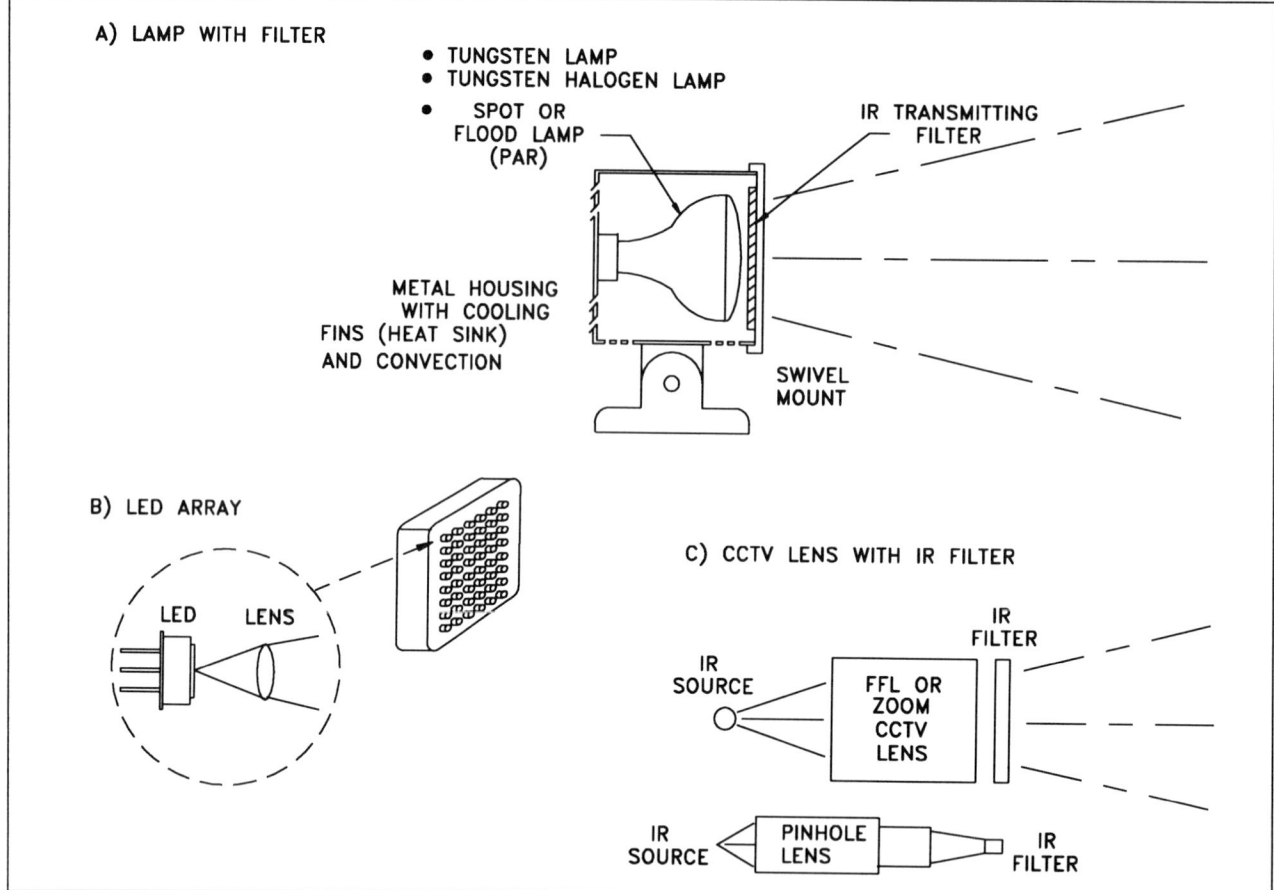

FIGURE 14-21 The principle of IR illumination

the sensitivity of the camera to the IR radiation. When a CCD camera is used, this last factor depends on whether an IR cut filter is used in the camera, and on the CCD sensor sensitivity to the IR energy.

14.5.1 Means for Concealment

Sources that emit both visible and IR light (tungsten, tungsten-hologen, xenon lamps, and others) can be optically filtered so that only the IR radiation leaves the source and irradiates the scene. High-efficiency, low-power LED semiconductors produce sufficient IR energy to illuminate an area suitable for covert operation while being invisible to the eye. Figure 14-21 illustrates the principle and several techniques of producing IR illumination.

The thermal lamp or LED source emits IR radiation, which reflects off the scene and off objects in it. The CCTV lens and camera collect the reflected IR energy to produce a normal television image signal.

The IR-emitting source is often concealed by installing it behind an opaque (tinted) plastic or one-way (partially aluminized) window. Another technique is to use a spectral beam-splitting window, which transmits the invisible IR radiation and blocks the visible radiation. Another tech-

nique is to conceal the IR-emitting source just as the pinhole lens is concealed, by locating the source at the focal plane of a pinhole lens and directing the energy at the same target the pinhole lens is viewing. Usually the beam from the pinhole lens IR source is made slightly larger than the FOV of the pinhole lens-camera combination. Alignment is necessary between the camera and IR source, since the IR beam must illuminate the same scene the pinhole lens is looking at. When the application is to perform covert surveillance at short distances and in small rooms (10 to 15 square feet or so), a wide-area IR illuminator is used, since the alignment is not critical.

14.5.2 IR Sources

There are numerous commercially available thermal lamp and LED IR sources for covert CCTV applications. They vary from short-range, low-power, wide-angle beams to long-range, high-power, narrow-angle beam types. Figure 14-22 illustrates two IR LED and thermal IR source illuminators.

A single IR LED emits enough IR energy to produce a useful picture at ranges up to a few feet with a CCD camera. By stacking many (several hundred) LEDs in an array,

silicon intensified target (SIT), intensified CCD (ICCD) or intensified SIT (ISIT) must be used (see Chapter 15). These intensified cameras operate at significantly lower light levels than the best tube or solid-state cameras. The newer ICCD camera has a sensitivity matching that of the SIT camera, with the SIT tube camera resolution still superior to that of the ICCD. All this increased sensitivity comes at a cost. Any intensified camera is expensive and should be considered only for critical security applications. Figure 14-23 shows an LLL covert pinhole lens camera that can see in a room with only a single candle or lighted match for illumination. The covert LLL camera uses a fast (f/2.0) pinhole lens coupled to a third-generation (GEN 3) microchannel plate image intensifier.

14.7 SPECIAL CONFIGURATIONS

CCTV cameras and lenses are concealed in many different objects and locations, including overhead track lighting fixtures, emergency lighting fixtures, exit signs, tabletop radios, table lamps, wall or desk clocks, shoulder bags, and attaché cases (Figure 14-24). Figure 14-24a shows a popular emergency light that was modified to house a CCTV camera and mini-lens system to view from behind the front bezel.

The emergency lighting fixture operates normally, can be tested for operation periodically, and its operation is in no way affected by the installation of the CCTV camera. The housing has an angled extension, which points downward by about 15 degrees so that the lens points downward and optimally views the area. Alternatively, an off-axis mini-lens could be used instead of the on-axis mini-lens to make the camera look downward. The lens views through the smoked (tinted) plastic front window and cannot be seen even at close range.

The exit light fixture is another convenient form for camouflaging a covert CCTV camera system (Figure 14-24b). The right-angle pinhole lens and CCD camera are located inside the unit and view out of either arrow on the exit sign. The right-to-left reversal of the image by the right-angle pinhole lens can be reversed using an electronic scan-reversal unit (Section 13.5) or a camera equipped with this feature. Alternatively, an off-axis mini-lens on a small PC-board camera could be used instead of the right-angle pinhole lens if a smaller enclosure is required.

A large wall-mounted clock is an ideal location for camouflaging the covert CCTV camera/lens combination (Figure 14-24c). The lens views out through one of the black numerals. In this case, the flat camera (approximately 7/8 inch deep) and right-angle mini-lens are mounted directly behind the numeral 11 on the clock. The camera uses offset optics (see inset in Figure 14-8) so that the camera views downward at approximately a 15-degree angle even though the clock is mounted vertically on the wall.

Figure 14-25 shows a portable covert CCTV system mounted in an attaché case containing the lens and camera

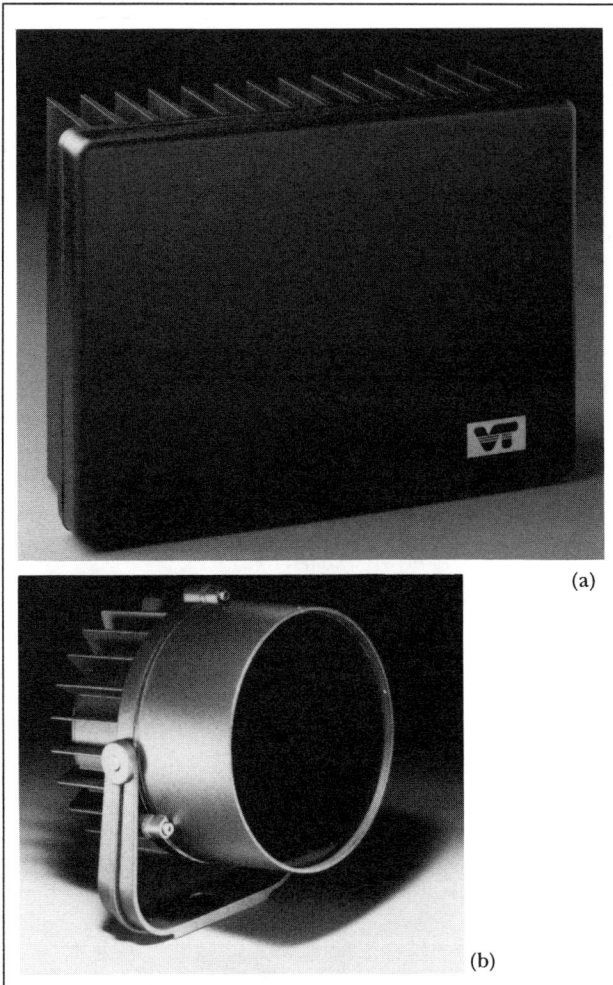

(a)

(b)

FIGURE 14-22 Covert IR source illuminators: (a) IR LED array, (b) IR thermal lamp

higher IR power is directed toward the scene and a larger area at distances up to 50–100 feet may be viewed.

14.6 LOW-LIGHT-LEVEL CAMERAS

The camera parameter most critical to the successful viewing of a scene under LLL conditions with a covert system is the camera sensor sensitivity. Most silicon-tube and CCD cameras have sensitivities of approximately 0.2 to 1 fc (0.1 lux), which does not result in satisfactory CCTV picture quality under dawn, dusk, nighttime, or poorly lighted indoor conditions. A few special CCD cameras produce better sensitivity than this. One is a high-sensitivity, monochrome, 1/2-inch format CCD camera using an interline transfer sensor having a faceplate sensitivity (full video) of 0.023 fc (0.0023 lux), which substantially increases its usefulness at low light levels. It also boasts a resolution of 570 TV lines.

When CCD camera sensitivity is not sufficient and additional lighting cannot be added, an LLL camera such as a

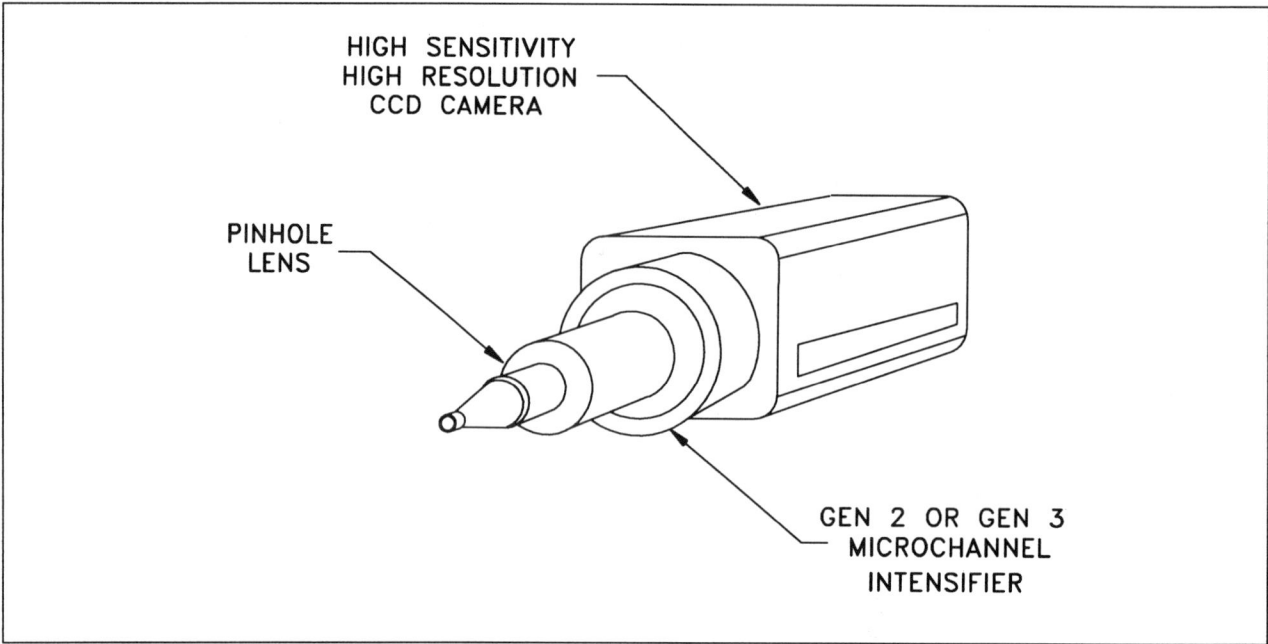

FIGURE 14-23 Low-light-level pinhole lens camera

CAMERA LENS VIEWS
THROUGH 1/4" DIA.
TINTED PLASTIC

CAMERA VIEWS
THROUGH HOLE
IN EITHER ARROW

CAMERA LENS VIEWS
THROUGH HOLE AT
NUMERAL "11"

FIGURE 14-24 Covert CCTV in emergency light, exit sign, and clock

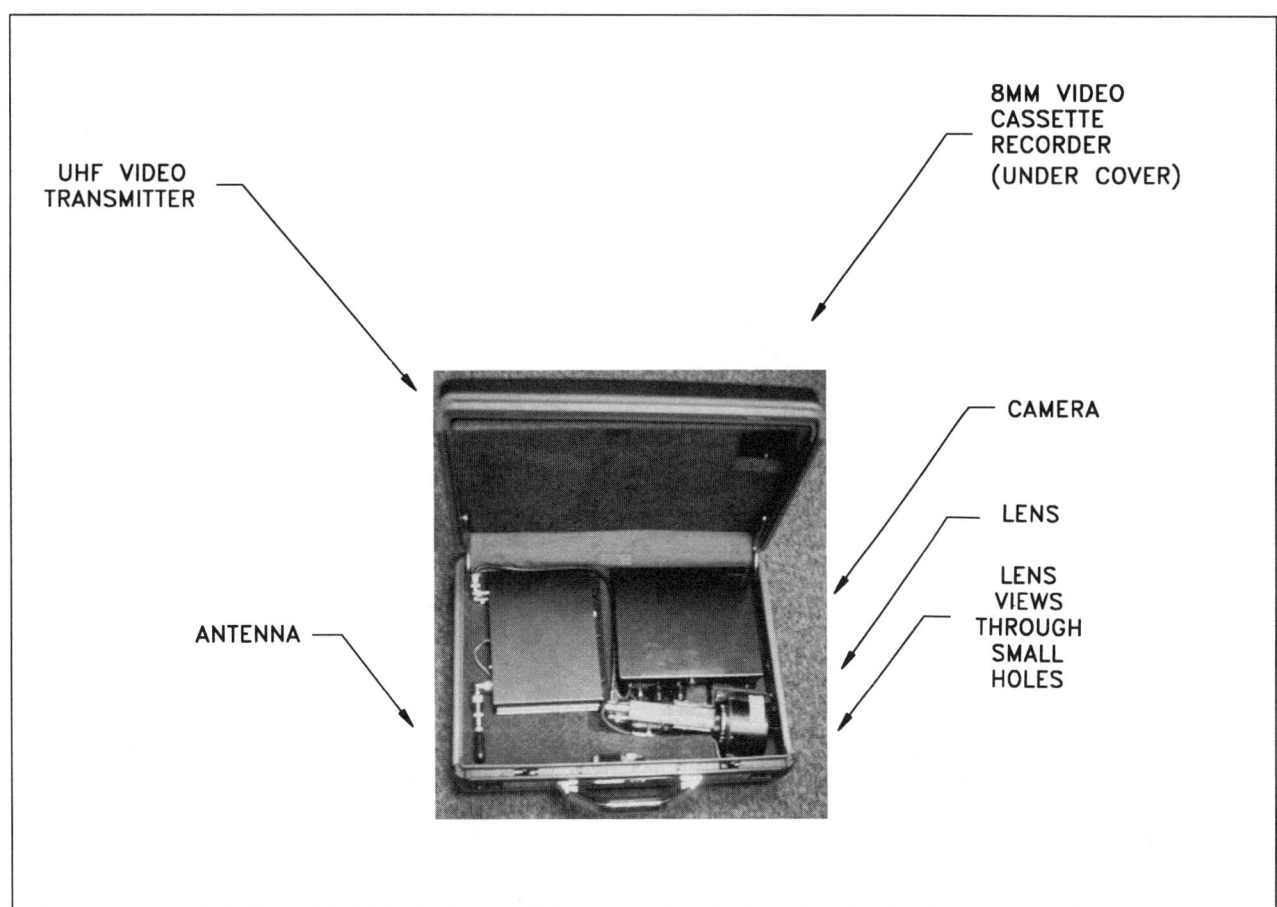

UHF VIDEO
TRANSMITTER

8MM VIDEO
CASSETTE
RECORDER
(UNDER COVER)

CAMERA

LENS

LENS
VIEWS
THROUGH
SMALL
HOLES

ANTENNA

FIGURE 14-25 Covert CCTV in attaché case

and a VCR. The operator activates the system via external switch controls. The covert lens views through small ¹⁄₁₆-inch-diameter hole(s) at the front end of the case and is inconspicuous to an observer. One system option is a wireless RF transmitter. The items in which covert cameras can be installed are limited only by the human imagination.

14.8 WIRELESS TRANSMISSION

For best results at lowest cost, the video signal from most covert CCTV installations is transmitted from the camera to the monitor (or VCR) via RG59/U, 75-ohm coaxial cable. If a dedicated telephone-grade line (two-wire) is available, a special line driver and receiver pair permit fair-to-good transmission of a real-time video signal over several thousand feet of continuous telephone wire (Chapter 6). This two-wire transmission requires that there be no break in the pair of wires: for example, they cannot go through a telephone switching station.

Covert CCTV applications often require that the camera/lens system be installed and removed quickly, or that it remain installed on location for only short periods of time. This may mean that a wired transmission link (such as coaxial cable or fiber optic) cannot be installed and a wireless transmission link from camera to monitor (or VCR) is required. This takes the form of a VHF or UHF RF, microwave, or light-wave (IR) video transmitter of low power mounted near the television camera. A description of these transmitters is given in Chapter 6, but those specifically applicable to covert applications are summarized here. The RF transmitters are of low power—from under 100 milliwatts to several watts - and transmit the video picture over ranges from 100 feet to several miles. In the United States, the FCC restricts the use of the higher-power transmitters to federal or government agencies and allows only low-power units for commercial or industrial use. Figure 14-26a shows a low-power, 100-mw transmitter and receiver operating at 920 MHz, which transmits an excellent monochrome or color video picture over a distance of 120 feet.

Figure 14-26b shows a 920-MHz transmitter that transmits an excellent monochrome or color picture and audio over a distance of 300 feet. Using a directional receiver antenna increases the range to 700 feet.

When considering RF frequencies, recognize that the VHF/UHF bands (150 to 950 MHz) may not produce optimum results if the transmission path is through a steel

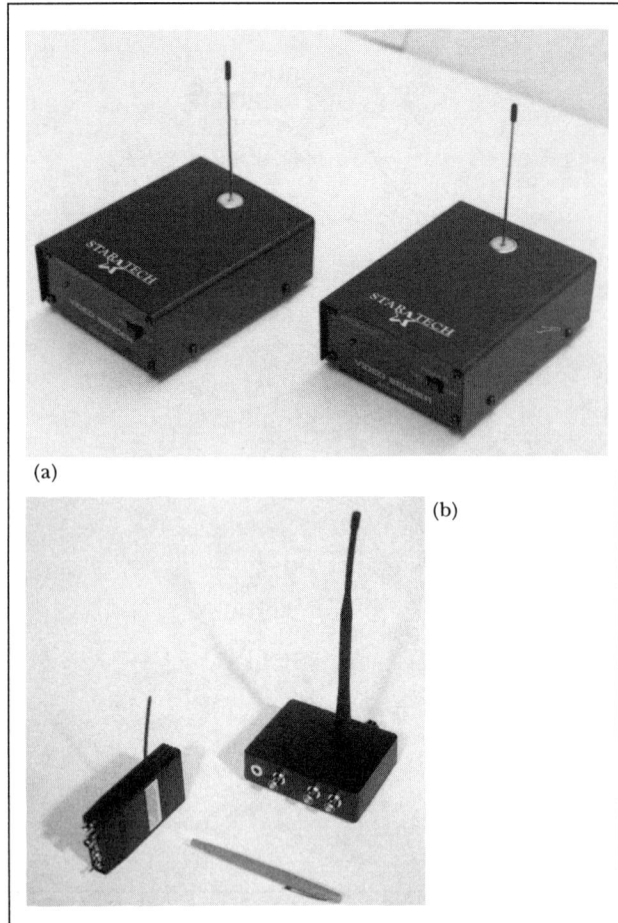

(a)

(b)

FIGURE 14-26   RF transmitters for covert CCTV transmission: (a) 920 MHz, 100-foot range, clear line of sight, (b) 920 MHz, 300-foot range, clear line of sight

building or near other metallic or reinforced-concrete structures. While the transmitter may have suitable range under outdoor, unobstructed conditions, when used indoors or between two points with obstructions, the only way to determine the useful range of the link is to put the system into operation. The deleterious effects most readily observed are (1) reduction in range; (2) ghost images, that is, multiple images produced by reflections of the VHF/UHF signal from metallic objects; and (3) unsynchronized pictures, that is, the picture breaks up. Repositioning the transmitter or receiver equipment often substantially improves or eliminates such problems.

Microwave transmission systems operate in the 2-to-22-gigahertz range and require FCC licensing and approval but can be used by government agencies and some by commercial customers. One condition in obtaining approval is to have a frequency search performed to ensure the system causes no interference to existing equipment in the area. Most microwave systems have a more directional transmitting pattern than RF transmitters. This means the antenna

directs the energy toward the receiver and therefore alignment between transmitter and receiver is more critical. Most microwave installations are line-of-sight but the microwave energy can be reflected off objects in the path between the transmitter and the receiver to direct the energy to the receiver, at a sacrifice in range. The higher frequency of operation and directionality make microwave installation and alignment more critical than the RF transmitters (see Chapter 6).

Another line-of-sight system requiring no FCC approval is a wireless gallium arsenide (GaAs) IR optical transmission system. This light-wave system requires no cable connection between the transmitter and the receiver and achieves ranges of hundreds to several thousands of feet (see Chapter 6). Its major limitation is the severe reduction in range under fog or heavy precipitation conditions.

14.9   COVERT CHECKLIST

- Optical speed or f-number is probably the most important reason for choosing one pinhole lens over another. The lower the f-number the better. An f/2 lens transmits four times more light than an f/4. This can mean the difference between using a standard vidicon, Newvicon, or silicon tube or CCD camera and using an LLL intensified CCD or tube camera.
- Most pinhole lenses have a FL between 3.8 mm and 22 mm and are designed for ⅓-, ½-, or ⅔-inch format cameras. Tables 14-1, 14-3, 14-4, and 14-5 show the FOVs obtained with these lenses. For example, using these tables or the Lens Finder Kit (Chapter 4), the FOV seen with the 11-mm lens on a ⅔-inch camera format at a distance of 15 feet is an area 12 feet wide by 9 feet high displayed on the television monitor. Note that the FOV is independent of the hole size through which the lens views, providing a hole produces no tunneling. When viewing through a wall with a wide-angle pinhole lens or mini-lens (3.8-, 5.5-, or 8-mm), the lens may require a cone-shaped hole or an array of small holes to prevent tunneling (vignetting) of the scene image.
- A short FL lens (5.5 mm) has a wide FOV and low magnification. A long FL lens (50 mm) has a narrow FOV and has high magnification.
- Medium FL lenses produce FOVs wide enough to see much of the action and still have enough resolution to identify the persons or actions in the scene. A short FL lens sees a wide FOV and objects are not well resolved. Long FL lenses see a narrow FOV with objects well resolved (clear).
- Under most conditions, the small-barrel, slow-taper pinhole lens is easier to install and is the preferred type over the wide-barrel, fast-taper shape. The user must weigh the pros and cons of both types.
- The use of a straight or right-angle pinhole lens depends on the space available behind the barrier for mounting

the lens and camera, and on the pointing direction of the lens.

- The fastest pinhole CCTV system is a mini-lens coupled to the camera. This is the best choice where the lowest cost and highest light efficiency are desired.

- A manual-iris lens is sufficient in applications where there are no large variations in light level, or where the light level can be controlled. Depending on the camera used, where there is more than a 50:1 change in light level, an automatic-iris pinhole lens or an electronically shuttered camera is needed.

- Most applications are solved using an "all-lens" system. In special cases where a thick barrier exists between a surface and the camera location, a rigid coherent fiber-optic bundle lens or borescope is used. If sufficient light is available, an "all-lens" borescope type should be used to obtain the cleanest picture. A second alternative is a remote-head camera.

- AC power is preferred for permanent covert camera installations. Either 117 or 24 volt AC wall-mounted converters are used. However, 24 volt AC is preferred over 117 volt AC, since it eliminates any fire or shock hazard and can be installed by security personnel without the help of electricians. Since most small cameras operate from 12 volt DC, a 117 volt AC to 12 volt DC converter is most popular. For temporary installations, 12 volt DC battery operation is used, with rechargeable or nonre-

chargeable batteries, depending on the application (see Chapter 16).

14.10 SUMMARY

Pinhole lenses are used for surveillance problems that cannot be solved adequately using standard FFL or zoom lenses. The fast f-numbers of some of these pinhole lenses make it possible to provide covert surveillance under normal or dimly lighted conditions. The small size of the front lens and barrel permit them to be covertly installed for surveillance applications.

A large variety of mini-, pinhole, fiber-optic and borescope lenses are available for use in covert security applications. These lenses have FL ranges from 3.8 to 22 mm, covering FOVs from 12 to 78 degrees. Variations, including manual- and automatic-iris, standard pinhole, mini- and off-axis-mini, fiber-optic and borescope, provide the user with a large selection.

Equipment is available to provide covert surveillance under lighted or unlighted conditions. Through the use of IR illumination, scenes can be viewed in total darkness. Compact lenses, small and low-power cameras, wireless RF, microwave, and IR transmission systems make the covert system portable.

Chapter 15

Low-Light-Level Cameras

CONTENTS

15.1 OVERVIEW

Many security surveillance applications require low-light-level (LLL) cameras to observe dimly lighted parking lots, warehouses, shopping malls, streets, back alleys, and so on. Often the high cost of installing additional lighting to accommodate conventional cameras would be easily offset by the extra cost of LLL cameras. Other applications include areas where it is impossible to install lighting because it is out of the control of the security force. Examples include (1) looking through darkened windows, (2) reading license plates or identifying vehicles in a darkened building, and (3) viewing (and videotaping) facial characteristics or witnessing an action in a dark area. Many of these surveillance problems are extremely difficult if not totally impossible to solve with normal tube or charge-coupled-device (CCD) cameras but are easily accomplished with an image-intensified camera system.

During the 1940s, the need for military LLL imaging devices accelerated the development of compact portable image-intensifying equipment. The silicon intensified target (SIT) and intensified SIT (ISIT) camera systems were already developed; though they had excellent LLL capabilities, they were not rugged or portable and could not be used for military applications. Several generations of image intensifiers have evolved since this period, beginning with generation zero (GEN 0) to present-day generation three (GEN 3). The GEN 0 device consists of an infrared (IR) light source illuminating a scene, and the reflected IR energy is detected by an IR-to-visible light converter tube. The visible scene image on the tube is relayed to a CCTV camera, thereby producing "night vision." This active system has the disadvantage of being detectable by an adversary with a simple IR viewer. The present state of the art are the GEN 2 and GEN 3 microchannel plate (MCP) image intensifiers. The GEN 3 intensifier has three times the sensitivity of the GEN 2, better resolution, and nearly four times the tube life. This chapter briefly describes GEN 0 and GEN 1 and covers in more detail GEN 2, GEN 2 Plus, and GEN 3 devices, which are the current state of the art. The combination of single-stage GEN 1, GEN 2, and GEN 3 image intensifiers with tube or CCD solid-state cameras results in state-of-the-art LLL devices.

LLL cameras are used in applications where the available illumination is not adequate for standard CCTV cameras. In a true LLL camera, the image viewed by the television image tube or solid-state sensor has been "intensified" or amplified before it reaches the sensor. The intensification technique does not add additional illumination to the scene but amplifies the reflected scene illumination. After the light image has reached the CCTV sensor, it is processed in the normal way and transmitted to the monitor. The three primary parameters determining the performance of LLL cameras are (1) the specific spectral content of the light, (2) the level of the scene illumination reaching the intensifier tube faceplate, and (3) the specific type of intensifier/camera used.

The name "night vision" is sometimes used to describe LLL cameras and indicates a device that sees in total darkness. By definition, however, it is impossible to see in total darkness with a nonthermal imaging system if the meaning of *seeing* is some sort of information transfer by means of light. Image-intensifying cameras are devices that can extend the threshold of seeing far below normal human limits. Most LLL cameras described in this chapter are passive cameras, that is, they do not emit any IR or other radiation and therefore only respond to residual visible and near-IR lighting from such faint objects as stars, moonlight (reflected sunlight), or other artificial lighting.

By definition, an image intensifier is any device that produces an observable image that is brighter at its output than at the input image. These devices are generically different from television cameras and SIT and ISIT tubes in that all of the points in the image are operated upon simultaneously, rather than sequentially as in scanning systems that move across the image point by point. Like the human eye or photographic film, the image intensifier operates on all points in the image *simultaneously*.

The following reasons explain why image-intensified CCTV cameras help us see at low illumination levels:

- Intensifier photocathode efficiency is better than that of the eye by a factor of between 3 and 5.
- Intensifiers can use the near-IR radiation from the scene, which the eye is not sensitive to.

- Intensifiers use lenses that are larger than the human eye lens without reducing the angular FOV.

15.2 HISTORY AND BACKGROUND

Interest in the development of image-intensifier tubes began shortly after the formalization of electron optics in the 1920s. Researchers worked toward improving the sensitivity of television camera tubes and converting IR radiation to visible light in the sensor. Both activities met with some success, resulting in two types of devices: (1) single-stage image intensifiers having a gain of approximately 50 and (2) IR-to-visible-image converters using active IR sources to illuminate the scene. Following the development of the single-stage devices, various multiple-stage (two- and three-stage) intensifiers were developed, resulting in what is generally referred to as GEN 1 image intensifiers. After early coupling problems associated with two- and three-stage intensifiers were solved by using high-resolution fiber-optic coupling plates, thousands of three-stage image intensifiers were fabricated and used in military applications. These military three-stage intensifiers had gains of approximately 35,000 and a limiting resolution of approximately 30 line pairs per millimeter, or approximately 360-TV-line resolution on a 1-inch (25-mm) tube—considered only fair resolution by normal CCTV television standards.

GEN 1 image-intensifier tubes were developed by U.S. Army scientists for military night vision use in the late 1950s, based on the knowledge and technology developed for the GEN 0 systems. The first GEN 1 intensifiers were large and inefficient and not very practical for security or military use. With additional development, the army reduced the size and increased the reliability of GEN 1 devices and developed the first single- and three-stage GEN 1 image intensifiers. Eventually, large quantities were produced for the military.

After some basic innovations and the invention of the MCP image intensifier, the GEN 2 intensifier emerged and is now in widespread use in the military and law enforcement. A further improved GEN 2 that uses a gallium arsenide (GaAs) photocathode is called a GEN 3 intensifier.

The human eye and many artificial devices such as television receivers and night vision devices operate in different wavelength regions within the visible and near-IR spectral range called the spectrum of electromagnetic radiation (Figure 15-1).

The human eye and these devices are tuned to be sensitive to specific wavelength bands within this spectrum. Each is sensitive to only a particular portion of the spectrum. Specifically, the human eye is sensitive to visible light, the spectral band of wavelengths from purple to deep red, covering the wavelength range from 0.4 to 0.7 microns (a micron equals one millionth of a meter) or 400 to 700

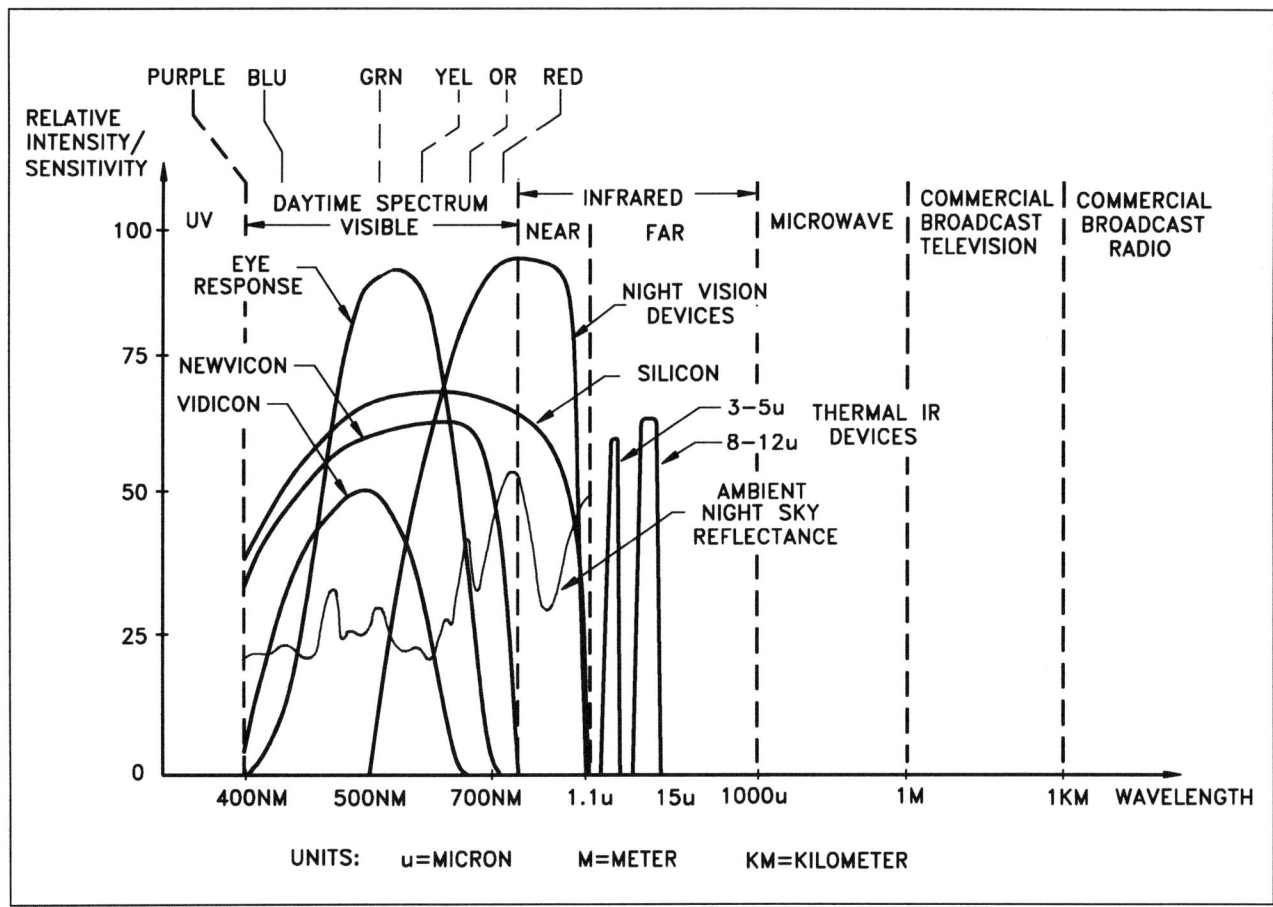

FIGURE 15-1 Optical electromagnetic spectrum

nanometers (nm). Most night vision devices are sensitive to electromagnetic radiation and wavelengths inside and outside the visible band. This includes the visible spectrum, from 400 to 700 nm, and the near-IR range, from approximately 700 to 1100 nm. Such devices are sensitive at light levels below the light threshold level of the human eye. The function of the night vision device is to detect electromagnetic energy at wavelengths or frequencies emitted by the scene that are within the visible and near-IR spectral range or below the threshold of human vision. These devices convert the electromagnetic energy into electrical energy, which is amplified and converted into visible light via the device's output phosphor screen. This visible light image on the screen is then relayed to a CCTV sensor, forming the LLL camera.

15.3 INTENSIFYING MECHANISMS

Image-intensifier tubes extend human visibility beyond that of the unaided eye by using (1) a photocathode (sensor) that has a higher quantum efficiency than the eye, (2) a photocathode with a broader spectral response (response to near-IR in addition to visible light) than the human eye,

and (3) an objective lens that is larger and can gather more light than the human eye.

Ambient light from sky illumination or residual ground-level illumination is reflected from the scene, collected by the objective lens, transmitted through the fiber-optics faceplate, and focused onto a light-sensitive photocathode. The image-intensifier tube electronically/optically amplifies an image of the scene being viewed and displays it directly on a luminescent (phosphor) screen (Figure 15-2).

The image produced by the objective lens is focused onto the input faceplate of the intensifier. The basic intensifier consists of a photocathode on which the image is focused by an objective lens, an electron amplifier lens, and an output phosphor screen upon which the final image is displayed. Image intensification occurs when the electrons emitted by the photocathode strike the phosphor screen after being accelerated by a high voltage, and produce a visual image on the phosphor screen. Each photon from each point in the scene is amplified by the intensifier and produces tens to thousands of photons at the cathode screen; by this method the image is greatly intensified. Luminance gains in single-stage GEN 1 image tubes can be on the order of 50 to 100. By coupling several image tubes, it is possible to obtain gains of 100,000 to 200,000. The

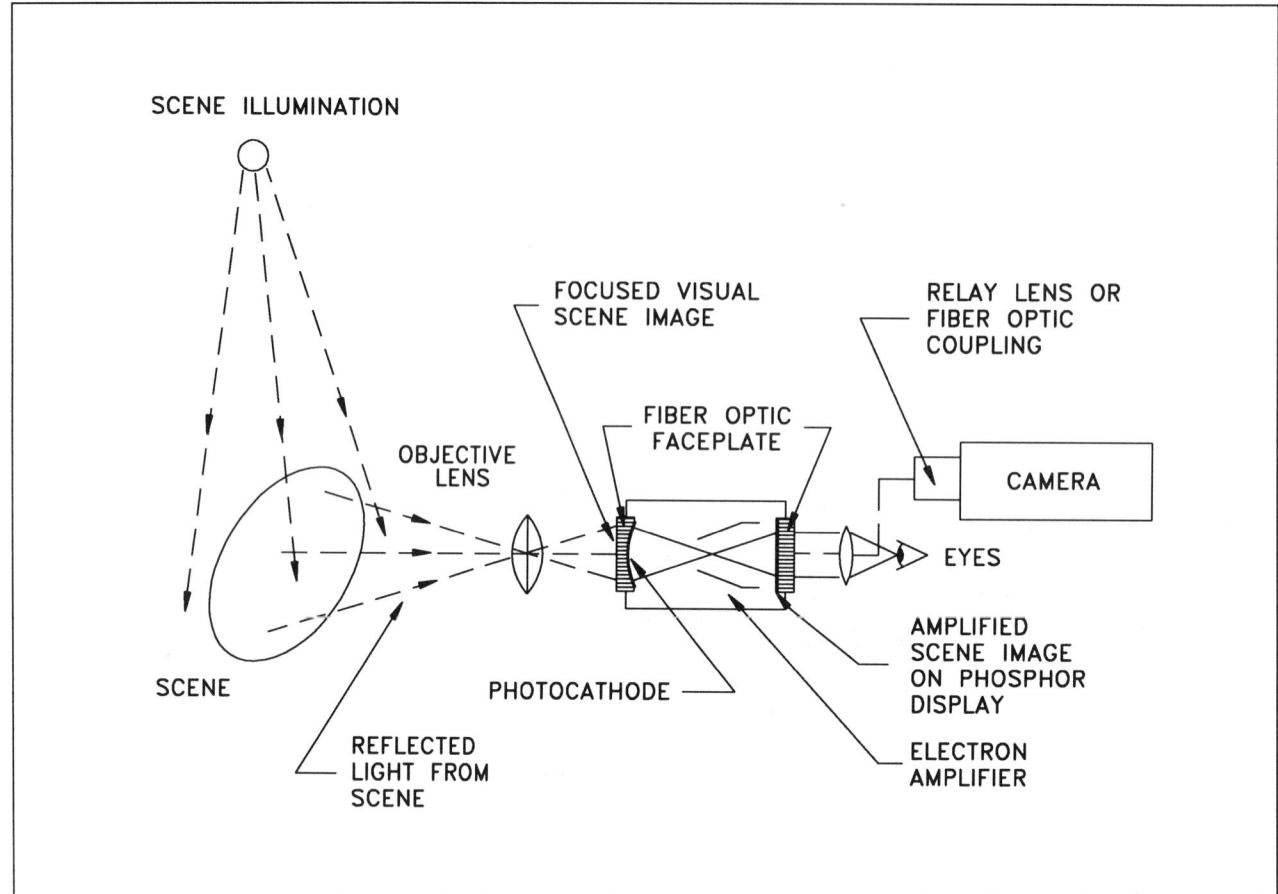

FIGURE 15-2 Image-intensifier components

image intensifier is coupled to a CCTV camera, forming an LLL CCTV system. There are several different generation tube types, categorized by the type of electron-optical focusing mechanism used.

15.4 ACTIVE VERSUS PASSIVE

There are basically two types of night viewing systems: (1) passive systems, which operate by intensifying available light, and (2) active systems, which require an IR light source to provide illumination. In the GEN 0 active system, the ability to see down to lower light levels—below the levels that standard television can see—is accomplished by illuminating the scene with an artificial IR source. This IR source takes the form of a filtered visible source, or an IR semiconductor LED or laser diode (LD).

15.4.1 Active Image Converter

To view scenes at practical distances, GEN 0 image converters are used in active systems requiring the scene to be artificially illuminated by IR radiation to obtain a usable image. The IR light source is either a thermal lamp with an optical filter, which transmits the near-IR light from the

light source but blocks or absorbs the visibly emitted light, or a semiconductor GaAs LED or LD. Since near-IR radiation is easily detected by using IR viewers, this technique is not applicable for many security applications; consequently, the military and industrial applications for GEN 0 devices represent only a small part of the night vision equipment used. The GEN 0 device converts the reflected IR into an intensified image suitable for direct viewing. Figure 15-3 shows an active IR source illuminating the scene to be observed and the LLL device detecting the reflected IR light from the scene via the lens.

The lens focuses the IR scene image onto the GEN 0 night vision device converter tube. GEN 0 tubes use a photocathode with an S-1 spectral response to convert IR to visible radiation. In this simple near-IR viewing system, invisible near-IR light reflected from the scene is focused onto a photoemissive surface (photocathode) sensitive to the 700-to-1200-nm near-IR radiation. This near-IR energy striking the photocathode surface causes electrons to be emitted from it, which are accelerated by a high voltage until they strike a phosphor screen at the opposite end of the tube, forming an image similar to the display of a miniature television picture tube. An eyepiece fitted with a lens allows this image to be viewed, or a relay lens coupling this phosphor screen to a CCTV camera allows viewing the scene on a remote monitor.

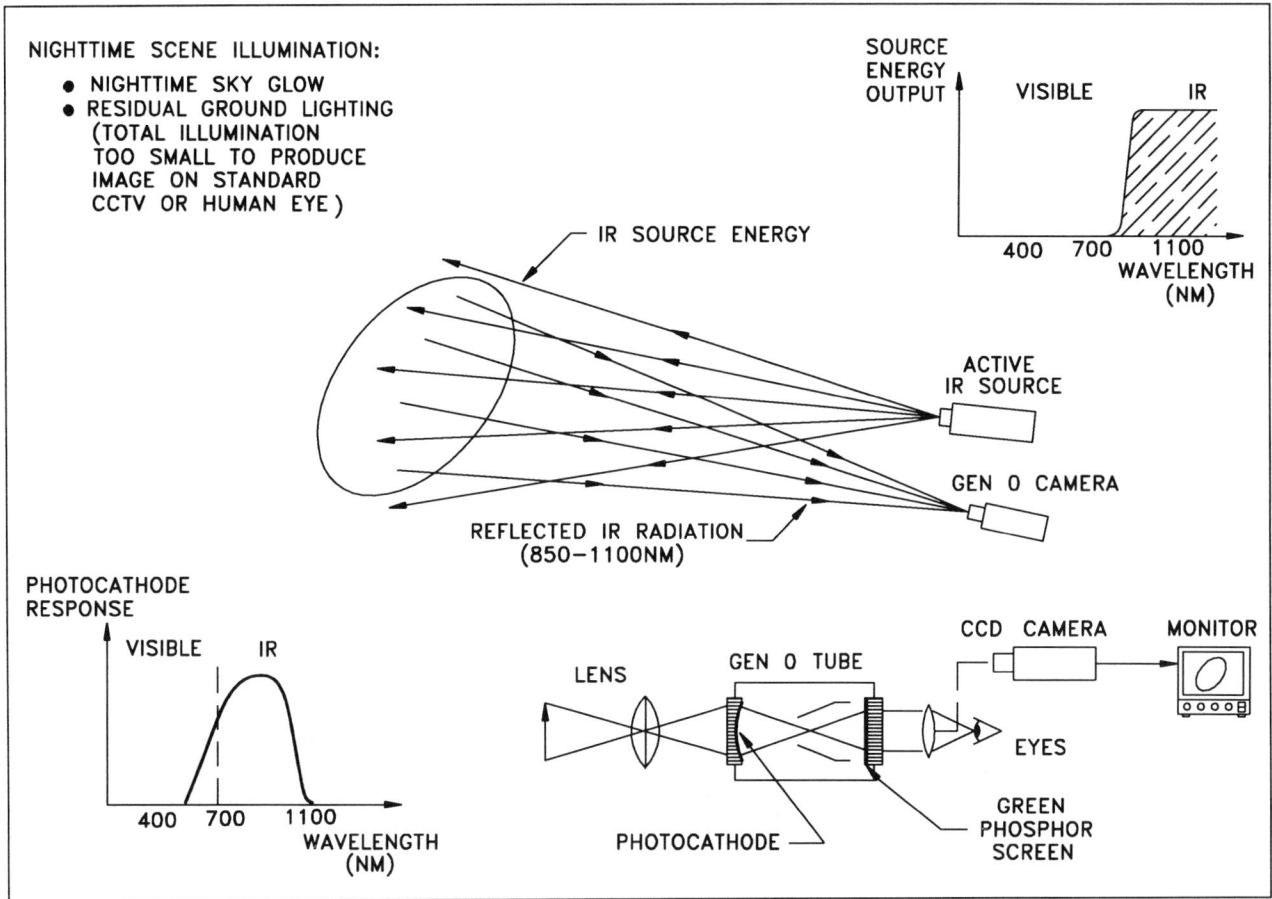

FIGURE 15-3   Active GEN 0 image converter

15.4.2 Passive Image Intensifier

Passive image intensifiers amplify visible and near-IR radiation and are "passive" in that they operate with available light and require no additional illumination, enabling clear vision in conditions of near-total darkness. The objective lens collects available light, however faint, and focuses it onto the GEN 1 photocathode sensor, which converts the light into electrical energy and directs it toward the phosphor screen at the output end of the tube (Figure 15-4).

In two- or three-stage GEN 1 intensifiers, each stage is coupled to the next via a short fiber-optic faceplate. The final, greatly intensified image is formed on the phosphor output screen and viewed through an eyepiece or coupled to a CCTV camera for eventual viewing on a monitor.

15.5   GEN 1 IMAGE INTENSIFIER

The GEN 1 intensifier originally developed for military night vision was coupled to an objective lens at the front end and an eyepiece at the rear end. The image intensifier tubes had an S-1 or S-20 spectral response (Figure 15-8) or some other variation of the multi-alkali photocathode. In order to increase amplification (gain), the logical develop-

ment was to cascade single-stage GEN 1 tubes to form two- and three-stage devices. The result was the GEN 1 system with three stages and a gain of about 30,000. Figure 15-5 shows the internal diagram of a single- and a three-stage GEN 1 image-intensifier tube system.

The GEN 1 intensification system operates by focusing the faint visible and near-IR scene image collected by the objective lens onto the front surface of the first-stage fiber-optic plate. The thin coherent fiber-optic faceplate transfers the scene image onto the light-sensitive photocathode. In the single-stage GEN 1 intensifier, the light energy is converted to electrons in the photocathode and accelerated to a phosphor screen suitable for viewing by the human eye, or coupled to a CCTV tube or solid-state sensor and displayed on a monitor.

The three-stage GEN 1 image intensifier is assembled from three single-stage intensifier tubes, fiber-optically coupled into one assembly. Electrons emitted from the input photocathode are accelerated and focused onto the first-stage phosphor screen as in the single-stage tube. These electrons are absorbed by the phosphor and their energy is radiated into a visible (green) image via the phosphor screen. This image is then transmitted via the input fiber optics on the second stage and the process repeated again. Light gains (amplification) of up to 100,000 have been

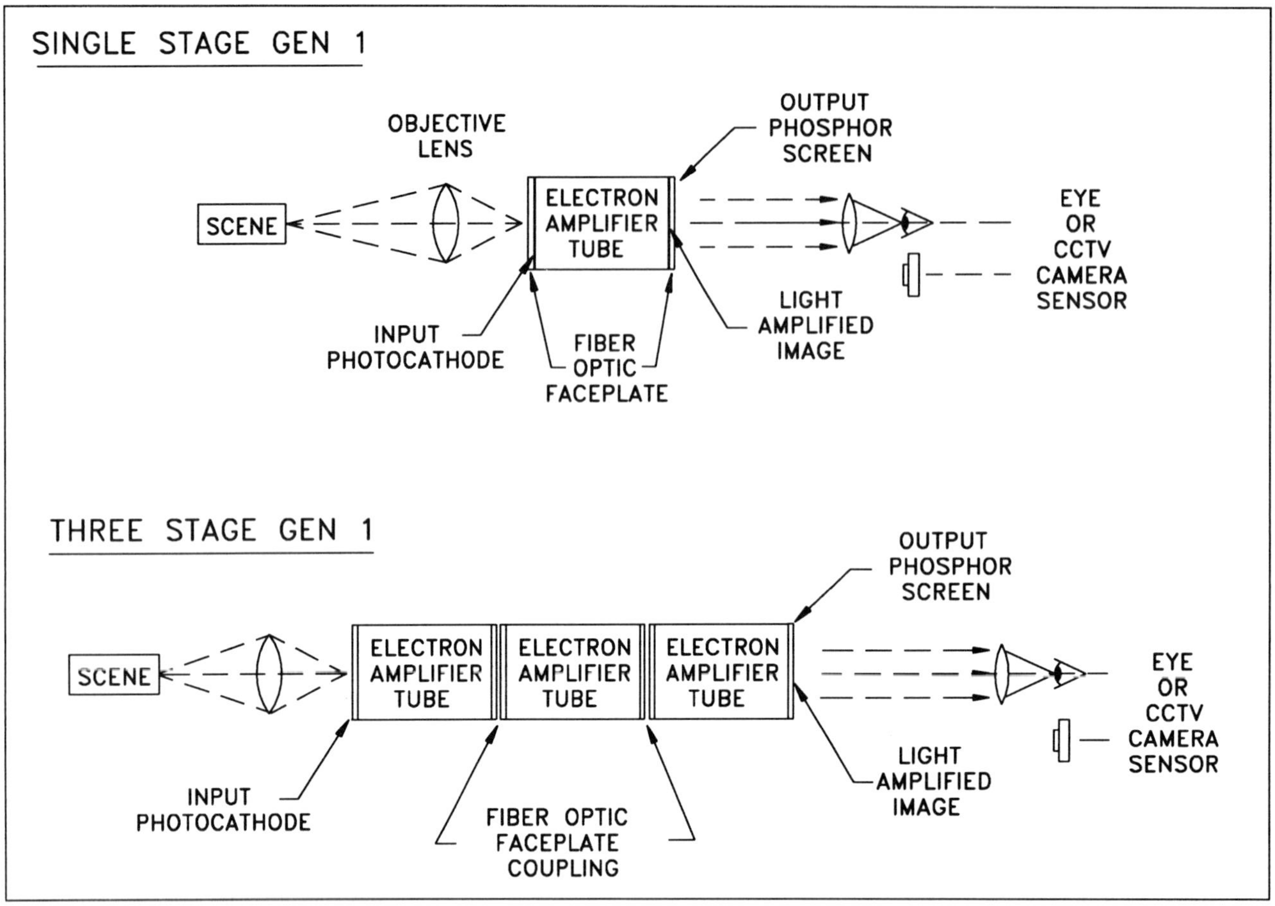

FIGURE 15-4 Passive GEN 1 intensifiers

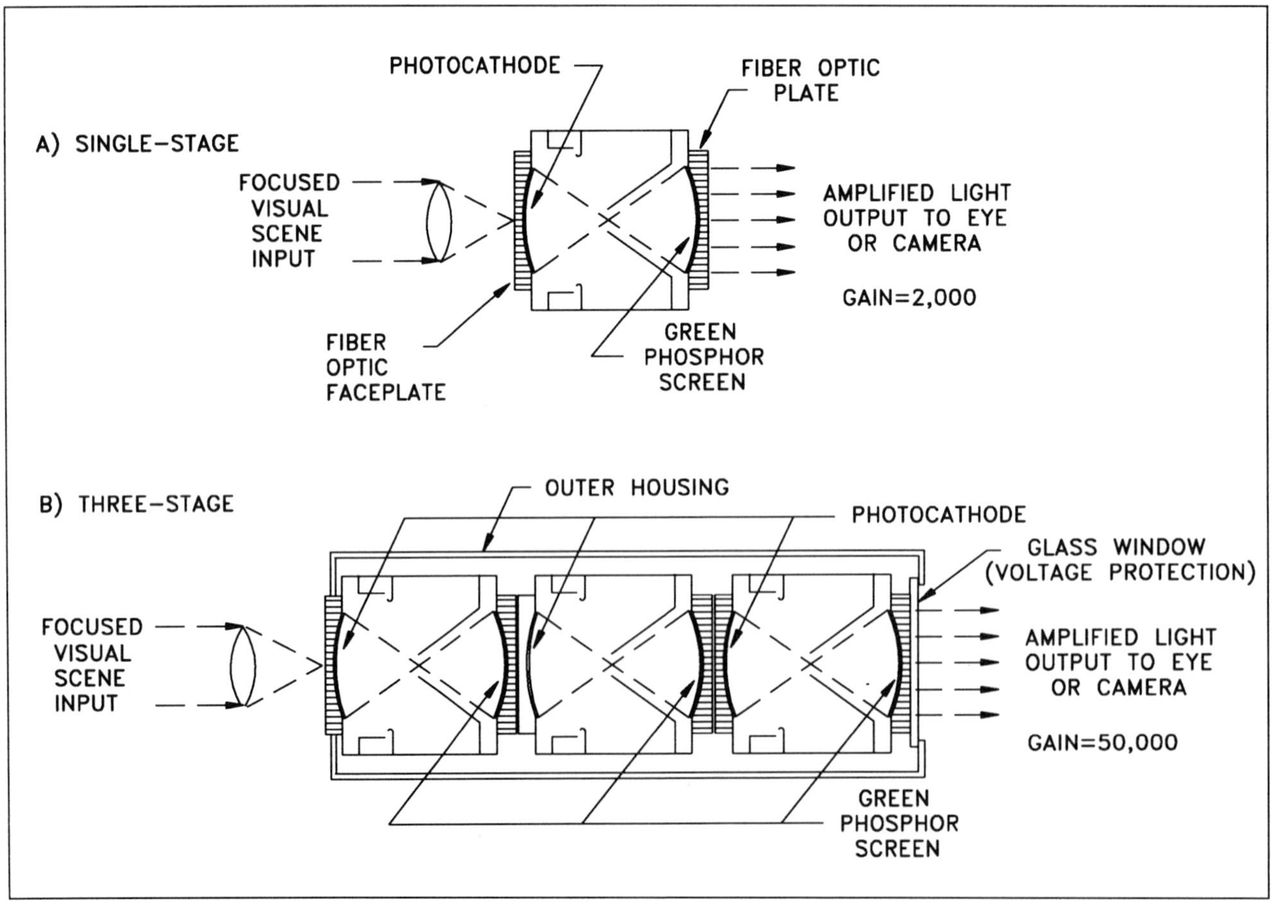

FIGURE 15-5 Diagram of single- and three-stage GEN 1 intensifiers

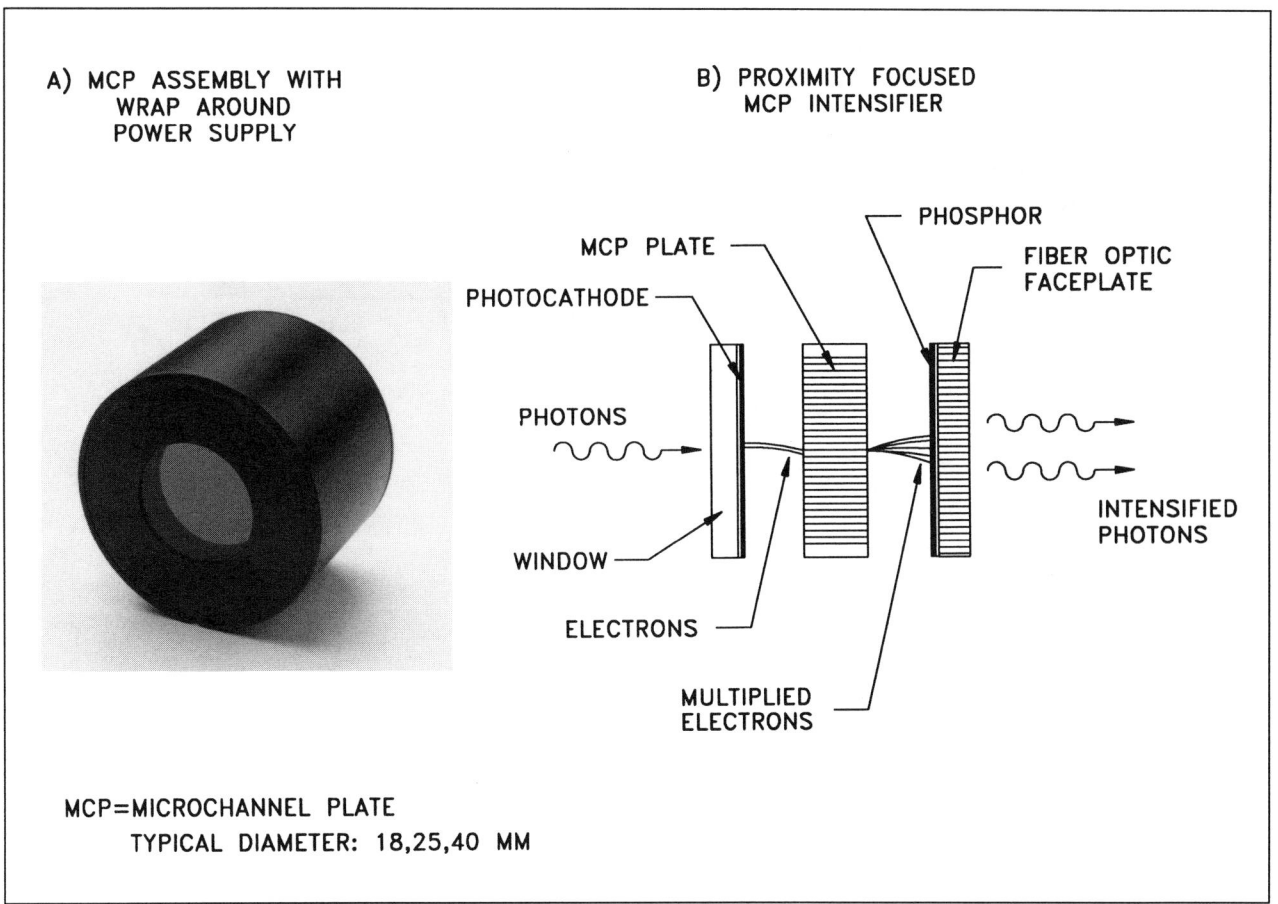

A) MCP ASSEMBLY WITH
   WRAP AROUND
   POWER SUPPLY

B) PROXIMITY FOCUSED
   MCP INTENSIFIER

PHOSPHOR

FIBER OPTIC
FACEPLATE

MCP PLATE

PHOTOCATHODE

PHOTONS

WINDOW

ELECTRONS

MULTIPLIED
ELECTRONS

INTENSIFIED
PHOTONS

MCP=MICROCHANNEL PLATE
   TYPICAL DIAMETER: 18,25,40 MM

FIGURE 15-6   GEN 2, 3 microchannel plate intensifier

achieved in these three-stage systems; however, a gain of 35,000 to 40,000 is typical. Expected single-stage gain of 50 × 50 × 50 = 125,000 is not achieved, due to coupling losses. One such loss occurs because the light available for the first and second stages is only green light from the phosphor, which constitutes a narrower spectral range of energy than the broadband scene illumination. Moreover, distortion is multiplied because of the three stages, and resolution reduced from about 50 line pairs per millimeter for a single stage to about 30 line pairs per millimeter for the three stages. Three-stage GEN 1 systems proved to be useful devices and thousands were manufactured and used by the armed forces and law-enforcement agencies.

The GEN 1 photocathode is sensitive to the 400-to-850-nm electromagnetic band covering the visible and the near-IR spectrum. Since these tubes can respond to very near IR radiation, they are able to detect whether someone is using a near-IR illuminator or active LLL system somewhere in the scene they are viewing, while at the same time remaining undetected themselves.

15.6 GEN 2 MICROCHANNEL PLATE IMAGE INTENSIFIER

Perhaps the most significant innovation in image intensifiers was made with the invention of the MCP intensifier, originally developed by Bendix Corporation as a single-stage channel multiplier in the 1950s. The GEN 2 MCP image intensifier replaced the GEN 1 electron tube multiplier (Figure 15-6).

The GEN 2 MCP intensifier uses secondary emission of photoelectrons as the gain mechanism and produces electron gains of many tens of thousands, depending on the voltage across the plate. The electrons leaving the MCP are proximity-focused onto the phosphor, but since most of the gain is in the MCP, lower voltages (approximately 5000 volts), compared with the 30,000 volts needed by three-stage GEN 1 intensifiers, are used, thus significantly reducing voltage-breakdown problems. The use of MCPs allows image tube size and weight to be decreased significantly as compared with GEN 1 three-stage intensifiers. They offer

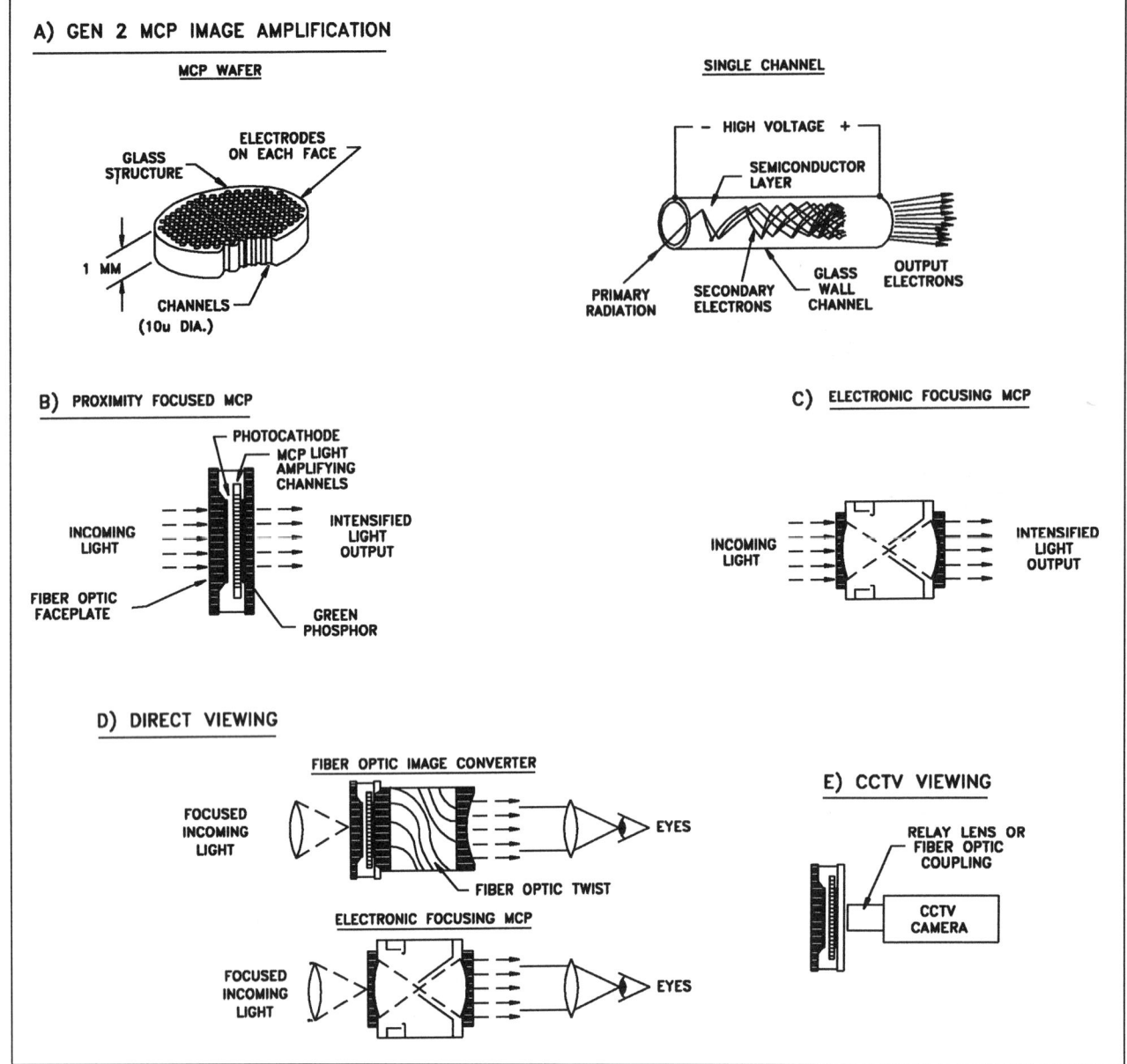

FIGURE 15-7   GEN 2, 3 microchannel intensifier gain mechanism

an advantage over the GEN 1 cascaded tubes with respect to overloading by bright lights. This superior MCP system performance is achieved due to the combined effect of inherent MCP saturation and a "smart" power supply. Figure 15-7 illustrates the GEN 2 device showing the input photocathode, amplifying MCP, and output phosphor screen.

GEN 2 image amplification (Figure 15-7a) is achieved through the use of an MCP amplifying device, a wafer-thin slice of hollow glass tubes that have been fused into a mosaic of many hundred thousands of tubes. Each glass tube has a conductive inside surface with secondary emission characteristics that cause several electrons to be emitted when struck by a single electron, thereby producing gain of the input light signal, to output electron signal. When a dim

nighttime scene is focused onto the GEN 2 photocathode, electrons are dislodged and accelerated toward the inside of each glass tube. This effect causes an electron image (considering all the hollow glass tubes together) to pass through the MCP. Each electron striking the MCP inside tube wall causes emission of secondary electrons (gain). Thousands of these secondary electrons impinge on the MCP phosphor output screen, converting the electron image into a visible display on the screen. Standard MCP amplification (gain) ranges from 35,000 to 100,000.

The GEN 2 intensifier assembly with power supply and image inverter is much smaller than the GEN 1 (1.5 versus 6 inches long), since the MCP is only the thickness of a credit card (typically 0.06 inch). It is much lighter in weight than the earlier GEN 1 three-stage tubes (2 ounces versus

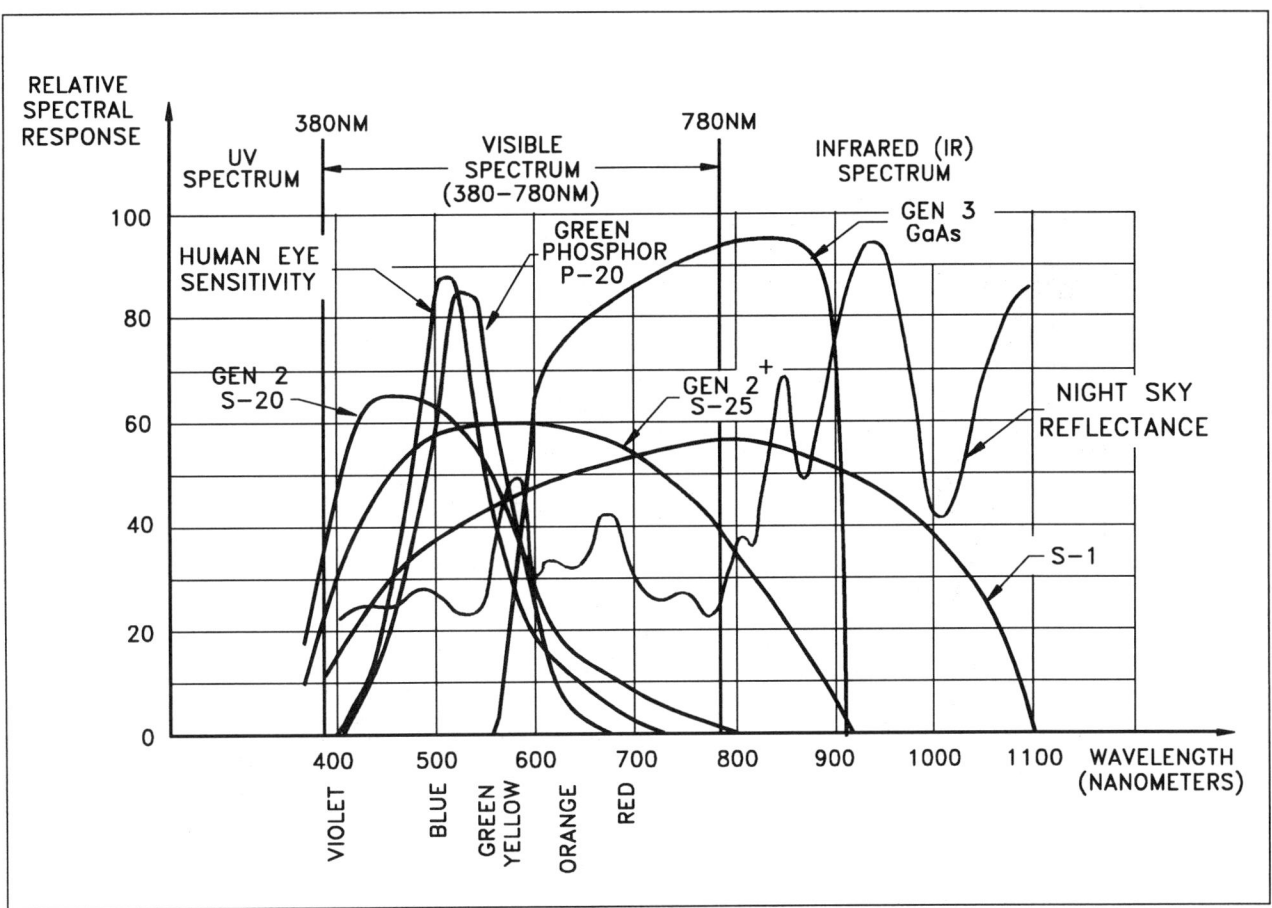

FIGURE 15-8 GEN 2 and GEN 3 MCP intensifier photocathode and phosphor screen spectral sensitivity

16 ounces). The GEN 2 device overcomes the problems of image flare and light overload caused by bright objects such as headlights, and the image distortion from the GEN 2 device is much lower than that of the GEN 1 device. Resolution for a 25-mm-diameter GEN 2 MCP image intensifier is about 28 to 35 line pairs per millimeter. The GEN 2 Plus is an improved GEN 2 with a spectral sensitivity extended further into the near-IR region, providing higher sensitivity and resolution.

There are two generic types of GEN 2 image intensifiers, distinguished chiefly by different focusing arrangements between the photocathode and the MCP (Figure 15-7b). The two types have many characteristics in common, since they both use the MCP to multiply the electron output of the photocathode to produce light gain. The type is chosen based on the expected use: for direct viewing by eye or for coupling to a CCTV camera. The electron-focusing type inverts the image, making it suitable for direct viewing, as it compensates for the inverted image produced by the objective lens. Although the proximity-focused type produces an inverted output image, directly suitable for television sensor applications, many systems use the electron-focusing type for CCTV when the CCTV camera is inverted. The proximity-focused output type, with the addi-

tion of a fiber-optic output coupler inverter, produces a right-side-up image if direct viewing is necessary. This fiber-optic twist increases the device length to approximately the same size as the electron-focusing version. In the proximity-focused device, the screen is placed close to the MCP so that the multiplied electrons travel only a short distance before they impact the screen. This minimizes spreading of the electrons, which would otherwise produce a subsequent loss of resolution.

Since most GEN 2 intensifiers have lifetimes of about 2000 hours, tube life is an important consideration. The most significant advantage of the focusing-electron structure over the proximity-wafer structure is its larger volume, which allows residual contaminants in the tube to spread out over a relatively large physical volume, thereby increasing the tube life.

15.6.1 Sensitivity

Image-intensifier photocathode sensitivity is expressed in two ways: (1) luminous sensitivity (sensitivity compared with white light) and (2) radiant sensitivity (sensitivity to near-IR radiation).

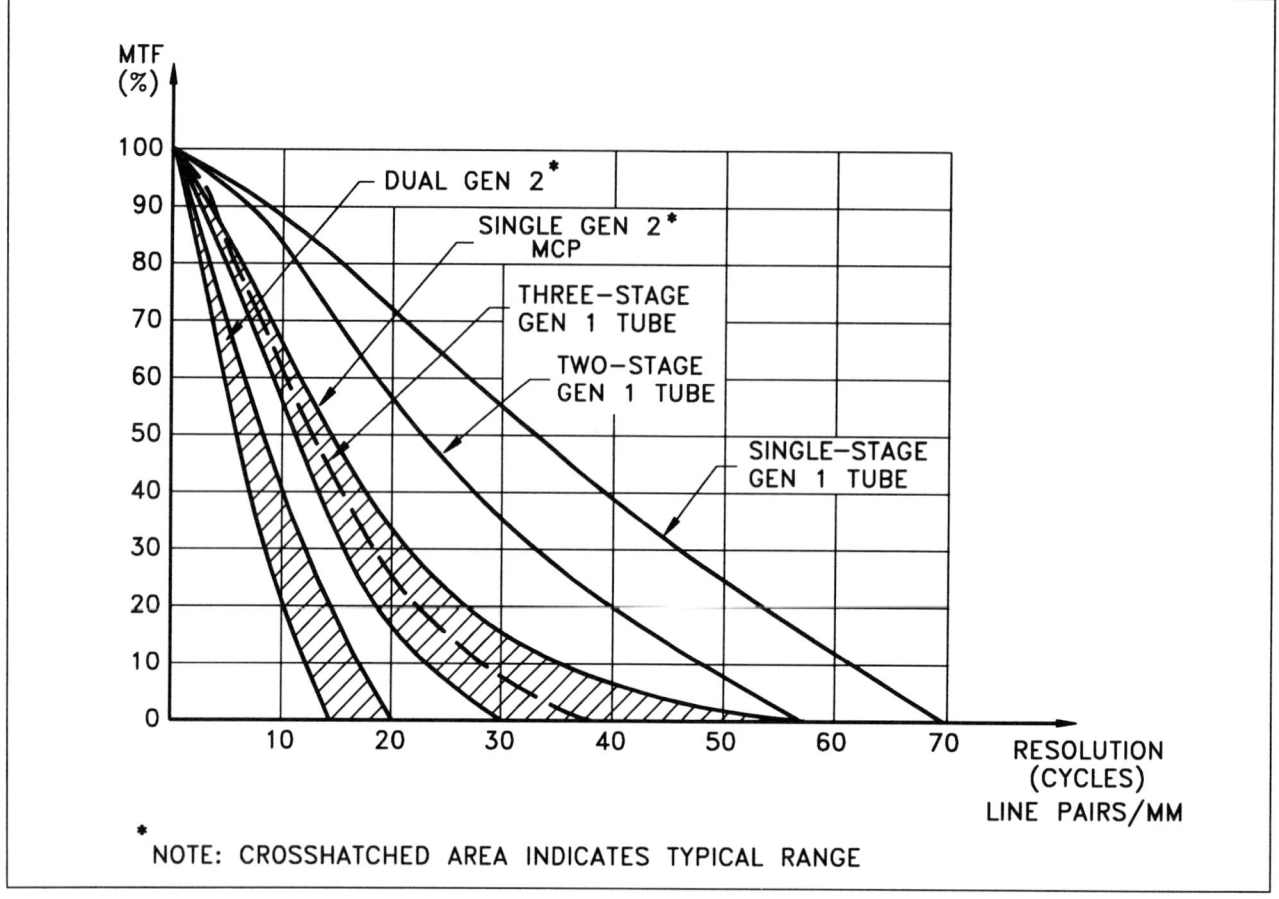

FIGURE 15-9 Intensifier modulation transfer function

GEN 2 intensifiers use S-25 (S-20ER, extended range) photocathodes, and GEN 3 (Section 15.7) use GaAs photocathodes for increased IR response and higher tube sensitivity. It can be seen that the GaAs photocathode has maximum response (sensitivity) in the red end of the visible and near-IR spectral region (850–920 nm), where the night sky illumination peaks. This higher efficiency of GaAs and spectral peaking account for the higher sensitivity of the GEN 3, GaAs intensifier. Figure 15-8 shows the spectral sensitivity of the different photocathode surfaces available for image intensifiers.

Figure 15-8 also shows the spectral output characteristics of the image-intensifier phosphor screen used in GEN 2 and 3 intensifiers. A typical night sky reflectance is shown, indicating the improved spectral match to the GEN 3 photocathode sensitivity.

The P-20 phosphor material chosen for the image-intensifier output phosphor screen optimizes the energy coupling of the light output to the spectral response of the eye or the CCTV sensor. The P-20 phosphor, which is yellow-green, is most prominently used for direct viewing applications because the spectral output is highest at the maximum sensitivity of the human eye. P-20 is also used to match the spectral sensitivity of the CCTV tube or solid-state sensor.

15.6.2 Resolution

Television system resolution is measured in TV lines. The resolution of photographic film, image intensifiers, and other nonscanning imaging tubes is measured by its modulation transfer function (MTF). An intensified system uses an image intensifier (nonscanned) and a CCTV camera (scanned); since the output is a television signal, the resolution is measured in TV lines. Manufacturers' data sheets may specify MTF and/or TV lines. The limiting resolution value commonly specified in image-intensifier data provides one single point on the MTF characteristic curve. Figure 15-9 gives some typical MTF characteristics for several image-intensifier tube types.

These data are for GEN 1 single-stage (brightest resolution and MTF), two-stage, and three-stage electrostatically focused image tubes with 18-mm-diameter size, and GEN 2 and GEN 3 MCP. As can be seen, coupling three stages of

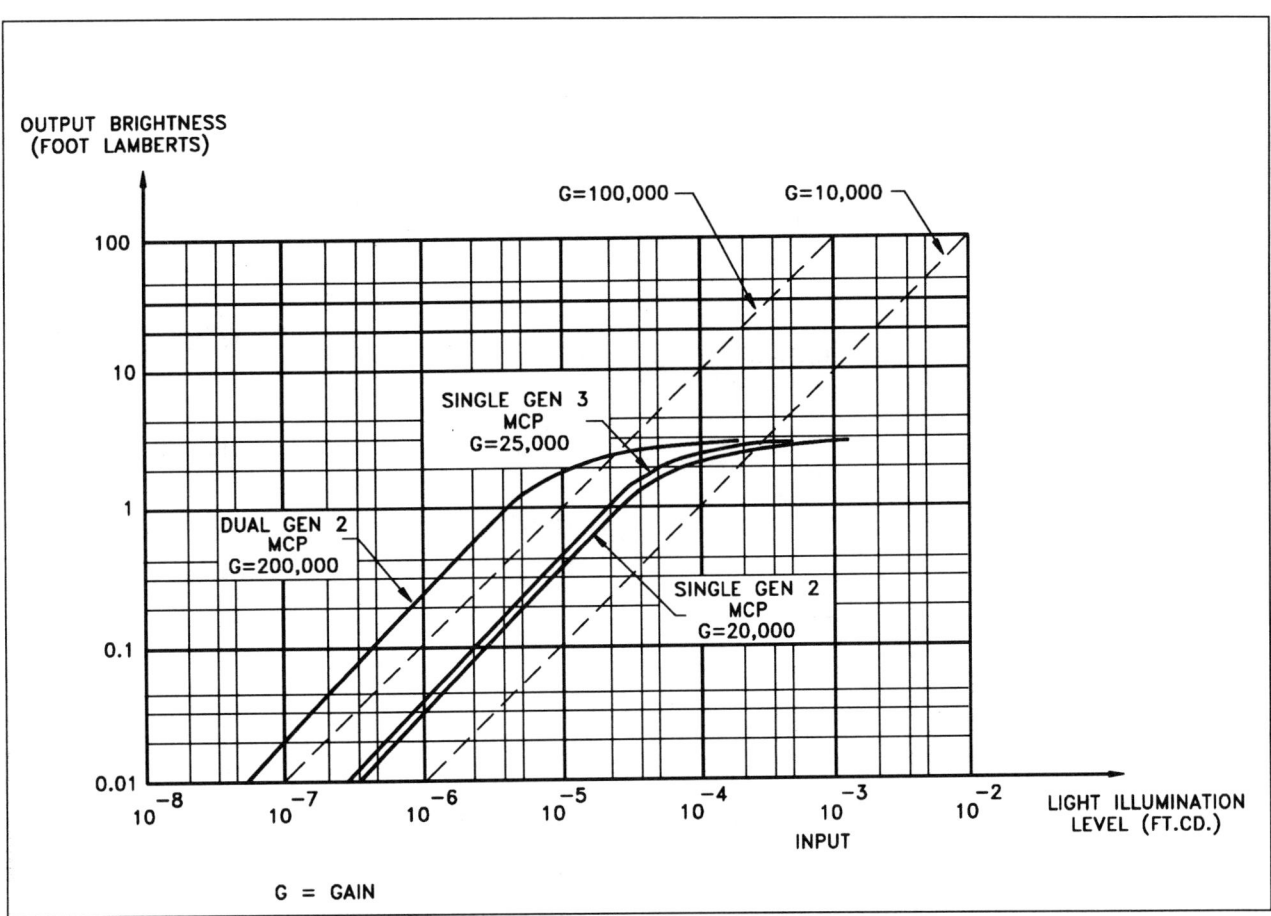

FIGURE 15-10   MCP intensifier gain saturation curve

intensification decreases the MTF and the resolution of the tube approximately to the level of a GEN 2 MCP. Notice that the GEN 2 and GEN 3 MCPs have noticeably lower MTF than the GEN 1 single-stage tube.

15.6.3 Light Overload

The typical real-world night environment is filled with light sources: flares, flashes, explosions, vehicle headlights, street lights, and others. GEN 1 devices exhibit considerable image persistence that results in image smear and often total obliteration of a scene when there are moving sources of light in the scene. Image blooming from point sources of light is common. Considerable effort was devoted to overcoming the deficiencies of GEN 1 devices, but with limited success. The result was the development of the GEN 2 MCP intensifier, which has far superior light-overload characteristics and has rapidly replaced the GEN 1 devices in military and commercial applications.

The MCP intensifier improves the scene contrast ratio and suppresses highlights caused by sun-glint or bright lights. The GEN 2 exhibits an inherent highlight saturation effect—highlights are suppressed before they reach the camera sensor. The MCP saturation is a localized effect in that individual channels operate essentially independently of each other. Although there is a small amount of crosstalk between channels, this localized saturation greatly aids in improving "white-out" image blooming that severely affects the three-stage GEN 1 device. The resulting GEN 2 and GEN 3 pictures have a good contrast ratio, enabling fine detail to be seen in the shadow areas of a bright source.

The MCP intensifier operates from 10^{-5} fc (10^{-4} lux) up to 0.1 fc (1 lux) with no additional automatic light control system required for the CCD sensor. Figure 15-10 shows the gain saturation curve for the MCP intensifier.

In the range from 10^{-7} fc (10^{-6} lux) to 3×10^{-5} fc (3×10^{-4} lux) faceplate illumination, the device has constant gain, but from 10^{-4} fc (10^{-3} lux) to approximately 10^{-1} fc (1 lux), the tube saturates: the output screen luminance remains constant even though the faceplate illumination increases. This nonlinear gain characteristic from low-light to highlight level accomplishes the highlight suppression, resulting in an improved image.

The GEN 2 intensifiers have a remarkable lack of persistence or image smear compared with the three-stage GEN 1 devices. Although the screen phosphors used in both generations have a rapid decay of light output to a low level

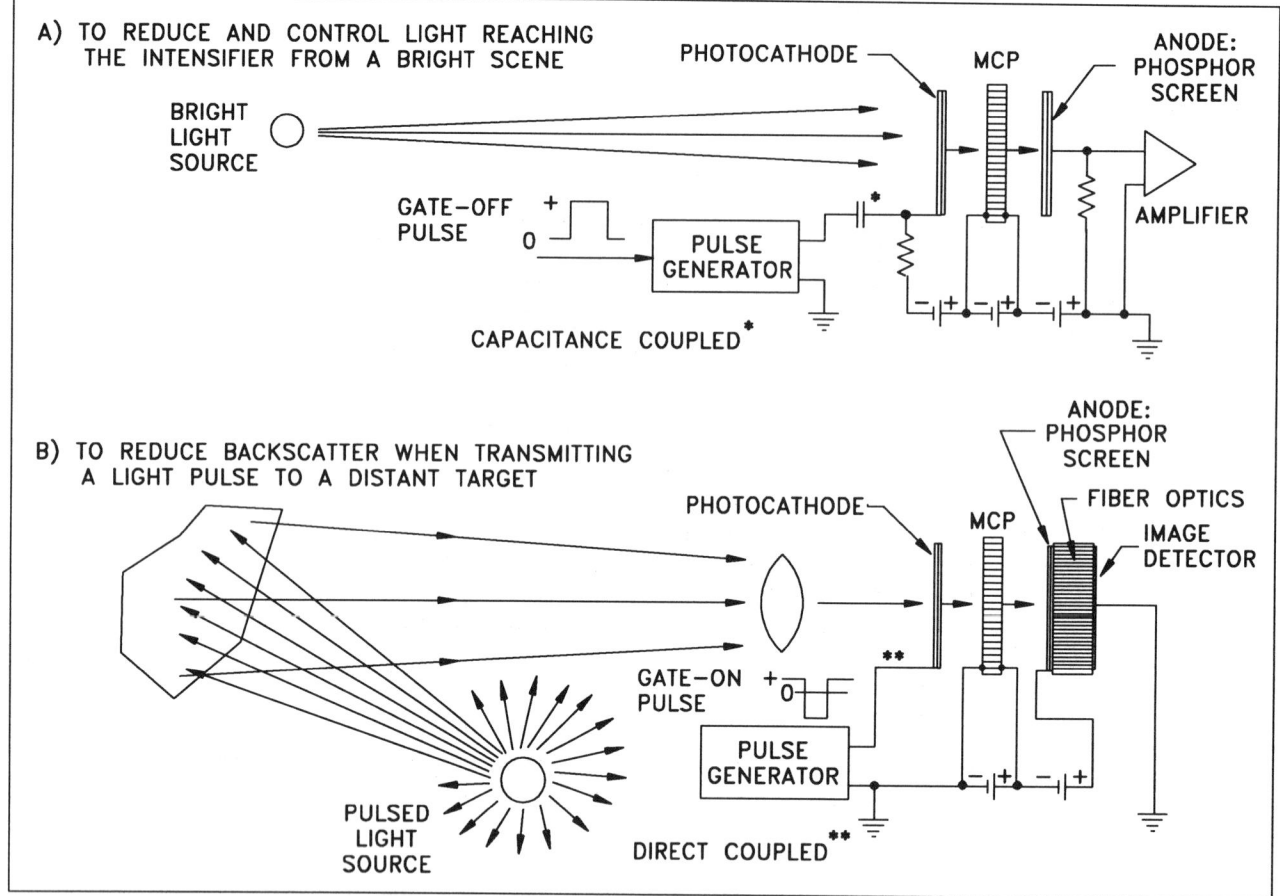

FIGURE 15-11 Range-gated intensifier system

when light excitation is removed, they have long persistence to final extinction. In GEN 1 intensifiers, this "tail" of the first-stage phosphor is amplified by the second and third stages, with the result that the output image smears, or "comet-tails" on moving bright sources. Since GEN 2 tubes require only one phosphor, no persistence is apparent.

15.6.4 Gating MCP Image Intensifier

Image intensifiers can be operated under a condition called gating, for the following reasons: (1) to reduce and control light reaching the passive intensifier system from a bright scene and (2) to actively transmit a light pulse to a distant target and detect light returned from it (active system) without being overloaded from near-field backscatter, which seriously degrades the picture. Figure 15-11a shows the arrangement to protect the intensifier and improve picture quality.

To control the amount of light (photons) being amplified, the MCP intensifier is electrically gated on and off for short, controlled periods of time. The higher the scene illumination, the shorter the period of time the MCP is

amplifying. This time-controlled automatic gain control keeps the brightness of the MCP output phosphor screen relatively constant. It also has the beneficial effects of (1) protecting the device from high-light-level damage, (2) extending tube life, and (3) improving sensitivity. For shuttering applications, the intensifier gain can be turned off by simply sensing a high light condition, say daylight condition, to turn off the MCP. This gating technique increases tube life substantially, and most LLL cameras have this protective feature.

LLL cameras can be used in an active gated mode by using a pulsed light source (filtered xenon lamp, LED, or laser) and gating the GEN 2 intensifier (Figure 15-11b). In the application of range-gating with a pulsed laser light source under conditions of fog, if the approximate range of the target is known, the intensifier is gated off electrically while the light pulse is being transmitted; the intensifier is turned on only when the light pulse reaches the target. The intensifier responds only to the reflected light from the target. This significantly improves the target-return-to-foreground-clutter ratio as compared with a continuous mode of operation. Range-gating reduces backscatter from near-range reflection caused by fog, precipitation, or other atmospheric aerosols or particles. However, range-gating makes the system active, which is a disadvantage in covert applications.

15.7 GEN 3 IMAGE INTENSIFIER

Military requirements and funding brought about the development of the GEN 3 intensifier. The primary improvements sought and achieved were (1) increased GEN 2 device sensitivity to IR illumination, which becomes more available as the natural light outdoor illumination decreases (Figure 15-1), (2) increased resolution, and (3) increased tube life. GEN 3 image intensifiers are structurally similar to GEN 2 except that the photocathode is constructed of high-quantum-efficiency GaAs. This material increases the intensifier sensitivity to IR radiation due to its high responsivity in the near-IR. Figure 15-8 shows the differences in spectral sensitivity and response of GEN 2 photocathodes constructed from S-20, S-25 (S-20ER), improved GEN 2 Plus multi-alkali, and GEN 3 gallium arsenide–cesium oxide (GaAsCsO) materials. The GEN 3 has a maximum responsivity in the 600-to-900-nm wavelength region, a better match to the night sky reflectance spectrum.

Several manufacturers produce a GEN 3 intensifier with an 18-mm photocathode and a P-20 phosphor screen with noninverting fiber-optic output window. The intensifier is combined with CCTV cameras either by lens coupling or with a fiber-optic faceplate and fiber-optic image taper to a fiber-optic faceplate silicon tube or CCD sensor (Section 15-8).

The GEN 3 image intensifier using state-of-the-art GaAs technology delivers significantly improved performance over GEN 2 by providing greater sensitivity in the near-IR spectrum where night illumination is more abundant. The GEN 3 device triples the photocathode sensitivity compared with GEN 2 and permits it to operate down to starlight illumination levels. The increased IR sensitivity is significant because scene target reflectivity from a variety of materials increases or is higher in the IR region compared with the visible spectrum. Likewise, the natural sky spectral radiation on the earth's surface increases in the higher IR wavelengths. GEN 3 intensifiers nearly quadruple tube life compared with GEN 2—from 2000 hours to approximately 7500 hours. GEN 2 and GEN 3 tubes incorporate automatic brightness control to provide constant image brightness under varying light-level conditions and bright source tube protection during exposure to high light levels.

15.8 INTENSIFIED CCD CAMERA

A system currently receiving much attention is the intensified CCD (ICCD), a GEN 1, 2, or 3 image intensifier coupled to a CCD or other solid-state sensor. The combination of the GEN 2 or GEN 3 image intensifier with the small solid-state camera provides an extremely small, sensitive, lightweight, low-power LLL device. The ICCD sensor with GEN 2 or GEN 3 compares favorably with the SIT tube in sensitivity, with somewhat lower resolution. One system uses the high-resolution capabilities of the GEN 1 single-stage tube intensifier coupled to a high-resolution CCD image sensor. In a more sensitive configuration, the GEN 2 MCP intensifier coupled with a CCD sensor produces an LLL camera system with a sensitivity comparable to that of the SIT camera. There are several technologies to couple the intensifier output to the CCD sensor (Figure 15-12).

The two coupling techniques used are optical relay lenses and fiber optics (1:1 magnification or reduced).

The following sections describe a GEN 1 intensifier coupled to a CCD sensor, and techniques for coupling the GEN 2 and GEN 3 intensifiers to a CCD sensor.

15.8.1 GEN 1 Coupled to CCD Camera

A single-stage GEN 1 electrostatically focused intensifier is used in two ways with CCTV cameras: (1) as a low-gain image intensifier and (2) to demagnify the GEN 2 or GEN 3 output image without resolution loss. While the GEN 1 is not as sensitive as a GEN 2 or GEN 3 intensifier tube, because of its excellent resolution the GEN 1–coupled, CCD-intensified camera is an excellent replacement for SIT cameras when resolution is of prime importance. This combination produces an increase in low light sensitivity by a factor of approximately 10 to 100 over that of the CCD camera, and approaches the sensitivity of a SIT tube camera. High resolution is achieved with the use of a $\frac{2}{3}$-inch format CCD having 800 (H) × 490 (V) pixels.

Figure 15-13 shows a diagram of the GEN 1–intensified CCD camera using fiber-optic coupling.

15.8.2 GEN 2 Coupled to CCD Camera

Higher gains and sensitivities can be achieved by substituting a GEN 2 MCP image intensifier for the GEN 1 intensifier. With a GEN 2 device having a gain between 20,000 or 30,000 times, and using a 4-to-1 tapered fiber-optic bundle reducer (Section 15.8.2.2), the overall gain improves over a CCD camera by a factor of 5000 to 7500. Several examples of this new-generation LLL camera are described next.

15.8.2.1 Relay Lens Coupling

The simplest technique for coupling the output image from the GEN 1, GEN 2, or GEN 3 image intensifier to the CCD television sensor is to use an optical relay lens having a demagnification of about 0.44 and 0.60 for the $\frac{1}{2}$- and $\frac{2}{3}$-inch format cameras, respectively (Figure 15-14).

The amount of demagnification is determined by the image-intensifier output diameter and the CCD sensor size.

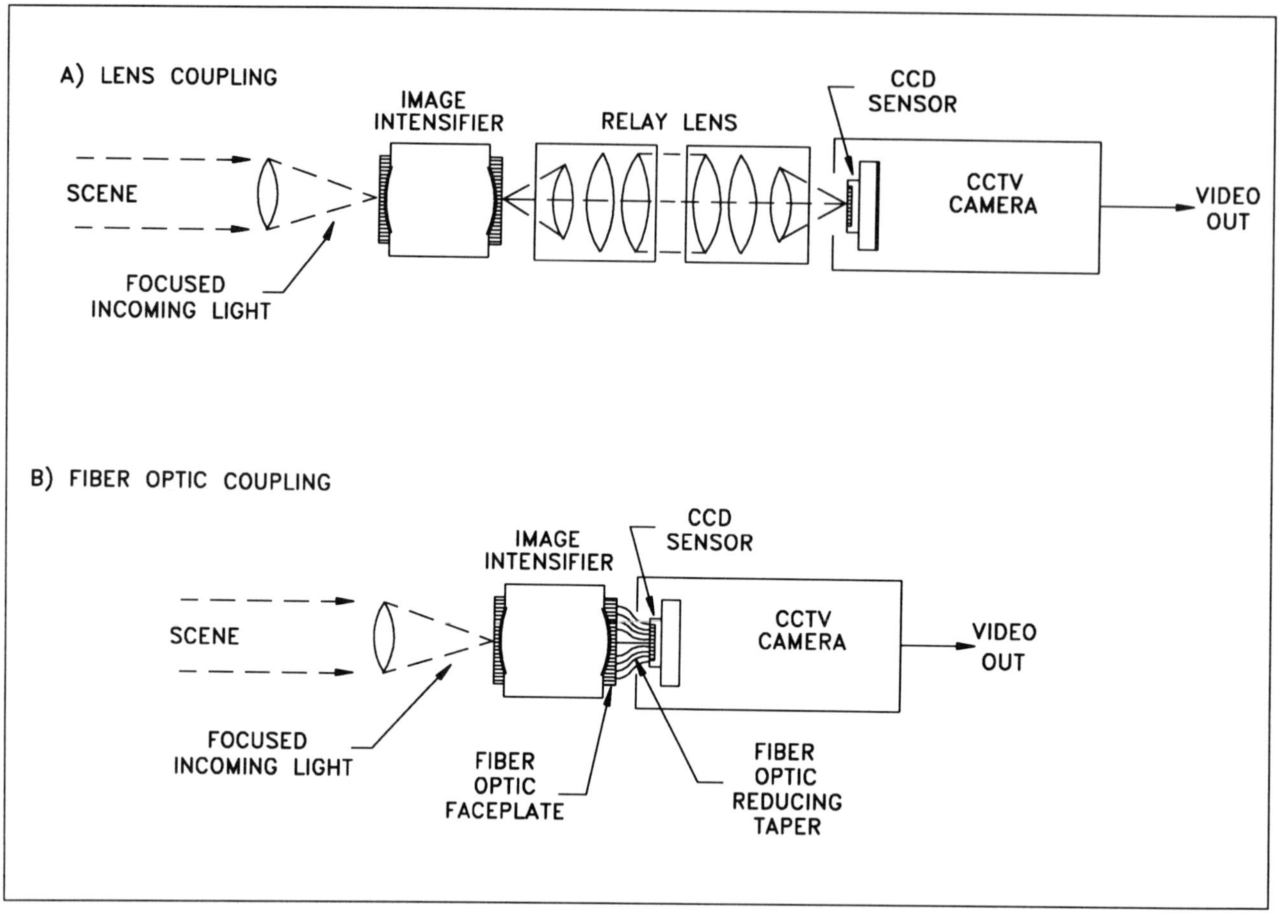

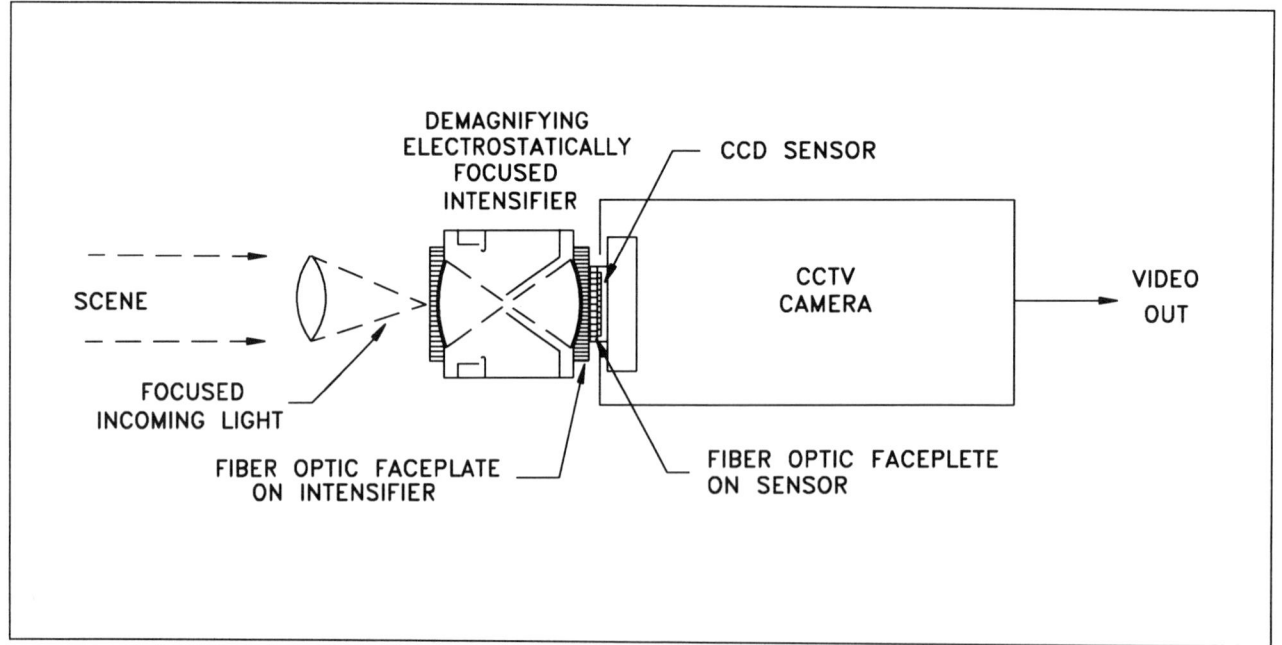

FIGURE 15-12   Methods for coupling intensifiers to CCTV sensors

FIGURE 15-13   GEN 1 intensifier fiber-optically coupled to CCD

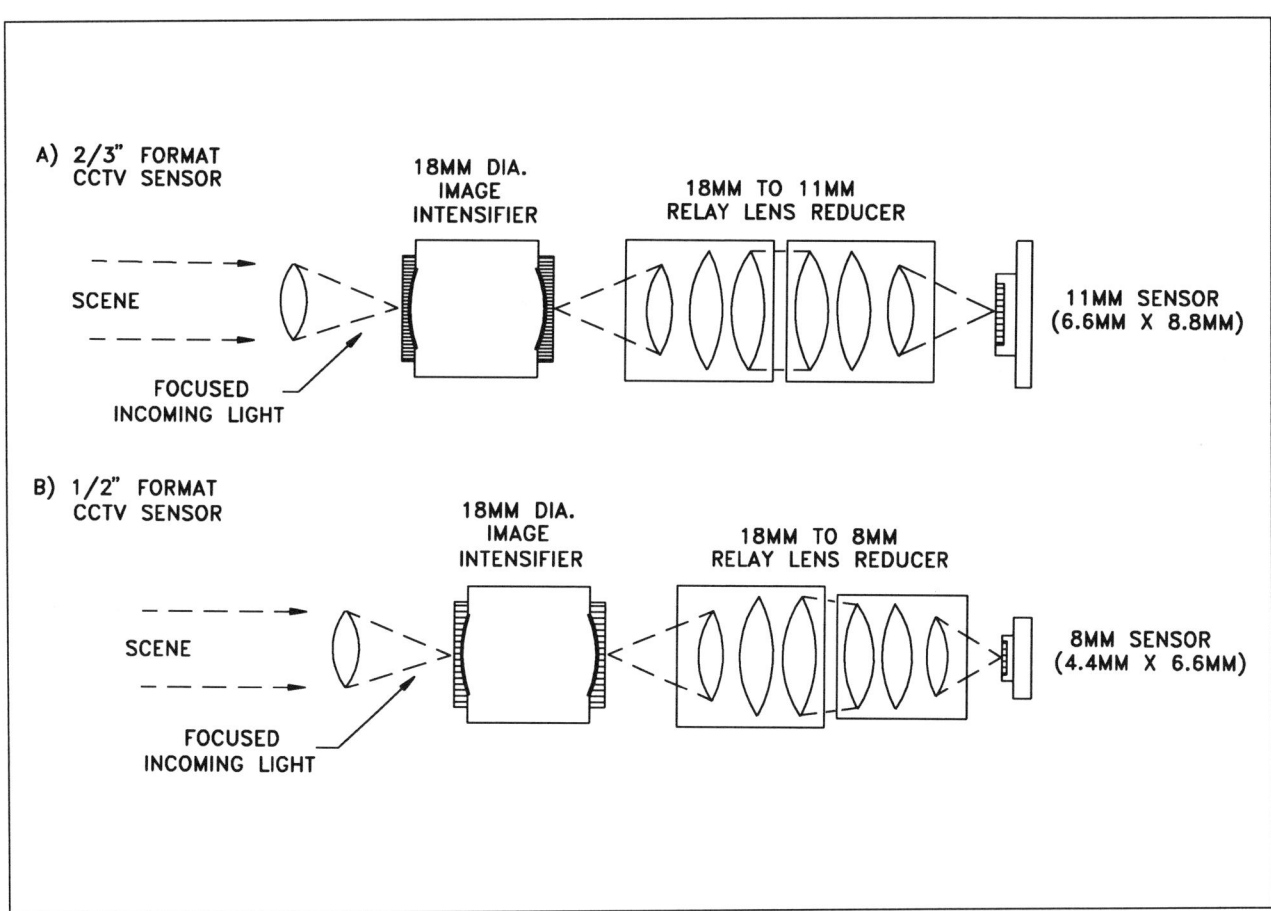

FIGURE 15-14 Relay lens coupling technique

For a standard 18-mm-diameter intensifier screen, the optimum image demagnification to a ⅔-inch (11-mm diagonal) format CCD (6.6 × 8.8 mm) is 0.6 and for the ½-inch (8-mm diagonal) format (4.8 × 6.4 mm) is 0.44. While lens coupling is simple to achieve, relay lenses are large in size, and optical coupling is not as efficient as fiber-optic coupling. A typical relay lens system uses an optically fast 25-mm lens in tandem with a 50- or 75-mm lens to demagnify the intensifier phosphor screen and refocus it onto the CCD sensor. The mechanical lens assembly maintains the critical alignment between the intensifier tube and CCD imager so that a sharp image of the intensifier phosphor screen is imaged onto the CCD sensor. An advantage of the lens coupling technique is the ability to replace any of the three components—independently—should one of them fail.

15.8.2.2 Fiber-Optic Coupling

A smaller and more light-efficient technique for coupling a GEN 1, GEN 2, or GEN 3 intensifier output screen to the CCD imager is to use a coherent fiber-optic bundle between the intensifier phosphor screen and the CCD sensor (Figure 15-15).

A coherent fiber-optic bundle is a closely packed (fused) group of several hundred thousand very fine glass fibers (one-tenth the diameter of a human hair) forming an image-carrying conduit. A tapered fiber-optic conduit is a fused bundle in which the entire bundle and fiber diameters change size from one end to the other. The tapered bundle introduces the additional feature of minifying the 18-mm transmitted intensifier image to match the smaller ½- or ⅔-inch CCD sensor format. Figure 15-15 shows two different minifying fiber-optic bundles having reducing ratios of 18 to 11 and 18 to 8. In this design, the coupler input side matches the diameter of the intensifier screen and the output side matches the CCD sensor format size. Tapered fiber optics have the following advantages over lens systems: (1) they are typically five times shorter in length than an equivalent lens system, (2) they result in a sturdy compact construction, and (3) their more efficient coupling permits operating the intensifier at a lower gain and therefore a higher signal-to-noise (S/N) ratio. A disadvantage is that if either the intensifier or the CCD fails, the entire assembly must be replaced.

Although each fiber transmits its portion of the scene from the intensifier phosphor to the CCD array with a loss

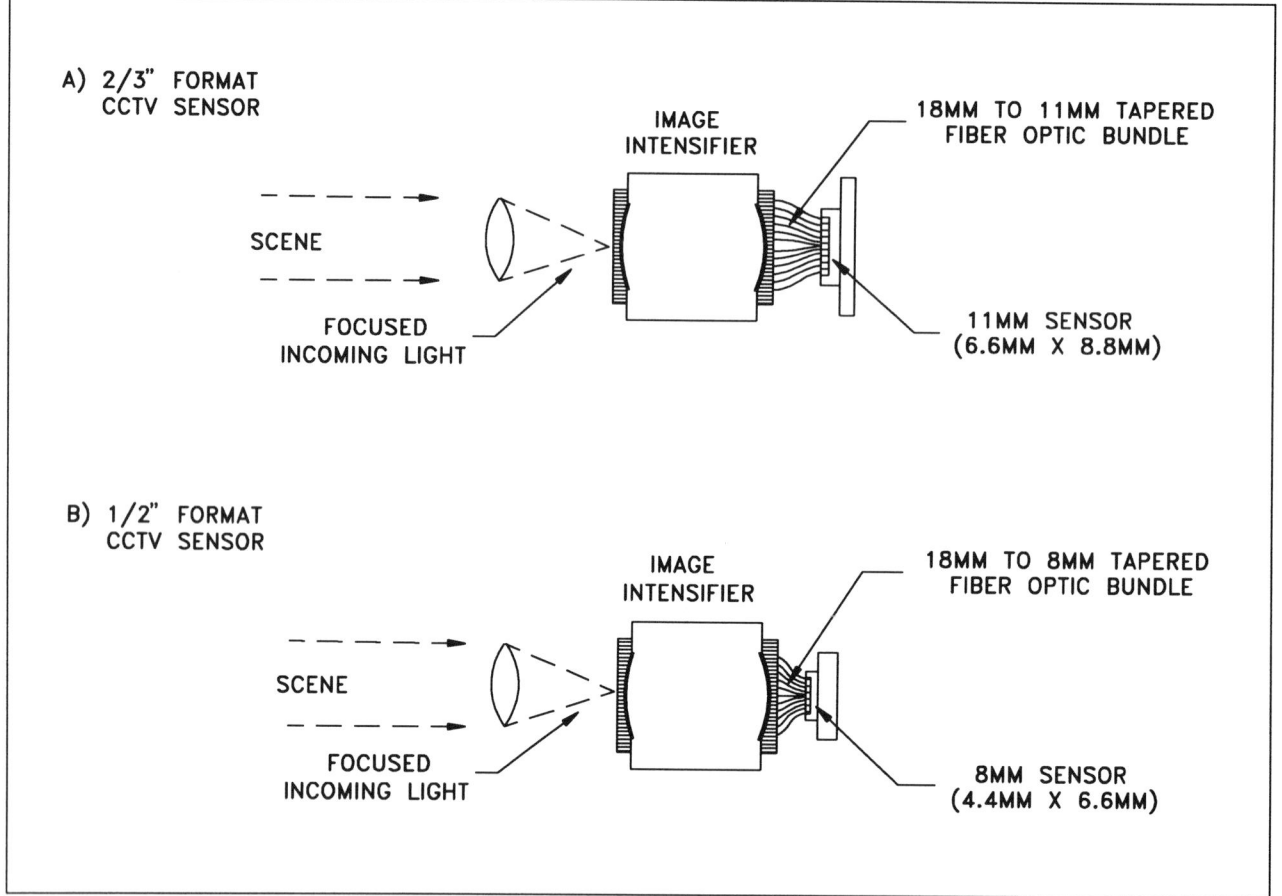

FIGURE 15-15 Fiber-optic-coupled intensifier

factor of about 4, this is substantially better than the relay lens coupling technique. Therefore, if a single-stage GEN 1 intensifier produces a gain of 100, the ICCD light output is 25 times the light intensity of the original scene. This is a significant improvement over the standard CCD imager and provides a means for increasing the sensitivity of this device with a lag-free image.

As an example, a GEN 2 intensifier fiber-optically coupled to a frame transfer (FT) device CCD has a sensitivity of 3.7×10^{-6} fc (4×10^{-5} lux) for full video output and 1.39×10^{-6} fc (1.5×10^{-5} lux) for minimum observable image. The resolution for this configuration is 260 TV lines. With a GEN 2 type and a 16-mm/7-mm fiber-optic reducer, a resolution of 425 TV lines is obtained.

Summarizing the relay lens and fiber-optic coupling methods, the latter permits a reduced package size because of the short fiber length as compared with the longer lens system. The straight or tapered fiber-optic coupler is approximately 20 millimeters long compared with a typical relay lens of 100 millimeters. The advantage of the lens coupling method is reduced cost, and the ability to replace the intensifier or CCTV camera in the event of failure, replacements, or upgrades, as it does not require sacrificing the entire assembly.

15.8.3 Comparison of Intensifier Sensitivities

Figure 15-16 shows the available light level under different conditions, from bright sunlight to starlight, and the sensitivities of SIT, ISIT, and ICCD cameras.

Across the top of the figure are the types and levels of illumination sources available, ranging from full sunlight to clear starlight conditions.

15.9 SIT AND INTENSIFIED SIT CAMERAS

Since the appearance of the first television camera tube, the trend has been to develop tubes with greater sensitivity to operate at lower and lower levels of faceplate illumination, yet maintain or improve image resolution and quality. From the first image orthocon in 1944 to tubes with built-in image intensifiers in the 1950s, and with fiber-optic coupling in the 1960s, evolved the first generation of practical LLL security cameras in the form of the silicon diode tube and the SIT, based on the silicon tube in the 1970s. The development of the SIT tube made other devices such as the intensified SIT and intensified CCD possible. The ISIT is still considered the highest-resolution, lowest-light-level im-

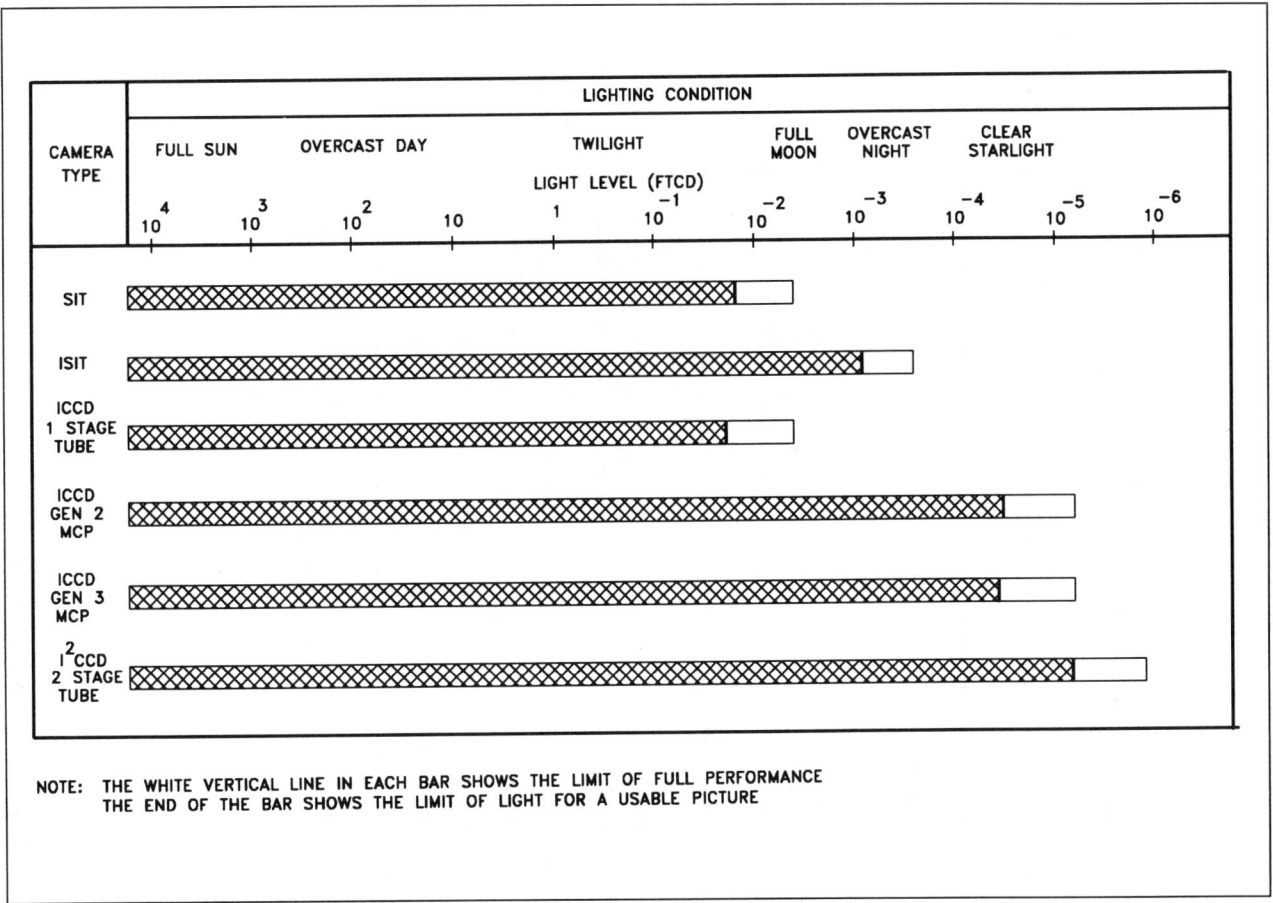

FIGURE 15-16 Illumination levels and intensified camera sensitivities

aging device available. The SIT and ISIT cameras offer high static resolution (nonmoving targets) under very LLL conditions and presently play a vital role in the U.S. government's ground border interdiction electron-optical surveillance programs, as well as in the important function of producing imaging for airport X-ray security equipment to detect contraband and other illegal materials.

SIT and ISIT cameras offer 100 to 1000 times improvement over the imaging power of the best CCD and tube (silicon, Newvicon) cameras. LLL cameras intensify light, while tube and CCDs only detect it. GEN 1, GEN 2, and GEN 3 intensifiers coupled to CCD sensors are rapidly replacing the SIT cameras.

15.9.1 Sensitivity/Gain Mechanism

The SIT image-intensifier tube has a photocathode, an electron-optic image-forming system (electrostatic or magnetic), and a phosphor screen that is excited by the photoelectrons (Figure 15-17).

The light output from the image tube is fiber-optically coupled to the fiber-optic faceplate input of the silicon camera tube photocathode.

The SIT camera tube uses a photocathode as the prime sensor, and a silicon-diode array target to produce gain by impact of the photoelectrons and a scanning beam to produce an output signal at the target. The most unusual feature of this tube is the silicon target, which produces electron gains of 3000 or more. The target consists of a two-dimensional PN-junction diode array with each diode facing the electron scanning beam of the tube, similar to the vidicon and silicon tubes. The silicon diodes constitute the storage mechanism required for holding the image while the electron beam is scanning the array. The ISIT combination (fiber-optic coupled image intensifier tube as input to the SIT tube) brings the sensitivity very close to the theoretical limit. Table 15-1 summarizes the sensitivity, gain, and resolution characteristics of commercially available tube and MCP intensifier devices.

15.9.2 Resolution and Image Lag

The SIT and ISIT camera resolution is limited by the 1-inch silicon diode tube, which is about 700 TV lines.

Resolution is only one of the important parameters in an LLL system. The second is image lag, which becomes worse

TYPE	AMPLIFYING MECHANISM	TYPICAL GAIN [3]	INTENSIFIER ACTIVE AREA DIAMETER (MM)	PHOTOCATHODE MATERIAL	TUBE LIFE MTBF [6] (HOURS)	RESOLUTION LINE PAIRS/MM (TV LINES)	WAVELENGTH SENSITIVITY (NM)	FACEPLATE SENSITIVITY (FTCD) MINIMUM	FACEPLATE SENSITIVITY (FTCD) FULL VIDEO
GEN 0	NONE (REFERENCE)	CONVERTS IR TO VISIBLE	25	S-1	—	50	700-1200	—	—
GEN 1 [1] 1-STAGE	ELECTRON TUBE	75-125	25	S-1, S-20 MULTI-ALKALI	50,000	50	400-850	—	—
2-STAGE	ELECTRON TUBE	2,500	25			40			
3-STAGE	ELECTRON TUBE	65,000	25			30			
GEN 2 [1]	MCP [2]	20,000	25	S-25 (S-20ER)[5] TRI-ALKALI	2,000	32-35	350-900	—	—
GEN 3 [1]	MCP	25,000	25	GALLIUM ARSENIDE (GaAs)	7,500	34-38	550-900	1×10^{-6} [7]	—
SIT	SILICON INTENSIFIED TARGET	1,600 [4]	16	MULTI-ALKALI	20,000	(700)	350-650	1×10^{-5}	—
ISIT	INTENSIFIED SIT	90,000 [4]	16	MULTI-ALKALI		(600)	350-650	1×10^{-6}	—
ICCD	GEN 2 MCP / GEN 3 MCP	3,000-6,000	18	S-25 / GaAs	2,000 / 7,500	(400) / (500)	350-900	1×10^{-6}	1×10^{-6}
ICCD	GEN 1 TUBE	100	18	S-20 S-25	50,000	50-85 (500)	400-850	1×10^{-5} TO 2.5×10^{-6}	1×10^{-4} TO 2.5×10^{-5}
I²ICCD	GEN 1 TUBE PLUS DEMAGNIFIED GEN 1 TUBE	—	18	S-20, S-25	50,000	(500)	400-850	3×10^{-6}	6×10^{-6}

1. DIRECT VIEWING. INTENSIFIER ONLY–NO CAMERA
2. MICROCHANNEL PLATE
3. LIGHT OUTPUT FROM PHOSPHOR SCREEN/LIGHT OUTPUT AT PHOTOCATHODE
4. CURRENT GAIN
5. ER=EXTENDED RANGE
6. MEAN TIME BETWEEN FAILURE
7. REFERENCE 10^{-6} FTCD EQUIVALENT TO STARLIGHT CONDITION

NOTE: 3 STAGE GEN 1, SIT, AND ISIT CAMERAS PRODUCE MORE PICTURE BLOOMING AND LAG THAN MCP-ICCD COUNTERPARTS

Table 15-1 LLL Intensifier and Intensifier-Camera Characteristics

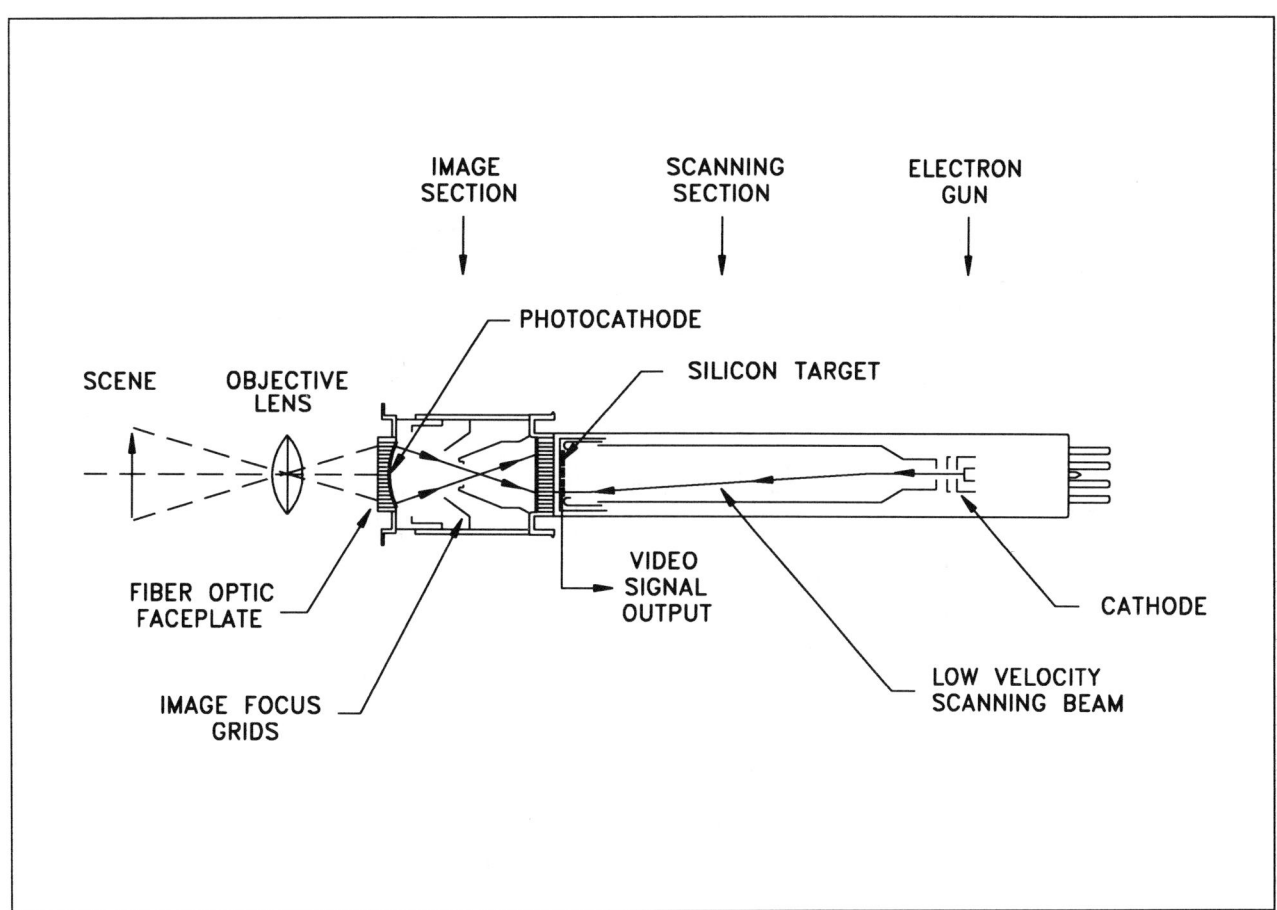

FIGURE 15-17   SIT image-intensifier diagram

as the signal level decreases. Lag is the residual picture scene or signal measured in the dark and is expressed as the percentage of the original signal present after three television fields ($3 \times \frac{1}{60}$th sec = $\frac{1}{20}$ sec) of scanning in the dark. SIT tubes exhibit significant "third" field lag. ICCD cameras show little or none of this image lag.

15.9.3 Light-Level Overload

Intensifiers are nighttime surveillance devices designed to work in near-dark conditions and are not intended for use in brightly illuminated park or street scenes at night, but rather poorly illuminated scenes, such as wooded areas, nonilluminated exterior perimeters, dark interior premises, or scenes at long ranges.

SIT and ISIT intensifiers are subject to the same precautions as other photosensitive tube pickup devices. Prolonged exposure to point sources of bright light results in permanent image retention and/or photocathode surface damage. Optimum intensifier use is obtained when scenes have a reasonably consistent light level throughout the scene. A shoreline under a cloudy, moonlit night or an unlit wooded area are excellent examples of intensified camera

applications. A parking lot or loading dock containing bright spotlights in some areas and total darkness in others can result in poor scene images unless the intensified camera views only the dark areas.

As the scene light level is reduced, scene blooming (the expansion of the white overload area due a bright light source at a location in the scene) caused by bright lights becomes a problem and is a phenomenon associated with most CCTV camera systems. It is particularly objectionable in LLL applications, where the scene contrast range is very small except for an occasional bright light or flash in the scene. When bright light blooming increases in the display, it obscures the picture information in it, causing a loss of intelligence. When the light intensity exceeds normal operating levels by a factor of approximately 1000 in SIT tubes, blooming becomes a problem. To reduce this objectionable overload, most SIT and ISIT cameras are equipped with automatic-iris lenses to compensate over the full dynamic range over which the cameras will be used. To protect the camera from dangerously bright objects, a shutter mechanism automatically closes the path of light reaching the sensor and protects the tube.

Imaging devices designed to operate under LLL conditions are not usually capable of imaging points of high

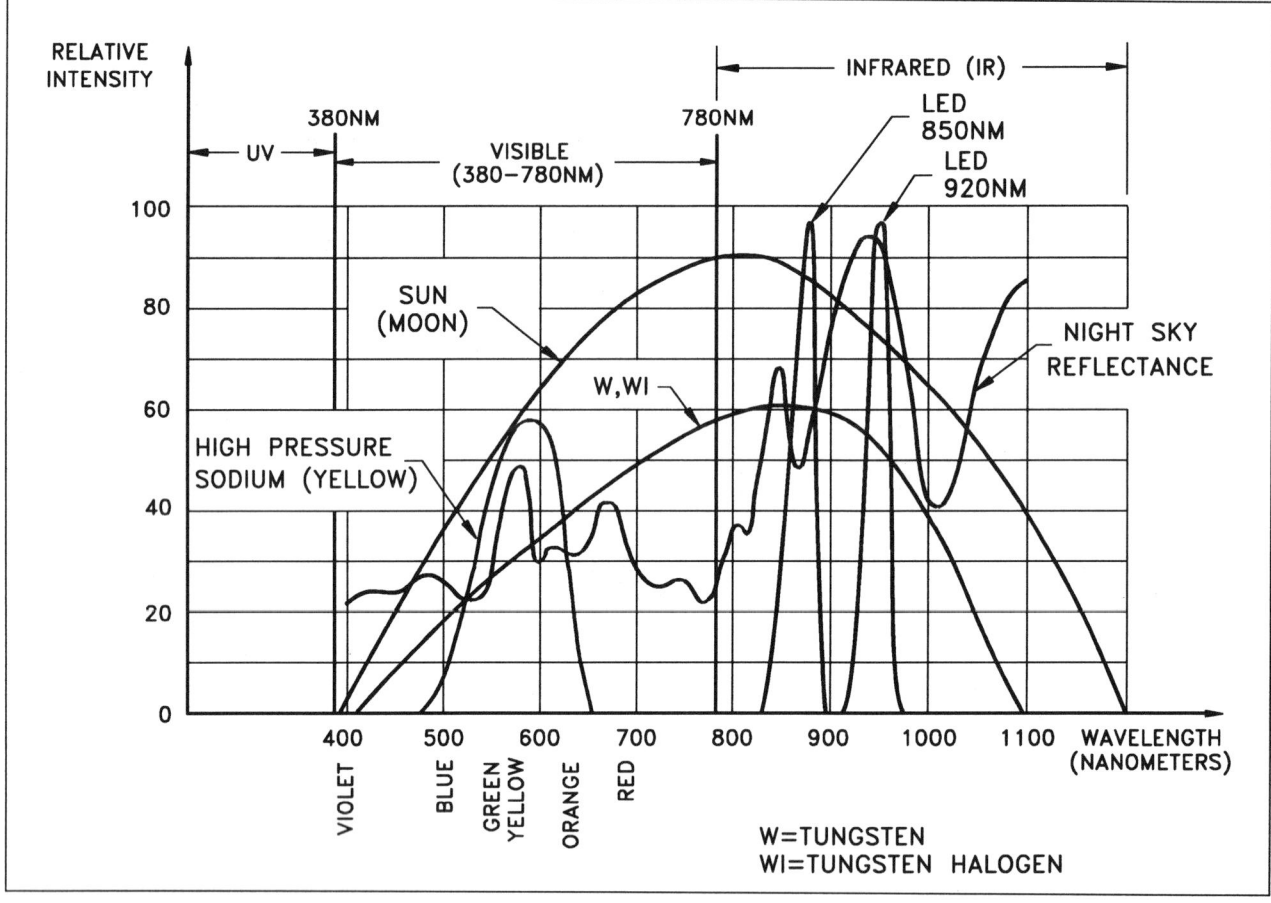

FIGURE 15-18   Infrared radiation from common sources

intensity within the LLL scene without severe spreading of the point source image into adjacent scene areas. Different LLL systems are capable of handling highlight objects in the scenes such as ground fires, flares, vehicle lights, or runway lights with different degrees of success. These bright light sources often contribute light levels much higher than collected from the remainder of the scene, while providing little illumination to the objects under surveillance. The energy emitted by these sources is spread out by the atmosphere and the intensifier optics. The result is a concentration of light spread over an area many times larger than the image of the point source in the scene, having the effect of obliterating detail over a large portion of the picture area. GEN 1 tube devices suffer from this phenomenon. GEN 2 devices have a built-in self-limiting saturation capability and therefore exhibit much less blooming.

GEN 2 intensifiers provide significantly improved overload immunity to unexpected bright lights from car headlights, street lights, and so on. The difference between three-stage GEN 1 and GEN 2 and GEN 3 intensifiers is best seen by actual viewing through these devices. The GEN 2 and GEN 3 devices exhibit superior performance over GEN 1 devices with respect to bright lights.

15.10 MULTISENSOR CAMERAS

Multisensor cameras are used in applications requiring daytime, dawn/dusk, and nighttime operation (Figure 15-18). During daytime operation, a standard monochrome or color camera is used. During dawn and dusk, a sensitive monochrome CCD camera is used. And during nighttime operation, an optically fast lens and LLL image intensified camera is used. An efficient use of a fast lens and image sensor switcher couples the appropriate camera to the single lens. The switchover is accomplished via remote control or by an automatic photocell. Above a predetermined light level, the daytime camera operates. When the light level drops below this level, the dawn/dusk camera operates, and at nighttime light level, the intensified camera operates. To ensure that damage does not occur to the intensifier photocathode, a neutral-density filter or automatic iris and a shutter mechanism are used. The multisensor system is designed to operate under moonlight, dawn/dusk, and bright sunlight conditions. The light level can increase from 10^{-5} to 10^{-1} lux during the night, from 0.1 lux to 100 lux at dawn, and up to 10^5 lux at midday, a total dynamic light range of 10 orders of magnitude.

CAMERA TYPE	SENSOR TYPE	PIXELS	OPTICS	COOLING	SENSITIVITY NETD** T (°C)	GRAY LEVELS	POWER CONSUMPTION (WATTS)	SPECTRAL RANGE (MICRONS)	VERTICAL TV LINES	FOV TOTAL (DEG)	IFOV ++ (MILLIRADIAN)
STARING FPA* SOLID STATE AREA ARRAY	PLATINUM SILICIDE	512x512	GERMANIUM FIXED FOCAL LENGTH OR ZOOM	STERLING* CYCLE	0.15	256 (8 BITS)		3–5	525	14x11 (50MM LENS)	
SCANNING MIRROR SOLID STATE LINEAR ARRAY	MERCURY CADMIUM TELLURIDE		GERMANIUM	STERLING CYCLE	0.18	4096 (12 BITS)	60	8–12	525	DUAL 20x13.2, 5x3.3	
INFRARED VIDICON	THERMAL VIDICON		GERMANIUM FIXED FOCAL LENGTH OR ZOOM	NONE, ROOM TEMPERATURE					525		
STARING FPA AREA ARRAY	MICROBOLOMETER	336x240 (80,000)	GERMANIUM FIXED FOCAL LENGTH OR ZOOM	NONE,+ ROOM TEMPERATURE	0.1		LOW	8–12	525	DUAL 15x9, 5x3	1, 0.33

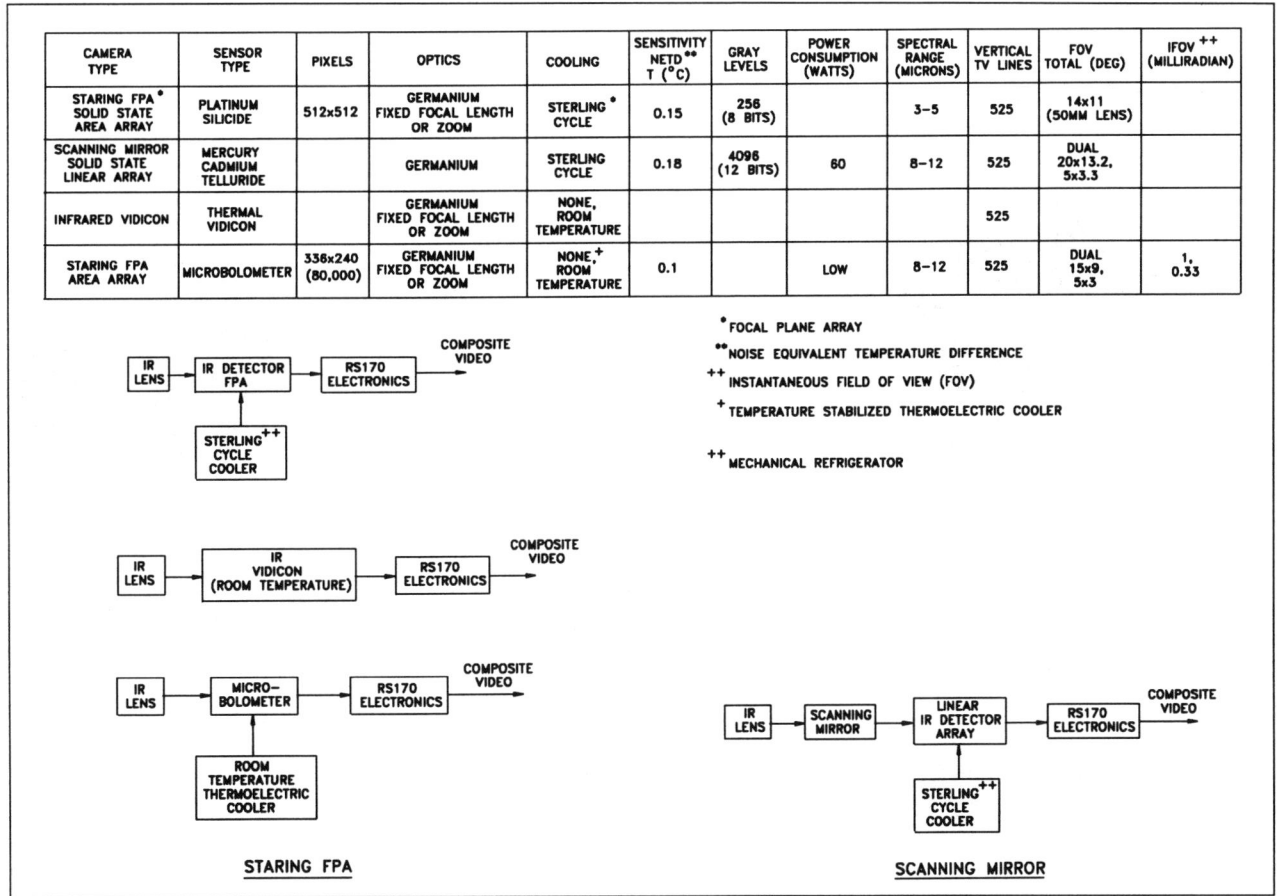

* FOCAL PLANE ARRAY

** NOISE EQUIVALENT TEMPERATURE DIFFERENCE

++ INSTANTANEOUS FIELD OF VIEW (FOV)

+ TEMPERATURE STABILIZED THERMOELECTRIC COOLER

++ MECHANICAL REFRIGERATOR

STARING FPA

SCANNING MIRROR

FIGURE 15-19 Thermal infrared video cameras

15.11 THERMAL-IR CAMERAS

The thermal-IR viewer is a night vision device that uses the difference in temperature of scene objects to produce a scene image. These passive devices require no light whatsoever and produce an image based solely on the *thermal temperature* difference of the objects in the scene. Unlike the image intensifier, thermal viewers can "see" in total darkness by detecting and displaying small temperature differences between objects and their backgrounds. Many objects (such as the human body, an automobile, a motor) emit considerable thermal radiation (heat) in the mid-IR wavelength region, which easily propagates through the atmosphere in the 3-to-5-micron and 8-to-12-micron electromagnetic spectrum (Figure 15-1).

Even objects of relatively low temperature (70°F) are "hot" enough with respect to a background of, say, 50°F to be detected by these sensitive IR devices. The lower resolution and very high cost of these devices have limited their use to military and other scientific applications. The thermal CCTV viewer uses an IR-transmitting germanium lens to collect and focus far-IR thermal images onto a special thermal detector array or thermal (pyroelectric) television tube. Figure 15-19 summarizes the parameters of four generic types of IR cameras.

The two most popular types are the staring focal plane array (FPA) and the scanning mirror. The FPA uses an IR lens, a cooled linear or area sensor (for example, a CCD), processing electronics, and a thermal cooler, making for an expensive, complex system. The platinum-silicide area sensor requires cooling to 77 degrees Kelvin (°K) for proper operation. The mercury-cadmium-telluride linear sensor requires the scene image to be scanned across it and cooling to 77°K. Both systems are very expensive, used in military and other government projects. The IR vidicon, as its name implies, is an IR-sensitive vidicon. It exhibits low resolution and has found limited use in surveillance. A newcomer is the room-temperature microbolometer staring FPA, which uses no cooler but only a temperature-stabilizing mount. This IR camera, when commercially available, will be a practical solution to many IR CCTV surveillance problems.

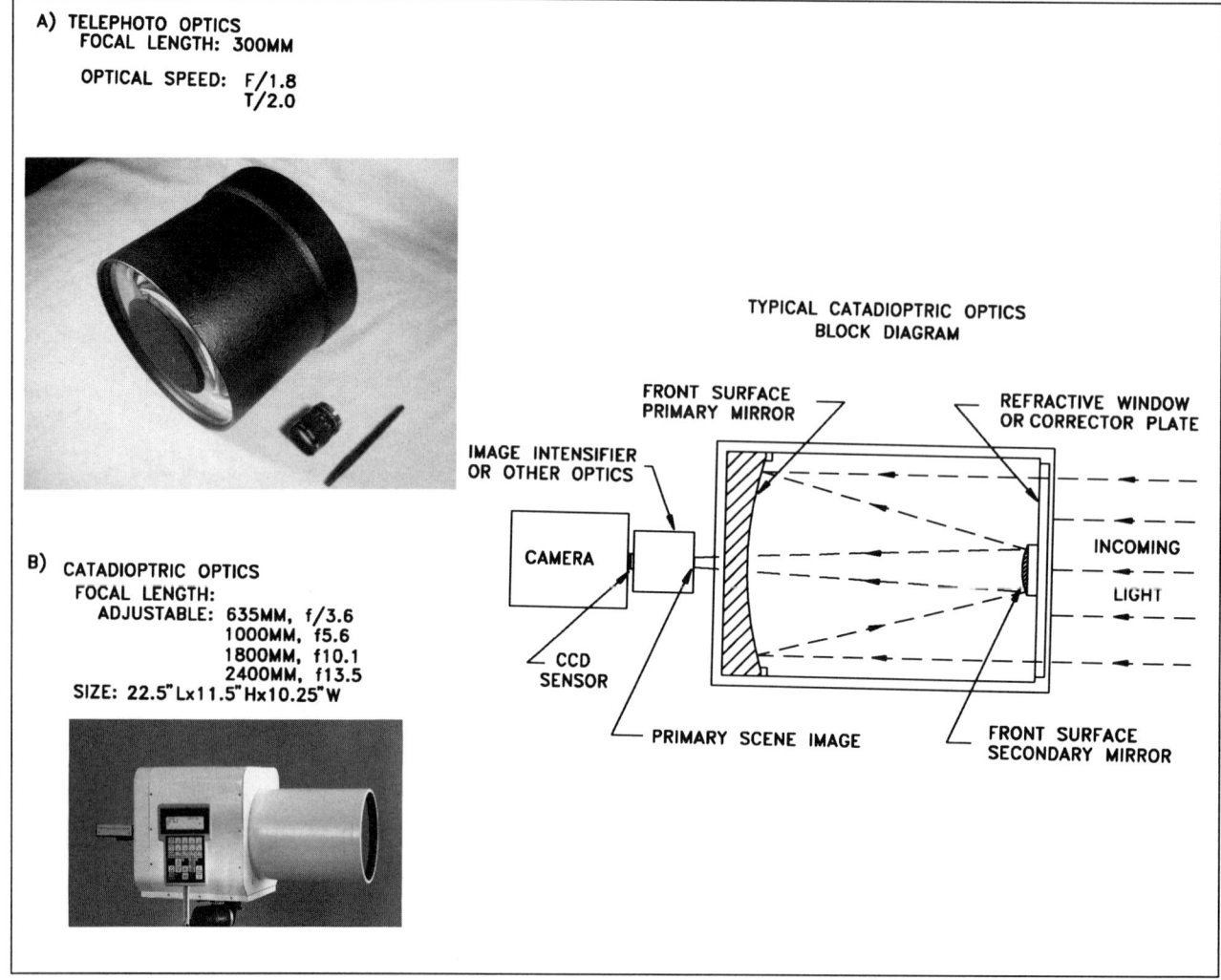

A) TELEPHOTO OPTICS
   FOCAL LENGTH: 300MM

   OPTICAL SPEED: F/1.8
                  T/2.0

B) CATADIOPTRIC OPTICS
   FOCAL LENGTH:
     ADJUSTABLE: 635MM, f/3.6
                 1000MM, f5.6
                 1800MM, f10.1
                 2400MM, f13.5
   SIZE: 22.5"Lx11.5"Hx10.25"W

TYPICAL CATADIOPTRIC OPTICS
BLOCK DIAGRAM

FRONT SURFACE
PRIMARY MIRROR

REFRACTIVE WINDOW
OR CORRECTOR PLATE

IMAGE INTENSIFIER
OR OTHER OPTICS

CAMERA

INCOMING

LIGHT

CCD
SENSOR

PRIMARY SCENE IMAGE

FRONT SURFACE
SECONDARY MIRROR

FIGURE 15-20   Long-range and catadioptric lens configurations

15.12 LOW-LIGHT-LEVEL OPTICS

The ability to use large optics with LLL devices accounts for a significant part of the improved visual capability of these devices over that of the eye.

15.12.1 The Eye versus LLL Camera Optics

As the light level is reduced, seeing is limited by the number of available photons in the visible and near-IR spectral range. The eye is handicapped by its small optical aperture. The pupil (lens) of the dark-adapted human eye is approximately 7 mm in diameter. Any optical system must concentrate collected light into this diameter in order that all of the collected light be useful to the eye. As an example, a standard 7 × 50 night binocular is a 7-power (magnification) binocular with a 50-mm-diameter objective lens. The

diameter of the exiting bundle of light is determined by dividing the objective diameter by the power. The light is thus concentrated into about a 7-mm-diameter bundle, all of which can be collected by the dark-adapted eye. Increasing the size of the light-collecting objective lens for the same magnification results in an exit diameter larger than that of the eye without any increase in useful light. An LLL camera benefits by using a larger collection aperture (objective lens diameter), since the intensifier tube can be 18, 25, or 40 mm in diameter, or 2.6, 3.6, or 5.7 times larger, respectively, than the human eye. Since the light gathered is a function of the area, these improvement factors are 6.76, 12.96, and 32.49, respectively.

The increased quantum efficiency, and the extension of the photocathode spectral sensitivity range of the intensifier into the near-IR where the night sky and other artificial light sources provide a greater number of photons and energy, likewise mean higher sensitivity over that of the eye.

LENS TYPE	FOCAL LENGTH (MM)	OPTICAL SPEED f/#	OPTICAL SPEED T/#	IMAGE FORMAT (MM)*	FIELD OF VIEW (FOV) (DEGREES)	MAGNIFICATION**	DIAMETER (INCHES)	WEIGHT LBS (KG)
CATADIOPTRIC: (REFRACTIVE AND/OR REFLECTIVE)	95	1.2	—	—	10.8	3.8	—	4.8 (2.2)
	135	1.6	—	25	10.6	5.0	3.8	4.4 (2.0)
	155	1.2	—	—	6.6	6.2	—	10.8 (4.9)
	170	1.5	—	25	—	—	4.9	6.6 (3.0)
	238	1.7	—	—	4.3	9.5	—	10.8 (4.9)
	300	1.4	2.0	40	—	—	9.0	23 (10.5)
	410	1.5	—	—	—	15.5	11.25	38 (17.3)
	500	1.6	2.0	16	1.84	—	13.4	56 (25.5)
	1725	5.6	—	—	0.6	70.0	—	59.4 (27)
CATADIOPTRIC (REFRACTIVE AND/OR REFLECTIVE) MULTI FOCAL LENGTH	635	3.6					22.5" LONG 11.5" HIGH 10.25" WIDE	—
	1000	5.6						
	1800	10.1						
	2400	13.5						

*MAXIMUM IMAGE (SENSOR) SIZE

**WITH 26 MM FOCAL LENGTH EYEPIECE

Table 15-2 Long-Range LLL Lens Parameters

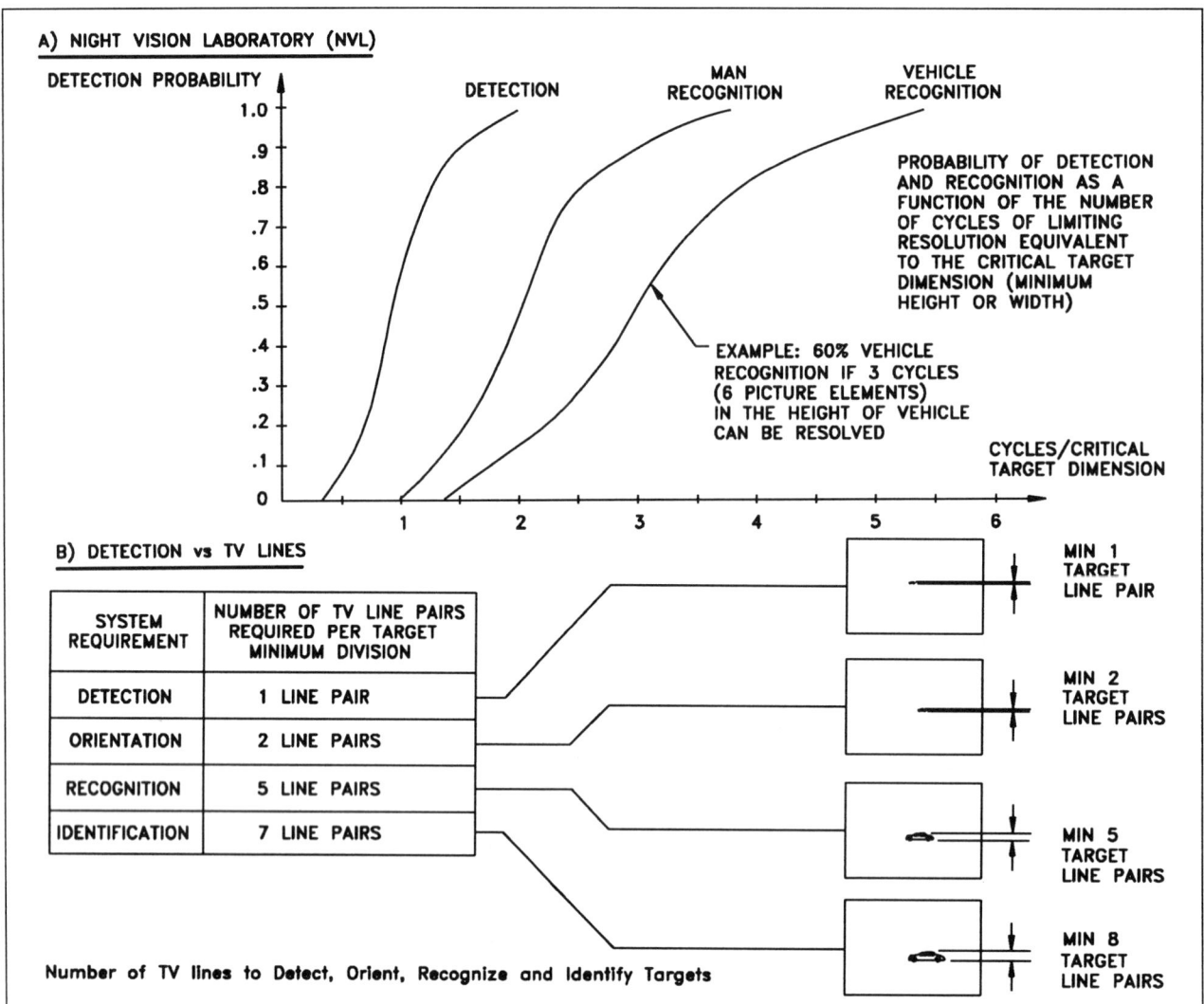

FIGURE 15-21 Target detection and recognition versus system resolution and target size

15.12.2 LLL Objective Lenses

Objective lenses for LLL television systems vary greatly depending on the application. Long-range systems use Cassegrain and catadioptric lenses fabricated from mirror or glass/mirror combinations. These 4- to 12-inch-diameter lenses weigh less than all-glass lenses and gather maximum visible and IR radiation. Figure 15-20 shows several examples of catadioptric lenses used with night vision systems.

Table 15-2 lists some commercially available lenses used in long-range nighttime surveillance applications.

15.13 TARGET DETECTION AND RECOGNITION PARAMETERS

To illustrate the capability of LLL systems, a typical application might be to view a scene with natural illumination during a clear, starlit night or, worse yet, cloudy starlit night

when the average illumination is down to a level of 10^{-5} to 10^{-6} fc.

The probability of detection and recognition of specific objects has been determined experimentally by the Night Vision Laboratory of the U.S. Army Electronics Command. Figure 15-21a shows the probability of detection and recognition as a function of limiting resolution and critical target dimension.

As an example, it can be seen that there is a 60% probability of recognizing a vehicle occupying 3 TV lines or if 3 cycles of spatial information are resolvable (6 picture elements) in the vehicle height. Likewise, there is a 90% probability of recognizing a man with the same number of picture elements resolvable.

How many line pairs of resolution at a target are required to detect, recognize, or identify? Numerous experiments have been performed, which indicate the following (Figure 15-21b):

1. One to 2 line pairs across the minimum target dimension are required for target detection (that is, in order to determine that a discontinuity exists in the field).
2. Two to 3 line pairs are necessary to determine target orientation.
3. Five to 8 line pairs are required for recognition (that is, to be able to estimate a rough outline and estimate aspect ratio).
4. Eight to 12 line pairs are required for target identification (that is, to allow the observer to perceive sufficient detail to classify the target as to a particular type).

These rules apply for targets with aspect ratio of about 1:1 to 5:1. Targets with higher aspect ratios are generally easier to detect: a telephone wire or pole is easier to detect than a round object.

15.14 SUMMARY

The SIT, ISIT, and ICCD image intensifier cameras provide the ability to "see" during periods of low light by amplifying the faint glow of the moon, starlight, or other artificial light. CCD cameras can barely produce usable pictures under nighttime, dawn, or dusk illumination conditions. The higher sensitivities of SIT and ICCD cameras allow them to function under light levels of a quarter moon (only 0.001 fc illumination). ISIT cameras detect images from a scene illuminated with 0.00001 fc light level (the light available from stars on a moonless night). Unlike near-IR active devices, these passive image intensifiers do not require additional light sources other than ambient light reflected from the scene. The SIT, ICCD, and ISIT cameras produce a video output that is displayed directly on a monitor.

Two disadvantages in using LLL systems as compared with standard solid-state CCTV cameras are cost and lifetime. Image-intensifying systems can cost 10 to 20 times more than a standard CCTV camera. Most GEN 2 or GEN 3 image-intensifier systems have lifetimes of about 2000 and 8000 hours, respectively. Compare this to standard CCTV systems, which may last many years (solid-state cameras, CCD, MOS, and others have no wear-out mechanism).

Devices have been developed for night vision by exploiting three basic imaging techniques: (1) near-IR image converters (GEN 0) used with searchlights or IR emitters; (2) image intensifiers (GEN 1, GEN 2, and GEN 3); and (3) mid-IR thermal-imaging devices.

All LLL viewing systems require a sensitive photocathode and a large lens to collect visible and near-IR radiation.

The unaided human eye can be compared with CCTV cameras and image-intensified cameras with respect to sensitivity and resolution. Since the lens diameter of the dark-adapted eye is about 7 mm, a good catadioptric lens used with night vision devices having an effective diameter of from 100 to 200 mm improves sensitivity over the unaided eye by about 500 to 1000 times. This translates into improvement in resolvable picture elements in a scene by a factor of approximately 1.5 to 3 times when compared with the human eye. When the human eye is compared with a night vision device using a 200-mm lens in addition to an LLL sensor, the advantage is approximately 125 times in the reduction factor in picture element size as compared with the unaided eye.

Chapter 16
System Power Sources

CONTENTS

16.1 OVERVIEW

CCTV security equipment operates on either alternating current (AC) or direct current (DC). The type (AC or DC) and voltage level used depend on the equipment chosen and the application intended. Most CCTV security systems use the AC power available from utility companies in standard 117-volt AC (VAC) outlets or at 24 VAC provided by step-down transformers. By far the most convenient power available for most security equipment is 117-VAC power. When it is necessary to locate the equipment remotely from the power source, it is convenient to use 24-VAC equipment and reduce the voltage to 24 VAC via a transformer. If the security equipment is installed for temporary surveillance, 12-volt DC (VDC) power obtained from batteries or other sources is a convenient solution. Another power source available in remote areas or for portable applications is solar power, either as the prime power source or as a battery charger.

An important factor to consider in powering any CCTV equipment is the backup equipment immediately available in the event the primary AC or DC source fails or is degraded. This backup power usually takes the form of an uninterruptible power supply (UPS). To protect the CCTV equipment from external, deleterious electrical noise, surge protectors or power conditioning equipment is used.

While the subject of security system power sources, backup power, and power-conditioning equipment may not seem exciting or important, it can be absolutely crucial in the event the power source for some or all of the security equipment for an installation fails. When this happens, the

COUNTRY	AC VOLTAGE (RMS)	FREQUENCY (HERTZ)	STANDARD	COLOR SYSTEM	NUMBER OF TV LINES	CHANNEL BANDWIDTH (MHZ)	VIDEO BANDWIDTH (MHZ)
ARGENTINA	220	50	N	PAL	625	6	4.2
AUSTRALIA	240	50	B	PAL	625	7	5
AUSTRIA	220	50	B,G	PAL	625	7,8	5
BELGIUM	220	50	B,G	PAL	625	7,8	5
BERMUDA	120	60	M	NTSC	525	6	4.2
CANADA	120	60	M	NTSC	525	6	4.2
CHINA	220	50	D	PAL	625	8	6
COLOMBIA	110	60	M	NTSC	525	6	4.2
DENMARK	220	50	B,G	PAL	625	7,8	5, 8
EGYPT	220	50	B	SECAM (V)	625	7	5
FINLAND	220	50	B,G	PAL	625	7,8	5
FRANCE	220	50	E,L	SECAM (V)	625	14,8	10, 6
GERMANY	220	50	B,G	PAL	625	7,8	5
(FORMER EAST)	220	50	D,K	SECAM (Y)	625	7,8	5
GREECE	220	50	B	SECAM (H)	625	7	5
INDIA	230	50	B	PAL	625	7	5
ISRAEL	230	50	B,G	PAL	625	7,8	5
ITALY	220	50	B,G	PAL	625	7,8	5
JAPAN	100	60	M	NTSC	525	6	4.2
KOREA	100	60	M	NTSC	525	6	4.2
NETHERLANDS	220	50	B,G	PAL	625	7,8	5
NEW ZEALAND	230	50	B	PAL	625	7	5
NORWAY	230	50	B	PAL	625	7	5
POLAND	220	50	D,K	SECAM (V)	625	8	8
PORTUGAL	220	50	B,G	PAL	625	7,8	M
PUERTO RICO	120	60	M	NTSC	525	6	5
SAUDI ARABIA	220	50	B,G	SECAM (H)	625	7,8	5
SOUTH AFRICA	220	50	I	PAL	625	8	5.5
SPAIN	220	50	B,G	PAL	625	7,8	5
SWEDEN	220	50	B,G	PAL	625	7,8	5
SWITZERLAND	220	50	B,G	PAL	625	7,8	5
TURKEY	220	50	B,G	PAL	625	7,8	5
UAE	220	50	B,G	PAL	625	7,8	5
UK	240	50	A,I	PAL	625	5,8	5, 5.5
USA	120	60	M	NTSC	525	6	4.2
RUSSIA	220	50	D,K	SECAM (V)	625	8	8
VENEZUELA	120	60	M	NTSC	525	6	6

THE FIELD FREQUENCY IS 50HZ EXCEPT FOR SYSTEM M WHICH USES 60HZ.

THE FM SOUND DEVIATION IS ± 50KHZ EXCEPT FOR SYSTEM M WHICH USES ±25KHZ.

SYSTEMS C,E,F AND H ARE BECOMING OBSOLETE AND ARE BEING REPLACED BY B, G OR L.

TABLE 16-1 International TV Standards in Selected Countries

result is equivalent to having no security at all. Therefore, it is absolutely necessary when installing any CCTV security system to decide whether part or all of the CCTV equipment is critical to the security operation, and how to implement the backup power in the event of primary power failure.

16.2 AC POWER

Most security equipment, including CCTV surveillance equipment, is powered from an AC power source conveniently provided from outlets in the facility. In the United States, the standard power is single-phase 117 VAC, 60 hertz (Hz). Table 16-1 summarizes the utility power sources available in selected countries around the world, indicating the differences in voltages available, from 100 VAC in Japan to 240 VAC in many European countries.

The alternating frequency is either 50 or 60 Hz, depending on the country. Most CCTV suppliers provide equipment suitable for operation on almost any of the combinations of voltages available around the world. The rated voltage listed is the nominal or ideal voltage; utility companies usually supply this voltage within a tolerance of plus or minus 10% of the voltage, and a frequency plus or minus 1 or 2 Hz. These numbers vary depending on the particular utility and the country.

For safety reasons, many CCTV installations use equipment operating from 24 rather than 117 VAC. The 117-VAC power can give an electrical shock if the wires are exposed and a person touches the live wire (117 VAC) and the ground wire or grounded equipment. When the voltage is reduced to 24 VAC using an electrical transformer, the voltage is too low to give a person an electrical shock and the equipment is therefore safer to install. A second attribute is that Class II transformers are available, which provide a 24-VAC output or lower (18, 16, and 10 VAC are also common); the designation Class II indicates that if the output terminals or wires are inadvertently shorted (touch each other) continuously, the transformer will not overheat or start a fire. For this reason, a Class II transformer can be installed by anyone, not just a professional electrician. When a 117- or 24-VAC system is installed, the voltage drop occurring along the cable from the power source to the equipment location must be considered. Table 16-2 summarizes the voltage drop occurring in different diameter (size) electrical conductors used to power equipment (check with local electrical codes)

This table gives an indication of the wire size required to power equipments of various power consumption. For example, if a camera requires 25 watts to operate and is located a distance of 400 feet from the 24-VAC source and requires a minimum of 22.5 volts to operate normally, what wire size is required? From the fundamental electrical relationship between power (P), voltage (V), and current (I):

$$\text{Power} = \text{Voltage} \times \text{current}$$

$$P = VI$$

Power is measured in watts (or volt-amperes), voltage in volts, and current in amps. The current drawn by the equipment is as follows:

$$I = \frac{P}{V} = \frac{25 \text{ watts}}{24 \text{ volts}} \approx 1 \text{ amp}$$

The maximum voltage drop is 24–22.5 volts, or 1.5 volts.

From Ohm's Law, the voltage drop in a wire is equal to the current in the wire times the resistance (measured in ohms) of the wire.

$$\text{Voltage} = \text{Current} \times \text{Resistance}$$

$$V = IR$$

For a 1.5-volt drop and a 1-amp current,

$$R = \frac{V}{I} = \frac{1.6 \text{ volts}}{1 \text{ amp}} = 1.5 \text{ ohms}$$

From Table 16-2, the 18 American wire gauge (AWG) size would match the requirements. If there is a question of which wire size to choose, always choose the next larger wire size, to provide an extra safety factor.

There are many manufacturers of 117-to-24-VAC plug-in wall-mounted transformers with Class II specification for powering CCTV equipment. Figure 16-1 shows several Class II transformers used for security systems.

To choose the correct Class II transformer size, it is necessary to know the CCTV equipment power consumption. The manufacturer specifies this in terms of watts or volt-amps (VA) consumed by the equipment. Transformers are also rated in watts or volt-amps. For example, if a system requires two cameras at a location, each requiring 24 VAC and consuming 8 watts each (16 watts total), use a Class II transformer with a 20-watt capacity.

Most cameras operate from 117 VAC or 24 VAC. If a 117-VAC outlet is available at the camera location, use it. If power is to be run from a remote location to the camera, a 117-to-24-VAC Class II step-down transformer powering a 24-VAC camera has an advantage—any technician can install it. Step-down transformers with a 10, 20, or 50 VA power rating are readily available and adequate to power the camera.

16.3 DC POWER

Many CCTV security equipments operate from DC power. This can be derived from an AC source using an AC-to-DC converter or a power supply, or from a battery. DC voltages in common use are 6, 9, 12, 24, and 28 VDC; 12 volts is the most common.

CONDUCTOR SIZE AWG*	SOLID WIRE DIAMETER (INCHES)	WIRE CMA**	CURRENT CARRYING CAPACITY+	RESISTANCE (OHMS/1000FT)	TWO CONDUCTOR CABLE LENGTH VS CURRENT DRAW++							
					(MA)					(AMPS)		
					100	300	500	700	900	1.1	1.3	1.5
28	0.013	159	0.16	66.20	—	—	—	—	—	—	—	—
26	0.016	253	0.254	41.60	—	—	—	—	—	—	—	—
24	0.020	404	0.404	26.20	467	156	93	67	52	42	36	31
22	0.025	640	0.642	16.50	745	248	149	106	83	68	57	50
20	0.032	1020	1.02	10.40	1,200	400	240	171	133	109	92	80
19	0.036	1290	1.29	8.21	—	—	—	—	—	—	—	—
18	0.040	1620	1.62	6.51	1,875	625	375	268	208	170	144	125
16	0.051	2580	2.58	4.09	3,000	1,000	600	429	333	273	231	200
14	0.064	4110	4.11	2.58	4,800	1,600	960	686	533	436	369	320
12	0.08	6530	6.53	1.62	7,500	2,500	1,500	1,071	833	682	571	500
10	0.102	10380	10.38	1.04	12,000	4,000	2,400	1,715	1,333	1,090	923	800

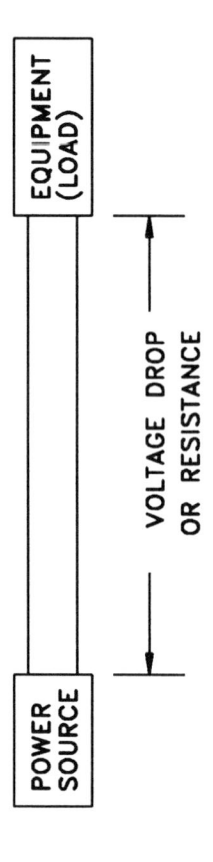

*   AMERICAN WIRE GAUGE
** CIRCULAR MIL AREA.

FOR STRANDED WIRE, MEASURE DIAMETER OF ONE STRAND IN MILS (0.001 IN).
SQUARE THE DIAMETER, AND MULTIPLY BY THE TOTAL NUMBER OF STRANDS TO CALCULATE CMA.

+   BASED ON 1000 CIRCULAR MILS PER AMP.

++ CABLE LENGTHS (FT) FOR A 10% VOLTAGE DROP.

Table 16-2   Wire Size versus Current Capacity and Voltage Drop

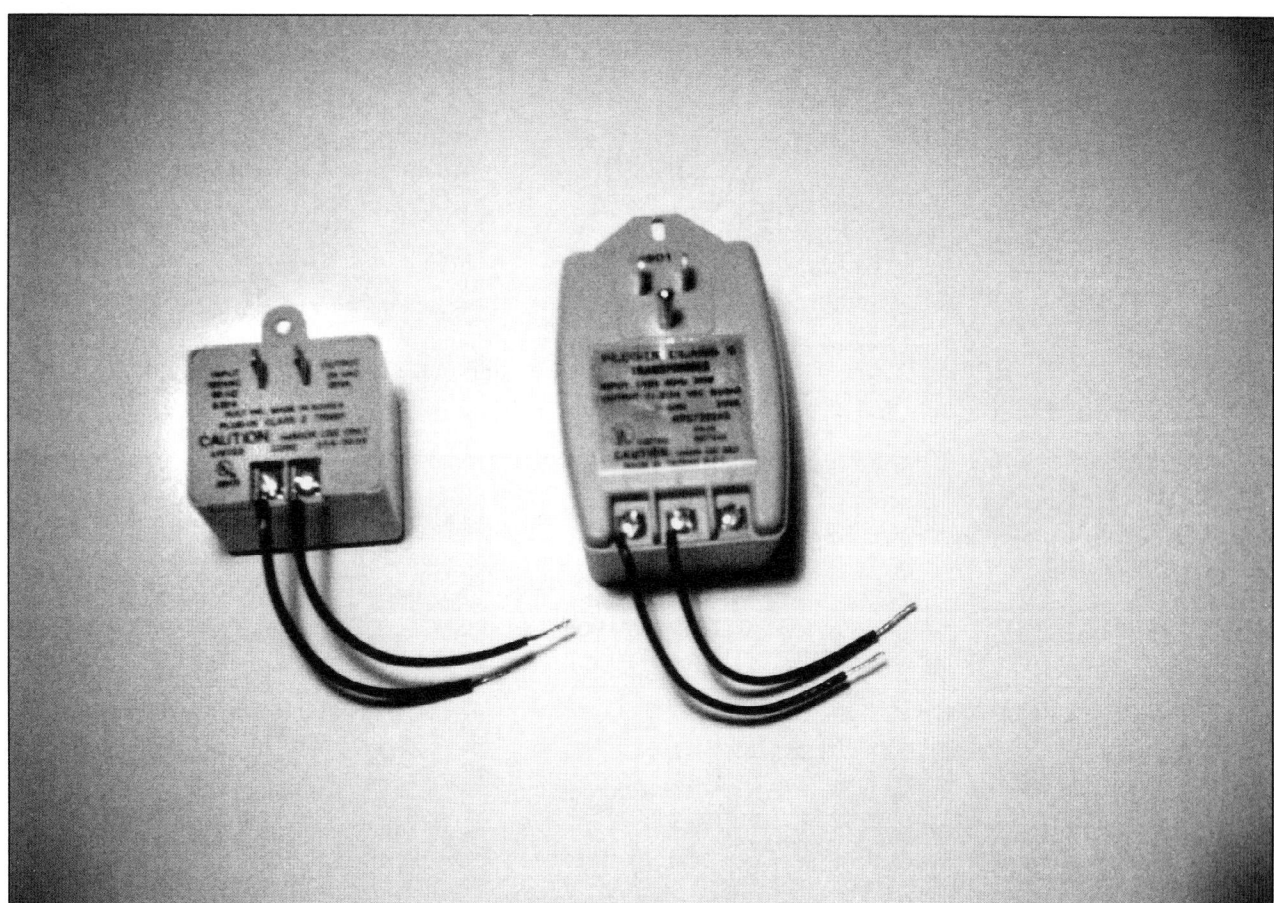

FIGURE 16-1 Class II AC transformers suitable for security equipment

16.3.1 DC Power Supply

Many CCTV components operate from 12-VDC power (normally 12 to 13.5) supplied by wall-mounted power converters that convert 117 VAC to 12 VDC. When larger amounts of current or better voltage regulation is required, small power supplies housed in metal electronic enclosures and having a standard 117-VAC power cord and 12-VDC output terminals are available. These supplies should be fused or have a circuit breaker for safety and to protect the equipment in case of electrical failure. If a 12-VDC CCD solid-state camera is used, either a 12-VDC power source or a wall-plug-mounted 117-VAC-to-12-VDC power converter is used. Figure 16-2 shows examples of a wall-outlet-mounted unit and the chassis-mounted power supply.

As in the case of the AC-powered transformers, it is necessary to determine the equipment current requirements and to size the DC power supply accordingly. As an example, if the equipment consists of three cameras and an infrared illuminator operating at 12 VDC and requiring 4 watts for each camera (333 milliamps) and 24 watts (2 amps) for the illuminator (2.333 amps total), choose a 117-VAC-to-12-VDC power supply with a minimum capacity of 2.5 amps. If the next higher value power or current rating of power supply is also a choice, choose it.

16.3.2 Batteries

A very convenient source of DC power is a battery. Batteries come in many shapes, sizes, and types, suitable for portable, temporary, or permanent applications. The endurance or power-delivering capability of a battery is measured in ampere-hours and is the product of its current capacity (amperes) and operation time (hours). As an example, a 5-ampere-hour (5-AH) battery is one that is capable of supplying 5 amperes of current for 1 hour. This current and time for a particular battery is an example; the battery doesn't have to deliver that current for that period of time. It can also deliver 1 amp for 5 hours or some other combination of current and time (10 amps for 30 minutes, ½ amp for 10 hours, and so on) within its range of operation. As the battery runs down, its voltage generally does not remain exactly constant but also decreases. The same battery can last longer if it is used intermittently rather than continuously, and as batteries get older, their ampere-hour rating

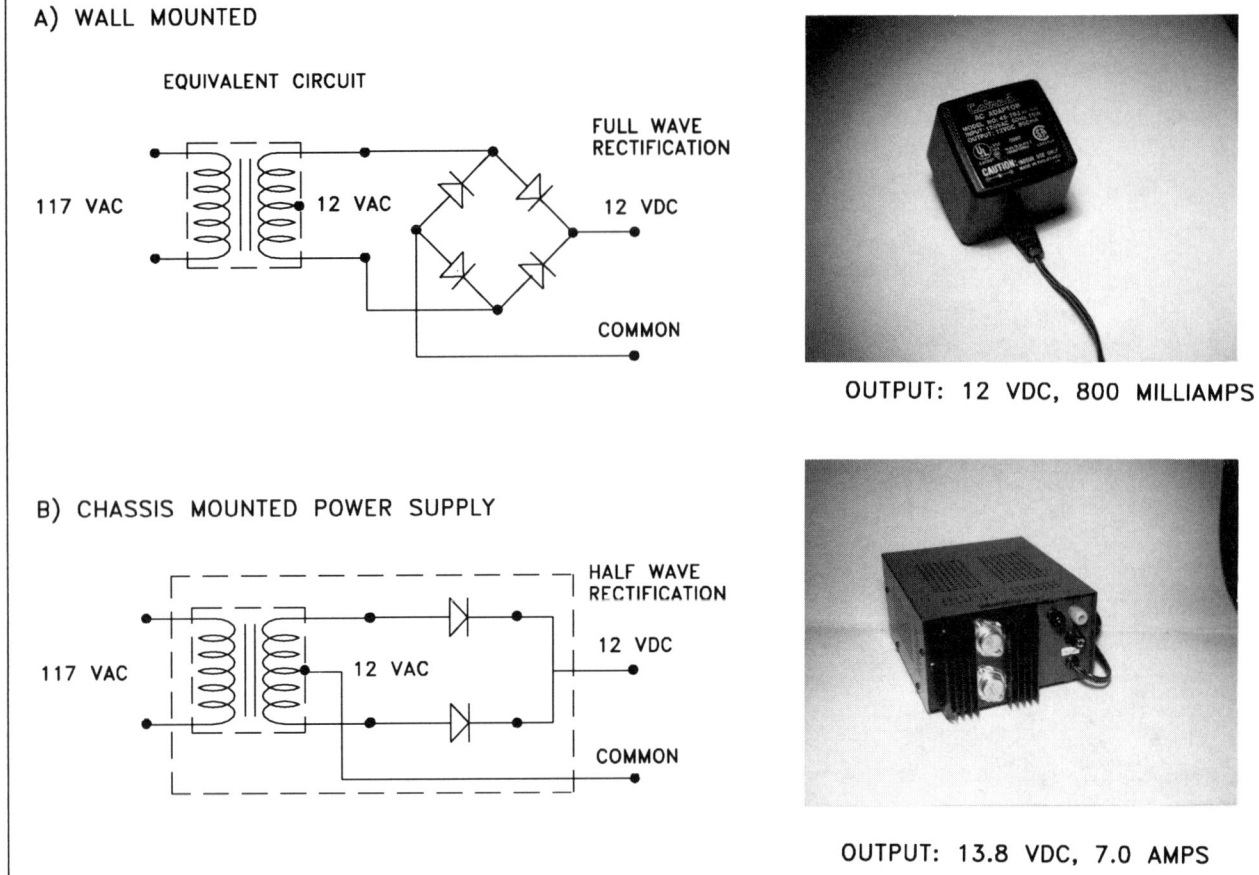

FIGURE 16-2   DC wall- and chassis-mounted power converters

decreases. Trying to extend a battery's life by using it as seldom as possible is not the answer either, since most batteries have a rated shelf life.

16.3.2.1 Lead-Acid

A common battery used for portable CCTV security applications is the gel-cell lead-acid automobile battery. It is probably the most often used battery for security systems in portable applications or for backup operation because it is readily available, provides reliable operation, and has the lowest initial cost. The lead-acid battery takes the form of a car battery or a smaller, more portable version having less amp-hour capacity and packaged in a safer container. This smaller type is called a gel-cell lead-acid battery. The lead-acid solution in this battery is contained in a "jell," which prevents spilling of the lead-acid solution in the event the battery case ruptures, and permits using the battery in any orientation.

The lead-acid battery cell has a voltage (potential) of approximately 2.15 VDC per cell. For the so-called 12-volt battery, 6 cells are connected in series to produce a total terminal voltage of 12.9 VDC. In the modern lead-acid cell, the acid electrolyte is sealed. The gel-cell battery charging rate is categorized as rapid, quick, standard, or trickle.

Table 16-3 gives typical values for lead-acid battery sizes, weights, and amp-hour capacity.

The lead-acid battery is a secondary battery and can therefore be recharged.

16.3.2.2 Carbon-Zinc

The carbon-zinc battery is the most popular type used in small, low-power-drain, portable electronic equipment. All such batteries provide the standard voltage of 1.50 VDC and are available in the common sizes; AAA, AA, C, and D. Table 16-4 summarizes the amp-hour capacity, sizes, and weights.

The carbon-zinc battery is a primary battery and cannot be recharged. Its popularity derives from its low cost, standardization, and availability.

16.3.2.3 Nickel-Cadmium

The nickel-cadmium battery, popularly referred to as the Ni-Cad, is available as a packaged unit designed to power many electronic security devices. The Ni-Cad battery ranges in size from the commonly available AAA, AA, C, and D up to sizes providing several amp-hours in many standard and custom configurations. The letter-size designations are

NOMINAL VOLTAGE (VDC)	NORMAL CAPACITY (AH) *			WEIGHT LB (KG)	DIMENSIONS LxWxH (INCHES)
	5HR	10HR	20HR		
6	0.85	0.91	1	0.61 (.275)	2x1.65x2
6	4.2	4.6	5	2.2 (.98)	2.63x2.63x3.78
6	10	10.9	12	4.4 (2)	6x2x3.7
12	1.7	1.8	2	1.8 (.83)	1.34x2.36x2.6
12	3.4	3.7	4	3.4 (1.53)	3.54x2.75x4
12	6.0	6.4	7	5.4 (2.45)	6x2.56x3.7
12	20.4	22	24	19.1 (8.7)	6.9x6.9x4.92
12	80	91	100	70.4 (32)	6.85x8.42x9.41
12	160	182	200	118.8 (54)	20.1x10.6x9.2

* AMP HOURS

ATTRIBUTES:

SEALED, MAINTENANCE FREE, LOW COST
6–12 MONTH SHELF LIFE
SECONDARY BATTERY–RECHARGEABLE

Table 16-3 Lead-Acid and Gel-Cell Battery Characteristics

slightly shorter than their carbon-zinc counterparts and in many applications are interchangeable with them. The output voltage of the Ni-Cad is slightly lower than the carbon-zinc, between 1.25 and 1.4 VDC. Ni-Cad batteries are secondary cells, are rechargeable, and in this respect are more closely related to lead-acid storage cells. A Ni-Cad battery is about one-third lighter than a common lead-acid battery of equal power rating, and smaller as well. Table

BATTERY DESIGNATION	SIZE DIA x H (INCHES)	NOMINAL VOLTAGE	AMP–HOUR CAPACITY (AH)	WEIGHT OZ. (GRAMS)
AAA	0.413x1.75	1.5		0.34 (9.7)
AA	0.57x1.99	1.5		0.67 (19)
C	1.03x1.97	1.5		1.80 (51.5)
D	1.34x2.42	1.5		3.52 (100)
	L x W x D			
	1.9x1.04x0.69	9.0		

Table 16-4 Carbon-Zinc Battery Characteristics

BATTERY DESIGNATION	SIZE DIA x H	NOMINAL VOLTAGE (VDC)	AMP HOUR CAPACITY (mAH) C/5 *	CHARGING RATE				WEIGHT (OZ)
				STANDARD		QUICK		
				CURRENT (mA)	TIME (HR)	CURRENT (mA)	TIME (HR)	
1/3 AA	0.55x0.65	1.2	110	11	15	27.5	6	0.23
AAA	0.39x1.73	1.2	220	22	15	66	5	0.35
AA	0.55x1.95	1.2	500	50	15	150	5	0.78
2/3 C	1.0x1.2	1.2	1000	100	15	333	4.5	1.59
C	1.0x1.94	1.2	2200	220	15	730	4.5	2.65
D	1.27x2.36	1.2	4000	400	15			4.8

ATTRIBUTES:

MAINTENANCE FREE, 3–6 MONTHS SHELF LIFE
EXCELLENT LOW TEMPERATURE PERFORMANCE
SECONDARY BATTERY–RECHARGEABLE; MUST BE FULLY DISCHARGED BEFORE RECHARGING
TO PREVENT LOWERING VOLTAGE AND AMP–HOUR CAPACITY

* DISCHARGE RATE OF 1/5 BATTERY mAH CAPACITY

Table 16-5   Nickel-Cadmium Battery Characteristics

BATTERY DESIGNATION	SIZE DIA x H (INCHES)	NOMINAL * VOLTAGE	AMP–HOUR CAPACITY (AH)	RATED LOAD (mA)	WEIGHT (OZ)
AA	1.98x0.56	2.8	1.1	46	0.49
C	1.0x1.56	2.8	3.4	125	1.55
D	1.31x3.19	2.8	8.3	175	2.82
—	1.96x1.64	2.8	10	420	3.7
—	4.59x1.64	2.8	28	1000	8.11
—	5.5x1.64	2.8	35	1250	9.87

ATTRIBUTES:

SEALED, HIGH ENERGY DENSITY
EXCELLENT HIGH–LOW TEMPERATURE PERFORMANCE
EXCELLENT SHELF LIFE: 5–10 YEARS
FLAT VOLTAGE DISCHARGE
PRIMARY BATTERY–NON RECHARGEABLE

* 5 CELLS IN SERIES=14.0 VDC

Table 16-6 Lithium Battery Characteristics

16-5 shows sizes, amp-hour capacity, and the charging rates for typical Ni-Cad batteries.

The discharge voltage curve for a Ni-Cad battery is quite flat (voltage remains constant) for about the first 3 or 4 hours, after which the voltage drops faster. Although Ni-Cad batteries are initially more expensive than their equivalent nonrechargeable type, they are more economical in the long run since they can be recharged as many as 1000 times before having to be replaced. One disadvantage of the Ni-Cad battery is that it has a "memory": when it is recharged, it may not recharge to its previous maximum charge. If this occurs, the battery will not work at its full capacity. To prevent this condition, the Ni-Cad should be allowed to discharge almost completely prior to recharging, and it should be recharged to its full condition, not partially recharged.

As a precaution, a Ni-Cad or any battery should never be discharged by shorting the two output (plus and minus) terminals or leads. This dangerous procedure produces excess heat and will damage or destroy the battery.

16.3.2.4 Alkaline

Alkaline batteries are often used in place of carbon-zinc to provide longer operation on the same battery. The alkaline battery is a primary battery and cannot be recharged.

16.3.2.5 Mercury

Mercury batteries are used in applications where long shelf life is required. The open circuit voltage remains relatively constant over the useful life of the battery. The mercury battery is a primary battery and cannot be recharged.

16.3.2.6 Lithium

The lithium battery is used when the smallest, lightest battery is required and cost is not a primary concern (Table 16-6).

The standard lithium battery is a primary battery and cannot be recharged; however, a special version is rechargeable. The lithium battery weighs approximately half as much as the Ni-Cad battery and approximately one-third as much as the lead-acid battery. The lithium battery is available in sizes AAA, AA, C, D, and larger.

A recent innovation has produced a rechargeable lithium battery based on a variation of the lithium technology. One convenient feature of the rechargeable lithium battery is that it does not exhibit any "memory" effect and retains its charge over 5 years. Its output voltage is a direct indication of the charge in the battery: as the battery discharges, its output voltage slowly decreases. Therefore, monitoring this voltage provides an accurate measure of the available capacity remaining in the battery. This change is accurate

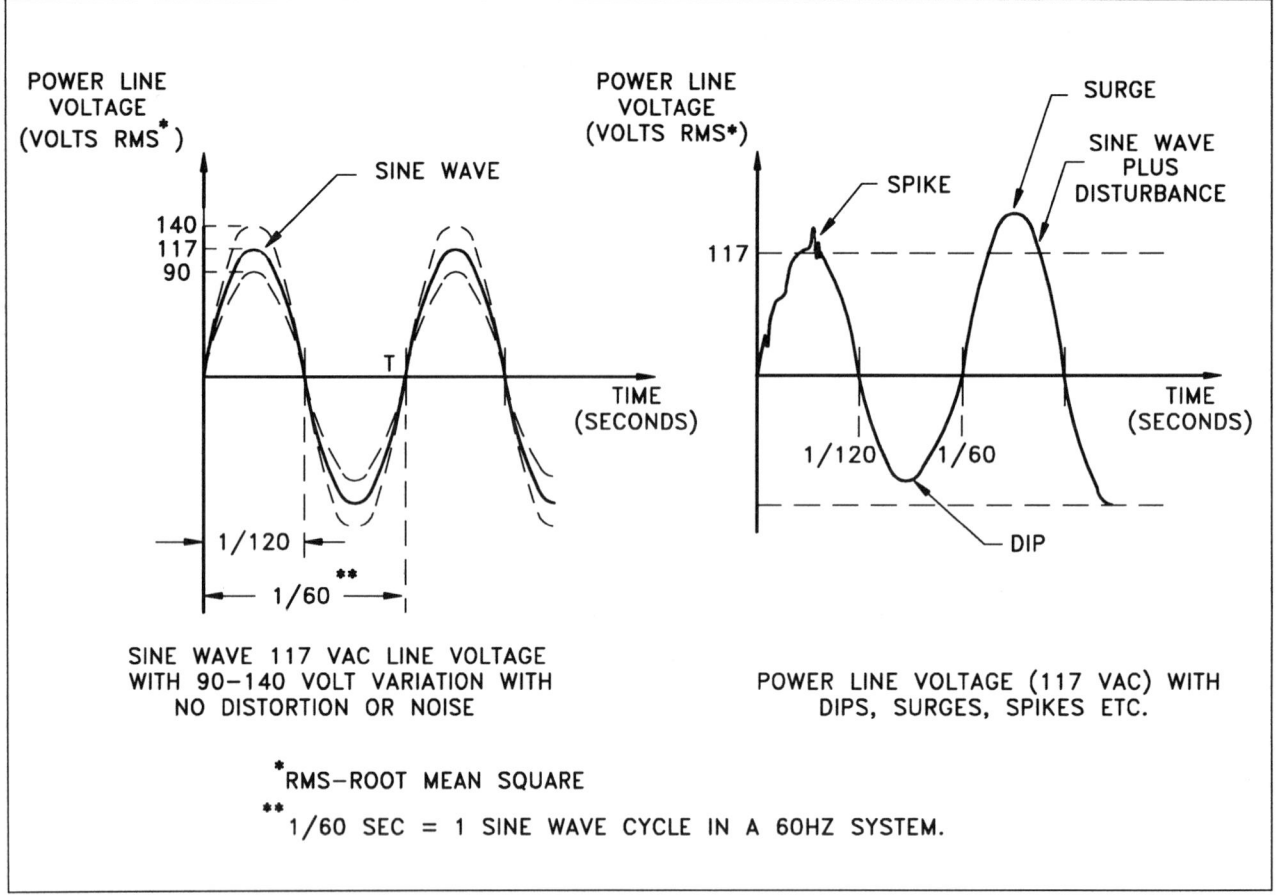

FIGURE 16-3 Typical power-line voltage variations and surges

over the entire battery life cycle irrespective of the voltage range chosen.

16.4 BACKUP POWER-LINE CONDITIONERS

Power problems are defined as any irregularity, disturbance, or interruption induced upon or occurring along a power line. These problems range from microsecond faults, voltage dips and surges, line spikes, and brownouts to common electrical noise and line-frequency harmonics.

Although everyone refers to power outlets as having a single voltage (such as 117 VAC in the United States), actually the voltage at the outlet is constantly changing 60 times a second (in the United States) and increases to approximately +170 volts, then plunges to approximately −170 volts. The effect of this rapid variation of voltage is a similar pulsation of electrical current (the flow of electrons) following at exactly the same rate in the wire. Since the term *power* is the combination (product) of voltage and current, it too is alternating or pulsing. The effect of this varying voltage at 60 Hz is to produce an "effective" or average 117 VAC. This is referred to as the root mean square (RMS) voltage. This discussion assumes that there

are no external influences to change the ideal power, current, and voltage being supplied. Add on to the standard power the line-voltage variations described previously (line dip, surges, spikes, brownouts, and so on) and the voltage, current, and power being delivered to a system look like that shown in Figure 16-3.

The shape of the waveform delivered by a utility company is a sine wave. When the utility company is doing its job correctly and there are no other external causes of electrical disturbances in the power line, the waveform at the output of every power outlet should be a perfect sine wave. In the real world, this situation does not occur. Power utility standards vary around the world and within single countries. In countries supplying nominal 117 volts RMS, the voltage varies from 90 volts to 130 volts (depending on the country), while the frequency is nominally 60 plus or minus 1 Hz. In countries supplying 220-VAC, 50-Hz power, the voltage may vary from 200 to 250 VAC and 50 plus or minus 1 Hz. Most electronic systems are designed to operate from "clean" sine waves. The power generated by most utility companies is closely regulated in both amplitude and frequency so that it meets the requirements of the millions of devices and equipments connected to the system. Unfortunately, there is little anyone can do to stop the voltage

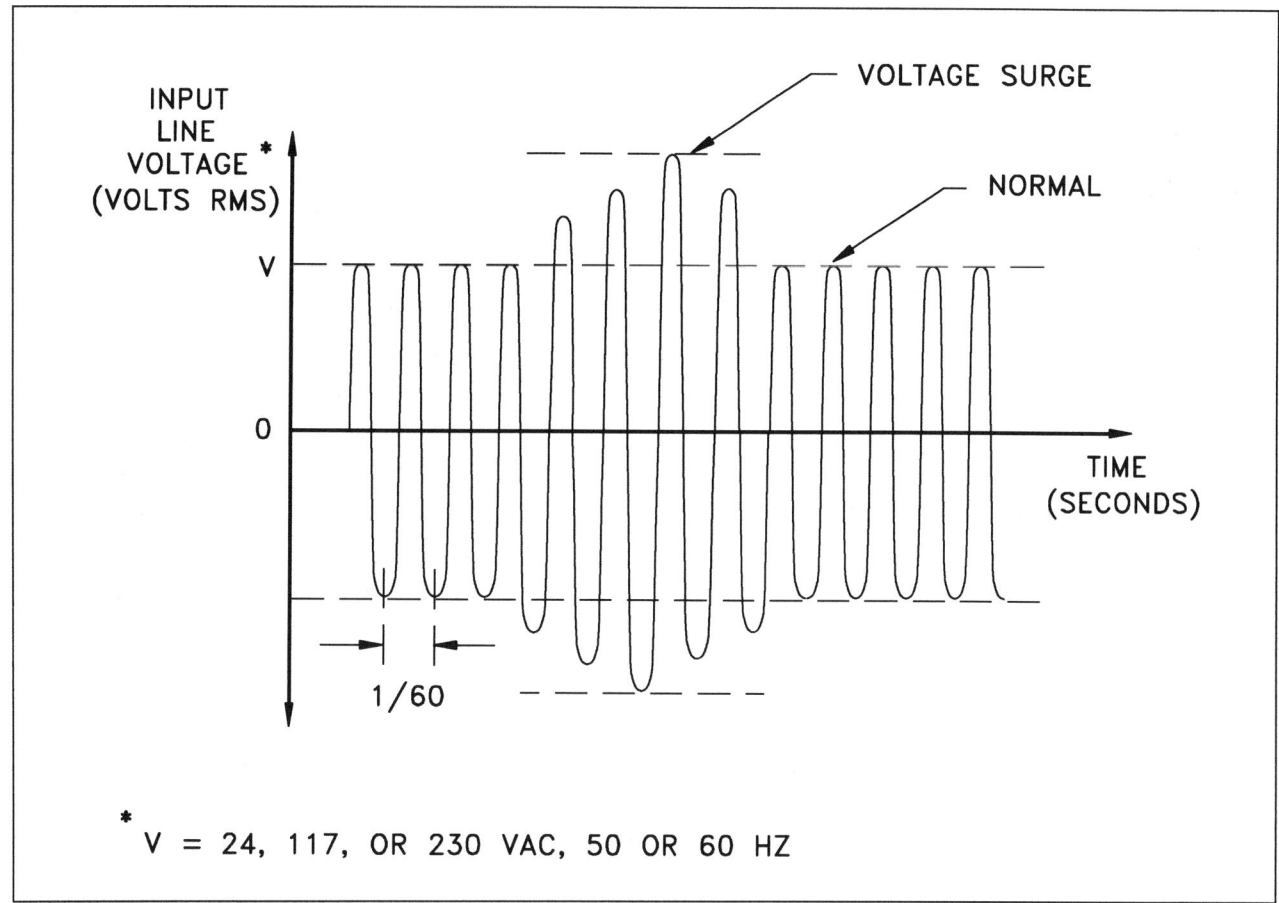

FIGURE 16-4 Voltage surge characteristics

distortions caused by storms, heavy machinery, and other disturbances from entering the power-line grid and creating problems. The following paragraphs describe six different deleterious conditions that alter and degrade equipment power.

16.4.1 Voltage Surges

A voltage surge is a temporary rise in amplitude (Figure 16-4). Unlike short voltage spikes, a voltage surge is defined as lasting at least one-half cycle ($\frac{1}{120}$ second). In many cases this type of disturbance is caused by switching off high-powered electric motors or other electrical equipment (causing a reduced current load and corresponding increase in voltage). Even something as commonplace as an air-conditioning system can momentarily boost the line voltage thousands of volts. Most everyone has witnessed the short change in brightness of lights coinciding with the activity of nearby electrical motors (such as air conditioners, refrigerators, tools). This is caused by an uncontrolled voltage and current surge, resulting in a matching increase in power dissipation in the light.

16.4.2 Voltage Spikes

When lightning strikes a power line or the ground nearby, a large, damaging voltage pulse can enter electronic equipment and destroy everything in its path. While the spike may last only a few milliseconds (a few thousandths of a second), it may reach thousands of volts. Storm-induced voltage spikes are responsible for huge equipment losses every year. The best course of action during a storm is to keep the equipment unplugged and thereby totally removed from the source of spikes. Unfortunately this is not possible in security applications. Security cannot shut down just because of a storm, and this critical service must remain in operation. The only practical answer is to have a high-quality power-line protection system.

16.4.3 Voltage Dips

A voltage dip is the opposite of a voltage surge. During a voltage dip, the line voltage decreases for a short period lasting at least one-half cycle ($\frac{1}{120}$ second) (Figure 16-5).

Dips are usually caused by a nearby sudden increase in the electrical load or power dissipated, such as turning on

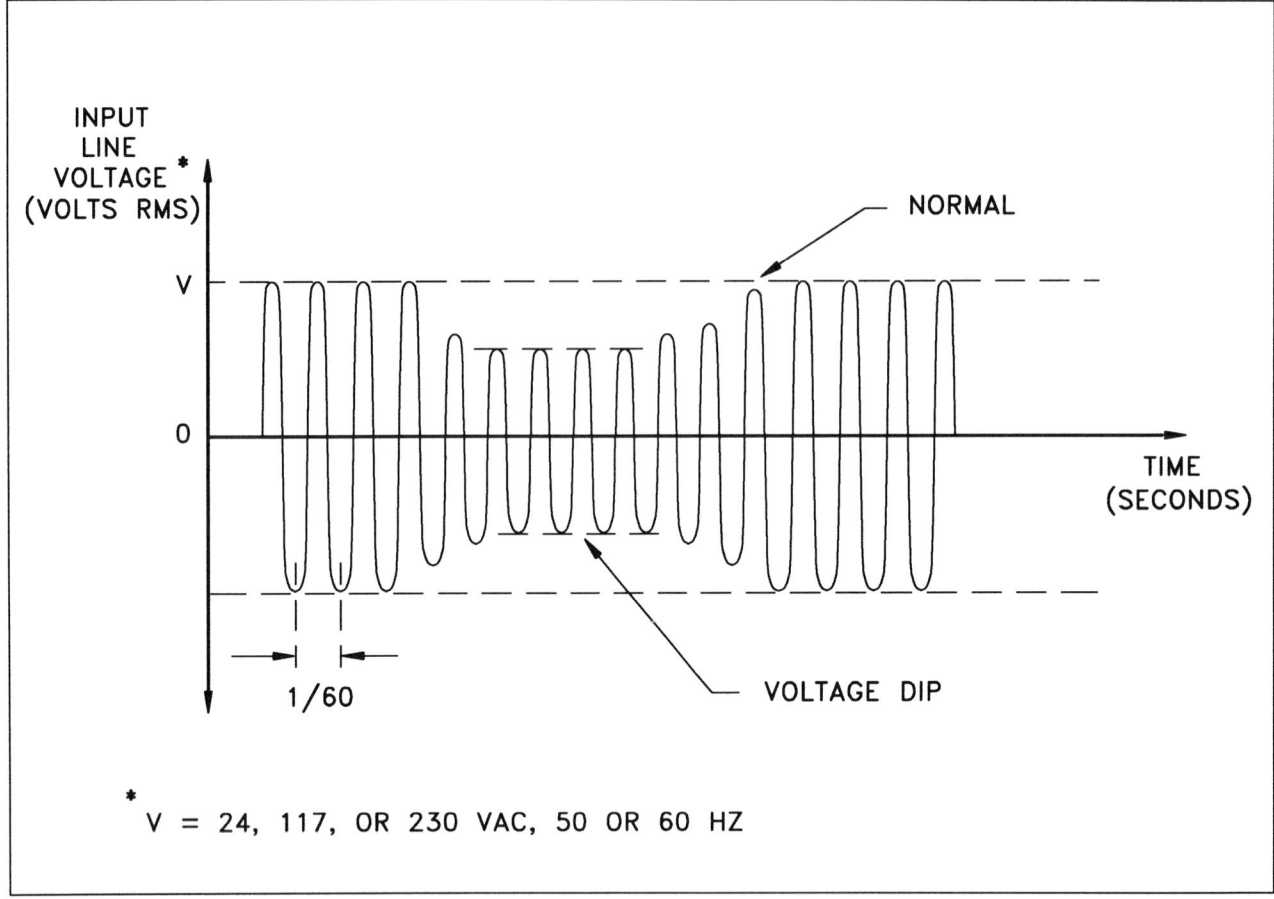

FIGURE 16-5 Voltage dip characteristics

a high-powered electrical motor, such as a refrigerator. Depending on the time required by the motor to come up to speed, the accompanying voltage dip can last for several seconds. During this time, all other electric equipment is forced to make due with reduced voltage, current, and power. The longer the situation continues, the more likely it is to cause problems elsewhere in the security system.

16.4.4 Brownouts

During periods of unusually high power demand (during hot summer days with many air conditioners operating), the power utility may not have enough generating capacity and might intentionally reduce the line voltage by up to 15%. This brownout can last from several hours to several days, depending on the condition. The brownout has the same effect on equipment as a prolonged voltage dip. Since most electronic equipment is designed to operate with its normal level of electrical power, brownouts can create many forms of abnormal behavior and failure. The only way to safeguard the operation of critical security equipment is to use auxiliary equipment to bring the power voltage back up to normal level.

16.4.5 Blackouts

The ultimate power problem is a blackout, when the power goes off and the devices that rely on it cease to operate. The risk usually associated with blackouts is a result of their unpredictability. One minute a computer or video system is humming along, the next minute it is suddenly down and everything is lost (or erroneous data are produced). The only way to keep critical electrical security equipment running during a blackout is to bring on-line a new supply of power, either from batteries to generate the AC voltage or from a mechanical generator (run on gasoline, diesel, or other fuel) to replace the original electrical power source.

16.4.6 Electrical Noise

Noise is often defined as anything we don't want (Figure 16-6). One type of noise we don't want in an electrical system is power-line noise. Lightning, radio transmitters, welding equipment, electrical switching equipment, poor brush contacts on DC motors, and many other electronic devices having switching power supplies are all sources of unwanted electrical noise. When these noise producers are

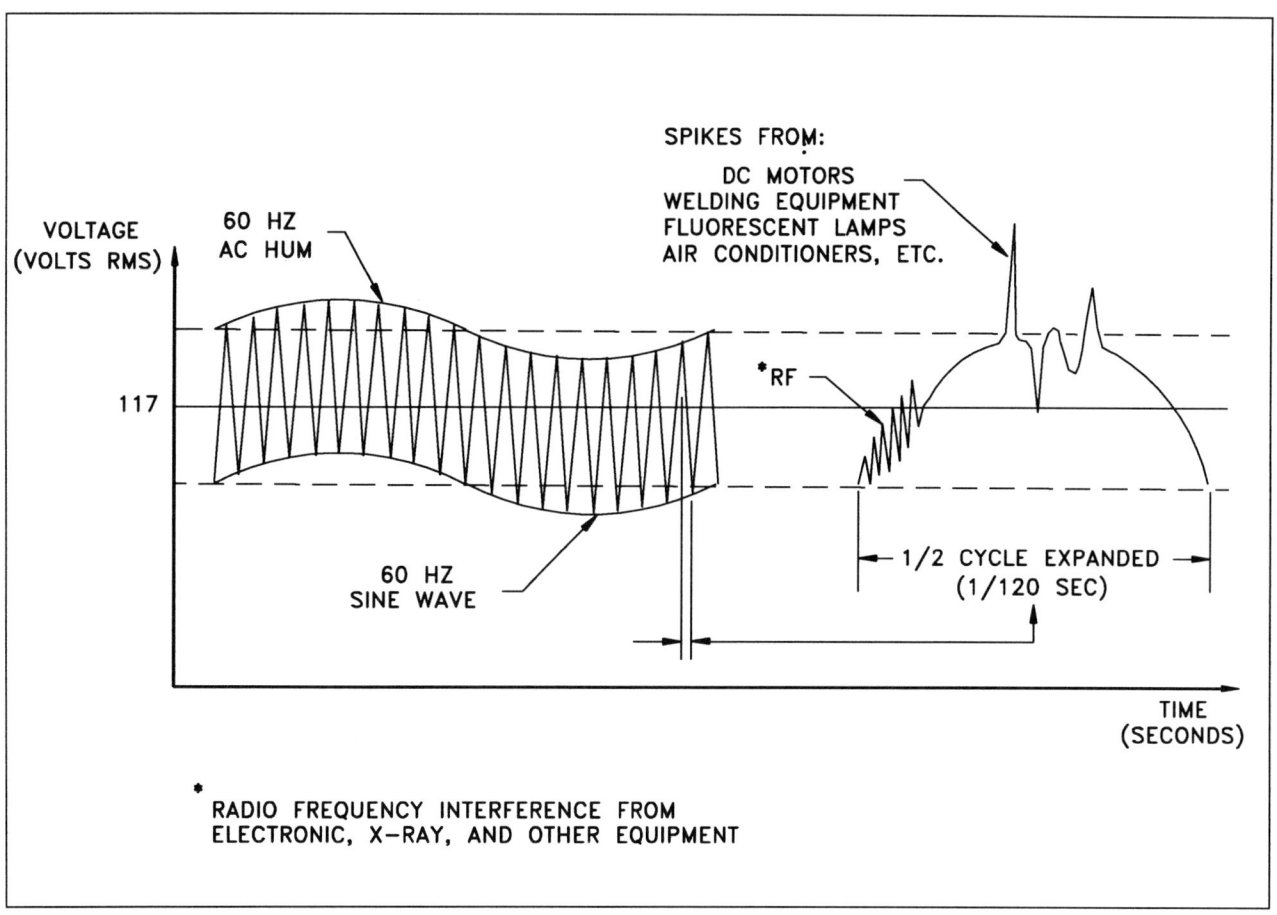

FIGURE 16-6   Electrical noise disturbances

operating, they disrupt the smooth sinusoidal power-line voltage and add many sharp-edged, high-frequency voltage changes. The noise may be repetitive or transient (that is, occurring intermittently or only once), but the effect on the equipment is the same.

Noise entering the equipment may be able to get past various safeguards (filters, surge protectors) against noise suppression and into the internal electronics. Since most modern electronic circuits are designed to operate with small high-frequency voltage variations, the noise has a tendency to randomly disrupt the normal operation and create havoc. Computer "bugs," "glitches," and errors are often the only outward symptoms that noise is causing a problem. The only real solution is to prevent noise from entering equipment by filtering the power-line input.

16.4.7 Solving the Problem

Because of all of the potential problems just mentioned, computers, telecommunications equipment, video equipment, and other sensitive electronic devices should not connect to raw line power. This type of equipment requires line conditioning and filtering to provide a source of continuous uninterrupted power at all times, including complete blackouts.

To gain independence from the less-than-perfect AC power services available, engineers and designers are increasingly turning to the uninterruptible power supply. Numerous manufacturers have UPS designs that satisfactorily eliminate most of the problems mentioned. Most UPS systems operate over a maximum period of several minutes to several hours—more in strategic applications. Some UPSs designed for the needs of large systems have power capacities from a few kilowatts to hundreds of kilowatts. The majority of systems have an output power range of 1 kilowatt or less.

UPS sources are categorized by output power capacity and are generally configured for continuous duty or standby operation. In the first type, also called on-line, power is supplied continuously to the load by a device called a static converter, which gets its own power from a storage battery. The battery charges itself continuously through the incoming utility power line.

The following sections describe the traditional UPS versus the line conditioner UPS.

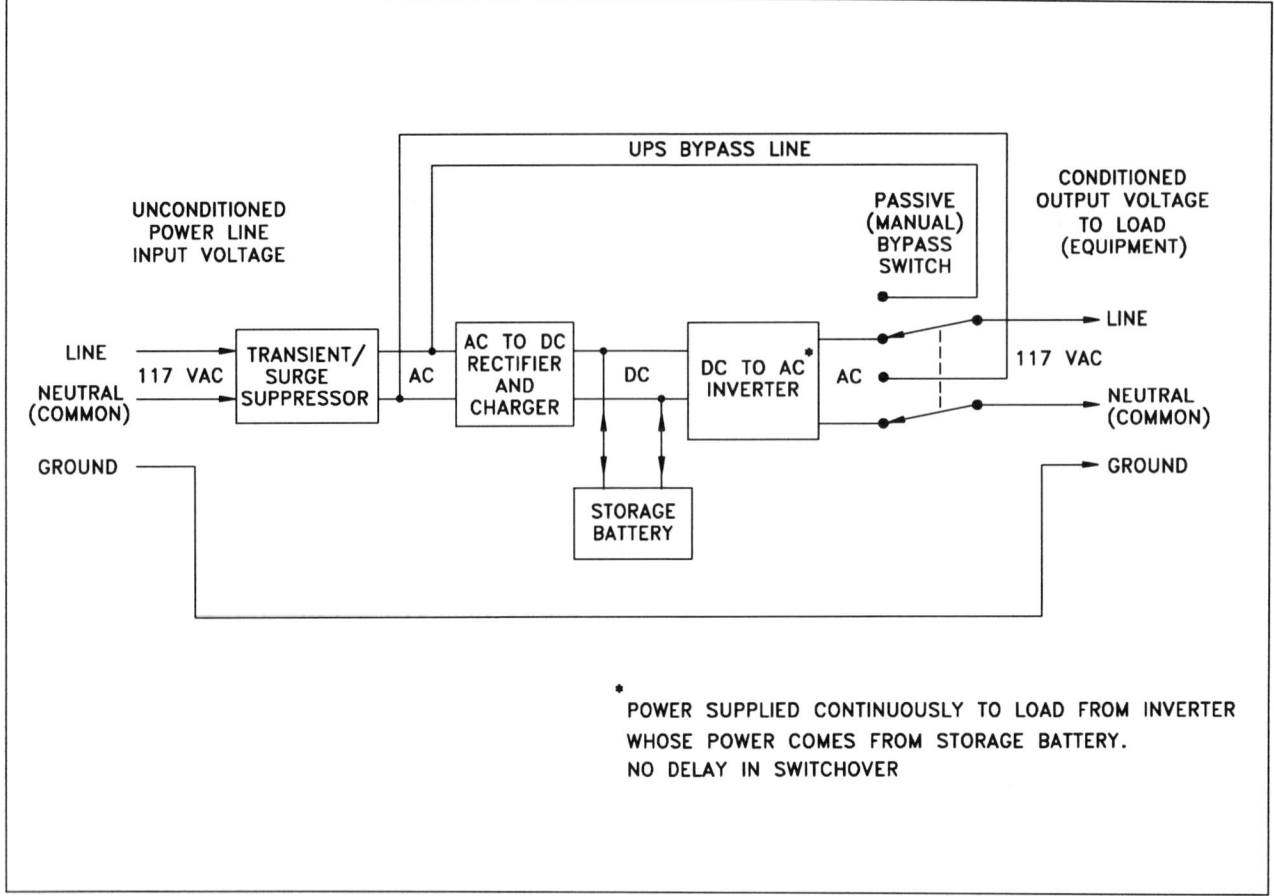

FIGURE 16-7 On-line UPS system block diagram

16.4.7.1 Traditional UPS System

The UPS system was first introduced in the 1950s when the silicon-controlled rectifier (SCR) was invented (Figure 16-7). The SCR is an electronic switch that either conducts or blocks electrical power and can continuously vary the amount of power reaching a load, in this case the security equipment.

The first type of UPS technology to evolve, and one that is still commonly used, uses a double-conversion technology, in which the incoming raw AC power is converted to DC in a large rectifier and charger system. The direct current generated is used to charge a battery bank and provide power to a DC-to-AC inverter. The inverter changes the current back into AC current, which is then used by the electronic equipment. In all these systems, there is a static or manual bypass switch to allow the incoming line power to be passed directly to the load if the UPS fails.

While this device provides good protection against almost all power variations and particularly against outages and deep brownouts, it does have one major drawback: its vulnerability to failure. That is the reason for inclusion of the bypass switch. The probability of failure exists because all components are constantly in use and under electrical and thermal stress. A second weakness is its low efficiency;

in some cases as low as 60% of the original power input is delivered to the electrical equipment being driven. The remaining power is converted to heat, which often calls for additional air conditioning. Also because of the noisiness of larger rated units, they are rarely placed in the worker environment. Remoting this equipment incurs additional expense in wiring cable between the UPS and the security equipment location.

16.4.7.2 Line-Conditioner UPS System

In 1983 a major breakthrough in the UPS design occurred with the introduction of the ferro-resonant UPS system (Figure 16-8). The ferro-resonant UPS system solves many problems at once by conditioning the power line at all times, whether the load is being powered by the line or by the UPS internal batteries. The previously described double-conversion technology was superseded in effect by a technology incorporating a constant-voltage ferro-resonant transformer and sophisticated line-sensing circuitry.

When the unit is operating in a normal mode, that is, when the incoming power line is stable, the UPS transformer cleans up the raw line power, eliminating spikes, dips, surges, noise, and minor brownouts or drops in voltage.

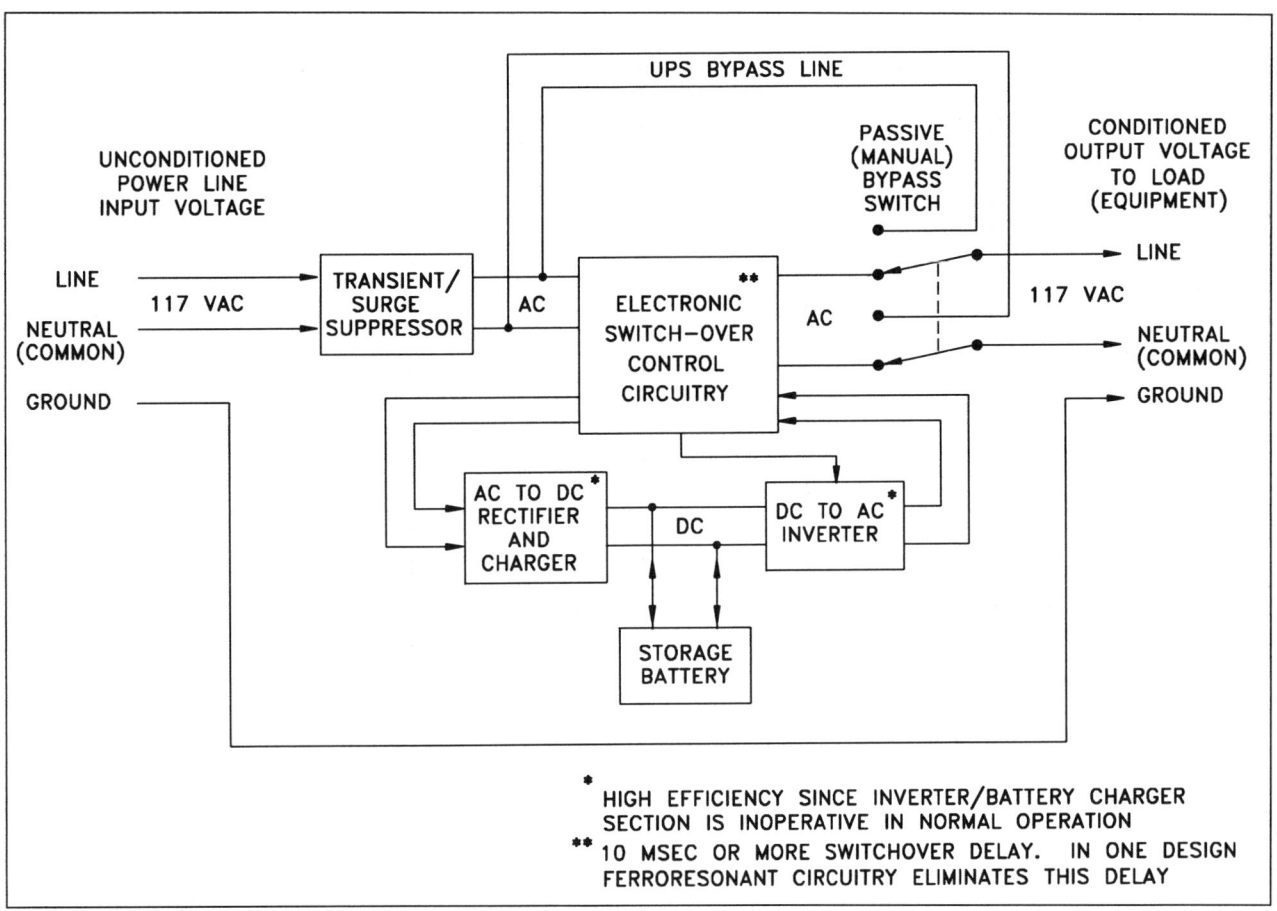

UPS BYPASS LINE

UNCONDITIONED POWER LINE INPUT VOLTAGE

PASSIVE (MANUAL) BYPASS SWITCH

CONDITIONED OUTPUT VOLTAGE TO LOAD (EQUIPMENT)

LINE
117 VAC
NEUTRAL (COMMON)
GROUND

TRANSIENT/ SURGE SUPPRESSOR

AC

ELECTRONIC SWITCH–OVER CONTROL CIRCUITRY **

AC

LINE
117 VAC
NEUTRAL (COMMON)
GROUND

AC TO DC * RECTIFIER AND CHARGER

DC

DC TO AC * INVERTER

STORAGE BATTERY

* HIGH EFFICIENCY SINCE INVERTER/BATTERY CHARGER SECTION IS INOPERATIVE IN NORMAL OPERATION
** 10 MSEC OR MORE SWITCHOVER DELAY. IN ONE DESIGN FERRORESONANT CIRCUITRY ELIMINATES THIS DELAY

FIGURE 16-8   Standby UPS system block diagram

In the event of power-line failure, the ferro-resonant transformer continues to provide energy from its magnetic field and capacitor circuits up to 16 milliseconds before a significant drop in output voltage occurs. This is often called the transformer's flywheel effect. During the 16 milliseconds, the electronic line-loss circuits detect the power-line problem and energize its battery-operated DC-to-AC inverter section, before the transformer output voltage drops noticeably. The security equipment being driven is unaware that the power being supplied is now coming from batteries rather than the electric line. Thus uninterruptible power is being delivered to the load because there is no break in the voltage or current supplied to the equipment.

There are additional benefits in the use of ferro-resonant transformer technology in the UPS, including 90% power efficiency, since double conversion has been eliminated. There is also higher reliability, since the double-conversion rectifier is eliminated entirely and the inverter is on only when the batteries are supplying power, rather than all the time. The UPS provides continuous power conditioning, with the ferro-resonant transformer eliminating the need for separate line filters and conditioners. This second-generation ferro-resonant transformer UPS costs less, since the older-technology hardware is eliminated, and the de-

sign is simpler, more reliable, smaller, requires no air conditioning, and is quieter.

While UPS systems might seem like an extra cost burden on the security system budget, they actually save money and provide a cost-effective solution to computer and CCTV security system power backup problem. Studies by Bell Laboratories, IBM, and the Institute of Electrical and Electronic Engineers over the past ten years have revealed that 87% of power problems affecting computers and video equipment are caused by power sags or outages, and 13% are caused by surges and spikes. These studies also revealed a tenfold increase in power problems affecting computers between 1978 and 1987. With 95% of computer failures caused by the AC power, it makes sense to protect security computers from such disturbances. Likewise, with 48% of software problems caused by the AC power source, it is prudent to install these protective systems for most security systems.

16.4.8 Equipment That Should Have Backup Power

In a small security system, the entire CCTV system and computer equipment can have backup power provided at reasonable cost. In medium-to-large systems, it is often

EQUIPMENT	QUANTITY	POWER (WATTS)	TOTAL VA** (VOLT−AMPS)
CAMERA	10	10	100
PAN/TILT	1	10	10
QUAD	1	30	30
MONITOR	2	25	50
P/T CONTROLLER	1	5	5
VMD *	1	50	50
VCR	1	60	60
SWITCHER	1	10	10
PRINTER	1	20	20
MISC.		25	25
	TOTAL	245	360

* VIDEO MOTION DETECTOR

** ASSUMING RESISTANCE LOAD (POWER FACTOR=1) VA=WATTS

Table 16-7 Backup Power Requirements for Ten-Camera CCTV System

difficult to justify 100% backup power for all CCTV and security equipment. A judgment must be made regarding which equipment will have backup power, as determined by its strategic value in the overall installation and security plan. Any security system has certain essential camera, monitor, and recording equipment that must continue operating in the event of a power disturbance or blackout. There are other categories of equipment that, while useful in normal security functions, can afford to be inoperative without jeopardizing the primary assets or safety of the facility. Essential items might include access control and identification equipment at critical points around the facility, internal CCTV surveillance cameras in areas critical to the operation of the facility or protection of valuable assets, and any monitoring and recording equipment for these scenes.

Once the equipment that must have backup power has been identified, the location and power consumption of this equipment must be determined and calculated. The next step is to outline the procedure for which this equipment will be transferred from the primary AC power source to the short-term UPS backup equipment and eventually to a temporary power source. The most practical device for the temporary transfer and temporary backup is the UPS. Figure 16-9 shows a block diagram of a ten-camera system with the equipment to be backed up, and how the UPS equipment fits into the scenario.

The following example illustrates how to determine the size and type of UPS system to choose for an installation having ten CCTV cameras, two monitors, auxiliary console equipment, VCR, and video printer (Table 16-7).

The total backup power required is 365 VA (watts). Since UPS systems are usually available in 250-, 500-, and 1000-VA ratings, choose the 500-VA unit. This will provide a good margin of safety.

16.5 SOLAR POWER

Solar photovoltaic (PV) power generators produce electricity directly from sunlight or overcast sky. In case sunlight is not available for long periods of time, the system includes rechargeable batteries so that the solar panels charge the batteries during sunlight and the batteries provide the power to the CCTV equipment on a continuous basis. In this latter case, the solar panels are used to charge the batteries, which then act as the source of power to the equipment around the clock. The PV generator is used in situations where an independent source of electrical power is required.

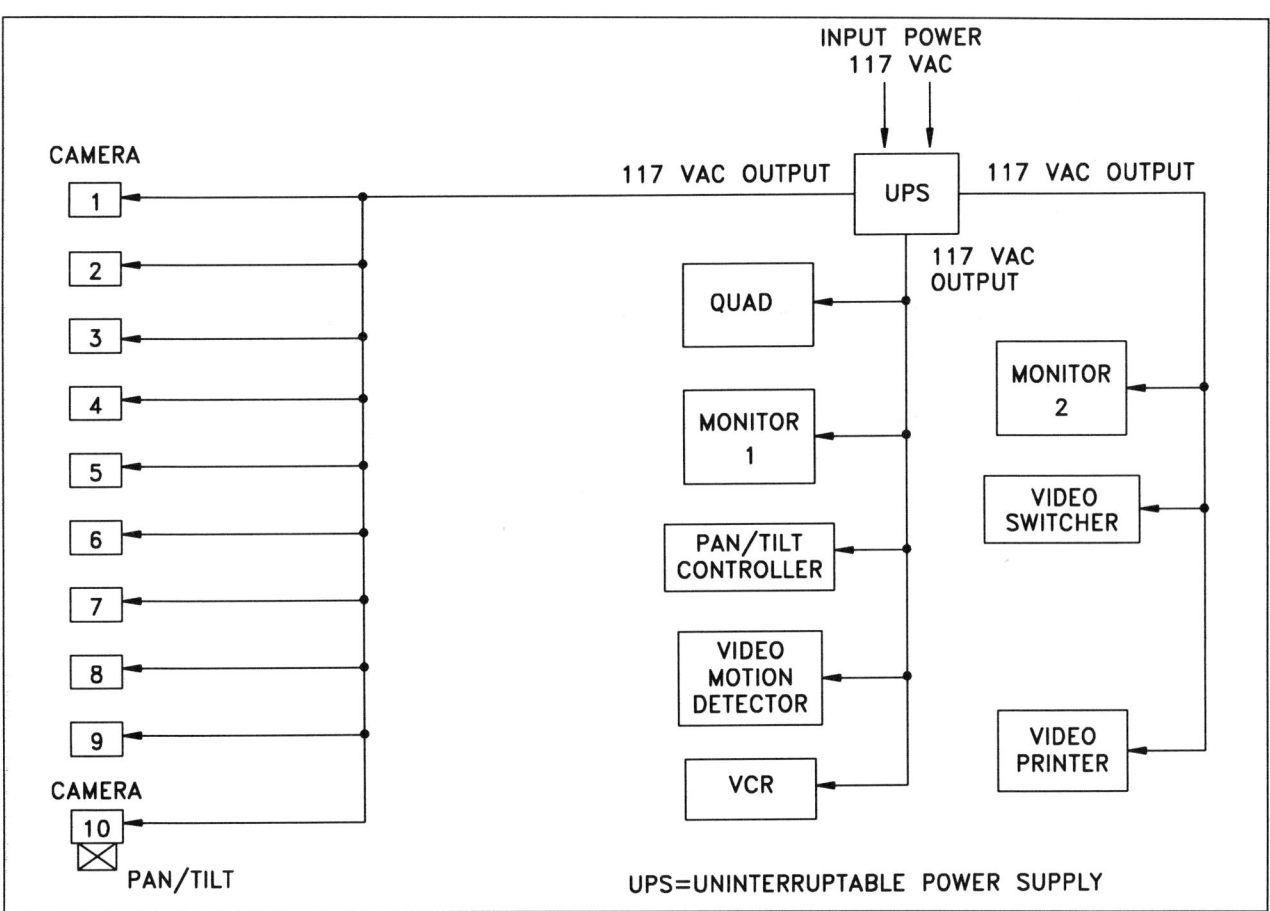

FIGURE 16-9 UPS backup system for ten-camera CCTV system

A wide range of PV modules are available, from small panels delivering 1 watt of power to large ones producing up to 60 watts of power. They are used alone or in multiple arrays when many kilowatts of power are required. These panels use third-generation photovoltaic technology and are constructed from thin-film amorphous silicon modules capable of long life (up to ten years). They are available with standard output voltages ranging from 6 to 30 VDC, capable of providing power for most security equipment operating at 12 and 28 VDC. The panels are rugged, easy to use, and weatherproof for continuous outdoor use. In some designs the solar cells are laminated between sheets of ethylene vinyl acetate and protected with either tempered glass or a polycarbonate plastic. Depending on the material, the panels are extremely resistant to mechanical stress, including impact of hail, and special types are available for protection against projectiles. Modules subject to rigorous environmental testing meet or exceed repetitive cycling between −40°C to 90°C at 85% relative humidity with no performance degradation. Some are also capable of high wind loading up to 125 miles per hour.

16.5.1 Silicon Solar Cells

Figure 16-10 shows an 11.5-watt solar panel module designed to power 6- or 12-VDC loads or rechargeable batteries. The panel consists of forty 2.5 by 10-centimeter semi-crystalline silicon solar cells. The dual voltage capability is enabled by movable jumper leads in the module's junction box. By connecting the 40 cells in a series string, 12 VDC power is achieved. When connected as two 20-cell series strings, with outputs in parallel, 6 VDC power is achieved.

The electrical output characteristics of these two configurations are shown in Figure 16-10. The output voltage available to a load at various load currents is a function of the load current. Some form of voltage regulation is required at the input of the electronic equipment being powered from them. As shown in Figure 16-10, the panel generates a minimum of 10 watts power at 6 or 12 VDC. The current at rated load is approximately 680 mA (0.68 amps) at 12 VDC and 1360 mA (1.36 amps) at 6 VDC (Table 16-8).

The solar panel current and power output are proportional to the illumination intensity (from the sun or other

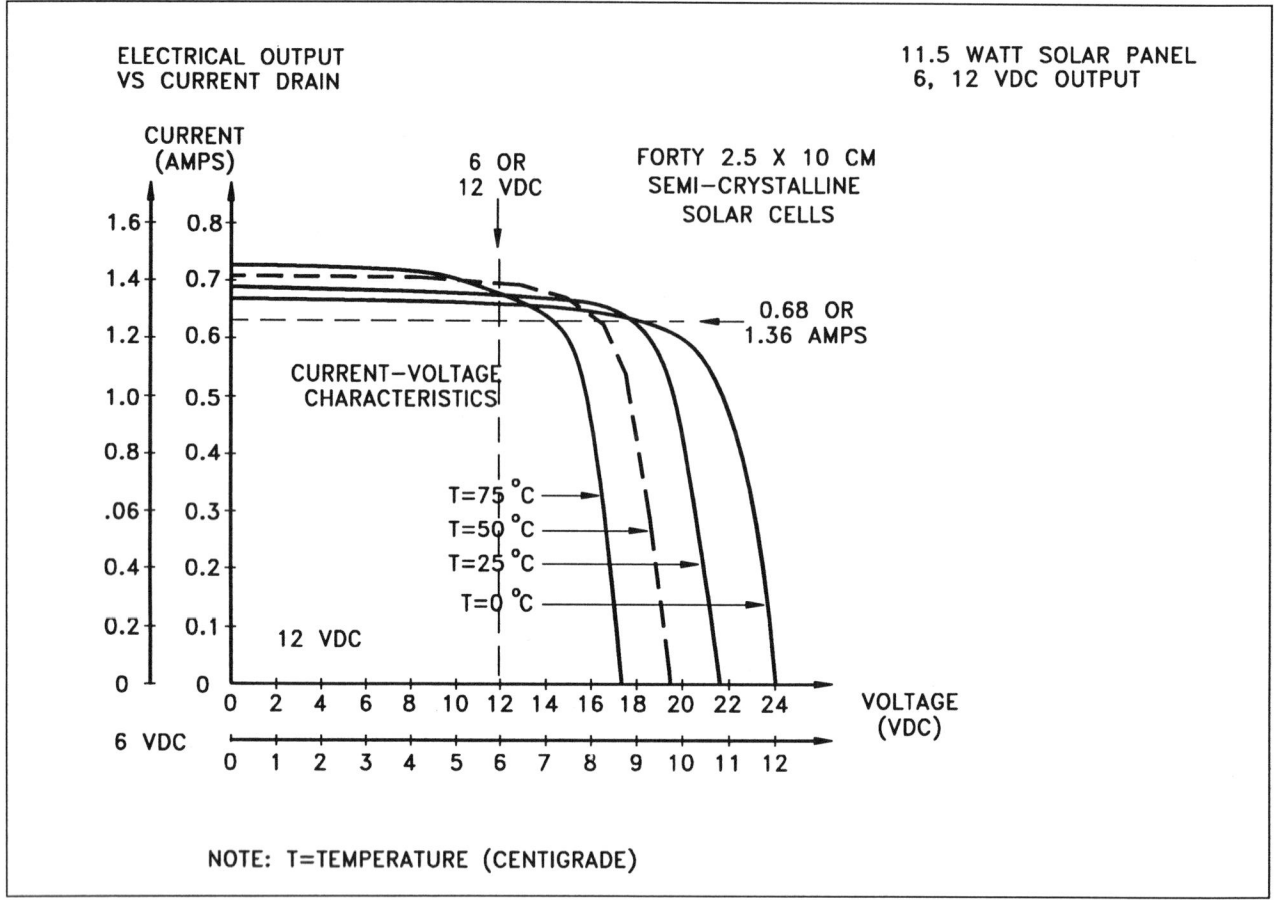

FIGURE 16-10 Solar panel module

source). At a given light intensity, a module output current is determined by the operating voltage: as voltage decreases, current decreases in conformance with the curve (Figure 16-10). Temperature also affects the performance, as shown by the electrical performance data in the figure and the table.

16.5.2 Flexible Solar Panel Array

A flexible, amorphous solar panel array is available for use in applications where a rigid structure is not suitable. These panels are used as primary DC power sources for equipment or for recharging batteries that are then used as the source of power. These flexible panels convert daylight sunlight into electricity and provide power from 3 to 280 watts, with charging levels of 1 to 80 ampere-hours per day. They are capable of charging Ni-Cad, gel-cell, and heavy-duty vehicular batteries. They are available in compact, lightweight, stowable packages that can be deployed rapidly for temporary or permanent use. The panels are completely weatherproof and are designed to operate under water if necessary. The units are rugged, shatterproof, and are designed to resist damage. In most cases they can continue to operate

even after being pierced by a hail of bullets. Many different sizes are available. A portable compact unit that provides from 3 to 10 watts is designed to charge batteries for portable CCTV cameras, transmitters, and recorders (Figure 16-11a).

An intermediate-size module powers from 15 to 100 watts. A 100-watt unit can power multiple cameras and transmitters and a low-power IR source at a remote location.

A vehicular battery-charging unit can maintain a full charge on two 100-amp-hour 12-volt batteries connected in series by supplying over 300 mA at 28.5 VDC (Figure 16-11b).

Figure 16-11c is a photograph of one of the flexible panels, which can be formed around a circular or other uneven shape without any harm caused to the solar panel.

Figure 16-12 shows a portable solar electric generator/battery charger designed primarily for military application. It is flexible, stowable for outdoor use, and can eliminate the need for rechargeable batteries in the field and eliminate the need for a spare primary battery or other power source. It provides 30 VDC with up to 1 amp current. It is designed to charge batteries such as the VB-542 and VB-590 used in many military operations for remote CCTV surveillance. This module takes advantage of the amor-

TYPE	SIZE (INCHES) HxWxT*	PEAK POWER (WATTS) 12 VDC	PEAK POWER (WATTS) 6 VDC	PEAK CURRENT (mA) 12 VDC	PEAK CURRENT (mA) 6 VDC	NUMBER OF CELLS	WEIGHT LBS (KG)	COMMENTS
SEMI CRYSTALLINE SILICON	9.5x7.25x0.9		2.5		300	20	2 (0.9)	
	13.5x10.25x0.8	5	5	260	520	40	4 (1.8)	
	17.5x12x2.0	10	10	680	1360	40	6 (2.7)	
	16.6x18.4x2.1	18.5	18.5	1060		36	6.5 (2.95)	
	23.3x19.8x2.1	30	30	1680			8.5 (3.86)	
AMORPHOUS SILICON	13.1x5.5x1.1	1.75		100			1.1 (0.5)	LOW POWER
	14.1x7.1x1.1		2.4		210		1.4 (0.65)	LOW POWER
	10.6x9.8x0.9	4.5		260			1.7 (0.77)	LOW POWER
	43.7x19.8x2.1	60		3400			1.59 (7.2)	HIGH POWER

24–28 VDC

TYPE	SIZE (INCHES)	PEAK POWER	PEAK CURRENT	WEIGHT LBS (KG)	COMMENTS
AMORPHOUS SILICON	17.5x12.4X0.13	8.5	250	2 (0.9)	MILITARY–MED POWER
	12.4x10.1x3**	28	800	8 (3.6)	MILITARY–HIGH POWER
	44.4x39.0x2.0	120	7000	30.8 (14)	HIGH POWER

*HEIGHT, WIDTH, THICKNESS

**PANEL FOLDED IN HALF

Table 16-8 Solar Panel Load Characteristics

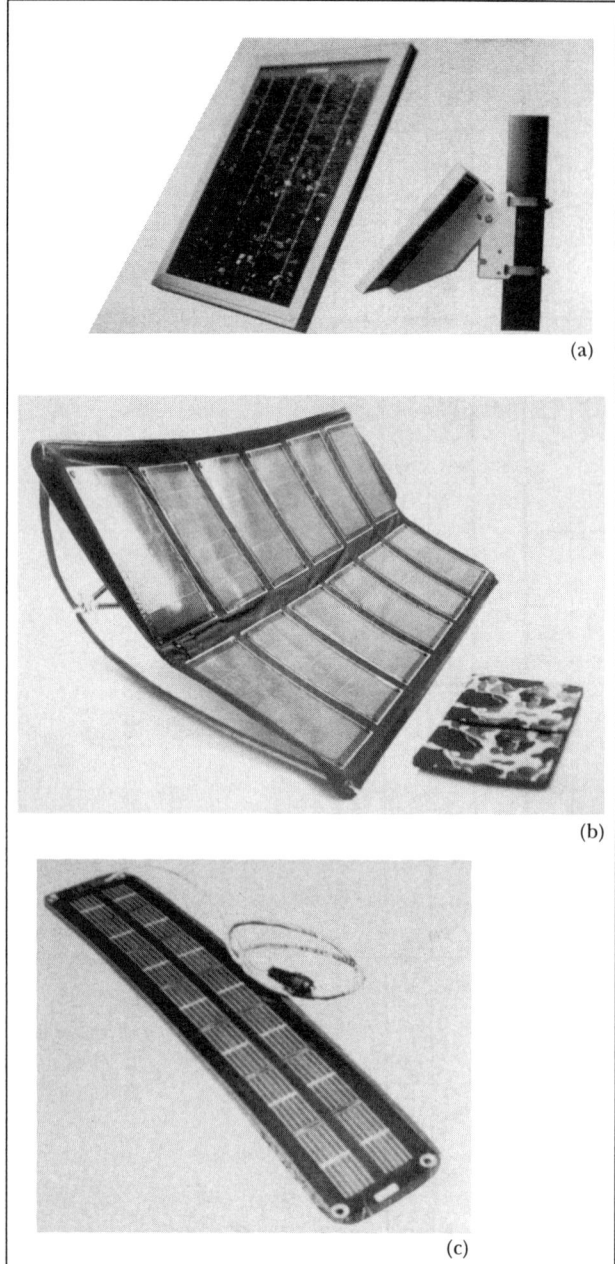

(a)

(b)

(c)

FIGURE 16-11 Portable and vehicular flexible solar panel arrays: (a) portable (3–10 watts), (b) vehicle (28.5 volts, 300 mA), (c) flexible

phous silicon alloy technology, in which the cell material is deposited in thin film layers onto a flexible substrate. The unit is lightweight, tough, nonbreakable, and easily transported and deployed when needed. The units are easily camouflaged when deployed, and since they emit no noise, heat, or other characteristic signature, they are self-camouflaged. The panel front surface is a dark color with a nonreflective finish. The outer covering is a weather-resistant nylon fabric.

16.6 POWER/SIGNAL DISTRIBUTION SYSTEM

Selecting the best power supply for security applications is only half the battle. There is still the problem of getting the power to the active electronic systems—camera, pan/tilt, switcher, monitor, and so on. A large percentage of security system equipment degradation can be traced to improper power distribution and/or ground loops. The symptoms of these inadequate designs are excessive noise in the picture or recorded image caused by (1) voltage spikes or other radiated noise into the system, (2) crosstalk between different signals, (3) AC power-line noise pickup (such as hum), or (4) poor load voltage regulation, causing erratic or unpredictable picture image degradation or equipment malfunctioning.

The primary function of any power distribution system is to provide a path for utility power (or other power) to reach the security equipment. Aside from meeting the electrical requirements of the security equipment, any wires or cables must meet the fire and safety codes of the locality in which they are used. Wire and cable must have insulation, voltage breakdown, and physical characteristics suitable for the voltage, current, power, and environment in which they will be used.

Regarding equipment operation, when a single security unit is connected to the utility or power supply output, the conductor sizes, that is, the wire diameter, must be sufficiently large so that there are no significant voltage drops on the conductors between the power supply and the equipment locations that would result in too low a voltage at the equipment. By specifying the proper conductor size based on the distance from the power source to the equipment

FIGURE 16-12 Portable solar electric generator/battery charger: 30 volts DC output at 1 amp (30 watts)

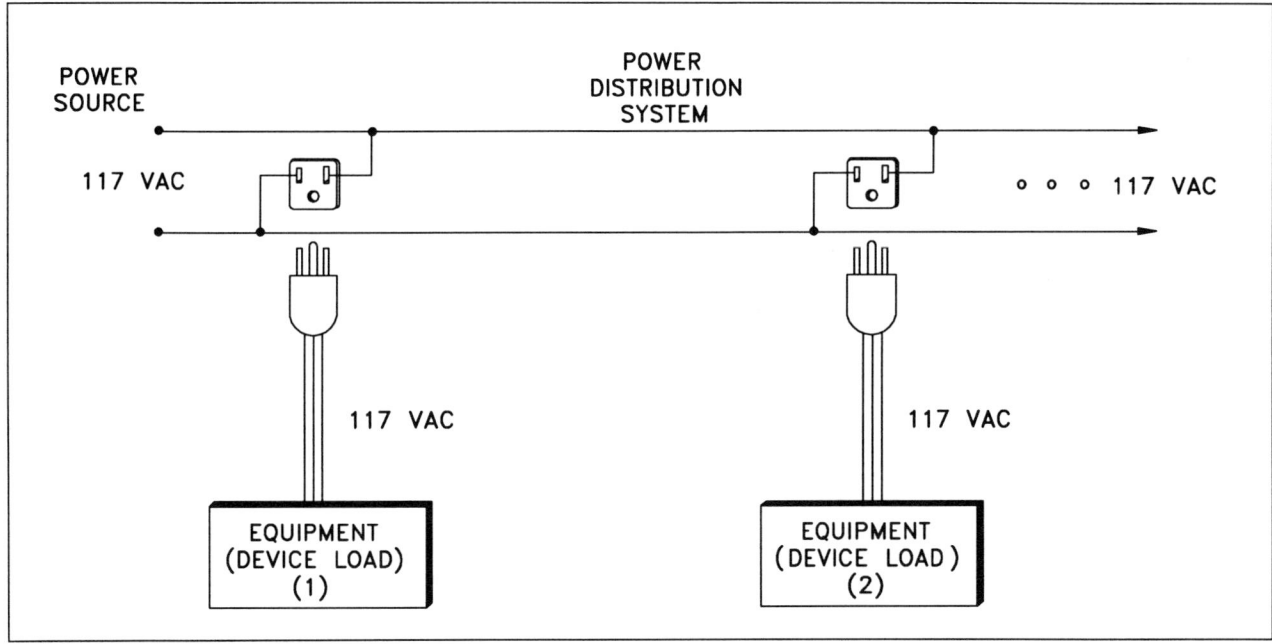

FIGURE 16-13 Parallel power source distribution system

and the amount of current the equipment requires, the voltage drop will be minimized. Unfortunately, when there are multiple loads connected to the same power source and the loads are at various distances from it, the calculations are more complex.

16.6.1 Distribution Techniques

In most security installations, the equipment is connected in parallel across the power source, as shown in Figure 16-13. Depending on the location at which each equipment is connected to the power conductors along the power line, each will receive a slightly different (lower) voltage, with the lowest voltage occurring at the last equipment on the line. For reliable operation, all equipments, including the one at the end of the line, must receive the minimum rated voltage for that equipment, usually 105 VAC in a 117-volt U.S. system. It is important that this voltage be available even during the normal line voltage variations provided by the utility company. As described previously, all equipments should receive their rated voltage during brownouts or other unusual circumstances, with a UPS used for this purpose.

Most electronic equipment functions with two generic cable types: signal (and control) and power. For safety purposes and to keep extraneous noise from contaminating signals, these two cable types are not routed in close proximity to each other. For safety reasons, most local electrical codes require that the electrical power cables be run in a separate conduit, not the same conduit as the signal cables. The reason for this is that the power cables are manufactured with insulation sufficient to withstand the higher voltages used in powering systems (120 volt, 240 volt), whereas the signal cables generally have insulation sufficient for voltages up to perhaps only 50 or 100 volts. To prevent accidental shorting and sparks and perhaps fire due to a voltage breakdown between the higher-voltage power cable and the low-voltage signal cable, these two cables must be in separate conduits.

With respect to contamination of the signal cable by power-line voltage transient surges, the signal and power cables use separate shielded cables or conduits to avoid this problem. A general practice is to shield the signal cables with solid foil or stranded shielding to prevent external electrical noise from reaching the signal cable conductors.

In larger CCTV security installations, the security equipment is powered with a parallel network. In more complex installations where several parallel power-line paths are used, or where equipment is located in different buildings or receiving power from different parts of the electric utility grid, the voltages may vary appreciably between different power distribution lines (Figure 16-14).

In these installations, several factors must be checked: (1) Are all voltages within the range required by the equipment? (2) Are there significant potential differences (more than a few volts) between the ground levels of each of the distribution lines? More often than not, there may be several volts—up to 25 volts—between what should be ground points on the ground line of each distribution line. If this condition exists, when the video equipment grounds are connected, the result will be damage to the equipment and/or degradation of the equipment's performance. As described in Chapter 6, hum bars appearing on the video image are a very apparent effect of a difference in ground voltage when multiple distribution lines are used. In more

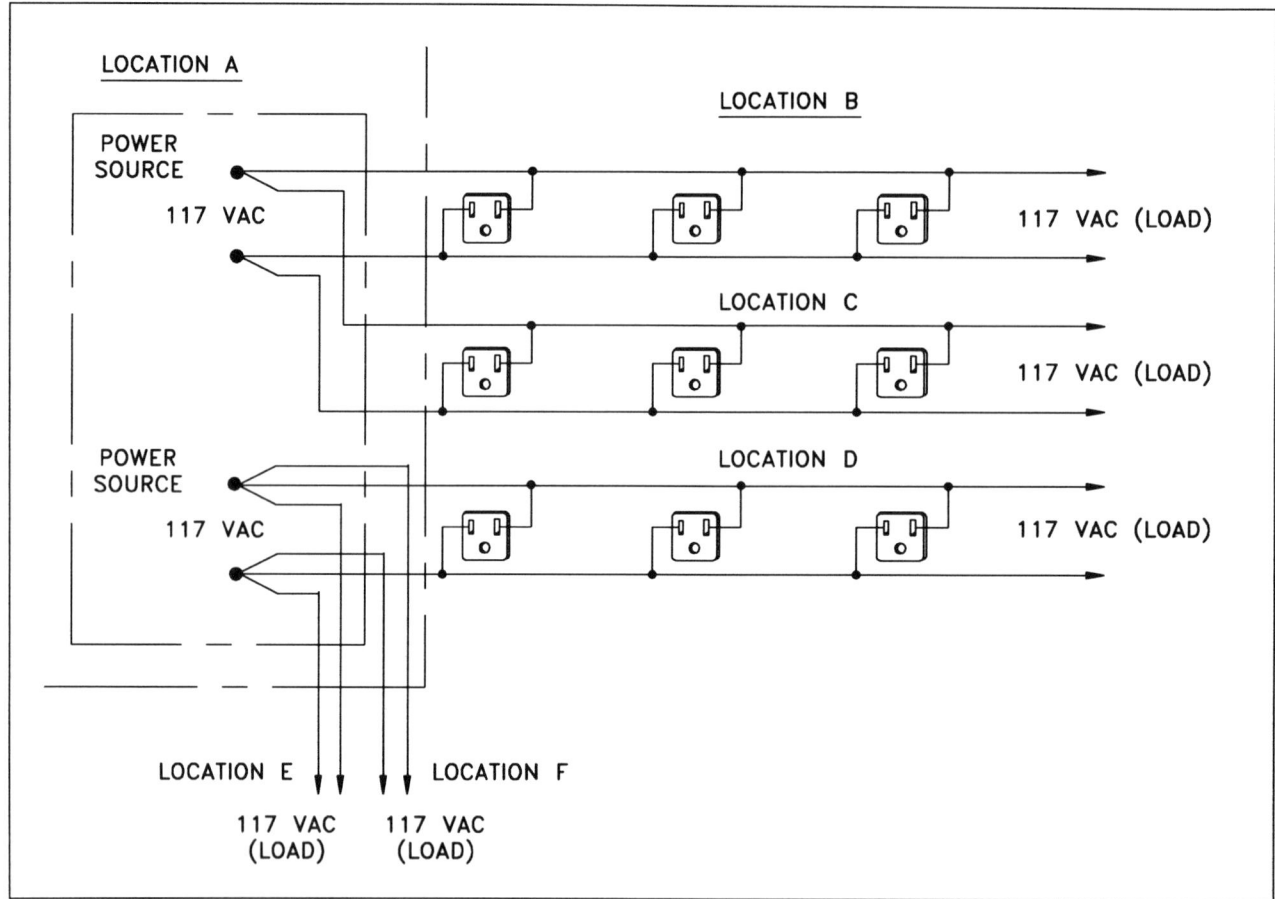

FIGURE 16-14   Multiple parallel power source grid

serious cases, equipment damage or personnel injury can occur when equipments at different ground levels are connected together. These problems (see Chapter 6) are eliminated when fiber-optic data and video links are used between the equipments, and/or electrical isolation transformers or optical couplers.

A second technique for connecting security equipment to power lines is called the radial system, in which each equipment is connected to the power supply terminals by a separate line (Figure 16-15).

Although there will still be different voltages at each equipment caused by unequal conductor lengths, these differences can be minimized by careful selection of conductor size. This technique eliminates some of the noise that may be induced on the power lines of sensitive equipment caused by heavy equipment on the line. Although the advantages of radial distribution over the parallel feed are evident on paper, they may be difficult to obtain in practice because of the many extra conductors required in the distribution system. In practice, a practical technique is a compromise, in which some equipments requiring only small amounts of current are combined in a parallel grid structure and the radial elements are used to power equipments with high current requirements. Figure 16-16 illustrates the combined use of the parallel and radial

distribution system, combining these two techniques. In this example, light loads are powered by parallel distribution lines while heavy loads are powered radially.

16.6.2 Ground Loops

An important aspect of the electrical distribution system is the use of good grounding practices and the elimination or avoidance of ground loops in the system. The optimum configuration can be determined only after a thorough analysis of the distribution system, as well as chassis and power-line grounding. Most electronic equipment is designed such that electronic circuitry is grounded at one or multiple points on the equipment chassis. The system should be designed so that these ground points are connected electrically with a good conductive path from each equipment to the next, and to the chassis or earth ground (green wire) of the power distribution grid. In the case of battery or solar systems, one point in the system is chosen and all equipment grounds brought back to this one point. In the installation of a UPS into the system, an appropriate chassis ground from the backup UPS or other voltage-transient suppression equipment must go back to the chassis ground. For safety purposes, all equipment chassis grounds

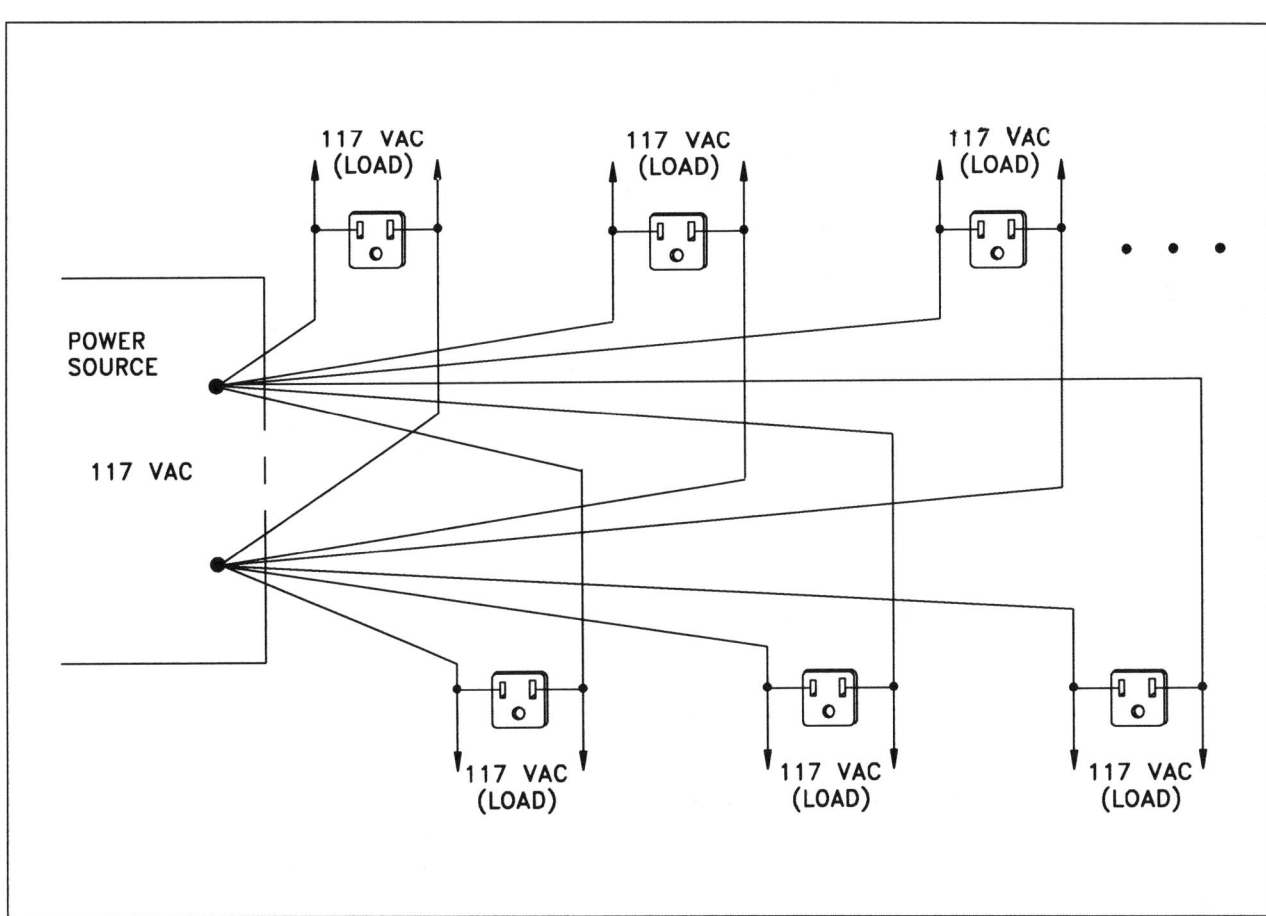

FIGURE 16-15   Radial power distribution systems

should be connected with an electrical wire to the mechanical frame housing the equipment and eventually the conductor going from this housing to earth ground through the third wire (green wire) in the system.

Ground loops are undesirable current paths coupling earthed points, equipment-ground points, or signal-return connections (Figure 16-17).

The finite resistance, inductance, and capacitance of conductors that carry these ground-loop currents provide a common cross-coupling impedance to inject unwanted signals, noise, and crosstalk into sensitive input circuits. Obviously, resistance and inductance can be minimized with heavy cable and short lengths. Also, when connecting several facility equipments to earth, approximately equal cable lengths and sizes should be used. To reduce or eliminate electrical interference from heavy machinery (motors, lights, air conditioner), use a radial power distribution system (Figure 16-18).

Equal impedances to a single earthed point help equalize potentials between different units and reduce both the intercoupling of interference and shock hazards. Independent cables to a single earthed point eliminate common impedance paths among different equipment units. A single earthed point avoids the ground-current circulation that can occur when several earthed points are used. Signals

from different locations should be isolated using fiber optics or isolation transformers.

Shielded signal cables are needed for interference-free signal handling. Cables carrying substantial power, which can induce interference into signal cables, also should be shielded—especially when both signals and power must flow in close proximity. Available types of shielded cable include single-wire, multi-wire, twisted-pair, coaxial, and multiple-shield.

Shields for all these types are made of braided metal, solid conduit, or metal foil. Braided shielding is light and easy to handle; its shielding effectiveness decreases with increasing frequency because of discontinuities and openings inherent in the weave. Shielding effectiveness, of course, depends on the type and thickness of the material used for the braid and the tightness of the weave. Both solid and foil shields are very effective, especially at high frequencies; the solid material has no discontinuities or openings.

The following guideline lists recommended practices to achieve noise-free CCTV picture performance; proper equipment grounding is a start.

• Coaxial cables should be terminated at both ends.
• Equipment chassis and cable shields should never be used for signal-return paths (with the exception of coaxial cable).

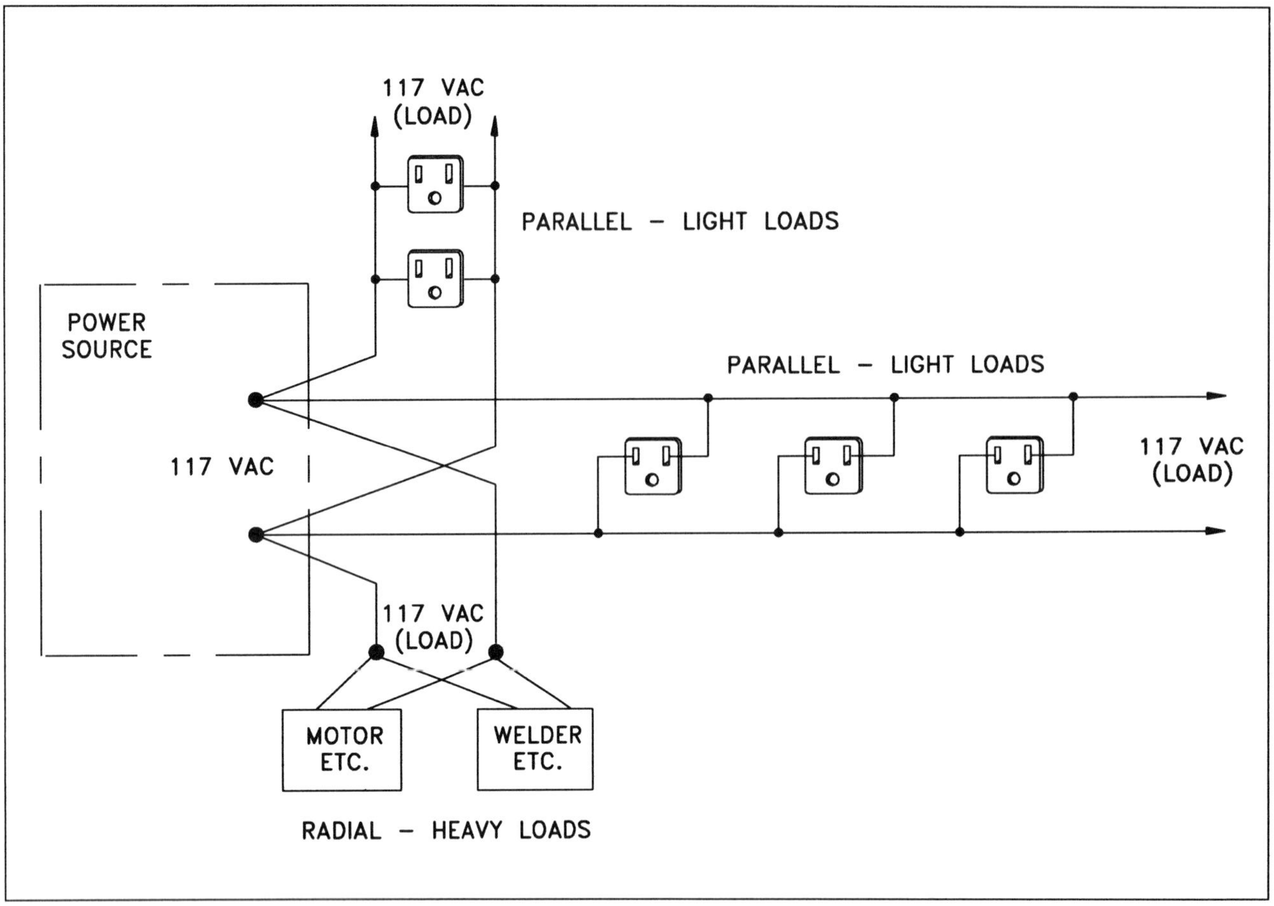

FIGURE 16-16 Parallel and radial combined power grid

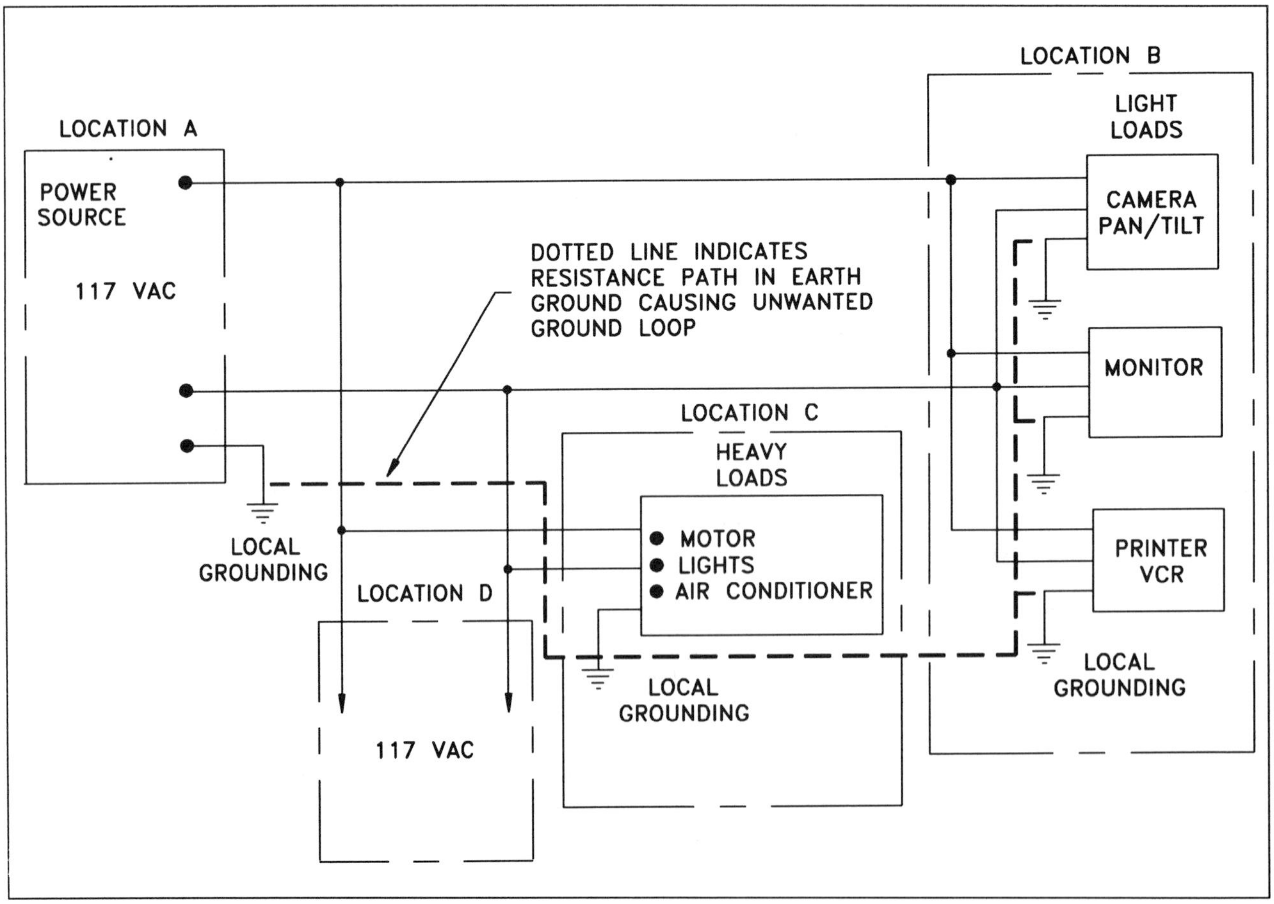

FIGURE 16-17 Potential ground loop sources

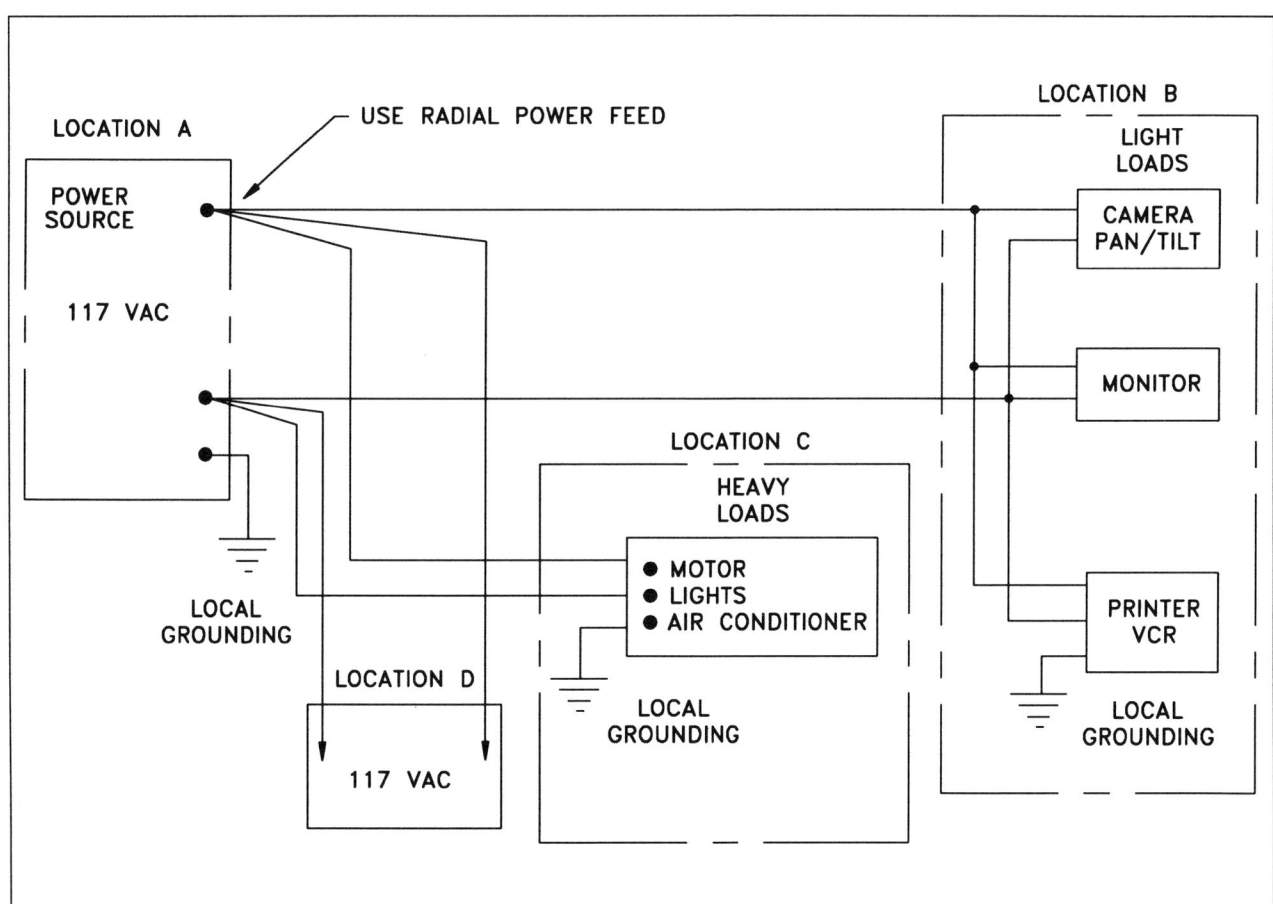

FIGURE 16-18 Practices for eliminating ground loops

- Signal circuits should use balanced cable, such as twisted-pair or shielded or balanced coaxial lines.
- Individual shields of balanced coaxial cables, when contained by a common shield, should be insulated from one another and from the common shield.
- Cables that carry high-level signals should not be bundled with cables carrying low-level signals, whether shielded or not.
- Each signal line should have its own independent return line running as close as possible to the signal line. In this way, the loop area is reduced, and common impedances with other signals avoided.
- Use twisted-pair cable (signal and return lines together) to ensure a minimum pickup-loop area. The voltage induced in one twist tends to cancel an oppositely induced voltage in the adjacent twist.

16.7 SUMMARY

Backup electrical power and power-line conditions are an essential consideration in any CCTV security system. UPS equipment is available and is economically prudent for all strategic security and safety equipment. The proper power-line filtering and conditioning saves time and money and increases security.

There is a choice of AC or DC power in the form of power line, power converters, batteries, and solar panels for powering all equipment necessary for any CCTV security application.

Distribution and ground loop problems should not be attacked with "black magic" fixes because these solutions tend to be temporary, and often create other chronic problems. A thoughtful examination of each current-demanding element in the system and its effect on the overall distribution network is far more valuable than a dozen quick fixes.

Improper grounding and shielding of power and signal lines is a major cause of noise interference in sensitive electronic equipment. Improper grounding occurs primarily because equipment designers often forget that every conductor to ground has resistance, inductance, and shunt capacitance. Moreover, when the ground conductors form so-called ground loops, they can inject, radiate, or pick up both low- and high-frequency interference.

PART 3

Chapter 17
Applications and Solutions

CONTENTS

17.1 OVERVIEW

Security and safety requirements encompass all the disciplines described in this book. The effectiveness of the CCTV system compared with its low investment cost attests to its widespread use. This chapter provides design guidelines and hardware information, specific case studies, and a checklist for representative security applications. The institutions, facilities, and surveillance areas cover a wide range: (1) government/industrial/business agencies, (2) retail stores, (3) correctional institutions, (4) banking facilities, and (5) lodging and casino establishments.

All the previous chapters have served as a basis for understanding the design requirements and hardware available to implement a practical CCTV security system. This chapter defines the hardware required and questions to ask (which must be answered) for several real cases. Each case states the problem and provides the information leading to a solution. A layout of the security problem identifies equipment locations and system requirements. A detailed block diagram serves to identify the functions, define the hardware requirements, and uncover potential problems. Each case solution includes a bill of materials (BOM) to define and choose the hardware. The BOM also serves as a basis for a request for quotation or a quotation from a vendor.

The BOM forms part of a checklist during the design and final system checkout after installation.

Within the limited space available in this chapter, several specific applications are analyzed to teach the user and practitioner the methodology used in designing a CCTV system. A layout, a block diagram, and BOM are provided for the following examples:

1. Three-Camera Lobby Surveillance System
2. Six-Camera Elevator Cab and Lobby System
3. Six-Camera Office Covert System
4. Twelve-Camera Parking Lot and Perimeter System
5. Eight-Camera Showroom Floor System

An overview of the application of CCTV to access control and how it relates to CCTV surveillance and integrates with electronic access control and other security systems is given. The use of CCTV access control is dealt with only briefly in this chapter since this topic is covered comprehensively in another book, *CCTV Access Control*.

CCTV equipment is used extensively to train security personnel in all aspects of security. It is a convenient, cost-effective, and powerful visual tool to acquaint new personnel with the physical facilities, the management, security, and safety procedures. Section 17.8 gives some examples of how CCTV can be used to train security guard personnel, and how to train installing technicians. An ingredient in the successful implementation and effective operation of any security system is a professional installer or installing company, as well as continued maintenance of the system. Section 17.9 outlines criteria to assist in choosing a reliable security installer.

When designing a CCTV security system, the equipment chosen depends on whether it will be used indoors or outdoors. In indoor applications such as lobbies, stairwells, stockrooms, elevators, or computer rooms, minimum environmental protection is required. The equipment may be required to operate under wide variations in light level if it is to be used for daytime and nighttime surveillance.

Outdoor equipment is subjected to environmental factors including extreme temperature and humidity, high winds, precipitation (rain, sleet, snow), dirt, dust, chemicals, and sand. The outdoor equipment must be designed to withstand and be serviceable under all these adverse environmental conditions. To maintain proper operation of lenses and cameras in outdoor environments, thermostatically controlled heaters and/or fans must be incorporated to maintain the interior of the housing within the temperature range of the camera and lens equipment. Periodic servicing of pan/tilt mechanisms must be performed, including lubrication of moving parts and checks for wear or deterioration of flexing or exposed wires.

CCTV equipment in outdoor applications operates under extreme variations in light level, ranging from low levels produced by artificial lighting for nighttime use to bright sunlight and sand or snow sun-reflected scenes. This often represents a one-million-to-one change in light level for which the camera system must compensate.

CCTV cameras and lenses are installed in indoor and outdoor environments using a simple camera bracket, a camera housing and bracket, or a recessed mounting in the ceiling or wall. Camera brackets serve to fix the camera and lens at a location but do not protect them from vandalism or the environment.

Indoor and outdoor CCTV equipment consists of camera mountings, camera housings, cameras, lenses, pan/tilt mechanisms, visible or IR illuminators, and the cable runs required to transmit the power and control signals to the equipment and to transmit the video signal and any other communication back to the console room. The security room equipment consists of monitors, VCRs, switchers, splitters, combiners, time/date generators, video printers, and other equipment.

17.2 GOVERNMENT/INDUSTRIAL/BUSINESS PROPERTY SURVEILLANCE

Applications covered in this section apply to facilities used by governmental agencies, manufacturers and industrial companies, and business services. CCTV surveillance applications include lobbies, elevators, offices, rear doors, shipping areas, hallways, and warehouses. Temporary and permanent covert CCTV systems are used in installations for viewing the general public entering a facility as well as monitoring employees. Wide-FOV optics and zoom lenses are used for parking lots and perimeter-fence-line environments.

17.2.1 Lobby Surveillance

A common surveillance area is an entrance lobby to a facility. The lobby has a front entrance door and one or more internal doors and is occupied by a receptionist, possibly a guard, visitors (business or public), and employees of the facility. The security functions to be performed

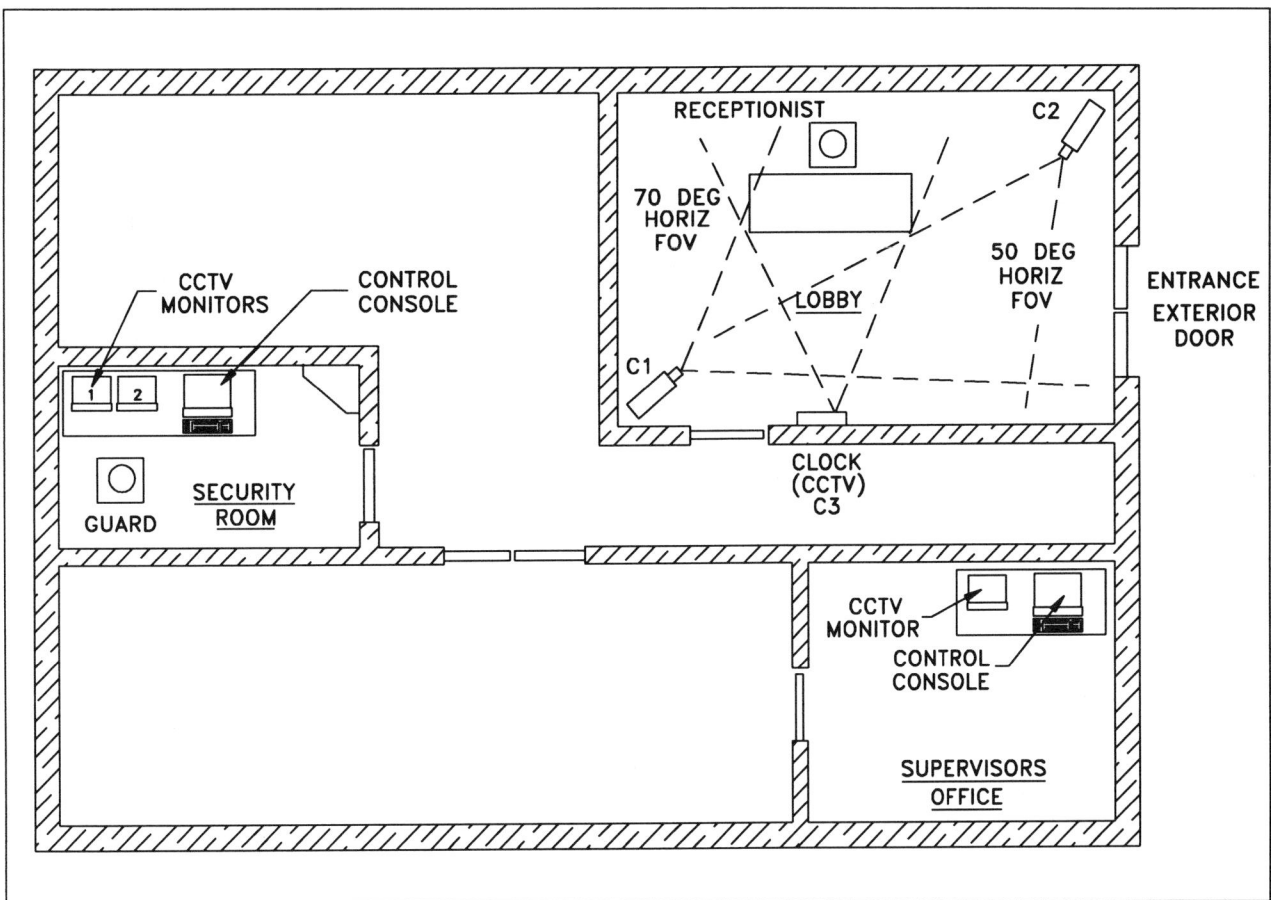

FIGURE 17-1 Three-camera lobby surveillance requirements

by the CCTV system, receptionist, and security staff include (1) viewing the lobby area to determine that order prevails, (2) monitoring and controlling entry and exit through the internal doors, (3) monitoring and controlling material movement in and out of the main entrance, and (4) guarding the receptionist for safety purposes. The following case describes a CCTV lobby surveillance system.

17.2.1.1 Case: Three-Camera Lobby Surveillance System

Figure 17-1 shows a simple but common CCTV problem: monitoring people entering and leaving the front lobby of a building, and the surveillance of the reception area.

One camera is located in the lobby close to the ceiling and looks at the entrance door and lobby with a lens having an FOV wide enough to see at least half of the lobby, the receptionist, and the front door. The second camera on the opposite wall views the other half of the lobby and the internal access door. The third camera is covertly mounted in a wall-mounted clock. Aside from choosing the equipment, choosing the camera locations is most important. The camera/lens should not be pointed in the direction of a bright light or the sun or toward an outside door or window if possible. If the sunlight enters the camera lens directly, blooming (white areas in scene) can obliterate a part of the scene. Newvicon, silicon tube, and CCD solid-state cameras using automatic-iris lenses or CCD cameras using shuttering will compensate for light-level changes but do not always compensate well for a very bright spot in a scene. A person under surveillance who is back-illuminated by light from an exterior window or door is usually seen on the monitor as a black silhouette and cannot be identified.

The video signals from the two overt CCTV cameras and covert camera in the lobby with the receptionist are transmitted to one or two monitors located at the remote security guard station. The two overt cameras view people entering and exiting through the front entrance door, most of the lobby, the receptionist, and the internal access door. The third covert camera provides backup in the event the overt cameras are rendered inoperative. Figure 17-2 shows the block diagram for this system.

What would a BOM for such a three-camera overt CCTV system look like? Table 17-1 lists the hardware necessary to complete such a system.

With each component is a list of parameters that must be specified in order to define it. A logical starting point is the camera type. The cameras can be monochrome or color, $1/3$-, $1/2$-, or $2/3$-inch format, C or CS lens mount, AC (24 or 117 VAC) or DC (12 VDC) powered. Choose a monochrome, $1/2$-inch format, C-mount, solid-state, 24-

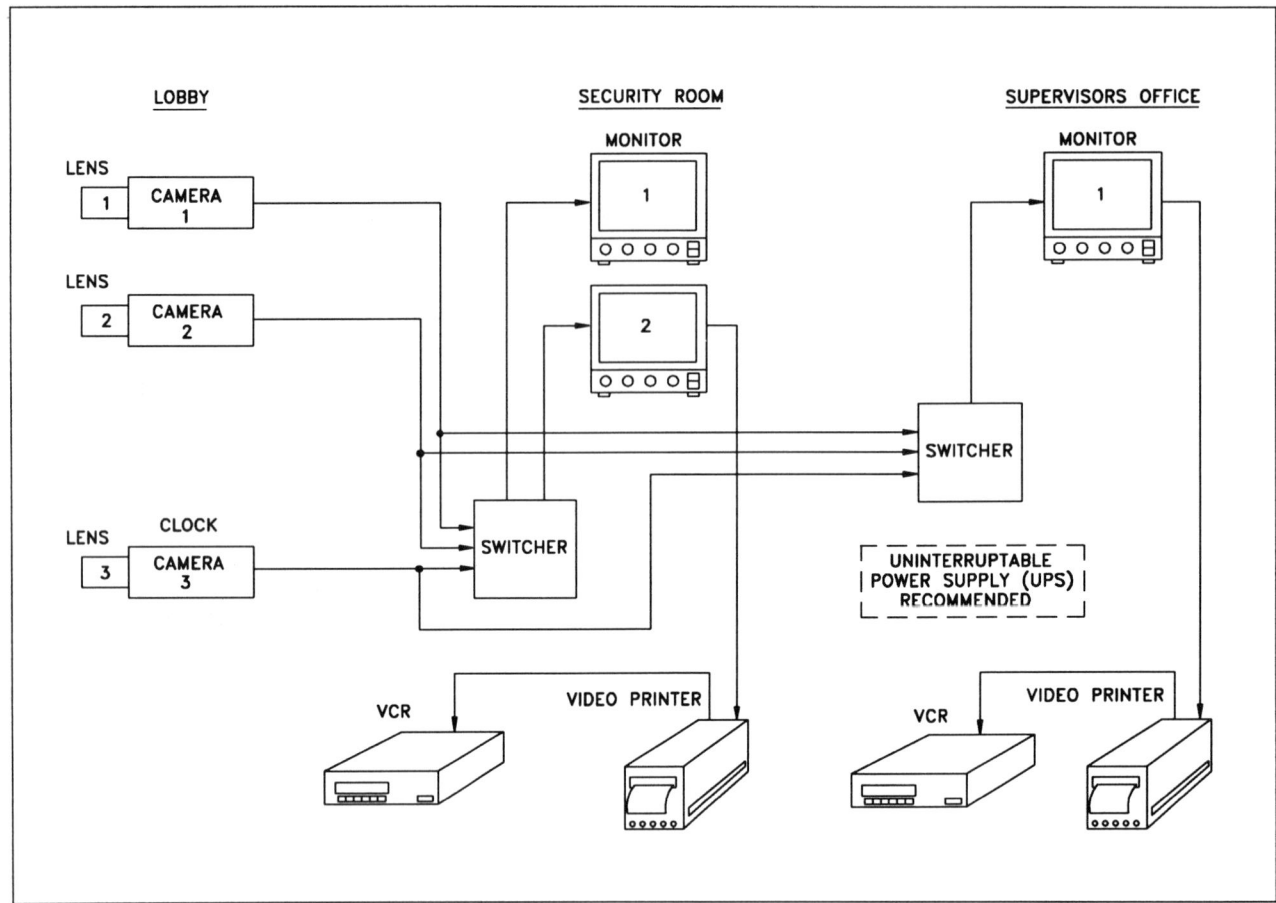

FIGURE 17-2   Three-camera CCTV lobby block diagram

VAC camera. From Figure 17-1, the lens horizontal FOV for each camera should be 50 degrees. Using Table 4-4 (or the Lens Finder Kit), a 6.5-mm FL automatic-iris lens capable of covering a ½-inch format, with a C mount, will do. An automatic-iris lens is suggested because there are outside windows and a door. Use a good quality RG59/U coaxial cable since the distance between camera and security room is less than 600 feet. Either three monitors or one monitor and a switcher can be used, depending on whether 100% full-time coverage or time-shared (switched) coverage is adequate.

The two overt cameras can be mounted directly on brackets or installed in housings and then bracket-mounted, depending on the vandalism/environmental factor. For a typical console, the monitor(s) should be 9-inch-diagonal for optimum viewing. The switcher is a four-position, alarming sequential one. The VCR is programmed to record the pictures sequentially, in time-lapse mode from cameras 1 and 2. In the event the receptionist (or other alarm input) sounds an alarm, the VCR automatically switches to real-time recording. The video printer is available to make hard-copy prints of any video scene from the monitor or VCR.

17.2.1.2   Lobby CCTV Checklist

- How many cameras are required to cover all pertinent areas?
- Will the light level remain relatively constant or will cameras view direct sunlight? Manual or automatic-iris lens?
- Should a covert camera be installed for backup?
- Should monochrome or color cameras and monitors be used?
- Should the VCR be time-lapse or should it record on manual demand or automatic (alarm input)?

17.2.2   Elevator Surveillance

Crimes against elevator passengers have increased significantly in recent years; the threat represents a potentially high liability to the owner of the elevator's building. The use of CCTV in elevators reduces the risk of harm being perpetrated on passengers in elevators. From an asset-protection point of view, elevators represent a valuable capital investment and should be under surveillance to deter vio-

Item	Qty.	Description	Location	Comments
1	2	Cameras 1, 2	Lobby	Type: Color or Monochrome; Sensor Type: CCD, MOS Sensor Format: $1/3$-, $1/2$-, or $2/3$-inch Lens Mount: C or CS (Depends on Camera) Automatic Iris or Electronic Shutter Input Voltage: 117 VAC or 24 VAC
2	2	Lenses 1, 2	Lobby	Iris: Manual or Automatic f/#: 1.4 to 1.8; Mount: C or CS (Depends on Camera) See Tables 4-1, 4-2, Lens Finder Kit
3	1	Camera 3	Lobby Small Camera in Wall Clock	Type: Monochrome; Sensor Type: CCD Sensor Format: $1/3$- or $1/2$-inch Lens Mount: Nonstandard Voltage: 12 VDC
4	1	Lens 3	Lobby	Iris: No Iris Mini-lens Focal Length: 8, 11 mm f/#: 1.6 to 1.8 Mount: Nonstandard
5	1	Housing: Wall Clock	Lobby	
6	2	Brackets: Cameras 1, 2	Lobby	Type: Wall or Ceiling Mounted
7	2	Housings: Cameras 1, 2	Lobby	Type: Wall or Ceiling Mounted Mounting: Bracket or Flush
8	1–4	Monitors	Security Room	Type: Monochrome or Color Screen Size: 9-inch Mounting: Desktop, Rack, Hanging Bracket Audio: With or Without
9	1	Switcher*	Security Room	Type: Manual or Automatic, Sequential, Homing, Looping, or Alarming Number of Camera Inputs: 4
10	1	Videocassette Recorder	Security Room, in Locked Compartment	Type: Real-time, Time-Lapse Format: VHS, 8 mm, VHS-C, SVHS, Hi-8
11	1	Video Printer	Security Room	Type: Thermal, Monochrome or Color Print size: 3.5×4.0 inches
12	—	Transmission Cable	Ceiling, Wall, Floor	Type: Coaxial, Fiber Optic, Twisted Pair Installation: Free Air, Conduit, Tray, Plenum
13	1	Uninterruptible Power Supply	Security Room	Recommended

Table 17-1 Three-Camera Lobby Surveillance Bill of Materials

*If switcher is used and there is only one viewing location, use one monitor. With no switcher, use three monitors.

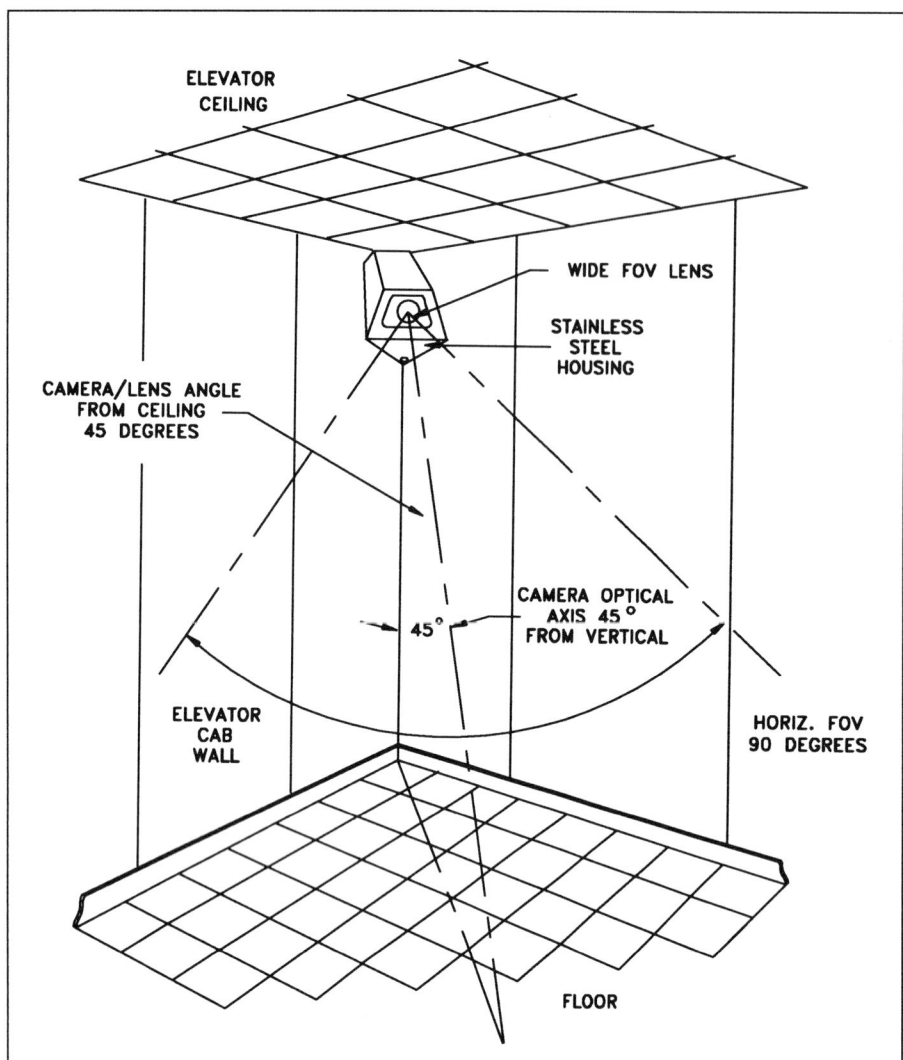

ELEVATOR
CEILING

WIDE FOV LENS

STAINLESS
STEEL
HOUSING

CAMERA/LENS ANGLE
FROM CEILING
45 DEGREES

CAMERA OPTICAL
AXIS 45°
FROM VERTICAL

45°

ELEVATOR
CAB
WALL

HORIZ. FOV
90 DEGREES

FLOOR

FIGURE 17-3 Elevator cab CCTV requirement

lators from defacing and vandalizing them. This application describes the equipment to monitor elevator cabs and the lobby.

A passenger entering an elevator with a stranger is temporarily "locked in" with the person until the elevator stops and the door opens on another floor. This serious problem has come about because almost all elevators are automatically controlled and there are no elevator operators. Years ago the elevator operator performed the function (whether consciously or not) of visual surveillance in the "locked" elevator cab. No one was ever alone in the elevator. The operator was a deterrent who could also assist in preventing molestations, robberies, and vandalism. This inherent protection now absent on most elevators is returned via the use of overt CCTV and audio intercom systems.

Repair costs resulting from vandalism in elevators are well known. Building owners expend thousands of dollars to "face-lift" elevator cab interiors that have been defaced and vandalized. Studies and actual practice have shown that the installation of television monitoring systems in public facilities has significantly decreased elevator cab vandalism, as well as crimes against individuals.

17.2.2.1 How to View 100% of the Elevator Cab

The essential requirement of an elevator CCTV surveillance system is to see a picture of the entire elevator interior, on a remote CCTV monitor. The picture should maximize the facial view of elevator cab occupants and have sufficient image quality and resolution to identify them.

A wide-FOV lens/camera system in an unobtrusive CCTV housing provides 100% visual coverage of an elevator interior. The picture is displayed remotely in the building lobby and/or manager's office for real-time surveillance. The potential elevator passenger entering the building lobby can view the lobby monitor and determine that the elevator is safe to enter. Likewise the passenger knows that everyone in the elevator is under surveillance by the security guard or other people in the lobby.

A study of the elevator cab and camera/lens FOV geometry shows that a wide-FOV camera located in a ceiling corner of the cab optimizes the elevator cab surveillance (Figure 17-3).

The camera system should have a 90-degree horizontal FOV, about a 70-degree vertical FOV, and a camera/lens

(a)

(b)

FIGURE 17-4   Wide-FOV elevator CCTV systems: (a) stainless-steel/painted steel corner mount, (b) polycarbonate corner mount

features for an elevator CCTV camera system include (1) a locked, hinged cover, (2) easy access to the lens and camera by maintenance personnel, (3) a removable, unbreakable window port made of polycarbonate material, and (4) an easily removable camera/lens assembly for servicing. The removable viewing window is usually 0.25-inch thick mar-resistant polycarbonate material. Tempered glass can be used in place of polycarbonate to provide a nonscratchable, chemical-resistant window.

Several other camera/lens/housing configurations are available to mount directly into the ceiling corner of the elevator cab.

The rears of the camera/lens housings are contoured to fit directly into the ceiling corner of the elevator cab, with the front of the housing projecting as little as possible into the cab interior.

These systems use wide-angle lenses, a 4.8-mm FL for a $\frac{2}{3}$-inch camera format, and a 3.5-mm FL for a $\frac{1}{2}$-inch camera. With either combination, approximately a 90-degree horizontal by 70-degree vertical FOV is obtained.

Figure 17-5 shows a photograph taken from a CCTV monitor of the camera scene in a 5 by 7-foot elevator.

An installation requiring a covert elevator CCTV is accomplished by using a small CCTV camera, a small remote-head camera, or right-angle optics (Figure 17-6).

A small camera ($1\frac{1}{4} \times 1\frac{1}{4} \times 2\frac{1}{8}$ inches) can be installed inside an elevator cab wall or ceiling with only the lens showing. The remote-head camera is mounted in the elevator cab wall or ceiling with only the wide-FOV lens in view in the cab.

The right-angle optical system bends the wide-field image by 90 degrees without vignetting (loss of light at the edge of the image) so that the camera can be mounted parallel to the cab ceiling. The camera right-angle adapter and wide-FOV lens are completely above the cab ceiling,

optical axis directed 45 degrees from each wall and 45 degrees down from the horizontal (ceiling). This viewing geometry results in 100% coverage of the elevator volume and provides excellent probability of occupant and activity identification. Another essential requirement for the cab CCTV camera system is that it be as unobtrusive and as small as possible, since the elevator cab is a confined area and space is at a premium.

As with other security-related equipment, it is essential that the camera system in the cab—which is exposed to abuse by potential vandals—be constructed to be vandal-proof, and to adequately protect the lens and camera inside. One durable housing enclosure material is stainless steel (Figure 17-4a).

A brushed, rippled, or textured finish hides the vandal's attempts to deface the surface. A second choice of material is polycarbonate plastic (Figure 17-4b). Other desirable

FIGURE 17-5   Elevator cab CCTV picture: a man trying to hide in corner under camera

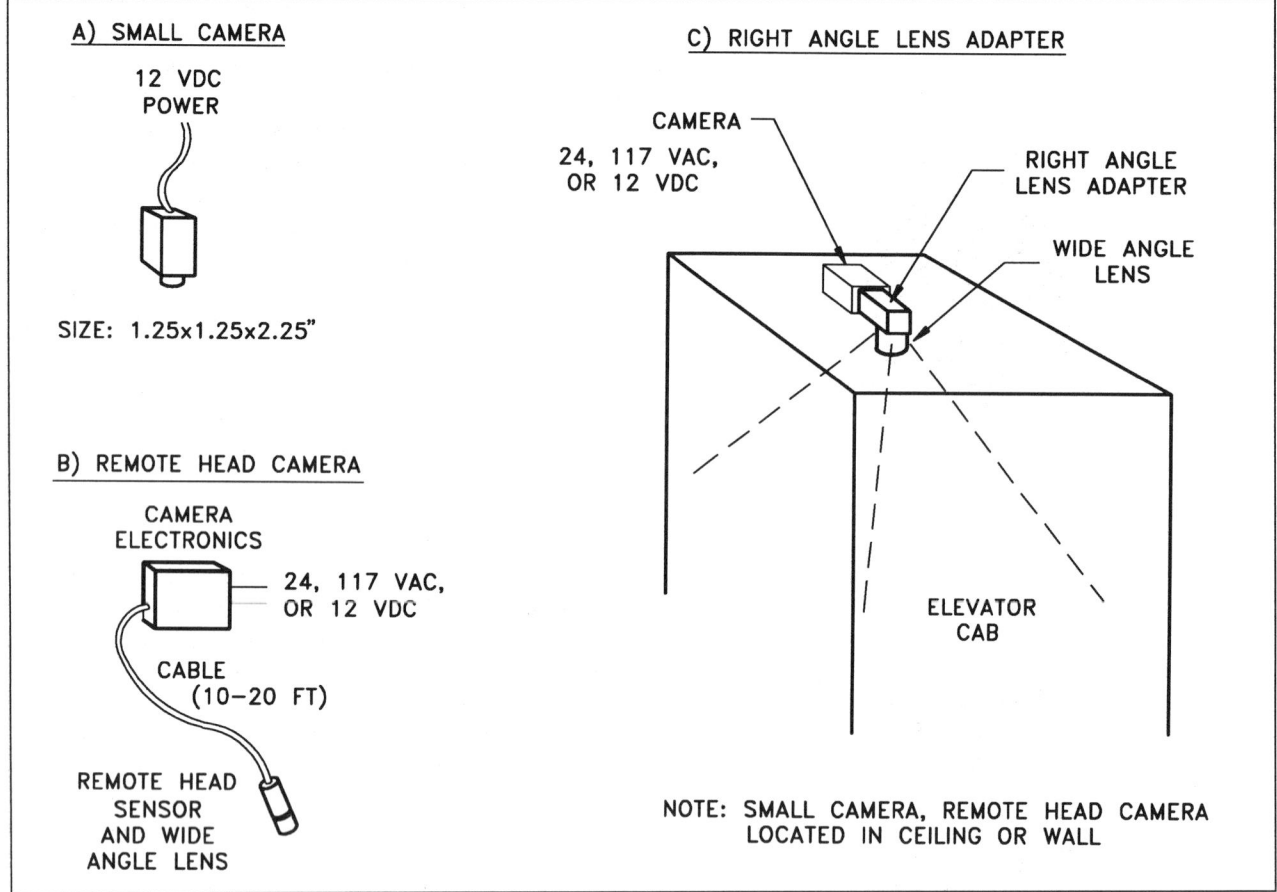

A) SMALL CAMERA

12 VDC
POWER

SIZE: 1.25x1.25x2.25"

B) REMOTE HEAD CAMERA

CAMERA
ELECTRONICS

24, 117 VAC,
OR 12 VDC

CABLE
(10-20 FT)

REMOTE HEAD
SENSOR
AND WIDE
ANGLE LENS

C) RIGHT ANGLE LENS ADAPTER

CAMERA

24, 117 VAC,
OR 12 VDC

RIGHT ANGLE
LENS ADAPTER

WIDE ANGLE
LENS

ELEVATOR
CAB

NOTE: SMALL CAMERA, REMOTE HEAD CAMERA
LOCATED IN CEILING OR WALL

FIGURE 17-6   Covert elevator cab CCTV systems

with only the wide-FOV lens protruding into the cab. The ceiling camera location is least conspicuous to passengers but has the disadvantage of viewing all occupants from overhead, so that fewer facial views are obtained.

In all of these installations, depending on the elevator cab construction and aesthetics required, the lenses can be hidden behind a semitransparent window to make them completely covert.

17.2.2.2 Case: Six-Camera Elevator Cab and Lobby System

Figure 17-7 shows the six-camera elevator cab and lobby CCTV surveillance system. The system consists of (1) one CCTV camera/lens and housing unit installed in each of the elevator cabs (four total); (2) two cameras and lenses in the lobby; (3) four video-transmission cables from the cab cameras to the console room; (4) a video switcher or quad splitter; (5) three monitors, one in the console room and two in the lobby; (6) a time-lapse/alarm VCR; and (7) a video hard-copy printer.

Two monitors in the main floor elevator lobby display a quad picture showing the interior of the four elevator cabs. The scenes from the elevator cabs and the two cameras in the lobby are displayed on the two monitors in the security

console. A time-lapse VCR in the console room records the elevator or lobby scenes, and a video printer prints out hard copy on demand. Figure 17-8 shows the block diagram for the six-camera elevator cab and lobby surveillance system.

In operation, each CCTV camera in the elevator cab (1, 2, 3, and 4) views the interior of its respective cab and transmits the video picture via the traveling coaxial cable in the elevator shaft to a stationary junction box halfway up the elevator shaft. The video signals are then transmitted conventionally over coaxial cable to the quad splitter/combiner and video switcher, and then on to the monitors, VCR, and video printer.

The camera power is obtained from an existing 117-VAC (or 24-VAC) power outlet in the cab ceiling on top of the cab. The video signal is transmitted from the elevator cab to the security console via coaxial or fiber-optic cable. This is not a routine television cable installation, and elevator maintenance personnel should be contacted for information pertinent to the specific installation. The video output is transmitted via a very flexible coaxial cable along the existing traveling cables in the elevator shaft, down to the lobby. The video cable is laced in with the existing cable run, with one end starting at the elevator cab and ending at a fixed junction terminal box halfway up the elevator shaft. From the junction terminal box, standard coaxial or

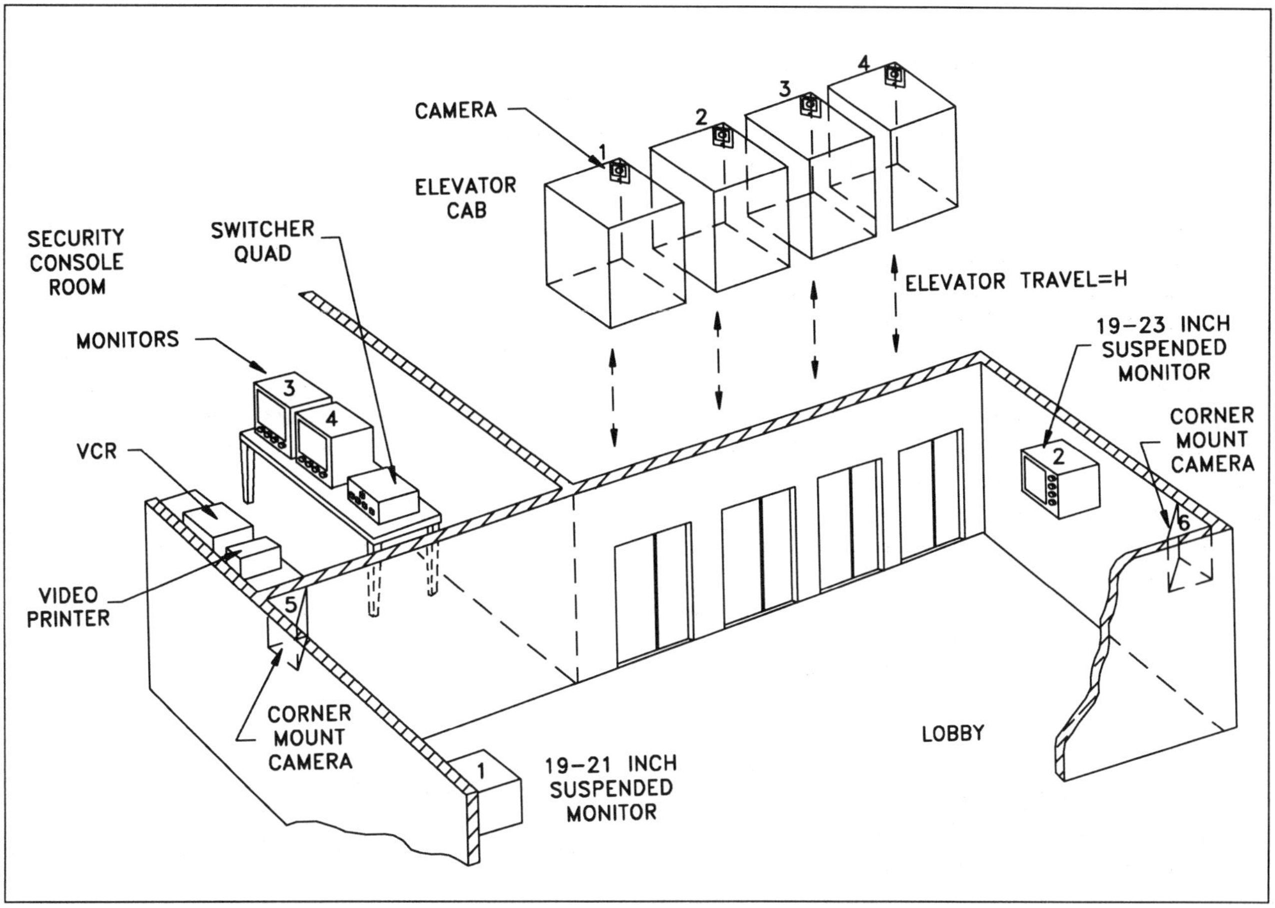

FIGURE 17-7   Six-camera elevator cab and lobby requirements

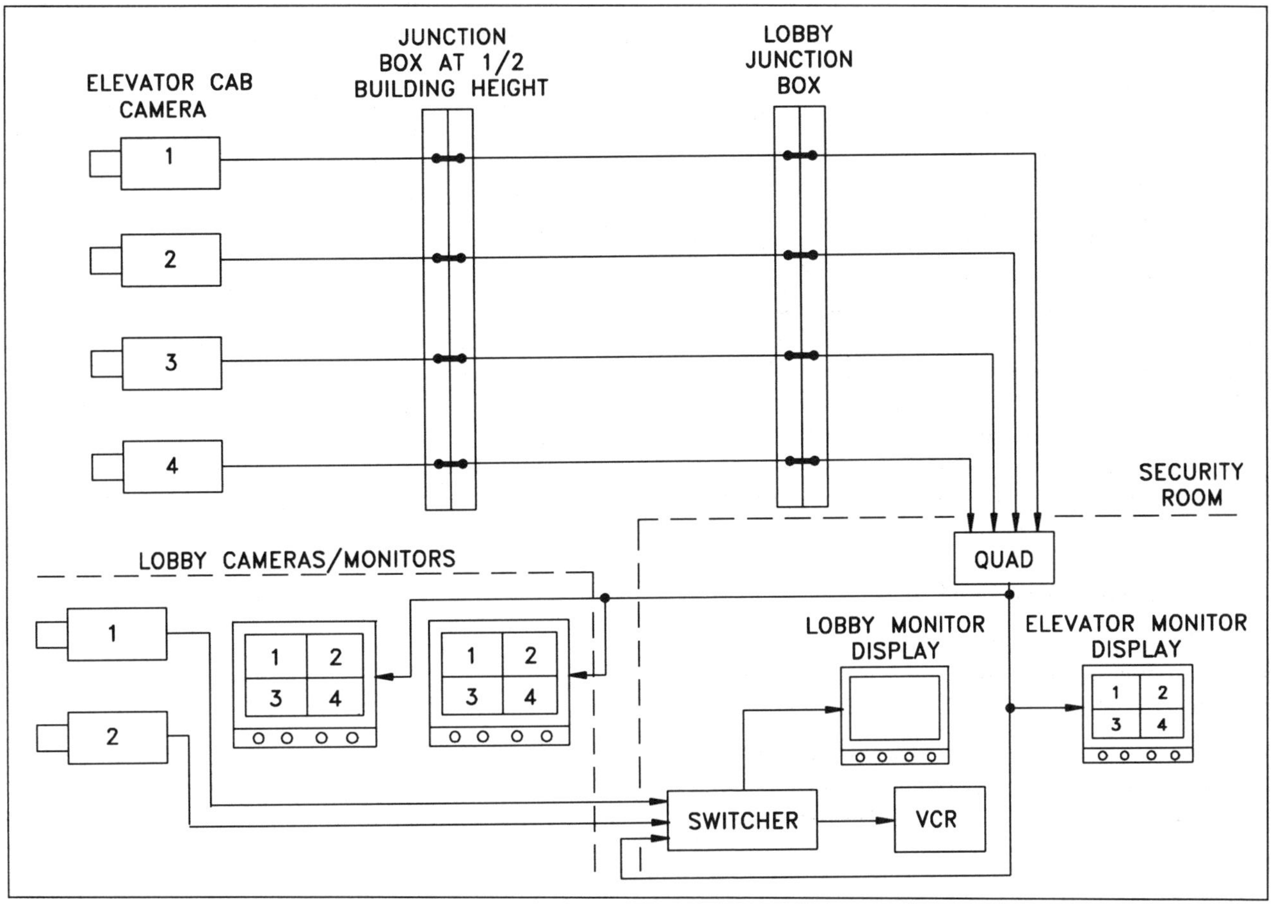

FIGURE 17-8   Six-camera elevator cab and lobby surveillance block diagram

fiber-optic cable techniques are used. The length of coaxial cable needed is equal to the height of the elevator travel (H). The location of the junction box is usually midway up the elevator shaft ($\frac{1}{2}$ H), minimizing the cable length.

Installing the traveling cable uses techniques familiar to experienced elevator installers and maintenance personnel. Since the cable flexes, bends, and twists each time the elevator cab moves, the coaxial cable chosen must be flexible. Stranded-type RG59/U is the most common coaxial cable used.

Fiber-optic cable offers the advantages of electrical noise immunity for the transmission of the video signal from the elevator cab to the junction box (and on to the monitor). Since the elevator shaft is an electrically "noisy" environment (electronics, relays, motors, switches, and so on in constant use), the use of fiber optics completely eliminates the possibility of electrical interference (see Chapter 6).

Table 17-2 is a BOM for the six-camera elevator cab surveillance system. The elevator camera can be a $\frac{1}{2}$- or $\frac{2}{3}$-inch format CCD with a C or CS lens mount. Since the lighting from floor to floor varies (whether from windows or varying types of fixtures), an automatic-iris lens or a manual-iris lens and a shuttered camera is chosen. From Tables 4-1, 4-2, and 4-4, a 3.5-mm FL will see 88 degrees horizontal on a $\frac{1}{2}$-inch format sensor and a 4.8-mm FL lens will see 95 degrees horizontal on a $\frac{2}{3}$-inch format. To obtain the 60-degree FOV required for the lobby lenses, a 12.5-mm FL lens on a $\frac{1}{2}$-inch or a 16-mm FL on a $\frac{2}{3}$-inch are used.

The most durable housing for an elevator is the stainless-steel design in Figure 17-4. Electrical access holes are drilled in the elevator cab, the housing is mounted, and the camera/lens is installed.

The lobby monitors should be large—17- to 23-inch diagonal—depending on lobby size and viewing distance, and mounted from ceiling or wall brackets. The faces and activities of persons in the elevator cab as seen from these monitors should be visible by anyone in the lobby. The two monitors in the security room are 9-inch, for maximum resolution for the guard. For recording, there are four VCR formats to choose from. The most common is VHS. The 8-mm is newer and provides a smaller cassette. The S-VHS and Hi-8 should be used when highest resolution is required. The video printer provides a convenient means to issue a hard-copy picture printout to dispatch a guard or for later apprehension or prosecution. For court prosecution, annotated time and date on the video frame are necessary.

17.2.2.3 Elevator CCTV Checklist

The following points should be considered for any elevator security requirement to provide maximum deterrence of crime with minimum intimidation of regular passengers.

- Should the cab CCTV be covertly or overtly installed? An overt system with warning labels can deter a would-be offender from committing a crime. The covert system has

the advantage that the offender is not aware that he is being viewed and immediate apprehension by a guard is more probable.
- Some passengers may be intimidated by overt CCTV; a concealed camera system may be less offensive.
- The elevator cab camera/lens should have a wide FOV. The optimum system has a 90-degree horizontal FOV (wall to wall) and about a 70-degree vertical FOV.
- No one should be able to hide anywhere in the elevator.
- The camera housing and window should be vandal-proof and be constructed to withstand direct impact from destructive blows. It must have a vandal-proof keylock.
- The elevator lighting should be sufficient to use a standard CCD camera.
- Does the situation warrant a monitor on each floor for added protection? As a minimum, most installations use a monitor in the lobby and the security room.
- Is a permanent VCR record of the monitor picture required?

17.2.3 Office Surveillance

Numerous government, industrial, and business facilities suffer millions of dollars in lost assets removed from office facilities by dishonest employees or maintenance contractors. Business hardware such as calculators, telephones, computers, and facsimile and copying machines are lost regularly. Software programs and company data files are physically or electronically copied and removed.

The most effective measure an institution can take to reduce or eliminate these losses is the installation of covert CCTV in those affected areas. The following section describes concealed CCTV camera/lens equipment for an office environment.

17.2.3.1 Case: Six-Camera Office Covert System

The purpose for covert television surveillance is to observe persons carrying out normal behavior as well as those carrying out unlawful acts. The CCTV system should assist in the apprehension and prosecution of individuals carrying out unlawful acts. The covert system can also act as a backup to an overt CCTV system in the event the overt system is sabotaged or has become inoperative in some way.

Figure 17-9 shows a layout of an office environment and covert cameras concealed judiciously in objects found in the office and the ceiling. The small cameras and lenses permit the covert CCTV to be easily installed in a ceiling, wall, or other article or fixture in the facility. Lenses and cameras can be installed in lamps, pictures, exit signs, emergency lights, radios, file cabinets, computers, and so on, providing windows of opportunity to view activities in a facility.

Item	Qty.	Description	Location	Comments
1	6	Camera	One in Each Elevator Cab (4 Total) Two in Lobby	Type: Monochrome Sensor Type: CCD Sensor Format: ⅓-, ½-, or ⅔-inch Lens Mount: C or CS Automatic Iris or Electronic Shutter Input Voltage: 117 VAC or 24 VAC
2	4	Lens	Elevator Cab	Iris: Manual or Automatic Focal Length: 3.5 mm on ½-inch or 4.8 mm on ⅔-inch f/#: 1.8 Mount: C or CS See Tables 4-1, 4-2, Lens Finder Kit
3	2	Lens	Lobby	Iris: Manual or Automatic Focal Length: 8 mm on ½-inch or 12.5 mm on ⅔-inch f/#: 1.4 to 1.8 Mount: C or CS See Tables 4-1, 4-2, Lens Finder Kit
4	4	Housing (if Overt)	Elevator Cab	Type: Indoor, Triangular, Vandalproof Material: Stainless Steel, Polycarbonate
5	2	Monitor	Lobby	Type: Monochrome Screen Size: 19, 21, or 23-inch Mounting: Hanging Bracket Audio: With or Without
6	2	Monitor	Security Room	Type: Monochrome Screen Size: 9-inch Mounting: Desktop, Rack Audio: With or Without
7	1	Videocassette Recorder	Security Room, in Locked Compartment	Type: Real-time, Time-lapse Format: VHS, 8 mm, VHS-C, S-VHS, Hi-8
8	1	Video Printer	Security Room	Type: Thermal, Monochrome or Color Print Size: 3.5 × 4.0 inches
9	1	Time/Date Generator	Security Room	
10	—	Transmission Cable*	Laced to Traveling Junction Box Termination in Shaft	Type: Coaxial, Fiber Optic, or Twisted Pair Installation: Plenum, Conduit in Wall

Table 17-2 Six-Camera Elevator Cab and Lobby Bill of Materials

*Option: IR atmospheric LED transmitter from elevator cab to receiver in junction box at top of elevator shaft. Coaxial or fiber-optic cable from box to security room.

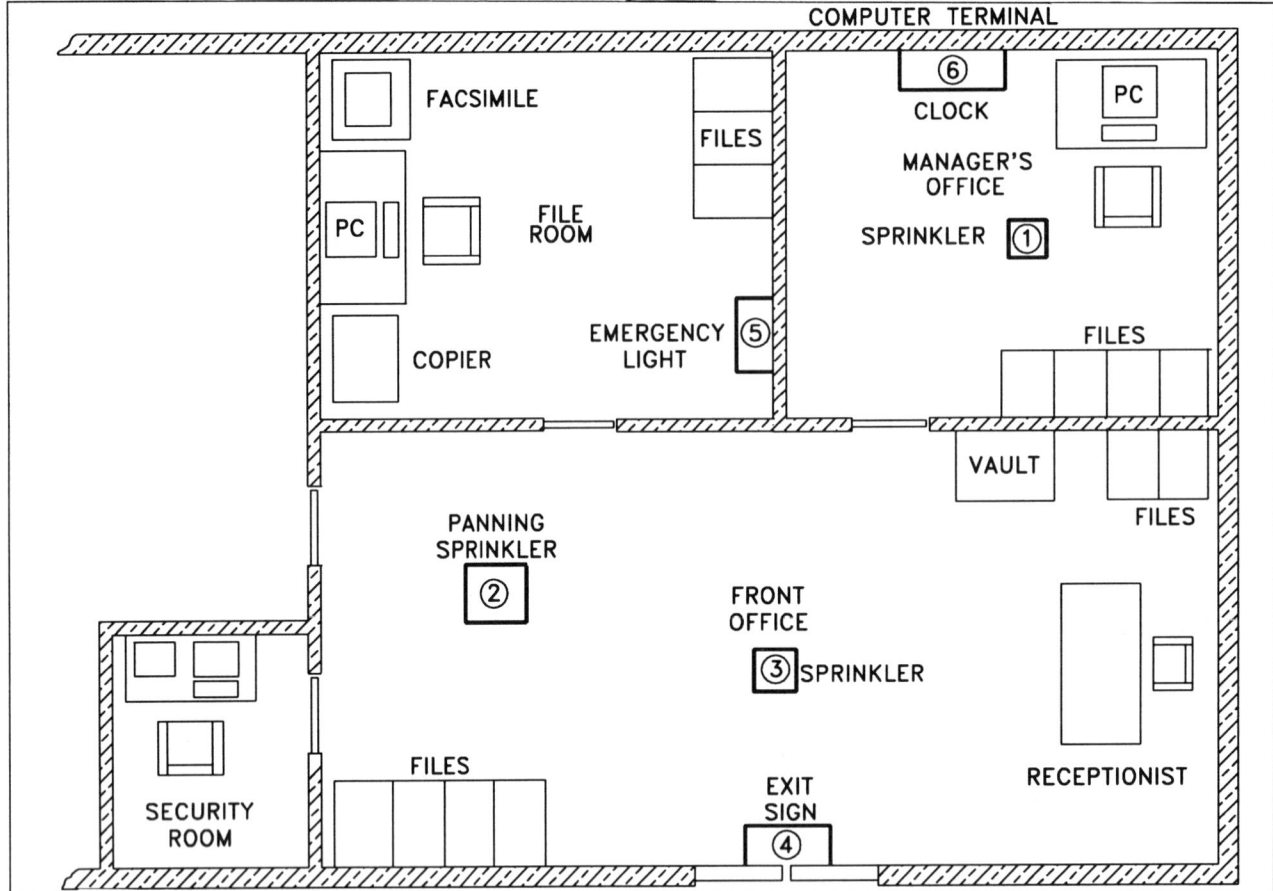

FIGURE 17-9 Six-camera covert office requirements

The pinhole lens disguised as a water sprinkler system mounted in the office ceiling provides an excellent view of office activities. This unique surveillance system is unobtrusively disguised as a conventional sprinkler head, as used in countless industrial and office premises. A straight or right-angle pinhole lens on a camera has a sprinkler head fixture mounted to the front of the lens with a beam-directing mirror to aim the camera FOV at the desired location. Only the sprinkler head and small mirror assembly protrude below the ceiling. The mirror can be adjusted over a range of angles to provide effective covert surveillance from a horizontal view, down to approximately minus 60 degrees. One having a straight down view is also available. At normal viewing distances of 10 to 20 feet, it is unlikely that any observer will recognize or be able to detect the mirror in the sprinkler head.

As with any other CCTV camera, power is obtained from (1) 117 or 24 VAC or (2) 12 VDC, with the video coaxial cable run to a monitor or recorder, or (3) with power up the coax (vidiplex). Low-power RF, microwave, or IR transmitters can provide transmission from the camera location to the monitoring location if it is impractical to install coaxial cable, other hard wire, or fiber optics between the two points.

Areas and objects of particular interest for theft in an office environment (Figure 17-9) include a vault, file cabinets, computers, facsimiles and copiers, software, and company records. The figure shows many suitable covert camera/lens locations. Three cameras and lenses are ceiling-mounted and disguised as water sprinkler heads. One of the sprinkler heads has a 360-degree panning capability to increase surveillance coverage and follow a moving person or activity in the scene. The remaining three cameras are hidden in an emergency lighting fixture, a wall clock, and an exit sign.

Figure 17-10 is a block diagram of the hardware and electrical interconnections required for the system.

The signals to and from the cameras, lenses, and panning units are transmitted to the security room via hard-wire, fiber-optic, or wireless means. The console equipment includes monitors, VCR(s), and a hard-copy video printer.

Table 17-3 is a bill of materials for the six-camera, covert office surveillance system.

The camera/lens in the clock in the manager's office uses a very small CCD camera and offset mini-lens optics (Chapter 4) to view the office door and files. The stationary ceiling sprinkler monitors the manager's desk and computer terminal. For the file room, an operating wall-

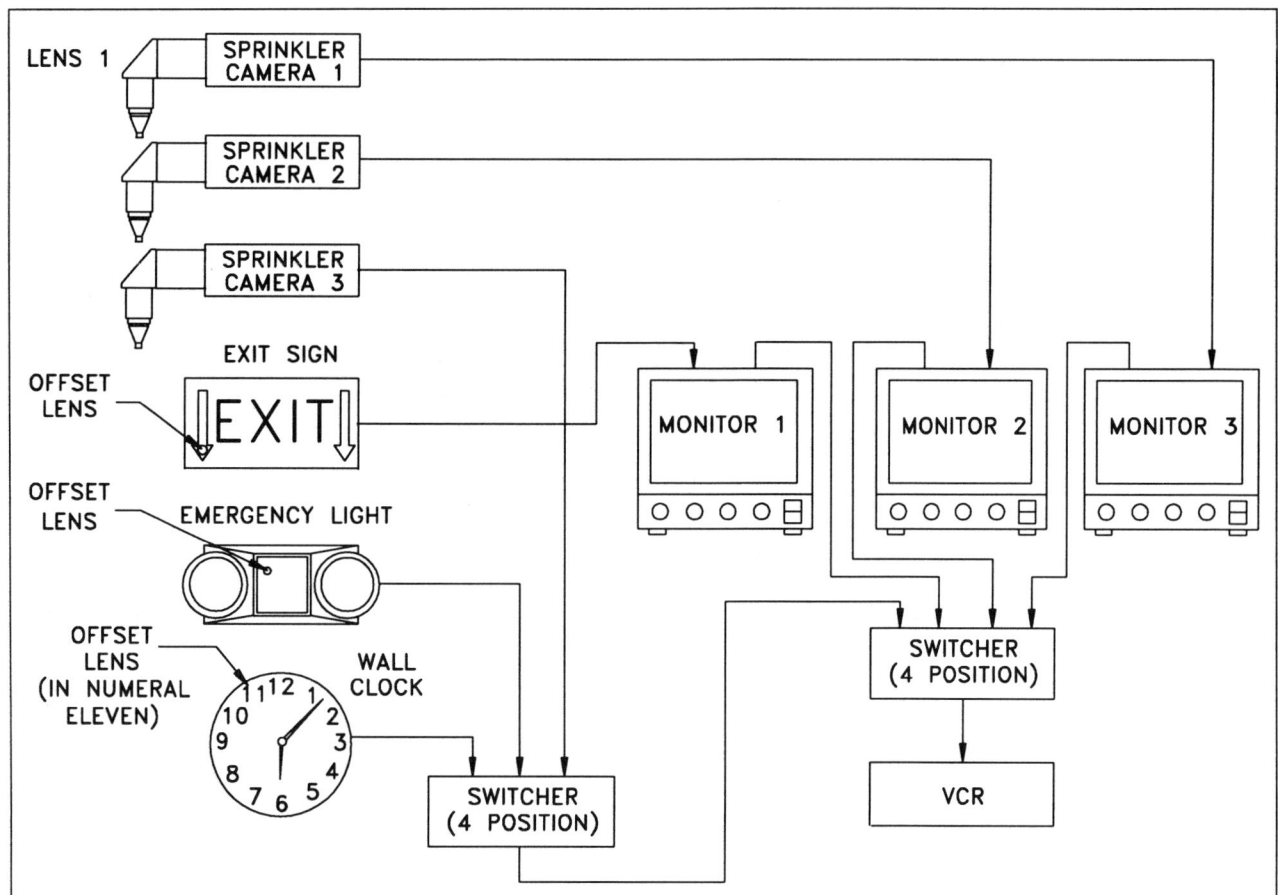

FIGURE 17-10 Six-camera covert office block diagram

mounted emergency lighting fixture monitors the computer, copier, facsimile, and files. The front office personnel, office entrance, vault, and files are monitored with two ceiling-mounted sprinkler cameras (one fixed, one panning) and an exit sign camera. The panning sprinkler can pan 360 degrees and uses an 11-mm FL lens and $\frac{1}{2}$-inch format camera. The exit sign views anyone leaving the manager's office or accessing the vault.

17.2.4 Outdoor Parking Lot and Perimeter Surveillance

Overall CCTV surveillance of an entire facility including a perimeter fence line, parking lot, building exterior, and loading dock area can range from a modest installation to a large comprehensive system with hundreds of cameras and dozens of monitors controlled by a microcomputer. Figure 17-11 illustrates the types of CCTV surveillance problems encountered in these applications.

Several fencepost or pedestal-mounted cameras are needed along the perimeter to detect intrusion or activity. These cameras, augmented by building-mounted cameras,

are used to view parking lots, entrances, and loading dock areas.

Requirements for perimeter parking lot surveillance include viewing a wide FOV (up to 360 degrees) while also having the ability to "home in on" and view a narrow FOV with high resolution. The lens/camera must be able to tilt up and down—usually 60 to 90 degrees to view an area far away or close in. Some techniques for solving the wide-versus narrow-FOV CCTV requirements are (1) using a camera on a pan/tilt platform with an FFL or zoom lens, (2) using multiple cameras to split the wide FOV into narrower FOVs, and (3) using a camera on a pan/tilt with two FFL lenses, one having a wide FOV and the other a narrow FOV.

The pan/tilt system (Chapter 10) permits the CCTV camera and lens to rotate horizontally and vertically so that the camera can look at scenes substantially outside the lens FOV. An FFL lens and camera system mounted on a stationary platform looks at only those parts of the scene as determined by the lens FOV and the camera sensor format (Tables 4-1, 4-2, and 4-4).

A zoom lens on a pan/tilt platform has a primary advantage over fixed-position, FFL lenses in that it can look at any

Item	Qty.	Description	Location	Comments
1	2	Camera	Main Office	Type: Monochrome; Sensor Type: CCD Sensor Format: $1/3$-, $1/2$-, or $2/3$-inch; Lens Mount: C or CS Input Voltage AC: 117 V or 24 V Input Voltage DC: 117-VAC-to-12-VDC Converter
2	2	Lens	Manager's Office, Front Office	Type: Right-Angle Sprinkler-Head Pinhole Iris: Manual; Focal Length: 11 mm f/#: 2.5
3	1	Panning Sprinkler Head	Front Office	Camera Type: Monochrome; Sensor Type: CCD Sensor Format: $1/2$-inch Lens Mount: C Lens Type: Sprinkler Head Focal Length: 11 mm; f/#: 2.5 Housing: Above the Ceiling, Fully Enclosed Pan Mechanism—Rotation: 360 Degrees, Variable Speed
4	1	Emergency Lighting Fixture	File Room	Camera Type: Monochrome; Sensor Type: CCD Sensor Format: $1/3$- or $1/2$-inch Lens Mount: Nonstandard; Lens Type: Mini Focal Length: 8 mm; f/#: 1.6
5	1	Clock Camera	Manager's Office	Camera Type: Monochrome; Sensor Type: CCD Sensor Format: $1/3$- or $1/2$-inch Lens Mount: Nonstandard; Lens Type: Mini Focal Length: 11 mm; f/#: 1.6
6	1	Fixed Sprinkler Head	Front Office	Camera Type: Monochrome; Sensor Type: CCD Sensor Format: $1/3$- or $1/2$-inch; Lens Mount: C, CS Lens Type: Pinhole Sprinkler Head; Focal Length: 11 mm f/#: 1.8
7	1	Monitor	Security Room	Type: Monochrome; Screen Size: 9-inch Mounting: Desktop, Rack; Audio: With or Without
8	1	Videocassette Recorder	Security Room, in Locked Compartment	Type: Real-time, Time-lapse Format: VHS, 8 mm, VHS-C, S-VHS, Hi-8
9	1	Video Printer	Security Room	Type: Thermal, Monochrome Print Size: 3.5 × 4.0 inches
10	2	Sequential Switcher	Security Room	Type: 4 Position
11	1	Time/Date Generator	Security Room	Optional
12	–	Transmission Cable	Wall, Ceiling, or Floor	Type: Coaxial or Fiber Optic
13	–	Covert Camera Installation		Temporary or Permanent

Table 17-3 Six-Camera Office Covert Bill of Materials

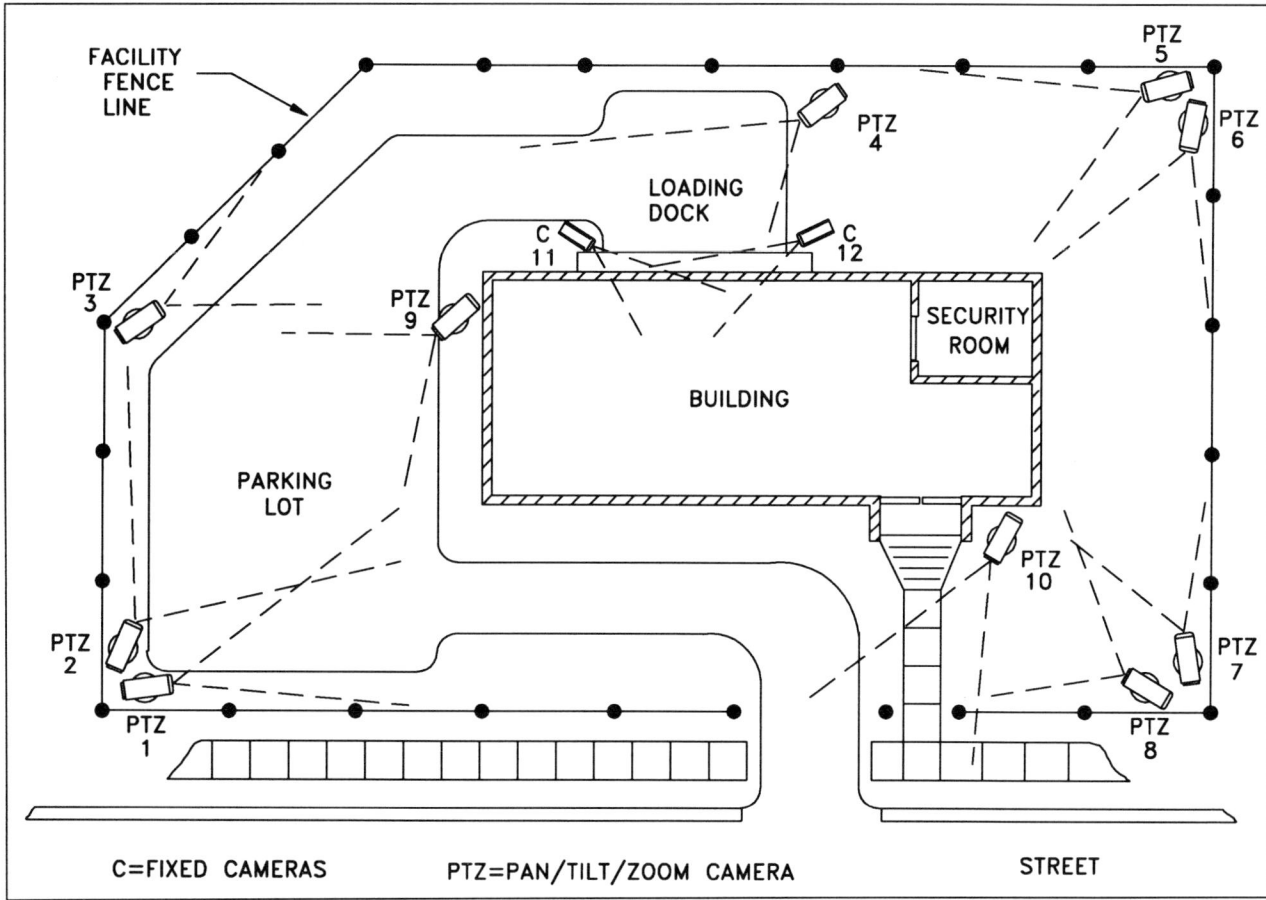

FIGURE 17-11 Outdoor parking lot and perimeter surveillance system

part of the scene by changing the lens/camera pointing direction via the pan/tilt mechanism. The zoom lens with its variable FL and therefore variable FOV provides wide-area coverage in the low-magnification, wide-angle mode and excellent resolution in the high-magnification, narrow-angle mode. The main disadvantage of the zoom pan/tilt installation is the inherent optical viewing dead zone, and operator time and dexterity required to manipulate pan/tilt and zoom functions. The inherent optical dead zone of the pan/tilt system originates because the system cannot be looking at all places at the same time. When it is pointing in a particular direction, all areas outside the FOV of the lens are not under surveillance and hence there is effectively no surveillance in those areas part of the time. Systems with preprogrammed presets eliminate some of this dead time.

Figure 17-12 shows the static versus dynamic FOV of the stationary and pan/tilt systems. The highlighted lines show the instantaneous FOV of the camera with a zoom lens (narrow- and wide-FOV extremes). The dashed lines show the total pan/tilt dynamic FOV and lens FOV, and represent the total angular coverage that the pan/tilt camera system can view. At any time, most of the area for which surveillance is desired is not being displayed on the television monitor. To partially overcome this shortcoming,

when the person hiding from the camera is out of the instantaneous FOV of the CCTV, the camera and pan/tilt mechanism are sometimes hidden, so that the person does not know where the camera is pointing at any particular instant.

The brute force approach to provide CCTV coverage of a perimeter and parking lot is to provide multiple cameras with fixed FOVs. Each camera/lens installation views a part of the entire scene. If the area covered by each camera is small, adequate resolution can be obtained. The primary advantages of the fixed CCTV camera installation over the pan/tilt type are its low initial installation costs, low maintenance costs, and lack of optical dead zones. The primary disadvantage is that only a small FOV can be viewed with good resolution: the wider the FOV, the smaller the amount of detail that can be seen.

The bifocal optical image splitting lens (Figure 4-31) offers a unique solution to the problem of displaying a wide and narrow FOV (telephoto) scene simultaneously on one monitor with one camera. It is particularly advantageous when expensive LLL SIT and ISIT tube and ICCD solid-state intensified cameras are used, since only one camera is used. When the system operates on a pan/tilt platform, wide-area coverage is always displayed simultaneously with a close-up of the area of interest. A pan/tilt system, with or

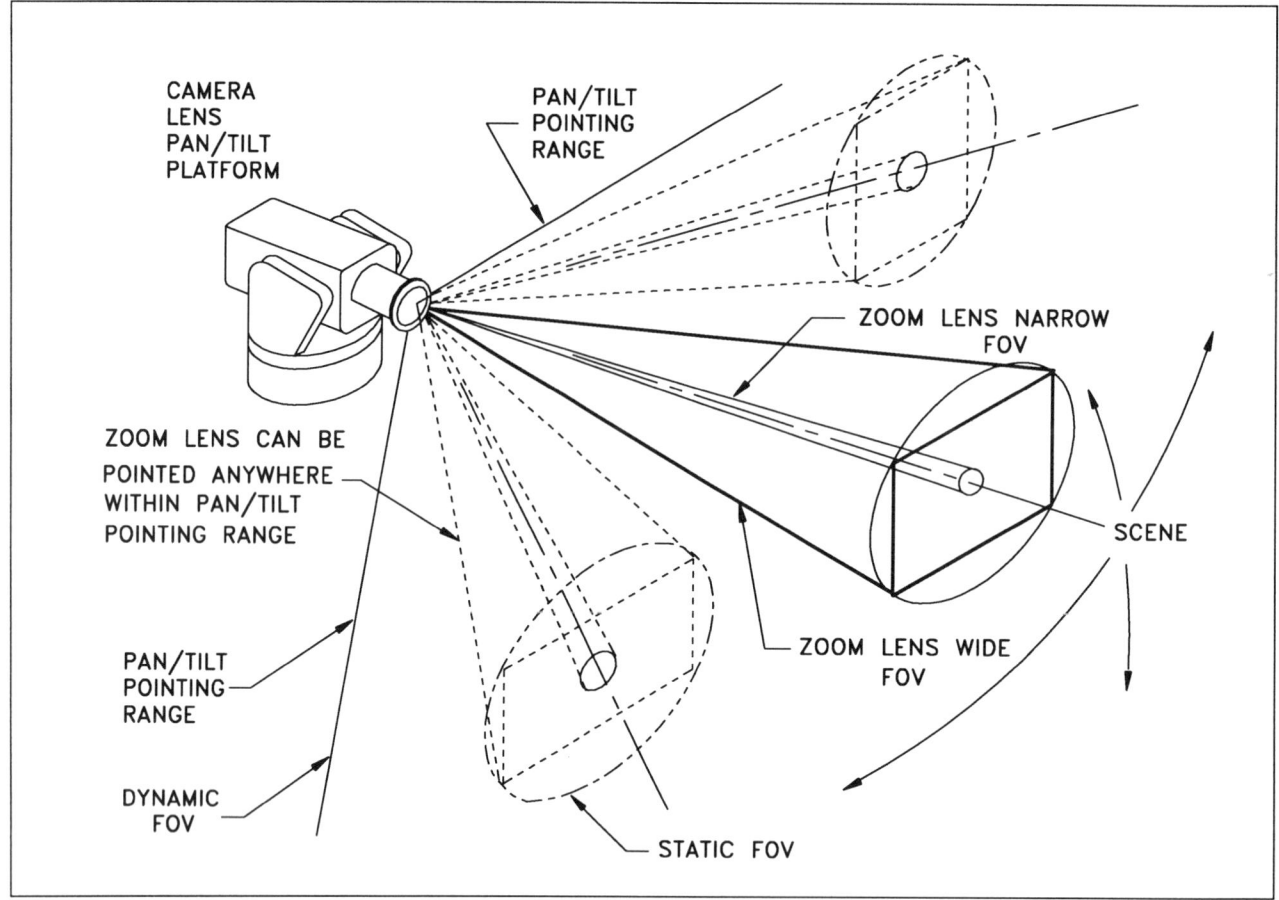

FIGURE 17-12   Static versus dynamic CCTV surveillance

without a zoom lens, cannot accomplish this. To obtain maximum flexibility and wide-angle coverage, a bifocal lens with a zoom lens and a narrow-angle (75 or 150-mm FL) lens is an excellent solution. The zoom lens permits a variable-FOV coverage from a wide angle of 40 degrees to a narrow angle of 4 degrees.

This combination is particularly good for outdoor parking lot and fence line (perimeter) applications. In the fence line application, if the pan/tilt in its normal condition is left pointing so that the telephoto FOV is looking at a perimeter gate or along the fence line, a video motion sensor programmed to respond only to the narrow-FOV scene could be used to activate an alarm or a VCR or alert a guard. Simultaneously, the wide-FOV zoom lens scene assures no dead zone in the scene.

17.2.4.1   Case: Twelve-Camera Parking Lot and Perimeter System

Figure 17-13 shows the block diagram for a twelve-camera outdoor parking lot and perimeter system.

Eight pedestal-mounted cameras (1 to 8) have pan/tilt/zoom capability to provide full area coverage of the perimeter, building, and facility grounds. The two building cameras (9, 10) have pan/tilt/zoom and provide

additional coverage of the parking lot and facility entrance. Cameras 11 and 12 are on stationary mounts with FFL lenses and monitor the loading dock areas.

Video signals from all cameras terminate in the security room. Transmission is via fiber optic to eliminate any external electrical interference (from machinery, lightning, or other causes) and environmental degradation (such as moisture, water, or rodents). The console room has a 16-position looping, homing, alarming switcher with its output going to monitors, a VCR, and a video printer. The three quad-camera combiners display the 12 cameras on three 9-inch monitors for 100% viewing of all cameras simultaneously.

Table 17-4 is a bill of materials for the 12-camera surveillance system. The perimeter cameras are monochrome and have $\frac{1}{2}$- or $\frac{2}{3}$-inch formats with an automatic-iris lens input. The monochrome CCD camera can view a lower light level than the CCD color camera. If viewing at low light levels is required, a more expensive ICCD, SIT, or ISIT camera can be substituted (Chapter 15). The zoom lenses chosen have FLs from 8 to 48 mm on a $\frac{1}{2}$-inch sensor or 11 to 110 mm on a $\frac{2}{3}$-inch sensor (see Table 4-5). Most outdoor housings should have a heater and/or fan, depending on climate. A window washer ensures a better

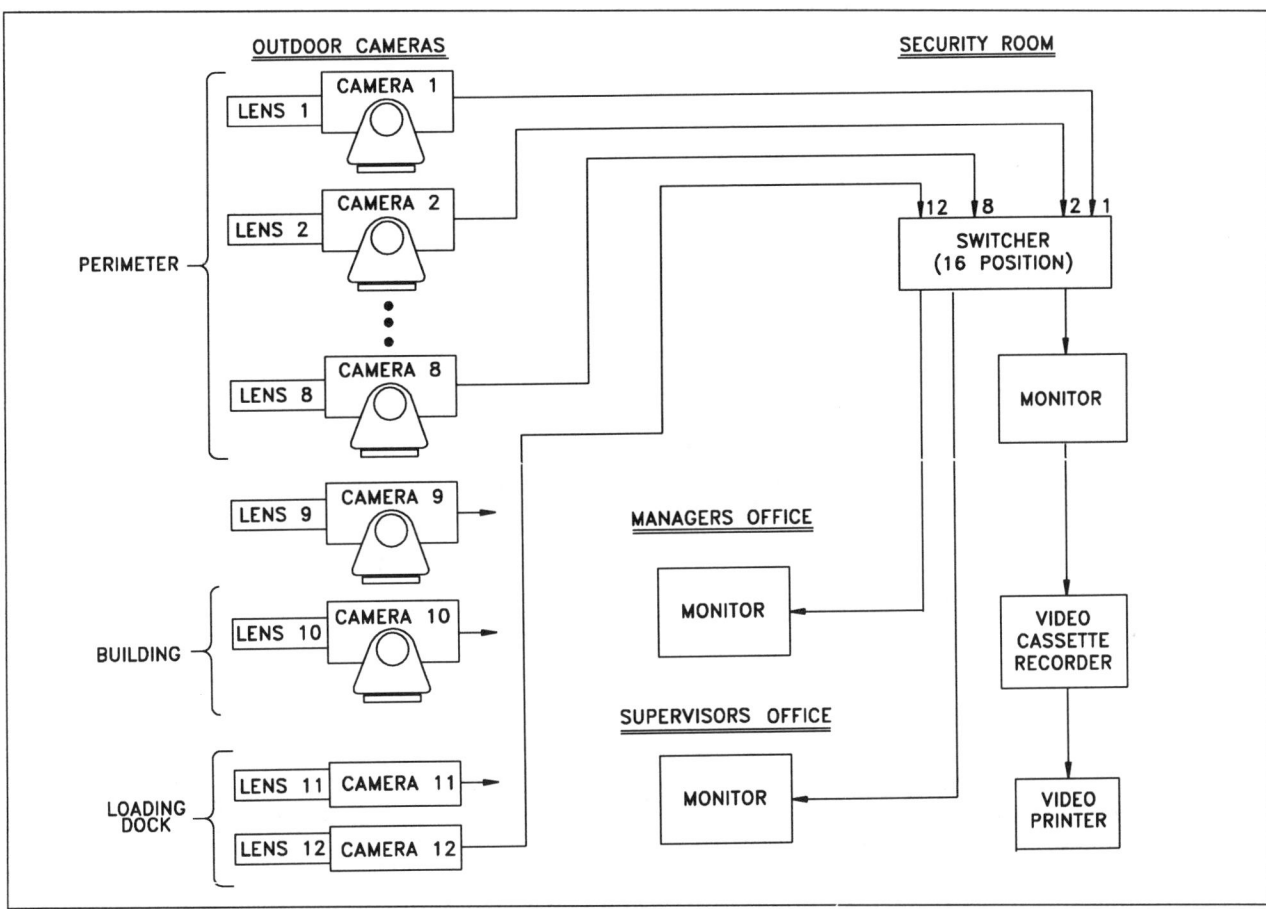

FIGURE 17-13 Twelve-camera outdoor parking lot and perimeter system block diagram

picture in most environments; in cold climates, antifreeze in the washer solution is necessary.

The pan/tilt platform (Chapter 10) should be chosen carefully to be sure its rating exceeds the weight load of the camera, lens, housing, accessories, wind loading and safety factor (10%). The preferred video transmission medium is fiber-optic cable, either direct-buried or in conduit (Chapter 6). Coaxial cable should not be used in geographical areas having thunderstorms.

The building-mounted pan/tilt/zoom cameras can be color or monochrome. Color makes identification of cars, equipment, and people easier.

The loading dock cameras are monochrome and use FFL lenses, since the area of interest is small and well defined. The combination is chosen so that personnel, actions, and material can be identified by the security guard.

If there is insufficient lighting in any of the areas, it should be augmented with auxiliary lighting (Chapter 3), or an intensified camera should be used.

17.2.4.2 Parking Lot/Perimeter CCTV Checklist

- Are FFL lenses adequate, or are zoom or bifocal lenses necessary?
- Are "dead zones" in the surveillance areas acceptable?

- Should monochrome or color cameras be used?
- What type of camera/lens housing (or concealment) should be used?
- What accessories are needed: heater? fan? window washer?
- What transmission medium should be used: coaxial cable? fiber optics? wireless? microwave? IR? RF?

17.3 RETAIL STORE SURVEILLANCE

The role of CCTV is to provide the remote "eyes" for the security officer; CCTV should be part of an overall retail store security system. The security operator(s) and management should be able to view many locations within the store from the security console and corporate offices. The viewed scenes should be of point-of-sale terminals, cashiers and merchandise, expensive products, and access and exit locations. Cameras can be overt and/or covert and have provision for panning and tilting where necessary. The system should automatically alert the security guard via video motion detectors (VMDs), alarm sensors, or signals from security or sales personnel on the floor. VCRs should record the viewed camera scenes in real time or time-lapse. Time-lapse VCRs should automatically change to the real-time mode

Item	Qty.	Description	Location	Comments
1	10	Camera	Perimeter, Building Mounted	Type: Monochrome or Color Sensor Type: CCD, MOS Sensor Format: $\frac{1}{2}$, $\frac{2}{3}$-inch Lens Mount: C or CS; Automatic Iris Input Voltage: 117 VAC or 24 VAC
2	2	Camera	Loading Dock	Type: Monochrome; Sensor Type: CCD, MOS Sensor Format: $\frac{1}{2}$, $\frac{2}{3}$-inch; Lens Mount: C or CS Automatic Iris; Input Voltage: 117 VAC or 24 VAC
3	10	Lens	Perimeter Line	Type: Zoom Focal Length Range: 17–90 mm f/#: Depends on Lens; Mount: C or CS Lens Controller Functions: Zoom, Focus, Iris (Manual and/or Automatic) See Tables 4-1, 4-2, Lens Finder Kit Presets: Depends on Application
4	2	Lens	Building Mounted	Type: Zoom Focal Length Range: 17–90 mm f/#: Depends on Lens; Mount: C or CS Lens Controller Functions: Zoom, Focus, Iris (Manual and/or Automatic) See Tables 4-1, 4-2, Lens Finder Kit Presets: Depends on Application
5	2	Lens	Loading Dock	Type: Fixed Focal Length Focal Length: 17–90 mm f/#: Depends on Lens Mount: C or CS See Tables 4-1, 4-2, Lens Finder Kit
6	12	Housing	Perimeter Line Building Mounted	Type: Outdoor, Environmental (Aluminum, Painted Steel, or Plastic) Shape: Rectangular, Dome Size: Large Enough to Contain Camera, Lens, and Accessories Accessories: Heater and/or Fan with Thermostats, Window Washer
7	10	Pan/Tilt	Perimeter Line Building Mounted	Type: Outdoor, Side or Overhead Load Capacity: To Suit Combined Weight of Housing, Lens, Camera, Accessories, Wind Loading, Safety Factor Range: Azimuth: 0 to 355 degrees or Continuous Elevation: 0 to –60 degrees; Scan Speed: Depends on Requirements Standard-Pan: 6 degrees/sec; Tilt: 6 degrees/sec High Speed-Pan: 60 degrees/sec; Tilt: 30 degrees/sec
8	1	Pan/tilt Controller	Security Room	Functions: Azimuth/Elevation Joystick, Speed Presets: Depends on Application Mount: Desktop, Rack
9	1	Switcher: 16-Position, Automatic	Security Room	

Table 17-4 Twelve-Camera Parking Lot and Perimeter Bill of Materials

Item	Qty.	Description	Location	Comments
10	3	Quad Combiner: 4 Cameras In, 1, 2, 3, 4, or Quad out	Security Room	
11	2	Monitor	Security Room	Type: Monochrome Screen Size: 9-inch Mounting: Desktop, Rack
12	1	Monitor	Security Room	Type: Monochrome Screen Size: 23-inch Mounting: Ceiling Bracket
13	1	Monitor	Security Room	Type: Color Screen Size: 23-inch Mounting: Ceiling Bracket
14	1	Videocassette Recorder	Security Room, in Locked Compartment	Type: Real-time, Time-lapse Format: VHS, 8 mm, VHS-C, S-VHS, Hi-8
15	1	Video Printer	Security Room	Type: Thermal, Monochrome Print Size: 3.5×4.0 inches
16	1	Video Printer	Security Room	Type: Thermal, Color Print Size: 3.5×4.0 inches
17	1	Time/Date Generator	Security Room	
18	—	Transmission Indoor	Perimeter Cameras, Junction Box, Building Cameras, Loading Dock Cameras to Console	Type: Cable-Coaxial or Fiber Optic Installation: Plenum, Tray, Conduit, Junction Box
	—	Transmission Outdoor	Perimeter Cameras to Building	Type: Cable-Coaxial, Fiber Optic, Wireless (RF, Microwave, IR Atmospheric) Installation: Underground Conduit

Table 17-4 Twelve-Camera Parking Lot and Perimeter Bill of Materials (*continued*)

when activated by these alarms. A video printer can provide hard copy when immediate action is required.

State-of-the-art CCTV equipment is available for almost any application at reasonable cost. Monochrome cameras provide the highest resolution and sensitivity, but the use of color television makes identification of persons and objects quicker and more accurate, providing there is sufficient light available. CCTV provides a significant cost advantage as compared with the equivalent security personnel required to do the same job.

17.3.1 Small Retail Stores

In the application of CCTV to retail stores, retail industry experts report that inventory shrinkage typically originates at the point of sale. It is estimated that three times as much shrinkage occurs at the cash register than anywhere else in the store. Shrinkage drops dramatically when point-of-sale data are recorded with a complete picture of exactly what was purchased, by whom, and through which salesperson. For cash register displays, the data include date, time, items purchased with quantities and prices, total cost, the amount tendered, the amount of change, transaction number, register number, and operator name.

At convenience stores, the number one cause of inventory shrinkage is employee theft at the cash register; the use of CCTV can significantly reduce this loss. When the video picture of the cashier, register, customer, and merchandise is combined with the cash register amount, the store manager sees what merchandise was purchased, what items and amounts were entered at the cash register, how payments were made, where they were deposited, and what change was given. CCTV surveillance in small retail stores is accomplished with equipment similar to the lobby case in Section 17.2.1. To connect the register transaction with the cashier, customer, and merchandise, a cash register interface unit is used. This can take the form of an electronic or electro-optical interface.

17.3.2 Large Retail Stores

Successful retail store security systems require the integration of alarm sensors, intrusion detectors, CCTV, and the coordination of personnel to carry out preplanned response procedures. Security personnel and employees need to work as a team. Two-way radio communication between floor security officers, the sales force, and control console personnel provides an effective means for maximizing CCTV capabilities. The security console operator (and management) view the selling floor via the CCTV system and look for system and procedural violations. When the operator sees a violation by an employee or customer on the floor, he radios an officer on the floor, who goes to the area under CCTV surveillance and assists or apprehends in

whatever way necessary. If an officer sees someone come into the store and thinks that person looks suspicious, the officer will radio the console operator that there is a suspicious person and give a description, which allows security personnel at the CCTV console to follow the person and his actions with the television camera. It is a cooperative effort, with the floor officers directly involved with the console operator and management.

17.3.2.1 Overt Showroom Floor Surveillance

The primary objectives for overt CCTV cameras on the sales floor are the deterrence of external and internal theft by shoplifting. Overt cameras located in areas having high visibility to customers and employees can watch areas of expensive merchandise. Monochrome or color cameras at ceiling level with pan and tilt pointing and zoom lenses view the main selling floor area. Fixed cameras with pan, tilt, and zoom (P/T/Z) can view large areas and are used effectively by security console operators to identify and watch suspicious actions by customers or potential thieves. An excellent P/T/Z camera module and vantage point is a ceiling-mounted color television camera. The optical magnification can be sufficient to identify a person as well as the merchandise being displayed on a counter. In the case of small jewelry, which is easily concealed, the use of a color camera makes it possible to detect a characteristic color feature unique to the item, which might not be apparent on a monochrome monitor. Fixed cameras in small boutique shops in a department store provide good views of the entire area.

Color cameras allow identification of jewelry and other articles, and the identification of an individual through differences in hair and skin color, clothing, and so on.

Cameras function as a deterrent and as a training/tracking system. In the training function, employees on the sales floor are monitored to detect procedural violations, such as leaving a showcase open, leaving a key in a lock, or showing too many pieces at one time.

Overt television cameras visible to the customer deter temptation of the casual shopper, customer, or thief. Professional thieves believe the selling floor is too large, there are too many people, and the security personnel can't see everything all the time. They see the cameras and know where the overt cameras are pointing. This suggests the use of optical domes to camouflage the cameras or the use of covert CCTV in the form of disguised sprinkler heads. The large plastic domes conceal the panning, tilting, and zooming CCTV system from the observer. The purpose for concealing the system is so the observer cannot see the direction in which the camera is pointing. By this subterfuge, one camera system scans and views a large area without the observer knowing at any instant whether he is under observation. Most domes in use are from 16 to 24 inches in diameter and drop below the ceiling by 8 to 12 inches. The requirement that the CCTV lens view through the dome results in light loss, typically 50%, and the image takes on some distortion because the optical quality of the dome is not perfect. Some domes can be "dummy" domes, which don't have cameras in them, but are wired so that cameras can be installed should the need arise.

To summarize, overt CCTV's first purpose in retail security is deterrence, the second is monitoring employee procedures and performance, and the third is aiding in the actual identification, apprehension, and prosecution of suspected individuals.

17.3.2.2 Covert Surveillance

Covert CCTV in large retail stores is used to spot and apprehend professional thieves and dishonest employees who are able to defeat the use of the overt CCTV system. The thief can't see (and thereby defeat) the covert cameras because they are cleverly concealed within the building structure and its fixtures. The availability of small covert cameras permits use in internal store investigations, where they can be installed and removed quickly.

Cameras and lenses can be concealed in many locations in a store. A unique combination of a pinhole lens with ceiling-mounted sprinkler head provides an extremely useful covert surveillance technique for retail stores. By law, most retail stores are required to have overhead sprinkler fire-suppression water systems. The covert surveillance sprinkler is in addition to and in no way affects the operation of the active fire-suppression sprinkler system. The sprinkler pinhole lens and camera is installed in the ceiling at a location to produce the required surveillance scene. At the customers' level on the showroom floor, the surveillance sprinkler unit goes unnoticed. The small adjustable mirror permits choosing the desired tilt (elevation) angle. A more sophisticated version of the sprinkler concept has a remote-control 360-degree panning capability. The most comprehensive system provides remote panning, tilting, and zooming but with a sacrifice in image quality and optical speed.

Two advantages of the moving-mirror sprinkler system over the dome are the absence of a large protruding dome suspended below the ceiling and easier installation, since only a small hole in the ceiling about $^3/_4$ inch in diameter is required to install the equipment. One limitation of the small mirror scanning system is that it cannot view the scene directly below its location.

17.3.2.3 Case: Eight-Camera Showroom Floor System

This retail store CCTV surveillance system incorporates overt and covert cameras to monitor showroom floor activity. The installation consists of four overt color and four covert monochrome cameras (Figure 17-14).

Four color cameras with zoom lenses on fast-scan pan/tilt platforms with preset capabilities provide excellent CCTV images of individuals, merchandise, and activities. The combination of fast-scan and preset allows the guard to track and follow a suspicious person or activity, or point and

zoom in quickly to a specific predetermined (memorized) location on the floor.

Four covert monochrome cameras hidden as ceiling-mounted panning sprinkler heads view central areas. One camera views the entrance and cashier; a second the jewelry department; a third coats and furs; and the fourth the 35-mm/video camera department. All panning sprinkler lenses scan 360 degrees. All control (pan, tilt, and lens functions) and video signals are transmitted to and from the camera and security room locations via hard wire or fiber-optic cable. Figure 17-15 shows the eight-camera retail showroom block diagram.

The camera video signals terminate in two quad combiners and an eight-position switcher in the security room. One monochrome monitor displays the four covert sprinkler scenes. Another color monitor displays the four pan/tilt/zoom scenes. All eight camera scenes are sequenced through a sequential switcher to a color monitor, VCR, or thermal video printer. A similar sequenced switcher, display, and recording/printing capability is located in a manager's office. The four color camera pan/tilt, camera functions, and presets are controlled from the controllers shown. The zoom lens and controller provide high-speed preset capability for the zoom (focal length), focus, and iris parameters. The pan/tilt platform and controller for the four color cameras provide high-speed camera pointing with presets.

Table 17-5 shows the BOM for the eight-camera showroom system.

17.3.2.4 Showroom Floor CCTV Checklist

- What areas must be under surveillance?
- What are the optimum camera locations to view customer and employee activity?
- Do some areas require 100% surveillance, eliminating the use of a P/T/Z camera?
- Where should monochrome and color cameras be used?
- Are overt or covert cameras needed, or both?
- Should the cameras be dome-enclosed, P/T/Z, or sprinkler/scanning small-mirror type?
- Is a register/point-of-sale transaction interface required? Electronic or electro-optical?

17.4 CORRECTIONAL FACILITY SURVEILLANCE

Large correctional facilities make widespread use of CCTV as a primary means for observing and controlling inmate activities, as well as monitoring access and exit of inmates, visitors, vendors, contractors, and employees. CCTV cameras are used to monitor inmates at initial incarceration, in cells, and during daily activities. Overt CCTV is particularly important in maintaining security along the perimeter of the facility through full-time visual monitoring. Some facilities have installed CCTV at the correctional facility to reduce

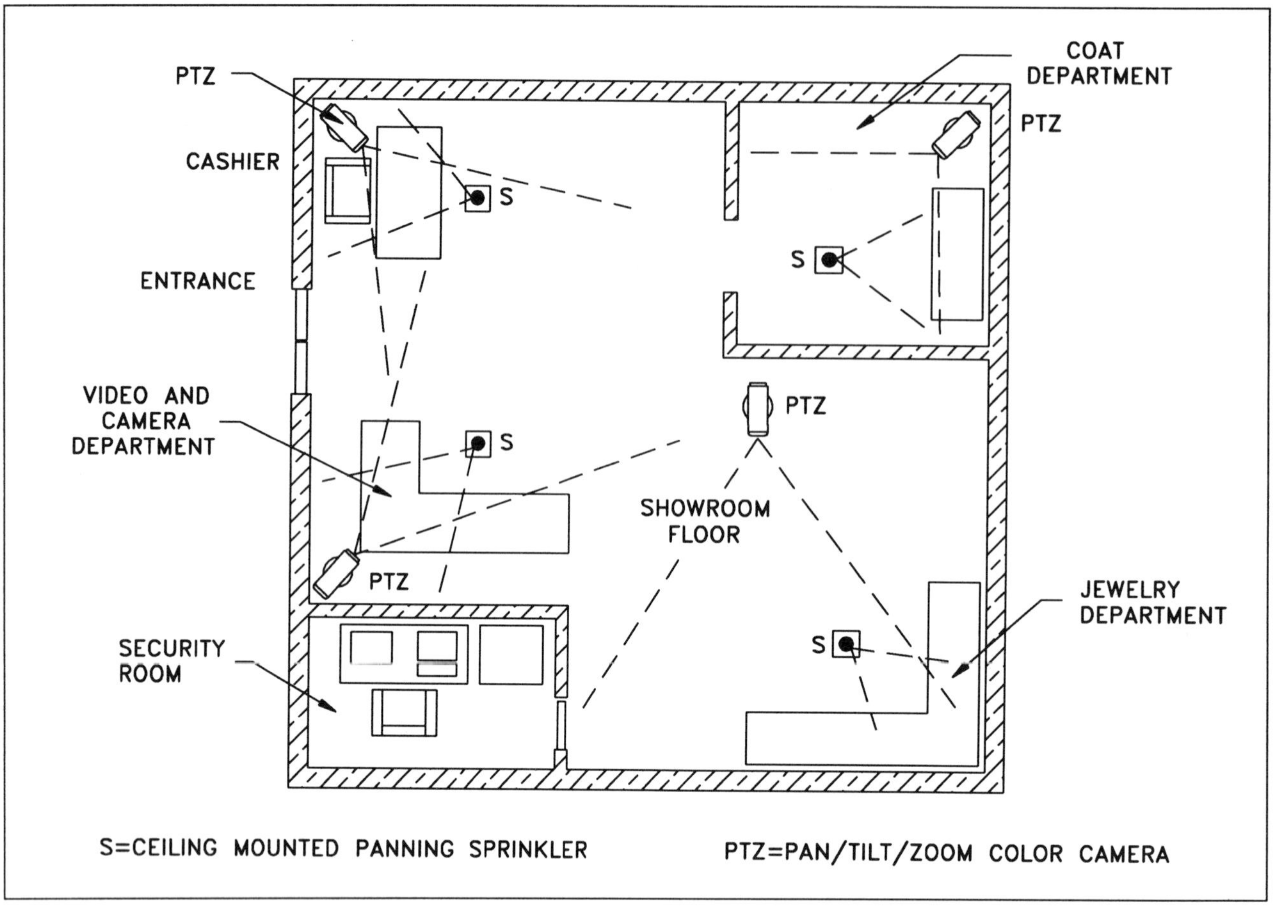

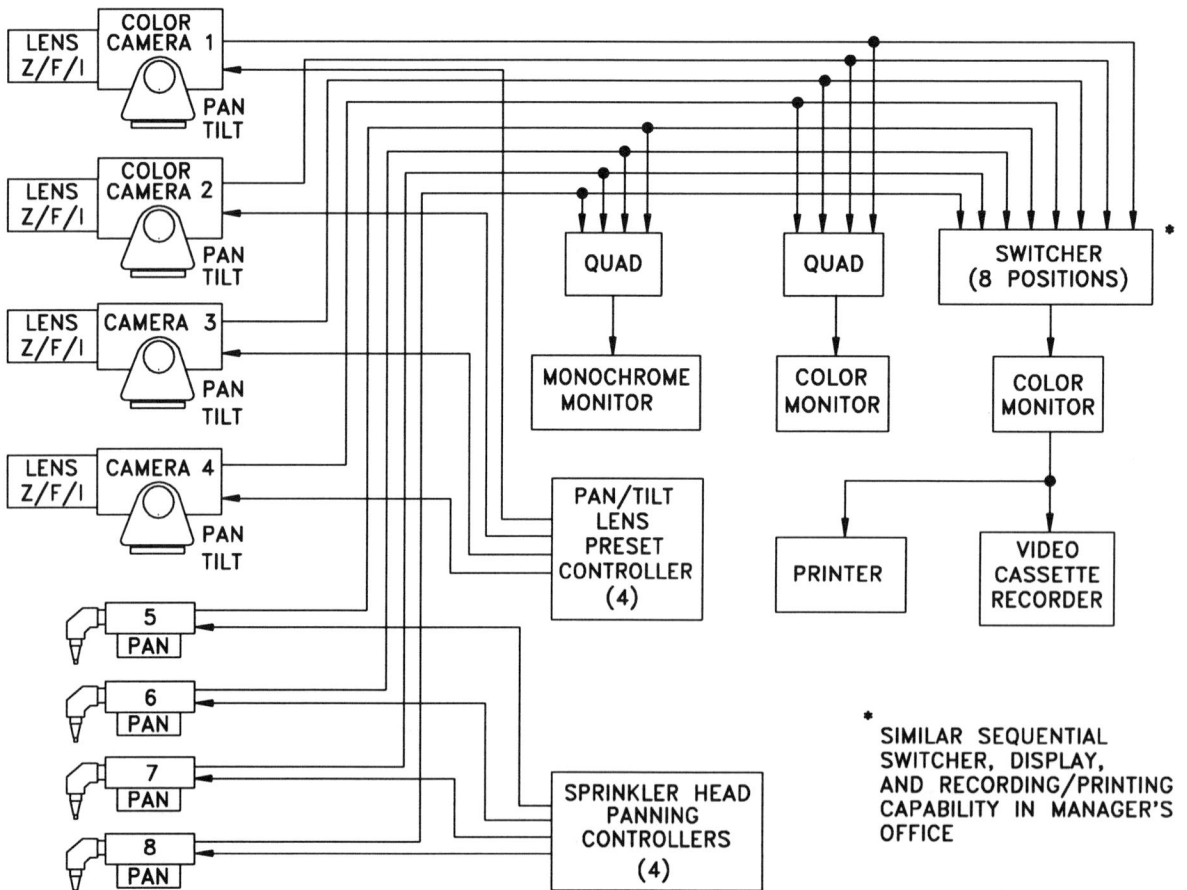

FIGURE 17-14 Eight-camera overt/covert retail showroom system requirements

S=CEILING MOUNTED PANNING SPRINKLER   PTZ=PAN/TILT/ZOOM COLOR CAMERA

FIGURE 17-15 Eight-camera retail showroom block diagram

Item	Qty.	Description	Location	Comments
1	4	Camera	Showroom	Type: Color Sensor Type: CCD, MOS Sensor Format: ½-, ⅔-inch Iris: Automatic Iris Lens Mount: C or CS; Input Voltage AC: 117 VAC or 24 VAC
2	4	Lens	Showroom	Type: Zoom Focal Length Range: 17–110 mm f/#: 1.8; Mount: C or CS Lens Controller Functions: Zoom, Focus, Iris (Manual and/or Automatic) Presets: Depends on Application See Tables 4-1, 4-2, Lens Finder Kit
3	4	Pan/tilt	Showroom	Overt P/T/Z to See Entrance, Floor Area, Showcases, Customers, Employees Type: Indoor, Side or Overhead Load Capacity: To Suit Combined Weight of Housing, Lens, Camera, Accessories, Safety Factor Range: Azimuth: 0 to 355 degrees or Continuous; Elevation: 0 to –90 degrees Scan Speed: Depends on Requirements; Standard-Pan: 6 degrees/sec. Tilt: 6 degrees/sec.; High Speed-Pan: 60 degrees/sec. Tilt: 30 degrees/sec.; Mount: Wall or Ceiling Bracket
4	1	Pan/tilt Controller	Security Room	Functions: Azimuth/Elevation Joystick, Speed Presets: Depends on Application
5	4	Camera Covert	Showroom	Covert Cameras in Sprinkler Head to Foil Professional Thieves, Employees Type: Monochrome Sensor Type: CCD; Sensor Format: ⅓-, ½-, or ⅔-inch Iris: Automatic Iris or Electronic Shuttered; Lens Mount: C or CS Input Voltage AC: 117 V or 24 V; Input Voltage DC: 117-VAC-to-12-VDC Converter
6	4	Lens	Showroom	Type: Right-Angle Sprinkler-Head Pinhole; Iris: Manual Focal Length: 22 mm; f/#: 3.9
7	2	Monitor	Security Room	Type: Color; Screen Size: 9-, 11-, or 13-inch Mounting: Desktop, Rack Audio: With or Without
8	1	Monitor	Security Room	Type: Monochrome Screen Size: 9-inch; Mounting: Desktop, Ceiling, Rack Audio: With or Without
9	1	Videocassette Recorder	Security Room, in Locked Compartment	Type: Real-time, Time-lapse Format: VHS, 8 mm, S-VHS, Hi-8
10	2	Video Printer	Security Room	Type: Thermal, Monochrome or Color Print Size: 3.5 × 4.0 inches
11	1	Time/Date Generator	Security Room	Transmission Type: Cable-Coaxial, Fiber Optic Installation: Plenum, Tray, Conduit

Table 17-5 Eight-Camera Retail Showroom Bill of Materials

the need for inmate transportation from the jail to the courtroom, to provide video arraignment of inmates.

To assist safety and security personnel during day-to-day and emergency situations, the CCTV systems include comprehensive graphics display capability. These large-screen color monitors integrate and display stored facility maps showing access and exit locations, alarm points, and camera locations. This information, combined with live video information, allows the guard to make on-the-spot decisions to open or close locked doors, call for reinforcements, and so on for the optimum safety and protection of all personnel.

Since the CCTV system is a vital security component in the overall personnel facility protection plan, it must have an electrical backup power system (Chapter 16). Critical cameras should also have an alternate transmission system (path)—preferably wireless—in the event the cabled video transmission path is severed or compromised.

17.4.1 Overt Surveillance

17.4.1.1 Perimeter

Perimeter surveillance is accomplished by using CCTV equipment similar to that described in Section 17.2.4, in conjunction with other intrusion-detection methods, including VMD, seismic, E-field, RF, and microwave technology. Outdoor VMD equipment (Chapter 12) should use digital motion-sensing to achieve the high probability of detection and low false-alarm rates required. Correctional perimeter surveillance also requires adequate nighttime lighting to obtain reliable VMD operation. Camera/lens housings and light luminaires should be rugged and capable of surviving attack.

17.4.1.2 Cell Blocks and Activity Areas

Cells, cell block, and activity areas are viewed with monochrome or color cameras—color is preferred for more positive article and inmate identification. Fixed camera/lens positions and P/T/Z equipment with presets should be chosen, depending on the location and requirement. Very wide angle CCTV systems with 90-degree horizontal FOV are available to monitor multiple cells with one camera.

Camera/lens housings with infrared illuminators are used in holding cells to view inmates 24 hours a day for their protection.

17.4.2 Covert Surveillance

Important areas for covert CCTV surveillance in correctional facilities include drug-dispensing rooms, disturbed-inmate holding cells, and in locations that assist staff security and law enforcement personnel during inmate uprising and hostage release. There are many suitable

locations and CCTV camera equipment available (Chapter 14) to conceal these devices.

17.5 BANKING SURVEILLANCE AREAS AND IDENTIFICATION

CCTV is in widespread use in banking and other financial institutions. More integrated security systems with CCTV as a key element are being implemented as a result of increased governmental regulations. Other reasons are the need for better protection of assets and personnel, and the ability to survive and recover from a disaster. In addition to the assets and personnel protection required, data, telecommunications, and computers must be controlled. Restriction of access to sensitive data processing, money-counting, vault storage, and communication rooms is controlled by CCTV/electronic access control systems. Automatic teller machines (ATMs) in widespread use require CCTV surveillance cameras to identify customers and tie the transaction with the person using the ATM. This reduces problems associated with money withdrawal and check cashing.

Security at automobile drive-in banking lanes can also benefit from CCTV surveillance.

17.5.1 Public Areas

The bank's main floor and safe deposit rooms require visual surveillance of customers and employees. CCTV cameras viewing teller/customer areas can record the customer for positive identification during check cashing. Usually one camera can view two teller stations. A clock and calendar located within the camera FOV document the transaction time and date.

A high-quality CCTV system must be used for the bank holdup cameras to permit personnel identification. The high-resolution S-VHS and Hi-8 VCR tape cassette formats produce the highest quality image—over 400 horizontal TV lines—producing approximately a 30% increase in resolution over the standard VHS, VHS-C, or 8-mm formats (Section 8.2.1.4).

17.5.2 Computer, Vault, and Money-Counting Rooms

Sensitive banking areas require CCTV access control and surveillance to maintain security. An integrated electronic and CCTV access control system provides full assets and data security. Electronic access control alone or with a personal identification number (PIN) does not provide full security and should be combined with some form of video identification, preferably a video image storage and retrieval system.

FIGURE 17-16 Customer check and face identification system

To identify holdup or internal theft suspects, overt CCTV should be augmented with covert CCTV. This may take the form of any of the equipments or applications described in Chapters 14 and 17.

17.5.3 Check Identification

An important function that reflects in bank profitability is customer check identification. This can take the form of two different requirements: (1) recording the picture of the customer, check cashed, and time/date and (2) positively identifying the customer cashing the check. In the first case, a device to video record the person's face and check annotated with the time/date on the same frame is required. Figure 17-16 shows a countertop system to accomplish these functions, which views the person, check, or other identification credential and annotates it with time and date.

Recording the person's face, credential, and data can be accomplished using a high-resolution VCR (S-VHS or Hi-8) or a video image storage and retrieval system (Figure 17-17).

The hard-disk video storage system (Section 8.5) accomplishes positive identification of the customer. In operation, the bank officer enrolls the customer into the system by recording the person's face, name, signature, ID number (PIN), and any other personal data onto a magnetic or optical storage medium. When the person requests a check to be cashed, the name, ID number, or bank card is entered into the keypad/reader and the image of the customer's face and all personal data are retrieved and displayed within a second on the monitor, for the teller's use. The video retrieval system communicates data and video over fiber-optic or coaxial local area network (LAN) or transmits over long distances via an electronic modem transmission system to branch offices.

17.5.4 Automatic Teller Machine

Security for the ATM requires that the face of the person accessing the machine, the transaction time, date, terminal number, and transaction number be recorded. The ATM text is typically presented as one line centered at the top of the screen. If the ATM transmits a message related to the transaction, the message is displayed on the second line.

When a customer disputes a deduction from his or her account for cash dispensed by an ATM, the burden of proof falls on the financial institution. With the message annotator, branch executives document the entire transaction in a single visual record, with all key financial data superimposed on a picture of the person performing the transaction.

17.5.5 Auto-Teller Drive-In

Many banks in suburban communities provide drive-in banking for their customers. Some drive-in lanes are situated so that the teller in the bank building can see the customer and no television is required. In most installations, however, the distance between the customer and teller precludes direct visual identification, and CCTV cameras and monitors are used. Systems are available to record the customer's face, check, and if required, the license plate of the automobile.

17.6 LODGING AND CASINO SURVEILLANCE AREAS

Lodging and casino facilities offer many opportunities for theft, vandalism, and personal injury to their employees, guests, and visitors. For this reason, there is a large variety of CCTV surveillance and access control equipment used at the facilities. Some of the areas requiring surveillance include (1) lobbies, (2) elevators, (3) hallways, (4) parking lots, (5) building entrances, (6) hotel gaming rooms, and (7) the casino floor. Locations requiring access control include money-counting rooms, sensitive data/equipment operations rooms, and security rooms. Because the casual

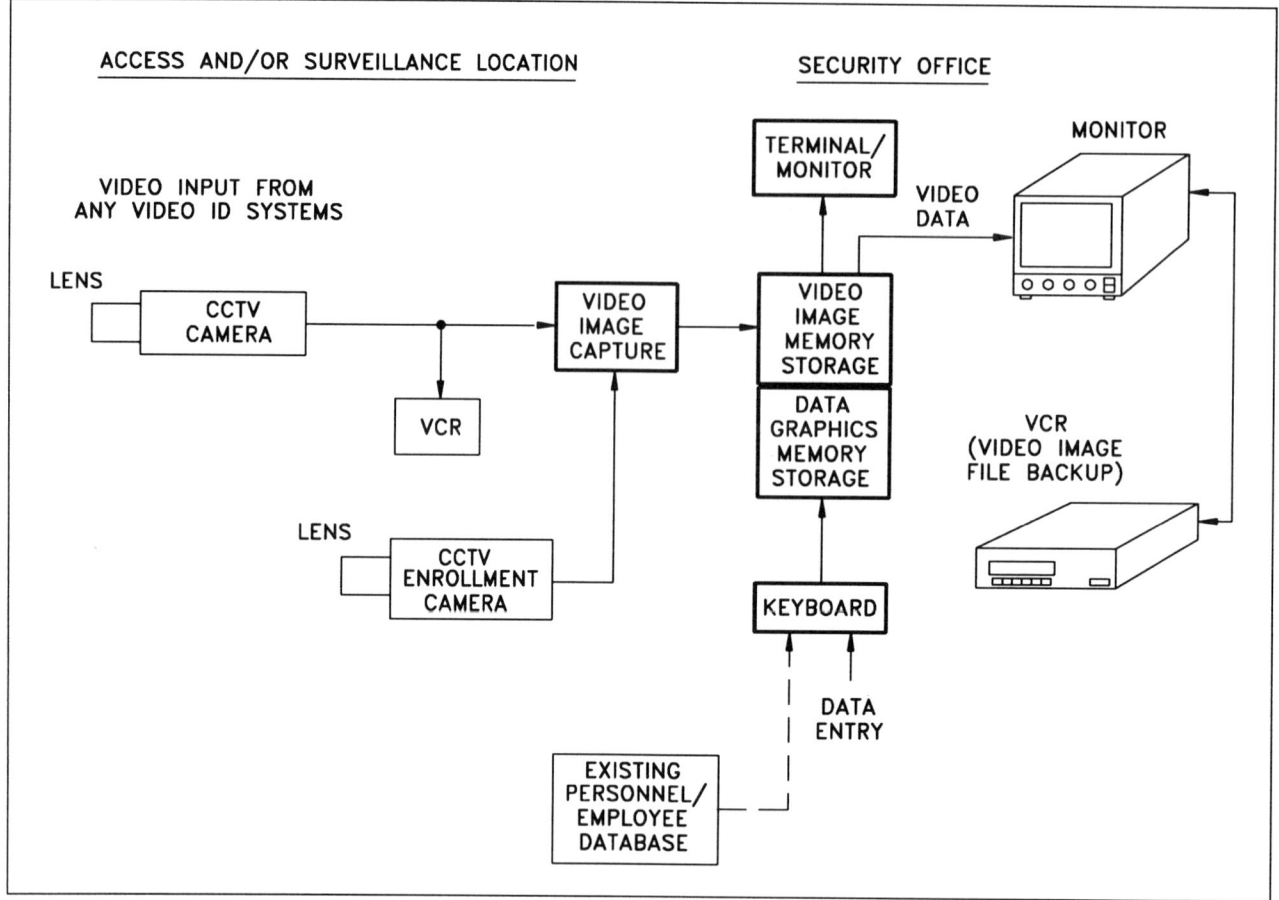

FIGURE 17-17   Video image storage and retrieval system

hotel/motel/casino atmosphere attracts many amateur and professional offenders, the use of overt and covert CCTV security is required.

17.6.1 Hallway or Corridor

The CCTV requirement is to view as many hallways as possible, showing personnel in the hallway, where they are going, and what they are doing. This can be accomplished using a combination of color and monochrome overt and covert cameras. To reduce the number of cameras at elevator lobby-hallway intersections, a tri-split lens (Chapter 4) can be used to advantage. Mounting this three-lens, one-camera system in a low-profile ceiling-mounted housing provides three pictures from one camera, on one monitor.

17.6.2 Casino Floor

A gambling casino floor requires many different types of CCTV security equipment and must satisfy and solve many different types of security requirements for the casino secu-

rity staff and management, as well as state and local gaming officials. CCTV equipments include monochrome and color, overt and covert systems, fixed-position and P/T/Z systems controlled by large computer-controlled switchers.

Cameras are located to view all gambling tables, all slot machines, bill- and coin-counting rooms, cage and chip rooms, money changing and cashier cages, the main vault, and bars. These cameras are located in the ceiling behind one-way glass or camouflaged in plastic domes or disguised sprinkler heads. The P/T/Z cameras are located and installed into the ceiling to be unobtrusive. All camera installations are aesthetically installed and integrated into the casino decor. The P/T/Z cameras have sufficient resolution to read the cards and count the number of chips in a stack. Many full-size casinos use video computer control systems capable of switching 128, 256, or 512 cameras and 16, 32, or 64 monitors and VCRs. This permits switching any camera to many different locations, monitoring, and recording devices. The combination of VMDs and other alarm sensors with the matrix switcher permits detecting, following, and recording the movements of suspicious persons. Monitors alarmed by these inputs and motion in the alarmed scenes cue the security guard.

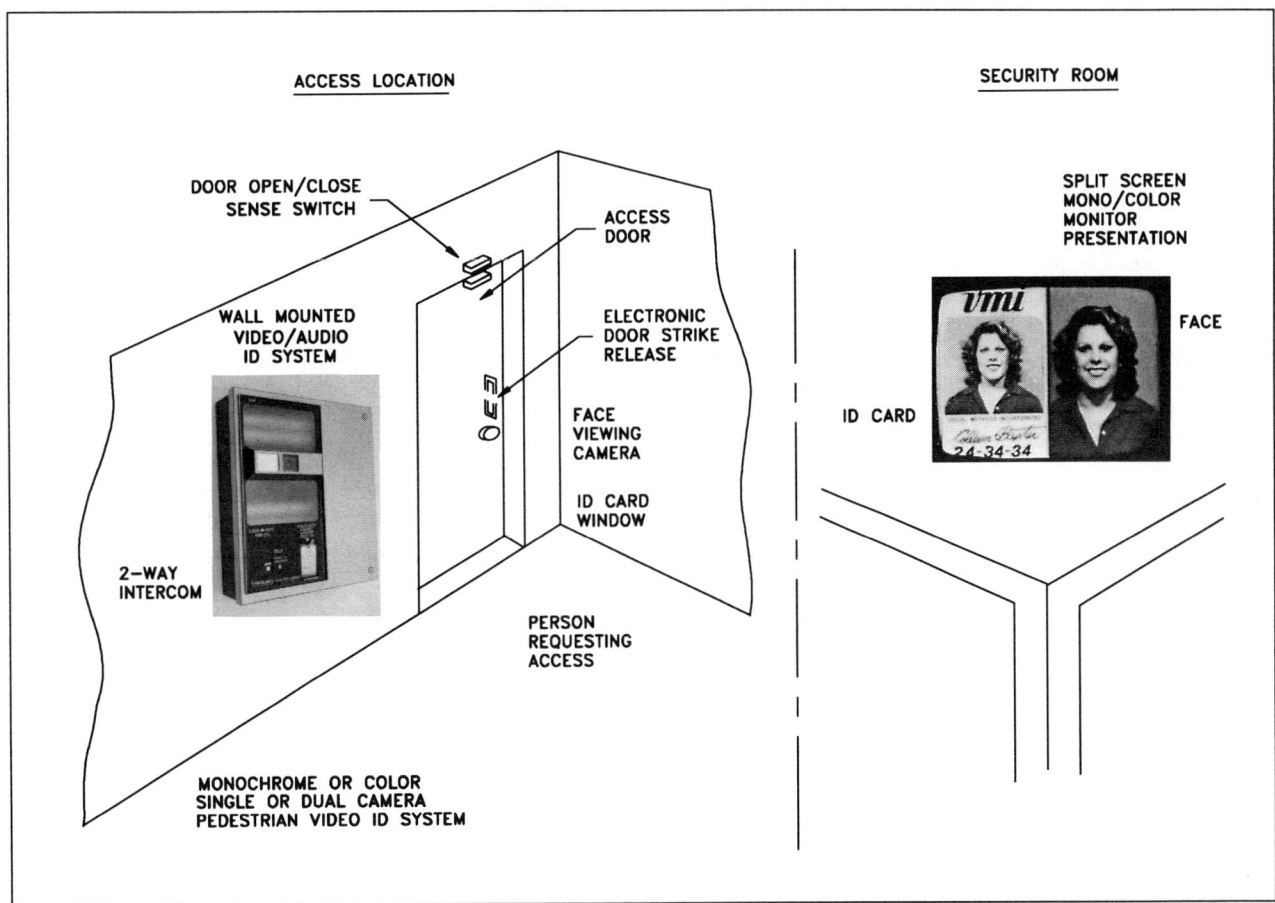

FIGURE 17-18   Photo ID video identification system

17.7 ACCESS CONTROL

CCTV plays an important role in access control. While most access control applications are solved using electronic access control systems, other means must be used in addition if positive identification of personnel is to be accomplished. An electronic access control card reader system identifies only the card, not the person. The addition of the PIN as an identification criterion increases the level of security. Unless the system is augmented with other identifying means such as photo ID or biometric systems (such as fingerprints, voiceprint, hand geometry, retinal pattern), positive identification does not result. Two forms of video identification are in common use: (1) photo ID badge and (2) image storage and retrieval.

17.7.1 Video ID System

The simple photo ID badge split-screen system shows the face of the person requesting access on one half of the screen and the person's photo ID card in the other half. This system, used with an electronic access control system with PIN, increases personnel identification significantly.

Figure 17-18 shows an example of a CCTV access control photo ID system.

The ID system is mounted immediately adjacent to an access door and contains lights and optics for the camera to view the person's face and the photo ID card presented by the person. A two-way audio intercom and call-button permit communication between the guard and person requesting access.

17.7.2 Video Storage and Retrieval System

When positive personnel identification from a remote site is required, the video image storage and retrieval method is used (see Figure 17-17).

With this system, a video image of each person requesting access to a facility is made at the time of enrollment into the system and is stored in a magnetic or optical storage medium for later retrieval. When a person who is enrolled in the system requests access (Figure 17-18), the system retrieves the stored video image of the person, and the guard compares the retrieved picture with the live image of the person obtained with a video camera at the access location. Since the stored and retrieved picture is always

FIGURE 17-19 Video turnstile access control system

FIGURE 17-20 Sheltered outdoor portal

under the control of the guard, positive identification results. For the convenience of the person requesting access, systems are offered with the ability of retrieving the image by name, ID number (PIN), or electronic ID card.

17.7.3 Video Turnstile

The video ID units described in Sections 17.7.1 and 17.7.2 can be mounted in a turnstile to eliminate the possibility of "tailgating," that is, an unauthorized person entering the facility immediately behind the person entering. The turnstile is equipped with an overhead CCTV camera that views the confined area. The guard ascertains that only one person is in the controlled/confined area, so that only one person is admitted at a time. The system can be integrated with an electronic access control system and video image storage and retrieval system to obtain positive identification of the person requesting access. Figure 17-19a shows an indoor single-rotor design, and Figure 17-19b an outdoor installation using a dual turnstile (for entry and exit or dual entry and dual exit). This outdoor installation has an environmental shelter enclosing the turnstile to protect personnel from inclement weather.

17.7.4 Video Portal

For installations not suitable for turnstiles, a portal (mantrap) is often used. This two-door portal provides the same anti-tailgating features as the turnstile, but with an environment more suitable for some office or architectural requirements. The larger space available in this portal allows room for other identification equipment, such as biometric or electronic, and makes for easier use by handicapped personnel. The portal can also be used with the video storage and retrieval system to provide positive personnel identification. Figure 17-20 shows an environmentally controlled outdoor portal enclosing the video ID system and overhead cameras.

17.7.5 Video Vehicle Control

Many personnel require access to a facility by means of a vehicle. For casual control, an electronic access control system using electronic ID cards or keys can be used. For positively identifying the vehicle driver, a video photo ID system or video image storage and retrieval system must be used. Figure 17-21 shows two video vehicle control systems for identifying (1) driver in a car, (2) driver in a car, van, or truck, (3) or ID credential (such as a photo ID card).

The pedestal-mounted video ID system is for vehicles of car height. The system houses two CCTV cameras to view the vehicle driver's face and a photo ID card (or other identification means) on two monitors at the remote guard room location. The unit has a dual lighting system for daytime and nighttime illumination requirements and a two-way audio intercom and call-button annunciator. The housing is environmentally designed with thermostatically controlled heaters and fans. The variable-height system has a movable carriage to raise and lower a dual camera and lighting system to that required for the vehicle present. The system can be raised or lowered by the vehicle operator (while remaining in the vehicle) or guard to accommodate a small car or a large tractor-trailer.

17.8 TRAINING SECURITY PERSONNEL

A CCTV security system can be used for on-the-job and new employee training. All new employees can visit the console room, where they witness the use of the television security

ENVIRONMENTAL PEDESTAL
SIZE:

ENVIRONMENTAL KIOSK
SIZE:

TRUCK HEIGHT

VAN HEIGHT

CAR HEIGHT

A) KIOSK ID UNIT
   CAR OCCUPANT
   IDENTIFICATION

B) MOVING CAMERA/LIGHTS ID UNIT
   OCCUPANT IDENTIFICATION
   CAR, VAN, TRUCK

FIGURE 17-21   Fixed- and variable-height video vehicle control systems

installation and are shown its capabilities and its purposes. Employees can view previously recorded videotapes, which is a valuable training tool for management. For example, in a retail store, employees are trained by showing them instances recorded on camera where a salesperson has left the merchandise vulnerable to theft and are instructed on how the salesperson should have watched a customer's actions. This is shown to new and existing employees to illustrate what happened, why it is wrong, what was lost as a result, and the proper action described. New employees can also see positive actions where someone has tried to operate against a salesperson. Incidents where the salesperson was alert and protected the merchandise are pointed out to new employees. Examples such as these teach a new trainee what to look for and how people operate.

17.9 CHOOSING A PROFESSIONAL SECURITY DESIGNER AND INSTALLER

To assemble an effective and successful security operation, management and the security director must know what needs to be protected, and what cost expenditure is justified to protect it. The security director must choose and collaborate with a reputable consultant, architect and engineer, and a security systems supplier to design an effective system together. High-quality equipment must be chosen that will provide reliable service over many years. The plan for response to a threat must be well organized, documented, explained, and agreed upon.

A successful security installation must include the use of a knowledgeable security dealer who understands the customer's security problems and the current technology available. The system requirements, design, and installation should be a joint effort between the requirements of the management/security staff and the capabilities of the security company. CCTV equipment is complex and requires the use of quality hardware and professional service and maintenance to produce continuous 24-hours-a-day, 365-days-a-year operation. It must be installed reliably and the dealer must provide prompt response time to an emergency requirement. The real key is knowing that the vendors used stand behind their products and that the dealer/installer is reputable and has a high commitment to service. The bottom line in any sales and service industry is that the service sells the product.

17.10 APPLICATIONS CHECKLIST

The following list enumerates some questions that should be asked when designing a CCTV surveillance system.

- What is to be protected? Assets? Personnel?
- What is the value of what is to be protected and what is the cost of the system needed to protect it?
- Are goods and/or personnel to be under surveillance?
- What type and how many cameras are required to view the personnel and articles to be protected?
- Where should cameras be for the best view?
- What should be the FOV of each camera?
- Are fixed-camera or P/T/Z cameras required?
- Which cameras should be overt?
- Which cameras should be covert?
- What monitoring equipment is needed at the console?
- Number of monitors? Should they be multiscene (split or combined screen) monitors? How many splits?
- Color or monochrome cameras and monitors?
- Number of VCRs? Real-time or time-lapse? VHS, S-VHS, 8-mm, or Hi-8?
- Video disk recorder: to store alarm frames, employee facial pictures, fast retrieve?
- Video switcher: manual, alarming, alarming with preset pan, tilt, and zoom parameters preprogrammed into the system? Salvo (gang) switching?
- Video printer: hard-copy printout of monitor or VCR? Monochrome or color?
- Daytime only? Daytime/nighttime—with added lighting or existing lighting? Intensified cameras required?
- Interfaced with facility security officers via two-way radio?
- Interfaced with sensors carried by roving guard personnel?
- Interfaced with door, window, or perimeter sensors? Intrusion alarms: dry-contact, microwave, infrared?
- Video motion detectors? Analog or digital?

17.11 SUMMARY

Successful CCTV surveillance systems are the result of a careful plan and a professional installation.

- Manufacturers have a wealth of information. Ask them for it.
- Attend educational seminars and exhibitor training courses to obtain professional, firsthand information. Attend manufacturers' exhibits and see real, functioning hardware.
- Write a system specification or have a consultant prepare one, and send it out to several vendors for quotation.
- Choose a well-established dealer/installer whom you think will be in business to service the installation 5 to 10 years hence. Choose one system integrator to make sure all system components and software are compatible.
- Include a training course for management and personnel who will operate the system.

Chapter 18
New Technology

CONTENTS

18.1 OVERVIEW

The CCTV industry has witnessed unprecedented growth and advancements in technology in the last decade. One factor responsible for this rapid growth has been the miniaturization and increased density of electronic circuits, resulting in smaller equipment having higher capabilities at competitive prices. Specific product areas benefiting from this technology advancement include solid-state cameras, VCRs, and computers and memories that manipulate and store video images. In the coming decade these and other technological advances will continue, further enhancing the capability of existing products and generating new ones. The following sections describe some of the advances that will come to fruition and find applications in the security field.

18.2 CAMERA TECHNOLOGY

Technology will continue to improve in solid-state sensors used in television cameras. Increased sensitivity and resolution, particularly with color cameras, whose development is driven by the consumer market, will make the use of color security systems more common than monochrome. Improvements in camera-sensor dynamic range should all but eliminate the need for automatic-iris lenses. The trend toward smaller sensor sizes will continue, along with the attendant smaller lenses. A technology showing high promise is IR thermal imaging. A breakthrough in technology will result in competitively priced cameras operating at room temperature that can see through fog.

18.2.1 LLL Cameras

With present technology, the ICCD is rapidly replacing tube-type SIT and ISIT LLL cameras. There will be further improvement in sensitivity, resolution, and reduced prices as volume increases.

18.2.2 Infrared Cameras

Present IR imaging cameras using cooled (mechanical refrigeration) forward-looking IR are prohibitively expensive. Thermal vidicon systems produce poor-quality pictures. New technology using room temperature focal plane arrays will provide good resolution at affordable prices.

18.3 MONITOR/DISPLAY TECHNOLOGY

The CRT monitor has remained the standard in the industry, but LCD technology is maturing and will have the resolution and cost advantage in the near future. LCD displays are available in standard sizes up to 10 inches, but they are expensive.

18.4 LENS TECHNOLOGY

Nothing can be done to change the laws of physics and optics; therefore, any changes or improvements in optics will come in the form of smaller size to match the smaller sensors and aspheric optics to improve the optical speed and image resolution. The availability and application of powerful computers and optical design programs will no doubt result in the design of a new generation of CCTV lenses, particularly zoom and multifocal length lenses, to match the small size and high resolution of the new small sensors. The use of automatic focusing lenses will increase. The development of moving-image tracking cameras will keep cameras pointed at the target of interest in the picture.

18.5 MAGNETIC AND OPTICAL RECORDER TECHNOLOGY

Recorder technology is driven by the consumer and computer industries. Higher recording densities, faster storage and access times, and new technologies will be forthcoming.

18.5.1 Magnetic

VHS and 8-mm format tape recording have been the standard in the security industry for many years, with S-VHS and Hi-8 formats making inroads for the higher resolution applications. New methods of recording and higher-density tapes to store more pictures and data are being developed. Digital storage of video on magnetic disks will become commonplace as storage media costs decrease.

18.5.2 Optical

Storage of video images in digital form on optical disks is in limited use and will continue to grow in popularity as a convenient storage media. The write-once, read-many (WORM) optical storage drive is now being used, and the new erasable drive will be used in many security applications as prices become competitive.

18.5.3 Semiconductor

If semiconductor storage technology continues its trend of significant increase in storage each year, there will be an increase in its use for temporary storage of video information.

18.6 INTEGRATED SYSTEMS

With CCTV playing a more important role in security applications and its use with other security components (alarms, fire protection, and so on), the key phrases will be "integrated security systems" and system "interoperability."

18.6.1 Computer-Controlled

The availability of powerful computers at competitive prices and integrated, multitasking software (for example, Microsoft Windows) will add significant security capabilities to otherwise limited installations.

18.6.2 Fiber-Optic-Linked

The availability of fiber-optic video and computer data networks covering long distances with real-time video capabilities will increase security capabilities significantly. Combining satellite transmission and global security from one security console location will become commonplace.

18.7 MULTITASKING

The application of multitasking with computers is commonplace for manipulating and communicating data and graphics. This capability will extend into video applications in the future.

18.7.1 Microsoft Windows

Software programs such as Microsoft Windows offer the capability to combine multiple tasks on one monitor screen to be controlled by one security guard. This environment will become commonplace in security in the near future.

18.7.2 Interactive

The use of two-way interactive communication with data, graphics, video, audio, and commands will significantly enhance the effectiveness of the guard in the security loop in the future.

18.8 DIGITAL IMAGE PROCESSING

Some solid-state cameras have a digital output to represent the video image signal. These cameras have been developed primarily for the industrial and desktop publishing fields, where a video image of a person is digitally stored for later retrieval. This digital format will provide benefits for the security-surveillance industry in the future. By using the output from the chip in its digital form rather than converting it to the standard analog signal used by the television monitor, an increase in the signal-to-noise ratio is obtained, which transfers into an improved overall sensitivity of the camera when operating in LLL environments. There is also a better use of the cameras' inherent resolution due to the direct one-to-one picture cell-to-digitizer correlation. For example a $1/2$-inch format CCD camera with a 739 (H) × 484 (V) frame-transfer array is degraded to a 560 horizontal TV lines picture. If the 739 (H) digital output was displayed on a digital monitor, the full 739-pixel resolution would obtain. The camera also exhibits 256 shades of gray, as compared with the typical ten shades of gray obtained with analog solid-state and tube cameras.

18.8.1 Image Enhancement

Presently, video camera outputs are sent directly to a video monitor for direct viewing or to a VCR for recording. Applying digital processing to the video signal can improve the quality of the picture by factors of five to ten, with some sacrifice in time delay. Using digital signal averaging and edge enhancement techniques as digital memory costs decline, further improvements will result. Image enhance-ment will allow cameras to see at lower light levels or at longer ranges.

18.8.2 Image Compression

Video images contain large amounts of information. A good-resolution monochrome picture with 740 horizontal × 480 vertical pixels and a gray scale of 16 shades (4-bit) contains 740 × 480 × 4 = 1,420,800 bytes of information. The real-time CCTV frame is processed in $1/30$ second. In order to transmit the video signal or store it efficiently in a computer magnetic or optical storage media, it is advantageous to compress the signal. This compression takes the form of removing any unnecessary information and/or reconstructing it into a simpler one requiring less bandwidth to transmit or less storage medium to store. Future systems will contain many variations of this process.

An area of development within slow-scan systems is that of reducing the bandwidth requirements to send a picture, thereby reducing the transmit time or increasing the number of pictures sent per second. This consists of electronically analyzing the television scene that is to be transmitted and subtracting out portions of the scene that are not changing in time or that have no motion. By having to transmit only the parts of the scene that are changing, the amount of information needed to be transmitted can decrease significantly, thereby allowing scenes to be sent over the telephone line at a faster rate. This scheme is an attempt to bring the snapshot mode to a faster rate, bringing it closer to real-time television transmission. Depending on the motion in the scene and the amount of detail (high resolution) in the scene, significant increases have been achieved and further improvements will continue.

18.9 HIGH-DEFINITION TELEVISION

The international consumer-broadcast video industry has several designs in the testing phase to bring to market a high-definition television (HDTV) system. This will bring superior picture quality and a larger viewing screen to the security console. A new aspect ratio of 16 to 9 (horizontal to vertical) will improve the picture information for scenes having a wide format (such as parking lots). There are presently several systems under consideration. The proposed European HDTV system has 1250 horizontal lines and a 50-field/second interlaced scan. The U.S system has 1125 horizontal lines at 60 fields/second. The higher resolution of these systems will put them on par with 35-mm film resolution.

gather more information. Future security systems will use some form of 3D for certain aspects of the security function.

18.10 THREE-DIMENSIONAL TELEVISION

There are several three-dimensional (3D) CCTV system techniques in use or being developed. 3D can significantly increase the intelligence in a video image and can enhance the ability of the guard to make a better judgment call than if made from conventional two-dimensional, flat-screen monitors. The 3D systems use two cameras and two video channels and some optical means to display the two images for the viewer. By viewing a 3D illusion, the guard can

18.11 SUMMARY

Each decade brings new and innovative products that improve the quality of CCTV security. Some come from new ideas, some from refinements of existing technology, and others from adaptations of ideas and equipments already in use in other fields. Communications between the manufacturer and the equipment user can be a catalyst to bring a new concept to market.

Chapter 19
CCTV Checklist

CONTENTS

19.1 OVERVIEW

The previous chapters have analyzed CCTV technology and the essential components of CCTV security systems. Chapter 17 analyzed five CCTV applications at different facilities. This chapter is a checklist that summarizes the salient points for the overall CCTV surveillance system. At the outset of any system design, these basic questions must be answered. The checklist concludes with a short form, which can be used as a startup questionnaire for the CCTV project.

19.2 CHECKLIST

The following list enumerates questions that should be asked when designing a CCTV surveillance security system.

1. What is the purpose of the CCTV system: crime deterrence, vandalism detection, offender identification? What is to be protected? What is the value of what is to be protected and what is the cost of the system needed to protect it? Are goods and/or personnel to be under surveillance?

2. Should the system be overt, covert, or both? Which cameras should be overt? Which cameras should be covert? Where should cameras be for the best view?

3. Is the surveillance indoor, outdoor, or both?

4. What type of housings are needed: conventional rectangular, dome, other, indoor or outdoor?

5. Will there be added lighting or existing lighting? Is the lighting natural or artificial? Will the system operate daytime and nighttime?

6. Is additional lighting required? What kind?

7. Will FFL, zoom, pinhole, or other lenses be used? Manual or automatic iris? Presets? What should be the FOV of each camera?

8. Should the system be monochrome or color? How many cameras are required to view the personnel and articles to be protected?

9. Will a video motion detector be used? Analog or digital? Will the system be interfaced with other alarms?

10. Is a pan/tilt mechanism required to cover all areas? Pan/tilt/zoom for which cameras? Limited or 360-degree continuous rotation?

11. Are standard or LLL cameras required?

12. What kind of switcher is needed? Sequential, homing, looping, alarming? How many cameras/monitors/VCRs must be switched?

13. What monitoring equipment is needed at the console? How many console monitoring locations?

14. Is a microprocessor system necessary? How many cameras/monitors/VCRs must be switched?

15. Installation environment: Building structure material? Single or multiple floors? Single or multiple buildings? Outdoor pedestal or fence line?

16. Number of monitors? Should they be multiscene (split-screen) monitors? Should a quad (or other) screen combiner be used?

17. Transmission environment: Should coaxial cable, fiber optic, wireless, two-wire, real-time, slow-scan be used?

18. Is backup power required to ensure 100% operation, zero down-time? If not 100%, what cameras, monitors, and other equipment must operate as a minimum to maintain adequate security and recovery?

19. What power conditioning and uninterruptible power supply equipment should be used?

20. Should the system be powered from 117 or 24 VAC, or 12 VDC, or a combination? Will solar power and/or batteries be used as a backup?

21. Security console: What is the level of the personnel monitoring the system? How many are there? What are the duties of monitoring personnel?

22. What recorded information is required? What kind of video recording will be used? VHS, S-VHS, 8-mm, Hi-8? Number of VCRs—real-time or time-lapse?

23. Is a hard-copy video printer necessary? If so, monochrome or color?

24. Will non-video alarms be used to alert guards and activate VCRs and printers?

25. Is there a requirement for video access control?

26. What local electrical codes apply?

27. Must any or all of the equipment meet UL, Canadian Safety Association, or other code approval?

28. Is equipment compatible? Does it follow EIA and Closed Circuit Television Manufacturers Association design recommendations?

19.3 SHORT FORM CHECKLIST

1. CCTV system purpose?
2. Overt or covert system?
3. Indoor or outdoor?
4. Housing type?
5. Natural or artificial lighting, additional required?
6. Lens type?
7. Monochrome or color system?
8. Standard or LLL cameras?
9. Motion detection?
10. Pan/tilt mechanism?
11. Video switcher type or microprocessor?
12. Installation environment
13. Transmission type and environment?
14. Security console: personnel required, duties, equipment?
15. Quad, screen combiner, or screen splitter?
16. VCR?
17. Video printer?
18. System power?
19. Power conditioning and backup power?
20. Other interfaces: audio, alarm inputs, paging, law enforcement?

Chapter 20
Education, Standards, and Bids

CONTENTS

20.1 OVERVIEW

The successful implementation of a CCTV surveillance system integrated into an overall security plan requires a thorough understanding of the technology, and its interoperability with the overall security system. The designer/installer must attend professional seminars and see manufacturers' demonstrations to remain informed of the latest techniques and technologies in the industry. The end user must likewise attend security assets and protection seminars and manufacturers' exhibits to remain current.

Crucial to the implementation of any successful CCTV surveillance system is the assemblage of a detailed, well-conceived, professional specification/bid package. Writing this package forces management, security personnel, consultants, and architect/engineers (A&E) to itemize what functions the system should perform and how it should function with the other parts of the security system. The written plan should be detailed and include suggested manufacturers, dealers, and installers. All electrical equipment must meet local electrical and fire codes. To design for maximum reliability, compatibility and safety, all system hardware and installation techniques should conform to the EIA, Closed Circuit Television Manufacturers Association (CCTMA), and UL safety recommendations.

20.2 PROFESSIONAL SEMINARS AND MANUFACTURERS' EXHIBITS

Numerous organizations conduct educational seminars and manufacturers' exhibitions in the United States and abroad. To ensure a successful CCTV surveillance system, it is necessary to attend these conferences periodically. One reason for attending is to learn the role of CCTV in the overall security plan. A second reason is to stay abreast of the rapid advancements being made in CCTV technology and security methods. The rapid deployment of new solid-state technology from the computer industry in the security industry requires regular attendance at these seminars and expositions.

The organization providing the most comprehensive educational seminars and manufacturers' exhibits in the security industry is the American Society for Industrial Security (ASIS). This professional organization conducts educational seminars directed toward end users in all aspects of assets protection.

The organization providing the most comprehensive educational seminars for dealer/installers and end users in CCTV technology is the CCTMA (Section 20.3).

LTC Training Center provides comprehensive seminars for most CCTV applications.

Seminar attendance and fulfillment of course requirements given by these organizations can lead to professional certification.

20.3 ELECTRONIC INDUSTRIES ASSOCIATION

The EIA is a trade organization composed of electronic equipment manufacturers. This organization provides a national forum to establish technical standards, educate practitioners, and create an influential voice in governmental legislative circles. The EIA television standards (RS-170, RS-330, and so on) form the basis of all U.S. CCTV standards. The EIA RS-232, RS-422, and RS-485 serve as the communications protocol for almost all security applications.

The CCTMA was formed in 1986 under the auspices of the EIA to promote the CCTV industry and project a positive image to equipment end users and the public at large.

The CCTMA actively encourages its member companies and the industry at large to establish standard goals that will result in excellence in equipment design and manufacture, professional installation and service, and maximum equipment interoperability with other security equipment.

The CCTMA sponsors comprehensive technical seminars to educate CCTV manufacturers, dealers, installers, consultants, architects, engineers, and end users through seminars, reports, and videotapes.

CCTMA members are dedicated to designing equipment that provides maximum compatibility and interchangeability among manufacturers. This commitment is manifest in many areas of equipment design, including: (1) common video input and output signal parameters for cameras, monitors, switchers, VCRs, printers, and so on; (2) common pin and terminal connections and mechanical configurations for power, video, automatic-iris lens, pan/tilt and lens controls, switcher functions, and others; and (3) common mechanical mounting configurations, hole pattern, location, and fastening hardware for camera mounts, pan/tilt mechanisms, monitor mounts, and electrical equipment mounting racks (standard 19-inch rack mounting panel).

New standards in process include a comprehensive equipment installation guide, including recommended wire and cable types, terminating procedures, and testing.

20.4 UNDERWRITERS LABORATORIES

Underwriters Laboratories is a corporation that tests equipment submitted by manufacturers and issues a UL seal of approval if the equipment meets its standards. The UL standards have resulted from industry requests to satisfy their needs for reliable and safe equipment. The criteria for approval are listed in various standards written by UL, and designers of CCTV security systems should be knowledgeable of the documents applicable to the industry.

Glossary

Many terms and definitions used in the security industry are unique to CCTV surveillance; others derive from the electro-optical industry. This comprehensive glossary will help the reader better understand the literature, interpret manufacturers' specifications, and write bid specifications and requests for quotation. These terms encompass the CCTV industry, basic physics, electricity, mechanics, and optics.

Aberration   Failure of an optical lens to produce exact point-to-point correspondence between an object and its image.

Achromatic lens (achromat)   A lens consisting of two or more elements, usually of crown and flint glass, that has been corrected for chromatic aberration with respect to two selected colors or light wavelengths.

Alarming switcher   See **Switcher**.

Ambient temperature   The temperature of the environment; that is, the temperature of the surrounding medium, such as gas or liquid, that comes into contact with the apparatus.

Amplifier   A device whose output is essentially an enlarged reproduction of the input which does not draw power from the input.

Amplifier, distribution   A device that provides several isolated outputs from one looping or bridging input. The amplifier has sufficiently high input impedance and input-to-output isolation to prevent loading of the input source.

Angle of view   The maximum scene angle that can be seen through a lens or optical assembly. Usually described in degrees, for horizontal, vertical, or circular dimension.

Aperture   An opening that will pass light, electrons, or other forms of radiation. In an electron gun, the aperture determines the size of, and has an effect on, the shape of the electron beam. In television optics, the aperture is the effective diameter of the lens that controls the amount of light reaching the image sensor.

Aperture, clear   See **Clear aperture**.

Aperture, numerical   See **Numerical aperture**.

Aperture stop   An optical opening or hole that defines or limits the amount of light passing through a lens system. The aperture stop takes the form of the front lens diameter in a pinhole lens, an iris diaphragm, a neutral density or spot filter.

Arc lamp   An electric-discharge lamp with an electric arc between two electrodes to produce illumination. The illumination results from the incandescence of the positive electrode and from the heated, luminous, ionized gases that surround the arc.

ASIS   American Society for Industrial Security.

Aspect ratio   The ratio of width to height for the frame of the video picture in CCTV or broadcast television. The NTSC standard is 4:3. The aspect ratio for proposed high-definition television is 16:9.

Aspheric   An optical element having one or more surfaces that are not spherical. The spherical surface of the lens is slightly altered to reduce spherical aberration.

Astigmatism   A lens aberration that causes an object point to be imaged as a pair of short lines at right angles to each other.

Attenuation   A reduction in light or electrical signal or energy strength. In electrical systems attenuation is often measured in decibels or decibel per unit distance. In optical systems the units of measure are f-number or optical density. See also **Decibel, Density**.

Audio frequency   Any frequency corresponding to a normally audible sound wave—roughly from 15 to 15,000 Hz.

Auto balance   A system for detecting errors in color balance in the white and black areas of the picture and automatically adjusting the white and black levels of both the red and blue signals as needed.

Auto light range   The range of light—such as sunlight to moonlight or starlight—over which a TV camera is capable of automatically operating at specified output.

Automatic brightness control (ABC)   In display devices, the self-acting mechanism that controls brightness as a function of ambient light.

Automatic frequency control (AFC)   A feature whereby the frequency of an oscillator is automatically maintained within specified limits.

Automatic gain control (AGC)   A process by which gain is automatically adjusted as a function of input or other specified parameter.

Automatic iris   A diaphragm device in the lens that self-adjusts optically to light level changes via the video signal from the television camera. The iris diaphragm opens or closes the aperture to control the light transmitted through the lens. Typical compensation ranges are 10,000–300,000 to 1. Automatic irises are used on Newvicon, silicon, Ultracon, SIT, ISIT tubes, and solid-state CCD and MOS cameras.

Automatic iris control   An electro-optic accessory to a lens that measures the video level of the camera and opens and closes the iris to compensate for light changes.

Automatic light compensation   The degree to which a CCTV camera can adapt to varying light conditions.

Automatic light control   The process by which the illumination incident upon the face of a pickup device is automatically adjusted as a function of scene brightness.

Automatic pedestal control   A process by which the pedestal height in a video signal is automatically adjusted as a function of input level or other specified parameter.

Automatic sensitivity control   The self-acting mechanism that varies system sensitivity as a function of the specified control parameters. This may include automatic target control, automatic light control, and so on, or any combination thereof.

Automatic target control   The self-acting mechanism that controls the image pickup tube target potential as a function of the scene brightness.

Axis, optical   See **Optical axis**.

Back focus   The distance from the last glass surface of a lens to the focused image.

Back porch   That portion of a composite video signal which lies between the trailing edge of a horizontal sync pulse and the trailing edge of the corresponding blanking pulse. The color burst, if present, is not considered part of the back porch.

Bandpass   A specific range of frequencies that will be passed through a device or system.

Bandwidth   The number of hertz (cycles per second) expressing the difference between the lower and upper limiting frequencies of a frequency band; also, the width of a band of frequencies.

Bandwidth limited gain control   A control that adjusts the gain of an amplifier while varying the bandwidth. An increase in gain reduces the bandwidth.

Barrel distortion   A distortion in television that makes the televised image appear to bulge outward on all sides like a barrel.

Beam   A concentrated, unidirectional flow of electrons, photons, or other energy: (1) A shaft or column of light; a bundle of rays consisting of parallel, converging, or diverging rays. (2) A concentrated stream of particles that is unidirectional. (3) A unidirectional concentrated flow of electromagnetic waves.

Beam splitter   An optical device for dividing a light beam into two or more separate beams. The splitting can be done in the parallel (collimated) beam or in the focused image plane. The splitting can be done spectrally.

Beam width (angular beam width)   The angular beam width of a conical beam of light, the vertex angle of the cone, which determines the rate at which a beam of energy diverges or converges. Lasers produce very narrow or very nearly parallel beams. Thermal light sources (floodlight) produce wide-angle beams.

Beta format   A ½-inch video cassette recorder format not compatible with the VHS format.

Bifocal lens   A lens system having two different focal length lenses that image two identical or different scenes onto a single camera sensor. The two scenes appear as a split image on the monitor.

Blackbody   A thermally heated body that radiates energy at all wavelengths according to specific physical laws.

Black clamp   An electronic circuit that automatically maintains the black video level (no light) at a constant voltage.

Black compression   Also called black saturation, the reduction in gain applied to a picture signal at those levels corresponding to dark areas in a picture with respect to the gain at that level corresponding to the midrange light value in the picture.

Black level   The picture signal level corresponding to a specified maximum limit for black peaks.

Black negative   The television picture signal in which the polarity of the voltage corresponding to black is negative with respect to that which corresponds to the white area of the picture signal.

Blanking   The process whereby the beam in an image pickup or cathode ray display tube is cut off during the retrace period.

Blanking level   The level of a composite picture signal that separates the range containing picture information from the range containing synchronizing information; also called pedestal, or blacker-than-black. The setup region is regarded as picture information.

Blooming   The defocusing of regions of the picture where the brightness is at an excessive level, due to enlargement of spot size and halation of the fluorescent screen of the cathode ray picture tube.

Borescope   An optical device used for the internal inspection of mechanical and other parts. The long tube contains a multiple lens telescope system that usually has a high f-number (low amount of light transmitted).

Boresight   An optical instrument used to check alignment or pointing direction. A small telescope mounted on a gun or television camera so that the optical axis of the telescope and the mechanical axis of the gun or camera coincide. The term also applies to the process of aligning other equipment, such as cameras and lasers.

Bounce   Sudden variation in picture presentation (brightness, size, and so on) independent of scene illumination.

Breezeway   In NTSC color, that portion of the back porch between the trailing edge of the sync pulse and the start of the color burst.

Bridging   Connecting two electrical circuits in parallel. Usually the input impedances are large enough so as not to affect the signal level.

Bridging amplifier   An amplifier for bridging an electrical circuit without introducing an apparent change in the performance of that circuit.

Brightness   The attribute of visual perception in accordance with which an area appears to emit more or less light. Luminance is the recommended name for this photometric quantity, which has also been called brightness.

Brightness control   The manual bias control on a cathode ray tube or other display device that determines both the average brightness and the contrast of a picture.

Burn-in   Also called burn. An image that persists in a fixed position in the output signal of a camera tube after the camera has been pointed toward a different scene. An image that persists on the face of a CRT monitor with no input video signal present.

C-mount   An industry standard for lens mounting. The C-mount has a thread with a 1-inch diameter and 32 threads per inch. The distance from the lens mounting surface to the sensor surface is 0.69 inches (17.526 mm).

Cable   A number of electrical conductors (wires) bound in a common sheath. These may be control cables, coaxial cables (for video signals) and/or remote control cables.

Camera control unit (CCU)   Remote module that provides control of camera electronic circuitry such as deflection and video amplification.

Camera format   Standard C- and CS-mount television cameras are made with $\frac{1}{3}$-, $\frac{1}{2}$-, $\frac{2}{3}$-, and 1-inch sensor image formats. The actual target areas used (scanned) on the sensors are 4.4 mm horizontal × 3.3 mm vertical for the $\frac{1}{3}$-inch, 6.4 mm horizontal × 4.8 mm vertical for the $\frac{1}{2}$-inch, 8.8 mm horizontal × 6.6 mm vertical for the $\frac{2}{3}$-inch, and 12.8 mm horizontal × 9.6 mm vertical for the 1-inch.

Camera housing   An enclosure designed to protect the CCTV camera from undue environmental exposure when placed outdoors, and from tampering or theft when outdoors or indoors.

Camera, television   An electronic device containing an electronic image tube or solid-state sensor. The image formed by a lens on the sensor is scanned rapidly by a moving electron beam or clocked out for a solid-state sensor. The sensor output varies with the local brightness of the image. These variations are transmitted to a display device, where the brightness of the scanning spot in a cathode ray tube is controlled. The scanning location at the camera and the scanned spot at the viewing tube are accurately synchronized.

Camera tube   An electron tube that converts an optical image into an electrical current by a scanning process. Also called a pickup tube or a television camera tube.

Candela (cd)   Unit of measurement of luminous intensity. The candela is the international unit that replaces the candle.

Candle power (cp)   Light intensity expressed in candles. One foot-candle (fc) is the amount of light emitted by a standard candle at 1-foot distance.

Catadioptric system   A telephoto optical system embodying both lenses and image-forming mirrors. Examples are the Schmidt, Maksutov, and Cassegrain telescope.

Cathode ray tube (CRT)   A vacuum tube in which electrons emitted by a heated cathode are focused into a beam and directed toward a phosphor-coated surface, which then becomes luminescent at the point where the electron beam strikes. Prior to striking the phosphor, the focused electron beam is deflected by two pairs of electrostatically charged plates located between the "gun" and the screen. Electromagnets are often used in place of the deflector plates.

CATV   Community antenna television. A television distribution system primarily used for consumer TV broadcast programming.

CCIR   International Radio Consultative Committee. The CCIR format uses 625 lines per picture frame, with a 50-Hz power line frequency.

CCTMA   Closed Circuit Television Manufacturers Association. A division of the EIA, the CCTMA is a full-service national trade organization. It promotes a positive image for the CCTV industry and promotes the interests of its members.

CCTV   Closed-circuit television. A television system that does not broadcast TV signals but transmits them over a closed circuit via an electrically conducting or fiber-optic cable.

CCTV camera   That part of the CCTV system which captures and transmits the picture.

CCTV monitor   That part of the CCTV system which receives the picture from the CCTV camera and displays it.

Ceiling mount   A bracket fastened to a ceiling to which a CCTV camera or monitor can be attached.

Charge coupled device (CCD)   A solid-state semiconductor imaging device. A self-scanning semiconductor array that uses charge coupling in a metal oxide semiconductor technology, surface storage, and information transfer by digital shift register techniques.

Chromatic aberration   A design flaw in a lens or lens system that causes the lens to have different focal lengths for radiation of different wavelengths. The dispersive power of a simple positive lens focuses light from the blue end of the spectrum at a shorter distance than light from the red end. This deficiency produces an image that is not sharp.

Clamping   The process and circuitry that establishes a fixed level for the television picture level at the beginning of each scanning line.

Clear aperture   The opening in the mount of an optical system or its components that restricts the extent of the bundle of rays incident on the given surface. It is usually circular and specified by its diameter.

Clipping   The shearing off of the peaks of a signal. For a picture signal, clipping may affect either the positive (white) or negative (black) peaks. For a composite video signal, the sync signal may be affected.

Close-up lens   A low-magnification (power) accessory lens that permits focusing on objects closer to the lens than it has been designed for.

Coaxial cable   A cable capable of carrying a wide range of frequencies with very low signal loss. In its simplest form it consists of a hollow metallic shield with a single wire accurately placed along the center of the shield and isolated from the shield by an insulator.

Color bar test pattern   A special test pattern for adjusting color TV receivers or color encoders. The upper portion consists of vertical bars of saturated colors and white. The lower horizontal bars have black-and-white areas and I and Q signals.

Color saturation   The degree of mixture of a color and white. When a color is mixed with little or no white, it is said to have a high saturation. Low saturation denotes the addition of a great amount of white, as in pastel colors.

Color temperature   The term used to denote the temperature of a blackbody light source that produces the same color as the light under consideration.

Composite video signal   The combined signals in a television transmission, including the picture signal, blanking signal, and vertical and horizontal synchronizing signals.

Compression   The reduction in gain at one level of a picture signal with respect to the gain at another level of the same signal. In video signal transmission or storage, the removal of redundant information to decrease the digital transmission or storage requirements.

Concave   A term describing a hollow curved surface of a lens or mirror; curved inward.

Contrast   The range of difference between light and dark values in a picture, usually expressed as contrast ratio (the ratio between the maximum and minimum brightness values).

Control panel   A rack at the monitor location containing a number of controls governing camera selection, pan and tilt controls, focus and lens controls, and so on.

Convergence   The crossover of the three electron beams of a three-gun tricolor picture tube. This normally occurs at the plane of the aperture mask.

Convex   A term denoting a spherically shaped optical surface of a lens or mirror; curved outward.

Cooler kit   An accessory unit used to cool the camera and lens within a camera housing in extreme hot weather conditions.

Corner reflector or corner cube prism   A corner reflector having three mutually perpendicular surfaces and a hypotenuse face. Light entering through the hypotenuse is totally internally reflected by each of the three surfaces in turn, and emerges through the hypotenuse face parallel to the entering beam and returns entering beams to the source. It may be constructed from a prism or three mutually perpendicular front surface mirrors.

Covert surveillance   In television security, the use of camouflaged (hidden) lenses and cameras for the purpose of viewing a scene without being seen.

Cross-talk   Interference between adjacent video, audio, or optical channels.

CS-mount   An industry standard for lens mounting. The CS-mount has a thread with a 1-inch diameter and 32 threads per inch. The distance from the lens mounting surface to the sensor surface is 0.492 inches (12.497 mm).

Cutoff frequency   That frequency beyond which no appreciable energy is transmitted. It may refer to either an upper or lower limit of a frequency band.

Dark current   The current that flows in a photoconductor when it is placed in total darkness.

Dark current compensation   A circuit that compensates the dark current level change with temperature.

DC restoration   The re-establishment by a sampling process of the DC and low-frequency components of a video signal that has been suppressed by AC transmission.

DC transmission   A form of transmission in which the DC components of the video signal are transmitted.

Decibel (dB)   A measure of the voltage or power ratio of two signals. In system use, a measure of the voltage or power ratio of two signals, provided they are measured across a common impedance. Decibel gain or loss is 20 times log 10 of the voltage or current ratio (Voutput/Vinput), and 10 times log 10 of the power ratio (Poutput/Pinput).

Decoder   The circuitry in a receiver that transforms the detected signal into a form suitable to extract the original modulation or intelligence.

Definition   The fidelity of a television system with respect to the original scene.

Delay distortion   Distortion resulting from the nonuniform speed of transmission of the various frequency components of a signal, caused when various frequency components of the signal have different times of travel (delay) between the input and the output of a circuit.

Delay line   A continuous or periodic structure designed to delay the arrival of an electrical or acoustical signal by a predetermined amount.

Density   A measure of the light-transmitting or -reflecting properties of an optical material. It is expressed by the common logarithm of the ratio of incident to transmitted light flux. A material having a density of 1 transmits 10% of the light, 2 transmits 1%, 3 transmits 0.1%, and so on. See also **Neutral density filter**.

Depth of field   For a lens, the area along the line of sight in which objects are in reasonable focus. It is measured from the distance behind an object to the distance in front of the object when the viewing lens shows the object to be in focus. Depth of field increases with smaller lens aperture (higher f-numbers), shorter focal lengths, and greater distances from the lens.

Depth of focus   The range of detector-to-lens distance for which the image formed by the lens is clearly focused.

Detail contrast   The ratio of the amplitude of video signal representing high-frequency components with the amplitude representing the reference low-frequency component, usually expressed as a percentage at a particular line number.

Detail enhancement   Also called image enhancement. A system in which each element of a picture is analyzed in relation to adjacent horizontal and vertical elements. When differences are detected, a detail signal is generated and added to the luminance signal to enhance it.

Detection, image   In television, the criterion used to determine whether an object or person is observed (detected) in the scene. Detection requires the activation of only 1 TV line pair.

Diaphragm   See **Iris diaphragm**.

Differential gain   The amplitude change, usually of the 3.58-MHz color subcarrier, introduced by the overall circuit, measured in dB or percent, as the picture signal on which it rides is varied from blanking to white level.

Differential phase   The phase change of the 3.58-MHz color subcarrier introduced by the overall circuit, measured in degrees, as the picture signal on which it rides is varied from blanking to white level.

Diopter   A term describing the optical power of a lens. It is the reciprocal of the focal length in meters. For example, a lens with a focal length of 25 cm (0.25 m) has a power of 4 diopters.

Distance, image   See **Image distance**.

Distance, object   See **Object distance**.

Distance, working   See **Working distance**.

Distortion, electrical   An undesired change in the waveform from that of the original signal.

Distortion, optical   A general term referring to the situation in which an image is not a true reproduction of an object. There are many types of distortion.

Distribution amplifier   See **Amplifier, distribution**.

Dot bar generator   A device that generates a specified output pattern of dots and bars, used for measuring scan linearity and geometric distortion of TV cameras and video monitors. Also used for converging cathode ray tubes.

Drive pulses   Sync and blanking pulses.

Dynamic range   In television, the useful camera operating light range, from highlight to shadow, in which detail can be observed in a static scene when both highlights and shadows are present. The voltage or power difference between the maximum allowable signal level and the minimum acceptable signal level.

Echo   A signal that has been reflected at one or more points during transmission with sufficient magnitude and time difference as to be detected as a signal distinct from that of the primary signal. Echoes can be either leading or lagging the primary signal and appear as reflections or "ghosts."

EIA interface   A standardized set of signal characteristics (time duration, waveform, voltage, current) specified by the Electronic Industries Association.

EIA sync signal   The signal used for the synchronizing of scanning specified in the Electrical Industry Association standards RS-170 (for monochrome), RS-170A (for color), RS-312, RS-330, RS-420, or subsequent specifications.

Electromagnetic focusing   A method of focusing a cathode ray beam to a fine spot by application of electromagnetic fields to one or more deflection coils of an electron lens system.

Electronic viewfinder   See **Viewfinder, electronic**.

Electrostatic focusing   A method of focusing a cathode ray beam to a fine spot by application of electrostatic potentials to one or more elements of an electron lens system.

Endoscope   An optical instrument resembling a long, thin periscope used to examine the inside of objects by inserting one end of the instrument into an opening in the object. Endoscopes comprise a coherent fiber-optic bundle with a small objective lens to form an image of the object onto one end of the bundle, and a relay magnifier lens at the sensor end to focus the fiber bundle image onto the sensor. In an illuminated version, light from an external lamp is piped down to the object by a second set of thicker fibers surrounding the image-forming bundle. See also **Fiberscope**.

Equalizer   An electronic circuit that introduces compensation for frequency discrimination effects of elements within the television system.

Fader   A control and associated circuitry for effecting fade-in and fade-out of video or audio signals.

Fiber-optic bundle, coherent   An optical component consisting of many thousands of hair-like fibers coherently assembled so that an image is transferred from one end of the bundle to the other. The length of each fiber is much greater than its diameter. The fiber bundle transmits a picture from one of its surfaces to the other, around curves and into otherwise inaccessible places by a process of total internal reflection. The positions of all fibers at both ends are located in an exact one-to-one relationship with each other.

Fiber-optics transmission   The process whereby light is transmitted through a long, transparent, flexible fiber, such as glass or plastic, by a series of internal reflections: (1) for video, audio, or data transmission over long distances (thousands of feet, many miles), single fibers in protective insulating jackets are used; (2) bundles of these fibers can transmit an entire image (coherent) where each single fiber transmits but one component of the whole image.

Fiberscope   A bundle of systematically arranged fibers that transmits a monochrome or full-color image which remains undisturbed when the bundle is bent. By mounting an objective lens on one end of the bundle and a relay or magnifying lens on the other, the system images remote objects onto a sensor. See **Endoscope**.

Field   One of the two equal parts into which a television frame is divided in an interlaced system of scanning. There are 60 fields per second in the NTSC system. The NTSC field contains 262 $\frac{1}{2}$ horizontal lines.

Field frequency   The number of fields transmitted per second in a television system. Also called field repetition rate. The U.S. standard is 60 fields per second (60-Hz power source). The European standard is 50 fields per second (50-Hz power source).

Field lens   Lens used to effect the transfer of the image formed by an optical system to a following lens system with minimum vignetting.

Field of view (FOV)   The width, height, or diameter of a scene to be monitored, determined by the lens focal length, the sensor size, and the lens-to-subject distance. The maximum angle of view that can be seen through a lens or optical assembly. Usually described in degrees, for a horizontal, vertical, or circular dimension.

Filter   An optically transparent material characterized by selective absorption of light with respect to wavelength (color). Electrical network of components to limit the transmission of frequencies to a special range (bandwidth).

Fixed focal length lens (FFL lens)   A lens having one or more elements producing a singular focal length. Units of measure: millimeters or inches.

Flatness of field   Appearance of the image to be flat. The object is imaged as a plane.

Fluorescent lamp   A high-efficiency, low-wattage arc lamp used in general lighting. A tube containing mercury vapor and lined with a phosphor. When current is passed through the vapor, the strong ultraviolet emission excites the phosphor, which emits visible light. The ultraviolet energy cannot emerge from the lamp as it is absorbed by the glass.

f-number The optical speed or ability of a lens to pass light. The f-number (f/#) denotes the ratio of the equivalent focal length (FL) of an objective lens to the diameter (D) of its entrance pupil (f/# = FL/D). The f-number is directly proportional to the focal length and inversely proportional to the lens diameter. A smaller f-number indicates a faster lens.

Focal length (FL) The distance from the lens center, or second principal plane to a location (plane) in space where the image of a distant scene or object is focused. FL is expressed in millimeters or inches.

Focal length, back The distance from the rear vertex of the lens to the lens focal plane.

Focal plane A plane (through the focal point) at right angles to the principal axis of a lens or mirror; that surface on which the best image is formed.

Focal point The point at which a lens or mirror will focus parallel incident radiation from a distant point source of light.

Focus (1) The focal point. (2) The adjustment of the eyepiece or objective of a visual optical device so that the image is clearly seen by the observer. (3) The adjustment of a camera lens, image sensor, plate, or film holder so that the image is sharp. (4) The point at which light rays or an electron beam form a minimum-size spot. Also the action of bringing light or electron beams to a fine spot.

Focus control, electronic A manual electric adjustment for bringing the electron beam of an image sensor tube or picture tube to a minimum size spot, producing the sharpest image.

Focus control, mechanical A manual mechanical adjustment for moving the television sensor toward or away from the focal point of the objective lens to produce the sharpest image.

Foot-candle (fc) A unit of illuminance on a surface 1 square foot in area on which there is incident light of 1 lumen. The illuminance of a surface placed 1 foot from a light source that has a luminous intensity of 1 candle.

Foot-lambert A unit of luminance equal to 1 candela per square foot or to the uniform luminance at a perfectly diffusing surface emitting or reflecting light at the rate of 1 lumen per square foot.

Frame The total picture area scanned while the picture signal is not blanked. In the standard U.S. NTSC 525-line system, the frame time is $1/30$ second. In the European 625-line system, the frame time is $1/25$ second.

Frame frequency The number of times per second that the frame is scanned. The U.S. NTSC standard is 30 times per second. The European standard is 25 times per second.

Frequency interlace The method by which color and black-and-white sideband signals are interwoven within the same channel bandwidth.

Frequency response The range or band of frequencies to which a unit of electronic equipment will offer essentially the same characteristics.

Front porch That portion of a composite picture signal which lies between the leading edge of the horizontal blanking pulse and the leading edge of the corresponding sync pulse.

Front surface mirror An optical mirror with the reflecting surface applied to the front surface of the glass instead of to the back. The reflecting material is usually aluminum with a silicon monoxide protective overcoat. A front surface mirror exhibits no secondary image or ghost.

f-stop See **f-number**.

Gain An increase in voltage or power, usually expressed in decibels.

Gallium arsenide diode A light-emitting diode semiconductor device that emits low-power infrared radiation. Used in television systems for small area illumination or with fiber optics for signal transmission. The radiation is incoherent and has a beam spread of typically 10 to 50 degrees and radiates at 850 nanometers in the IR spectrum.

Gallium arsenide laser A narrow-band, narrow-beam IR radiation device. The radiation is coherent and has a very narrow beam pattern, typically $1/2$ to 2 degrees, and radiates in the IR spectrum.

Gamma A numerical value of the degree of contrast in a television picture that is used to approximate the curve of output magnitude versus input magnitude over the region of interest. Gamma values range from 0.6 to 1.0.

Gamma correction To provide for a linear transfer characteristic from input to output device by adjusting the gamma.

Genlock An electronic device used to lock the frequency of an internal sync generator to an external source.

Geometric distortion Any aberration that causes the reproduced picture to be geometrically dissimilar to the original scene.

Ghost A spurious image resulting from an echo (electrical) or a second or multiple reflection (optical). A front surface mirror produces no ghost, while a rear surface mirror produces a ghost. RF and microwaves reaching a receiver after reflecting from multiple paths produce ghosts.

Gray scale Variations in value from white, through shades of gray, to black, on a television screen. The gradations approximate the tonal values of the original image picked up by the TV camera. Most CCTV cameras produce at least 10 shades of gray.

Halo   A glow or diffusion that surrounds a bright spot on a television picture tube screen.

Hertz (Hz)   The frequency of an alternating signal. Formerly called cycles per second. The U.S. power-line frequency is 60 Hz.

High-contrast image   A picture in which strong contrast between light and dark areas is visible. Intermediate values, however, may be missing, as in a laser print or photocopy.

High-frequency distortion   Distortion effects that occur at high frequency. In television, generally considered as any frequency above 15.75 kHz.

Highlights   The maximum brightness of the TV picture occurring in regions of highest illumination.

Horizontal (hum) bars   Relatively broad horizontal bars, alternately black and white, which extend over the entire picture. They may be stationary or may move up or down. Sometimes referred to as a "venetian-blind" effect. In 60-Hz systems, hum bars are caused by approximate 60-Hz interfering frequency or one of its harmonic frequencies (such as 120 Hz).

Horizontal blanking   Blanking of the picture during the period of horizontal retrace.

Horizontal resolution   See **Resolution, horizontal**.

Horizontal retrace   The return of the electron beam from the right to the left side of the raster after the scanning of one horizontal line.

Hue   Corresponds to colors such as red, blue, and so on. Black, gray, and white do not have hue.

Hum   Electrical disturbance at the power supply frequency or harmonics thereof.

Hum modulation   Modulation of a radio frequency, or detected signal, by hum.

Identification, image   In television, the criterion used to determine whether an object or person can be identified in the scene. It requires approximately 7 TV-line pairs to identify an object or person.

Illuminance   Luminous flux incident per unit area of a surface; luminous incidence. (The use of the term *illuminance* for this quantity conflicts with its more general meaning.)

Illumination, direct   The lighting produced by visible radiation that travels from the light source to the object without reflection.

Illumination, indirect   The light formed by visible radiation that, in traveling from the light source to the object, undergoes one or more reflections.

Image   A reproduction of an object produced by light rays. An image-forming optical system collects light diverging from an object point and transforms it into a beam that converges toward another point, thus producing an image.

Image distance   The axial distance measured from the image to the second principal point of a lens.

Image format   In television, the size of the area of the image at the focal plane of a lens, which is scanned by the television sensor.

Image intensifier   A class of electronic imaging tubes equipped with a light-sensitive photocathode—electron emitter at one end and a phosphor screen at the other end. An electron tube or microchannel plate amplifying (intensifying) mechanism produces an image at its output brighter than the input. A device coupled by fiber optics to a TV image pickup tube to increase sensitivity. Can be single stage or multistage.

Image pickup tube   An electron tube that reproduces on its fluorescent screen an image of an irradiation pattern incident on its input photosensitive surface.

Image plane   The plane at right angles to the optical axis at the image point.

Impedance   The input or output characteristic of an electrical system or component. For maximum power and signal transfer, a cable used to connect two systems or components must have the same characteristic impedance as the system or component. Impedance is expressed in ohms. Video distribution systems have standardized on 75-ohm unbalanced and 124-ohm balanced coaxial cable.

Incident light   The light that falls directly onto an object.

Infrared radiation   The invisible portion of the electromagnetic spectrum that lies beyond about 750 nanometers (red end of the visible spectrum) and extends out to the microwave spectrum.

Intensified CCD (ICCD)   A charge coupled device sensor camera fiber optically coupled to an image intensifier. The intensifier is a tube or microchannel plate.

Intensified silicon intensified target (ISIT)   An SIT tube with an additional intensifier, fiber-optically coupled to provide increased sensitivity. Two intensifiers are stacked in series to yield a gain of about 2,000 times that of a standard vidicon.

Intensified vidicon (IV)   A standard vidicon-type TV image pickup tube of the direct readout type coupled with fiber optics to an intensifier to increase sensitivity.

Interference   Extraneous energy that tends to interfere with the desired signal.

Interlace, 2 to 1   A scanning technique used in CCTV systems in which the two fields making up the frame are synchronized precisely in a 2 to 1 ratio, and where the time or phase relationship between adjacent lines in successive fields is fixed.

Interlace, random   A scanning technique used in CCTV systems in which the two fields making up the frame are not synchronized, and where there is no fixed time or phase relationship between adjacent lines in successive fields.

Interlaced scanning   A scanning process in which the distance from center to center of successively scanned lines is two or more times the nominal line width, and in which the adjacent lines belong to different fields. It is a scanning process used to reduce image flicker and is 2:1 in the NTSC system.

Iris   An adjustable opto-mechanical aperture built into a camera lens to permit control of the amount of light passing through the lens.

Iris diaphragm   A mechanical device within a lens used to control the size of the aperture through which light passes. A device for opening and closing the lens aperture to adjust the f-stop of a lens.

Isolation amplifier   An amplifier with input and output circuitry designed to eliminate the effects of changes made by either upon the other.

Jitter   Instability of a signal in either its amplitude, phase, or both, due to mechanical disturbances or to changes in supply voltage, component characteristics, and so on.

Lag   In a television pickup tube, the persistence of the electrical charge image for two or more frames after excitation is removed.

Laser   An acronym for light amplification by simulated emission of radiation. A laser is an optical cavity, with plane or spherical mirrors at the ends, that is filled with light-amplifying material, and an electrical or optical means of stimulating (energizing) the material. The light produced by the atoms of the material generates a brilliant beam of light, which is emitted through one of the mirrors. The output beam is highly monochromatic (pure color) and coherent.

Laser diode   See **Gallium arsenide laser**.

Leading edge   The major portion of the rise of a pulse, taken from the 10 to 90% level of total amplitude.

Lens   A transparent optical component consisting of one or more pieces of optical glass with surfaces so curved (usually spherical) that they serve to converge or diverge the transmitted rays of an object, thus forming a real or virtual image of that object.

Lens, fresnel   A lens that is cut into narrow rings and flattened out. The lens has narrow concentric rings or steps, each acting to focus radiation into an image.

Lens speed (f-number)   Refers to the ability of a lens to transmit light, represented as the ratio of the focal length to the diameter of the lens. A fast lens would be rated f/1.4. A much slower lens might be designated as f/8. The larger the f-number, the slower the lens.

Lens system   Two or more lenses so arranged as to act in conjunction with one another.

Light   Electromagnetic radiation detectable by the eye, ranging in wavelength from about 400 nm (blue) to 750 nm (red).

Limiting resolution   A measure of resolution usually expressed in terms of the maximum number of TV lines per TV picture height discernible on a test chart.

Line amplifier   An amplifier for audio or video signals that drive a transmission line. An amplifier (generally broadband) installed at an intermediate position connected to a main cable run to compensate for loss.

Linearity   The state of an output that incrementally changes directly or proportionally as the input changes.

Line pairs   Term used in defining television resolution. One TV line pair constitutes one black line and one white line. The 525 NTSC has 485 line pairs.

Load   That component which receives the output energy of a device.

Loss   A reduction in signal level or strength, usually expressed in dB. A power dissipation serving no useful purpose.

Low-frequency distortion   Distortion effects that occur at low frequency. In television, generally considered as any frequency below 15.75 kHz.

Low light level (LLL) camera and television   CCTV systems capable of operating below normal visual response. A CCTV camera capable of operating in extremely poorly lighted areas.

Lumen (lm)   The unit of luminous flux, equal to the flux through a unit solid angle (steradian) from a uniform point source of 1 candela or to the flux on a unit surface of which all points are at a unit distance from a uniform point source of 1 candela.

Luminance   Luminous intensity (photometric brightness) of any surface in a given direction per unit of projected area of the surface as viewed from that direction, measured in foot-lamberts.

Luminance signal   That portion of the NTSC color television signal which contains the scene luminance or brightness information.

Luminous flux   The time rate of flow of light.

Lux   International System unit of illumination in which the meter is the unit of length. One lux equals 1 lumen per square meter.

Magnetic focusing   A method of focusing an electron beam by the action of a magnetic field.

Magnification   A number expressing the change in object to image size. Usually expressed with a 1-inch focal length lens and a 1-inch format sensor as a reference (magnification = M = 1). A lens with a 2-inch focal length is said to have a magnification of M = 2.

Matching   The obtaining of like electrical impedances to provide a reflection-free transfer of signal.

Matrix switcher   A combination or array of electro-mechanical or electronic switches that route a number of signal sources to one or more designations.

Maximum aperture   The largest size to which the iris diaphragm of the lens can be opened (the lowest f-number).

Megahertz (MHz)   Unit of frequency equal to 1 million Hz.

Mercury arc lamp   An intense electric arc lamp that generates blue-white light when electric current flows through mercury vapor.

Metal arc lamp   An intense arc lamp that generates a white light when an electric current flows through the multimetal vapor.

Microcomputer   A tabletop or portable digital computer composed of a microprocessor, active memory storage, and permanent memory storage (disk) and which computes and controls functions via a software operating system and applications program.

Micron   Unit of length: one millionth of a meter.

Microphonics   Audio-frequency noise caused by the mechanical vibration of elements within a system or component.

Microprocessor   The brain of the microcomputer. A very large scale integrated circuit comprising the computing engine of a microcomputer. The electronic chip (circuit) that does all the calculations and control of data. In larger machines it is called the central processing unit.

Microwave transmission   In television, a transmission means that converts the camera television signal to a modulated (AM or FM) microwave signal via a transmitter, and a receiver that demodulates the received microwave signal to the baseband CCTV signal for display on a monitor.

Mirror, first or front surface   An optical component on which the reflecting surface is applied to the front of the glass instead of the back, the front being the first surface of incidence and reflectance. It produces a single image with no ghost.

Mirror, rear surface   An optical component on which the reflecting surface is applied to the rear of the glass. It produces a secondary or ghost image.

Modulation   The process, or results of the process, whereby some characteristic of one signal is varied in accordance with another signal. The modulated signal is called the carrier. The carrier may be modulated in three fundamental ways: by varying the amplitude, called amplitude modulation (AM); by varying the frequency, called frequency modulation (FM); or by varying the phase, called phase modulation (PM).

Moire   In television, the spurious pattern in the reproduced picture resulting from interference beats between two sets of periodic structures in the image. Caused by tweed or checkerboard patterns in the scene.

Monitor   A device for viewing a television picture from a camera output, which does not incorporate channel selector or audio components. The monitor displays the composite video signal directly from the camera, videotape recorder, or special-effects generator.

Monochrome signal   Black and white with all shades of gray. In monochrome television, a signal for controlling the brightness values in the picture. In color television, that part of the signal which has major control of the brightness values of the picture, whether displayed in color or in monochrome. The minimum number of shades of gray for good image rendition is 10.

Monochrome transmission   The transmission of a signal wave that represents the brightness values in the picture, not the color (chrominance) values.

Motion detector   A device used in security systems that reacts to any movement on a CCTV camera/monitor by automatically setting off an alarm and/or indicates the movement on the monitor.

Motorized lens   A camera lens fitted with small electric motors that by remote control can focus the lens, open or close the iris diaphragm, or in the case of the zoom lens, change the focal length.

Multiplex   Putting two or more signals into a single channel.

NAB   National Association of Broadcasters.

NAEB   National Association of Educational Broadcasters.

Nanometer (nm)   Unit of length: one billionth of a meter.

Negative image   A picture signal having a polarity that is opposite to normal polarity and that results in a picture in which the white and black areas are reversed.

Neutral density filter   An optical attenuating device. A light filter that reduces the intensity of light without changing the spectral distribution of the light. See **Density.**

Newvicon A television pickup tube with a cadmium and zinc telluride target with sensitivity about 20 times that of a vidicon target. It has a spectral response of 470 to 850 nm, is relatively free from burn-in, has good resolution, and needs an automatic iris.

Noise The word noise originated in audio practice and refers to random spurts of acoustical or electrical energy, or interference. In television, it produces a "salt-and-pepper" pattern over the televised picture. Heavy noise is sometimes referred to as "snow."

Nonbrowning A term used in connection with lens glass, faceplate glass, fiber optics, used in radiation-tolerant television cameras. Nonbrowning glass does not discolor (turn brown) when irradiated with atomic particles and waves.

Noncomposite video A video signal containing all information except sync.

Notch filter A special filter designed to reject a very narrow band of electrical frequencies or optical wavelengths.

NTSC National Television Systems Committee. A committeemittee that worked with the FCC in formulating standards for the present-day U.S. color television system. NTSC has 525 horizontal scan lines, 60 frames per second ($^{525}\!/_{60}$). Commonly used in the United States and Japan.

Numerical aperture The sine of the half-angle of the widest bundle of rays capable of entering a lens, multiplied by the refractive index of the medium containing that bundle. In air, the refractive index n = 1.

Object distance The distance between the object and the cornea of the eye, or the first principal point of the objective in an optical device.

Objective lens The optical element that receives light from an object scene and forms the first or primary image. In cameras, the image produced by the objective is the final image. In telescopes and microscopes, when used visually, the image formed by the objective is magnified by an eyepiece.

Optical axis The line passing through the centers of curvatures of the optical surfaces of a lens or the geometric center of a mirror or window; the optical centerline.

Optical splitter An optical lens–prism and/or mirror system that combines two or more scenes and images them onto one television camera. Optical components are used to combine the scene.

Orientation, image In television, the criterion used to determine the angular orientation of a target (object, person) in an image. At least 2 TV-line pairs are required.

Overshoot The initial transient response to a unidirectional change in input, which exceeds the steady-state response.

Overt surveillance In television, the use of any openly displayed television lenses or cameras to view a scene.

PAL Phase alternating line system. A color television system in which the subcarrier derived from the color burst is inverted in phase from one line to the next in order to minimize errors in hue that may occur in color transmission. The PAL format is used in Western Europe, Australia, parts of Africa, and the Middle East. It has 625 horizontal scan lines and 25 frames per second.

Pan and tilt Camera-mounting device that allows movement in both the azimuth (pan) and the elevation (tilt) planes.

Pan, panning Rotating or scanning a camera around a vertical axis to view an area in a horizontal direction.

Pan/tilt/zoom Three terms associated with television cameras to indicate the horizontal (pan), vertical (tilt), and magnification (zoom) they are capable of producing.

Passive Incapable of generating power or amplification. A nonpowered device that generally presents some loss to a system.

Peak-to-peak The amplitude (voltage) difference between the most positive and the most negative excursions (peaks) of an electrical signal.

Pedestal level See **Blanking level**.

Persistence In a cathode ray tube, the period of time a phosphor continues to glow after excitation is removed.

Phosphor A substance capable of luminescence. It is used in fluorescent lamps, television monitors, viewfinders, and image intensifier screens.

Phosphor-dot faceplate A glass plate in a tricolor picture tube. May be the front face of the tube or a separate internal plate. In either case, its rear surface is covered with an orderly array of tricolor lines or tricolor phosphor dots. When excited by electron beams in proper sequence, the phosphors glow in red, green, and blue to produce a full-color picture.

Photocathode An electrode used for obtaining photoelectric emission.

Photoconductivity The changes in the electrical conductivity (reciprocal of resistance) of a material as a result of absorption of photons.

Photoconductor A material whose electrical resistance varies in relationship with exposure to light.

Photoelectric emission The phenomenon of emission of electrons by certain materials upon exposure to radiation in and near the visible region of the spectrum.

Photon-limited sensitivity When the quantity of available light is the limiting factor in the sensitivity of a device.

Photopic vision   Vision that occurs at moderate and high levels of luminance and permits distinction of colors. This light-adapted vision is attributed to the retinal cones in the eye; in contrast, twilight or scotopic vision uses primarily the rods.

Pickup tube   A television camera image pickup tube. See also **Image pickup tube**.

Picture element (pixel)   Any segment of a scanning line, the dimension of which along the line is exactly equal to the nominal line width.

Picture size   The useful area of a picture tube. In the standard NTSC format, the horizontal to vertical to diagonal ratios are 4:3:5, respectively.

Picture tube   See **Cathode ray tube**.

Pin-cushion distortion   Distortion in a television picture that makes all sides appear to bulge inward.

Pinhole lens   A lens designed to have a relatively small (0.06 inch to 0.375 inch) front lens diameter to permit its use in covert (hidden) camera applications.

Pixel time   The time required to scan a single pixel.

Plumbicon   Trade name for a direct-readout type of lead-oxide television image pickup tube manufactured by N.V. Philips used in some color video cameras. The Plumbicon has very low picture lag and excellent red color sensitivity.

Preamplifier   An amplifier used to increase the output of a low-level source so that the signal can be further processed without additional deterioration of the signal-to-noise ratio.

Preset   A term used in television pointing systems (pan/tilt/zoom). A computer stores pre-entered azimuth, elevation, zoom (magnification), focus, and iris combinations, which are later accessed when commanded by an operator or an alarm.

Primary colors   Three colors wherein no mixture of any two can produce the third. In color television these are the additive primary colors red, green, and blue (RGB).

Pulse   A variation of a quantity whose value is normally constant. This variation is characterized by a rise and a decay, and has finite amplitude and duration.

Pulse rise-time   Time interval between upper and lower limits of instantaneous amplitude; specifically, 10 and 90% of the peak-pulse amplitude, unless otherwise stated.

Radio frequency (RF)   A frequency at which coherent electromagnetic radiation of energy is useful for communication purposes. The entire range of such frequencies, including the AM and FM radio spectrum and the VHF and UHF television spectrum.

Random interlace   See **Interlace, random**.

Raster   A predetermined pattern of scanning lines that provides substantially uniform coverage of an area. The area of a camera or CRT tube scanned by the electron beam.

Raster burn   See **Burn-in**.

Recognition, image   In television, the criterion used to determine whether an object or person can be recognized in a television scene. A minimum of 5 TV-line pairs are required to recognize a person or object.

Reference black level   The picture signal level corresponding to a specified maximum limit for black peaks.

Reference white level   The picture signal level corresponding to a specified maximum limit for white peaks.

Resolution, horizontal   The amount of resolvable detail in the horizontal direction in a picture; the maximum number of individual picture elements that can be distinguished. It is usually expressed as a number of distinct vertical lines, alternately black and white, which can be seen in a distance equal to picture height. 500 to 600 TV lines are typical with the standard 4.2-MHz CCTV bandwidth.

Resolution, limiting   Picture details that can be distinguished on the television screen. Vertical resolution refers to the number of horizontal black-and-white lines that can be resolved in the picture height. Horizontal resolution refers to the number of vertical black-and-white lines that can be resolved in a width equal to the picture height.

Resolution, vertical   The amount of resolvable detail in the vertical direction in a picture. It is usually expressed as the number of distinct horizontal lines, alternately black and white, which can be seen in a picture. 350 TV lines are typical in the 525 NTSC system.

Retained image   Also called image burn. A change produced in or on the sensor target that remains on the output device (such as a CRT) for a large number of frames after the removal of a previously stationary light image.

Right-angle lens   A multi-element optical component that causes the optical axis of the incoming radiation (from a scene or image focal plane) to be redirected by 90 degrees.

Ringing   In electronic circuits, an oscillatory transient occurring in the output of a system as a result of a sudden change in input.

Ripple   Amplitude variations in the output voltage of a power supply caused by insufficient filtering.

Roll   A loss of vertical synchronization that causes the picture to move up or down on a television receiver or CCTV monitor.

Rolloff   A gradual decrease in attenuation of a signal voltage.

Saturation In color, the degree to which a color is undiluted with white light, or is pure. The vividness of a color, described by such terms as bright, deep, pastel, pale, and so on. Saturation is directly related to the amplitude of the television chrominance signal.

Scanning Moving the electron beam of an image pickup or a CRT picture tube horizontally across and slowly down the target or screen area respectively. Moving the charge packets in a charge coupled device out of the sensor.

Scotopic vision Vision that occurs in faint light, or in dark adaptation. It is attributed to the operation of the retinal rods in the eye. Contrast with daylight or photopic vision, using primarily the cones.

SECAM Sequential Couleur A'Memorie, a color television system developed in France and used in some countries that do not use either the NTSC or PAL systems. Like PAL, SECAM has 625 horizontal scan lines and 25 frames per second but differs significantly in the method of producing color signals.

Sensitivity In television, a factor expressing the incident illumination upon a specified scene to produce a specified picture signal at the television camera output.

Shutter In an optical system, an opaque material placed in front of a lens, optical system, or sensor for the purpose of protecting the sensor from bright light sources, or for timing the length of time the light source reaches the sensor or film.

Shuttering An electronic technique used in solid-state cameras to reduce the charge accumulation from scene illumination to increase the dynamic range of the camera sensor. Analogous to the shutter in a film camera.

Signal strength The intensity of the television signal measured in volts, millivolts, microvolts, or decibels. Using 0 dB as the standard reference is equal to 1000 microvolts in RF systems and 1 volt in video systems.

Signal-to-noise ratio The ratio of the peak value of the video signal to the value of the noise. Usually expressed in decibels. The ratio between useful television signal and disturbing noise or snow.

Silicon intensified target (SIT) An image intensifier fiber-optically coupled to a silicon faceplate vidicon resulting in a sensitivity 500 times that of a standard vidicon.

Silicon monoxide A thin-film dielectric (insulator) used as a protective layer on aluminized mirrors. It is evaporated on the mirror as a thin layer, and after exposure to the air the monoxide tends to become silicon dioxide or quartz, which is very hard and completely transparent.

Silicon target tube A high-sensitivity television image pickup tube of the direct-readout type using a silicon diode made up of a mosaic of light-sensitive silicon material. It has a sensitivity between 10 and 100 times more than a sulfide vidicon and has high resistance to image burn-in.

Slow-scan An electronic television transmission system consisting of a transmitter and receiver that transmit single-frame television pictures at rates slower than the normal NTSC frame rate of 30 per second. The CCTV frames are modulated and transmitted to a distant receiver-demodulator to be displayed on a CCTV monitor. The slow-scan process periodically sends "snapshots" of the scene. Typical sending rates are from 2 to 72 frames per second.

Snow Heavy random noise manifest on a phosphor screen as a changing black and white "peppered" random noise. See **Noise**.

Sodium lamp A low- or high-pressure discharge metal vapor arc lamp using sodium as the luminous radiation source. The lamp produces a yellow light and has the highest electrical-to-light output efficiency (efficacy) of any lamp. Because of its poor color balance it is not recommended for color CCTV systems.

Spike A transient of short duration, comprising part of a pulse, during which the amplitude considerably exceeds the average amplitude of the pulse.

Switcher A video electronic device that connects one of many input cameras to one or several output monitors, recorders, and so on, by means of a panel switch or electronic input signal.

Switcher, alarming An automatic switcher that is activated by a variety of sensing devices. Once activated, the switcher connects the camera to the output device (such as a monitor or recorder).

Switcher, bridging A sequential switcher with separate outputs for two monitors, one for programmed sequence and the second for extended display of a single area.

Switcher, fader A control device that permits each of two or more cameras to be selectively fed into the distribution or display system. The "fader" permits gradual transition from one camera to another.

Switcher, homing A switcher in which (1) the outputs of multiple cameras can be switched sequentially to a monitor, (2) one or more cameras can be bypassed (not displayed), or (3) any one of the cameras can be selected for continuous display on the monitor (homing). The switcher has three front-panel controllable modes: (1) Skip, (2) Automatic (sequential) and (3) Select (display one camera continuously). The lengths of time each camera output is displayed are independently selectable by the operator.

Switcher, manual A switcher in which the individual cameras are chosen by the operator manually by pushing the switch for the camera output signal chosen to be displayed, recorded, or printed.

Switcher, sequential   A generic switcher type that allows the video signals from multiple cameras to be displayed, recorded, or printed one at a time in sequence.

Sync   A contraction of *synchronous* or *synchronization*.

Sync generator   A device for generating a synchronizing signal.

Synchronizing   Maintaining two or more scanning processes in phase.

Sync level   The level of the peaks of the synchronizing signal.

Sync signal   The signal employed for the synchronizing of scanning.

Talk-back   A voice intercommunicator; an intercom.

Target   (1) In image pickup tubes, a structure using a storage surface that is scanned by an electron beam to generate a signal output current corresponding to a charge-density pattern stored on it. (2) In solid-state sensors, a semiconductor structure using picture elements to accumulate the picture charge, and a scanning readout mechanism to generate the video signal. (3) In surveillance, an object (person, vehicle, and so on) or activity of interest present in an image of the scene under observation.

Target voltage   In a camera tube with low-velocity scanning, the potential difference between the cathode and the backplate.

Tearing   A term used to describe a picture condition in which groups of horizontal lines are displaced in an irregular manner.

Test pattern   A chart especially prepared for checking overall performance of a television system. It contains combinations of lines and geometric shapes of specific sizes and spacings. In use, the camera is focused on the chart, and the pattern is viewed at the monitor for image fidelity (resolution). The chart most commonly used is the EIA resolution chart.

Tilt   A deviation from the ideal low-frequency response; unsatisfactory low-frequency response.

Transient   An unwanted signal existing for a brief period of time that is superimposed on a signal or power line voltage.

Tri-split lens   A multi-element optical assembly that combines one-third of each of three scenes and brings them to focus (adjacent to one another) at the focal plane of a television camera sensor. Three separate objective lenses are used to focus the scenes onto the splitter assembly.

T-stop   A system of rating the light throughput of lenses for sensitivity purposes. It provides an equivalent aperture of a lens having 100% transmission efficiency. This system is based on actual light transmission and is considered a more realistic test than the f-stop system.

Tungsten-halogen (halide) lamp   Type of lamp once called quartz-iodine, which contains a filament (tungsten) and a gas (halogen) in a fused quartz enclosure. The iodine or bromine added to the fill produces a tungsten-halogen cycle, which provides self-cleaning and an extended lifetime. The higher filament temperature produces more light and substantially longer lifetime.

Tungsten lamp   An optical radiation source composed of a tungsten filament surrounded by an inert gas (nitrogen, xenon) enclosed in a glass or quartz envelope (bulb). An AC or DC electric current passing though the filament causes it to heat to incandescence, producing visible and infrared radiation.

UHF (ultra-high frequency)   In television, a term used to designate the part of the RF spectrum in which channels 14 through 83 are transmitted.

UL certified   A certification given by Underwriters Laboratory to certain items that are impractical to list and which the manufacturer can use to identify the material.

UL listed   A label that signifies that a product meets the safety requirements as set forth by UL safety testing standards.

Ultracon tube   The trade name (RCA/Burle Industries) of a special form of silicon tube.

Ultraviolet (UV)   An invisible region of the optical spectrum located immediately beyond the violet end of the visible spectrum, and between the wavelengths of approximately 100 and 380 nanometers. Radiation just beyond the visible spectrum (at the blue end of the visible spectrum) ordinarily filtered or blocked to prevent eye damage.

Underwriters Laboratory (UL)   A testing laboratory that writes safety standards used by manufacturers when designing products. A private research and testing laboratory that tests and approves manufactured items for certification or listing providing they meet required safety standards.

Vectorscope   A special oscilloscope used in color TV camera system calibration. The vectorscope graphically indicates on a CRT the absolute angle between the different color signals, with respect to a reference signal, and to each other. These angles represent the phase differences between the signals.

Vertical resolution   The number of horizontal lines that can be seen in the reproduced image of a television pattern. The 525 TV line NTSC system limits the vertical resolution to 350 TV lines maximum.

Vertical retrace   The return of the electron beam to the top of the picture tube screen or the pickup tube target at the completion of the field scan. The retrace is not displayed on the monitor.

VHF (very high frequency) In television, a term used to designate the part of the RF spectrum in which channels 2 through 13 are transmitted.

Video A term pertaining to the bandwidth and spectrum position of the signal resulting from television scanning. In current CCTV usage video means a bandwidth of 4.2 MHz.

Video amplifier A wideband amplifier used for passing video picture signals.

Video band The frequency band used to transmit a composite video signal.

Videocassette recorder (VCR) A device that accepts signals from a video camera and a microphone and records images and sound on magnetic tape in a cassette. The VCR can play back the recorded images and sound for viewing on a television receiver or CCTV monitor or printing out on a video printer.

Video database In television systems, a computer and permanent magnetic or optical memory that is capable of storing and retrieving many analog or digital video images. Each video image has a unique ID number (address) and location on the media.

Video file See **Video database**.

Video signal (noncomposite) The picture signal. A signal containing visual information without the horizontal and vertical synchronization and blanking pulses. See also Composite video signal.

Vidicon tube An electron tube used to convert light into an electrical signal. The spectral response covers most of the visible light range and most closely approximates the human eye response (400–700 nm).

Viewfinder A small electronic or optical viewing device attached to a television camera so that the operator can view the scene the camera sees.

Viewfinder, electronic A small CCTV monitor attached to the video camera that allows the operator to view the exact scene being viewed by the camera.

Viewfinder, optical A small optical sighting device attached to a television camera or other device so that the operator can see the scene as the camera (device) sees it.

Vignetting The loss of light through a lens or optical system at the edges of the field due to inadequate lens design or an internal obstruction. Most well-designed lenses minimize vignetting.

Visible spectrum That portion of the electromagnetic spectrum to which the human eye is sensitive. The range covers from 400 to 700 nanometers.

Wavelength The length of an electromagnetic energy wave measured from any point on one wave to the corresponding point on the next wave, usually measured from crest to crest. Wavelength defines the nature of the various forms of radiant energy that compose the electromagnetic spectrum and determines the color of light. Common units for measurement are the nanometer (1/10,000 micron), micron, millimicron, and the Angstrom.

White clipper A nonlinear electronic circuit providing linear amplification up to a predetermined voltage and then unity amplification for signals above the predetermined voltage.

White compression Amplitude compression of the signals corresponding to the white regions of the picture.

White level The picture signal level corresponding to a specified maximum limit for white peaks.

White peak The maximum excursion of the picture signal in the white direction.

White peak clipping Limiting the amplitude of the picture signal to a preselected maximum white level.

Working distance The distance between the front surface of an objective lens and the object being viewed.

Xenon arc lamp An arc lamp containing the rare gas xenon, which is excited electrically to emit a brilliant white light. The lamps are available in short-arc (high-pressure) and long-arc (low-pressure).

Zoom (1) Optical: to enlarge or reduce the size of a CCTV sensor image on a continuously variable basis. (2) Electronic: to electronically amplify (magnify) or attenuate (demagnify) the electronic image size of a CCTV picture.

Zoom lens An optical system of continuously variable focal length, the focal plane remaining in a fixed position. Groups of lens components are moved to change their relative physical positions, thereby varying the focal length and angle of view through a specified range of magnifications.

Bibliography

Active Night Vision Devices. National Institute of Law Enforcement and Criminal Justice, Standard NILECJ, June 1975.

Anderson, John Eric. *Fiber Optics: Multi-Mode Transmissions.* Technical memorandum. Galileo Electro-Optics Corp., 1990.

Bell, Trudy E. "TheNew Television: Looking Behind the Tube." *IEEE Spectrum,* August 1984, 48–56.

Bose, Keith W. *Video Security Systems,* 2nd ed. Stoneham, Mass.: Butterworth, 1982.

Chryssis, George. *Power Supply Distribution and Grounding.* Datel-Intersil, September 1979.

Clifford, Martin. *The Camcorder.* 1989.

Cook, Jack. "Making Low-Loss Single-Mode Connectors." *Laser Focus,* Dorran Photonics, 1983.

Dubois, Paul A. "The Design and Application of Security Lighting." *Security Management,* September 1985.

Electro-Optics Handbook. Burle Industries Inc., 1974.

Fay, John J., ed. *Encyclopedia of Security Management.* Stoneham, Mass.: Butterworth, 1993.

Fennelly, Lawrence J. *Effective Physical Security.* Stoneham, Mass.: Butterworth, 1992.

Fennelly, Lawrence J., ed. "Museum, Archive, and Library Security." Chapter 16 in Herman Kruegle, *Museum Television Security.* Stoneham, Mass.: Butterworth, 1983, 427–93.

———. "Controlling Cargo Theft." Chapter 15 in Herman Kruegle, *CCTV.* Stoneham, Mass.: Butterworth, 1983, 327–402.

———. "Handbook of Loss Prevention and Crime Prevention, 2nd ed." Chapter 18 in Herman Kruegle, *Closed Circuit Television Security.* Stoneham, Mass.: Butterworth, 1989, 320–63.

Fiber Optics: Theory and Applications. Technical memorandum. Galileo Electro-Optics Corp., 1990.

Freese, Robert P. "Optical Disks Become Erasable." *IEEE Spectrum,* February 1988, 41–45.

Goldstein, Herbert. "Consider the Options When Choosing Your Uninterruptible Power Supply." *Security Technology & Design,* March/April 1993.

Grant, Joseph C., ed. *Nickel-Cadmium Battery Application Engineering Handbook,* 2nd ed. General Electric, 1975.

Inglis, Andrew F. *Video Engineering.* New York: McGraw-Hill, 1993.

Jaeger, R. E. *Optical Communication Fiber Fabrications.* Galileo Electro-Optics, 1976.

Jefferson, Robert. "Shedding Light on Security Problems." *Security Management,* December 1992.

Kincaid, B. E., S. Cowen, and D. Campbell. "Application of GaAs Lasers and Silicon Avalanche Detectors to Optical Communications." *Optical Engineering,* September/October 1974.

Kruegle, Herman. "The Basics of CCTV." *Security Management,* January 1981.

———. "Elevator CCTV Security." *Elevator World,* March 1978.

Laurin, Teddi C. *Photonics Design and Applications Handbook,* 38th ed. Laurin Publishing, 1992.

———. *Photonics Dictionary,* 38th ed. Laurin Publishing, 1992.

Levin, Robert E., and Thomas M. Lemons. "High-Intensity Discharge Lamps and Their Environment." *IEEE Transactions on Industry and General Applications,* vol. IGA-7, no. 2, March/April 1971.

Lighting Design. American Electric Co. Pub. No. LSBC, 1991, 168–84.

Luther, Arch C. *Digital Video in the PC Environment,* 2nd ed. New York: McGraw-Hill, 1991.

McHale, John J. "Tungsten Transport in Quartz-Iodine Lamps." *Illuminating Engineering*, April 1971.

McPartland, Joseph F. *National Electrical Code Handbook*, 18th ed. New York: McGraw-Hill, 1984.

Martzloff, Francois. "Protecting Computer Systems Against Power Transients." *IEEE Spectrum*, April 1990, 37–40.

O'Brien, James T. "Intensified Cameras for Low Light Situations." *Advanced Imaging*, June 1989.

NILECJ. U.S. Government Printing Office Stock No.270000257. July 1974.

Philips Lighting. *Security Lighting*. Publication P-3368, North American Philips Corp., 1992.

Raia, Salvatore L., and Raymond W. Payne. "Security Equipment and Technology." *Security World*, February 1978.

Reis, Charles. "Your Range of Choices in Color Hard Copy Devices." *Advanced Imaging*, September 1990, 54–59.

Richmond, Joseph C. *Test Procedures for Night Vision Devices*. Report. National Institute of Law Enforcement and Criminal Justice.

Rintz, Carlton L. "Designing with Image Tubes." *Photonics Spectra*, December 1989.

Rintz, Carlton L., and George A. Robinson. "A High-Resolution Intensified CCD Imager."

Robins, G. A. "The Silicon Intensifier Target Tube: Seeing in the Dark." *SMPTE Journal* 86 (June 1977).

Rogowitz, Bernice E. "Displays: The Human Factor." *Byte*, July 1992, 195–200.

Sampat, Nitin. "The RS-170 Video Standard and Scientific Imaging: The Problems." *Advanced Imaging*, February 1991.

Schumacher, William. "Test Conditions for Fiber Optic Connectors and Splices." AMP Inc. 1982.

Smith, Cecil. "The Resolution Solution." *AV Video Journal*, November 1991.

Smith, Coleman Cecil. *Mastering Television Technology*. NewmanSmith, 1991.

Smith, Warren J. *Modern Optical Engineering*, 2nd ed. New York: McGraw-Hill, 1990.

Technician's Guide to Fiber Optics, 2nd ed. AMP Inc. 1993.

Walsh, Timothy J. *Protection of Assets*. The Merritt Co., 1985.

Watkins, Harry S., and John S. Moore. "A Survey of Color Graphics Printing." *IEEE Spectrum*, July 1984, 26–37.

Weiler, Harold D. "Monograph 1: Methods of Adjusting and Evaluating Video Equipment." Video Concepts Corp., 1968.

Westman, H. P., ed. *Reference Data for Radio Engineers*, 5th ed. Howard Sams and Co., Inc., 1974.

Wiza, Joseph L. "Microchannel Plate Detectors." *Nuclear Instruments and Methods* 162 (1979): 587–601.

Index